Genera of Hyphomycetes

J.W.Carmichael
The University of Alberta Mold Herbarium

W.Bryce Kendrick
Biology Department,
The University of Waterloo

I.L.Conners
Mycology Section, The Biosystematics Research Institute, Ottawa (retired)

Lynne Sigler
The University of Alberta Mold Herbarium

Genera of Hyphomycetes

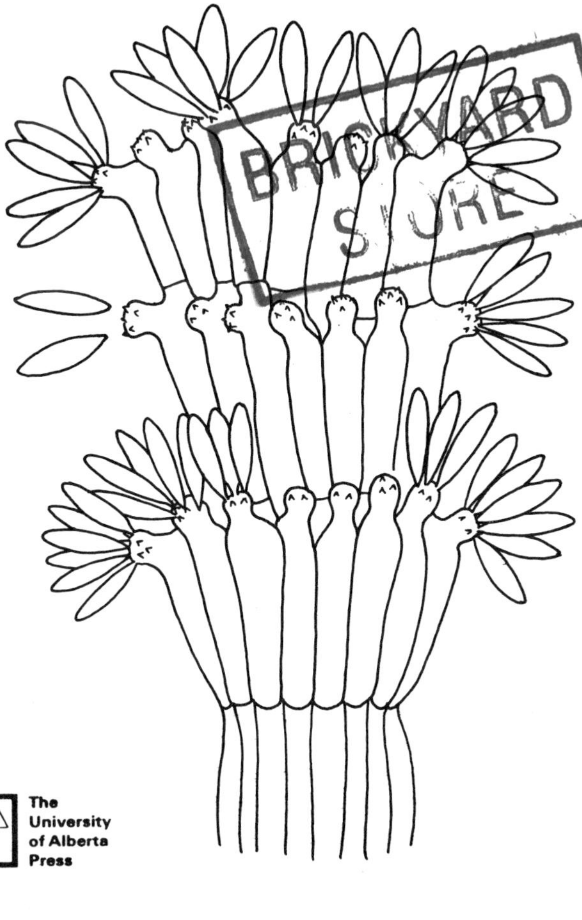

The University of Alberta Press

First published by
The University of Alberta Press
Edmonton, Alberta, Canada
1980

ISBN 0-88864-063-3 (coil bound)
ISBN 0-88864-038-2 (hard bound)

Canadian Cataloguing in Publication Data

Main entry under title:
Genera of Hyphomycetes
Bibliography: p.
Includes index.
ISBN 0-88864-063-3 (coil bound). --ISBN 0-88864-038-2 (hard bound.)

1. Moniliales. 2. Fungi — Classification I. Carmichael, J. W.
QK625.M74G45 589.2'4 C80-091031-1

Printed by
Hignell Printing Limited
Winnipeg, Manitoba

This book is dedicated to the memory of
Luella K. Weresub
who died in Ottawa on 12 October 1979.
Luella was not only an ever-helpful colleague,
but also a friend, conscience, and adviser to us all.

Contents

Acknowledgments

The production of this book, in progress since 1972, would not have been possible without the efforts of many people in addition to the authors. We are pleased to specifically acknowledge the following.

Gareth Morgan-Jones, Bruce Peterman, and Odette Rose provided invaluable assistance with the drawings.

Luella Weresub, Frank DiCosmo, T.R. Nag Raj, Joanna Mott, Shannon Berch, and Laurie Kendrick helped at various junctures in the preparation of the plates.

The Mycology Section of the Biosystematics Research Institute, Canada Agriculture, Ottawa, provided working space and facilities for Conners. Their library, herbarium, and Index of Fungus Names were made available on several occasions to Kendrick and Carmichael. Luella Weresub, Dave Malloch, Stan Hughes, and other members of the Mycology Section gave freely of their time for discussions on nomenclatural and taxonomic problems.

The Universities of Alberta and Waterloo provided salaries, space, and equipment for three of us (see addresses on title page).

The Natural Sciences and Engineering Research Council of Canada and the Medical Research Council of Canada provided operating grant funds to Kendrick and Carmichael during this project.

Debbie Reinhart, Ed Grams, and Dave Holberton gave advice on photo-typesetting.

Introduction

I. Purpose and History

A) THE PURPOSE AND CONTENTS OF THIS BOOK

This book was produced to meet three main needs: first, to permit the identification of unknown Hyphomycetes to form-genus; second, to indicate the current status and application of all the generic names proposed for Hyphomycetes; and, third, to provide access to the important taxonomic literature on each form-genus.

To achieve these objectives, the book contains the following main parts: 1) an extensively annotated list of all the names proposed for Hyphomycete form-genera, 2) line drawings showing the characteristic morphology of one or more species for each accepted form-genus, and 3) key-lists to assist in identification.

In addition, we have: a) indicated the ecological importance of each form-genus by listing the substrates or hosts from which the included species have been reported, and b) listed for each anamorph genus the genera of Higher Fungi to which species have been reported to be connected. We have provided a cross-index under the names of the genera of Higher Fungi, but no cross-index under host or substrate.

B) A BRIEF HISTORY OF COMPILATIONS OF HYPHOMYCETES

Taxonomic studies draw their data from three different sources: the field, the museum, and the library. The results of these endeavors appear in the literature as: 1) descriptions and records of species,

2) monographs and revisions of genera and families, and 3) compilations and revisions of the accumulated knowledge published in 1) and 2). Publication of a major compilation usually results in a flurry of renewed activity in basic taxonomic studies.

The earliest compilations of Hyphomycetes which are still widely consulted are those of Persoon (1801) and Fries (1821-32). These works are used as the starting point for priority of taxonomic names of fungi (see Part IIIB). Both Persoon and Fries combined all three aspects of taxonomic study and they treated a wide variety of fungi, not just the Hyphomycetes. By far the most influential of the later nineteenth century compilers was P.A. Saccardo. He published descriptions of all the known species of Hyphomycetes in 1886 as Volume IV of his *Sylloge Fungorum*. The current usefulness of these early works is limited primarily by their lack of illustrations, but also by the fact that they lack keys, are written in Latin, and their references to the earlier literature are often cryptic.

Early in the 20th century a number of multi-volume, multi-author floras appeared. Perhaps the most useful treatment of the Hyphomycetes was that of Lindau in *Rabenhorst's Kryptogamen-Flora von Deutschland, Oesterreich and der Schweiz* (Zweite auflage, Erster Band, VIII Abteilung, 1907 and IX Abt., 1910). Not only are the descriptions of the species in a modern language (German), but there are many illustrations, most of which were carefully redrawn from the original descriptions. Lindau also took pains to provide unambiguous references to the earlier literature.

In 1909, F.E. Clements published *The Genera of Fungi: A key to Saccardo's Sylloge Fungorum*. This was followed in 1931 by Clements and Shear's *The Genera of Fungi*. This book treated more than 5,000 generic names, giving for each the type species and a reference to the original publication. There were descriptions of the orders and families of fungi, keys to the genera, and 58 plates of illustrations by E.S. Clements. Clements and Shear's work was the standard text for identifying fungi for forty years, with a fourth printing in 1964.

An interesting compilation of the Fungi Imperfecti was presented as a thesis at Yale University in 1931 by H.B. Bender. This work was not illustrated, but it contained descriptions and keys for the genera, plus a host index. The thesis was never published as a book, presumably because Clements and Shear's book was published the same year.

One of the first modern attempts to provide keys and illustrations for the genera of Hyphomycetes was in Barnett's (1955) *Illustrated Genera of Fungi Imperfecti.* This useful work had many original drawings as well as those traced from previously published works. Revised editions were published in 1960 and 1972.

One of the problems encountered by Barnett and by Clements and Shear was uncertainty as to the identity of the type species of many classical Hyphomycete genera. This problem was largely remedied when S.J. Hughes (1958) examined type and authentic specimens from the herbaria of Persoon, Link, Nees, and other early mycologists (see also Part IIIB).

About the same time, Hughes (1953) focused attention on the exact nature of conidiogenesis as an important character for the classification of conidial fungi. A quarter-century of progress in this area has been summed up in a beautifully illustrated book by Cole and Samson (1979).

Also in recent years, a number of monographs and floras have appeared with excellent original illustrations of Hyphomycetes. Those that deal with large numbers of species include von Arx (1970, 1974), Barron (1968), Ellis (1971, 1976), and Matsushima (1971, 1975). In addition, there have been hundreds of papers and books with new or revised descriptions and illustrations of species and genera. Indeed, the last two decades have witnessed an almost explosive increase in our knowledge of the Hyphomycetes and their teleomorphic connections.

In 1973, Volumes IVA and IVB of *The Fungi* were published under the editorship of Ainsworth, Sparrow, and Sussman. These volumes attempted a taxonomic review of all the currently accepted genera of fungi, with keys to all and references and illustrations for some. Although 31 authors labored on the different groups of fungi, the treatments were, in some respects, less complete than those of Clements and Shear. Based on an estimate from the index, approximately 4,000 genera were accepted by the authors. There are undoubtedly at least an equal number of generic names in the literature which were rejected for one reason or another.

The treatment of the Hyphomycetes, in Volume IVA of *The Fungi* (Kendrick and Carmichael, 1973) was the precursor of this book. For this new version, the 1973 compilation has been very substantially revised and enlarged. First, the literature has been covered for the period 1972 to 1978 inclusive, with partial coverage for 1979. Second, most of the doubtful older names have been carefully rechecked and their status established more firmly. Third, two additional sets of key characters are provided. Fourth, the plates have been reorganized, many of the old drawings revised and many new ones added (853 compared with 566). Fifth, reported substrates and hosts are now listed. Sixth, hundreds of new references have been added. Seventh, an index of perfect state connections has been added.

In addition, advances in text processing technology have enabled us to produce a much more readable format for the list of generic names. Different type sizes, bold face type, and italics are used to separate the different kinds of information in the list. These changes were accomplished by writing computer programs to act on the original data file. This file was begun in 1970 using an electro-mechanical paper-tape punch. It has been revised and added to at frequent intervals since then. Unfortunately, it was impractical to convert the original upper case text to the normal upper and lower case format now available on large computing and text processing systems.

II. Classification

A) KINGDOMS

Traditionally the fungi, slime molds, and bacteria were classified as members of the division Thallophyta of the plant kingdom (cf. Wilson, Loomis, and Steeves, 1971). From about the middle of the 19th century proposals have been made to exclude the fungi from the plant kingdom (e.g. Haeckel, 1866) and to exclude from the fungi the slime molds (e.g. de Bary, 1858) and bacteria (e.g. Cohn, 1875). Despite various proposals for revision, it was not until the development of the electron microscope (in the early 1950s) that the majority of biologists were at last forced to abandon their classical perspective. The textbook *The Microbial World* by Stanier, Doudoroff, and Adelberg (1957) popularized Haeckel's Protista as an independent kingdom and revised it by resurrecting Haeckel's Monera as the Lower Protists (bacteria and procaryotic algae) and renaming the rest of Hogg's Protoctista (1860) as the Higher Protists (fungi, protozoa, and eucaryotic algae). Since then, there has been an increasing tendency to regard the Lower Protists, the Fungi, and the Protozoa each as a separate Kingdom (Whittaker, 1969). We think this is the most satisfactory approach for these groups. The eucaryotic algae have been variously treated— as plants, as an independent kingdom, or partly as plants (the green algae) and partly as one or more independent kingdoms (red algae, brown algae, etc. See Whittaker, 1969). The slime molds have mostly been accepted as protozoa by protozoologists (e.g. Kudo, 1946) ever since de Bary (1858) first named them Mycetozoa. However, mycologists seem loath to give them up and Ainsworth (1973) still divided the Fungi into two divisions—Myxomycota and Eumycota. We prefer to call the plasmodial protists Mycetozoa rather than Myxomycota and to refer them to the Protozoa rather than the Fungi. Von Arx (1974) proposed an interesting arrangement with the Mycetozoa and Mastigomycota (see below) included in phyla of the kingdom Protista and the rest of the fungi in the kingdom Mycota.

B) DIVISIONS OF FUNGI

We recognize four Divisions of the Fungi (corresponding to four of Ainsworth's subdivisions) plus the anamorph division Fungi Imperfecti:

Mastigomycota	Lower Fungi
Zygomycota	Lower Fungi
Ascomycota	Higher Fungi
Basidiomycota	Higher Fungi
Fungi Imperfecti	anamorphs of Higher Fungi.

The Fungi Imperfecti include asexual states (anamorphs) of Ascomycota and Basidiomycota. The name Deuteromycota has been used for this group (Saccardo, 1902; Ainsworth, 1973), but we prefer not to use it, since it implies a division equivalent to the other four.

It should be noted in passing that the anamorphic states of Mastigomycota, Zygomycota, and the Teleomycetes of the Basidiomycota can be recognized as such and referred to the appropriate taxa in these divisions (see appropriate chapters in *The Fungi*, Vol. IVB, 1973). Thus they are not usually included in the anamorph division Fungi Imperfecti, and we have followed this practical arrangement.

C) CLASSES OF FUNGI IMPERFECTI

Within the Fungi Imperfecti we recognize the three anamorph classes given by Ainsworth (1971,1973):

Blastomycetes (yeasts)
Hyphomycetes (molds)
Coelomycetes (pycnidial fungi).

Like the anamorph division Fungi Imperfecti, these three classes are morphodemes, not taxa (cf. Heslop-Harrison, 1962, pp. 33-34). Even though they are given taxonomic names, they are not mutually exclusive groups, and a single species may have anamorphs referable to two or even all three classes, plus a teleomorph (sexual state) in the Ascomycota or Basidiomycota.

The distinctions between the Hyphomycetes and the other two anamorph classes are not clear cut. Form-genera such as *Geotrichum, Hormonema, Aureobasidium,* and *Collecephalus* are usually placed in the Hyphomycetes but could almost equally well go in the Blastomycetes. Pycnidial and acervular anamorphs are usually considered to belong to the Coelomycetes while sporodochial anamorphs are classified in the Hyphomycetes. However, intergrades occur between these three types of conidiomata (see Kendrick and Nag Raj, 1979). In addition, acervuli are defined partly on the basis of host tissue and thus cannot be formed in agar cultures (cf. Barron, 1968, pp. vii and 6). In the List of Form-genera, we have included a few names based on acervuli when we thought this might be useful to the reader attempting to identify molds. However, we did not attempt to cover all of this group. A compilation of Coelomycete nomenclature has been published by Sutton (1977) and illustrations of many species have appeared in the *Icones Generum Coelomycetum* (see Morgan-Jones, Nag Raj, and Kendrick, 1972 for *Icones* I). Form-genera based on sterile mycelia and on molds producing only bulbils or sclerotia are also disposed in the Hyphomycetes.

D) SUB-GROUPS OF HYPHOMYCETES

In the past, the Hyphomycetes have been divided into so-called orders and/or families on the basis of their conidiomata and color (see Saccardo, 1886, p. 2; Clements and Shear, 1931, p. 29). Within the major groups, form-genera were arranged according to the prominence of the conidiophores or mycelium and the shape and septation of the conidia (for a brief summary, see Hughes, 1953, pp. 577-580, or Smith, 1962, pp. 112-113). The Mycelia Sterilia were usually added as an appendix for form-genera based on sclerotia, bulbils, or mycelium, rather than conidia.

More recently, proposals have been made to recognize "families," or sections or series of the Hyphomycetes based on slimy or dry conidia (Wakefield and Bisby, 1941) or the ontogeny of the conidia (Vuillemin, 1910, 1911; Hughes, 1953; and many others since). Barron (1968) summarized a number of these systems and proposed his own system of 10 named series based on conidium ontogeny, plus a miscellaneous group and the Mycelia Sterilia. He provided a key to the series and to the form-genera within each series, but the descriptions and illustrations of the form-genera are given in alphabetic order. Ellis (1971, p. 11) recognized six groups of dematiaceous Hyphomycetes based on conidium ontogeny and arranged his descriptions according to the groups. Von Arx (1974) keyed out seven series of Moniliales and treated the form-genera by series, but he cautioned the reader that the keys were not intended to "give a natural classifiation at all; they rather are prepared for practical use." Kendrick and Carmichael (1973) provided 53 key-lists of form-genera divided according to Saccardoan spore group, arrangement of conidia, and color of conidia. Within each list, the form-genera were arranged according to the type of conidiogenous cells. The key-lists were not mutually exclusive— one form-genus could appear in two or more lists if any of its key characters were variable or doubtful. A separate list was added at the end for sterile forms. Bibliographic treatment of the form-genera was alphabetic, but the plates of illustrations were arranged to bring similar fungi together as much as possible. In this book, the same system is followed, except that the key-lists are also divided by type of conidiogenous cells, and the presence of sporodochia, synnemata, setae, hyphopodia, and a few other characters are noted. Form-genera based on bulbils or sclerotia are separated from those based on mycelium.

A recent attempt to re-define orders and families in the Fungi Imperfecti was made by Luttrell (1979). The discussion following his presentation (pp. 255-264) demonstrates that there is still a wide diversity of opinion on how best to overcome the taxonomic problems created by anamorphic reproduction in the Higher Fungi. This subject has also been reviewed recently by Nicot (1977).

E) FORM-GENERA

The "genera" used for anamorphs of Higher Fungi are form-genera or pseudo-taxa (Hennebert and Weresub, 1979; Weresub and Pirozynski, 1979b, p. 673). Application of form-generic names for anamorphs of Higher Fungi is regulated by Article 3, Note 1, and Article 59 of the Botanical Code (Stafleu et al., 1972). Under these Articles, four special restrictions are placed on the use of form-generic names:

1) Form-generic names apply only to the state or morph represented by their type. They do not apply to the whole organism if it has more than one state or morph.

2) Form-genera are not mutually exclusive. A single species may have anamorphs referable to two or more form-genera, and also a teleomorph classified and named in the Higher Fungi.
3) Form-genera are not assignable to the families of the Higher Fungi.
4) The nomenclature of the Higher Fungi is independent of the nomenclature of anamorph genera (except that homonyms are not allowed). Transfer of specific epithets between form-genera and true genera is not allowed.

Since the names of form-genera have exactly the same form as the names of genera, the distinction is often overlooked or misunderstood. For this reason, we always use the term "form-genus" whenever it is applicable, even though this practice is more cumbersome than using the term "genus" to refer to both true genera and form-genera. Schroeter (1897, as quoted by Weresub and Pirozynski, 1979a, p. 19) went even further and prefixed the letters FG (Formgattung) to the name of every anamorph genus.

Although all anamorph genera are currently treated as form-genera, it is quite possible that some of them may eventually be incorporated into the families of the Higher Fungi beside their teleomorph-producing relatives. They will then still be anamorph genera but will no longer be form-genera. Therefore, we have not accepted Pirozynski and Weresub's recommendation (1979, pp. 675-676) to reserve the term form-genus exclusively for fossil pseudo-taxa.

Both genera and form-genera house collections of species that are separated from each other by more or less arbitrary decisions of taxonomists. The criteria for recognizing separate genera vary from one group of organisms to another. However, taxonomists usually hope and assume that the species included in a taxonomic genus are related by ancestry. Since form-genera are pseudo-taxa, created primarily for convenience in identification, and are not necessarily mutually exclusive, there seems to be no good reason to require or expect that the included species will be related in any way except by similarity of their key characters. However, mycologists still disagree as to whether the species in an anamorph form-genus should be phylogenetically related.

Contemporary mycologists do agree that Hyphomycete form-genera should be uniform as to type of conidiogenesis. Even though the exact method of conidium ontogeny is obscure or variable in some species, the principle is not in dispute. Also, slimy and dry-spored species are rarely lumped together, although a few species are variable in this respect. For almost all other characters, at least some accepted form-genera are variable. For example, both dark and hyaline species are accepted in *Stachybotrys* and *Monilia*; mononematous and synnematous forms are accepted in *Beauveria* and *Penicillium*; and amero- and didymosporous species are accepted in *Cladosporium, Wardomyces*, and *Periconiella*.

There are a number of areas in the Hyphomycetes where the form-genera are not separated from each other on any explicit basis. For example, the differences between *Chrysosporium, Myceliophthora*, and *Sporotrichum* are not satisfactorily elucidated. Agreement is also lacking as to the limits of the genera in the *Cercospora* complex, the *Nodulisporium* complex, the *Dactylaria-Scolecobasidium-Ochroconis* group, and many other areas. As compilers, we have simply had to do the best we could with these inadequately known groups. The uncertainty is indicated in the List of Form-genera by the ?= symbol or by the suggestion to *see* or *compare with* another form-genus.

A few form-genera based on two kinds of conidia have been proposed. By and large, mycologists have rejected these "double form-genera," or have restricted the application of the name to one of the types of conidia. For example, see the entries under *Chalaropsis, Gamsia, Thielaviopsis,* and *Triadelphia*.

F) SPECIES

The species is the basic unit of biological taxonomy (I.C.B.N., Art. 2, Stafleu et al., 1972). Nevertheless, there is no fully satisfactory definition for species. This is probably because there are different mechanisms for maintaining "species" populations in different groups of organisms. In any event, taxonomists assume that there are self-perpetuating genetic lines which propagate themselves more or less independently of other such lines. The individuals embodying one such line are the members of a species. Thus, species have an independent existence, while all higher taxa are simply constructs of the taxonomist.

Since the "genera" in the Hyphomycetes are form-genera, it has been suggested that the species in this group are necessarily form-species or pseudo-species (Hennebert and Weresub, 1979, p. 230; Weresub and Pirozynski, 1979a, p. 25). We cannot accept this viewpoint, unless form-species is arbitrarily defined to include any species named in a form-genus (cf. Ainsworth, 1971). The Code makes no such restriction— see Article 3, Note 1. We would prefer to define a form-species as a subgeneric category for two or more morphologically similar species (see Simmons, 1952). There is no provision in the Code for giving binomials to such series of species. However, it is common practice to designate them by a subgeneric or section name or to use the binomial for the best known species followed by a term such as series, group, or aggregrate.

A number of terms have been proposed for species of Fungi Imperfecti which have apparently dropped the teleomorphic stage from their life-cycle. They have been called "imperfect species" (Hennebert, 1971, p. 207), "autonomous Fungi Imperfecti" (Hennebert and Weresub, 1979, p. 29, footnote), "autonomous conidial fungi" (Weresub and Pirozynski, 1979a, p. 21), and "anamorphic holomorphs" (see discussion in Kendrick, 1979, pp. 38-39). Any of these terms seems appropriate to convey the desired meaning.

III. Nomenclature

A) CROSS-REFERENCE NAMES

Because it is not possible to allot anamorph genera to appropriate places among the families of Higher Fungi, their classification and nomenclature have remained separate (see Weresub and Pirozynski, 1979a). For this reason, a single fungus species may have one binomial for the teleomorph and a second binomial for the conidial anamorph. In fewer instances, two or more binomials have been proposed for different anamorphs of one species. Once the connection between the different morphs has been established, it would seem desirable to use a single binomial to refer to the species (ICBN, Principle IV and Article 2). The need to refer to different morphs of a species and to allot anamorphs to their appropriate form-genera can be met by using cross-reference names instead of additional binomials (see Carmichael, 1962; Kendrick and Carmichael, 1973). Cross-reference names consist of the name of an anamorph genus coupled to a species name (binomen) by the words *state of* or *anamorph of* (or abbreviations therefore, or their equivalent in other languages). For example, *Monilia* state of *Neurospora sitophila* is the cross-reference name for the conidial anamorph previously given the separate species name *Monilia sitophila*. Cross-reference names may also be written with the species name first, for example, *Neurospora sitophila* anam. *Monilia*. The use of cross-reference names for anamorphs with known teleomorphs is now fairly widely accepted. Recently, it has been suggested that cross-reference names should be used also to refer to different anamorphs of a species with no known teleomorph (Sigler and

Carmichael, 1976; Carmichael, 1979). We have used this system occasionally in the list of genera— see, for example, the type species listed under *Echinobotryum* and *Gamsia*. An extensive discussion of the use of cross-reference names and other systems for naming pleomorphic anamorphs is given by Weresub (1979).

B) STARTING POINT DATE

In 1958, S.J. Hughes published his remarkable "*Revisiones Hyphomycetum Aliquot,*" based on an examination of more than 1,000 type and authentic collections preserved in European herbaria. This work marked the beginning of a modern reassessment of classical Hyphomycete genera, which is now approaching completion. During his study, Hughes became convinced that the starting point for priority of Hyphomycete names, as defined in Article 13f of the International Code of Botanical Nomenclature was confusing and difficult to use. He therefore chose to use Persoon's *Synopsis Methodica Fungorum* as the starting point work for Hyphomycetes, with the publication date being taken as 31:XII:1801. In 1959 Hughes formally proposed that Article 13 be altered to change the starting point for Hyphomycete nomenclature. Despite his cogent arguments in favor of the change (Taxon 8:96-103, 1959), the proposal was defeated at the next International Botanical Congress.

When we (Kendrick and Carmichael, 1973) came to compile the genera of Hyphomycetes for Chapter 10 in *The Fungi*, Vol. IVA, we also found that using Fries's works as a starting point made it difficult or impossible to be sure of the proper citation and application of some names. Therefore we followed Hughes's proposal and used Persoon's *Synopsis* as the starting point.

Since then, R.H. Petersen has reviewed the literature on fungi published during the year 1821, as well as the pre-1821 source literature. In a series of papers (see *Mycotaxon* 6:78,1977) he provided lists of the species' names used in 1821 publications and their pre-1821 sources. He also discussed some of the nomenclatural problems connected with the choice and use of various starting points. Petersen's lists are of limited help with the generic names for Hyphomycetes, because generic names were often "validated" by Fries in 1821 (in the introduction to the *Systema*) without mention of any species. Indeed, this is one of the problems associated with using Fries's work as a starting point for the Hyphomycetes (see Hughes 1959, pp. 98-99, points 3A and 4A).

Pre-1821 Name List

Pre-1821 Names	Status, or 'Validation' Using 1821 Start
Achitonium Kunze 1819	?, never illustrated
Acladium Link 1809	Persoon 1822 Mycol. Eur. 1:28
Acremonium Link 1809	Fries 1821 Syst. Mycol. 1:XLIV
Acrosporium Nees 1817	=*nom. rej.*
Acrotamnium Nees 1817	=*nom. dub.*
Actinocladium Ehrenberg 1819	Persoon 1822 Mycol. Eur. 1:31
Aegerita Persoon 1801	Fries 1821 Syst. Mycol. 1:XLI
Aleurisma Link 1809	=*Trichoderma* Persoon 1801
Alternaria Nees 1817	Wallroth 1833 Fl. Crypt. German. 2:148 *nom. cons.*
Alysidium Kunze 1817	Steudel 1824 Nomencl. Bot., p 54
Alytosporium (Link) Ehrenberg 1818	=*nom. dub.*
Amphitrichum Nees & Nees 1818	=*nom. dub.*
Antennaria Link 1809	=*nom. illegit.*
Anthina Fries 1819	=sterile mycelium of ?*Peniophora* (B)
Apiosporium Kunze 1817	=*Sclerotium* Tode *ex* Persoon 1801
Arthrinium Kunze 1817	Fries 1821 Syst. Mycol. 1:XLIV
Aspergillus Micheli *ex* Link 1809	Fries 1821 Syst. Mycol. 1:XLV
Asterophora Ditmar 1809	Fries 1821 Syst. Mycol. 1:LI
Asterosporium Kunze 1819	=Coelomycetes
Atractium Link 1809	Fries 1821 Syst. Mycol. 1:XLI
Bactridium Kunze 1817	Fries 1821 Syst. Mycol. 1:XL
Botrytis Micheli *ex* Persoon 1801	Fries 1821 Syst. Mycol. 1:XLV
Byssocladium Link 1815	?=sterile mycelium

Pre-1821 Name List

Pre-1821 Names	Status, or 'Validation' Using 1821 Start
Campsotrichum Ehrenberg 1819	?, never illustrated
Cephalotrichum Link 1809	Fries 1821 Syst. Mycol. 1:XLVIII
Chloridium Link 1809	Fries 1821 Syst. Mycol. 1:XLVI
Circinotrichum Nees 1817	Persoon 1822 Mycolog. Europ. 1:19
Cladobotryum Nees 1817	Brongniart 1824 In Cuvier, Dictionaire 33:535
Cladosporium Link 1815	Fries 1821 Syst. Mycol. 1:XLVI
Coniosporium Link 1809	Fries 1821 Syst. Mycol. 1:XL
Conoplea Persoon 1801	Fries 1832 Syst. Mycol. 3:491
Coremium Link 1809	=*Penicillium* Link 1809
Coryne Nees 1817	S. F. Gray 1821 Nat. Arr. Brit. Pl. 1:595
Coryneum Nees 1817	Link 1825 Linn. Spec. Plant. IV 6(2):124
Dactylium Nees 1817	=*nom. dub.* non *Dactylium* Saccardo 1886 *nom. cons.*
Dematium Persoon 1801	=*nom. confusum*
Dermosporium Link 1815	?=Aegerita Persoon 1801
Epicoccum Link 1815	Schlechtendal 1824 Synop. Pl. Crypt. P 136
Epochnium Link 1809	=*Monilia* Persoon 1801
Exosporium Link 1809	Fries 1821 Syst. Mycol. 1:XL
Fibrillaria Sowerby 1803	=sterile mycelium or immature fruit bodies of (B)
Fusarium Link 1809	Fries 1821 Syst. Mycol. 1:XLI
Fusidium Link 1809	Fries 1821 Syst. Mycol. 1:XL
Fusisporium Link 1809	?, never illustrated
Geotrichum Link 1809	Persoon 1822 Mycolog. Europ. 1:26
Gemmularia Rafinesque 1819	Steudel 1824 Nomencl. Bot. Pl. Crypt. 183

Pre-1821 Name List

Pre-1821 Names	Status, or 'Validation' Using 1821 Start
Gonytrichum Nees & Nees 1818	Fries 1832 Syst. Mycol. 3:346-348 (as a synonym of *Myxotrichum*)
Haplaria Link 1809	=*Botrytis* Persoon 1801
Helicomyces Link 1809	Wallroth 1833 Fl. Crypt. German. 2:147
Helicosporium Nees 1817	Fries 1821 Syst. Mycol. 1:XLVI
Helicotrichum Nees & Nees 1818	=*Helicosporium* Nees 1817
Helminthosporium Link 1809	Fries 1832 Syst. Mycol. 3:354-362
Himantia Persoon 1801	=sterile mycelium of *Corticium* (B), *etc.*
Hormiscium Kunze 1817	=*Torula* (Persoon) Link 1809
Hyphasma Rebentisch 1804	=*nom. dub.*
Hypha Persoon 1818	=*nom. dub.*
Illosporium Martius 1817	?=*nom. dub.*
Isaria Persoon 1801	=*nom. dub.*
Melanconium Link 1809	Fries 1821 Syst. Mycol. 1:XL
Monilia Persoon 1801	see Donk 1963 Taxon 12:266-271
Mycobanche Persoon 1818	=*Sepedonium* Link 1809
Mycogone Link 1809	Persoon 1822 Mycolog. Europ. 1:26
Myrothecium Tode *ex* Link 1809	Fries 1825 Syst. Orb. Veg. 1:152
Oidium Link 1809	=*Sporotrichum* Link 1809, non *Oidium* Link 1824
Ozonium Link 1809	=sterile mycelium of *Coprinus* (B)
Penicillium Link 1809	S. F. Gray 1821 Nat. Arr. Brit. Pl. 1:554
Periconia Tode *ex* Persoon 1801	Fries 1832 Syst. Mycol. 3:307-308
Polyactis Link 1809	=*Botrytis* Persoon 1801
Polythrincium Kunze & Schmidt 1817	Fries 1832 Syst. Mycol. 3:367-368

Pre-1821 Name List

Pre-1821 Names	Status, or 'Validation' Using 1821 Start
Pyrenium Tode *ex* Persoon 1801	=*nom. dub.*
Racodium Persoon 1801	=sterile mycelium of lichens
Rhizoctonia De Candolle 1815	Fries 1821 Syst. Mycol. 1:LV
Rhizomorpha Roth *ex* Persoon 1801	=sterile, for rhizomorphs of *Armillaria* (B)
Sarcopodium Ehrenberg 1818	Schlectendal 1824 Synop. Pl. Crypt. 2:101
Sclerococcum Fries 1819	Fries 1825 Syst. Orb. Veg. P 172-173
Sclerotium Tode *ex* Persoon 1801	S. F. Gray 1821 Nat. Arr. Brit. Pl. 1:590-592
Scolicotrichum Kunze 1817	=*nom. confusum*
Sepedonium Link 1809	Fries 1832 Syst. Mycol. 3:438
Spermodermia Tode 1790	=immature *Hypoxylon*
Spondylocladium Martius 1817	=*Stachylidium* Link 1809
Sporidesmium Link 1809	Fries 1821 Syst. Mycol. 1:XL
Sporotrichum Link 1809	Fries 1821 Syst. Mycol. 1:XLIV
Stachylidium Link 1809	S. F. Gray 1821 Nat. Arr. Brit. Pl. 1:553
Stilbum Persoon 1801	S. F. Gray 1821 Nat. Arr. Brit. Pl. 1:563
Thanatophytum Nees 1817	=*Rhizoctonia* De Candolle 1815
Torula (Persoon) Link 1809	Fries 1821 Syst. Mycol. 1:XLVI
Trichoderma Persoon 1801	Fries 1821 Syst. Mycol. 1:XLV & 3:214,1829
Trichothecium Link 1809	S. F. Gray 1821 Nat. Arr. Brit. Pl. 1:550
Tubercularia Tode *ex* Persoon 1801	Fries 1821 Syst. Mycol. 1:XLI *nom. cons.*
Verticillium Nees 1817	Link 1824 In Linnaeus, Spec. Plant., IV 6(1):76
Virgaria Nees 1816	Saccardo 1886 Syll. Fung. 4:280
Volutella Tode 1790	Fries 1832 Syst. Mycol. 3:467 *nom. cons.*
Xylostroma Tode 1790	=sheets of mycelium

For this book, we have again used 31:XII:1801 as the starting point. Thus, some of our citations are not in line with the current code. However, as a contribution to solving the starting point problem, we have extracted from our main list and set forth here all the generic names published before 1821. There are 93 of them, 39 of which are not in current use. These 39 are names which are: synonyms (13), rejected, confused, illegitimate, or dubious (11), apply to sterile mycelium or immature fruit bodies (8), of uncertain application, i.e., type or authentic material is not known (6), or apply to Coelomycetes (1). The bases for these dispositions will be found in the main list.

It is interesting to note that only 21 of the names in our pre-1821 list were in use before 1801. These include the 15 names listed as Persoon 1801 (9 proposed earlier by Persoon, 5 by Tode, and 1 by Micheli), plus 3 from Link 1809 (1 by Micheli, 1 by Tode, and 1 by Persoon), and 3 from Tode in 1790 which were not dealt with by Persoon or Link. Fourteen of those very early names are in current use, 3 are nomina dubia or confusa and 4 refer to sterile mycelium. Of the 72 names first used between 1802 and 1820, 33 were proposed by Link (22 current), 15 by one or both Nees (8 current), 9 by Kunze (5 current), 4 by Ehrenberg (2 current), and 11 by eight other authors (4 current). For the 54 generic names in current use, we have listed the earliest post-1820 validation we could find. Some of these are earlier validations than those usually given in the literature.

The problems of determining the type species for these early names, based on the 1821 starting point, are such that we did not think it worthwhile to try to sort them out and list them. For an interesting example, the case of *Monilia fructigena*, see Korf and Kohn (1979). In the List of Form-genera, the type species given are based on the 1801 starting point.

C) AUTHORITIES FOR SYNONYMS

The establishment of facultative synonyms, especially above the species level, is a matter of taxonomic judgement or opinion. We have indicated the source of these opinions in a variety of ways:

1) Where the synonym is indicated by an = followed by the notation *fide* plus an author and date, it means that the author established or accepted the synonymy in a publication in that year. It does not necessarily mean that he was the first person to establish the synonymy, it merely indicates the source from which we took the opinion. The place of publication (if indicated only by the name and date) will be found in the list of references given for one or the other of

the two form-genera or in the References. Where the word *see* is used instead of fide, it means that we have based the synonymy on information published in the reference indicated, but the author of the publication did not explicitly propose the synonymy.
2) Where the synonym is indicated by an = followed by the notation fide plus an author with no date, it indicates a previously unpublished opinion.
3) Where the synonymy is indicated by ?=, not followed by a reference, it indicates possible synonymy, suggested by the authors of this compilation, based on published descriptions. These questionable synonyms are included to indicate the need for their confirmation or rejection based on type material or for clarification of the distinction between the two genera.

D) AUTHORITIES FOR LECTOTYPES

The ICBN provides that the earliest designation of a lectotype has priority and must be followed "... unless the choice was based on a misinterpretation of the original description" (Article 8). In citing the type species of anamorph genera we have indicated when lectotypification is involved and have given a reference where possible. However, it was sometimes difficult to determine where the lectotype was originally proposed. Thus, these references may not always be to the earliest designation.

E) AUTHORITIES FOR CONNECTIONS

The evidence for declaring that an anamorph is part of the life cycle of an ascal or basidial fungus varies in reliability from one case to another (see Kendrick, Di Cosmo et al., in Kendrick, 1979, pp. 283-284). We have listed these connections as ASCO STATE = or BASIDIO STATE = followed by one or more generic names and references to the publications from which we took the information. As in the case of synonyms, the cited reference is not necessarily the first to make the connection. Where the connection is listed as ?=, there is doubt about the validity of the connection. Where the connection is listed as =?, there is doubt about the name applied to the teleomorph. Frequently, an unnamed anamorph is described as part of a species of the Higher Fungi. We have sometimes tentatively allotted these anamorphs to a form-genus and have given the reference there. Where the anamorph did not exactly fit the form-genus, we have added the suffix *-like* to the form-genus name in the cross-index. In other cases, we have simply called the anamorph *phialoconidia* or *arthroconidia*.

There are entries under these terms in the List of Form-genera.

Since we were compiling information about genera, we did not list the species involved in the connections. It should be realized, however, that a connection listed by generic names means only that one species placed in the anamorph genus has been connected to one species in the teleomorph genus. Thus, when more than one teleomorph genus is mentioned, each one is connected to a different species in the anamorph genus. The names of the species involved can be found in the references listed. Many connections between species are reported in Volumes 1 and 2 of *The Whole Fungus* (Kendrick, 1979). In the Index of Perfect State Connections, we provide an alphabetic cross-index to the teleomorph names cited as connections in the List of Form-genera.

Since anamorph form-genera are monomorphic, we have also listed connections between different anamorphs of pleomorphic species. These are indicated as OTHER CONIDIAL STATES = or OTHER STATES =. The **single letter in brackets** following the generic names in such cases indicates the group to which the genus belongs (see Abbreviations section for definitions).

IV. Notes on the List of Form-genera

A) SCOPE AND LIMITATIONS OF THE LIST OF FORM-GENERA

This compilation is intended to include all of the generic names proposed for Hyphomycetes since 1801 (see Part IIIB for a discussion of the starting date and Part IIC for our circumscription of the Hyphomycetes). We believe our coverage is complete to the end of 1978. We have also cited as many 1979 names as we could find before the book went to press, but we were unable to include illustrations for them. Earlier compilers (e.g., Lindau 1907; Clements and Shear, 1931) complained of the difficulty of finding descriptions of new fungi scattered in a variety of periodicals and books. We have had the advantage of *The Index of Fungi* published by the Commonwealth Mycological Institute and *Abstracts of Mycology* published by BioSciences Information Service. These periodicals survey a large portion of the world's literature and give references to descriptions of new taxa. This is especially fortunate because there are now so many journals that even large university libraries can only afford a limited selection of them. While our coverage of new generic names is, we hope, reasonably complete, we will have missed some publications with important revisions, keys, illustrations, or other material of interest to those trying to identify or classify Hyphomycetes. We would appreciate learning of any errors or omissions in our list of names or in the References cited.

This compilation does not include many of the names proposed for fossil Hyphomycetes. The recent proposals of Pirozynski and Weresub (1979) suggest that a substantial revision in the system for naming fossil fungi is likely to occur.

The List of Form-genera now includes 2032 names (compared with 1541 in 1973). Of these, we have accepted 885 as applying to different, recognizable anamorphs. New combinations have been proposed in *Beauveria* and *Trichocladium*.

B) FORMAT OF ENTRIES

1) Genus and Type Species

The first part of each entry consists of the form-genus name, its author(s), date of publication, and place of publication, then a diagonal slash followed by the type species and the author(s) and date of publication of the type species. Where the author and date for the type species are identical to those for the genus, they have not been repeated. When the name was a new combination, the author of the epithet is given in brackets but is followed by the author of the combination only when different from the author of the genus. When the first part of the entry required more than one line, the continuation line is indented four spaces. Names of accepted form-genera are given in large, bold-face italic type.

2) Status or Key-words and Figure Number

For **rejected names**, the status is given either as ?, =, or ?=. A question mark means that the name is validly published, but the characters of the genus or the identity of the type species are not definitely known. For a few recent names, the status is given as ? because we have not been able to obtain a copy of the original description. In these cases, the notation *(not seen)* appears after the reference. The status designation following an equals sign is usually self-explanatory, as either a synonym, nomen dubium, etc. If no reference is given to support the synonymy, or if only a name and date are given, the full reference will be found under the name of the accepted genus or in the References. Occasionally, the equals sign is followed by a group designation such as =BLASTOMYCETES or

=COELOMYCETES. In these cases, the name is "rejected" only for use as a form-genus of the Hyphomycetes.

The ?= designation is used for probable, facultative synonyms established from published descriptions, rather than from the examination of type specimens. Where no authority is given for these synonyms, they are suggested by the compilers.

For a few long-disused names, the usual format is abbreviated so that the entire entry consists of a simple statement of the form *genus-name = disposition*, usually followed by a reference to the authority for the disposition.

For **accepted names**, the characters of the form-genus are indicated by a series of four or more key-words (in bold-face type) and a figure number. The key-words form an abbreviated description of the conidia, conidiogenous cells (and conidiomata, if present) of the type and other species currently included in the form-genus. Thus, some of the included species may lack some of the characters indicated by the key-words. Furthermore, some of the characters indicated by key-words (e.g., acervuli) are dependent on growth conditions for their expression. Key-words joined by OR or AND are alternative states for the same character.

Normally, the **first key-word** indicates the Saccardoan spore group for the conidia. The alternatives are AMERO, DIDYMO, PHRAGMO, DICTYO, SCOLECO, HELICO, and STAURO. These spore types are defined and illustrated in the *Dictionary of the Fungi* (Ainsworth, 1961 or 1971, Fig. 1) and in more detail in *The Whole Fungus* (see Kendrick and Nag Raj, 1979, pp. 43-48). An additional conidial type, the cheirospore, is indicated by the key-word **CHEIRO**. This key-word is used for multicellular conidia shaped rather like a hand with the fingers held together (cf. Ellis, 1971, p. 26). When the form-genus is not based on a conidial anamorph, the **first key-word** will be BULBILS, SCLEROTIA, or STERILE. The key-word **BULBILS** is used in a broad sense for multicellular propagative structures. The key-word **SCLEROTIA** is used for multi-cellular resistant structures whose primary role seems to be endurance rather than multiplication or dispersal. The key-word **STERILE** indicates a form-genus based on a mycelial anamorph without conidia, bulbils, or sclerotia. In many cases, we have listed form-genera based only on hyphae as questionable, since they do not seem to have any characters differentiating them from other such form-genera.

The **second key-word** indicates the arrangement of the conidia as they are borne on the conidiogenous cells. Only five general types of conidium arrangement were specified because more specific information is lacking for some genera and because even these types are somewhat intergrading. The five types recognized are as follows:

ARTHRO (arthrocatenate)— conidia are formed in chains by the simultaneous or random fragmentation of a hypha, which may be branched.

BLASTO (blastocatenate)— conidia formed in chains with the youngest spore at the tip or distal end of the chain. These chains may branch.

BASO (basocatenate)— conidia formed in chains with the youngest conidium at the basal or proximal end of the chain. Included here are phialoconidia in chains, meristem arthroconidia, retrogressive holoblastic conidia, and chains produced from annellated conidiogenous cells. These chains do not branch.

GLOIO (gloiosporae)— conidia produced in slimy heads or masses. Most commonly these are phialoconidia.

CETERI (the rest)— conidia not in chains or slime. This includes solitary conidia and conidia produced from inflated or sympodially proliferating conidiogenous cells.

The **third key-word** indicates the cell wall color of the conidia only— **HYALO** for light or bright colored and **PHAEO** for dark colored. The color of the hyphae and conidiophores is not taken into account. Where the color is *pale olivaceous* or *hyaline, becoming dark at maturity,* we have listed both HYALO and PHAEO as key-words. Similarly, where a genus was considered to be variable or borderline in any of the key characters, then more than one key-word is given.

The **fourth key-word** indicates the type of conidiogenus cell (or type of conidiogenous axis when several conidiogenous cells form a morphological entity). Unfortunately, for some genera the conidiogenous cells have not been carefully examined and their nature remains in doubt. We have indicated these cases with a question mark. Other genera produce a plurality of conidia by a combination of methods or by methods which are intermediate or variable with respect to our classification. Nevertheless, the nature of the conidiogenous cells was the most stable and useful character we could find for further subdivision. The types of conidiogenous cells recognized are as follows:

NONSP (nonspecialized)— the conidiogenous cells resemble vegetative hyphae. They are not morphologically specialized for the simultaneous or successive production of many conidia, e.g.,

Chrysosporium, Helicoma, Oncopodium, Mammaria, Wardomyces, Pithomyces.

PHIAL (phialiform)— the conidiogenous cell does not elongate or enlarge during the production of a succession of conidia from a fixed growing point, e.g., *Phialophora, Chalara, Penicillium, Fusarium.*

ANNEL (annelliform)— the conidiogenous axis proliferates percurrently during the production of a succession of conidia so that it becomes annellated or ringed, e.g., *Annellophora, Scopulariopsis, Gyrothrix, Ceratocladium, Spilocaea.*

RACHI (rachiform)— the elongating conidiogenous axis is relatively narrow compared to the width of the conidial attachment and the axis is pushed first to one side and then the other as it grows past a succession of terminal conidia. This sympodial development produces a geniculate or zig-zag rachis, e.g., *Tritirachium, Beauveria, Pyricularia, Drechslera, Spiropes guareicola.* In a few fungi (e.g., *Polythrincium, Costantinella*) the axis usually grows to the same side of each terminal conidium and is not really rachiform.

RADUL (raduliform)— the elongating conidiogenous axis is relatively wide compared to the width of the conidial attachment and the axis tends to become clavate or somewhat inflated rather than zig-zag. In some genera the conidia are initially lateral, as well as (or instead of) terminal, e.g., *Rhinocladiella, Virgaria, Scolecobasidium, Spiropes clavatus, Haplariopsis, Arthrobotrys, Dactylaria, Spadicoides.* The prefix radul is derived from Mason (1933, p. 9).

AMPUL (ampulliform)— the condidiogenous cell swells to form an ampulliform, clavate, or lobed fertile area which then bears many conidia simultaneously over its surface, e.g., *Oedecephalum, Gonatobotrys, Botrytis, Ostracoderma, Botryosporium.*

MISC (miscellaneous)— a plurality of conidia is achieved in some manner other than the ones above, e.g., *Arthrinium, Basipetospora, Cladobotryum, Trichothecium, Piricauda,* or else the conidiogenous cells are morphologically specialized but do not produce a succession of conidia, e.g., *Nigrospora, Spegazzinia, Fusicladiella, Zygosporium.*

One or more **additional key-words** are given for some form-genera. These key-words may occur in any order and specify a variety of features pertaining to the hyphae, conidia, or conidiomata. Definitions for the additional key-words are as follows:

ACERVULI— indicates that one or more species in the form-genus produces acervuli when growing on a host plant. In culture, the same organism may produce its conidia diffusely on the hyphae, on sporodochium-like structures, or it may fail to sporulate. See Part

IIC for references and a discussion of acervuli.

ALEURI (aleuric dehiscence of conidia)— the conidia are released by fracture or dissolution of the supporting cell or hypha (see Carmichael, 1971). Cole and Samson (1979, p. 25) called this kind of dehiscence "rhexolytic."

APPEND (appendages on conidia)— indicates appendages of some kind on the conidia. See Kendrick and Nag Raj (1979) for a discussion about the occasional difficulty in deciding whether a protruberance from the conidium is an arm of the conidium or an appendage (they use the term "setulae" for appendages).

AQUATIC— this indicates that the fungus has conidia adapted for aquatic dispersal, usually scoleco- or stauro-conidia. Although this group of fungi was originally discovered in surface scum of ponds or streams, it is now realized that they are essentially litter fungi with conidia adapted for aquatic dispersal (Bandoni, 1972). The group was recently monographed by Ingold (1975,1979).

CLAMPS— indicates that clamp connections may be present on the hyphae and sometimes also on the conidia. This feature is taken as evidence that the anamorph is related to the Basidiomycota.

PODIA (hyphopodia or stromatopodia)— indicates that the fungus produces short lateral branches from the hyphae that serve as appressoria which adhere firmly to the host plant. From each appressorium, a narrow filament penetrates the plant cuticle (see Ellis, 1971, pp. 8 and 24).

PYCNIDIA— see comments under ACERVULI.

SETAE— indicates that some species in the form-genus produce specialized, stiff, often pointed hyphae in association with the conidiophores or conidiomata. "Setae" which grow on the conidia are referred to by the key-word APPEND. The setae formed by fungi with slimy conidia often form a "basket" which supports the slime drop at the top of the conidiophore.

SPORODOCHIA— indicates that some species in the form-genus form sporodochia, at least under some conditions of growth. See the comments under ACERVULI.

STROMA— indicates that loose or fasciculate conidiophores are formed from or on a compact mass of hyphae, usually when growing on a higher plant host. When the conidiophores are short and closely crowded, the entire structure is called a sporodochium.

SYNNEMATA— indicates that some species in the form-genus, when growing under appropriate conditions, have their conidiophores aggregated into a compound stalk. The stalk may be completely covered with conidiogenous cells or the synnema may be differentiated into a sterile basal part and a fertile distal part or head.

3) Substrates

In many form-genera the species are related and have a common habitat. However, the species in a form-genus are not necessarily related, and they may be quite diverse in their metabolism and ecological niches. We have simply compiled the substrates reported in the literature for any of the species included in each form-genus. Our task of compiling substrates was greatly facilitated by the substrate records published by Martin Ellis in his two books on dematiaceous Hyphomycetes (Ellis, 1971,1976). The list of substrates follows the figure number and is introduced by the word ON. The general types of substrates are given first, followed by specific examples or comments in brackets. The most common substrate for Hyphomycetes is material from green plants, living or dead. We have tried to distinguish between the saprophytic and parasitic modes of growth by referring to all shed plant parts as LITTER. When the substrate is given as leaves, stems, or other plant parts, the implication is intended that they are the leaves or stems of living plants. Since the original reports are sometimes vague as to whether the fungus grows on living, senescent, or shed plant parts, we have been forced to make the best possible guess in such cases. This is particularly the case for fungi recorded on wood and bark. The original reports frequently do not make clear whether the tree was living or dead. Where the substrate is given as *on trees* or *on twigs*, living trees are implied.

Fungi recorded as *on soil* have been isolated from soils, but their actual substrate may be some specific soil component such as a particular type of litter, the remains of animals or insects, or some other specific substrate.

Fungi which are described by the key-word AQUATIC do not have a substrate listed. All of these fungi are presumed to grow on shed leaves and other litter in wet places (see discussion above under the key-word AQUATIC).

Fungi which are described as *on air* have only been isolated as contaminants or airborne spores and their substrate is not known.

4) Comments and Connections

Immediately following the substrate list, there may be comments about the status of the form-genus and suggestions to SEE or COMPARE WITH similar form-genera. Reported connections between a species in the form-genus and other form-genera or genera of Higher

Fungi are given following an equals sign, e.g. ASCO STATE=, BASIDIO STATE=, or OTHER CONIDIAL STATE=. References are given for each reported connection to the Higher Fungi. A discussion of the evidence for accepting connections is given in Part IIIE. A cross-index under the names of the genera of Higher Fungi is provided in the Index of Perfect State Connections.

5) References

References to the taxonomic literature on the form-genus are preceded by diagonal slashes. We have tried to list all monographs or other major contributions and any papers with exceptionally good illustrations or illustrations of little-known species. For small or recently described form-genera, we have given references for new species and new combinations. For genera with a voluminous literature, we have selected only those papers that we thought would be most useful to the would-be identifier of an unknown fungus.

The title of journal papers is not given, but the reason for including the reference is indicated by abbreviated descriptive terms in brackets. Some of the commonest terms are ILLUS for illustrations of microscopic characters, MONOGR for monographs, REVIS for papers revising either the generic concept or the included species, KEY for keys to some or all of the species, NEW SP for new species (singular), NEW SPP for new species (plural), and REFS to indicate that the paper contains references to other literature on the form-genus.

In many cases, we have abbreviated the most-often-cited references in a non-standard fashion. Where only the author's name and a date are given, the full reference may be found in the References. Where a book title has been abbreviated, the full title can be found via the list of Abbreviations and the References.

6) Synonyms

Accepted and probable synonyms are listed in alphabetic order at the end of the entry for each genus. Explanation of the = and ?= signs and the authorities cited are given in Part IIIC. We did not have the congruency sign available to indicate obligate synonyms, but, since we have listed the type species for all form-genera, this is not a major failing.

7) Bullet

A large dot or bullet terminates the entry for each form-genus. When the last line on a page is not followed by a bullet, the entry for that form-genus is continued on the next page.

V. Notes on the Illustrations

A) SOURCES

Most of the illustrations in this book have been redrawn from photographs or drawings originally published by others. In some cases our drawings are composites, based on several original illustrations. In many cases we have modified the original illustrations, either slightly or substantially, in order to achieve as much uniformity in style as possible and in order to reflect our interpretation of the nature of the conidiogenous cells. We have not indicated the original source(s) of the drawings directly in the legends to the plates, but the references which contain illustrations are noted in the List of Form-genera. In addition to the illustrations provided by the authors of new species, and those published in scattered journal papers, we have drawn especially from the illustrations in recent books by von Arx (1970, 1971), Barron (1968), Ellis (1971, 1976), and Matsushima (1971, 1975) and from the illustrations published by various authors in *Mycological Papers of the Commonwealth Mycological Institute*; *Studies in Mycology*, published by the Centraalbureau voor Schimmelcultures; and *Fungi Canadenses,* published by the National Mycological Herbarium, Biosystematics Research Institute, Ottawa.

All of the plates in this book are newly assembled, but we have retained unaltered about half of the 566 drawings used in the 1973 version (Kendrick and Carmichael, 1973). The other half have been improved, replaced, or eliminated. Additional drawings illustrating new form-genera bring the total illustrations to 853.

The illustrations are essentially complete for form-genera described before the end of 1978. Form-genera described during 1979 have been included in the List, but we were not able to include illustrations for the most recent ones.

B) DRAWING CONVENTIONS

Two drawing conventions have been used more consistently in the new plates (although there are still exceptions). First, stippling is used to indicate cell-wall color. The darker the stippling, the darker the cell-walls. Occasionally, very light stippling is used to indicate the contours of hyaline fungi. Second, a dotted outline, when used, indicates a slimy mass of conidia (key-word GLOIO).

The figures are drawn at various magnifications, but no scales are provided. We apologize for this defect. When conidiomata are illustrated, they are drawn at a much smaller scale than the conidia and conidiogenus cells.

Usually only the type species is illustrated for each form-genus. Occasionally a different species has been used when no suitable illustration of the type species was available. In a few instances, more than one species is illustrated, sometimes on separate plates.

C) ARRANGEMENT

The illustrations were grouped into plates with the objective of making it as easy as possible to compare an unknown fungus with the pictures. As far as possible, each plate contains form-genera that resemble each other.

The illustrations are arranged first by Saccardoan spore group. The distribution of the groups is as follows:

Spore Group	Plates	No.	(%)
Amerospores	1- 55	55	(43)
Didymospores	56- 68	13	(10)
Phragmospores	69- 93	25	(19)
Dictyospores	94-101	8	(6)
(Cheirospores, plate 98)			
Helicospores	102-106	5	(4)
Scolecospores	107-111	5	(4)
Staurospores	112-127	16	(12)
(Cheirospores, plate 116)			
Bulbils and misc.	128-129	2	(2)

The second criterion for arrangement was usually the kind of conidiogenous cell. For example, the first 17 plates depict form-genera with phialides. However, in some plates the second criterion for grouping illustrations was the color, shape, or ornamentation of the conidia. Plate 55 consists of form-genera that were grouped together because they all grow on insects.

Further criteria for grouping illustrations vary as dictated by the numbers and kinds of fungi in each group. Since the groupings in the plates are not the same as those in the key-lists, the plates form an additional pictorial key to assist with identification.

The rationale for the grouping in each plate is indicated by a heading of key-words and other descriptive terms.

VI. Notes on the Key-lists

The first four key-words describing accepted form-genera were used to sort the names into separate sub-lists. Within each sub-list, any additional key-words are enclosed in brackets and appended to the generic name. The plate and figure numbers are given at the right. Since the sorting was done in the same order that the key-words are given, the sub-lists are separated first by Saccardoan spore type, second by arrangement of the conidia, third by color, and fourth by the type of conidiogenous cell. Form-genera which are variable in any of the first four key-word characters appear in the key-lists in more than one place.

When an unknown fungus has been keyed out to a sub-list of form-genera with the same key-words, it is then necessary to compare the unknown with the illustrations of all the fungi in the sub-list. Once the most likely form-genus is found, confirmation of the identification may be sought in the references given for that form-genus. If there is no suitable form-genus listed, the fungus may be undescribed.

It would have been possible to provide additional key-lists, sorted first according to conidium arrangement, color, or conidiogenous cells. However, we did not feel that such permutations would provide enough additional assistance to make it worth the extra pages.

LIST OF FORM-GENERA

ABGLIOPHRAGMA ROY & GUJARATI 1966. TBMS 49:363 /A. SETOSUM
=WIESNERIOMYCES KOORDERS 1907 FIDE KENDRICK 1973. •

ACALYPTOSPORA DESMAZIERES 1848 (ANN. SCI. NAT. SER. 3, 16:243) =GLANDULAR LEAF HAIRS FIDE SACCARDO 1901 (SYLL. FUNG. 15:1) •

ACAROCYBE SYDOW 1937. ANN. MYCOL. BERLIN 35:285 /A. HANSFORDII
PHRAGMO,CETERI,PHAEO,MISC,SYNNEMATA,APPEND, 86C, ON LEAVES.
/ELLIS, CMI PAPERS 76:2-5,1960 (ILLUS,KEY) /ELLIS, DH P 384-385,1971 (ILLUS) •

ACAROCYBELLA ELLIS 1960. CMI PAPERS 76:5 (ILLUS) /A. JASMINICOLA (HANSFORD)
PHRAGMO OR SCOLECO,CETERI,PHAEO,NONSP,APPEND, 91F, ON LEAVES.
/ELLIS, DH P 370-372,1971 (ILLUS) •

ACAULIUM SOPP 1912. SKR. VIDENSK.-SELSK. CHRISTIANIA 11:42-53 /A. ALBONIGRESCENS
=SCOPULARIOPSIS BAINIER 1907 FIDE ELLIS 1971, OR =MICROASCUS SEE EMMONS & DODGE 1931 (MYCOLOGIA 23:329). •

ACAULOPAGE DRECHSLER 1935. MYCOLOGIA 27:185 (176-205,ILLUS) /FIVE ORIGINAL SPECIES, ALL NEW
STAURO,CETERI,HYALO,NONSP,APPEND, 112B, ON AMOEBAE. ?=ZYGOMYCETES.
/TUBAKI, BOT. MAG. TOKYO 71:131-132,1958 (ILLUS) /INGOLD, GUIDE P 88-89,1975 (ILLUS) /DUDDINGTON, THE FUNGI IV B, P 231-234,1973 (ILLUS,REFS) •

ACHITONIUM KUNZE 1819. FLORA 2:50 (NO ILLUS) /A. ACICOLA
?
/LINDAU, RABENH. KRYPT. FL. 2 AUFL. 1 BD. 9:456-457,1910 (REFS)
=*CHROOSTROMA* CORDA 1837 FIDE LINDAU 1910
=*LEUCOSPORIUM* CORDA 1837 FIDE LINDAU 1910
=*PACTILIA* FRIES FIDE FRIES (SEE SACCARDO, SYLLOGE FUNG. 4:673,1886)
=*PHYMATOSTROMA* CORDA 1837 FIDE LINDAU 1910 •

ACHORION REMAK 1845. DIAG. PATH. UNTERSUCH. P 193 /A. SCHOENLEINI (LEBERT)
=TRICHOPHYTON MALMSTEN 1845. •

ACICULARIELLA ARNAUD 1954. BULL. SOC. MYCOL. FRANCE 69:298 & 301 (ILLUS,NO LATIN DIAGNOSIS) /A. LASIOSPHAERIAE
?=MICRODOCHIUM SYDOW 1924 SEE MOUCHACCA & SAMSON 1973. •

ACICULOCONIDIUM KING & JONG 1976. MYCOTAXON 3:401-408 /TRICHOSPORON ACULEATUM PHAFF ET AL 1956 =A. ACULEATUM (PHAFF ET AL)
=BLASTOMYCETES. •

ACINULA FRIES 1822. SYSTEMA MYCOL. 2:267 /A. CANDICANS
?, STERILE. •

ACLADIUM LINK 1809. MAG. GES. NATURF. FREUNDE, BERLIN 3:11 /A. CONSPERSUM (LECTOTYPE FIDE HUGHES 1958, P 731) =A. STATE OF BOTRYOBASIDIUM CONSPERSUM ERIKSSON 1958
AMERO,BLASTO OR CETERI,HYALO,RADUL, 23H, ON WOOD AND BARK. BASIDIO STATE = BOTRYOBASIDIUM.
/LINDER, LLOYDIA 5:165-207,1942 (ILLUS, SUB OIDIUM) /HUGHES, CAN. J. BOT. 36:731-732,1958 (LIST) /ERIKSSON & RYVARDEN, THE CORTICIACEAE OF NORTHERN EUROPE, VOL. 2:145-175,1973 (ILLUS,KEY) /ELLIS, MDH P 52-54,1976 (ILLUS,NEW COMB)
?=*AMPHIBLISTRUM* CORDA 1837 FIDE DONK 1962 SEE OIDIUM
=*SPOROCEPHALIUM* CHEVALLIER 1826 FIDE HUGHES 1958 •

ACMOSPORIUM CORDA 1839. ICONES FUNGORUM 3:12 /A. BOTRYOIDEUM
=ASPERGILLUS LINK 1809 FIDE HUGHES 1958. •

ACONTIOPSIS NEGRU 1961. COMUN. ACAD. REPUB. POP. ROM. 11:839-842 (ILLUS) /A. CRATAEGI
=CYLINDROCLADIUM MORGAN 1892 FIDE KENDRICK 1973. •

ACONTIUM MORGAN 1902. J. MYCOL. 8:4-5 /THREE ORIGINAL SPECIES
?, NOT ILLUSTRATED BY AUTHOR. •

ACREMONIELLA (SACCARDO) SACCARDO 1886. SYLLOGE FUNGORUM 4:302 /ACREMONIUM ATRUM CORDA 1837 (ICON. FUNG. 1:11) =A. ATRA (CORDA) SACCARDO 1886 (LECTOTYPE, NO TYPE SPECIMEN AVAILABLE)
=NOMEN ILLEGITIMUM BECAUSE SACCARDO INCLUDED THE TYPE SPECIES OF MONOTOSPORA CORDA 1837 FIDE HOLUBOVA-JECHOVA 1974 (FOLIA GEOBOT. PHYTOTAX., PRAHA 9:315-316). SEE HARZIA. •

ACREMONIULA CIFERRI 1962. ATTI IST. BOT. UNIV. LAB. CRITTOGAM. PAVIA, SER. 5, 19:85 /A. SUPRAMELIOLA
AMERO,CETERI,PHAEO,NONSP,ALEURI?, 49H, ON FUNGI (CLYPEOLELLA & SCHIFFNERULA).
/DEIGHTON, CMI PAPERS 118:2-5,1969 (ILLUS) /ELLIS, DH P 57-58,1971 (ILLUS) /ELLIS, MDH P 36-37,1976 (ILLUS,NEW SP) •

ACREMONIULA ARNAUD 1954. BULL. SOC. MYCOL. FRANCE 69:268-269 /A. SARCINELLAE (PATOUILLARD & HAR.)
=NOMEN INVALIDUM, NO LATIN DIAGNOSIS, NON ACREMONIULA CIFERRI 1962. •

ACREMONIUM LINK 1809. MAG. GES. NATURF. FREUNDE, BERLIN 3:15 /A. ALTERNATUM (LECTOTYPE)
AMERO,BASO OR GLOIO,HYALO,PHIAL, 2A & 3B, ON MANY SUBSTRATES. ASCO STATE= EMERICELLOPSIS, NECTRIA, WALLROTHIELLA, MYCOCITRUS, PELORONECTRIELLA, HYPOCREA, & TRICHOSPHAERELLA FIDE GAMS 1971 (P 19-21), OR CERATOCYSTIS, SEE UPADHYAY & KENDRICK 1975 (MYCOLOGIA 67:798-805) OR SAGENOMA, SEE STOLK & ORR 1974 (MYCOLOGIA 66:677) OR PECKIELLA, NIGROSABULUM, NEOCOSMOSPORA OR CALONECTRIA FIDE VON ARX 1974 OR ?=CEPHALOTHECA SEE UDAGAWA & HORIE 1971 (J. GEN. APPL. MICROBIOL. 17:141-159) OR PSEUDEUROTIUM SEE UDAGAWA 1965 (TRANS. MYCOL. SOC. JAPAN 6:78-90, FIGS 3 & 4) OR HAPSIDOSPORA OR MYCOARACHIS FIDE DE HOOG 1977 (CBS STUDIES 15:137) OR LEVISPORA SEE ROUTIEN 1957 (MYCOLOGIA 49:189-191,ILLUS) OR PROTOCREA SEE MERCURI & RANALLI 1976 (PHYSIS BUENOS AIRES 35:303-317) OR THIELAVIA FIDE SAMSON ET AL 1977 (TBMS 69:417-423) OR CORDYCEPS FIDE PACIONI & FRIZZI 1978 (CAN. J. BOT. 56:391-394,ILLUS) OR CONIOCHAETA SEE HAWKSWORTH 1978 (NORW. J. BOT. 25:17). OTHER CONIDIAL STATE =VOLUTELLA.
/GAMS, CEPHALO. SCHIMMEL. 262 P,1971 (MONOGR) /SAMUELS, MYCOLOGIA 65:401-420,1973 (ILLUS,ASCO) /GAMS, TBMS 64:389-404,1975 (ILLUS,NEW SP) /SAMUELS, NEW ZEALAND J. BOT. 14:231-260,1976 (ILLUS,KEY,ASCO)
=*CEPHALOSPORIUM* AUCT. NON CORDA FIDE VON ARX 1970
?=*CULICINOMYCES* COUCH, ROMNEY & RAO 1974
?=*GABARNAUDIA* SAMSON & GAMS 1974
=*MASTIGOCLADIUM* MATRUCHOT 1911 FIDE GAMS 1971
?=*PSEUDOFUSIDIUM* DEIGHTON 1969
?=*TILACHLIDIUM* PREUSS 1851 •

ACROCLADIUM PETRAK 1949. SYDOWIA 3:263-264 /A. ANDINUM
?, NOT ILLUSTRATED BY AUTHOR.
/PETRAK, SYDOWIA 4:568-569,1950 (NEW SP,NO ILLUS) •

ACROCONIDIELLA LINDQUIST & ALIPPI 1964. DARWINIANA 13:612 /HETEROSPORIUM TROPAEOLI BOND 1947 =A. TROPAEOLI (BOND)
PHRAGMO,CETERI,PHAEO,RACHI, 82G, ON LEAVES (TROPAEOLUM & ESCHOLTZIA). COMPARE WITH CURVULARIA.
/ELLIS, DH P 461-462,1971 (ILLUS) /ELLIS, MDH P 406-408,1976 (ILLUS,NEW COMB) •

ACROCONIDIELLINA ELLIS 1971. CMI PAPERS 125:22-27 (ILLUS) /A. LOUDETIAE
DIDYMO OR PHRAGMO,CETERI,PHAEO,RACHI OR RADUL,SETAE, 61A, ON LEAVES. COMPARE WITH ACROCONIDIELLA. SCLEROTIAL STATE =SCLEROTIUM.
/ELLIS, MDH P 408-411,1976 (ILLUS,KEY) •

ACROCYLINDRIUM BONORDEN 1851. HANDB. MYK., P 97-98 (ILLUS) /THREE ORIGINAL SPECIES. NO TYPE MATERIAL AVAILABLE FIDE GAMS & HAWKSWORTH 1975 (KAVAKA 3:60)
=VERTICILLIUM NEES FIDE BARRON 1968. •

ACRODESMIS SYDOW 1926. ANN. MYCOL. 24:424-426 (NO ILLUS) /A. CESTRI
=PERICONIELLA SACCARDO 1885 FIDE ELLIS 1967. •

ACRODICTYS ELLIS 1961. CMI PAPERS 79:5-19 (ILLUS,KEY) /A. BAMBUSICOLA
DICTYO,CETERI,PHAEO,NONSP OR ANNEL,APPEND, 98F, ON STEMS, BARK & WOOD. OTHER CONIDIAL STATE =IDRIELLA.
/ELLIS, CMI PAPERS 103:33-35,1965 (ILLUS) /ELLIS, DH P 128-131,1971 (ILLUS) /CRANE & DUMONT, CAN. J. BOT. 53:843-851,1975 (ILLUS,NEW SP) /MATSUSHIMA, ICONES P 3-4 & PL 148-150,1975 (ILLUS) /ELLIS, MDH P 101-103,1976 (ILLUS,KEY) /LUNGHINI & RAMBELLI, GIORN. BOT. ITAL. 112:189-194,1978 (ILLUS) •

ACRODONTIUM DE HOOG 1972. CBS STUDIES 1:23-27 (ILLUS) /CHLORIDIUM CRATERIFORME VAN BEYMA 1933 =A. CRATERIFORME (VAN BEYMA)
AMERO,CETERI,HYALO OR PHAEO,RACHI, 23G, ON WOOD, BARK & LEAVES. SIMILAR TO TRITIRACHIUM, BUT WITH LESS SINUOUS RACHIDS. COMPARE WITH PUCIOLA. ASCO STATE =ASCOCORTICIUM FIDE DE HOOG 1972
/OBERWINKLER, CASAGRANDE & MULLER, NOVA HEDWIGIA 14:283-289,1967 (ILLUS SUB RHINOTRICHELLA) /DE HOOG & RAO, PERSOONIA 8:207-209,1975 (ILLUS,NEW SP) /KUSHWAHA & AGRAWAL, MYKOSEN 20:97-100,1977 (NEW SP,ILLUS) /COLE & SAMSON, CONIDIAL FUNGI P 36-37,1979 (ILLUS,CONIDIUM ONTOGENY) •

ACROGENOSPORA ELLIS 1971. DH P 114-115 (ILLUS) /MONOTOSPORA SPHAEROCEPHALA BERKELEY & BROOME =A. SPHAEROCEPHALA (BERKELEY & BROOME)
AMERO,CETERI,PHAEO,ANNEL, 18A, ON WOOD & BRANCHES. ASCO STATE = FARLOWIELLA FIDE MASON 1941 (SUB MONOTOSPORA) AND ELLIS 1976.
/MASON, CMI PAPERS 5:103-113,1941 (ILLUS,REVIS) /ELLIS, CMI PAPERS 131:5-6,1972 (ILLUS) /MATSUSHIMA, ICONES P 4 & PL 114,1975 (ILLUS) /ELLIS, MDH P 74,1976 (ILLUS) /HUGHES, NEW ZEALAND J. BOT. 16:312-317,1978 (ILLUS,NEW SPP) /COLE & SAMSON, CONIDIAL FUNGI P 76-78,1979 (ILLUS)
=*MONOTOSPORA* SACCARDO 1880 NON CORDA, SEE MASON 1941 •

ACROPHIALOPHORA EDWARD 1961. MYCOLOGIA 51:784 (782-784,ILLUS,DATED 1959) /A. NAINIANA =A. FUSISPORA (SAKSENA) ELLIS 1971
AMERO,BASO,HYALO OR PHAEO,PHIAL, 4B, ON PLANTS & SOIL.
/VON ARX, GENERA P 179-181,1970 (ILLUS) /SAMSON & MAHMOOD, ACTA. BOT. NEERL. 19:804-808,1970 (ILLUS,KEY) /ELLIS, DH P 533-534,1971 (ILLUS) •

ACROPHRAGMIS KIFFER & REISINGER 1970. REV. ECOL. BIOL. SOL 7:16 (12-16,ILLUS) /A. CORONATA
PHRAGMO OR STAURO,CETERI,PHAEO,ANNEL,APPEND, 70B, ON BRANCHES & LITTER.
/ELLIS, MDH P 107-108,1976 (ILLUS) /HUGHES, FUNGI CANADENSES 143,1979 (NEW SP,ILLUS) •

ACROSPEIRA BERKELEY & BROOME 1857. IN BERKELEY, INTRO. TO CRYPTOGAM. BOT. P 305 /A. MIRABILIS
DICTYO OR HELICO,CETERI,PHAEO,NONSP,ALEURI, 96B, ON "SPANISH CHESTNUTS". OTHER CONIDIAL STATE =HARZIELLA.
/BARRON, HYPHOMYCETES P 87-88,1968 (ILLUS,REFS) /VON ARX, GENERA P 237-238,1970 (ILLUS)
=*SPIROSPORA* MANGIN & VINC. 1920 FIDE HUGHES 1958 •

ACROSPIRA MONTAGNE 1857. ANN. SCI. NAT. 4, 8:299-301 /A. CROUANII
?, NOT ILLUSTRATED BY AUTHOR. •

ACROSPORELLA RIEDL & ERSHAD 1976/77. SYDOWIA 29:166-169 (ILLUS) /A. SARMENTORUM
?=CERCOSPORA. •

ACROSPORIUM NEES 1817. DAS SYSTEM ..., P 53 /A. MONILIOIDES =ACROSPORIUM STATE OF ERYSIPHE GRAMINIS DE CANDOLLE EX MERAT 1821
=NOMEN REJ. 1975, SEE OIDIUM LINK 1824 NOM. CONS. 1975. •

ACROSTALAGMUS CORDA 1838. ICONES FUNGORUM 2:15 /A. CINNABARINUS (LECTOTYPE)
=VERTICILLIUM NEES FIDE HUGHES 1958. •

ACROSTAPHYLUS ARNAUD EX SUBRAMANIAN 1956. J. INDIAN BOT. SOC. 35:483 (NO LATIN DIAGNOSIS FOR TYPE SPECIES) /A. HYPOXYLI ARNAUD 1954 (BULL. SOC. MYCOL. FRANCE 69:272, FIG 1,N, P 269)
=NODULISPORIUM PREUSS FIDE BARRON 1968. •

ACROSTAURUS DEIGHTON & PIROZYNSKI 1972. CMI PAPERS 128:94-96 (ILLUS) /A. TURNERI
STAURO,CETERI,PHAEO,ANNEL, 125B, ON FUNGI (ASTERINA & CIRSOSIA).
/ELLIS, MDH P 108-109,1976 (ILLUS) •

ACROTAMNIUM NEES 1817. DAS SYSTEM P 75 /A. VIOLACEUM (LECTOTYPE FIDE DONK 1962)
=NOMEN DUBIUM FIDE DONK 1962 (TAXON 11:103-104). •

ACROTHECA FUCKEL 1860. JAHRB. VER. NATURK., NASSAU 15:43 /A. GEI
=RAMULARIA UNGER FIDE HUGHES 1958. A. ACUTA GROVE IS NOW DISPOSED AS PLEUROPHRAGMIUM ACUTUM (GROVE) ELLIS 1976 (MDH P 164). •

ACROTHECIELLA KOORDERS 1907. VERH. AKAD. AMSTERDAM 2:13:250-251 (ILLUS) /A. JAVANICA
?=CURVULARIA BOEDIJN 1933. •

ACROTHECIUM CORDA 1851. IN PREUSS, LINNAEA 24:110-111 (NO ILLUS) /A. MULTISPORUM PREUSS 1851
=NOMEN DUBIUM FIDE HUGHES 1958. FOR A. GLOBOSUM SEE HUGHES 1978 (NEW ZEALAND J. BOT. 16:352-353,SUB STERIGMATOBOTRYS). •

ACTINICEPS BERKELEY & BROOME 1876. J. LINN. SOC., LONDON 15:85 /A. THWAITESI
=DIMORPHOCYSTIS, A BASIDIOMYCETE, FIDE BOEDIJN 1959. •

ACTINOCHAETE FERRO 1907. NUOV. GIORN. BOT. ITAL. 14:232 /A. ARACHNOIDEA
=NOMEN CONFUSUM FIDE M. B. ELLIS. •

ACTINOCLADIUM EHRENBERG 1819. JAHRB. GEWACHSK. 1:52 /A. RHODOSPORUM
STAURO,CETERI,PHAEO,NONSP, 126E, ON WOOD & BARK.
/ELLIS, DH P 137-138,1971 (ILLUS) /HUGHES, NEW ZEALAND J. BOT. 16:317-318,1978 (ILLUS) •

ACTINODOCHIUM SYDOW 1927. ANN. MYCOL. 25:146-148 (ILLUS) /A. CONCINNUM
?=CYLINDROCOLLA BONORDEN 1851. •

ACTINOMMA SACCARDO 1884. ATTI. IST. VENETO, 6 SER., 2:462-463 (NO ILLUS) /A. GASTONIS
=ATICHIA FLOTOW 1850 FIDE MEEKER 1975. •

ACTINONEMA PERSOON 1822. MYCOL. EUROP. 1:51-52 /A. CAULINICOLA
?, ON STEMS. STERILE FIDE HUGHES 1958, BUT SEE VON ARX 1974 (GENERA P 169) FOR ACTINONEMA FRIES 1849 SENSU GROVE 1937.
/SUTTON, CMI PAPERS 141:5-6,1977 (REFS) •

ACTINONEMELLA HOHNEL 1923. MYKOL. UNTERSUCH. 1:338 =NOMEN NUDUM •

ACTINOPELTE SACCARDO 1913. ANN. MYCOL. 11:315 /A. JAPONICA
=TUBAKIA SUTTON 1973, NON ACTINOPELTE SITZENB. 1861. •

ACTINOSPORA INGOLD 1952. TBMS 35:68-69 (ILLUS) /A. MEGALOSPORA
STAURO,CETERI,HYALO,NONSP,APPEND,AQUATIC, 120C, ASCO STATE =MILADINA FIDE DESCALS & WEBSTER 1978 (TBMS 70:466-472).
/GOOS, TBMS 55:335-337,1970 (ILLUS) /CRANE & DUMONT, CAN. J. BOT. 53:843-851,1975 (ILLUS,NEW SP, PROBABLY NOT CONGENERIC WITH A. MEGALOSPORA) /DESCALS, NAWAWI & WEBSTER, TBMS 67:207-222,1976 (ILLUS) •

ACTINOSTILBE PETCH 1925. ANN. BOT. GARD. PERADENIYA 9:327 /A. VANILLAE
?, NOT ILLUSTRATED BY AUTHOR. •

ACURTIS FRIES 1849. SUMMA VEG. SCAND. 2:337 /CLAVARIA GIGANTEA SCHWEINITZ =A. GIGANTEA (SCHWEINITZ)
?, STERILE SEE DONK 1962 (TAXON 11:76). BASIDIO STATE =ENTOLOMA FIDE DONK 1962. •

ADHOGAMINA SUBRAMANIAN & LODHA 1964. ANTONIE VAN LEEUWENHOEK 30:319-323 (ILLUS) /A. RUCHIRA
=GILMANIELLA BARRON 1964 FIDE HENNEBERT 1968. •

AEGERITA PERSOON 1801. SYN. METH. FUNG. 684 /A. CANDIDA =AEGERITA STATE OF PENIOPHORA CANDIDA LYMAN 1907 (?=METULODONTIA FARINOSA =BULBILLOMYCES FARINOSUS)
BULBILS,CETERI OR GLOIO,HYALO,NONSP,CLAMPS, 129E, ON WOOD & BARK. BASIDIO STATE =PENIOPHORA FIDE HUGHES 1958. COMPARE WITH CRISTULARIELLA.
/TUBAKI, J. HATTORI BOT. LAB. NO. 20, P 145-146,1958 (ILLUS) /DONK, TAXON 11:77,1962 (REFS) /MATSUSHIMA, ICONES P 4-5 & PL 373-374,1975 (ILLUS,NEW SP, & P 106 & PL 370,ILLUS SUB PAPULASPORA VIRIDIS) /FUNK, CAN. J. BOT. 51:1645 & PL II,1973 (ILLUS SUB PHYMATOTRICHUM FUNGICOLA) /HENNEBERT, PERSOONIA 7:191,1973 (NEW COMB)
=*CROCYSPORIUM* CORDA 1837 FIDE HUGHES 1958
=*DERMOSPORIUM* LINK 1815 FIDE FRIES 1829
=*EGERITA* AUCT. -ORTHOGRAPHIC VARIANT
?=*SPINULOSPORA* DEIGHTON 1973
?=*TERMITOSPHAERA* CIFERRI 1935 •

AEGERITELLA BALAZY & WISNIEWSKI 1974. POZN. TOW. PRZYJ. NAUK, PR., KOM. NAUK. ROL. LES. 38:3-15 (NOT SEEN) /A. SUPERFICIALIS
?, ON INSECTS (FORMICA, CAMPONOTUS).
/BALAZY & WISNIEWSKI, ACTA MYCOL. 13:271-274,1977 (NEW SP,ILLUS) •

AEGERITOPSIS HOHNEL 1903. ANN. MYCOL., BERLIN 1:532-533 /A. NULLIPORIOIDES
?, NOT ILLUSTRATED BY AUTHOR. •

AEROPHYTON ESCHWEILER 1824. SYLL. PLANT. SOC. BOT. RATISB. 1:163-165 /A. PRINCIPIS
?, ORIGINAL ILLUSTRATION NOT DIAGNOSTIC. •

AGARICOSTILBUM WRIGHT 1970. MYCOLOGIA 62:679-682 (ILLUS) /A. PALMICOLA
AMERO,CETERI,HYALO,RADUL,SYNNEMATA,SETAE, 29D, ON LITTER (PHOENIX, COCOS). ASCO STATE ?=ANTHOSTOMA-LIKE SEE WRIGHT 1970 (P 682).
/SUBRAMANIAN & NATARAJAN, MYCOLOGIA 69:1224-1226,1977 (ILLUS,REVIS)
=*AMEROBOTRYUM* SUBRAMANIAN & NATARAJAN 1976 FIDE SUBR. & NATARAJ. 1977 •

AGARWALIA TIWARI & AGRAWAL 1974. J. INDIAN BOT. SOC. 52:134 (133-138,DATED 1973,ILLUS) /A. TERRICOLA
AMERO,BLASTO,PHAEO,NONSP, 39B, ON SOIL. CLOSE TO PERICONIA. •

AGYRIELLA ELLIS & EVERHART 1897. BULL. TORREY BOT. CLUB 24:470 /A. BETHELI
=NOMEN ILLEGITIMUM, NON AGYRIELLA SACCARDO 1884. •

AGYRIELLA SACCARDO 1884. ATTI IST. VENET. SCI., LETT. ED ARTI, SER. 6, 2:454-455 /A. NITIDA (LIBERT)
AMERO,GLOIO,HYALO,PHIAL,SPORODOCHIA, 14B, ON BRANCHES (RIBES & RUBUS).
/ELLIS, DH P 525-526,1971 (ILLUS) •

AKANTHOMYCES LEBERT 1858. ZEITSCH. WISSEN. ZOOLOGIE 9:447 /A. ACULEATA
AMERO,BASO,HYALO,PHIAL,SYNNEMATA, 10F, ON INSECTS. ASCO STATE =CORDYCEPS LINK FIDE MAINS 1950.
/MAINS, MYCOLOGIA 42:566-589,1950 (ILLUS) /SAMSON & EVANS, ACTA BOT. NEERL. 23:28-35,1974 (ILLUS,REVIS)
=*INSECTICOLA* MAINS 1950 FIDE KENDRICK 1973 •

AKENOMYCES ARNAUD 1954. ANN. EC. AGRIC. MONTPELLIER 29:26 (ILLUS) /A. ENIGMATICUS
SCLEROTIA, ON LITTER (CAREX). PYRIFORM SCLEROTIA. •

ALATOSPORA INGOLD 1942. TBMS 25:384 (381-384,ILLUS) /A. ACUMINATA
STAURO,CETERI,HYALO,PHIAL,APPEND,AQUATIC, 118D.
/TUBAKI, BULL. NAT. SCI. MUSEUM, TOKYO 3:250-252,1957 (ILLUS) /INGOLD, GUIDE P 68-69,1975 (ILLUS) /MARVANOVA, ARCH. PROTISTENK. BD. 119:68-74,1977 (ILLUS,NEW SP) /DYKO, TBMS 70:409-411,1978 (ILLUS,NEW SP) •

ALBOSYNNEMA MORRIS 1967. MYCOPATHOLOGIA 33:179-181 (ILLUS) /A. ELEGANS
PHRAGMO,GLOIO,PHAEO,PHIAL?,SYNNEMATA, 74B, ON LITTER.
/SHERWOOD, MYCOTAXON 1:117-119,1974 (ILLUS,NEW SP) •

"ALEURIOCONIDIA" REPORTED AS ANAMORPHS OF GENERA OF GYMNOASCACEAE (A), PEZIZA (A), TRICHOPHAEA (A) SEE PADEN 1972 (PERSOONIA 6:405), CONIOCHAETIDIUM (A) SEE MALLOCH & CAIN 1971 (CAN. J. BOT. 49:879), & DIADEMOSPORA (H). •

ALEURISMA LINK 1809. MAG. GES. NATURF. FREUNDE, BERLIN 3:18-19 /A. SPORULOSUM =TRICHODERMA PERSOON 1801 FIDE HUGHES 1958. •

ALEURODOMYCES BUCHNER 1912. ARCHIV PROTISTENK., JENA 26:100-114 & TAF 4 /A. SIGNORETII.
?, A SEPTOBASIDIUM-LIKE FUNGUS IN ALEURODES (WHITE FLY) LARVAE. •

ALEUROPHORA MAGALHAES 1916. BRASIL-MEDICO 30(47):369-370 /A. BENIGNA
?=NOMEN DUBIUM, NO ILLUSTRATION, PROBABLY NO TYPE SPECIMEN. •

ALEUROSPORIA GRIGORAKIS 1925. COMP. REND. 179:1425,1924 & ANN. SCI. NAT. BOT. SER. 10, 7:413,1925 /A. ACUMINATA (BODIN) (LECTOTYPE FIDE CLEMENTS AND SHEAR 1931, P 410)
=TRICHOPHYTON MALMSTEN PRO PARTE FIDE DODGE 1935. •

ALLANTOSPORA WAKKER 1896. MEDED. PROEFST. OOST-JAVA, SER. 2, 28:8-9 /A. RADICICOLA
=CYLINDROCARPON WOLLENWEBER 1913 FIDE BOOTH 1966. •

ALLELOCHAETA PETRAK 1955. SYDOWIA 9:464-465 /A. GAUBAE
?, NOT ILLUSTRATED BY AUTHOR. PROBABLY NOT A HYPHOMYCETE. •

ALLESCHERIA HARTIG 1899. SEE HARTIGELLA SYDOW 1900. •

ALLESCHERIELLA HENNINGS 1897. HEDWIGIA 36:244 /A. UREDINOIDES =A. CROCEA (MONT.) HUGHES 1950 =A STATE OF BOTRYOBASIDIUM CROCEUM LENTZ 1967
AMERO,CETERI,PHAEO,NONSP,ALEURI,SPORODOCHIA?, 51B, ON WOOD. COMPARE WITH SPOROTRICHUM. BASIDIO STATE =BOTRYOBASIDIUM SEE LENTZ 1967 (MYCOPATHOLOGIA 32:1-25).
/HUGHES, CMI PAPERS 41:1-8,1951 (ILLUS) /ELLIS, DH P 60-62,1971 (ILLUS) /KOHLMEYER, REV. MYCOL. 41:199-202,1977 (ILLUS,NEW SP) •

ALLIOSPORA PIM 1883. JOUR. BOT., LONDON 21:234-235 /A. SAPUCAYAE
?, NOT ILLUSTRATED BY AUTHOR, PROBABLY BASED ON AN ASPERGILLUS. •

ALLONEMA SYDOW 1934. ANNAL. MYCOL. 32:283-285 /GEOTRICHUM ROSEUM GROVE 1886 (J. BOT. 24:198 & TAB 266, FIG 8) =A. ROSEUM (GROVE)
?, NOT ILLUSTRATED BY SYDOW. GROVE SHOWS A FEW CATENATE, BACILLIFORM SPORES, 9-10 X 16-30, ON JUNCUS. VON ARX 1974 (GENERA P 182-183) LISTS ALLONEMA AS A SYNONYM OF SPORENDONEMA. •

ALLOSPHAERIUM LINK 1826 (ABHANDL. K. AKAD. WISS. BERLIN 1824:181,NO ILLUS,NO SPECIES MENTIONED) •

ALPHITOMYCES REISSEK 1856. S. B. AKAD. WISS. WIEN 21:323-327 & PL /A. SCHROETERI
?, LOOKS LIKE A COREMIAL PAECILOMYCES, SEE SAMSON 1974 (CBS STUDIES 6:10). •

ALTERNARIA NEES 1817. DAS SYSTEM ... P 72 /A. TENUIS
DICTYO OR PHRAGMO,BLASTO,PHAEO,RACHI,APPEND,SYNNEMATA, 101A, ON PLANTS (MANY KINDS). ASCO STATE =CLATHROSPORA OR LEPTOSPHAERIA FIDE SIMMONS 1952, OR PLEOSPORA FIDE ELLIS 1971 OR COMOCLATHRIS SEE DE HOOG 1977 (CBS STUDIES 15:139). SCLEROTIAL STATE =SCLEROTIUM.
/SIMMONS, MYCOLOGIA 59:67-92,1967 (ILLUS) /ELLIS, CMI DESCRIPTIONS 162-164,1968 (ILLUS) /ELLIS, DH P 464-497,1971 (ILLUS,NEW COMB) /ELLIS & HOLLIDAY, CMI DESCRIPTIONS 241-250,1970 & 345-346,1972 (ILLUS) /ELLIS, MDH P 411-427,1976 (ILLUS,NEW COMBS) /ANAHOSUR, CMI DESCRIPTIONS 581-583,1978 (ILLUS)
?=*DICTYOCEPHALA* MEDEIROS 1962
=*ELOSIA* PERSOON 1822 FIDE HUGHES & SIMMONS 1958
=*MACROSPORIUM* FRIES 1832 FIDE HUGHES 1958
?=*PRATHODA* SUBRAMANIAN 1956 SEE CEJP & DEIGHTON 1969
=*RHOPALIDIUM* MONTAGNE 1846 FIDE ELLIS 1971
?=*ULOCLADIUM* PREUSS 1851 •

ALYSIDIOPSIS SUTTON 1973. CMI PAPERS 132:5-8, FIG 1 /A. PIPSISSEWAE
AMERO OR DIDYMO,BLASTO,PHAEO,NONSP, 39C, ON PLANTS (CHIMAPHILA). COMPARE WITH CLADOSPORIUM.
/ELLIS, MDH P 157-158,1976 (ILLUS) •

ALYSIDIUM KUNZE 1817. IN KUNZE & SCHMIDT, MYKOL. HEFTE 1:11 /A. FULVUM =TRICHODERMA DUBIUM PERSOON 1801 (SYN. METH. FUNG. P 233) =ALYSIDIUM DUBIUM (PERSOON) ELLIS 1971 =ALYSIDIUM STATE OF BOTRYOBASIDIUM AUREUM PARMASTO 1965
AMERO,BLASTO,HYALO OR PHAEO,NONSP, 36E, ON WOOD. COMPARE WITH ACLADIUM & SOROCYBE. BASIDIO STATE =BOTRYOBASIDIUM DONK FIDE ERIKSSON & RYVARDEN 1973 (CORTICIACEAE OF NORTHERN EUROPE 2:145-175,ILLUS)
/LINDER, LLOYDIA 5:165-207,1942 (ILLUS SUB OIDIUM) /HUGHES, CAN. J. BOT. 36:731-732,1958 (REFS) /DONK, TAXON 11:78,1962 (REFS) /ELLIS, DH PP 89-90,1971 (ILLUS,REVIS) /ELLIS, MDH P 51-52,1976 (ILLUS,NEW COMB) /ELLIS & GRIFFITHS, CAN. J. BOT. 55:676-684,1977 (ILLUS,CONIDIUM ONTOGENY)
=*OIDIUM* AUCT. NON LINK 1824 NOM. CONS. 1975
=*HAPLOTRICHUM* LINK 1824 FIDE HUGHES 1958 •

ALYTOSPORIUM (LINK) EHRENBERG 1818. SYLVAE MYC. BEROL. P 11 /A. BOMBACINUM (LECTOTYPE FIDE HUGHES 1958)
=NOMEN DUBIUM FIDE HUGHES 1958 (P 833) OR =TOMENTELLA FIDE STALPERS 1975 (REV. MYCOL. 29:97-101). •

AMALLOSPORA PENZIG 1897. MALPIGHIA 11:461-464 & TAV X /A. DACRYDION
STAURO,GLOIO,HYALO,NONSP?, 116C, ON WOOD.
/TUBAKI, TRANS. MYCOL. SOC. JAPAN 6:44-46,1965 (ILLUS) •

AMASTIGIS CLEMENTS & SHEAR 1931, ORTHOGRAPHIC VARIANT OF AMASTIGOSPORIUM •

AMASTIGOSPORIUM BONDARZEVA-MONTEVERDE 1921. MAT. MYKOL. OBSLED. ROSS. 5(1):1-3 /A. GRAMINICOLA
=MASTIGOSPORIUM RIESS 1852 FIDE HUGHES 1951.
=*AMASTIGIS* CLEMENTS & SHEAR 1931, ORTHOGRAPHIC VARIANT •

AMBIVINA KATZ 1974. NOVA HEDWIGIA 25:811-814 (ILLUS) /A. FILOBASIDIA
AMERO,CETERI,HYALO,AMPUL, 35A, ON LITTER (QUERCUS & CORNUS). A BASIDIOMYCETE. INCLUDED HERE BECAUSE IT LOOKS LIKE A HYPHOMYCETE AND HAS NO CLAMP CONNECTIONS. COMPARE WITH DEXHOWARDIA, NIA, DIGITATISPORA & BOTRYOCONIS. •

AMBLYOSPORIOPSIS FAIRMAN 1922. PROC. ROCHESTER ACAD. SCI. 6:132 & PL 21 /A. PARASPHENOIDES
?, ORIGINAL ILLUSTRATION NOT DIAGNOSTIC, ASPERGILLUS?, RHIZOPUS?, OEDOCEPHALUM? •

AMBLYOSPORIUM FRESENIUS 1863. BEITR. ZUR. MYKOL. 3:99 & PL 12, FIG 17-22 /A. BOTRYTIS
AMERO,ARTHRO OR CETERI,HYALO,NONSP,ALEURI, 44C-D, ON FUNGI (LACTARIUS & OTHERS), WOOD, LITTER, DUNG & SOIL. OTHER STATE ="SCLEROTIA".
/PIROZYNSKI, CAN. J. BOT. 47:325-334,1969 (REVIS,ILLUS) /MATSUSHIMA, ICONES P 7 & PL 19,1975 (ILLUS) /COLE & SAMSON, CONIDIAL FUNGI P 113-115,1979 (ILLUS)
=*GUEGUENIA* BAINIER FIDE PIROZYNSKI 1969
=*SEARCHOMYCES* MEHROTRA & MEHROTRA 1965 FIDE PIROZYNSKI 1969 •

AMBROSIAEMYCES TROTTER 1934. ANN. ROY. INST. SUP. AGRIC. PORTICI III, 6:262-275 /A. ZEYLANICUS
=NOMEN CONFUSUM FIDE VON ARX & HENNEBERT 1965 (MYCOPATHOLOGIA 25:309-315). •

AMBROSIELLA BRADER 1964. MEDED. LANDB-HOOGESCH. WAGENINGEN 64:41 /A. XYLEBORI BRADER EX VON ARX & HENNEBERT 1965
AMERO,BLASTO OR BASO,HYALO,NONSP OR PHIAL?, 47B, ON WOOD & INSECTS (BEETLES). COMPARE WITH PHIALOPHOROPSIS.
/VON ARX & HENNEBERT, MYCOPATHOLOGIA 25:312-315,1965 (ILLUS) /BATRA, MYCOLOGIA 59:976-1017,1967 (REVIS,KEY,ILLUS) •

AMEROBOTRYUM SUBRAMANIAN & NATARAJAN 1976. MYCOLOGIA 67:1211-1215 (DATED 1975,ILLUS) /A. INDICUM

=AGARICOSTILBUM FIDE SUBRAMANIAN & NATARAJAN 1977. •

AMEROSPORIELLA HOHNEL 1923. MYKOL. UNTERSUCH. U. BERICHTE 1(3):348 /SINE SPECIES NOMINE
=NOMEN ILLEGITIMUM.
=*AMEROSPORIS* CLEMENTS & SHEAR 1931 FOR AMEROSPORIELLA •

AMEROSPORIS CLEMENTS & SHEAR 1931 = NOM. NOV. FOR AMEROSPORIELLA •

AMPHIBLISTRUM CORDA 1837. ICONES FUNGORUM 1:11, TAB III, FIG 166 /A. HYPOCHNOIDES
?=ACLADIUM LINK 1809 SEE DONK 1962 (TAXON 11:78). •

AMPHICHAETE KLEBAHN 1914. MYCOL. ZBL. 4:16-17 (FIG 36) /A. ECHINATA
=AMPHICHAETELLA HOHNEL 1916, NON AMPHICHAETA MCALPINE 1904. •

AMPHICHAETELLA HOHNEL 1916. S. B. AKAD. WISS. WIEN 125:92 (NO ILLUS) /A. ECHINATA (KLEBAHN)
AMERO,CETERI,HYALO,NONSP,SPORODOCHIA,APPEND, 54E, ON LEAVES (LICUALA).
/MORGAN-JONES, CAN. J. BOT. 51:1431-1433,1973 (ILLUS) •

AMPHICHORDA FRIES 1825 (SYST. ORB. VEG. P 170) ='ISARIA' FIDE FRIES 1832 (SYST. MYCOL. 3:270) •

AMPHITIAROSPORA AGNIHOTHRUDU 1963. SYDOWIA 16:74-76, FIGS 1-6 /A. NEOTTIOSPOROIDES
=DINEMASPORIUM (COELOMYCETE) FIDE SUTTON 1977 (CMI PAPERS 141:10-11). •

AMPHITRICHUM NEES & NEES 1818. NOVA ACTA ACAD. CAES. LEOP. 9:249 /A. EFFUSUM
=NOMEN DUBIUM, HUGHES (1958, P 735) DID NOT FIND TYPE. •

AMPHOBOTRYS HENNEBERT 1973. PERSOONIA 7:192 & FIG 3, P 186 /A. RICINI (BUCHWALD 1949) = A. STATE OF BOTRYOTINIA RICINI (GODFREY) WHETZEL 1945
AMERO,CETERI,PHAEO,AMPUL,ALEURI, 33B, ON PLANTS (RICINUS). CLOSE TO BOTRYTIS. OTHER STATES = BOTRYOTINIA (A) (=SCLEROTINIA), SCLEROTIUM (SCLEROTIAL STATE) & MYRIOCONIUM (SPERMATIAL STATE) ALL FIDE HENNEBERT 1973. •

AMPHOROMORPHA THAXTER 1914. BOT. GAZ. 58:249 (235-253, ILLUS) /A. ENTOMOPHILA ?, ON INSECTS. PERHAPS NOT A HYPHOMYCETE. •

AMPULLIFERA DEIGHTON 1960. CMI PAPERS 78:36-42 (ILLUS) /A. FOLIICOLA
AMERO OR DIDYMO OR PHRAGMO,CETERI OR BLASTO,PHAEO,NONSP,PODIA, 36A, ON LICHENS.
/DEIGHTON, CMI PAPERS 101:28-31,1965 (ILLUS,EMEND) •

AMPULLIFERELLA BATISTA & CAVALCANTI 1964. PORT. ACTA BIOL. SER. B, 7:348-350 & FIG 1 /A. AMOEBOIDES
?=AMPULLIFERA, BUT SEE SUTTON 1969 (CAN. J. BOT. 47:611-616). •

AMPULLIFERINA SUTTON 1969. CAN. J. BOT. 47:609 (609-616,ILLUS) /A. PERSIMPLEX
DIDYMO OR PHRAGMO,ARTHRO,PHAEO,NONSP,PODIA, 67B, ON LEAVES OR LITTER.
/ELLIS, DH P 30,1971 (ILLUS) /ELLIS, MDH P 19,1976 (ILLUS,NEW SP) •

AMPULLIFEROPSIS BATISTA & CAVALCANTI 1964. PORT. ACTA BIOL., SER. B, 7:349 (349-355, ILLUS) /A. MYRIAPODA
?=AMPULLIFERA SEE SUTTON 1969 (CAN. J. BOT. 47:611-616). SEE 90E. •

ANAVIRGA SUTTON 1975. TBMS 64:406-409 (ILLUS) /A. LAXA
STAURO,CETERI,PHAEO,NONSP OR RACHI OR RADUL, 127A, ON LITTER. OTHER STATE =PHIALOCEPHALA (H).
/DESCALS & SUTTON, TBMS 67:269-274,1976 (ILLUS,NEW SP) •

ANCYLOSPORA SAWADA 1944. REPT. GOV. RES. INST. FORMOSA 87:77-78 (NO ILLUS, NO LATIN DIAGNOSIS) /A. COSTI
=PSEUDOCERCOSPORA FIDE DEIGHTON 1976. •

ANDREAEA PALM & JOCHEMS 1923 (DELI. PROEF. MEDAN-SUMATRA BULL. 19:1-21 & PL III)
SEE ANDREAEANA •

ANDREAEANA PALM & JOCHEMS 1924. RECTIFICUTIE VAN BULL. 19 DELI. PROEF. MEDAN-SUMATRA, P 1 /A. DELIENSIS (PALM & JOCHEMS)
?, APPEARS TO BE AN ACREMONIUM. ORIGINAL ILLUSTRATION NOT DIAGNOSTIC.
=*ANDREAEA* PALM & JOCHEMS 1923 NON EHRHART
=*PALMOMYCES* MAIRE 1926, NOM. NOV. FOR ANDREAEA •

ANEMATIDIUM GRONCHI 1931. BOLL. IST. SIEROTER. MILANO 10:242, FIGS 1-10 /A. OXIPHILUM
?, ORIGINAL ILLUSTRATION & DESCRIPTION NOT DIAGNOSTIC. •

ANGIOPOMA LEVEILLE 1841. ANN. SCI. NAT. 2 SER., 16:235-242 /A. CAMPANULATUM
=DRECHSLERA STATE OF PYRENOPHORA SEMIPERDA
=NOM. REJ. PROP., SEE SUTTON 1976 (MYCOTAXON 3:377-380). •

ANGUILLOSPORA INGOLD 1942. TBMS 25:402 /A. LONGISSIMA (SACC. & SYDOW)
SCOLECO,CETERI,HYALO,NONSP,AQUATIC, 108G, ASCO STATE =MOLLISIA FIDE WEBSTER 1961, OR MASSARINA FIDE WILLOUGHBY & ARCHER 1973 (FRESHWATER BIOL. 3:219-239). OTHER CONIDIAL STATE =PHIALOCEPHALA-LIKE SEE WEBSTER 1961.
/TUBAKI, BULL. NAT. SCI. MUSEUM, TOKYO 3:251-252,1957 (ILLUS) /WEBSTER, TBMS 44:559-564,1961 (ILLUS) /IQBAL, TBMS 59:301-307,1972 (ILLUS,NEW SP) /INGOLD, GUIDE P 78-79,1975 (ILLUS) •

ANGULIMAYA SUBRAMANIAN & LODHA 1969. ANTONIE VAN LEEUWENHOEK 30:327 (317-330, ILLUS) /A. SUNDARA
AMERO,BASO,HYALO,PHIAL,SETAE, 15C, ON DUNG. COMPARE WITH PHIALOPHORA.
/ELLIS, DH P 532-533,1971 (ILLUS) •

ANGULOSPORA NILSSON 1962. SVENSK. BOT. TIDSKR. 56:354-355 (ILLUS) /A. AQUATICA
SCOLECO OR HELICO,CETERI,HYALO,NONSP,AQUATIC, 107A. •

ANKISTROCLADIUM PERROTT 1960. TBMS 43:557-558 (ILLUS) /A. FUSCUM
=CASARESIA FRAGOSO 1920 FIDE KENDRICK 1973. •

ANNELLODOCHIUM DEIGHTON 1969. CMI PAPERS 118:28-30 (ILLUS) /A. RAMULISPORUM
AMERO OR DIDYMO,BLASTO OR BASO,PHAEO,ANNEL,SPORODOCHIA, 60D, ON FUNGI (DIATRYPE) ON BRANCHES.
/CRANE & DUMONT, CAN. J. BOT. 53:849-851,1975 (ILLUS) /ELLIS, MDH P 137-138,1976 (ILLUS) •

ANNELLOPHORA HUGHES 1951. TBMS 34:544-546 /A. SOLANI (SYDOW)
PHRAGMO,CETERI OR BLASTO,PHAEO,ANNEL, 75A, ON LEAVES & TWIGS. COMPARE WITH SPORIDESMIUM.
/ELLIS, CMI PAPERS 70:84-89,1958 (ILLUS,EMEND,KEY) /ELLIS, DH P 125-127,1971 (ILLUS)
=*CHAETOTRICHUM* SYDOW NON RABENHORST FIDE HUGHES 1951 •

ANNELLOPHORELLA SUBRAMANIAN 1962. PROC. INDIAN ACAD. SCI. SECT B, 55:6 /A. DENSUM (SYDOW & SYDOW) =A. FAUREAE (HENNINGS) ELLIS 1963
DICTYO OR PHRAGMO,CETERI,PHAEO,ANNEL, 98D, ON LEAVES & WOOD.
/ELLIS, CMI PAPERS 87:11-13,1963 (ILLUS) /ELLIS, DH P 87-89,1971 (ILLUS) /ELLIS, MDH P 50-51,1976 (ILLUS) •

ANNELLOPHRAGMIA SUBRAMANIAN 1963. PROC. INDIAN ACAD. SCI., SECT. B, 58:349 /A. COONOORENSIS (SUBRAMANIAN)
PHRAGMO,CETERI,PHAEO,ANNEL OR RADUL,SYNNEMATA,SPORODOCHIA, 87D, ON LEAVES (THYSANOLAENA). MUCH LIKE SPIROPES.
/SUBRAMANIAN, PROC. INDIAN ACAD. SCI. SECT. B, 42:283-292,1955 (ILLUS, SUB ARTHROBOTRYUM) /ELLIS, DH P 260-261,1971 (ILLUS) •

ANODOTRICHUM (CORDA) RABENHORST 1844. KRYPT.-FL. 1:86 /A. OLIGOCARPUM (CORDA) CORDA 1838
=BLASTOTRICHUM CORDA 1838 FIDE LINDAU 1907. •

ANOMOMYCES HOHNEL 1919. BERL. DEUTSCHL. BOT. GES. 37:153 (NO ILLUS) /SPHAERIA ARBUTICOLA SOWERBY 1802 =A. ARBUTICOLUS (SOWERBY)
?, SEE DISCUSSION BY SUTTON 1975 (CMI PAPERS 138:58-61). •

ANSATOSPORA NEWHALL 1944. PHYTOPATHOLOGY 34:98 /A. MACROSPORA (OSTERWALDER)
=NOMEN NUDUM, SEE UNDER MYCOCENTROSPORA. •

ANTENAGLIUM ALBUQUERQUE 1969. IN ALBUQUERQUE ET AL., ANTENAGLIUM NOVO GENERO DE FUNGO IMPERFECTO ENCONTRADO UN SOLOS DO AMBIENTE NATIVO DA AMAZONIA (INST. PESQ. E EXPER. AGROPEC., PARA), P 4-7 (ILLUS) /TWO ORIGINAL SPECIES, TYPE SPECIES NOT SPECIFIED
=GLIOCEPHALOTRICHUM ELLIS & HESSELTINE 1962 FIDE KENDRICK. •

ANTENNARIA LINK 1809. LINN. SPEC. PLANT IV 6(1):118 /A. ERICOPHILA (LINK) LINK 1824
=NOMEN ILLEGITIMUM FIDE HUGHES 1958. •

ANTENNATARIA -SEE ANTENNULARIA •

ANTENNATULA FRIES EX STRAUSS 1850. FLORA 33 (BEILAGE):99 /A. PINOPHILA (NEES) STRAUSS 1850
PHRAGMO,CETERI,PHAEO,NONSP, 91E, ON TWIGS (ABIES). ASCO STATE =EUANTENNARIA FIDE HUGHES 1972.
/HUGHES, NEW ZEALAND J. BOT. 8:178-197,1970 (ILLUS,EMEND) /HUGHES, NEW ZEALAND J. BOT. 10:225-242,1972 (ILLUS) /HUGHES, NEW ZEALAND J. BOT. 12:299-356,1974 (ILLUS,NEW SPP) /HUGHES, MYCOLOGIA 68:703-707 & 772-773,1976 (ILLUS,REVIS) /ELLIS, MDH P 32,1976 (ILLUS)
=*HORMISCIELLA* BATISTA 1956 FIDE HUGHES 1970
?=*TRICHOTHALLUS* STEVENS 1925 SEE HUGHES 1976 (P 773) •

ANTENNINA FRIES 1849 (SUMMA VEG. SCAND. 2:445) =ANTENNULARIA REICHENBACH 1828 FIDE HUGHES 1958 •

ANTENNOPSIS HEIM 1952. BULL. SOC. MYCOL. FRANCE 67:347 /A. GALLICA HEIM & BUCHLI 1952
PHRAGMO,CETERI,PHAEO,ANNEL, 76C, ON INSECTS (RETICULOTERMES). •

ANTENNULARIA REICHENBACH 1828. CONSPECT. REGNI VEGET. 1:5 /A. ERICOPHILA (LINK) HOHNEL 1909
=NOMEN DUBIUM FIDE HUGHES 1970, BUT SEE VON ARX 1975 (CBS STUDIES 8:101).
/HUGHES, NEW ZEALAND J. BOT. 8:156,1970 (REVIS)
=*ANTENNARIA* LINK 1809 FIDE HUGHES 1970
=*ANTENNATARIA* REICHENBACH 1841
=*ANTENNINA* FRIES 1849
=*GIBBERA* FRIES SUBGENUS ANTENNULARIA (REICHENBACH) PETRAK 1947 •

ANTHINA FRIES 1819. NOVIT. FL. SUEC. PART. V, CONT. P 79 /CERATONEMA DILATATUM ROTH 1800 =A. FLAMMEA (JUNGHUHN) FRIES 1832 (SYST. MYCOL. 3:283)
?, ON LITTER. STERILE, BASIDIO STATE ?=PENIOPHORA FIDE DONK 1962.
/DONK, TAXON 11:78-79,1962 (REFS) •

ANTHOPSIS FILIPELLO MARCHISIO, FONTANA & LUPPI MOSCA 1977. CAN. J. BOT. 55:115-117 (ILLUS) /A. DELTOIDEA
AMERO,GLOIO,PHAEO,PHIAL, 6E, OTHER CONIDIAL STATE =PHIALOPHORA. •

ANTIPODIUM PIROZYNSKI 1974. CAN. J. BOT. 52:1143-1146, FIG 1 /A. SPECTABILE =A. STATE OF OPHIONECTRIA TRICHOSPORA
PHRAGMO,BASO,HYALO,PHIAL, 73E, ON ROTTEN WOOD. ASCO STATE =OPHIONECTRIA FIDE PIROZYNSKI 1974.
/ROSSMAN, MYCOLOGIA 69:355-360,1977 (ILLUS OF ASCO STATE) •

ANTROMYCES FRESENIUS 1850. BEITRAGE ZUR MYKOL. 1:37-38 (ILLUS) /A. COPRIDIS
AMERO OR DIDYMO,ARTHRO,HYALO,NONSP,SYNNEMATA, 67A, ON INSECT COCCOONS (COPRIS).
/LINDAU, RABENH. KRYPT. FL. 2 AUFL. 1 BD. P 390-391,1910 (ILUS) •

ANTROMYCOPSIS PATOUILLARD & TRABUT 1897. IN PATOUILLARD, BULL. SOC. MYCOL. FRANCE 13:215-216 (ILLUS) /A. BROUSSONETIAE PATOUILLARD 1897 =A. STATE OF PLEUROTUS CYSTIDIOSUS FIDE POLLACK & MILLER 1976
AMERO,GLOIO,PHAEO,MISC?,CLAMPS,SYNNEMATA, 45B, ON WOOD. BASIDIO STATE =PLEUROTUS.
/MORRIS, WESTERN ILLINOIS UNIV. SER. BIOL. SCI. 3:20-21,1963 (ILLUS) /JONG & PENG, MYCOLOGIA 67:1235-1238,1975 (ILLUS) /POLLACK & MILLER, MEM. N.Y. BOT. GARD. 28:174-178,1976 (ILLUS) /MOORE, CAN. J. BOT. 55:1251-1256,1977 (ILLUS OF TYPE) /GHOSH, PATHAK & SINGH, NORWEGIAN J. BOT. 24:79,1977 (NEW VAR,ILLUS) •

ANULOHYPHA CIFERRI 1962. ATTI IST. BOT. UNIV. PAVIA, SER. 5, 19:88-89 (ILLUS) /A. SERPENS
?, STERILE. •

ANULOSPORIUM SHERBAKOFF 1933. MYCOLOGIA 25:262 & PL 35 /A. NEMATOGENUM
?, STERILE, BASED ON NEMATODE TRAPPING RINGS. •

ANUNGITEA SUTTON 1973. CMI PAPERS 132:10-13 (ILLUS) /A. FRAGILIS
AMERO OR DIDYMO OR PHRAGMO,BLASTO,HYALO OR PHAEO,RADUL, 65E, ON BARK. COMPARE WITH POLYSCYTALINA, POLYSCYTALUM & HORMIACTIS.
/MATSUSHIMA, ICONES P 7-9 & PL 59-61,1975 (ILLUS,REVIS,NEW SPP) /SUTTON & HODGES, NOVA HEDWIGIA 29:593,1978 (ILLUS,NEW SP) •

AORATE SYDOW 1929. IN SYDOW & PETRAK, ANN. MYCOL. 27:84-85 (ILLUS) /A. COSTARICANA
=TITAEA SACC. 1876 FIDE AINSWORTH 1961. •

APHANOCLADIUM GAMS 1971. CEPHALO. SCHIMMEL. P 196-198 (ILLUS, KEY) /ACREMONIUM ALBUM PREUSS =A. ALBUM (PREUSS)
AMERO,CETERI,HYALO,MISC,ALEURI, 54D, ON FUNGI, SPIDERS, DUNG & LITTER.
/VON ARX, GENERA P 182-183,1970 (ILLUS) /GAMS, PERSOONIA 7:161-169,1973 (REVIS,ILLUS) /MATSUSHIMA, ICONES P 9 & PL 7,1975 (ILLUS)
?=*ENGYODONTIUM* DE HOOG 1978 •

APHOTISTIS HUMBOLDT 1793 =NOMEN DUBIUM FIDE DONK 1962 (TAXON 11:79).
?=RHIZOMORPHA SEE AINSWORTH 1971. •

APIOSPORIUM KUNZE 1817. IN KUNZE & SCHMIDT, MYKOL. HEFTE 1:8-9 /A. SALICIS KUNZE & SCHMIDT 1817
=SCLEROTIUM, FIDE HOHNEL FROM AINSWORTH 1971. •

APOSPORELLA THAXTER 1920. BOT. GAZ. 69:11-13 & PL III /A. ELEGANS
AMERO,CETERI,HYALO,PHIAL? OR ARTHRO?, 55B, ON INSECTS. •

APOTEMNOUM CORDA 1833. STURM'S DEUTSCHL. FLORA, BD. 3, HEFT 13:77-78 & TAB 39 /A. MACULANS
?, SACCARDO TREATED AS CLASTEROSPORIUM, BUT CORDA SAID SLIMY. •

APPELIA "(SACC.) SACCARDO" =TRICHOCONIS SEE SACCARDO 1906 (SYLL. FUNG. 18:747) •

ARACHNOPHORA HENNEBERT 1963. CAN. J. BOT. 41:1165-1169 (ILLUS) /A. FAGICOLA
DICTYO OR STAURO,CETERI,PHAEO,ANNEL,APPEND, 115D, ON LITTER (FAGUS).
/ELLIS, DH P 139,1971 (ILLUS) /HUGHES, FUNGI CANADENSES 164,1980 (ILLUS) •

ARANEOMYCES HOHNEL 1909. S. B. AKAD. WISS. WIEN 118:894-896 (ILLUS) /A. ACARIFERUS
=TITAEA SACCARDO 1876 FIDE DAMON 1952 (J. WASH. ACAD. SCI. 42:365-367). •

ARBUSCULA BATISTA & PERES 1965. MYCOPATHOLOGIA 25:162-165 (ILLUS) /A. EUGENIAE
BULBILS,BLASTO,PHAEO,NONSP, 129D, ON LITTER (LEAVES).
/SUBRAMANIAN & VITTAL, CAN. J. BOT. 51:977-981,1973 (ILLUS) •

ARCUADENDRON SIGLER & CARMICHAEL 1976. MYCOTAXON 4:355-361 (ILLUS) /A. OVATUM
AMERO,BLASTO,HYALO,NONSP,ALEURI, 41D, ON SOIL & MAMMALS. •

ARDHACHANDRA SUBRAMANIAN & SUDHA 1978. CAN. J. BOT. 56:729-731 (ILLUS) /PSEUDOBELTRANIA SELENOIDES DE HOOG 1977 =A. SELENOIDES (DE HOOG)

AMERO,CETERI,PHAEO,RADUL, 21C, ON LITTER (BAPHIA?, IXORA & BROAD LEAF TREES). /MATSUSHIMA, MICROFUNGI P 49,1971 (ILLUS SUB RHINOCLADIELLA CRISTASPORA) /DE HOOG & HERMANIDES-NIJHOF, CBS STUDIES 15:199,1977 (NEW SP,NEW COMB FOR R. CRISTASPORA,NO ILLUS) •

ARNAUDINA TROTTER 1931. IN SACCARDO, SYLLOGE FUNGORUM 25:986 /A. MANAOSENSIS (ARNAUD) TROTTER 1925 (AS 'MANAOENSIS')
?=SEPTOIDIUM FIDE VON ARX 1975 (CBS STUDIES 8:50). NEW NAME FOR EXOSPORINA ARNAUD NON OUDEMANS, SEE HUGHES & PIROZYNSKI 1972 (CAN. J. BOT. 50:2531-2532). ASCO STATE =PARODIELLINA FIDE ARNAUD 1921.
/ARNAUD, ANN. EPIPHYT. 7:105-106,1921 (ORIG DESCR)
=*EXOSPORINELLA* BENDER 1932, MYCOLOGIA 24:410-411 •

ARTHRINIUM KUNZE 1817. MYKOL. HEFTE 1:9-10 /A. CARICICOLA
AMERO,CETERI,PHAEO,MISC,ALEURI, 46A-B, ON GRASSES (TUNCUS, CAREX, SCIRPUS) & OTHER PLANTS. ASCO STATE =APIOSPORA FIDE HUDSON 1963 (TBMS 46:19-23,ILLUS SUB PAPULARIA) OR PHYSALOSPORA FIDE VON ARX 1974 (GENERA P 246) OR PSEUDOGUIGNARDIA FIDE ELLIS 1971 (DH P 574-575). COMPARE WITH CORDELLA & PTEROCONIUM.
/ELLIS, CMI PAPERS 103:1-33,1965 (MONOGR,KEY,ILLUS) /ELLIS, DH P 567-575,1971 (ILLUS,KEY) /ELLIS, MDH P 477-478,1976 (ILLUS,NEW SP) /HUDSON & MCKENZIE, TBMS 66:359-362,1976 (CONIDIAL & ASCOSPORE SIZES) /COLE & SAMSON, CONIDIAL FUNGI P 99-102,1979 (ILLUS)
=*CAMPTOUM* LINK 1824 FIDE HUGHES 1958
=*GONATOSPORIUM* CORDA 1839 FIDE HUGHES 1958
=*GONIOSPORIUM* LINK 1824 FIDE HUGHES 1958
=*INNATOSPORA* VAN BEYMA 1919 FIDE ELLIS 1971
=*MICROTYPHA* SPEGAZZINI 1910 FIDE ELLIS 1971
=*PAPULARIA* FRIES 1825 FIDE ELLIS 1965
=*PHAEOHARZIELLA* LOUBIERE 1924 FIDE ELLIS 1965
=*PSEUDOBASIDIUM* TENGWALL 1924 FIDE ELLIS 1971
=*RACEMOSPORIUM* MOREAU & MOREAU 1941 FIDE ELLIS 1965
?=*RHINOCEPHALUM* KAMYSCHKO 1961
=*SPOROPHLEUM* NEES EX LINK 1824 FIDE HUGHES 1958
=*TUREENIA* HALL 1915 FIDE ELLIS 1971 •

ARTHROBOTRYELLA SIBILIA 1928. BOL. STAZ. PAT. ROME, N. S. 8:448, FIGS 1 & 2 /A. HERNICA
?=CORDANA PREUSS 1851. NO TYPE MATERIAL EXISTS SEE HUGHES 1978 (NEW ZEALAND J. BOT. 16:326). •

ARTHROBOTRYOMYCES BATISTA & BEZERRA 1961. INST. MYCOL. UNIV. RECIFE PUBL. NO. 321:9-12 (ILLUS) /A. AMAZONENSIS
?=ARTHROBOTRYUM CESATI 1854. •

ARTHROBOTRYS CORDA 1839. PRACHTFLORA P 43 /A. SUPERBA
AMERO OR DIDYMO OR PHRAGMO,CETERI,HYALO,RADUL OR AMPUL, 62G, ON NEMATODES. DISTINCTION FROM DACTYLELLA NOT CLEAR.
/DRECHSLER, MYCOLOGIA 29:447-552,1937 (ILLUS,NEW SPP) /HAARD, MYCOLOGIA 60:1140-1159,1968(ILLUS,KEY,REVIS) /MATSUSHIMA, ICONES P 9-11 & PL 205-209,1975 (ILLUS,NEW SP) /ROXON & JONG, MYCOTAXON 3:162-164,1975 (CULTURE OF A. ENTOMOPAGA) /MCCULLOCH, TBMS 68:176-177,1977 (ILLUS,NEW SP) /SCHENCK, KENDRICK & PRAMER, CAN. J. BOT. 55:977-985,1977 (ILLUS,NEW SP,NEW COMBS) /BARRON, THE NEMATODE-DESTROYING FUNGI 140P,1977 (ILLUS,ECOL,REFS) /BARRON, CAN. J. BOT. 57:1371-1373,1979 (NEW SP,ILLUS)
=*CANDELABRELLA* RIFAI & COOKE 1966 FIDE BARRON 1968
?=*DACTYLELLA* GROVE 1884
=*DIDYMOZOOPHAGA* SAPRUNOV & GALIULINA 1951 (NOMEN ILLEGITIMUM)
?=*DUDDINGTONIA* COOKE 1969
?=*GENICULARIA* RIFAI & COOKE 1966
?=*GENICULIFERA* RIFAI 1975
?=*GONYELLA* SYDOW & SYDOW 1919
?=*NEMATOPHAGUS* MEKHTIEVA 1975 •

ARTHROBOTRYUM CESATI 1854. HEDWIGIA 1:TAB 4, FIG 1, ERKLARUNG DER TAB 4 & 5 /A. STILBOIDEUM
PHRAGMO,GLOIO,HYALO OR PHAEO,ANNEL,SYNNEMATA, 77D, ON WOOD (QUERCUS) &

BRANCHES (KALMIA), NON ARTHROBOTRYUM ROSTRUP 1916 = GONYELLA SYDOW 1919.
/HUGHES, THE NATURALIST, OCTOBER-DECEMBER:171-172,1951 (ILLUS) /ELLIS, DH P 335-336,1971 (ILLUS) /ELLIS, MDH P 350-351,1976 (ILLUS)
?=*ARTHROBOTRYOMYCES* BATISTA & BEZERRA 1961
=*ARTHROGRAPHIUM* CESATI FIDE SACCARDO FROM AINSWORTH 1971
=*LINDAUOMYCES* KOORDERS 1907 FIDE CLEMENTS & SHEAR 1931
=*WETTSTEINIELLA* KUNTZE FIDE AINSWORTH 1971 •

ARTHROBOTRYUM ROSTRUP 1916 -SEE GONYELLA •

ARTHROCLADIUM PAPENDORF 1969. TBMS 52:483-489 (ILLUS) /A. CAUDATUM
PHRAGMO OR SCOLECO,CETERI,PHAEO,NONSP, 71B, ON SOIL. •

"ARTHROCONIDIA" REPORTED AS CONIDIAL STATES OF ARACHNIOTUS (A) & OTHER GYMNOASCACEAE SEE APINIS 1964 (CMI PAPERS 96:1-56,ILLUS) & MULLER & PACHA-AUE 1968 (NOVA HEDWIGIA 15:552 & TAB 77-E, SUB ARACHNIOTUS FLAVOLUTEUS) & MOUSTAFA 1973 (TBMS 61:392-393) OR ASCOCALVATIA OR XYNOPHILA SEE MALLOCH & CAIN 1971 (CAN. J. BOT. 49:839) OR FAURELINA VON ARX 1978 (PERSOONIA 10:27-28). BASIDIO STATE = COPRINUS OR PHLEBIA SEE SIGLER & CARMICHAEL 1976 (MYCOTAXON 4:368-372), OR PENIOPHORA, GLOEOPHYLLUM, ?TYROMYCES, PORIA, MYCOACIA, SPONGIPELLIS, GLOBIFOMES, TRAMETES, PERENNIPORIA, CORIOLOPSIS, DICHOMITUS, AMYLOSTEREUM, BJERKANDERA, PHELLINUS GLOEOCYSTIDIELLUM, RIGIDOPORUS SEE STALPERS 1978 (CBS STUDIES 16:39) OR PHLEBOPUS SEE PANTIDOU 1961 (CAN. J. BOT. 39:1163-1167,ILLUS) OR PLEUROTUS SEE WATLING & KENDRICK 1979 (NATURALIST 104:1-4) OR MELANOTUS SEE AMMIRATI ET AL 1979 (MYCOLOGIA 71:310-321). SEE ALSO ENTRIES UNDER GEOTRICHUM, MALBRANCHEA, PTYCHOGASTER, ARTHROGRAPHIS, SCYTALIDIUM & SPORENDONEMA. OTHER STATES = PAPULASPORA SEE GAMS & DOMSCH 1969 (NOVA HEDWIGIA 18:10-11,ILLUS), EPIDERMOPHYTON, MICROSPORUM, TRICHOPHYTON, CHRYSOSPORIUM, OR HARPOSPORIUM SEE BARRON 1979 (CAN. J. BOT. 57:886-889). •

ARTHROGRAPHIS COCHET EX SIGLER & CARMICHAEL 1976. MYCOTAXON 4:359-367 (ILLUS)
/OIDIODENDRON KALRAE TEWARI & MACPHERSON 1971 =A. KALRAE (TEWARI & MACPHERSON) SIGLER & CARMICHAEL 1976 =A. STATE OF PITHOASCUS LANGERONII VON ARX 1978 =A. LANGERONI COCHET 1939 NOMEN NUDUM
AMERO,ARTHRO,HYALO,NONSP, 42C, ON MAN & SOIL. ASCO STATE =PITHOASCUS FIDE VON ARX 1978 OR ?ARACHNOTHECA SEE MULLER & PACHA-AUE 1966 (NOVA HEDWIGIA 15:554 & TAB 77-N, SUB ARACHNIOTUS GLOMERATUS).
/COCHET, ANN. PARASIT. HUM. COMP. 17:97-101 & PL III,1939 (ILLUS) /VON ARX, PERSOONIA 10:23-26,1978 (ILLUS SUB PITHOASCUS) •

ARTHROGRAPHIUM CESATI 1854 (IN RABENHORST, HERB. MYCOL. NO. 1821; BOT. ZEIT. 1854:186) =ARTHROBOTRYUM CESATI FIDE SACCARDO 1901 (SYLL. FUNG. 15:48) •

ARTHROSPORIA GRIGORAKIS 1925. ANN. SCI. NAT. BOT. X,7:414 /SEVEN ORIGINAL SPECIES =NOMEN ILLEGITIMUM FIDE DODGE 1935. •

ARTHROSPORIUM SACCARDO 1880. MICHELIA 2:32 /A. ALBICANS (SACCARDO)
PHRAGMO,CETERI,HYALO,RACHI,SYNNEMATA, 86B, ON WOOD.
/MORRIS, WESTERN ILLINOIS UNIV. SER. BIOL. SCI. 3:24-25,1963 (ILLUS) /WANG, MYCOLOGIA 64:1175-1179,1972 (ILLUS)
=*PHRAGMOSTILBE* SUBRAMANIAN FIDE WANG 1972
?=*STILBOMYCES* ELLIS & EVERHART 1896 •

ARTICULARIA HOHNEL 1909. S. B. AKAD. WISS. WIEN 118:407-410 (ILLUS) /A. QUERCINA
AMERO,GLOIO,HYALO,AMPUL?,SYNNEMATA, ON FUNGI (ASCOMYCETELLA QUERCINA). COMPARE WITH MICROSTROMA AND SEE FIG 32D.
/PECK, BULL. TORREY BOT. CLUB 8:49-51 & PL VII,1881 (ILLUS) /CHARLES, MYCOLOGIA 27:74-82,1935 (ILLUS) /HAWKSWORTH & MIELKE, PHYTOPATH. 52:451-454,1962 (ECOL) /COX, MYCOLOGIA 68:500,1976 (DESCR OF TYPE SPECIES) •

ARTICULARIELLA HOHNEL 1909. S. B. AKAD. WISS. WIEN 118:410 /A. AURANTIACA =MICROSTROMA NIESSL FIDE VON ARX 1970 (P 71). •

ARTICULOSPORA INGOLD 1942. TBMS 25:372-376 (ILLUS) /A. TETRACLADIA
STAURO,CETERI,HYALO,NONSP,AQUATIC, 119B.
/TUBAKI, BULL. NAT. SCI. MUSEUM, TOKYO 3:251-253,1957 (ILLUS) /NOLAN, MYCOLOGIA

64:1169-1174,1972 (ILLUS) /MATSUSHIMA, ICONES P 11-12 & PL,1975 (ILLUS,NEW SP) /INGOLD, GUIDE P 22-23,1975 (ILLUS)
=*GENICULOSPORA* NILSSON EX MARVANOVA & NILSSON 1971 FIDE INGOLD 1975 •

ARTOCREAS BERKELEY & BROOME 1873 =MICHENERA BERKELEY & CURTIS 1868 SEE DONK 1962 (TAXON 11:89) •

ARTOTROGUS MONTAGNE 1845 ?=PYTHIUM FIDE MIDDLETON FROM DONK 1962 (TAXON 11:79-81) •

ARXIELLA PAPENDORF 1967. TBMS 50:73-75 (ILLUS) /A. TERRESTRIS
DIDYMO,BLASTO,HYALO,RADUL, 65B, ON SOIL.
/BARRON, HYPHOMYCETES P 92-93,1968 (ILLUS) •

ASCHIZOTRICHUM RIEUF 1962. CAH. RECH. AGRON. RABAT 15:61-62 (ILLUS) /A. GRISEUM =WIESNERIOMYCES KOORDERS 1907 FIDE KENDRICK 1973. •

ASCOCHYTITES BABAYUN & TASLAKHCHYAN 1973 (MIKOL. FITOPATHOL. 7:180) FOSSIL FUNGI IMPERFECTI. NOT VALIDLY PUBLISHED SEE SUTTON 1977 (CMI PAPERS 141:17). •

ASCOCONIDIUM SEAVER 1942. MYCOLOGIA 34:412-414 (ILLUS) /A. CASTANEAE
PHRAGMO OR DICTYO,CETERI,HYALO,PHIAL, 97G, ON BRANCHES (TSUGA & CASTANEA).
ASCO STATE =SAGERIA FIDE FUNK 1975.
/FUNK, CAN. J. BOT. 44:219-222,1966 (ILLUS) /FUNK, CAN. J. BOT. 53:1196-1199,1975 (ILLUS) /MATSUSHIMA, ICONES P 12 & PL,1975 (ILLUS) /NAG RAJ & KENDRICK, CHALARA P 40-41 & 164-166,1975 (KEY,ILLUS) •

ASPERGILLOPSIS SPEGAZZINI 1910. AN. MUS. NAC. BUENOS AIRES 20(SER.3,13):434-436 (ILLUS) /A. NIGRA (VAN TIEGHEM)
=ASPERGILLUS MICHELI EX LINK 1809 (NIGER SERIES). •

ASPERGILLOPSIS SOPP 1912. VID.-SELSK. SKR. I. M.-N. KL. 11:201-204 & TAFEL 20 /A. FUMOSIS
=NOMEN ILLEGITIMUM, NON ASPERGILLOPSIS SPEG. 1910. •

ASPERGILLUS MICHELI EX LINK 1809. MAG. GES. NATURF. FREUNDE, BERLIN 3:16 /A. GLAUCUS LINK 1809 (LECTOTYPE FIDE CLEMENTS & SHEAR 1931)
AMERO,BASO,HYALO OR PHAEO,PHIAL, 8C, ON SOIL, PLANTS, DUNG & OTHER SUBSTRATES. ASCO STATE =EUROTIUM, DICHLAENA, EMERICELLA, EDYUILLIA, NEOSARTORYA, PETROMYCES, FIDE MALLOCH & CAIN 1972, OR FENNELLIA, WARCUPIELLA, CHAETOSARTORYA, HEMICARPENTELES FIDE VON ARX 1974 (GENERA P 91-94 & 195-196) OR SARTORYA FIDE UDAGAWA & KAWASAKI 1968 (TRANS. MYCOL. SOC. JAPAN 8:115-121,ILLUS), OR SCLEROCLEISTA SEE MALLOCH & CAIN 1973.
/BULLER, TRANS. R. SOC. CANADA, SECT. IV, SER III, 9:1-25 & PL I-IV,1915 (MICHELI'S ILLUS & DESCR) /RAPER & FENNELL, THE GENUS ASPERGILLUS 686 P,1965 (MONOGR) /ONIONS, CMI DESCRIPTIONS 91-95,1966 (ILLUS) /ELLIS, DH P 547-548,1971 (ILLUS) /SUBRAMANIAN, CURR. SCI. 41:755-761,1972 (ASCO STATES,NOT SEEN) /MALLOCH & CAIN, CAN. J. BOT. 50:2613-2628,1972 (ILLUS,ASCO STATES) /MALLOCH & CAIN, CAN. J. BOT. 51:1647-1648,1973 (REVIS OF ASCO STATES) /SAMSON & MOUCHACCA, ANTONIE VAN LEEUWENHOEK 41:343-351,1975 (ILLUS, NEW SPP) /CHRISTENSEN & RAPER, MYCOLOGIA 70:200-205,1978 (ILLUS,NEW SP) /CHRISTENSEN, RAPER & STATES, MYCOLOGIA 70:332-342,1978 (ILLUS,NEW SPP) /CHRISTENSEN & RAPER, TBMS 71:177-191,1978 (KEY) /BARTOLI & MAGGI, TBMS 71:383-394,1978 (ILLUS,NEW SP) /KOZAKIEWICZ, TBMS 70:175-186,1978 (ILLUS,CONIDIUM ONTOGENY) /SAMSON, CBS STUDIES 18:1-38,1979 (ILLUS,COMPILATION)
?=*ALLIOSPORA* PIM 1883
=*ASPERGILLOPSIS* SPEGAZZINI 1910
=*CLADASPERGILLUS* RITG. FIDE AINSWORTH 1971
?=*CLADOSARUM* YUILL & YUILL 1938 FIDE AINSWORTH 1971
=*EUASPERGILLUS* LUDWIG FIDE AINSWORTH 1971
?=*GUTTUROMYCES* RIVOLTA FIDE AINSWORTH 1971
=*RAPERIA* SUBRAMANIAN & RAJENDRAN 1976 FIDE SAMSON 1979
=*REDAELLIA* CIFERRI 1930 FIDE CARMICHAEL & FENNELL
=*RHOPALOCYSTIS* GROVE 1911 FIDE RAPER & FENNELL 1965
?=*SCEPTROMYCES* CORDA 1831 SEE LINDAU 1907
?=*SPERMATOLONCHA* SPEGAZZINI 1909
?=*SPHAEROMYCES* MONTAGNE 1845

=*STERIGMATOCYSTIS* CRAMER 1859 FIDE CLEMENTS & SHEAR 1931
?=*STILBOTHAMNIUM* HENNINGS 1897 •

ASPERISPORIUM MAUBLANC 1913. BULL. SOC. MYCOL. FRANCE 29:357-358 /A. CARICAE (SPEG.) = A. STATE OF SPHAERELLA CARICAE MAUBL. 1913
DIDYMO,CETERI,PHAEO,RADUL,SPORODOCHIA, 63D, ON LEAVES. ASCO STATE =SPHAERELLA FIDE MAUBLANC 1913. COMPARE WITH PASSALORA.
/HUGHES, CAN. J. BOT. 31:575,1953 (ILLUS) /ELLIS, DH P 273-274,1971 (ILLUS) /SUTTON, CMI PAPERS 138:182-185,1975 (ILLUS) /MATSUSHIMA, ICONES P 12 & PL 180,1975 (ILLUS,NEW SP) /DEIGHTON, TBMS 66:549-550,1976 (ILLUS,NEW COMB) /ELLIS, MDH P 240-243,1976 (ILLUS,KEY,NEW COMBS) •

ASTELECHIA CIFERRI 1962. ATTI IST. BOT. UNIV. PAVIA SER 5,19:90 & 139 & PL V, FIGS 4-6 /A. RADIANS
?, ORIGINAL ILLUSTRATION NOT DIAGNOSTIC. •

ASTEROBOLUS REDHEAD & PERRIN 1972. CAN. J. BOT. 50:409-412 /A. GAULTHERIAE
=VALDENSIA PEYRONEL 1923, SEE REDHEAD & PERRIN 1972 (CAN. J. BOT. 50:2083-2084). •

ASTEROMYCES MOREAU & MOREAU EX HENNEBERT 1962. CAN. J. BOT. 40:1211-1213 (1203-1216,ILLUS) /A. CRUCIATUS
AMERO,CETERI,PHAEO,RADUL, 22D, ON SOIL (SAND) & ROOTS (AMMOPHILA).
/MOREAU & MOREAU, REV. MYCOL. 6:49-94,1941 (NO LATIN DIAGNOSIS) /ELLIS, DH P 78-79,1971 (ILLUS) /COLE, MARINE BIOL. 38:147-158,1976 (ILLUS) •

ASTEROPHORA DITMAR 1809. SCHRAD. J. BOT. 3:56 /A. LYCOPERDOIDES =A. STATE OF NYCTALIS ASTEROPHORA FIDE BREFELD FROM BULLER 1924
AMERO OR STAURO,CETERI,HYALO,NONSP,SYNNEMATA,CLAMPS,APPEND, 51E, ON FUNGI (RUSSULA). BASIDIO STATE =NYCTALIS FIDE BREFELD.
/BULLER, RESEARCHES ON FUNGI 3:446-463 1924 (ILLUS,REFS) /INGOLD, TBMS 24:29-32,1940 (ILLUS) /HUGHES, IN KENDRICK P 20-23,1971 (ILLUS) /DONK, TAXON 11:79-81,1962 (REFS) /DURAND & NICOT, REV. MYCOL. 33:96-107,1968 (ILLUS,REFS) /SINGER, AGARICALES IN MODERN TAXONOMY, CRAMER, P 69 & 224-225 & PL 11,1975 (ILLUS,DISCUSSION)
=*ASTEROSPERMA* PERSOON 1818 FIDE DONK 1962
=*ASTEROTRICHUM* BONORDEN 1851
=*STELLIFERA* LEMAN 1824 FIDE DONK 1962 •

ASTEROSPERMA PERSOON 1818 =ASTEROPHORA DITMAR 1809 FIDE DONK 1962 (TAXON 11:80) •

ASTEROSPORIUM KUNZE 1819 (FLORA 2(15):225) =COELOMYCETES SEE HUGHES 1958 (P 738) & SUTTON 1977 (CMI PAPERS 141:20). •

ASTEROTHECIUM WALLROTH 1836. IN ENDLICHER, GEN. PLANT. P 25 /A. STRIGOSUM (WALLROTH) RABENHORST 1844 (DEUTSCHL. KRYPT.-FL. 1:286)
=STEPHANOMA WALLROTH 1833 FIDE LINDAU IN RABENHORST 1907 (KRYPT.-FL. 2 AUFL., 1 BD., 8:225). •

ASTEROTRICHUM BONORDEN 1851. HANDB. ALLGEM. MYKOL. P 82-83 & FIG 224 /A. DITMARI, NOM. NOV. FOR ASTEROPHORA AGARICOIDES DITMAR
=ASTEROPHORA DITMAR 1809 FIDE DONK 1962 (TAXON 11:80-81). •

ASTOMA GRAY 1821 =SCLEROTIUM FIDE RABENHORST 1844 (KRYPT.-FL. GERM. P 238) •

ASTRABOMYCES BATISTA 1961. PUBL. INST. MICOL. UNIV. RECIFE, 320:22-27 (ILLUS) /A. AMAZONENSIS BATISTA & CAVALCANTI 1961
?, ORIGINAL ILLUSTRATION DIFFICULT TO INTERPRET. •

ASTRAGOXYPHIUM BATISTA, NASCIMENTO & CIFERRI 1963 =LEPTOXYPHIUM SPEGAZZINI 1918 FIDE HUGHES 1976 (MYCOLOGIA 68:777 & 747-748) •

ASTRODOCHIUM ELLIS & EVERHART 1897. AM. NAT. 31:430 /A. COLORADENSE
?, NOT ILLUSTRATED BY AUTHORS. •

ATELEOTHYLAX OTA & LANGERON 1923. ANN. PARASIT. HUM. COMP. 1:333 /TRICHOPHYTON CURRII CHALMERS & MARSHALL 1914 =A. CURRII (CHALMERS & MARSHALL)
=NOMEN DUBIUM. •

ATICHIA FLOTOW 1850. LINNAEA 23:149-150 /A. MOSIGII =A. GLOMERULOSA (ACHARIUS) STEIN IN COHN 1879 FIDE MEEKER 1975
STAURO,GLOIO,CETERI,HYALO OR PHAEO,NONSP,SPORODOCHIA, 122G, ON LEAVES. ASCO STATE =SEURATIA FIDE MEEKER 1975.
/MEEKER, CAN. J. BOT. 53:2462-2482 & 2483-2496,1975 (ILLUS,REVIS)
=*ACTINOMMA* SACCARDO 1884 FIDE MEEKER 1975
=*HETEROBOTRYS* SACCARDO 1880 FIDE MEEKER 1975
=*HYPHODICTYON* MILLARDET 1866 FIDE MEEKER 1975
=*MORULARIA* NANNIZZI 1925 FIDE MEEKER 1975
?=*MYRIOPHYSA* FRIES 1849 FIDE HOHNEL FROM AINSWORTH 1971
=*MYRIOPHYSELLA* SPEGAZZINI 1910 FIDE SANTESSON FROM AINSWORTH 1971 •

ATRACTIELLA SACCARDO 1886. IN SACCARDO & MALBRANCHE, ATTI IST. VENET. SCI., LETT. ED ARTI, SER. 6, 1:1280 (NO ILLUS) /A. BRUNAUDIANA (SACCARDO)
? •

ATRACTILINA DEARNESS & BARTHOLOMEW 1924. IN DEARNESS, MYCOLOGIA 16:175 /A. CALLICARPAE =A. PARASITICA (WINTER) DEIGHTON & PIROZYNSKI 1972
PHRAGMO,CETERI,HYALO OR PHAEO,RACHI,SYNNEMATA, 72F, ON FUNGI (MELIOLA & OTHERS) ON LEAVES.
/DEIGHTON & PIROZYNSKI, CMI PAPERS 128:33-44,1972 (ILLUS) /ELLIS, MDH P 177-178,1976 (ILLUS) •

ATRACTINA HOHNEL 1904. HEDWIGIA 43:298 /A. BISEPTATA
=STERIGMATOBOTRYS OUDEMANS 1886 FIDE HUGHES 1958. •

ATRACTIUM LINK 1809. MAG. GES. NATURF. FREUNDE, BERLIN 3:10 /A. STILBASTER
PHRAGMO OR SCOLECO,GLOIO,HYALO,PHIAL,SYNNEMATA, 111E, ON SOIL & ROOTS (CITRUS & OTHERS). ASCO STATE =NECTRIA (SUB SPHAEROSTILBE) FIDE SEAVER 1909 (MYCOLOGIA 1:178). /F
/MORRIS, WESTERN ILLINOIS UNIV. SER. BIOL. SCI. 3:30-31,1963 (ILLUS) /BOOTH & HOLLIDAY, CMI DESCRIPTIONS 391,1973 (ILLUS SUB SPHAEROSTILBE)
?=*MICROCERA* DESMAZIERES 1848 •

ATRICHOPHYTON CASTELLANI & CHALMERS 1919. MANUAL OF TROPICAL MEDICINE, ED. 3, P 1008 /A. ALBISCANS (NIEUWENHUIS)
=CHRYSOSPORIUM CORDA 1833 FIDE CARMICHAEL. •

ATTAMYCES KREISEL 1972. Z. ALLG. MIKROBIOL. 12:648-649 (ILLUS) /A. BROMATIFICUS
STERILE, ON ANT NESTS. BASIDIO STATE ?=ROZITES FIDE KREISEL (=LEUCOAGARICUS). •

AUREOBASIDIUM VIALA & BOYER 1891. REV. GEN. BOT. 3:369 /A. VITIS =A. PULLULANS (DE BARY) ARNAUD 1918
AMERO,GLOIO,HYALO OR PHAEO,AMPUL, 31C, ON PLANTS AND OTHER SUBSTRATES. HAS BEEN CONFUSED WITH HORMONEMA, SEE HERMANIDES-NIJHOF 1976 & 1977 (VERH. KON. AKAD. WETENSCH. AFD. NAT. 66:5-6 & 67:7). OTHER CONIDIAL STATE =KABATIELLA SEE HERMANIDES-NIJHOF 1977. ASCO STATE =GUIGNARDIA OR POTEBNIAMYCES OR DOTHIDEA OR ?=MUELLERITES, SEE HERMANIDES-NIJHOF 1977.
/COOKE, MYCOPATHOLOGIA 12:1-45,1959 (REFS) /COOKE, MYCOPATHOLOGIA 21:225-271,1963 (ILLUS,PHYSIOL) /VON ARX, BIBLIO. MYCOL. (CRAMER) BD 24:70-82,1970 (ILLUS,REVIS) /HERMANIDES-NIJHOF, CBS STUDIES 15:141-177,1977 (MONOGR)
=*AUREOBASIS* CLEMENTS & SHEAR 1931, ORTHOGRAPHIC VARIANT
=*CHRYSOBASIDIUM* CLEMENTS 1909 FIDE DONK 1956
=*DEMATOIDIUM* STAUTZ 1931, FIDE HERMANIDES-NIJHOF 1977
=*PULLULARIA* BERKHOUT 1923 FIDE COOKE 1962 •

AUREOBASIS CLEMENTS & SHEAR 1931 FOR AUREOBASIDIUM •

AUREOMYCES RUOKOLA & SALONEN 1970. MYCOPATHOLOGIA 42:273-276 (ILLUS) /A. MIRABILIS
=CEPHALOASCUS HANAWA 1920 (A) FIDE VON ARX 1974 (?=CONIDIASCUS HOLTERMANN 1898 FIDE MALLOCH). •

AZOSMA CORDA 1831. STURM'S DEUTSCHL. FLORA, BD. 3, HEFT 12:35-36 & TAB 18 /A. HELMINTHOSPORIODES
?, ORIGINAL ILLUSTRATION & DESCRIPTION NOT DIAGNOSTIC. •

BACILLISPORA NILSSON 1962. BOT. NOTISER. 115:77-79 (ILLUS) /B. AQUATICA
PHRAGMO OR SCOLECO,CETERI,HYALO,NONSP,AQUATIC, 89C. •

BACTRIDIOPSIS HENNINGS 1904. HEDWIGIA 43:397-398 (ILLUS) /B. ULEI
AMERO,CETERI,PHAEO,NONSP, 47C, ON WOOD. NON BACTRIDIOPSIS FRAGOSO & CIFERRI 1927 (BOL. SOC. ESP. HIST. NAT. 27:330-331,ILLUS).
/DAMON & DOWNING, MYCOLOGIA 46:209-221,1954 (ILLUS) •

BACTRIDIUM KUNZE 1817. IN KUNZE AND SCHMIDT, MYKOL. HEFTE 1:5 /B. FLAVUM
PHRAGMO,CETERI,HYALO,NONSP,SPORODOCHIA, 89D, ON WOOD & BARK.
/HUGHES, NEW ZEALAND J. BOT. 4:522-532,1966 (REVIS,ILLUS)
=*DAMNOSPORIUM* CORDA FIDE AINSWORTH 1971
=*ERICIANELLA* BRONDEAU FIDE FRIES 1849
=*PODOBACTRIDIUM* PETCH FIDE HUGHES 1966 •

BACTRODESMIELLA ELLIS 1959. CMI PAPERS 72:14-15 (ILLUS) /B. MASONII (HUGHES)
PHRAGMO,BASO OR CETERI,PHAEO,ANNEL, 76F, ON LITTER (FAGUS).
/ELLIS, DH P 104-105,1971 (ILLUS) •

BACTRODESMIUM COOKE 1883. GREVILLEA 12:35 /B. ABRUPTUM (BERK. & BR.) MASON & HUGHES 1953 (LECTOTYPE)
PHRAGMO,CETERI,PHAEO,NONSP,ALEURI,SPORODOCHIA, 90B, ON WOOD & BARK. OTHER CONIDIAL STATE =TRIADELPHIA.
/ELLIS, CMI PAPERS 72:1-15,1959 (REVIS,KEY,ILLUS) /ELLIS, DH P 100-104,1971 (ILLUS,KEY) /HOLUBOVA-JECHOVA, FOLIA GEOBOT. PHYTOTAX., PRAHA 7:407-418,1972 (ILLUS,KEY,NEW SPP) /MATSUSHIMA, ICONES P 12-13 & PL,1975 (ILLUS) /ELLIS, MDH P 67-69,1976 (ILLUS,NEW SPP) /SUTTON, BOL. SOC. ARGENT. BOT. 18:155-156,1977 (ILLUS,NEW SP) /HUGHES, NEW ZEALAND J. BOT. 16:318,1978 (ILLUS) •

BAHUCHASHAKA SUBRAMANIAN 1977. KAVAKA 5:93-98 (NOT SEEN) /DWAYABEEJA GRAMINIS MATSUSHIMA 1975 =B. GRAMINIS (MATSUSHIMA)
?, SEE ALSO POLYSCHEMA.
/MATSHUSHIMA, ICONES P 66 & PL,1975 (ILLUS) •

BAHUKALASA SUBRAMANIAN & CHANDRASHEKARA 1978. KAVAKA 6:41-48 (NOT SEEN) /B. SAMALA
?, ON DUNG. •

BAHUPAATHRA SUBRAMANIAN & LODHA 1964. ANTONIE VAN LEEUWENHOEK 30:329 & FIGS 3-4, P 322 /B. SAMALA
?=CLADORRHINUM SACC. & MARCH., SEE VON ARX 1973 (VERH. KON. NEDERL. AKAD. WET. AFD. NAT. 61:63). •

BAHUSAGANDA SUBRAMANIAN 1971. HYPHOMYCETES (NEW DELHI) P 387-389 (ILLUS,NO LATIN DIAGNOSIS) /DEIGHTONIELLA INDICA SUBRAMANIAN 1958 =B. INDICA (SUBRAMANIAN)
?=TORULA (PERSOON) LINK 1809 (SEE 88E). •

BAHUSAKALA SUBRAMANIAN 1958. J. INDIAN BOT. SOC. 37:61-63 (ILLUS) /B. OLIVACEO-NIGRA (BERKELEY & BROOME)
DIDYMO OR PHRAGMO,ARTHRO,PHAEO,NONSP, 42B, ON PLANTS (YUCCA, AGAVE, SALIX). ASCO STATE = AULOGRAPHINA FIDE MULLER, HARR & SULMONT 1969 (REV. MYCOL. 33:369-378,ILLUS), OR XYLOGONE SEE VON ARX 1974 (GENERA P 103,ILLUS)
/VON ARX, GENERA P 168-169,1970 (ILLUS) /ELLIS, DH P 29,1971 (ILLUS) /SIGLER & CARMICHAEL, MYCOTAXON 4:366-368,1976 (ILLUS) /ELLIS, MDH P 17-18,1976 (ILLUS,NEW SP) •

BAHUSANDHIKA SUBRAMANIAN 1956. J. INDIAN BOT. SOC. 35:469 /B. INDICA (SUBRAMANIAN)
?=TORULA (PERSOON) LINK 1809 (SEE 88F).
/ELLIS, MDH P 369-370,1976 (ILLUS) •

BAHUSUTRABEEJA SUBRAMANIAN & BHAT 1977. CAN. J. BOT. 55:2202-2206 (ILLUS) /B. DWAYA
AMERO OR STAURO,GLOIO,HYALO,PHIAL,APPEND, 1E, ON LITTER (COFFEA). •

BAINIERIA ARNAUD 1952. BULL. SOC. MYCOL. FRANCE 68:188-189 (ILLUS,NO LATIN DIAGNOSIS) /B. HYALINA
?=SCOPULARIOPSIS BAINIER 1907 SEE GAMS 1969 (TBMS 51:798-802, SUB S. STATE OF WARDOMYCES DIMERUS). •

BALANIUM WALLROTH 1833. FLORA CRYPT. GERMAN. 2:159-160 /B. STYGIUM
AMERO OR DIDYMO,CETERI,PHAEO,AMPUL OR NONSP, 68B, ON WOOD & SOIL.
/ELLIS, CMI PAPERS 79:19-20,1961 (ILLUS) /HUGHES & HENNEBERT, CAN. J. BOT. 39:1505-1508,1961 (ILLUS) /ELLIS, DH P 186-188,1971 (ILLUS) /ELLIS, MDH P 156,1976 (ILLUS) •

BARGELLINIA BORZI 1888. MALPIGHIA 2:469 /B. MONOSPORA
=WALLEMIA JOHAN-OLSON 1887 FIDE VON ARX 1970. •

BARNETTELLA RAO & RAO 1964. MYCOPATHOLOGIA 22:55-57 (ILLUS) /B. SPECIOSA
DICTYO,BASO,PHAEO,MISC?,SPORODOCHIA, (COELOMYCETE), ON LITTER (GRASS).
/SATYANARAYANA & RAO, MYCOPATHOLOGIA 45:267-268,1971 (NEW COMB) /KENDRICK & CARMICHAEL, THE FUNGI VOL 4A, PL 35,1973 (ILLUS)
=*VINCULUM* ROY, DWIVEDI & KHANNA 1965 FIDE SATYANARAYANA & RAO 1971 •

BARTHELETIA ARNAUD 1954. BULL. SOC. MYCOL. FRANCE 69:300 & FIG 10 M-N (NO LATIN DIAGNOSIS) /B. PARADOXA
?, ON LEAVES (GINKGO). NO TYPE SPECIMEN, SEE NICOT & CHARPENTIE 1971 (BULL. SOC. MYCOL. FRANCE 87:28). •

BARYEIDAMIA KARSTEN 1888. HEDWIGIA 27:132-137 (ILLUS) /HELICOSPORANGIUM PARASITICUM EIDAM 1883 =B. PARASITICA (EIDAM)
=PAPULASPORA PREUSS 1851 FIDE SACCARDO. •

BASIASCUM CAVARA 1888. ATTI IST. BOT. PAVIA II, 1:433-434 (ILLUS) /B. ERIOBOTRYAE
=SPILOCAEA FRIES 1825 FIDE HUGHES 1958. •

BASIDIDYMA CIFERRI 1962. ATTI IST. BOT. UNIV. PAVIA, SER 5 19:94-95 & PLATE VIII, FIGS 1-3 /B. PEREXIGUA
?, ORIGINAL ILLUSTRATION NOT DIAGNOSTIC. •

BASIDIELLA COOKE 1878. GREVILLEA 6:118 /B. SPHAEROCARPA
?
/COOKE, J. QUE. MICRO. CLUB, SER 2, 2:PL IX, FIG 1A TO E •

BASIDIOBOTRYS HOHNEL 1909. S. B. AKAD. WISS. WIEN 118:418-421 /B. CLAUTRIAVII (PATOUILLARD)
=XYLOCLADIUM SYDOW 1900 BASED ON SAME TYPE. •

BASIFIMBRIA SUBRAMANIAN & LODHA 1968. CURR. SCI. 37:246-248 (ILLUS) /B. AUREA
AMERO,CETERI,HYALO,RADUL, 24A, ON DUNG (EQUUS). COMPARE WITH SPOROCEPHALUM. RHINOTRICHUM SUBALUTACEUM PECK MAY BELONG HERE SEE HUGHES 1978 (NEW ZEALAND J. BOT. 16:345).
/DE HOOG & HERMANIDES-NIJHOF, CBS STUDIES 15:211,1977 (NEW COMB) •

BASIPETOSPORA COLE & KENDRICK 1968. CAN. J. BOT. 46:991-992 (987-992,ILLUS) /B. RUBRA =BASIPETOSPORA STATE OF MONASCUS RUBER VAN TIEGHEM 1884
AMERO,BASO,HYALO,MISC, 40G, ON SEEDS & SILAGE (ZEA, ORYZA). ASCO STATE =MONASCUS VAN TIEGHEM FIDE COLE & KENDRICK 1968.
/MATSUSHIMA, ICONES P 13 & PL,1975 (ILLUS,NEW SPP THAT ARE NOT CONGENERIC WITH B. RUBRA) •

BASIPILUS SUBRAMANIAN 1961 (PROC. NAT. INST. SCI. INDIA 27:243) =SEIMATOSPORIUM (MELANCONIALES) FIDE VON ARX 1970 (P 158). •

BASISPORIUM MOLLIARD 1902. BULL. SOC. MYCOL. FRANCE 18:167-170 (ILLUS) /B. GALLARUM = NIGROSPORA ORYZAE
=NIGROSPORA ZIMM. 1902 FIDE MASON 1933. •

BASITORULA ARNAUD 1954. BULL. SOC. MYCOL. FR. 69:276 (NO LATIN DIAGNOSIS) /B. CINGULATA
=GLIOMASTIX GUEGUEN FIDE DICKINSON 1968. •

BATCHELOROMYCES MARASAS, VAN WYK & KNOX-DAVIES 1975. J. S. AFR. BOT. 41:41-45 (ILLUS) /B. PROTEAE
AMERO OR DIDYMO,BASO,PHAEO,ANNEL,SPORODOCHIA, 20C, ON LEAVES. •

BEAUVERIA VUILLEMIN 1912. BULL. BOT. SOC. FRANCE 59:40 /BOTRYTIS BASSIANA BALSAMO 1835 =B. BASSIANA (BALSAMO) =BEAUVERIA DENSA (LINK 1809) PICARD 1914 SEE DE HOOG 1972
AMERO,CETERI,HYALO,RACHI,SYNNEMATA,SPORODOCHIA, 30A, ON INSECTS, SOIL & DUNG. ISARIA FELINA (DE CANDOLLE 1815, FL. FR. 6:30) FRIES 1932 (SYST. MYCOL. 3:227) =**BEAUVERIA FELINA** (DE CANDOLLE) CARMICHAEL COMB. NOV.
/MACLEOD, CAN J. BOT. 32:818-890,1954 (REVIS,ILLUS) /MACLEOD, ANN. N.Y. ACAD. SCI. 60:58-70,1954 (ECOL,DESCR) /DE HOOG, CBS STUDIES 1:4-14,1972 (ILLUS,REVIS) /DE HOOG & RAO, PERSOONIA 8:209-210,1975 (ILLUS,NEW SP) /BRADY, CMI DESCRIPTIONS 602-603,1979 (ILLUS) /COLE & SAMSON, CONIDIAL FUNGI P 38,1979 (ILLUS)
=*ISARIA* AUCT., SEE DE HOOG 1972
?=*SUTRAVARANA* SUBRAMANIAN & CHANDRASHEKARA 1977
?=*BEEJASAMUHA* SUBRAMANIAN & CHANDRASHEKARA 1977 •

BEAUVERIPHORA MATSUSHIMA 1975. ICONES P 14 & PL 245:3-6, PL 246:1-5 /B. SPOROTRICHIS (AS 'SPOROTRICHE')
AMERO,GLOIO,CETERI,HYALO,RACHI,SYNNEMATA, 29C, ON LITTER (CASTANEA). OTHER CONIDIAL STATES =?SPOROTHRIX OR ?NODULISPORIUM SEE MATSUSHIMA 1975. •

BEEJADWAYA SUBRAMANIAN 1977. KAVAKA 5:93-98 (NOT SEEN) /CLADOSPORIUM BISPORUM MATSUSHIMA =B. BISPORA (MATSUSHIMA) =B. BISPORA (MATSUSHIMA)
DIDYMO,CETERI,PHAEO,RADUL, ON LITTER (MUSAE). OTHER STATE =DACTYLARIA (H). •

BEEJASAMUHA SUBRAMANIAN & CHANDRASHEKARA 1977. CAN. J. BOT. 55:245-248 (245-253,ILLUS) /B. SAMALA
?=BEAUVERIA VUILLEMIN 1912. •

BELTRANIA PENZIG 1882. NUOV. GIORN. BOT. ITAL. 14:72 /B. RHOMBICA
AMERO,CETERI,PHAEO,RADUL OR MISC,SETAE,STROMA,APPEND, 22F, ON LITTER.
/PIROZYNSKI, CMI PAPERS 90:1-16,1963 (REVIS,KEY,ILLUS) /BARRON, HYPHOMYCETES P 98-99,1968 (ILLUS) /PIROZYNSKI & PATIL, CAN. J. BOT. 48:567-571,1970 (ILLUS,NEW SPP) /ELLIS, DH P 237-239,1971 (ILLUS,KEY) /MATSUSHIMA, ICONES P ~~14-15 &~~ PL 168-172,1975 (ILLUS,NEW SP) /ELLIS, MDH P 200-203,1976 (ILLUS) •

BELTRANIELLA SUBRAMANIAN 1952. PROC. INDIAN ACAD. SCI. B,36:227 /B. ODINAE
AMERO,CETERI,PHAEO,RADUL,SETAE,STROMA, 22E, ON LITTER. ASCO STATE =PSEUDOMASSARIA FIDE HODGES & BARR 1971 (MYCOLOGIA 58:562).
/PIROZYNSKI, CMI PAPERS 90:26-28,1963 (ILLUS) /PIROZYNSKI & PATIL, CAN. J. BOT. 48:571-575,1970 (ILLUS,NEW SPP, NEW COMB) /ELLIS, DH P 240-243,1971 (ILLUS SUB BELTRANIELLA & ELLISIOPSIS) /ELLIS, CMI PAPERS 131:14-15,1972 (ILLUS,NEW SP) /MATSUSHIMA, ICONES P 15-16 & PL 171-172,1975 (ILLUS,NEW SP) /ELLIS, MDH P 204-206,1976 (ILLUS,KEY)
=*ELLISIOPSIS* BATISTA & NASCIMENTO 1956 FIDE PIROZYNSKI & PATIL 1970. •

BELTRANIOPSIS BATISTA & BEZERRA 1960. PUBL. INST. MIC. UNIV. RECIFE 296:4-9 /B. ESENBECKIAE
AMERO,CETERI,PHAEO,RADUL OR MISC,SETAE,STROMA, 22C, ON LITTER.
/PIROZYNSKI, CMI PAPERS 90:24-26,1963 (ILLUS) /ELLIS, DH P 240,1971 (ILLUS) /ELLIS, MDH P 203-204,1976 (ILLUS) •

BENIOWSKIA RACIBORSKI 1900. PARAS. ALGEN PILZE JAVAS, 2:37-38 /B. GRAMINIS =B. SPHAEROIDEA (KALCHBRENNER & COOKE) MASON 1928
AMERO,BLASTO OR CETERI,HYALO,NONSP,SPORODOCHIA, 53C, ON LEAVES. BASIDIO STATE ?=HAPALOPILUS OR ?GALZINIA SEE STALPERS 1978.
/MASON, CMI PAPERS 2:26-27,1928 (REFS) /BARRON, HYPHOMYCETES P 100,1968 (ILLUS) /STALPERS, CBS STUDIES 16:22,1978 (NO ILLUS) /COLE & SAMSON, CONIDIAL FUNGI P 34-35,1979 (ILLUS,CONIDIUM ONTOGENY) •

BERKELEYNA KUNTZE 1898. REVIS. GEN. PL. 3:447 /B. CURTA (BERKELEY)
=PERICONIA TODE EX PERSOON 1801 FIDE ELLIS 1971. •

BERKLEASMIUM ZOBEL 1854. IN CORDA, ICONES FUNGORUM 6:4 /B. CORDAEANUM =B. CONCINNUM (BERK.) HUGHES 1958
DICTYO,CETERI,PHAEO,NONSP,SPORODOCHIA, 101E, ON WOOD.
/GOOS, CAN. J. BOT. 47:503-504 & PL 1,1969 (ILLUS) /ELLIS, DH P 105,1971 (ILLUS) /ELLIS, MDH P 69-70,1976 (ILLUS,NEW SP) •

BERTEROMYCES CIFERRI 1954. SYDOWIA 8:267-269 /B. AENEUS
=CERCOSPORIDIUM EARLE FIDE DEIGHTON 1967. •

BEVERWYKELLA TUBAKI 1975. TRANS. MYCOL. SOC. JAPAN 16:138 (132-140,ILLUS) /PAPULASPORA PULMONARIA BEVERWIJK 1954 (ANTONIE VAN LEEUWENHOEK 20:1) =B. PULMONARIA (BEVERWIJK)
BULBILS OR DICTYO,CETERI,PHAEO,NONSP,AQUATIC, 129B. •

BIDENTICULA DEIGHTON 1972. TBMS 59:424-427 (ILLUS) /B. CANNAE
PHRAGMO,CETERI,HYALO,RACHI, 78A, ON LEAVES (CANNA, MUSA).
/ELLIS, MDH P 181-182,1976 (ILLUS) •

BIFLAGELLOSPORA MATSUSHIMA 1975. ICONES P 16 & PL 347:1-3, PL 365:6 /B. JAPONICA
STAURO,CETERI,HYALO OR PHAEO,NONSP OR ANNEL, 112F, ON LITTER. COMPARE WITH IYENGARINA. •

BIHARIA THIRUMALACHAR & MISHRA 1953. SYDOWIA 7:79-80 (ILLUS) /B. VANGUERIAE =S. VANGUERIAE (THIRUMALACHAR & MISHRA) DEIGHTON 1979 (CMI PAPERS 144:53)
=STENELLA SYDOW FIDE DEIGHTON 1979. •

BILBOQUE VIEGAS 1960. BRAGANTIA 19:903 & FIG 1, P 898 /B. MAGNIFICUM
?, ORIGINAL ILLUSTRATION & DESCRIPTION NOT DIAGNOSTIC. •

BILGRAMIA PANWAR, PUROHIT & CHAUHAN 1974. CURRENT SCIENCE, INDIA 43:552-553 (ILLUS) /B. INDICUM
=PEYRONELIA FRAGOSO & CIFERRI. ALTHOUGH PEYRONELIA WAS DESCRIBED AS TRANSVERSELY SEPTATE THE ILLUSTRATION SHOWS DICTYOSPORES & THE CLASSIFICATION WAS GIVEN AS DEMATIACEAE PHAEODICTYOSPORAE. •

BIOCONIOSPORIUM BATISTA & BEZERRA 1964. PUBL. INST. MICOL. UNIV. RECIFE 417:4-8 & FIG 1-2, P 13-14 /B. BACCHARIDIS
DICTYO OR STAURO,CETERI,PHAEO,NONSP,APPEND, 95A, ON LEAVES (BACCHARIS). COMPARE WITH PITHOMYCES.
/ELLIS, MDH P 34-35,1976 (ILLUS) •

BIPOLARIS SHOEMAKER 1959. CAN. J. BOT. 37:882 /B. MAYDIS (NISIKADO)
=DRECHSLERA ITO 1930 FIDE ELLIS 1971. •

BISPORA CORDA 1837. ICONES FUNGORUM 1:9 /B. MONILIOIDES =B. ANTENNATA (PERSOON) MASON 1953
DIDYMO,BLASTO,PHAEO,NONSP, 66A, ON WOOD & LEAVES (DECIDUOUS TREES).
/HUGHES, CAN. J. BOT. 36:740,1958 (LIST) /BARRON, HYPHOMYCETES P 102-103,1968 (ILLUS) /ELLIS, DH P 90-91,1971 (ILLUS) /ELLIS, MDH P 54-55,1976 (ILLUS) /SCHOKNECHT & CRANE, MYCOLOGIA 69:533-535,1977 (ILLUS,NEW COMB) /SHARMA, J. IND. BOT. SOC. 57:102-105,1978 (NEW SP) •

BISPOROMYCES VAN BEYMA 1940. ANTONIE VAN LEEUWENHOEK 6:276-278 /B. CHLAMYDOSPORIS
AMERO,GLOIO,HYALO,PHIAL OR MISC, 3J, ON FUNGI (SCLEROTINIA) & SOIL. SEE ALSO CACUMISPORIUM. ASCO STATE =CHAETOSPHAERIA SEE MATSUSHIMA 1975.
/MATSUSHIMA, ICONES P 28-29 & PL 273:1-4 & 274:3-4,1975 (ILLUS SUB CHLORIDIUM)
?=*CHLORIDIUM* LINK FIDE HUGHES 1958 •

BISPOROSTILBELLA BRANDSBERG & MORRIS 1971. MYCOLOGIA 63:1076-1080 (ILLUS) /B. FUSCA
DIDYMO,BASO,PHAEO,ANNEL,SYNNEMATA, ON SOIL. •

BITUNICOSTILBE MORELET 1971. BULL. SOC. NAT. ARCHEOL. TOULON VAR 195:7 (17-20 ? NOT SEEN) /ISARIOPSIS CLAVATA ELLIS & MARTIN =B. CLAVATA (ELLIS & MARTIN)
?. •

BIZZOZERIELLA SPEGAZZINI 1888. AN. SOC. ARG. 26:73-74 /B. PHYLLOGENA
?, NOT ILLUSTRATED BY AUTHOR. •

BLASTOBOTRYS KLOPOTEK 1967. ARCH. MIKROBIOL. 58:92-95 (ILLUS) /B. NIVEA
AMERO,CETERI,HYALO,AMPUL OR RACHI,SETAE, 31G, ON COMPOST & SEEDS (PIPER). ASCO STATE ?=PORONIA SEE JONG & ROGERS 1969 (MYCOLOGIA 61:853-862,ILLUS SUB PORONIA OEDIPUS) AND STIERS, ROGERS & RUSSELL 1973 (CAN. J. BOT. 51:481-484 & PL I-III,SUB PORONIA PUNCTATA).
/DE HOOG, CBS STUDIES 7:10-13,1974 (ILLUS) /MARVANOVA, TBMS 66:217-222,1976 (ILLUS,NEW SPP) /SESMA & RAMIREZ, MYCOPATHOLOGIA 63:41-45,1978 (ILLUS,NEW SP,KEY) /COLE & SAMSON, CONIDIAL FUNGI P 42-43,1979 (ILLUS) •

BLASTOCONIUM CIFERRI 1931. J. DEP. AGRIC. PORTO RICO 15:233-234 /B. TROPICUM
?, NOT ILLUSTRATED BY AUTHOR. BASED ON A DARK, YEAST-LIKE FUNGUS. •

BLASTODICTYS ELLIS 1976. MDH P 149-150 (ILLUS) /DACTYLOSPORIUM HIBISCI HANSFORD 1943 =B. HIBISCI (HANSFORD)
DICTYO,CETERI,PHAEO,AMPUL,APPEND, 94E, ON LEAVES (HIBISCUS). •

BLASTOMYCES COSTANTIN & ROLLAND 1888. BULL. SOC. MYCOL. FRANCE 4:156 & PL XXIII /B. LUTEUS COSTANTIN & ROLLAND =CHRYSOSPORIUM MERDARIUM (LINK) CARMICHAEL 1962 (CAN. J. BOT. 40:1143)
=CHRYSOSPORIUM CORDA 1833 FIDE CARMICHAEL 1962. •

BLASTOMYCES GILCHRIST & STOKES 1898. J. EXPER. MED. 3:76 (53-83 & PL IV-VIII) /B. DERMATITIDIS =B. STATE OF AJELLOMYCES DERMATITIDIS MCDONOUGH & LEWIS 1968 (MYCOLOGIA 60:76-83,1LLUS)
=BLASTOMYCETES SEE ORIGINAL DESCRIPTION. NEEDS CONSERVATION AGAINST BLASTOMYCES COST. & ROLL. ASCO STATE =AJELLOMYCES FIDE MCDONOUGH & LEWIS 1968. HYPHO STATE =CHRYSOSPORIUM SEE CARMICHAEL 1962 (CAN. J. BOT. 40:1143 & 1154-1156 & PL III). •

BLASTOMYCOIDES CASTELLANI 1928. AM. J. TROP. MED. 8:381 /COCCIDIOIDES IMMITIS STILES B. IMMITIS (STILES)
=NOMEN ILLEGITIMUM FIDE CARMICHAEL 1962. •

BLASTOPHORELLA BOEDIJN 1937. BLUMEA, SUPPL. 1:140-141 (ILLUS) /B. SMITHII
DIDYMO,CETERI,HYALO,AMPUL, 64B, ON LITTER (ARENGA).
/ELLIS, DH P 397-398,1971 (ILLUS) •

BLASTOPHORUM MATSUSHIMA 1971. MICROFUNGI P 8-9 & FIG 67,1-4 /B. TRUNCATUM
AMERO OR DIDYMO OR PHRAGMO,GLOIO,HYALO OR PHAEO,RADUL OR MISC, 81C, ON LEAVES & LITTER (CASTANOPSIS, QUERCUS). COMPARE WITH CACUMISPORIUM PREUSS 1851.
/MATSUSHIMA, ICONES P 16-17 & PL,1975 (ILLUS,NEW SP) •

BLASTOTRICHUM CORDA 1838. ICONES FUNGORUM 2:10 & TAB IX, FIG 50 /B. CONFERVOIDES
=NOMEN DUBIUM, SEE GAMS & HOOZEMANS 1970 (PERSOONIA 6:99). •

BLISTUM SUTTON 1973. CMI PAPERS 132:16-20,1973 (ILLUS) /B. TOMENTOSUM (SCHRADER 1799 EX PERSOON 1801)
AMERO,GLOIO,HYALO,PHIAL,SYNNEMATA, 11F, ON SLIME MOLDS (TRICHIA, ETC) & PLANTS (CAREX, JUNCUS, PHRAGMITES & OTHERS). SEGREGATED FROM STILBUM PERSOON ON THE BASIS OF SUBGLOBOSE, VERRUCOSE CELLS PROTRUDING FROM THE STIPE OF THE SYNNEMA.
/TULLOCH, CMI PAPERS 130:32-24,1972 (ILLUS SUB MYROTHECIUM MASONII) /RAO & SUTTON, KAVAKA 3:25-28,1975 (DISCUSSION) /SEIFERT, MYCOTAXON 10:233-240,1979 (ILLUS,NEW SP,EMEND) •

BLODGETTIA WRIGHT 1881. TRANS. IRISH ACAD. 28:25 /B. BORNETII
?, STERILE. •

BLOXAMIA BERKELEY & BROOME 1854. ANN. MAG. NAT. HIST. SER. 2, 13:468 /B. TRUNCATA
AMERO,BASO,HYALO,PHIAL,SPORODOCHIA OR SYNNEMATA, 12D, ON WOOD (FAGUS, SALIX, ULMUS).
/PIROZYNSKI & MORGAN-JONES, TBMS 51:185-187,1968 (ILLUS) /ELLIS, DH P 510-511,1971 (ILLUS) /NAG RAJ & KENDRICK, CHALARA P 41-43 & 167-169,1975 (KEY,ILLUS)
=*ENDOSPOROSTILBE* SUBRAMANIAN 1958 FIDE NAG RAJ & KENDRICK 1975
?=*EXOSPORINA* OUDEMANS 1904 •

BODINIA OTA & LANGERON 1923. ANN. PARAS. HOM. COMP. 1:329 /B. VIOLACEA (SABOURAUD) BODIN 1923
=TRICHOPHYTON MALMSTEN 1845. •

BOMPLANDIELLA SPEGAZZINI 1886. AN. SOC. ARG. 22:222 (NO ILLUS) /B. GUARANITICA
?, CORRECTED TO BONPLANDIELLA BY SACCARDO. •

BONORDENIELLA PENZIG & SACCARDO 1901. MALPIGHIA 15:259 & PL 80, FIG 4 /B. MEMORANDA
=CONIOSPORIUM LINK 1809 FIDE ELLIS 1971 (CMI PAPERS 125:2). •

BONPLANDIELLA, SEE BOMPLANDIELLA •

BORNETINA MANGIN & VIALA 1903. C. R. ACAD. SCI., PARIS 136:398 & 1699 /B. CORIUM ?=B. STATE OF DIACANTHODES PHILIPPINENSIS FIDE DONK 1962
?, BASIDIO STATE ?=DIACANTHODES FIDE DONK 1962.
/DONK, TAXON 11:81,1962 (REFS) •

BOSTRICHONEMA CESATI 1859. ERBAR. CRITT. ITAL. NO. 149 /B. ALPESTRE =B. POLYGONI (UNGER) SCHROETER ?=B. STATE OF DIPLOCARPON POLYGONI FIDE MULLER 1977
DIDYMO,CETERI,HYALO,RADUL, 67G, ON LEAVES (POLYGONUM). ASCO STATE ?=DIPLOCARPON FIDE MULLER 1977.
/LINDAU, IN RABENHORST KRYPT.-FLORA 2 ED, 1 BAND, 8:381-382,1907 (ILLUS)
/MULLER, BEITR. KRYPT.-FL. SCHWEIZ, BD 15, HEFT 1:39-40,1977 (ILLUS) •

BOSTRYCHONEMA -ORTHOGRAPHIC VARIANT OF BOSTRICHONEMA •

BOTRYDIELLA BADURA 1963. ALLIONIA 9:182-184 (ILLUS) /B. BICOLOR =STAPHYLOTRICHUM COCCOSPORUM
=STAPHYLOTRICHUM MEYER & NICOT 1956 FIDE VON ARX 1970. •

BOTRYOCLADIUM PREUSS 1851. LINNAEA 24:134 (ILLUS IN STURM'S DEUTSCHL. FL. TOME 1, BD. 6, HEFT 36, TAB. 65, 1862) /B. DELECTATUM
=NEMATOGONUM DESMAZIERES 1834 FIDE SACCARDO FROM AINSWORTH 1971. •

BOTRYOCONIS SYDOW & SYDOW 1906. ANN. MYCOL. 4:344-345 /B. SACCARDOI
AMERO,CETERI,PHAEO,MISC,SPORODOCHIA, 35E, ON LITTER. THIS IS A BASIDIOMYCETE.
/MALENCON, BULL. SOC. MYCOL. FRANCE 69:77-100,1953 (ILLUS) •

BOTRYODERMA PAPENDORF & UPADHYAY 1969. TBMS 52:257 /B. LATERITIUM
AMERO,CETERI,PHAEO,NONSP,ALEURI,SETAE, 48A, COMPARE WITH MYCELIOPHTHORA AND SPOROTRICHUM. •

BOTRYOMONILIA GOOS & PIROZYNSKI 1975. CAN. J. BOT. 53:2927-2932 (ILLUS) /B. SCHEELEAE
AMERO,ARTHRO,HYALO,NONSP,ALEURI?, 44B, ON LITTER (SCHEELEA). COMPARE WITH STAHELIELLA. OTHER STATE ="SCLEROTIA".
/COLE & SAMSON, CONIDIAL FUNGI P 113 & 115,1979 (ILLUS) •

BOTRYONIPHA PREUSS 1852. LINNAEA 25:79 /B. ALBA
?, STERILE BASED ON THE ACANTHOPHYSES OF AN HYMENOMYCETE.
/HUGHES, FRIESIA 9:61-63,1969 (ILLUS) •

BOTRYOPHIALOPHORA LINDER 1944. IN BARGHOORN & LINDER, FARLOWIA 1:403-404 /B. MARINA LINDER 1940
=MYRIOCONIUM SYDOW FIDE VON ARX 1970. •

BOTRYOSPORIUM CORDA 1831. STURM'S DEUTSCHL. FL. III:9-10 & TAB 5 /B. DIFFUSUM (ALBERTINI & SCHWEINITZ)
AMERO,CETERI,HYALO,AMPUL, 32B, ON PLANTS & LITTER. SCLEROTIAL STATE =SCLEROTIUM.
/MASON, CMI PAPERS 2:27-29,1928 (LIST) /BARRON, HYPHOMYCETES P 103-104,1968 (ILLUS) /HUGHES & CONWAY, CAN. J. BOT. 56:2405-2407,1978 (ILLUS)
=*PEYLIA* OPIZ FIDE AINSWORTH 1971
=*RADAISIELLA* BAINIER 1910 FIDE MAIRE FROM AINSWORTH 1971 •

BOTRYOTRICHUM SACCARDO & MARCHAL 1885. IN MARCHAL, BULL. SOC. ROY. BOT. BELG. 24:66-67 (ILLUS) /B. PILULIFERUM =BOTRYOTRICHUM STATE OF CHAETOMIUM PILULIFERUM DANIELS 1961
AMERO,CETERI,HYALO,NONSP,ALEURI,SETAE, 47G, ON CELLULOSE & SOIL. ASCO STATE =CHAETOMIUM FIDE DANIELS 1961 OR FARLOWIA SEE DE HOOG 1977 (CBS STUDIES 15:138). COMPARE WITH STAPHYLOTRICHUM & CHRYSOSPORIUM.
/DANIELS, TBMS 44:79-86,1961 (ILLUS) /UDAGAWA & CAIN, CAN. J. BOT. 47:1939-1951,1969 (ILLUS SUB CHAETOMIUM SEMISPIRALE) /UDAGAWA, FURUYA & HORIE, BULL. NAT. SCI. MUSEUM, TOKYO 16:510-511,1973 (ILLUS SUB CHAETOMIUM LONGIROSTRE) /UDAGAWA & HORIE, TRANS. MYC. SOC. JAPAN 16:337-339,1975 (ILLUS, SUB B. STATE OF CHAETOMIUM PILULIFEROIDES) /MATSUSHIMA, ICONES P 17 & PL 36,1975 (ILLUS,NEW SP) •

BOTRYOXYLON CIFERRI 1962. ATTI IST. BOT. UNIV. PAVIA, SER. 5, 19:97 /B. GENICULATUM (CORDA) =CONOPLEA GENICULATA (CORDA) HUGHES 1958
=CONOPLEA PERSOON 1801 FIDE HUGHES 1958 (CAN. J. BOT. 36:755). •

BOTRYPES PREUSS 1853. LINNAEA 25:740 /B. ROSEA
=CILICOPODIUM CORDA FIDE LINDAU IN RABENHORST 1910. •

BOTRYTIS MICHELI EX PERSOON 1801. SYNOPSIS METH. FUNG. P 690 /B. CINEREA PERSOON 1801 = B. STATE OF BOTRYOTINIA FUCKELIANA (DEBARY) WHETZEL 1945 (LECTOTYPE FIDE CLEMENTS & SHEAR 1931)
AMERO,CETERI,PHAEO,AMPUL,ALEURI, 34E, ON PLANTS (MANY KINDS). ASCO STATE =BOTRYOTINIA WHETZEL (=SCLEROTINIA), SPERMATIAL STATE =MYRIOCONIUM SYDOW, SCLEROTIAL STATE =SCLEROTIUM PERSOON, ALL FIDE HENNEBERT 1973, WHO SEGREGATES STREPTOBOTRYS, AMPHOBOTRYS, VERRUCOBOTRYS, DICHOBOTRYS & PHYMATOTRICHOPSIS FROM BOTRYTIS. PHILLIPSIA (A) HAS BOTRYTIS-LIKE AND MYRIOCONIUM-LIKE CONIDIAL STATES, SEE PADEN 1977 (CAN. J. BOT. 55:2685-2692,ILLUS).
/BULLER, TRANS. R. SOC. CANADA, SECT IV, SER III, 9:1-25 & PL I-IV,1915 (MICHELI'S ILLUS & DESCR) /ELLIS, DH P 178-184,1971 (ILLUS,KEY) /HENNEBERT, PERSOONIA 7:183-204,1973 (ILLUS,REVIS,REFS) /ELLIS & WALLER, CMI DESCRIPTIONS 431-433,1974 (ILLUS) /ELLIS, MDH P 149-150,1976 (ILLUS) /JARVIS, BOTRYOTINIA & BOTRYTIS SPECIES, CANADA DEPT. OF AGRICULTURE MONOGRAPH NO 15, 195 PP, 1977 (BIOLOGY,REFS) /COLE & SAMSON, CONIDIAL FUNGI P 44-45,1979 (ILLUS)
=*CEPHALOCLADIUM* REICHARDT FIDE AINSWORTH 1971
=*HAPLARIA* LINK 1809 FIDE HUGHES 1958
=*PHYMATOTRICHUM* BONORDEN 1851 SEE HENNEBERT 1973
=*POLYACTIS* LINK 1809 FIDE HUGHES 1958
?=*PTERODINIA* CHEVALLIER FIDE AINSWORTH 1971 •

BOTRYTOIDES MOORE & ALMEIDA 1937. ANN. MISSOURI BOT. GARD. 23:545-547 & PL 26, FIG 3 (DATED 1936) /B. MONOPHORA, NOMEN ILLEGITIMUM
=RHINOCLADIELLA NANNFELDT 1934 FIDE DE HOOG 1977 (CBS STUDIES 15:89).
/SCHOL-SCHWARZ, ANTONIE VAN LEEUWENHOEK 34:140,1968 (REFS) •

BRACHIOSPHAERA NAWAWI 1976. IN DESCALS, NAWAWI & WEBSTER, TBMS 67:213 (207-222,ILLUS) /B. TROPICALIS
STAURO,CETERI,HYALO OR PHAEO,RACHI,AQUATIC, 115B. •

BRACHYCLADIUM CORDA 1838. ICONES FUNGORUM 2:14 (ILLUS) /B. PENICILLATUM
=DENDRYPHION WALLROTH 1833 FIDE HUGHES 1958. •

BRACHYDESMIELLA ARNAUD EX HUGHES 1961. CAN. J. BOT. 39:1095-1097 (ILLUS) /B. BISEPTATA
PHRAGMO OR AMERO,CETERI,PHAEO,RADUL,APPEND, 93E, ON WOOD (FRAXINUS, FRAGUS).

/ELLIS, DH P 459-460,1971 (ILLUS) •

BRACHYDESMIUM (SACCARDO) COSTANTIN 1888. LES MUCEDINEES SIMPLES P 167-168 (ILLUS) /B. OVOIDEUM (CORDA) COSTANTIN 1888
? •

BRACHYHELICOON ARNAUD 1952. BULL. SOC. MYCOL. FRANCE 68:208-209, FIG 7U /B. XYLOGENUM
HELICO,GLOIO,HYALO,PHIAL?, 106D, ON WOOD. NO LATIN DIAGNOSIS. •

BRACHYSPORIELLA BATISTA 1952. BOL. SOC. AGRIC. PERNAMBUCO 19:108-109 (ILLUS) /B. GAYANA
PHRAGMO,CETERI,PHAEO,ANNEL, 77E, ON WOOD & LITTER.
/ELLIS, CMI PAPERS 72:15-19,1959 (ILLUS,KEY) /ELLIS, DH P 157-159,1971 (ILLUS,KEY) /ELLIS, MDH P 145-146,1976 (ILLUS) /CRANE & DUMONT, CAN. J. BOT. 56:2613-2616,1978 (ILLUS,NEW SP) /HUGHES, NEW ZEALAND J. BOT. 16:319,1978 (ILLUS)
=*MONOSPORELLA* HUGHES NON KEILIN FIDE ELLIS 1959
=*MONOTOSPORELLA* HUGHES 1958 FIDE ELLIS 1959 •

BRACHYSPORIUM SACCARDO 1880. MICHELIA 2:28 /B. OBOVATUM (BERKELEY) SACCARDO 1886 (SYLL. FUNG. 4:427)
PHRAGMO,CETERI,PHAEO,RADUL OR MISC, 80B, ON WOOD & BARK.
/ELLIS, CMI PAPERS 106:43-54,1966 (KEY,ILLUS) /ELLIS, DH P 221-224,1971 (ILLUS,KEY) /HOLUBOVA-JECHOVA, FOLIA GEOBOT. PHYTOTAX., PRAHA 7:214-224,1972 (ILLUS,KEY) /HUGHES & KOKKO, FUNGI CANADENSES NO. 69,1975 & NO. 108,1977 (ILLUS) /HUGHES, FUNGI CANADENSES NO. 107 & 109,1977 (ILLUS)
?=*RHYNCHOMYCES* WILLKOMM 1866 (NON SACCARDO 1885) •

BRIAREA CORDA 1831. STURM'S DEUTSCHL. FL. III (PILZE), BD 3, HEFT 11:11 /B. ELEGANS =ASPERGILLUS GLAUCUS GROUP
=ASPERGILLUS LINK 1809 FIDE THOM & RAPER 1945 (FROM HUGHES 1958). •

BRIOSIA CAVARA 1888. ATTI. IST. BOT. PAVIA 2:321 /B. AMPELOPHAGA
AMERO,BASO,PHAEO,MISC,SYNNEMATA, 43H, ON LEAVES. SIMILAR TO OJIBWAYA SUTTON.
/SUTTON, CMI PAPERS 132:82-86,1973 (ILLUS,REVIS) /SIGLER & CARMICHAEL, MYCOTAXON 4:372-379,1976 (ILLUS,REVIS) /ELLIS, MDH P 24-25,1976 (ILLUS) •

BROMICOLLA EICHWALD 1843 (FLORA 19:108-109,ILLUS) =SCLEROTIUM FIDE SACCARDO FROM AINSWORTH 1971. •

BROOMEOLA KUNTZE 1891. REV. GEN. PLANT. 2:845 /B. GLAUCA (BERKELEY & BROOME) ?, NOM. NOV. FOR ENDODESMIA BERK. & BROOME 1874 NON BENTH. 1862. •

BRYOCHYSIUM LINK 1833 =RHIZOCTONIA FIDE SACCARDO 1901 (SYLL. FUNG. 15:46) •

"BULBILS" HAVE BEEN REPORTED IN CULTURES OF MANY FUNGI, ESPECIALLY APHYLLOPHORALES, E.G. ATHELIA, CERATOBASIDIUM, SISTOTREMA, THANATEPHORUS FIDE STALPERS 1978 (CBS STUDIES 16:21, SUB SCLEROTIA,NO ILLUS). SEE ALSO ENTRIES UNDER BURGOA, MINIMEDUSA, & PAPULASPORA. •

BURGOA GOIDANICH 1937. BOLL. STAZ. PAT. VEG., ROMA N. S. 17:354 & 359-360 /B. VERZUOLIANA (LECTOTYPE FIDE WERESUB & LECLAIR 1971)
BULBILS,CETERI,HYALO OR PHAEO,NONSP,CLAMPS, 128F, ON WOOD. BASIDIO STATE =SISTOTREMA FIDE WERESUB & LECLAIR 1971. OTHER CONIDIAL STATE ="OIDIA".
/WERESUB & LECLAIR, CAN. J. BOT. 49:2203-2213,1971 (ILLUS,REVIS) •

BUSSEELIA HENNINGS 1902. ENGLER'S BOT. JAHRB. 33:40 (NO ILLUS) /B. CARYOPHYLLI ?, NOT ILLUSTRATED. •

BYSSOCLADIUM LINK 1815. MAG. GES. NATURF. FREUNDE, BERLIN 7:35 /B. CANDIDUM (LECTOTYPE FIDE HUGHES 1958)
=BASIDIOMYCETE FIDE HUGHES 1958, BUT SEE DONK 1962 (TAXON 11:81). •

BYSSUS MICHELI EX LINNAEUS 1753 (SP. PLANTARUM 2:1168). SEE FRIES 1849 (SUMMA VEG. SCAND. 2:122). USED FOR STERILE MYCELIUM, SEE HENNEBERT & KORF 1975 (MYCOLOGIA 67:221). SEE ALSO DONK 1962 (TAXON 11:81). •

CACUMISPORIUM PREUSS 1851. LINNAEA 24:130 /C. TENEBROSUM =C. CAPITULATUM (CORDA) HUGHES 1958
PHRAGMO,GLOIO,PHAEO,ANNEL,MISC, 76E, ON WOOD & BARK. HAS BEEN CONFUSED WITH PLEUROTHECIUM. COMPARE WITH CHLORIDIUM & BISPOROMYCES. SEE ALSO MATSUSHIMA 1975 (PL 239:3-4 & 369, SUB CORDANA & PL 275:1-3, SUB CYLINDROTRICHUM).
/MANGENOT, REV. MYCOL. 18:143-146,1953 (ILLUS) /GOOS, MYCOLOGIA 61:52-56,1969 (ILLUS) /SUTTON, CMI PAPERS 132:20-23,1973 (ILLUS OF TYPE,REVIS) /MATSUSHIMA, ICONES P 19 & PL 240,1975 (ILLUS) /ELLIS, MDH P 100,1976 (ILLUS) /COLE & SAMSON, CONIDIAL FUNGI P 68,1979 (ILLUS)
?=*BLASTOPHORUM* MATSUSHIMA 1971 •

CADOPHORA LAGERBERG & MELIN 1928. IN LAGERBERG & AL., SVENSKA SKOGSVFOREN. TIDSKR. 25:263 /C. FASTIGIATA =PHIALOPHORA FASTIGIATA (LAGERBERG & MELIN) CONANT
=PHIALOPHORA MEDLAR 1915 FIDE CONANT 1937. •

CALCARISPORA MARVANOVA & MARVAN 1963. CAS. SLEZSK. MUS. OPAVE, SER. A, 12:106-109 (ILLUS) /C. HIEMALIS
SCOLECO,CETERI,HYALO,PHIAL,AQUATIC, 109D.
/INGOLD, GUIDE P 80-81,1975 (ILLUS) •

CALCARISPORIELLA DE HOOG 1974. CBS STUDIES 7:68-70 (ILLUS) /CALCARISPORIUM THERMOPHILUM EVANS 1971 =C. THERMOPHILA (EVANS)
AMERO,CETERI,HYALO,RACHI, 24F, ON COAL. •

CALCARISPORIUM PREUSS 1851. LINNAEA 24:124 /C. ARBUSCULA
AMERO,CETERI,HYALO,RADUL OR RACHI, 26A, ON FUNGI & WOOD. OTHER STATE ="SCLEROTIA" SEE WANG IN KENDRICK 1979.
/HUGHES, CMI PAPERS 43:1-6,1951 (ILLUS) /DEIGHTON & PIROZYNSKI, CMI PAPERS 128:99-103,1972 (ILLUS,NEW SPP) /DE HOOG, CBS STUDIES 7:70-73,1974 (ILLUS,REVIS) /MATSUSHIMA, ICONES P 19 & PL,1975 (ILLUS,NEW SP) /DE HOOG, PERSOONIA 10:77,1978 (ILLUS,NEW COMB) •

CALCEISPORA MATSUSHIMA 1975. ICONES P 20 & PL 368 FIGS 1-4 /C. HACHIJOENSIS
AMERO,GLOIO,HYALO,PHIAL,APPEND, 2D, ON BARK (MACHILI). •

CALDARIOMYCES WORONICHIN 1926. ANN. MYCOL. 24:261-264 /C. FUMAGO ='FUMAGO' OF ZOPF 1878
=LEPTOXYPHIUM SPEGAZZINI 1918 FIDE HUGHES 1976 (MYCOLOGIA 68:777-782). •

CALOSTILBELLA HOHNEL 1919. BERL. DEUTSCHL. BOT. GES. 37:160 /C. CALOSTILBE
PHRAGMO,GLOIO,HYALO OR PHAEO,PHIAL?,SYNNEMATA, 74F, ON WOOD AND LITTER. ASCO STATE =NECTRIA FIDE SAMUELS 1976 (MEM. N. Y. BOT. GARD. 26:22).
/MORRIS, WESTERN ILLINOIS UNIV. SER. BIOL. SCI. 3:34-35,1963 (ILLUS) /BOOTH & HOLLIDAY, CMI DESCRIPTIONS 392,1973 (ILLUS) •

CAMAROSPORIUM HARKNESS IN CLEMENTS & SHEAR 1931 (P 396- IN ERROR FOR CAMPOSPORIUM. NON CAMAROSPORIUM SCHULZER (COELOMYCETES). •

CAMPOSPORIUM HARKNESS 1884. BULL. CALIF. ACAD. SCI. 1:37-38 /C. ANTENNATUM
PHRAGMO,CETERI,PHAEO OR HYALO,RADUL,APPEND, 69E, ON WOOD, BARK & LITTER. ASCO STATE =LEPTOSPHAERIA FIDE MULLER IN KENDRICK 1971 (P 190).
/HUGHES, CMI PAPERS 36:3-16,1951 (ILLUS) /ELLIS, DH P 226-228,1971 (ILLUS,KEY) /INGOLD, GUIDE P 86-87,1975 (ILLUS) /MATSUSHIMA, ICONES P 21 & PL 201-203,1975 (ILLUS) /ELLIS, MDH P 197-198,1976 (ILLUS,NEW SP) /HUGHES, NEW ZEALAND J. BOT. 16:319-320,1978 (ILLUS)
=*CAMAROSPORIUM* HARKNESS IN CLEMENTS & SHEAR •

CAMPSOTRICHUM EHRENBERG 1819. IN LINK, JAHRB. F. GEWACHSK. 1(2):55 /C. BICOLOR ?, HUGHES (1958) EXAMINED THE TYPE SPECIMEN BUT DID NOT ILLUSTRATE. •

CAMPTOMERIS SYDOW 1927. ANN. MYCOL. BERLIN 25:143-144 (ILLUS) /C. CALLIANDRAE
DIDYMO OR PHRAGMO,CETERI,PHAEO,RADUL,SPORODOCHIA, 85D, ON LEAVES.
/HUGHES, CMI PAPERS 49:14-19,1952 (ILLUS) /ELLIS, DH P 285-287,1971 (ILLUS,KEY) •

CAMPTOSPORIUM LINK EX DUBY 1830. BOT. GALLICUM 2:928 /C. GLAUCUM (PERSOON) DUBY AS "EHRENB." =MENISPORA GLAUCA PERSOON
=NOMEN ILLEGITIMUM FIDE HUGHES 1958. •

CAMPTOUM LINK 1824. LINN. SPEC. PLANT. IV, 6(1):44 /ARTHRINIUM CURVATUM KUNZE =C. CURVATUM (KUNZE)
=ARTHRINIUM KUNZE 1817 FIDE HUGHES 1958. •

CAMPYLOSPORA RANZONI 1953. FARLOWIA 4:373 & FIG 11 /C. CHAETOCLADIA
STAURO,CETERI,HYALO,NONSP,APPEND,AQUATIC, 118B.
/NAWAWI, TBMS 63:603-606,1974 (ILLUS,NEW SP) /INGOLD, GUIDE P 24-25,1975 (ILLUS) •

CANCELLIDIUM TUBAKI 1975. TRANS. MYCOL. SOC. JAPAN 16:357-360 (ILLUS) /C. APPLANATUM
BULBILS OR DICTYO,CETERI,PHAEO,NONSP, 129A, ON WOOD IN WATER. •

CANDELABRELLA RIFAI & COOKE 1966. TBMS 49:160-162 (ILLUS) /C. JAVANICA
=ARTHROBOTRYS CORDA 1839 FIDE BARRON 1968. •

CANDELABRUM VAN BEVERWIJK 1951. ANTONIE VAN LEEUWENHOEK 17:283 (278-285,ILLUS) /C. SPINULOSUM
STAURO,CETERI,HYALO,NONSP,APPEND,AQUATIC, 114A.
/TUBAKI, TRANS. MYCOL. SOC. JAPAN 16:132-140,1975 (ILLUS,NEW SP) /GAMUNDI, ARAMBARRI & GIAIOTTI, DARWINIANA 21:86-87 & FIGS 8-9,1977 (ILLUS)
=*DIONYSIA* ARNAUD 1952 FIDE KENDRICK 1973
?=*GLOMOPSIS* HENDERSON 1961 •

CANDELOSPORA REA & HAWLEY 1912. PROC. ROYAL IRISH ACAD. 31:11-12 /C. ILICICOLA
=CYLINDROCLADIUM MORGAN 1892 FIDE BOOTH 1960. •

CAPILLARIA PERSOON 1822. MYCOL. EUROP. 1:50 /?, SIX ORIGINAL SPECIES
=NOMEN ILLEGITIMUM (NON CAPILLARIA STACKH. 1809). •

CAPNOBOTRYS HUGHES 1970. NEW ZEALAND J. BOT. 8:205 /C. NEESII
DIDYMO OR PHRAGMO,CETERI,PHAEO,RADUL, 63G, ON BRANCHES (SOOTY MOLD). ASCO STATE =METACAPNODIUM FIDE HUGHES 1972.
/HUGHES, NEW ZEALAND J. BOT. 10:225-242,1972 (ILLUS) /HUGHES, MYCOTAXON 1:121-122,1974 (NEW SP,NO ILLUS) /HUGHES, MYCOLOGIA 68:707-712,1976 (ILLUS) /ELLIS, MDH P 346-349,1976 (ILLUS) /HUGHES, FUNGI CANADENSES 132,1979 (ILLUS) •

CAPNOCYBE HUGHES 1966. NEW ZEALAND J. BOT. 4:335 (333-353, ILLUS) /C. FRASERAE =CAPNOCYBE STATE OF METACAPNODIUM FRASERAE (HUGHES) HUGHES 1976
PHRAGMO,GLOIO,PHAEO,RACHI OR RADUL,SYNNEMATA, 87B, ON BRANCHES (SOOTY MOLD). ASCO STATE =METACAPNODIUM FIDE HUGHES 1976. OTHER STATES =CAPNOPHIALOPHORA (H), HORMIOKYRPSIS (H) & TORULOPSIELLA (H).
/HUGHES, NEW ZEALAND J. BOT. 10:225-242,1972 (ILLUS) /HUGHES, MYCOLOGIA 68:707-712 & 782,1976 (ILLUS,REVIS) •

CAPNODENDRON HUGHES 1976. MYCOLOGIA 68:750-751 (& 714-719, ILLUS) /C. TRICHOMERICOLA (CIFERRI, BATISTA & NASCIMENTO)
PHRAGMO,BLASTO,PHAEO,RADUL, 93B, ON BRANCHES & LEAVES (SOOTY MOLD). •

CAPNOKYMA HUGHES 1975. NEW ZEALAND J. BOT. 13:638 (637-644, ~~NO~~ ILLUS) /C. CORTICOLA
PHRAGMO,CETERI,PHAEO,NONSP, 92E, ON ~~BRANCHES~~ (SOOTY MOLD). /TREES
/HUGHES, MYCOLOGIA 68:706-707,1976 (ILLUS) •

CAPNOPHIALOPHORA HUGHES 1966. NEW ZEALAND J. BOT. 4:352 (333-353, ILLUS) /C. FRASERAE =C. STATE OF LIMACINIA FRASERAE HUGHES 1966
AMERO,GLOIO,HYALO,PHIAL, 17D, ON BRANCHES (SOOTY MOLD). ASCO STATE =METACAPNODIUM FIDE HUGHES 1972 OR STRIGOPODIA FIDE BARR 1972 (CONTRIB. UNIV. MICHIGAN HERB. 9:621), OR OPHIOCAPNOCOMA FIDE HUGHES 1976. OTHER STATES =CAPNOSPORIUM (H), HORMIOKRYPSIS (H), TORULOPSIELLA (H) & CAPNOCYBE (H) FIDE HUGHES 1972 & 1976.
/HUGHES, NEW ZEALAND J. BOT. 10:225-242,1972 (ILLUS) /HUGHES, MYCOLOGIA 68:707-712 & 787-788,1976 (ILLUS) /ELLIS, MDH P 455-456,1976 (ILLUS) •

CAPNOSPORIUM HUGHES 1976. MYCOLOGIA 68:752 & 708 (ILLUS) /C. MONILIFORME
PHRAGMO,CETERI,PHAEO,NONSP, 93A, ON BRANCHES & LEAVES (SOOTY MOLD). ASCO STATE =METACAPNODIUM FIDE HUGHES 1972. OTHER STATE =CAPNOPHIALOPHORA (Q.V.) FIDE HUGHES 1976.
/HUGHES, NEW ZEALAND J. BOT. 10:232,1972 (ILLUS) •

CAPNOSTYSANUS SPEGAZZINI 1918. PHYSIS 4(17):295 /C. STYSANOPHORUS (PENZIG & SACCARDO)
?, NOT ILLUSTRATED BY SPEGAZZINI. •

CARLOSIA ARNAUD 1954. BULL. SOC. MYCOL. FRANCE 69:294 /C. MELIOLAE (ZIMMERMAN)
=ISTHMOSPORA STEVENS 1918 FIDE KENDRICK 1973. •

CARRIONIA BRICENO-IRAGORRY 1938. REV. CLIN. LUIZ RAZETTI, CARACAS 1:121 (NO LATIN DIAGNOSIS) /HORMODENDRUM PEDROSOI BRUMPT =C. PEDROSOI (BRUMPT)
=FONSECAEA NEGRONI 1936, SEE ALSO DE HOOG AND HERMANIDES-NIJHOF 1977 (CBS STUDIES 15:182). •

CASARESIA GONZALEZ FRAGOSO 1920. BOL. SOC. ESPAN. HIST. NAT. 20:112-114 (ILLUS) /C. SPHAGNORUM
STAURO,CETERI,PHAEO,NONSP,APPEND,AQUATIC, 127E, OTHER STATE ="PHIALOCONIDIA" FIDE WEBSTER & DESCALS 1975 (TBMS 64:529-531,ILLUS).
/INGOLD, GUIDE P 86-87,1975 (ILLUS) /HUGHES, NEW ZEALAND J. BOT. 16:320-321,1978 (ILLUS)
=*ANKISTROCLADIUM* PERROTT 1960 FIDE KENDRICK 1973 •

CATENOPHORA LUTTRELL 1940. MYCOLOGIA 32:530-536 (ILLUS) /C. PRUNI
AMERO,GLOIO,HYALO OR PHAEO,MISC,ACERVULI, ON LEAVES (PRUNUS, YUCCA).
/MORGAN-JONES, ICONES GENERUM COELOMYCETUM IX:15-16,1977 (ILLUS) /NAG RAJ, CAN. J. BOT. 55:757-765,1977 (ILLUS,NEW SP) •

CATENOSUBULISPORA MATSUSHIMA 1971. MICROFUNGI P 10 & FIGS 6-8 & PL 3:1-2 /C. HONIARAENSIS
STAURO,CETERI OR BLASTO,HYALO,RADUL, 124C, ON WOOD (CELTIS). •

CATENULARIA GROVE 1886. IN SACCARDO, SYLLOGE FUNGORUM 4:303 /C. SIMPLEX =CATENULARIA STATE OF CHAETOSPHAERIA CUPULIFERA (BERKELEY & BROOME) SACCARDO 1883
AMERO,BASO OR GLOIO,PHAEO,PHIAL,SETAE, 3F, ON WOOD & BARK (MANY DECIDUOUS TREES). ASCO STATE =CHAETOSPHAERIA FIDE HUGHES 1965 OR ?=MELIOLINA SEE DE HOOG 1977 (CBS STUDIES 15:139). COMPARE WITH CHLORIDIUM.
/HUGHES, NEW ZEALAND J. BOT. 3:136-150,1965 (REVIS,ILLUS) /ELLIS, DH P 512-513,1971 (ILLUS) /ELLIS, MDH P 443-444,1976 (ILLUS,NEW SP)
=*PSILONIELLA* COSTANTIN 1888 FIDE ELLIS 1971
=*HAPLOCHALARA* LINDER 1933 FIDE MASON 1941 •

CATTANEA GAROVAGLIO 1875. RC. IST. LOMB. SCI. LETT., 2 SER., 8:125 /C. HEPTASPORA =D. HEPTOSPORUM (GAROVAGLIO) DAMON 1952
=DICTYOSPORIUM CORDA 1836 FIDE ELLIS 1971. •

CENOCOCCUM MOUGEOT & FRIES 1825. SYST. ORB. VEG. 1:364 /C. GEOPHILUM FRIES 1825
?, STERILE. •

CENTROSPORA NEERGAARD 1942. ZENTRALBL. BAKT. 2,54:411 /C. OHLSENII =C. ACERINA (HARTIG) NEWHALL 1946
=MYCOCENTROSPORA DEIGHTON 1972, NON CENTROSPORA TREVISAN 1845
/LINCZEVSKI, TAXON 19:948-949,1970 (NOM. CONS. PROP.) •

CEPHALIOPHORA THAXTER 1903. BOT. GAZ. 35:157-158 & PL V, FIG 11-13 /C. TROPICA (LECTOTYPE)
DIDYMO OR PHRAGMO,CETERI,PHAEO,AMPUL, 64A & 93D, ON LITTER, DUNG, SOIL & OTHER SUBSTRATES.
/BARRON, HYPHOMYCETES P 112-113,1968 (ILLUS) /ELLIS, DH P 176-178,1971 (ILLUS,KEY) /MATSUSHIMA, ICONES P 22-23 & PL 2-4,1975 (ILLUS)
=*CEPHALOMYCES* BAINIER 1907 FIDE CARMICHAEL 1973 •

CEPHALOCLADIUM REICHENBACH 1828 (CONSPECT. REGNI. VEG. P 6) =BOTRYTIS SACCARDO 1901 (SYLL. FUNG. 15:82) •

CEPHALODIPLOSPORIUM KAMYSCHKO 1961. BOT. MATER (NOT. SYST. SECT. CRYPT. INST. BOT. ACAD. SCI. USSR 14:221-222, ILLUS) /C. ELEGANS
=FUSARIUM FIDE GAMS 1971 (CEPHALO. SCHIMMEL. P 208). •

CEPHALODOCHIUM BONORDEN 1851. HANDB. MYKOL. P 135 & TAF. XI, FIG 227 /C. ALBUM
? •

CEPHALOMYCES BAINIER 1907. BULL. SOC. MYCOL. FRANCE 23:109-110 (ILLUS) /C. NIGRICANS
=CEPHALIOPHORA THAXTER 1903 FIDE CARMICHAEL 1973 (FROM THE DESCRIPTIONS, C. NIGRICANS APPEARS TO BE THE SAME SPECIES AS CEPHALIOPHORA IRREGULARIS THAXT.) •

CEPHALOPHORUM "NEES" =MISCITATION BY SACCARDO, SEE BENJAMIN 1968 (TAXON 17:524) •

CEPHALOSPORIOPSIS PEYRONEL 1916. MEM. R. ACC. SCI. TORINO, SER 2, 66:52 (ILLUS) /C. ALPINA (PEYRONEL), NO TYPE PRESERVED FIDE GAMS 1971
DIDYMO,GLOIO,HYALO,PHIAL, 57C, ON SOIL. GAMS WOULD ASSIGN SUCH FUNGI TO ACREMONIUM, SEPTOFUSIDIUM, FUSARIUM OR CYLINDROCARPON (CEPHALO. SCHIMMEL. P 18 & 205,1971)
/BARRON, HYPHOMYCETES P 113-114,1968 (ILLUS) •

CEPHALOSPORIUM CORDA 1839. ICONES FUNGORUM 3:11-12 & TAF. II, FIG 29 /C. ACREMONIUM CORDA 1842
?, SEE VON ARX 1970. FOR CEPHALOSPORIUM AUCT., NON CORDA, SEE ACREMONIUM LINK.
/GAMS, CEPHALO. SCHIMMEL. 262 P,1971 (MONOGR) •

CEPHALOTHECIUM CORDA 1838. ICONES FUNGORUM 2:14 /C. ROSEUM ?=TRICHOTHECIUM ROSEUM (PERSOON) LINK 1809
=TRICHOTHECIUM LINK 1809 (PROBABLY, BUT TYPE OF C. ROSEUM NOT SEEN) FIDE HUGHES 1958. •

CEPHALOTRICHUM LINK 1809. MAG. GES. NATURF. FREUNDE, BERLIN 3:20 & TAB 1, FIG 34 /PERICONIA STEMONITIS PERSOON 1801 =C. STEMONITIS (PERSOON) (LECTOTYPE FIDE HUGHES 1958)
AMERO,BASO,PHAEO,ANNEL,SYNNEMATA, 19E, ON LITTER, SOIL, DUNG & OTHER SUBSTRATES. OTHER CONIDIAL STATES =ECHINOBOTRYUM OR SCOPULARIOPSIS.
/MORTON & SMITH, CMI PAPERS 86:70-85,1963 (REVIS,KEY,ILLUS SUB DORATOMYCES) /HUGHES, CAN. J. BOT. 36:744,1958 (REFS) /ELLIS, DH P 328-331,1971 (ILLUS SUB DORATOMYCES,KEY) /CONWAY & KIMBROUGH, MYCOTAXON 2:127-131,1975 (ILLUS,NEW SP SUB DORATOMYCES) /VON ARX, GENERA P 208-209,1974 (REFS) /MATSUSHIMA, ICONES P 62-63 & PL,1975 (ILLUS SUB DORATOMYCES) /DABINETT & WELLMAN, CAN. J. BOT. 56:2031-2049,1978 (DISCUSSION OF GENERA,NO ILLUS)
=*DORATOMYCES* CORDA 1829 FIDE HUGHES 1958 & VON ARX 1974
?=*LEIGHTONIOMYCES* HAWKSWORTH & SUTTON 1977
?=*SAROPHORUM* SYDOW & SYDOW 1916
?=*STILBELLULA* BOEDIJN 1951
?=*STYSANOPSIS* FERRARIS 1909 SEE HUGHES 1958
=*STYSANUS* CORDA 1837 FIDE HUGHES 1958
=*SYNPENICILLLIUM* COSTANTIN 1888 FIDE MORTON & SMITH 1963
?=*TRICHURUS* CLEMENTS & SHEAR 1896 •

CERACEA CRAGIN 1885. BULL. WASHBURN COLL. LAB. NAT. HIST. 1(2):82 /C. VERNICOSA
?, TREMELLALES?
/MARTIN, MYCOLOGIA 41:77-86,1949 •

CERATOCLADIUM CORDA 1839. PRACHT-FLORA ... , P 41 /C. MICROSPERMUM
AMERO,CETERI,HYALO,ANNEL OR MISC,SETAE, 20E, ON WOOD & BARK (FAGUS).
/HUGHES, CMI PAPERS 47:5-8,1951 (ILLUS) /ELLIS, DH P 360-361,1971 (ILLUS) •

CERATOCLADIUM PATOUILLARD 1898. BULL. SOC. MYCOL. FRANCE 14:196-197 (NO ILLUS) /C. CLAUTRIAVII
=XYLOCLADIUM SYDOW 1900. •

CERATONEMA PERSOON 1822 (MYCOL. EUROP. 1:48) ?=SPHACELIA. CERATONEMA IS BASED ON A CONIDIAL STATE OF CORDYCEPS SPHECOCEPHALA FIDE DONK 1962 (TAXON 11:81). •

CERATONEMA ROTH 1800 ?=NOMEN DUBIUM SEE DONK 1962 (TAXON 11:78 & 81) •

CERATOPHORUM SACCARDO 1880. MICHELIA 2:22 /C. HELICOSPORUM (SACCARDO)
PHRAGMO OR HELICO,CETERI,PHAEO,NONSP,PODIA,APPEND, 70D, ON LEAVES (QUERCUS). COMPARE WITH ANTENNATULA.
/ELLIS, CMI PAPERS 70:13-16,1958 (ILLUS) /ELLIS, DH P 47-48,1971 (ILLUS,KEY) /SUTTON, CMI PAPERS 138:107,1975 (DESCR) •

CERATOPODIUM CORDA 1837. ICONES FUNGORUM 1:19-20, TAV V, FIG 264 /C. ALBUM ?=GRAPHIUM (FIDE SACCARDO 1886). •

CERATOSPORELLA HOHNEL 1919. BERL. DEUTSCHL. BOT. GES. 37:155 /C. BICORNIS (MORGAN) HOHNEL 1923
STAURO OR CHEIRO,CETERI,PHAEO,NONSP OR ANNEL,STROMA, 127D, ON LITTER & WOOD.
/HUGHES, CMI PAPERS 46:22-25,1951 (ILLUS) /HUGHES, NEW ZEALAND J. BOT. 9:351-354,1971 (ILLUS) /ELLIS, DH P 135-136,1971 (ILLUS) /MATSUSHIMA, ICONES P 23 & PL 350,1975 (ILLUS) /ELLIS, MDH P 106-107,1976 (ILLUS) •

CERATOSPORIUM SCHWEINITZ 1832. TRANS. AMER. PHIL. SOC. II, 4:300 /C. FUSCESCENS
STAURO,CETERI,PHAEO,NONSP, 126B, ON WOOD. OTHER CONIDIAL STATE =SELENOSPORELLA.
/HUGHES, CMI PAPERS 39:1-11,1951 (ILLUS) /HUGHES, NEW ZEALAND J. BOT. 2:305-309,1964 (ILLUS) /ELLIS, DH P 53-54,1971 (ILLUS) /MATSUSHIMA, ICONES P 23-24 & PL,1975 (ILLUS,NEW SP) •

CERCOCLADOSPORA AGARWAL & SINGH 1974. PROC. NAT. ACAD. SCI. INDIA 42:436-439 (ILLUS,DATED 1972) /C. ADINAE
=PSEUDOCERCOSPORA SPEGAZZINI 1910 FIDE DEIGHTON 1976 (CMI PAPERS 140:8-9). •

CERCODEUTEROSPORA CURZI 1932. BOL. STAZ. PAT. VEG. ROMA 12:149-151 (ILLUS) /C. TRICHOPHILA
=MYCOVELLOSIELLA RANGEL 1917 FIDE CASTELLANI & CIFERRI. •

CERCOSEPTORIA PETRAK 1925. ANN. MYCOL. 23:68-70 /C. CHAMAESYCEAE (STEVENS & DALBEY)
SCOLECO,CETERI,HYALO OR PHAEO,RACHI, 110B, ON LEAVES. NOM. NOV. FOR SEPTORIOPSIS STEVENS & DALBEY, NOT VERY DIFFERENT FROM CERCOSPORA. ASCO STATE =MYCOSPHAERELLA FIDE VON ARX 1975 (CBS STUDIES 8:91).
/DEIGHTON, CMI PAPERS 140:159-167,1976 (ILLUS,NEW COMB) •

CERCOSPERMA ARNAUD 1954. BULL. SOC. MYCOL. FRANCE 69:289-290 (ILLUS,NO LATIN DIAGNOSIS) /C. SUBSESSILE
?=SPORIDESMIUM. NO TYPE SPECIMEN -SEE NICOT & CHARPENTIE 1971 (BULL. SOC. MYCOL. FRANCE 87:28). •

CERCOSPORA FRESENIUS 1863. BEITR. ZUR. MYKOL. P 91-92 /C. APII (LECTOTYPE)
SCOLECO OR PHRAGMO,CETERI,PHAEO OR HYALO,RADUL OR RACHI,STROMA, 78D, ON LEAVES (MANY PLANTS). ASCO STATE =MYCOSPHAERELLA FIDE MULLER IN KENDRICK 1971, OR SPHAERULINA FIDE VON ARX 1975 (CBS STUDIES 8:92), OR MICROCYCLUS SEE DE HOOG 1977 (CBS STUDIES 15:139), OR LEPTOSPHAERIA SEE SIVANESAN 1976 (CMI DESCRIPTIONS 506). COMPARE WITH MYCOVELLOSIELLA & RAMULARIA. WE FIND IT DIFFICULT TO DISTINGUISH AMONG THE SEGREGATES PROPOSED FOR CERCOSPORA-LIKE FUNGI & HAVE LISTED THEM AS TENTATIVE SYNONYMS, ALTHOUGH SOME OF THEM ARE ALSO INCLUDED IN THE KEYS AND ILLUSTRATED SEPARATELY. FOR DISCUSSION OF THIS GROUP OF GENERA, SEE DEIGHTON 1976 (CMI PAPERS 140:4-7) & 1979 (144:1-56). FOR A DISCUSSION OF SPECIES CONCEPT IN CERCOSPORA SEE ELLIS 1976 (MDH P 244).
/CHUPP, MONOGR. OF THE GENUS CERCOSPORA,1954 (MONOGR) /BARRON,

HYPHOMYCETES P 116-117,1968 (ILLUS) /ELLIS, DH P 275-277,1971 (ILLUS,REFS) /DEIGHTON, CMI PAPERS 133:1-62,1973 & 137:1-75,1974 (ILLUS,REVIS) /BROWN & MORGAN-JONES, MYCOTAXON 4:299-306,1976 (ILLUS,NEW COMB) /PIROZYNSKI, FUNGI CANADENSES 22,1974 & 62,1975 (ILLUS) /MULDER & HOLLIDAY, CMI DESCRIPTIONS 411-420,1974 & & 462-466,1975 (ILLUS) /ELLIS, MDH P 243-293,1976 (ILLUS,KEYS) /BROWN & MORGAN-JONES, MYCOTAXON 6:261-276,1977 (ILLUS,EXCLUSIONS)

?=*CERCOSEPTORIA* PETRAK 1925
?=*CERCOSPORELLA* SACCARDO 1880
?=*CERCOSPORIDIUM* EARLE 1901
?=*CERCOSPORINA* SPEGAZZINI 1910
?=*ELLETEVERA* DEIGHTON 1969
?=*ERIOCERCOSPORA* DEIGHTON 1969
?=*ISARIOPSIS* FRESENIUS 1865 SEE HODGES & HAASIS 1962
?=*PARACERCOSPORA* DEIGHTON 1979
?=*PASSALORA* FRIES 1849
?=*PHAEOISARIOPSIS* FERRARIS 1909
?=*PSEUDOCERCOSPORA* SPEGAZZINI 1910
?=*PSEUDOCERCOSPORELLA* DEIGHTON 1973
?=*STENOSPORA* DEIGHTON 1969
=*VIRGASPORIUM* COOKE 1875 FIDE ELLIS 1971 •

CERCOSPORELLA SACCARDO 1880. MICHELIA 2:20 /C. CANA (SACCARDO 1876) (LECTOTYPE FIDE DEIGHTON 1973) =C. VIRGAUREAE (THUMEN) ALLESCHER 1895 FIDE DEIGHTON 1973

SCOLECO OR PHRAGMO,CETERI,HYALO,RACHI, 110D, ON LEAVES. FOR CERCOSPORA SPECIES WITH PROTRUBERANT SCARS. ASCO STATE =MYCOSPHAERELLA FIDE VON ARX 1975 (CBS STUDIES 8:91).

/BARRON, HYPHOMYCETES P 116-117,1968 (REFS) /DEIGHTON, CMI PAPERS 133:1-62,1973 (ILLUS,REVIS) /PIROZYNSKI, FUNGI CANADENSES NO. 61,1975 (ILLUS) •

CERCOSPORIDIUM EARLE 1901. MUHLENBERGIA 1:16 /C. EUPHORBIAE (TRACY & EARLE) INED. =C. CHAETOMIUM (COOKE) DEIGHTON 1967

DIDYMO OR PHRAGMO,CETERI,HYALO OR PHAEO,RACHI,STROMA, 85E, ON LEAVES. MUCH LIKE CERCOSPORA. ASCO STATE =MYCOSPHAERELLA FIDE ELLIS 1971 (P 281).

/DEIGHTON, CMI PAPERS 112:27-77,1967 (REVIS,ILLUS) /ELLIS, DH P 279-284,1971 (ILLUS,KEY) /SUTTON, CMI PAPERS 132:25-27,1973 (ILLUS) /ELLIS, MDH P 294-297,1976 (ILLUS,NEW COMBS)

=*BERTEROMYCES* CIFERRI 1954 FIDE DEIGHTON 1967
?=*PSEUDOCERCOSPORIDIUM* DEIGHTON 1973 •

CERCOSPORINA SPEGAZZINI 1910. AN. MUS. NAC. BUENOS AIRES 20(SER 3,13):424-429 /THIRTEEN ORIGINAL SPECIES

?, PROBABLY CERCOSPORA. •

CERCOSPORIOPSIS MIURA 1928. FLORA OF MANCHURIA & EAST MONGOLIA, 3, CRYPTOGAMS, P 527 /EIGHT ORIGINAL SPECIES

=NOMEN ILLEGITIMUM, INCLUDES TYPE SPECIES OF PSEUDOCERCOSPORA, Q.V. •

CERCOSPORULA ARNAUD 1954. BULL. SOC. MYCOL. FRANCE 69:289-290 (ILLUS, NO LATIN DIAGNOSIS) /C. CRASSIUSCULA

?, (SEE 107B) NO TYPE SPECIMEN -SEE NICOT & CHARPENTIE 1971 (BULL. SOC. MYCOL. FRANCE 87:28-29).

/MATSUSHIMA, ICONES P 24 & PL 156,1975 (ILLUS) •

CEREBELLA CESATI 1851. APUD RABENHORST, BOT. ZTG. 9:669 /C. ANDROPOGONIS =EPICOCCUM ANDROPOGONIS (CESATI) SCHOL-SCHWARZ 1959

=EPICOCCUM LINK 1815 FIDE SCHOL-SCHWARZ 1959 (SEE 100F).

/ELLIS, DH P 73-74,1971 (ILLUS) •

CERIOMYCES CORDA 1837. STURM'S DEUTSCHL. FLORA III (PILZE), HEFT 14-15:133 & PLATE 61 /C. FISCHERI

=PLANT GALL FIDE DONK FROM VON ARX 1973 (PERSOONIA 7:130). HAS BEEN USED FOR PTYCHOGASTER SPP. •

CHAETENDOPHRAGMIA MATSUSHIMA 1971. MICROFUNGI OF THE SOLOMON ISLANDS & PAPUA-NEW GUINEA, KOBE, P 12 & FIG 106 /C. TRIANGULARIA

PHRAGMO OR STAURO,CETERI,PHAEO,ANNEL,APPEND, 121A, ON LITTER.

/PIROZYNSKI, CMI PAPERS 129:42-44,1972 (ILLUS) /MATSUSHIMA, ICONES P 24-25 & PL,1975 (ILLUS,NEW SPP) /ELLIS, MDH P 98-99,1976 (ILLUS) /SUTTON & HODGES, NOVA HEDWIGIA 29:599-600,1978 (ILLUS,REVIS,NEW COMB) •

CHAETENDOPHRAGMIOPSIS SUTTON & HODGES 1978. NOVA HEDWIGIA 29:596-599 (ILLUS) /C. PULCHRA
PHRAGMO,CETERI,PHAEO,NONSP OR ANNEL,APPEND, ON LITTER (EUCALYPTUS). •

CHAETOBASIDIELLA HOHNEL 1918. BERL. DEUTSCHL. BOT. GES. 35:317 /C. VERMICULARIOIDEA, DESCRIBED BUT NOT ILLUSTRATED IN MITT. BOT. LAB. TECH. HOCHSCH. WIEN, BD. II, HEFT 2:35-37,1925
? •

CHAETOBLASTOPHORUM MORGAN-JONES 1977. MYCOTAXON 5:484-487 (ILLUS) /C. INGRAMII
AMERO,CETERI OR GLOIO,HYALO,MISC,SETAE, 73G, ON LITTER (QUERCUS). •

CHAETOCHALARA SUTTON & PIROZYNSKI 1965. TBMS 48:350-351 (ILLUS) /C. BULBOSA
AMERO,BASO OR GLOIO,PHAEO OR HYALO,PHIAL,SETAE, 2H, ON LITTER. DISTINGUISHED FROM CHALARA ONLY BY PRESENCE OF SETAE. ASCO STATE ?=CALYCELLINA & HYALOSCYPHA, SEE NAG RAJ & KENDRICK 1975 (P 55)
/ELLIS, DH P 509-510,1971 (ILLUS) /PIROZYNSKI & HODGES, CAN. J. BOT. 51:157-160,1973 (ILLUS) /NAG RAJ & KENDRICK, CHALARA P 39-40 & 148-156,1975 (KEY,ILLUS) /ELLIS, MDH P 442-443,1976 (ILLUS) •

CHAETOCONIDIUM ZUKAL 1887. VERH. ZOOL. BOT. GES. WIEN 37:45 & TAF. I, FIG 7 /C. ARACHNOIDEUM
PHRAGMO OR AMERO,CETERI,HYALO,NONSP,APPEND, 71A, ON MEAT EXTRACT.
/LINDAU, RABENHORST KRYPT. FL. 2 ED. 8:185-186,1907 (ILLUS) •

CHAETODOCHIS CLEMENTS 1909 ?='CHAETOSTROMA' FIDE AINSWORTH 1971 •

CHAETODOCHIUM HOHNEL 1932. MITT. BOT. INST. TECHN. HOCHSCH. WIEN 9:44 /C. BUXI (DE CANDOLLE EX FRIES)
=VOLUTELLA TODE EX FRIES 1832 FIDE TULLOCH 1972 (CMI PAPERS 130:36). •

CHAETOPSELLA HOHNEL 1930. MITT. BOT. LAB. TECH. HOCHSCH. WEIN 7:44 /C. GRISEA (EHRENB.)
=CHAETOPSIS GREVILLE 1885 FIDE HUGHES 1951. •

CHAETOPSINA RAMBELLI 1956. R. C. ACCAD. BOLOGNA (FIS.) 3:5 /C. FULVA
AMERO,GLOIO,HYALO,PHIAL,SETAE, 17G, ON LITTER.
/BARRON, HYPHOMYCETES P 117-120,1968 (ILLUS) /ELLIS, DH P 526-527,1971 (ILLUS) /RAMBELLI & LUNGHINI, GIORN. BOT. ITAL. 110:253-258,1976 (ILLUS,NEW SP) /MORGAN-JONES, MYCOTAXON 8:411-416,1979 (ILLUS,NEW SP) •

CHAETOPSIS GREVILLE 1825. EDINB. PHIL. J. 13:63-64 /C. WAUCHII =C. GRISEA (EHRENBERG) SACCARDO 1880 FIDE HUGHES 1951
AMERO OR DIDYMO,GLOIO,HYALO,PHIAL,SETAE, 57B, ON WOOD & BARK.
/HUGHES, TBMS 34:569-573,1951 (ILLUS) /ELLIS, DH P 563-564,1971 (ILLUS)
=*CHAETOPSELLA* HOHNEL 1930 FIDE HUGHES 1951 •

CHAETOSIRA CLEMENTS, IN CLEMENTS & SHEAR 1931, A SUPERFLUOUS NAME FOR WIESNERIOMYCES. •

CHAETOSPORIUM CORDA 1833. IN STURM'S DEUTSCHL. FL. 3(13):65-66 & TAB 33 /C. GRAMINIS
=NOMEN DUBIUM FIDE SACCARDO 1886 (SYLL. FUNG. 4:761).
=*CHAETOTRICHUM* RABENHORST 1844, PRO PARTE •

CHAETOSTROMA CORDA 1829. IN STURM'S DEUTSCHL. CRYPT. FLORA 2:123-124 (ILLUS) /?
=NOMEN DUBIUM FIDE TULLOCH 1972 (CMI PAPERS 130:36). •

CHAETOTRICHUM RABENHORST 1844 (DEUTSCHL. KRYPT.-FL. 1:68) WAS A SUPERFLUOUS REPLACEMENT FOR CHAETOSPORIUM CORDA & CHRYSOMYXA UNGER. •

CHAETOTRICHUM SYDOW 1927. ANN. MYCOL. 25:150 /C. SOLANI =ANNELOPHORA SOLANI (SYDOW) HUGHES 1951
=ANNELLOPHORA HUGHES 1951, NON CHAETOTRICHUM RABENHORST 1844. •

CHALARA (CORDA) RABENHORST 1844. DEUTSCHL. KRYPT.-FLORA 1:38 /C. FUSIDIOIDES
AMERO OR DIDYMO OR PHRAGMO,BASO OR GLOIO,HYALO OR PHAEO,PHIAL, 2C, ON LITTER & WOOD. ASCO STATE =CERATOCYSTIS FIDE ELLIS 1971 & RAYNER & HUDSON 1977 (TBMS 68:315-316) OR CRYPTENDOXYLA SEE NAG RAJ & KENDRICK 1975 OR SPHAERONAEMELLA SEE MALLOCH 1974 (FUNGI CANADENSES 53) OR PYXIDIOPHORA SEE MULLER & VON ARX 1973 (IN THE FUNGI IVA, P 127). OTHER CONIDIAL STATES =THIELAVIOPSIS, HUMICOLA-LIKE & TRICHOCLADIUM-LIKE. SEE ALSO CHAETOCHALARA.
/MORGAN-JONES, CMI DESCRIPTIONS 141-143,1967 (ILLUS) /BARRON, HYPHOMYCETES P 120-122,1968 (ILLUS) /NAG RAJ & KENDRICK, CAN. J. BOT. 49:2119-2122,1971 (ILLUS) /ELLIS, DH P 501-509,1971 (ILLUS) /NAG RAJ & KENDRICK, A MONOGRAPH OF CHALARA AND ALLIED GENERA 200 P,1975 (MONOGR) /ELLIS, MDH P 438-441,1976 (ILLUS,KEY) /HAWES & BECKETT, TBMS 68:304-307,1977 (ILLUS,CONIDIUM ONTOGENY SUB THIELAVIOPSIS BASICOLA)
=*CHALAROPSIS* PEYRONEL 1916, PRO PARTE, FIDE BARRON 1968
=*CYLINDROCEPHALUM* BONORDEN 1851 FIDE HUGHES 1958
=*EXCIOCONIDIUM* PLUNKETT 1975 FIDE NAG RAJ & KENDRICK 1975
=*THIELAVIOPSIS* WENT 1893, PRO PARTE, FIDE BARRON 1968 •

CHALAROPSIS PEYRONEL 1916. STAZ. SPER. AGRIC. ITAL. 49:595 /C. THIELAVIOIDES =CHALARA & HUMICOLA STATES OF CHALARA THIELAVIOIDES (PEYRONEL) NAG RAJ & KENDRICK 1975 (SEE CHALARA)
?, THE TYPE SPECIES HAS CHALARA AND HUMICOLA-LIKE CONIDIA.
/NAG RAJ & KENDRICK, CHALARA, 200P,1975 (MONOGR) •

CHANTRANSIOPSIS THAXTER 1914. BOT. GAZ. 58:246 (235-253, ILLUS) /C. DECUMBENS (LECTOTYPE)
AMERO,CETERI,HYALO,NONSP, 55E, ON INSECTS (BEETLES). •

CHARDONIA CIFERRI 1930. IN CHARDON & TORO, J. DEPT. AGRIC. PORTO RICO 14(4):295 /C. ROSEA
?, NOT ILLUSTRATED BY AUTHOR. •

CHEIROCONIUM HOHNEL 1910. S. B. AKAD. WISS. WIEN 119:665 /C. BEAUMONTII (BERK. & CURT. EX HOHNEL) =S. TINCTUM (PECK) HUGHES 1958
=SIROTHECIUM KARSTEN 1887 FIDE HUGHES 1958. •

CHEIROMYCELLA HOHNEL 1910. S. B. AKAD. WISS. WIEN 119:664 /C. SPEIROIDES (HOHNEL) =C. MICROSCOPICA (KARSTEN) HUGHES 1958
CHEIRO,GLOIO,PHAEO,NONSP,ALEURI?,SPORODOCHIA, 116B, ON WOOD (CONIFERS).
/ELLIS, DH P 325-326,1971 (ILLUS) /MATSUSHIMA, ICONES P 26-27 & PL,1975 (ILLUS,NEW SPP) •

CHEIROMYCES BERKELEY & CURTIS 1857. IN BERKELEY, INTROD. CRYPT. BOT. P 312-313 & FIG 70C /C. STELLATUS
CHEIRO OR STAURO,CETERI,PHAEO,NONSP?,SPORODOCHIA, 116A, ON LEAVES (SALVIA).
/DAMON, MYCOLOGIA 42:554-562,1950 (ILLUS,REVIS) /MOORE, MYCOLOGIA 50:682,1958 (REFS)
=*STIGMOPSIS* BUBAK 1914 FIDE MOORE 1958 •

CHEIROPODIUM SYDOW 1915. ANN. MYCOL. 13:42-43 (ILLUS) /C. FLAGELLATUM =CLASTEROSPORIUM FLAGELLATUM (SYDOW) ELLIS 1958
=CLASTEROSPORIUM SCHWEINITZ 1832 FIDE ELLIS 1958. •

CHEIROSPORA MOUGEOT & FRIES 1825. IN FRIES, SYST. ORB. VEG. 1:365 /STILBOSPORA CHEIROSPORA FRIES =C. BOTRYOSPORA (MONTAGNE) HUGHES 1958 (P 748)
DICTYO OR BULBILS,CETERI,PHAEO,NONSP,SPORODOCHIA, ON WOOD (ALNUS). COMPARE WITH STRUMELLA.
/HUGHES, CAN. J. BOT. 36:748,1958 (REFS)
=*THYRSIDIUM* MONTAGNE 1846 FIDE HUGHES 1958
=*RHABDOSPORIUM* CHEVALLIER 1826 FIDE HUGHES 1958 (AS ?=)
=*HYPEROMYXA* CORDA 1839 FIDE HUGHES 1958
=*MORONOPSIS* DELACROIX 1891 FIDE SACCARDO SUB THYRSIDIUM

=*MYRIOCEPHALUM* DE NOTARIS 1842 FIDE HUGHES 1958 •

CHELISPORIUM SPEGAZZINI 1911. AN. MUS. NAC. BUENOS AIRES 20(SER. 3,13):463-464 (ILLUS) /C. HYSTERIOIDES
?•

CHIKANEEA SUTTON 1973. CMI PAPERS 132:29-33 (ILLUS) /C. HOLLERONIAE
PHRAGMO,ARTHRO OR BLASTO?,HYALO,RADUL, 81B, ON BARK (POPULUS). •

CHIONOMYCES DEIGHTON & PIROZYNSKI 1972. CMI PAPERS 128:74-82 (ILLUS) /C. MELIOLICOLA (CIFERRI)
PHRAGMO OR SCOLECO,CETERI,HYALO,RACHI,APPEND, 71C, ON FUNGI (MELIOLA, IRENOPSIS, AMAZONICA). ASCO STATE =MELIOLIPHILA SEE DE HOOG 1977 (CBS STUDIES 15:140).
/MATSUSHIMA, ICONES P 27 & PL,1975 (ILLUS,NEW SP) •

CHLAMYDOALEUROSPORIA GRIGORAKIS 1924. COMP. REND. 179:1425 /C. GRANULOSA (SABOURAUD) GRIGORAKIS 1925 (ANN. SCI. NAT. BOT. SER. 10, 7:412) =TRICHOPHYTON MENTAGROPHYTES (LECTOTYPE)
=TRICHOPHYTON MALMSTEN 1845. •

CHLAMYDOMYCES BAINIER 1907. BULL. SOC. MYCOL. FRANCE 23:240-241 /C. DIFFUSUS =C. PALMARUM (COOKE) MASON 1928
DIDYMO OR AMERO,CETERI,HYALO,NONSP,ALEURI, 68F, ON PLANTS & LITTER. OTHER CONIDIAL STATE =HARZIELLA.
/MASON, CMI PAPERS 2:37,1928 (REFS) /BARRON, HYPHOMYCETES P 122-123,1968 (ILLUS) /MATSUSHIMA, ICONES P 27-28 & PL 35,1975 (ILLUS) /ELLIS, MDH P 45-46,1976 (ILLUS)
?=*DISSOACREMONIELLA* KIRILENKO 1970 •

CHLAMYDORUBRA DESHPANDE & DESHPANDE 1966. MYCOPATHOLOGIA 29:272 /C. VERRUCOSA
?, ORIGINAL DESCRIPTION & ILLUSTRATION NOT DIAGNOSTIC. •

"CHLAMYDOSPORES" ARE FOUND IN SO MANY DIFFERENT FUNGI THAT THEIR PRESENCE IS NOT ROUTINELY NOTED IN THIS LIST. DARK-COLORED CHLAMYDOSPORES MAY BE REFERRED TO FORM-GENERA SUCH AS HUMICOLA & TRICHOCLADIUM. LIGHT COLORED CHLAMYDOSPORES WITH DISTINCTIVE CHARACTERS MAY BE REFERRED TO FORM-GENERA SUCH AS ASTEROPHORA. SOME MAY BE FORMED ON COREMIUM-LIKE CARPOPHORES SEE SINGER 1975 (AGARICALES IN MODERN TAXONOMY P 69). •

CHLAMYDOSPORUM PEYRONEL 1913 (INAUG. DIS., PADORA P 18) =DUBIOUS FIDE CLEMENTS & SHEAR 1931 (P 413). •

CHLORIDIELLA ARNAUD 1954. BULL. SOC. MYCOL. FRANCE 69:270 & FIG 1F-G (NO LATIN DIAGNOSIS) /C. LEUCOPODA (BONORDEN)
=IDRIELLA NELSON & WILHELM 1956 FIDE NICOT & CHARPENTIE 1971
/NICOT & CHARPENTIE, BULL. SOC. MYCOL. FRANCE 87:421-424,1971 (ILLUS) /NICOT & MOUCHACCA, REV. MYCOL. 36:185-193,1972 (ILLUS,NEW SPP) •

CHLORIDIUM LINK 1809. MAG. GES. NATURF. FREUNDE, BERLIN 3:13 /C. VIRIDE
AMERO,GLOIO,HYALO OR PHAEO,PHIAL, 3H, ON WOOD & SOIL. ASCO STATE =CHAETOSPHAERIA FIDE MULLER IN KENDRICK 1971. COMPARE WITH CATENULARIA. OTHER CONIDIAL STATE =HUMICOLA.
/BARRON, HYPHOMYCETES P 123-125,1968 (ILLUS) /KIFFER, MANGENOT & REISINGER, REV. ECOL. BIOL. SOL 6:185-191,1969 (ILLUS) /ELLIS, DH P 558-559,1971 (ILLUS) /WILCOX, GANMORE-NEUMANN & WANG, CAN. J. BOT. 52:2279-2282,1974 (ILLUS) /SIVASITHAMPARAM, TBMS 64:335-338 & PL 27,1975 (ILLUS) /MATSUSHIMA, ICONES P 28-29 & PL,1975 (ILLUS,NEW SP) /GAMS & HOLUBOVA-JECHOVA, CBS STUDIES 13:1-48,1976 (MONOGR) /COLE & SAMSON, CONIDIAL FUNGI P 68-69,1979 (ILLUS,CONIDIUM ONTOGENY)
?=*BISPOROMYCES* VAN BEYMA 1940 FIDE HUGHES 1958
=*CIRRHOMYCES* HOHNEL 1903 FIDE HUGHES 1958
=*PSILOBOTRYS* SACCARDO 1879 FIDE HUGHES 1958
?=*SPHAEROMYCETELLA* ARNAUD 1954 •

CHMELIA SVOBODOVA 1966. BIOLOGIA, BRATISLAVA 21:81-85 /C. SLOVACA
?, FOR A BLACK YEAST-LIKE AGENT OF CHROMOMYCOSIS. ORIGINAL DESCRIPTION & ILLUSTRATION DIFFICULT TO INTERPRET.
/EMMONS ET AL., MEDICAL MYCOLOGY P 363,1970 (DESCR) /DE HOOG AND HERMANIDES-NIJHOF, CBS STUDIES 15:214,1977 (DESCR OF TYPE STRAIN) •

CHRISTIASTER KUNTZE =GONATOBOTRYUM SACCARDO FIDE AINSWORTH 1971 •

CHROMATIUM LINK 1826 (ABHANDL. K. AKAD. WISS. BERLIN 1824:180, NO ILLUS,NO SPECIES MENTIONED) ="EIN AUSGEBILDETES DEMATIUM" FIDE LINK 1826 •

CHROMELOSPORIUM CORDA 1833. STURM'S DEUTSCHL. FLORA III (PILZE), BD. 3, HEFT 13:81 & TAB 41 /C. OCHRACEUM
AMERO,CETERI,HYALO,AMPUL,SYNNEMATA, 33F, ON WOOD, LITTER & SOIL. ASCO STATE =PEZIZA FIDE HENNEBERT 1973.
/HUGHES & BISULPUTRA, CAN. J. BOT. 49:361-366,1970 (ILLUS) /HENNEBERT, PERSOONIA 7:195-198 & FIG 6, P 186,1973 (ILLUS,REVIS) /HENNEBERT & KORF, MYCOLOGIA 67:214-240,1975 (ILLUS) /O'DONNELL, FIELDS & HOOPER, CAN. J. BOT. 54:1084-1093,1976 (ILLUS SUB PEZIZA QUELEPIDOTA) /MATSUSHIMA, ICONES P 103 & PL 5,1975 (ILLUS SUB OSTRACODERMA DICHOTOMA) /ELLIS, MDH P 151-154,1976 (ILLUS,KEY,NEW COMB) •

CHROMOSPORIUM CORDA 1825. STURM'S DEUTSCHL. FLORA III (PILZE) BD. 2 HEFT 9:119 /C. ROSEUM
=NOMEN DUBIUM SEE HUGHES 1958. •

CHROMOSTYLIUM GIARD 1889 - SEE TULLOCH 1976 (TBMS 66:409,REFS) •

CHROOSTROMA CORDA 1837. ICONES FUNGORUM 1:5, FIG 48 /C. PINI
=ACHITONIUM KUNZE FIDE LINDAU IN RABENHORST 1910. •

CHRYSACHNE CIFERRI 1938. ANN. MYCOL. 36:239 /C. HYPHOCHORIA
? •

CHRYSOBASIDIUM CLEMENTS 1909 IS A SUPERFLUOUS NAME FOR AUREOBASIDIUM VIALA & BOYER 1891 FIDE DONK 1956 (REINWARDTIA 4:114) •

CHRYSOSPORIUM CORDA 1833. STURM'S DEUTSCHL. FLORA III, BD 2, HEFT 13:85-86 (ILLUS) /C. CORII =C. MERDARIUM (LINK) CARMICHAEL 1962
AMERO,CETERI,HYALO,NONSP,ALEURI, 54B, ON KERATIN, DUNG, SOIL & OTHER SUBSTRATES. ASCO STATE =ARTHRODERMA, CTENOMYCES OR THIELAVIA, FIDE CARMICHAEL 1962 OR AJELLOMYCES FIDE MCDONOUGH 1968 (SUB BLASTOMYCES, MYCOLOGIA 60:76-83), OR ECHINOPODOSPORA FIDE JONG & DAVIS 1974 (MYCOLOGIA 66:463-468,ILLUS), OR CORYNASCUS FIDE VON ARX 1975 (CBS STUDIES 8:21-22 & KAVAKA 3:33) OR GYMNOASCUS, KERATINOPHYTON OR ANIXIOPSIS SEE VON ARX 1974 OR ZOPFIELLA SEE FURUYA & UDAGAWA 1975 (JOUR. JAPANESE BOT. 50:249-251,ILLUS SUB Z. MARINA) OR EMMONSIELLA (SEE ENTRY UNDER HISTOPLASMA) OR ARACHNIOTUS SEE UDAGAWA & TAKADA 1968 (TRANS MYCOL. SOC. JAPAN 9:12-13,ILLUS) OR APINISIA FIDE APINIS & REES 1976 (TBMS 67:522-524,ILLUS) OR BETTSIA SEE SKOU 1975 OR RENISPORA FIDE SIGLER ET AL 1979. BASIDIO STATE =PHANEROCHAETE FIDE BURDSALL & ESLYN 1974 (MYCOTAXON 1:123-133) OR LAETIPORUS SEE VAN DER WESTHUIZEN 1975 (CMI DESCRIPTIONS 441,ILLUS). OTHER CONIDIAL STATES =MICROSPORUM, TRICHOPHYTON. BLASTO STATE =BLASTOMYCES GILCHRIST & STOKES Q.V. & HISTOPLASMA. COMPARE WITH SPOROTRICHUM, MYCELIOPHTHORA, GEOMYCES & MYRIODONTIUM.
/CARMICHAEL, CAN. J. BOT. 40:1137-1173,1962 (REVIS,KEY,ILLUS) /UDAGAWA & HORIE, BULL. NAT. SCI. MUSEUM, TOKYO 15:191-195,1972 (NEW SP,ILLUS SUB THIELAVIA NOVOGUINEENSIS) /MATSUSHIMA, ICONES P 29-31 & PL 15-18,1975 (ILLUS SUB SPOROTRICHUM OCHROKERATINOPHILUM MATS. SP. NOV. =CHRYSOSPORIUM STATE OF CTENOMYCES SERRATUS FIDE CARMICHAEL) /SKOU, FRIESIA 11:70-72,1975 (ILLUS,NEW SP) /OORSCHOT, VERH. KON. NEDERL. AKAD. WET. AFD. NAT. 67:8,1976 (REFS) /SIGLER & CARMICHAEL, MYCOTAXON 4:374,1976 (REVIS) /SIGLER, GAUR, LICHTWARDT & CARMICHAEL, MYCOTAXON 10:133-141,1979 (ILLUS,NEW SP SUB RENISPORA)
=*ALEURISMA* AUCT. NON LINK FIDE CARMICHAEL 1962
=*ATRICHOPHYTON* CASTEL. & CHALM. 1919 FIDE CARMICHAEL 1973
=*BLASTOMYCES* COSTANTIN & ROLLAND FIDE CARMICHAEL 1962
=*GILCHRISTIA* CIFERRI & REDAELLI FIDE CARMICHAEL 1962

=*GLENOSPORELLA* NANNIZZI FIDE CARMICHAEL 1962
=*RHINOCLADIOPSIS* KAMYSCHKO 1961 FIDE CARMICHAEL
=*ZYMONEMA* DE BEURMANN & GOUGEROT FIDE CARMICHAEL 1962 •

CHUPPIA DEIGHTON 1965. IN DEIGHTON & PIROZYNSKI, CMI PAPERS 101:32-34 (ILLUS) /C. SARCINIFERA
DICTYO,CETERI,PHAEO,NONSP, 100A, ON LEAVES (SOLANUM). ASCO STATE =NEOPARODIA FIDE VON ARX 1975 (CBS STUDIES 8:52). COMPARE WITH MONODICTYS & ACROSPEIRA.
/ELLIS, DH P 48-49,1971 (ILLUS) /MATSUSHIMA, ICONES P 31-32 & PL 30,1975 (ILLUS) •

CIFERRIOXYPHIUM BATISTA & MAIA 1963. IN BATISTA & CIFERRI, QUADERNO IST. BOT. UNIV. PAVIA 31:65 /C. GIGANTEUM =C. CHAETOMORPHUM (SPEGAZZINI) HUGHES 1976
PHRAGMO,GLOIO,HYALO OR PHAEO,PHIAL,SYNNEMATA, 74D, ON LEAVES (SOOTY MOLD). ASCO STATE =AITHALODERMA FIDE HUGHES 1976. SEE ALSO LEPTOXYPHIUM.
/HUGHES, MYCOLOGIA 68:762-764 & 731-738,1976 (ILLUS)
=*PHRAGMOXYPHIUM* BATISTA & CIFERRI 1963 FIDE HUGHES 1976 •

CILICIOPODIUM CORDA 1833. IN STURM'S DEUTSCHL. FLORA, PILZE 3:57-58 /C. VIOLACEUM
AMERO,GLOIO,HYALO,PHIAL,SYNNEMATA, 9B, ON PLANTS & LITTER. ASCO STATE =NECTRIA FIDE BOOTH 1959. DOUBTFULLY DISTINCT FROM TUBERCULARIA, CRINULA, ETC., SEE BOOTH 1959 (CMI PAPERS 73:26-32,ILLUS).
/BARRON, HYPHOMYCETES P 127-128,1968 (ILLUS)
=*BOTRYPES* PREUSS 1852 FIDE LINDAU IN RABENHORST 1910
=*CLAVULARIA* KARSTEN 1883 FIDE CLEMENTS & SHEAR 1931 •

CILIOFUSA CLEMENTS & SHEAR 1931 (P 404) =A SUPERFLUOUS NAME FOR CILIOFUSARIUM •

CILIOFUSARIUM ROSTRUP 1892. BOT. TIDSSKR. 18:77 /C. UMBROSUM =MENISPORA GLAUCA PERSOON 1822 FIDE HUGHES 1958
=MENISPORA PERSOON 1822 FIDE HUGHES 1958
=*CILIOFUSA* CLEMENTS & SHEAR 1931 •

CIRCINOCONIS BOEDIJN 1942. 150TH ANNIV. VOL. R. BOT. GARD. CALCUTTA P 209-211 (ILLUS) /C. PARADOXA
HELICO,CETERI,PHAEO,ANNEL, 104B, ON PLANTS.
/ELLIS, DH P 159-160,1971 (ILLUS) •

CIRCINOTRICHUM NEES 1817. DAS SYSTEM ... P 19 /C. MACULIFORME
SCOLECO,CETERI OR GLOIO,HYALO,RADUL OR MISC,SETAE,STROMA, 54F, ON LITTER. DIFFERS FROM GYROTHRIX ONLY IN HAVING UNBRANCHED SETAE.
/PIROZYNSKI, CMI PAPERS 84:3-8,1962 (ILLUS) /PIROZYNSKI & PATIL, CAN. J. BOT. 48:575-578,1970 (ILLUS,NEW SPP) /ELLIS, DH P 356-358,1971 (ILLUS) /PIROZYNSKI & HODGES, CAN. J. BOT. 51:159-162,1973 (ILLUS) /CUNNINGHAM, MYCOLOGIA 66:122-129,1974 (REVIS) /ELLIS, MDH P 361-363,1976 (ILLUS,KEY)
=*GYROTRICHUM* SPRENGEL 1827 FIDE HUGHES 1958
=*DEPHILIPPIA* RAMBELLI 1959 FIDE PIROZYNSKI 1962 •

CIRRENALIA MEYERS & MOORE 1960. AMER. J. BOT. 47:346-348 /C. MACROCEPHALA (KOHLMEYER)
HELICO,CETERI,PHAEO,NONSP,SPORODOCHIA,AQUATIC, 104D, ON WOOD & BARK IN WATER. COMPARE WITH ZALERION.
/SUTTON, CMI PAPERS 132:33-36,1973 (ILLUS,NEW SP) /ELLIS, MDH P 38-40,1976 (ILLUS,KEY) •

CIRRHOMYCES HOHNEL 1903. ANN. MYCOL. 1:529-530 /C. CAUDIGERUS =CHLORIDIUM LINK FIDE HUGHES 1958. •

CITROMYCES WEHMER 1893 (BER. DEUTSCH. BOT. GES. 11:333) =ASPERGILLUS FIDE THOM FROM AINSWORTH 1971. •

CLADASPERGILLUS RITGER 1831 (SCHR. MARB. GES. 2:89) =ASPERGILLUS FIDE AINSWORTH 1971 •

CLADOBOTRYUM NEES 1816. DAS SYSTEM ... P 56 /C. VARIUM =C. VARIOSPERMUM (LINK) HUGHES 1958 =C. STATE OF HYPOMYCES AURANTIUS
AMERO OR DIDYMO OR PHRAGMO,BASO OR CETERI,HYALO,MISC, 67F, ON FUNGI (RUSSULA, BOLETUS, POLYPORUS & OTHERS). ASCO STATE =HYPOMYCES TULASNE 1865 FIDE GAMS & HOOZEMANS 1970. OTHER STATE =?TRICHOCLADIUM & "CHLAMYDOSPORES". COMPARE WITH SIBIRINA.
/GAMS & HOOZEMANS, PERSOONIA 6:95-110,1970 (ILLUS,KEY,NEW COMBS) /COLE & KENDRICK, CAN. J. BOT. 49:595-599,1971 (ILLUS,CONIDIUM ONTOGENY) /GAMS, PERSOONIA 7:167-169,1973 (REFS) /MATSUSHIMA, ICONES P 32-33 & PL,1975 (ILLUS) /DE HOOG, PERSOONIA 10:34-47,1978 (ILLUS,NEW SPP,KEY) /COLE & SAMSON, CONIDIAL FUNGI P 84-89,1979 (ILLUS,CONIDIUM ONTOGENY)
=*DACTYLIUM* SACCARDO NON NEES, PRO PARTE
=*DIDYMOCLADIUM* SACCARDO 1886 FIDE HUGHES 1958
=*DIPLOCLADIUM* BONORDEN 1851 FIDE HUGHES 1958
?=*CYLINDROPHORA* BONORDEN 1851, SEE ALSO GAMS & HOOZEMANS P 99
=*MUCROSPORIUM* PREUSS 1851 FIDE DE HOOG 1978 •

CLADOBYSSUS RITGEN 1831 (SCHR. MARB. GES. 2:92) =HYPHA PERSOON FIDE SACCARDO 1901 (SYLL. FUNG. 15:89) •

CLADOGRAPHIUM PEYRONEL 1918. NUOV. GIORN. BOT. ITAL. 25:439-440 (ILLUS) /C. RIVULORUM
? •

CLADORRHINUM SACCARDO & MARCHAL 1885. IN MARCHAL, BULL. SOC. BOT. BELG. 24:64 /C. FOECUNDISSIMUM
AMERO,GLOIO,HYALO,PHIAL,SETAE, 15D, ON SOIL & DUNG. ASCO STATE =APIOSORDARIA FIDE JONG & DAVIS 1974 (MYCOLOGIA 66:467), OR ARNIUM SEE JENG AND KRUG 1977 (CAN. J. BOT. 55:84-85,ILLUS), OR ECHINOPODOSPORA SEE MORINAGA, MINOURA & UDAGAWA 1978 (TRANS. MYCOL. SOC. JAPAN 19:135-138,ILLUS). OTHER STATE =PAPULASPORA FIDE GAMS & DOMSCH 1969.
/VON ARX & GAMS, NOVA HEDWIGIA 13:199-208,1966 (ILLUS,EMEND) /UDAGAWA & FURUYA, TRANS. MYCOL. SOC. JAPAN 13:52-54,1972 & 16:215-216,1975 (ILLUS SUB C. STATE OF APIOSORDARIA VERRUCULOSA) /GAMS & DOMSCH, NOVA HEDWIGIA 18:6-8,1969 (ILLUS,NEW SPP)
?=*BAHUPAATHRA* SUBRAMANIAN & LODHA 1964, SEE VON ARX 1973 •

CLADOSARUM YUILL & YUILL 1938. TBMS 22:199 (ILLUS) /C. OLIVACEUM
?=ASPERGILLUS MUTANT FIDE RAPER FROM AINSWORTH 1971. •

CLADOSPORIELLA DEIGHTON 1965. IN DEIGHTON & PIROZYNSKI, CMI PAPERS 101:34-36 (ILLUS) /C. CERCOSPORICOLA
SCOLECO,BLASTO,PHAEO,RADUL, 78F, ON FUNGI (ON LEAVES).
/DEIGHTON, CMI PAPERS 118:33-39,1969 (ILLUS) /ELLIS, DH P 303,1971 (ILLUS) •

CLADOSPORIUM LINK 1815. MAG. GES. NATURF. FREUNDE, BERLIN 7:37 /C. HERBARUM (PERSOON) =C. STATE OF MYCOSPHAERELLA TASSIANA (DE NOTARIS) JOHANSON 1884 FIDE BARR 1958
AMERO OR DIDYMO OR PHRAGMO,BLASTO,PHAEO,RADUL,STROMA, 66E, ON LEAVES & LITTER. ASCO STATE =MYCOSPHAERELLA OR VENTURIA FIDE ELLIS 1971, APIOSPORINA FIDE CORLETT 1975 & 1976 (FUNGI CANADENSES NO. 76 & 84), OR ZOPFIA FIDE REDDY 1976 (INDIAN PHYTOPATH. 28:286-287,DATED 1975,NOT SEEN). COMPARE WITH RAMULARIA, HORMOCONIS & SOROCYBE.
/ELLIS, DH P 308-319,1971 (ILLUS,KEY) /VON ARX, VERH. KON. NEDERL. AKAD. WETENSCH. AFD. NAT. 61:62,1973 (SYNONYMS) /MATSUSHIMA, ICONES P 33-36 & PL,1975 (ILLUS,NEW SPP) /ELLIS, MDH P 325-344,1976 (ILLUS,KEY,NEW SPP,NEW COMBS,NEW NAME)
?=*BIHARIA* THIRUMALACHAR & MISHRA 1953 SEE VON ARX 1973
=*DIDYMOTRICHUM* BONORDEN 1851 FIDE HUGHES 1958
=*FONSECAEA* NEGRONI 1936 PRO PARTE
?=*FULVIA* CIFERRI 1954 SEE VON ARX 1973
=*FUSICLADIOPSIS* KARAK. & VASSIL. 1937 FIDE VON ARX 1973
=*GRAPHIOPSIS* TRAIL 1899 SEE MASON & ELLIS 1953
=*HETEROSPORIUM* KLOTZSCH 1877 FIDE HUGHES 1958
?=*HORMODENDRUM* BONORDEN 1853 FIDE HUGHES 1958
=*KARAKULINIA* GOLOVINA 1964 FIDE VON ARX 1973 SEE ALSO MYCOVELLOSIELLA RANGEL 1917 FIDE VON ARX 1973

=*MYDONOSPORIUM* CORDA 1833 FIDE LINDAU IN RABENHORST 1907
=*MYXOCLADIUM* CORDA 1837 FIDE HUGHES 1958
=*PHAEORAMULARIA* MUNTANOLA 1960 FIDE VON ARX 1973
?=*POLYRHIZIUM* GIARD 1889 FIDE PETCH FROM AINSWORTH 1971
=*SPOROCLADIUM* CHEVALLIER 1826 FIDE HUGHES 1958
?=*STENELLA* SYDOW 1930 SEE VON ARX 1973
?=*TANDONELLA* PRASAD & VERMA 1970 •

CLADOSTERIGMA PATOUILLARD 1892. IN PATOUILLARD & LAGERHEIM, BULL. SOC. MYCOL. FRANCE 8:138 & PL XII, FIG 3A TO F /C. FUSISPORA
?, ORIGINAL ILLUSTRATION DIFFICULT TO INTERPRET. •

CLADOTRICHUM CORDA 1832. STURM'S DEUTSCHL. FLORA III (PILZE), BD 3, 12:39-40 /C. POLYSPORUM
=OEDEMIUM LINK 1824 FIDE HUGHES 1958.
=*SPHINCTROSPORIUM* KUNZE EX FRIES FIDE SACCARDO 1886 (SYLL. FUNG. 4:374) •

CLASTEROSPORIUM SCHWEINITZ 1832. TRANS. AMER. PHIL. SOC. II, 4:300 /C. CARICINUM
PHRAGMO,CETERI,PHAEO,ANNEL OR NONSP,PODIA,SETAE 75G, ON LEAVES (CAREX, CYPERUS & OTHERS). ASCO STATE =CLYPEOLELLA, YAMAMOTOA, EUPELTE OR ASTERODOTHIS FIDE VON ARX 1975 (CBS STUDIES 8:44).
/ELLIS, CMI PAPERS 70:1-13,1958 (KEY,ILLUS) /ELLIS, DH P 122-125,1971 (ILLUS,KEY) /ELLIS, MDH P 100-101,1976 (ILLUS,NEW SP)
= *CHEIROPODIUM* SYDOW 1915 FIDE ELLIS 1958
= *CLASTERISPORIUM* (ORTHOGRAPHIC VARIANT)
= *COMETELLA* SCHWEINITZ 1835 FIDE ELLIS 1958
= *HYMENOPODIUM* CORDA 1837 FIDE ELLIS 1958
= *SPORHELMINTHIUM* SPEGAZZINI 1918 FIDE HUGHES 1958 •

CLATHROCOCCUM HOHNEL 1911. S. B. AKAD. WISS. WIEN 120:473 /C. COMPACTUM (BERKELEY & CURTIS) HOHNEL EX CLEMENTS & SHEAR 1931 (P 404)
?=EPICOCCUM FIDE AINSWORTH 1971. •

CLATHROSPHAERA ZALEWSKI 1888. ROZPRAWY I SPRAWY. Z. POSIEDZ.-WYDZIALU MAT. PRZYROD, KRAKOW 18:151-191 /C. SPIRIFERA
=NOMEN CONFUSUM FIDE VAN BEVERWIJK 1951. •

CLATHROSPHAERINA VAN BEVERWIJK 1951. TBMS 34:289 /C. ZALEWSKII
DICTYO OR BULBILS,CETERI,HYALO,NONSP,AQUATIC, 97H, ASCO STATE =HYALOSCYPHA FIDE DESCALS & WEBSTER 1976 (TBMS 67:525-528,ILLUS).
/TUBAKI, J. HATTORI BOT. LAB. NO. 20, P 152-154,1958 (ILLUS) /PUROHIT, PANWAR & VYAS, INDIAN PHYTOPATHOL. 26:757-759,1974 (ILLUS,NEW SP,NOT SEEN) •

CLATHROTRICHUM PATOUILLARD 1921. BULL. SOC. MYCOL. FRANCE 37:33-35 (ILLUS) /C. SUBCARNEUM
?=MARTINDALIA OR HEYDENIA. •

CLAVARIANA NAWAWI 1976. IN DESCALS, NAWAWI & WEBSTER, TBMS 67:217-219 (207-222,ILLUS) /C. AQUATICA
STAURO,CETERI,HYALO,RACHI,APPEND,AQUATIC, 118C. •

CLAVARIOPSIS DE WILDEMAN 1895. ANN. SOC. BELG. MICROSC. 19:200 /C. AQUATICA
STAURO,CETERI,HYALO,NONSP,APPEND,AQUATIC, 119E, ASCO STATE =COROLLOSPORA FIDE SHEARER & CRANE 1971 (MYCOLOGIA 63:253-256,ILLUS).
/ANASTASIOU, MYCOLOGIA 53:11-16,1961 (ILLUS) /INGOLD, GUIDE P 26-27,1975 (ILLUS) •

CLAVATOSPORA NILSSON EX MARVANOVA & NILSSON 1971. TBMS 57:531-532 /HELISCUS LONGIBRACHIATUS INGOLD 1942 (TBMS 25:362) =C. LONGIBRACHIATA (INGOLD) NILSSON 1971 IN MARVANOVA & NILSSON
STAURO,CETERI,HYALO,PHIAL,APPEND,AQUATIC, 119A, A SEGREGATE FROM HELISCUS.
/NILSSON, SYMB. BOT. UPSALIENS. 18(2):88,1964 (NOT SEEN) /NAWAWI, TBMS 61:390-393,1973 (ILLUS,NEW SP) /INGOLD, GUIDE P 70-71,1975 (ILLUS) •

CLAVULARIA KARSTEN 1883. MEDD. SOR. FAUNA & FL. FENNICA 9:67 (NO ILLUS) /C. FUSISPORA (LECTOTYPE FIDE CLEMENTS & SHEAR 1931)

=CILICIOPODIUM CORDA FIDE CLEMENTS & SHEAR 1931. •

CLINOCONIDIUM PATOUILLARD 1898. BULL. SOC. MYCOL. FRANCE 14:156 /UREDO FARINOSA HENNINGS 1897 =C. FARINOSUM (HENNINGS) PATOUILLARD EX SACCARDO & SYDOW 1902 (IN SACCARDO, SYLL. FUNG. 16:1093)
AMERO,CETERI,HYALO,AMPUL OR MISC,SPORODOCHIA, 35D, ON PLANTS. THIS IS A BASIDIOMYCETE.
/MALENCON, BULL. SOC. MYCOL. FRANCE 69:77-100,1953 (ILLUS) •

CLINOTRICHUM COOKE 1871. POP. SCI. REV. 10:30 & PL 68, FIG 1-3 /C. LANOSUM =NOMEN NUDUM
=NOMEN ILLEGITIMUM FIDE HUGHES 1958, THE FUNGUS BECOMES ACLADIUM LANOSUM (COOKE) CARMICHAEL 1973 (THE FUNGI, VOL IVA, P 355), BASIONYM- RHINOTRICHUM LANOSUM COOKE 1871 (HANDBOOK OF BRITISH FUNGI 2:591). FOR ILLUSTRATIONS OF A. LANOSUM SEE BALFOUR-BROWN 1952 (TBMS 35:273-276, FIG 1A). •

CLONOSTACHYOPSIS HOHNEL 1907. S. B. AKAD. WISS. WIEN 116:149 /C. POPULI (HARZ)
=CLONOSTACHYS CORDA 1839 FIDE CLEMENTS & SHEAR 1931. •

CLONOSTACHYS CORDA 1839. PRACHTFLORA P 31-32 & PL XV /C. ARAUCARIA
AMERO,BASO OR GLOIO,HYALO,PHIAL, 6A, ON SOIL. INTERMEDIATE BETWEEN GLIOCLADIUM & PAECILOMYCES.
/TUBAKI, TRANS. MYCOL. SOC. JAPAN 4:83-90,1963 (ILLUS) /BARRON, HYPHOMYCETES P 177-179,1968 (SUB GLIOCLADIUM,ILLUS,REFS) /HAWKSWORTH & PUNITHALINGAM, TBMS 64:89-99,1975 (ILLUS,REVIS)
=*CLONOSTACHYOPSIS* HOHNEL 1907 FIDE CLEMENTS & SHEAR 1931
?=*MARIANNAEA* ARNAUD EX SAMSON 1974 •

CLOSTEROALEUROSPORIA GRIGORAKIS 1924. COMP. REND. 179:1424 /C. AUDOUINI (GRUBY) GRIGORAKIS 1925 (ANN. SCI. NAT. BOT. SER. 10, 7:142,LECTOTYPE)
=NOMEN ILLEGITIMUM, AN UNNECESSARY NEW NAME FOR MICROSPORUM. •

CLOSTEROSPORIA GRIGORAKIS 1924. COMP. RENDU ACAD. SCI. 179:1425 /C. LANOSA (SABOURAUD) GRIGORAKIS 1925 (ANN. SCI. NAT., SER. 10 7:410-411,LECTOTYPE)
=MICROSPORUM GRUBY 1834. •

COCCIDIOIDES STILES 1896. IN RIXFORD & GILCHRIST, JOHNS HOPKINS HOSP. REP. 1:243 /C. IMMITIS RIXFORD & GILCHRIST 1896
AMERO,CETERI,HYALO,MISC, 54A, ON SOIL & ANIMALS (MAN, RODENTS). AS A FORM-GENUS BASED ON ENDOSPORULATING SPHERULES, COCCIDIOIDES FITS BETTER IN THE BLASTOMYCETES THAN THE HYPHOMYCETES. OTHER CONIDIAL STATE =MALBRANCHEA FIDE SIGLER & CARMICHAEL 1976 (MYCOTAXON 4:374-375 & 458-460,ILLUS).
/SUN & HUPPERT, SABOURAUDIA 14:185-198,1976 (ILLUS)
=*BLASTOMYCOIDES* CASTELLANI 1928 FIDE CARMICHAEL 1962
?=*POSADASIA* CANTON 1898 FIDE DODGE 1935 •

COCCOBOTRYS BOUDIER & PATOUILLARD 1900. BULL. SOC. MYCOL. FRANCE 16:141-144 & TAB V, FIG 1 /CENOCOCCUM XYLOPHILUM FRIES =C. XYLOPHILUS (FRIES)
SCLEROTIA, ON BARK (PALMS). COMPARE WITH AEGERITA. BASIDIO STATE =LEPIOTA FIDE DONK 1962 (TAXON 11:82). •

COCCOPLEUM EHRENBERG 1818 (SYLVA MYC. BERLIN P 27) =SCLEROTIUM FIDE LINDAU IN RABENHORST 1910 •

COCCOSPORA WALLROTH 1833. FLORA CRYPT. GERMAN. 2:176 /C. AURANTIACA
?=SCLEROTIUM SEE DAMON & DOWNING 1954 (MYCOLOGIA 46:209-211,ILLUS). •

COCCOSPORELLA KARSTEN 1893. ACTA SOC. F. F. FENNICA 9:11 /C. CALOSPORA
=MYCOGONE LINK 1809 FIDE HUGHES 1958. •

COCCOSPORIUM CORDA 1832. STURM'S DEUTSCHL. FL. III (PILZE) BD 3 HEFT 12:49-50 /C. MACULIFORME
=NOMEN DUBIUM SEE HUGHES 1958.
=*FREYNELLA* KUNTZE FIDE AINSWORTH 1971 •

COCCOTRICHUM LINK 1824. LINN. SPEC. PLANT. IV 6(1):25-26 /C. MARTII =ALEURISMA GRANULOSUM MARTIUS 1817
=NOMEN DUBIUM SEE HUGHES 1958. •

CODINAEA MAIRE 1937. PUBL. INST. BOT. BARCELONA 3:15-16 (ILLUS) /C. ARISTATA
AMERO OR DIDYMO OR PHRAGMO,GLOIO,HYALO,PHIAL,SETAE,APPEND,STROMA, 1A, ON LITTER & WOOD. ASCO STATE =CHAETOSPHAERIA FIDE HUGHES & KENDRICK 1968. MAY BE A SYNONYM OF DICTYOCHAETA SPEGAZZINI 1923. SEE GAMUNDI ET AL 1977.
/HUGHES & KENDRICK, NEW ZEALAND J. BOT. 6:323-375,1968 (REVIS,ILLUS) /PIROZYNSKI & PATIL, CAN. J. BOT. 48:578-579,1970 (ILLUS,NEW SP) /ELLIS, DH P 561-562,1971 (ILLUS) /SHEARER & CRANE, MYCOLOGIA 63:245-247,1971 (ILLUS,NEW SP) /SUTTON & HODGES, NOVA HEDWIGIA 26:517-522,1975 (ILLUS,NEW SP) /MATSUSHIMA, ICONES P 36-37,1975 (ILLUS,NEW SP) /GAMS & HERMANIDES-NIJHOF, CBS STUDIES 13:54-55,1976 (ILLUS,NEW SP) /ELLIS, MDH P 471-474,1976 (ILLUS,KEY,NEW SPP) /GAMUNDI, ARAMBARRI & GIAIOTTI, DARWINIANA 21:95-96 & FIG 6G-K,1977 (ILLUS SUB DICTYOCHAETA)
?=*CODINAEOPSIS* MORGAN-JONES 1976
?=*DICTYOCHAETA* SPEGAZZINI 1923
=*MENISPORELLA* AGNIHOTHRUDU 1962 FIDE HUGHES & KENDRICK 1968 •

CODINAEOPSIS MORGAN-JONES 1976. MYCOTAXON 4:166-170 (ILLUS) /CODINAEA GONYTRICHOIDES SHEARER & CRANE 1971 (MYCOLOGIA 63:245) =C. GONYTRICHOIDES (SHEARER & CRANE)
AMERO,GLOIO,HYALO,PHIAL,SETAE, 1F, COMPARE WITH CODINAEA MAIRE 1937. •

COELOANGUILLOSPORA DYKO & SUTTON 1978 =FILOSPORELLA NAWAWI 1976 FIDE DYKO & SUTTON 1978 (MYCOTAXON 7:323-326) •

COELOGRAPHIUM (SACCARDO) GAUMANN 1920. BULL. JARD. BOT. BUITENZ. SER 3, 2:13 & TAB 2 & 3 /STILBUM CAVIPES OUDEMANS 1883 (HEDWIGIA 22:62) =C. 'CAVICEPS' (OUDEMANS) GAUMANN 1920
?, TYPE SPECIES NOT ILLUSTRATED BY OUDEMANS OR GAUMANN. •

COELOSPORIUM LINK 1824. LINN. SPEC. PLANT. IV, 6(1):127 /C. FRUTICULOSUM (=DEMATIUM ARTICULATUM PERSOON)
=NOMEN DUBIUM FIDE HUGHES 1958. •

COLEMANIELLA AGNIHOTHRUDU 1974. J. COFFEE RES. 4:3-6 (ILLUS) /C. OSSOORII
DICTYO,CETERI,PHAEO,MISC,PODIA, 90D, ON LITTER (FICUS).
/ELLIS, MDH P 366-367,1976 (ILLUS) •

COLEODICTYOSPORA CHARLES 1929. PHYTOPATHOLOGY 19:1051-1053 (ILLUS) /C. CUBENSIS
DICTYO,CETERI,PHAEO,NONSP, 95G, ON SACCHARUM. •

COLEOMYCES MOREAU & MOREAU 1937. BULL. SOC. MYCOL. FRANCE 53:33-38 (ILLUS) /C. RUFUS
=CYLINDROCARPON WOLLENWEBER 1913 FIDE BOOTH 1966 (CMI PAPERS 104:1). •

COLLECEPHALUS SPENCER 1972. HEMEROCALLIS JOURNAL 26:12-16 (ILLUS) /C. HEMEROCALLIDIS (AS 'HEMEROCALLI')
AMERO,GLOIO,HYALO,AMPUL, ON LEAVES (HEMEROCALLIS). COMPARE WITH TRICHOSPORONOIDES.
/LUNSFORD & SPENCER, PHYTOPATHOLOGY 66:245-248,1976 (ILLUS,NEW VAR) •

COLLETOCONIS DE HOOG & VAN DER AA 1978. IN DE HOOG, PERSOONIA 10:48-51 (33-81,ILLUS) /GLOEOSPORIUM AECIDIOPHILUM SPEGAZZINI =C. AECIDIOPHILA (SPEGAZZINI) DE HOOG, VAN DER AA & SINGH 1978
?, COMPARE WITH MYROTHECIUM. •

COLLETOSPORIUM LINK 1824. SPEC. PLANT. ED. 4, 6(1):25 /C. UMBRINUM (LECTOTYPE FIDE HUGHES 1958)
?, STERILE FIDE HUGHES 1958 (CAN. J. BOT. 36:753). •

COLLETOTRICHOPSIS BUBAK 1904 =COLLETOTRICHUM FIDE VON ARX 1970 •

COLLETOTRICHUM CORDA 1831. STURM'S DEUTSCHL. KRYPT. FLORA III 3:41 /C. LINEOLA
AMERO,GLOIO,HYALO,PHIAL,SETAE,SPORODOCHIA,PODIA, ON PLANTS & LITTER. ASCO STATE =GLOMERELLA FIDE POLITIS 1975 (MYCOLOGIA 67:56-62,ILLUS).
/VON ARX, PHYTOPATH. ZEITSCHR. 29:413-508,1957 (REVIS) /BARRON, HYPHOMYCETES P 131-132,1968 (ILLUS) /VON ARX, BIBLIO. MYCOL (CRAMER) BD. 24:51-63,1970 (ILLUS,REVIS) /SUTTON & WATERSTON, CMI DESCRIPTIONS 222,1970 (ILLUS) /MORDUE, CMI DESCRIPTIONS 131-134,1967 & 315-317,1971 (ILLUS) /KOKKO & SHOEMAKER, FUNGI CANADENSES NO. 51,1974 (ILLUS) /VON ARX, GENERA P 167-168,1974 (ILLUS ON P 124) /FIRMAN & WALLER, CMI PHYTOPATH. PAPERS 20:1-53,1977 (ILLUS,COFFEE BERRY DISEASE)
=*COLLETOTRICHOPSIS* BUBAK 1904 FIDE VON ARX 1970
=*DICLADIUM* CESATI 1852 FIDE VON ARX 1970
=*ELLISIELLA* SACCARDO 1880 FIDE VON ARX 1970
=*FELLNERIA* FUCKEL FIDE VON ARX 1970
=*GLOEOSPORIOPSIS* SPEGAZZINI 1911 FIDE VON ARX 1970
=*SCHIZOTRICHELLA* MORRIS 1956 FIDE TULLOCH 1972
=*STEIROCHAETE* BRAUN & CASP. 1890 FIDE VON ARX 1970
=*VERMICULARIA* FRIES 1849 FIDE VON ARX 1970 •

COLLODOCHIUM HOHNEL 1902. S. B. AKAD. WISS. WIEN 111:1029 (NO ILLUS) /C. ATROVIOLACEUM
? •

COLUMNOPHORA BUBAK & VLEUGEL 1916. IN BUBAK, ANN. MYCOL. 14:349 /C. RHYTISMATIS (BRESADOLA) BUBAK 1916
=STIGMINA SACCARDO 1880 FIDE NAG RAJ & KENDRICK 1975 (CHALARA P 47-49). •

COMETELLA SCHWEINITZ 1835. IN FRIES, FLORAM SCAND. 1:362 /C. CARICINA (FRIES) FRIES 1849 (SUMMA VEG. SCAN. 2:506)
=CLASTEROSPORIUM SCHWEINITZ 1832 FIDE HUGHES 1958. •

CONDYLOSPORA NAWAWI 1976. TBMS 66:363-365,1976 (ILLUS) /C. SPUMIGENA
SCOLECO,CETERI,HYALO,RACHI,AQUATIC, 108F.
/MATSUSHIMA, ICONES P 167 & PL,1975 (ILLUS SUB MFC 1744)
?=*OBSTIPISPORA* SINCLAIR & MORGAN-JONES 1979 •

CONIODICTYUM HARIOT & PATOUILLARD 1909. BULL. SOC. MYCOL. FRANCE 25:13-14 (ILLUS) /C. CHEVALIERI
DICTYO,CETERI,HYALO,MISC,SPORODOCHIA, 97B, ON STEMS & FRUITS (ZIZIPHUS). THIS IS A BASIDIOMYCETE FIDE MALENCON 1953. OTHER CONIDIAL STATE =?HYALODENDRON -SEE ILLUS IN MALENCON 1953.
/MALENCON, BULL. SOC. MYCOL. FRANCE 69:77-100,1953 (ILLUS,REVIS) •

CONIOSCYPHA HOHNEL 1904. ANN. MYCOL. 2:58-59 (NO ILLUS) /C. LIGNICOLA
AMERO,BASO OR CETERI,PHAEO,PHIAL? OR MISC, 2G, ON WOOD IN WATER.
/SHEARER, MYCOLOGIA 65:128-136,1973 (REVIS,ILLUS) /SHEARER & MOTTA, CAN. J. BOT. 51:1747-1751,1973 (ILLUS) /MATSUSHIMA, ICONES P 38 & PL,1975 (ILLUS,NEW SP) /ELLIS, MDH P 444-445,1976 (ILLUS)
=*CYLICOGONE* VAN EMDEN & VEENBAAS-RIJKS 1973 FIDE JONG & ROXON 1975 •

CONIOSPORIOPSIS SPEGAZZINI 1918. PHYSIS 4:291 /C. FUMAGO (SCHWEINITZ)
=NOMEN CONFUSUM FIDE HUGHES 1958. •

CONIOSPORIUM LINK 1809. MAG. GES. NATURF. FREUNDE, BERLIN 3:8 /C. OLIVACEUM
DICTYO,BASO,PHAEO,MISC,SPORODOCHIA, 101C, ON WOOD & LITTER. ASCO STATE =HYSTERIUM FIDE HUGHES 1958.
/HUGHES, CMI PAPERS 37:10-17,1950 (ILLUS) /HUGHES, THE NATURALIST 93-98,1952 (ILLUS SUB SIRODESMIUM) /ELLIS, DH P 39-41,1971 (ILLUS) /ELLIS, MDH P 26-27,1976 (ILLUS) /SCHOKNECHT & CRANE, MYCOLOGIA 69:535-537,1977 (ILLUS,NEW COMB)
=*SIRODESMIUM* DE NOTARIS 1849 FIDE HUGHES 1958
=*BONORDENIELLA* PENZ. & SACC. 1901 FIDE HUGHES 1958
?=*DISCOSPORIUM* SACCARDO & SYDOW (NON HOHNEL) FIDE AINSWORTH 1971 •

CONIOTHECIELLA SPEGAZZINI 1918. PHYSIS 4:295 (NO ILLUS) /C. "PHYLLOGENA (DESMAZIERES)" (FROM SACCARDO, SYLL. FUNG. 1:79,1882- PROBABLY IN ERROR FOR CONIOTHECIUM PHYLLOPHILUM DESMAZIERES 1845)
? •

CONIOTHECIUM CORDA 1833. STURM'S DEUTSCHL. FLORA III (PILZE), BD. 3, HEFT 13:71-72 /C. ATRUM
=NOMEN DUBIUM, HUGHES 1958 DID NOT FIND TYPE. •

CONOPLEA PERSOON 1801. SYN. METH. FUNG. P 234 /C. SPHAERICA
AMERO,CETERI,PHAEO,RADUL,STROMA, 25C, ON LITTER & WOOD. ASCO STATE =URNULA FIDE HUGHES 1953 (CAN. J. BOT. 31:659) AND WOLF 1958 (MYCOLOGIA 50:843,SUB STRUMELLA), OR PLECTANIA FIDE PADEN 1972 (PERSOONIA 6:408), OR KORFIELLA FIDE DE HOOG 1977 (CBS STUDIES 15:137).
/HUGHES, CAN J. BOT. 38:659-696,1960 (REVIS,ILLUS,KEY) /ELLIS, DH P 234-237,1971 (ILLUS,KEY) /ELLIS, MDH P 200-201,1976 (ILLUS) /HUGHES, NEW ZEALAND J. BOT. 16:320-325,1978 (ILLUS,NEW SPP)
=*STREPTOTHRIX* CORDA 1839 FIDE HUGHES 1960
=*BOTRYOXYLON* CIFERRI 1962, FIDE HUGHES 1958
?=*STREBLOCAULIUM* CHEVALLIER 1837 •

COOKSONOMYCES SWART & GRIFFITHS 1974. TBMS 63:152 (151-155, ILLUS) /C. BANKSIAE
PHRAGMO,GLOIO OR CETERI,PHAEO,ANNEL,SPORODOCHIA, 77C, ON LITTER (BANKSIA). •

COPROTRICHUM BONORDEN 1851. HANDB. ALLGEM. MYKOL. P 76 & TAF. VI, FIG 132 /C. PURPURASCENS (LECTOTYPE) =SPORENDONEMA PURPURASCENS (BONORDEN) MASON & HUGHES 1957
=SPORENDONEMA DESMAZIERES 1827, BASED ON MASON & HUGHES APUD WOOD 1957 (NATURE, LONDON 179:328). •

CORALLINOPSIS LAGARDE 1917. ARCH. ZOOL. EXPER. GEN. 56:286-288 & PL X /C. PILULIFERA
?, A SYNNEMATOUS ACREMONIUM ON INSECTS. SEE ALSO SYNNEMATIUM. •

CORALLODENDRON JUNGHUHN 1838. BATAV. GENOOTSCH. VERHAND. 17:7 /C. LEUCOCEPHALUM
?
/MORRIS, WESTERN ILLINOIS UNIV. SER. BIOL. SCI. 3:36-37,1963 (ILLUS)
=*STILBODENDRON* SYDOW 1916 FIDE MORRIS 1963
=*HASSKARLINDA* KUNTZE 1891 BASED ON SAME TYPE •

CORDALIA GOBI 1885. MEM. ACAD. IMP. SCI. ST. PETERSBOURG 32(14):13 /C. PERSICINA (DITMAR)
=TUBERCULINA SACCARDO 1880 FIDE LINDAU IN RABENHORST 1910. •

CORDANA PREUSS 1851. LINNAEA 24:129-130 /C. PAUCISEPTATA (LECTOTYPE)
DIDYMO,CETERI,PHAEO,RADUL, 62B, ON WOOD, BARK & LEAVES.
/HUGHES, CAN. J. BOT. 33:259-263,1955 (ILLUS) /ELLIS, DH P 199-202,1971 (ILLUS) /HUGHES, FUNGI CANADENSES NO. 34,1974 (ILLUS) /MATSUSHIMA, ICONES P 39-40 & PL,1975 (ILLUS,NEW SPP WHICH MAY BELONG IN CACUMISPORIUM) /DE HOOG, ACTA BOT. NEERL. 22:209,1973 (ILLUS,NEW SP) /UDAGAWA & FURUYA, TRANS. MYCOL. SOC. JAPAN 16:218-221,1975 (ILLUS) /ELLIS, MDH P 162-163,1976 (ILLUS,NEW SP)
=*PREUSSIASTER* KUNTZE 1891 FIDE HUGHES 1958
?=*ARTHROBOTRYELLA* SIBILIA 1928 •

CORDELLA SPEGAZZINI 1886. AN. SOC. CIENT. ARGENTINA 22:210 /C. CONIOSPORIOIDES (LECTOTYPE)
AMERO,CETERI,PHAEO,MISC,ALEURI,SETAE, 46F, ON CULMS & LITTER (BAMBUSA, CAREX). COMPARE WITH PTEROCONIUM & ARTHRINIUM.
/ELLIS, CMI PAPERS 103:30-33,1965 (ILLUS) /ELLIS, DH P 575-577,1971 (ILLUS) /ELLIS, MDH P 478,1976 (ILLUS,NEW SP) •

COREMIELLA BUBAK & KRIEGER 1912. IN BUBAK, ANN. MYCOL. 10:52-53 /C. CYSTOPOIDES =C. CUBISPORA (BERK. & CURTIS) ELLIS 1971
AMERO,ARTHRO,PHAEO,NONSP,ALEURI,SYNNEMATA, 43F, ON LITTER.
/ELLIS, DH P 32-34,1971 (ILLUS) /SUTTON, CMI PAPERS 132:82-86,1973 (ILLUS,REVIS) /SIGLER & CARMICHAEL, MYCOTAXON 4:375-387,1976 (ILLUS) /COLE & SAMSON, CONIDIAL FUNGI P 120-122,1979 (ILLUS)
?=*GEOTRICHELLA* ARNAUD 1954 •

COREMIOPSIS SIZOVA & SUPRUN 1957. VESTN. MOSKOV. UNIV. 2:55-56 /C. ROSEA
?, ORIGINAL ILLUSTRATION INADEQUATE TO CHARACTERIZE, PROBABLY PAECILOMYCES. •

COREMIUM LINK 1809. MAG. GES. NATURF. FREUNDE, BERLIN 3:19 /C. GLAUCUM
=PENICILLIUM LINK 1809 FIDE AINSWORTH 1960. •

COREOMYCETOPSIS THAXTER 1920. BOT. GAZ. (LANCASTER) 69:13-14 (1-27) & PL IV, FIGS 32-36 /C. OEDIPUS
AMERO,CETERI,HYALO,PHIAL, 55F, ON INSECTS ~~(TERMITES)~~. / RETICULOTERMES
/BLACKWELL & KIMBROUGH, MYCOLOGIA 68:551-558,1976 (ILLUS) •

CORETHROPSIS CORDA 1839. PRACHTFLORA P 1 /C. PARADOXA
?, NO ONE SINCE CORDA HAS REPORTED SEEING THE TYPE OR A FUNGUS THAT MATCHES THE ORIGINAL ILLUSTRATION. THE NAME SHOULD BE CONSIDERED A NOMEN DUBIUM. •

COROLLIUM SOPP 1912. MONOGR. DER PILZGR. PENICILLIUM, P 98-100 & TAF. X, FIG 108 /C. DERMATOPHAGUM
=PAECILOMYCES BAINIER 1907 SEE CARMICHAEL 1973. •

CORONOSPORA ELLIS 1971. CMI PAPERS 125:16-17 (ILLUS) /C. DENDROCALAMI
STAURO,CETERI,PHAEO,RACHI, 117E, ON LEAVES (DENDROCALAMUS).
/MATSUSHIMA, ICONES P 40 & PL,1975 (ILLUS,NEW SP) /ELLIS, MDH P 228-229,1976 (ILLUS) •

CORYMBOMYCES APPEL & STRUNK 1904. CENT. BAKT. 2(11):632-634 (ILLUS) /C. ALBUS
=GLIOCLADIUM CORDA 1840 FIDE CARMICHAEL 1973. •

CORYNE NEES 1817. DAS SYSTEM DER PILZE UND SCHWAMME, P 157 & FIG 143 /ACROSPERMUM DUBIUM PERSOON 1797 (NEOTYPE TRTC 32215) =C. DUBIA (PERSOON) GRAY 1821 =C. SARCOIDES (JACQUIN) BONORDEN 1851 =C. STATE OF ASCOCORYNE SARCOIDES
AMERO,GLOIO,HYALO,PHIAL,SPORODOCHIA, 11C, ON WOOD. DIFFERENTIATION FROM TUBERCULARIA, CRINULA, ETC., NOT CLEAR. ASCO STATE =ASCOCORYNE FIDE GROVES & WILSON 1967.
/GROVES & WILSON, TAXON 16:35-41,1967 (REVIS,REFS) /TUBAKI, IN KOBAYASI ET AL., ANN. REP. INST. FERM., OSAKA 3:114-115,1967 (ILLUS SUB PIROBASIDIUM) /KORF & CANDOUSSAU, BULL. SOC. MYCOL. FRANCE 90:209-215,1974 (REVIS,ILLUS) /ROSS, CAN. J. BOT. 54:735-737,1976 (ILLUS)
=*PIROBASIDIUM* HOHNEL 1902 FIDE GROVES & WILSON 1967
=*ENDOSTILBUM* MALENCON 1964 FIDE KORF & CANDOUSSAU 1974 •

CORYNESPORA GUSSOW 1906. ZEITSCHR. PFLANZENKR. 16:13 /C. MAZEI =C. CASSIICOLA (BERK. & CURT.) WEI 1950
PHRAGMO OR SCOLECO,BLASTO OR CETERI,PHAEO,ANNEL,SYNNEMATA,STROMA, 75B, ON LEAVES, LITTER & WOOD. ASCO STATE =PYRENOPHORA SEE PAUL & PARBERY 1968 (TBMS 51:707-710 SUB HELMINTHOSPORIUM DICTYOIDES).
/ELLIS, CMI PAPERS 65:1-15,1957 (KEY,ILLUS) /ELLIS, CMI PAPERS 76:19-36,1960 (ILLUS) /ELLIS, DH P 372-378,1971 (ILLUS) /ELLIS, MDH P 376-379,1976 (ILLUS,KEY,NEW SP) /DYKO & SUTTON, MYCOTAXON 8:119-124,1979 (ILLUS,NEW SP) •

CORYNESPORELLA MUNJAL & GILL 1961. INDIAN PHYTOPATHOL. 14:7-9 /C. URTICAE
PHRAGMO OR SCOLECO,CETERI OR BLASTO,PHAEO,NONSP,STROMA, 91G, ON STEMS (URTICA).
/ELLIS, DH P 379-380,1971 (ILLUS) •

CORYNEUM NEES 1817. DAS SYSTEM ... P 34 /C. UMBONATUM =CORYNEUM STATE OF PSEUDOVALSA LONGIPES (TULASNE) SACCARDO 1879 FIDE HUGHES 1958
PHRAGMO,CETERI,PHAEO,ANNEL,ACERVULI, (MELANCONIALES) ON BARK & LEAVES (QUERCUS & MANY OTHER PLANTS). ASCO STATE =PSEUDOVALSA FIDE HUGHES 1958 OR MELANCONIS FIDE VON ARX 1974 (GENERA P 128), OR ASTEROMASSARIA SEE DE HOOG 1977 (CBS STUDIES 15:139).
/HUGHES, CAN. J. BOT. 33:343-349,1955 (ILLUS) /SUTTON, CMI PAPERS 138:1-224,1975 (MONOGR) •

CORYNODESMIUM WALLROTH EX REICHENBACH 1828. CONSP. REGNI VEGETAB., PARS PRIMA, LIPSIAE, P 4 /?

=NOMEN NUDUM, SEE HUGHES 1958 (CAN. J. BOT. 36:757-758). •

COSMARIOSPORA SACCARDO 1880. MICHELIA 2:34 & 172 (ILLUS IN FUNGI ITALICI, FIG 769,1881) /C. BIZZOZERIANA SACCARDO 1881
? •

COSTANTINELLA MATRUCHOT 1892. RECHERCHES SUR DEVEL. DE QUELQUES MUCEDIN. P 97 /C. CRISTATA =C. TERRESTRIS (LINK) HUGHES 1958
AMERO,CETERI,HYALO,RACHI OR MISC,SETAE, 26B, ON WOOD, LITTER & SOIL. ASCO STATE =MORCHELLA FIDE PADEN 1972 (PERSOONIA 6:411).
/BARRON, HYPHOMYCETES P 135-136,1968 (ILLUS) /ELLIS, DH P 364-365,1971 (ILLUS) •

CRASPEDODIDYMUM HOLUBOVA-JECHOVA 1972. CESKA MYKOL. 26:70-73 (ILLUS) /C. ELATUM
AMERO,CETERI,PHAEO,PHIAL?, 16B, ON LITTER (PHOENIX). CONIDIOGENESIS RESEMBLES CONIOSCYPHA.
/ELLIS, MDH P 457-458,1976 (ILLUS) •

CREMASTERIA MEYERS & MOORE 1960. AMER. J. BOT. 47:347-348 (ILLUS) /C. CYMATILIS
?=STERILE, ?=HUMICOLA. •

CRINULA FRIES 1821. SYSTEMA MYCOLOGICUM 1:493 /C. CALICIIFORMIS =CRINULA STATE OF HOLWAYA MUCIDA (SCHULZER) KORF & ABAWI 1971
AMERO,GLOIO,HYALO,PHIAL,SYNNEMATA, 9E, ON WOOD & BARK. ASCO STATE =HOLWAYA SACCARDO FIDE MORRIS 1963. CRINULA DOES NOT SEEM TO DIFFER FROM CILICIOPODIUM. SEE ALSO CORYNE.
/MORRIS, WESTERN ILLINOIS UNIV. SER. BIOL. SCI. 3:38-39,1963 (ILLUS) /KORF & ABAWI, CAN. J. BOT. 49:1879-1883,1971 (ILLUS,REVIS) /SUTTON, CMI PAPERS 132:45-47,1973 (ILLUS) •

CRISTULA CHENANTAIS 1920. BULL. SOC. MYCOL. FRANCE 35:208-209 & PL 18, FIGS 11-12 (DATED 1919) /C. INTEGRA
?, ORIGINAL ILLUSTRATION DIFFICULT TO INTERPRET. •

CRISTULARIA (SACCARDO) COSTANTIN 1888. LES MUCEDINEES SIMPLES P 17 & 134-135 (ILLUS) /BOTRYTIS GRANULIFORMIS SACCARDO 1881 (FUNGI ITAL. 691, ILLUS & SYLL. FUNG. 4:134,1886) =C. GRANULIFORMIS
? •

CRISTULARIELLA HOHNEL 1916. S. B. AKAD. WISS. WIEN 125:122-124 (NO ILLUS) /POLYACTIS DEPRAEDANS COOKE 1885 (J. QUEKETT MICROSCOP. CLUB, SER. 2, 2:141-142,NO ILLUS) =C. DEPRAEDANS (COOKE)
BULBILS,CETERI,HYALO OR PHAEO,MISC, 129C, ON LEAVES (ACER & OTHERS). OTHER CONIDIAL STATE =MYRIOCONIUM. COMPARE WITH AEGERITA (E.G., SEE MATSUSHIMA 1975, PL 370).
/WATERMAN & MARSHALL, MYCOLOGIA 39:690-698,1947 (ILLUS,NEW SP) /YOKOYAMA & TUBAKI, TRANS. MYCOL. SOC. JAPAN 15:189-195,1974 (ILLUS,REFS) /REDHEAD, CAN. J. BOT. 53:700-707,1975 (ILLUS,REVIS) /NIEDBALSKI, CRANE & NEELY, MYCOLOGIA 71:722-730,1979 (ILLUS,REFS) •

CROCYSPORIUM CORDA 1837. ICONES FUNGORUM 1:5 /C. AEGERITA =AEGERITA STATE OF PENIOPHORA CANDIDA LYMAN
=AEGERITA PERSOON 1801 FIDE HUGHES 1958. •

CROTALIA LIRO 1939. MYCOTHECA FENNICA, DIE ETIKETTEN, HELISNKI P 111 (NO ILLUS) /C. CINTRACTIAE-FISCHERI =C. STATE OF ANTRHACOIDEA FISCHERI KUKKONEN
?
/ZAMBETTAKIS, REV. MYCOL. 35:158-175,1970 (REFS,NO ILLUS) •

CRYPTOCORYNEUM FUCKEL 1869. JAHRB. NASSAU VER. 23-24 /C. FASCICULATUM =C. CONDENSATUM (WALLROTH) MASON & HUGHES 1953 EX HUGHES 1958
CHEIRO OR STAURO,CETERI,PHAEO,NONSP,SPORODOCHIA, 116G, ON WOOD & BARK.
/ELLIS, DH P 108-109,1971 (ILLUS) /ELLIS, CMI PAPERS 131:2-3 & PL 1-A,1972 (ILLUS,NEW SP) /ELLIS, MDH P 71-72,1976 (ILLUS) /SCHOKNECHT & CRANE, MYCOLOGIA 69:543-545,1977 (ILLUS) •

CRYPTOPHIALE PIROZYNSKI 1968. CAN. J. BOT. 46:1123-1127 (ILLUS) /C. KAKOMBENSIS
SCOLECO OR DIDYMO OR HELICO,GLOIO,HYALO,PHIAL,SETAE, 108A, ON LITTER (BAPHIA & QUERCUS).
/ELLIS, DH P 539-540,1971 (ILLUS) /MATSUSHIMA, ICONES P 40-41 & PL,1975 (ILLUS,NEW SP) /LANGE, CAN. J. BOT. 56:532-541,1978 (ILLUS) •

CRYPTOSTROMA GREGORY & WALLER 1951. TBMS 34:589 & 593-594 (ILLUS) /C. CORTICALE (ELLIS & EVERHART)
AMERO,BASO,PHAEO,PHIAL,STROMA,SETAE?, 17A, ON TREES & WOOD (ACER).
/ELLIS, DH P 523-524,1971 (ILLUS) •

CULCITALNA MEYERS & MOORE 1960. AMER. J. BOT. 47:346-349 (ILLUS) /C. ACHRASPORA =TRICHOCLADIUM ARACHASPORUM (MEYERS & MOORE) DIXON
=TRICHOCLADIUM HARZ 1871 FIDE HUGHES 1969. BUT SEE ELLIS 1976 (MDH P 38,ILLUS). •

CULICIDOSPORA PETERSEN 1960. BULL. TORREY BOT. CLUB 87:342-347 (ILLUS) /C. AQUATICA
STAURO,CETERI,HYALO,NONSP,APPEND,AQUATIC, 112D.
/INGOLD, GUIDE P 28-29,1975 (ILLUS) •

CULICINOMYCES COUCH, ROMNEY & RAO 1974. MYCOLOGIA 66:374-379 (ILLUS) /C. CLAVISPORUS
AMERO,GLOIO,HYALO,PHIAL, ON LARVAE OF INSECTS (DIPTERA). COMPARE WITH PAECILOMYCES.
/SWEENY, AUSTRAL. J. ZOOL. 23:49-57 & 59-64,1975 (ILLUS) /ONIONS, CMI DESCRIPTIONS 611,1979 (ILLUS) •

CURCULIOSPORA ARNAUD 1954. BULL. SOC. MYCOL. FRANCE 69:287 & FIG 3 R, P 273 /C. SYDOWII
HELICO,CETERI,PHAEO,NONSP, 104A, ON FUNGI (ASTERINELLA). NO LATIN DIAGNOSIS. NO TYPE SPECIMEN -SEE NICOT & CHARPENTIE 1971 (BULL. SOC. MYCOL. FRANCE 87:29). •

CURVIDIGITUS SAWADA 1943. REP. GOVT. RES. INST. FORMOSA 86:176 /C. DAPHNIPHYLLI
=NOMEN ILLEGITIMUM (JAPANESE DIAGNOSIS) SAME TYPE SPECIES AS TROCHOPHORA MOORE FIDE PIROZYNSKI 1972.
/PIROZYNSKI, CMI PAPERS 129:12,1972 (DISCUSSION) •

CURVISPORIUM CORBETTA 1963. RISO 12(3):28 /C. RIBALDII
=CURVULARIA BOEDIJN 1933 FIDE CARMICHAEL 1973.
=*CURVUSPORIUM* & CURVOSPORIUM (ORTHOGRAPHIC VARIANTS IN ORIGINAL PUBLICATION •

CURVULARIA BOEDIJN 1933. BULL. JARD. BOT. BUITENZ. III, 12:123-127 /C. LUNATA (WAKKER) =CURVULARIA STATE OF COCHLIOBOLUS LUNATA NELSON & HAASIS 1964
PHRAGMO,BLASTO OR CETERI,PHAEO,RACHI OR RADUL,STROMA, 79C, ON LEAVES & SEEDS (MANY PLANTS). ASCO STATE =PSEUDOCOCHLIOBOLUS FIDE TSUDA ET AL 1977. COMPARE WITH DRECHSLERA.
/ELLIS, CMI PAPERS 106:1-43,1966 (KEY,ILLUS) /LAUNDON, CMI DESCRIPTIONS 307,1971 (ILLUS) /ELLIS, DH P 452-459,1971 (ILLUS,KEY) /MATSUSHIMA, ICONES P 41-44 & PL,1975 (ILLUS,NEW SP) /ELLIS, MDH P 404-407,1976 (ILLUS,KEY) /TSUDA, UEYAMA & NISHIHARA, MYCOLOGIA 69:1109-1120,1977 (ILLUS)
=*CURVISPORIUM* CORBETTA 1963 FIDE CARMICHAEL 1973
?=*ACROTHECIELLA* KOORDERS 1907
=*MALUSTELLA* BATISTA & LIMA 1960 FIDE ELLIS 1971 •

CURVULARIOPSIS ELLIS 1961. CMI PAPERS 82:39-41 (ILLUS) /C. CYMBISPERMA (PATOUILLARD)
PHRAGMO OR HELICO,CETERI,PHAEO,RACHI,SYNNEMATA,PODIA,STROMA, 79A, ON LEAVES (RUBUS).
/ELLIS, DH P 204,1971 (ILLUS) •

CUSPIDOSPORIUM CIFERRI 1955. SYDOWIA 9:303 /C. CUSPIDATUM (SACCARDO) =CORYNESPORA PULVINIFORMIS
=EXOSPORIUM LINK 1809 FIDE ELLIS 1971. •

CUSTINGOPHORA STOLK, HENNEBERT & KLOPOTEK 1968. IN STOLK & HENNEBERT, PERSOONIA 5:195-197 /C. OLIVACEA
AMERO,GLOIO,PHAEO,PHIAL, 8G, ON COMPOST. ASCO STATE =CHAETOSPHAERIA FIDE BARR & CRANE 1979.
/ELLIS, DH P 548-549,1971 (ILLUS) /BARR & CRANE, CAN. J. BOT. 57:865-867,1979 (ILLUS,NEW SP SUB CHAETOSPHAERIA) •

CUTICULARIA DUCOMET 1907. ANN. AGR. RENNES 1:235 /C. ILICIS
?, STERILE. •

CYCLOCONIUM CASTAGNE 1845. CATAL. PLANTES . . . MARSEILLES P 220 /C. OLEAGINEUM =SPILOCAEA OLEAGINEA (CASTAGNE) HUGHES 1953
=SPILOCAEA FRIES 1825 FIDE HUGHES 1958. •

CYLICOGONE VAN EMDEN & VEENBAAS-RIJKS 1973. ACTA BOT. NEERL. 22:637-638 (ILLUS) /C. REGENERANS
=CONIOSCYPHA HOHNEL 1904 FIDE JONG & ROXON 1975 (MYCOLOGIA 67:424-425, NO ILLUS). •

CYLINDRIUM BONORDEN 1851. HANDB. MYKOL. P 34 /C. FLAVOVIRENS (DITMAR) (LECTOTYPE) =FUSIDIUM AERUGINOSUM LINK 1809 FIDE HUGHES 1958
=FUSIDIUM LINK 1809 FIDE HUGHES 1958. •

CYLINDROCARPON WOLLENWEBER 1913 HYPHOMYCETE NOM. CONS. 1969. PHYTOPATHOLOGY 3:225 /C. CYLINDROIDES WOLLENWEBER 1913 (CONSERVED XI INTERNATIONAL BOTANICAL CONGRESS 1969)
DIDYMO OR PHRAGMO,GLOIO,HYALO,PHIAL,SPORODOCHIA, 73I, ON TWIGS, LEAVES & FRUIT (MANY PLANTS). ASCO STATE =NECTRIA FRIES 1825 & CALONECTRIA DE NOTARIS 1864 FIDE BOOTH 1966. OTHER STATES =STEMPHYLIOPSIS (H) SEE BARRON & ONIONS 1966 & PAECILOMYCES (H) SEE BOOTH 1966 (P 11 & 12).
/BOOTH, CMI PAPERS 104:1-56,1966 (REVIS,KEY,ILLUS) /MATSUSHIMA, ICONES P 44-47 & PL,1975 (ILLUS,NEW SPP) /SAMUELS, NEW ZEALAND J. BOT. 16:73-82,1978 (NEW SPP) /ROSSMAN, MYCOTAXON 8:321-328 & 485-558,1979 (REVIS CALONECTRIA)
=*ALLANTOSPORA* WAKKER 1896 FIDE BOOTH 1966
=*COLEOMYCES* MOREAU & MOREAU 1937 FIDE BOOTH 1966
=*EURICOA* BATISTA & MAIA 1955 FIDE CARMICHAEL 1973
=*HYALOFLOREA* BATISTA & MAIA 1955 FIDE CARMICHAEL 1973
=*MOESZIA* BUBAK 1914 FIDE VON ARX 1970 •

CYLINDROCEPHALUM BONORDEN 1851. HANDB. MYK. P 103 /C. AUREUM (CORDA) =CHALARA AUREA (CORDA) HUGHES
=CHALARA (CORDA) RABENHORST 1844 FIDE HUGHES 1958. •

CYLINDROCLADIOPSIS YEN 1979. MYCOTAXON 8:233-237 (ILLUS) /C. LAGERSTRAEMIAE
PHRAGMO,ARTHRO OR CETERI,HYALO,NONSP, 93C, ON LEAVES (LAGERSTRAEMIA). COMPARE WITH SEPTOTRULLULA. •

CYLINDROCLADIUM MORGAN 1892. BOTANICAL GAZETTE 17:190-192 (ILLUS) /C. SCOPARIUM
DIDYMO OR PHRAGMO,GLOIO,HYALO,PHIAL,SETAE, 73F, ON PLANTS (PICEA, PRUNUS, CITRUS & MANY OTHERS). ASCO STATE =CALONECTRIA DE NOTARIS 1864 FIDE BARRON 1968. OTHER STATE =SCLEROTIA.
/BARRON, HYPHOMYCETES P 139-140,1968 (ILLUS,REFS) /BOOTH & GIBSON, CMI DESCRIPTIONS 362,1972 (ILLUS) /PEERALLY, CMI DESCRIPTIONS 421-430,1974 (ILLUS) /MATSUSHIMA, ICONES P 47 & PL,1975 (ILLUS) /HUNTER & BARNETT, MYCOLOGIA 70:614-635,1978 (COMPARATIVE PHYSIOLOGY,REFS) /DE HOOG, PERSOONIA 10:73-75,1978 (ILLUS,NEW COMB)
=*ACONTIOPSIS* NEGRU 1961 FIDE KENDRICK 1973
=*CANDELOSPORA* REA & HAWLEY 1912 FIDE BOOTH 1960
=*TETRACYTUM* VANDERWALLE 1945 FIDE WORMALD (SUB CANDELOSPORA) FROM AINSWORTH 1971 •

CYLINDROCOLLA BONORDEN 1851. HANDB. ALLGEM. MYKOL. P 149 /C. URTICAE (PERSOON) =CYLINDROCOLLA STATE OF CALLORIA FUSARIOIDES (BERK.) FRIES 1849 FIDE LINDAU 1907
AMERO,ARTHRO,HYALO,NONSP,SPORODOCHIA, 42F, ON LEAVES & LITTER. ASCO STATE

=CALLORIA FRIES 1849 FIDE LINDAU 1907
/LINDAU, RABENHORST KRYPT.-FL. 2 ED. 1 BD. 9 ABT. P 477-479,1907 (ILLUS,REFS)
/TUBAKI, IN KOBAYASI ET AL., ANN. REP. INST. FERM., OSAKA 3:114 & 116,1967 (ILLUS)
?=*ACTINODOCHIUM* SYDOW 1927 •

CYLINDRODENDRUM BONORDEN 1851. HANDB. ALLGEM. MYKOL. P 98 & FIG 127 /C. ALBUM (LECTOTYPE)
DIDYMO,GLOIO,HYALO,PHIAL, 57D, ON LITTER.
/PETCH, TBMS 27:91,1944 (NEW SP,NO ILLUS) /DE HOOG, PERSOONIA 10:70-71,1978 (A SPECIES MATCHING CYLINDRODENDRUM IS TREATED SUB PSEUDOMICRODOCHIUM CANDIDUM) •

CYLINDRODOCHIUM BONORDEN 1851 (HANDB. ALLG. MYKOL. P 132) IS AN OBLIGATE SYNONYM OF CYLINDROSPORIUM GREVILLE 1823 •

CYLINDRONEMA SCHULZER 1866. IN SCHULZER ET AL., VERHANDL. ZOOL.-BOT. GES. WIEN 16:34 (NO ILLUS) /"C. SOLANI TUBEROSI"?
=NOMEN DUBIUM •

CYLINDROPHORA BONORDEN 1851. HANDB. ALLGEM. MYKOL. P 92 & FIGS 110 & 131 /THREE ORIGINAL SPECIES
?=CLADOBOTRYUM SEE ALSO GAMS & HOOZEMANS 1970 (PERSOONIA 6:99) & DE HOOG 1978 (PERSOONIA 10:67). •

CYLINDROSPORA KUEHN 1877. HEDWIGIA 16:120-121 /TWO SPECIES MENTIONED
=NOMEN ILLEGITIMUM, PUBLICATION OF NEW GENUS APPARENTLY NOT INTENDED. •

CYLINDROSPORELLA HOHNEL 1916. SBR. K. ADAD. WISS. WIEN, MATH.-NAT. KL. 125:96 (NO ILLUS) /C. CARPINI (LIBERT)
=ASTEROMA DE CANDOLLE 1815 FIDE SUTTON 1977 (CMI PAPERS 141:19 & 56-57). ASCO STATE =GNOMONIA FIDE VON ARX 1974.
/VON ARX, GENERA P 167,1974 (SYNOS,REFS) •

CYLINDROSPORIUM GREVILLE 1823. SC. CRYPT. FL. 1, NO. 27 (ILLUS) /C. CONCENTRICUM
AMERO,GLOIO,HYALO,PHIAL,ACERVULI, (COELOMYCETE), ON LEAVES (BRASSICA). ASCO STATE =PYRENOPEZIZA.
/RAWLINSON, SUTTON & MUTHYALU, TBMS 71:425-439,1978 (ILLUS)
=*CYLINDRODOCHIUM* BONORDEN 1851, BASED ON SAME TYPE SPECIES •

CYLINDROTRICHUM BONORDEN 1851. HANDB. ALLGEM. MYKOL. P 88 /C. OLIGOSPERMUM (CORDA) (LECTOTYPE)
DIDYMO OR PHRAGMO,GLOIO,HYALO,PHIAL,SETAE, 57A, ON WOOD & LITTER. ASCO STATE =CHAETOSPHAERIA SEE BARR & CRANE 1979 (CAN. J. BOT. 57:835).
/BARRON, HYPHOMYCETES P 141-142,1968 (ILLUS) /ELLIS, DH P 560-561,1971 (ILLUS) /MATSUSHIMA, ICONES P 47-48 & PL,1975 (ILLUS,NEW SPP,REVIS) /ELLIS, MDH P 470,1976 (ILLUS,NEW SP) /GAMS & HERMANIDES-NIJHOF, CBS STUDIES 13:48-54,1977 (REVIS,KEY,ILLUS) /MORGAN-JONES, MYCOTAXON 5:487-490,1977 (ILLUS,NEW SP,NEW NAME) •

CYLOMYCES CLEMENTS 1931 (IN CLEMENTS & SHEAR), A SUPERFLUOUS NAME FOR LISTEROMYCES PENZIG & SACCARDO 1901 •

CYPHELLOPHORA DE VRIES 1962. MYCOPATHOLOGIA 16:47-50 (ILLUS) /C. LACINIATA
DIDYMO OR PHRAGMO,GLOIO,PHAEO,PHIAL, 57F, ON BARK (POPULUS). ASCO STATE ?=PARKERELLA SEE FUNK 1976 (CAN. J. BOT. 54:868) & DE HOOG 1977 (CBS STUDIES 15:138). •

CYSTODENDRON BUBAK 1914. ANN. MYCOL. 12:212 /C. DRYOPHILUM (PASSERINI)
AMERO,GLOIO,HYALO OR PHAEO,PHIAL, 6F, ON LEAVES (QUERCUS). NOT VERY DIFFERENT FROM PHIALOPHORA. ASCO STATE =BISPORELLA, SEE CARPENTER 1975 (MYCOTAXON 2:123-126).
/ELLIS, DH P 511,1971 (ILLUS) •

CYSTOPHORA RABENHORST 1844 NON AGARDH 1841, SEE VOGLINOANA •

DACRINA FRIES 1832 (NON FRIES 1825) =STRUMELLA SACCARDO FIDE SACCARDO 1886 (SYLL. FUNG. 4:742) •

DACRYMYCELLA BIZZOZERO 1885. ATTI IST. VENETO SCI. LETT. ARTI 6 SER. 3:308 & TAB III, FIG 8 /D. FERTILISSIMA
?, ORIGINAL ILLUSTRATION DIFFICULT TO INTERPRET. •

DACRYODOCHIUM KARSTEN 1896. HEDWIGIA 35:47 /D. FLUXILE
?, NOT ILLUSTRATED BY AUTHOR. •

DACTULIOPHORA LEAKY 1964. TBMS 47:341-350 (ILLUS) /D. TARRII
BULBILS,CETERI,PHAEO,MISC,SETAE,APPEND, 129G, ON LEAVES. HAS BULBILS WITH SETAE, NO CONIDIA. COMPARE WITH ECHINOCHONDRIUM. •

DACTYLARIA SACCARDO 1880. MICHELIA 2:20 /ACROTHECIUM PURPURELLUM SACCARDO 1877 =D. PURPURELLA (SACCARDO) SACCARDO 1886
DIDYMO OR PHRAGMO OR SCOLECO,CETERI,HYALO OR PHAEO,RADUL, 62F & 81F, ON WOOD, LEAVES & LITTER. FOR NEMATODE-TRAPPING SPECIES SEE ARTHROBOTRYS & DACTYLELLA. COMPARE WITH OCHROCONIS & SCOLECOBASIDIUM. OTHER CONIDIAL STATES =IDRIELLA & MONODICTYS SEE MATSUSHIMA 1975 (ICONES PL 182:1-2 & PL 187:1). SEE ALSO GIBELLULA. ASCO STATE =ACROSPERMUM SEE WEBSTER 1956 (TBMS 39:361-366,SUB DIPLORHINOTRICHUM).
/COOKE & GODFREY, TBMS 47:61-74,1964 (ILLUS,KEY) /BHATT & KENDRICK, CAN. J. BOT. 46:1253-1257,1968 (REVIS, ILLUS) /DE HOOG & VON ARX, KAVAKA 1:55-60,1973 (ILLUS,REVIS) /MATSUSHIMA, ICONES P 48-53 & PL,1975 (ILLUS,NEW SP,ALSO SUB PLEUROPHRAGMIUM P 114-117) /ELLIS, MDH P 170-174,1976 (ILLUS,KEY,NEW SP,NEW NAME,NEW COMB) /DE HOOG & HERMANIDES-NIJHOF, CBS STUDIES 15:212-213,1977 (NEW COMB)
=*DIPLORHINOTRICHUM* HOHNEL 1902 FIDE BHATT & KENDRICK 1968
?=*OCHROCONIS* DE HOOG & VON ARX 1973
?=*PLEUROPHRAGMIUM* COSTANTIN 1888 SEE DE HOOG & VON ARX 1973 •

DACTYLARIOPSIS MECHTIEVA 1967. MIKOL. I FITOPATOL. 1:278-279 /D. BROCHOPAGA (DRECHSLER)
?=DACTYLELLA GROVE 1884. •

DACTYLELLA GROVE 1884. JOUR. BOT. 22:199 /D. MINUTA
DIDYMO OR PHRAGMO OR SCOLECO,CETERI,HYALO,NONSP OR RACHI,APPEND, 89A, ON NEMATODES & NEMATODE EGGS. ASCO STATE =MASSARINA FIDE WEBSTER FROM INGOLD 1975 (GUIDE P 30-31,ILLUS). DISTINCTION FROM ARTHROBOTRYS NOT CLEAR.
/DRECHSLER, MYCOLOGIA 29:447-552,1937 (ILLUS,NEW SPP) /COOKE & GODFREY, TBMS 47:61-74,1964 (ILLUS,KEY) /COOKE & DICKINSON, TBMS 48:621-629,1965 (ILLUS) /MATSUSHIMA, ICONES P 53-56 & PL,1975 (ILLUS,NEW SP) /DOWSETT, REID & VAN CAESEELE, CAN. J. BOT. 55:2945-2970,1977 (ILLUS SUB DACTYLARIA) /BARRON, THE NEMATODE-DESTROYING FUNGI 140P,1977 (ILLUS,ECOL) /STIRLING & MANKAU, MYCOLOGIA 70:774-783,1978 (ILLUS,NEW SP)
?=*DACTYLARIOPSIS* MECHTIEVA 1967
?=*DACTYLELLINA* MORELET 1968
?=*DACTYLINA* ARNAUD EX SUBRAMANIAN 1964 (NOMEN ILLEGITIMUM)
?=*DRECHSLERELLA* SUBRAMANIAN 1964
?=*GOLOVINIA* MECHTIEVA 1967
?=*MONACROSPORIUM* OUDEMANS 1884 SEE BARRON 1968 •

DACTYLELLINA MORELET 1968. BULL. SOC. SCI. NAT. ARCHEOL. TOULON VAR. 178:6 /DACTYLELLA LEPTOSPORA DRECHSLER 1937 (MYCOLOGIA 29:504-508,ILLUS) =D. LEPTOSPORA (DRECHSLER)
?=DACTYLELLA. SEGREGATED BECAUSE CONIDIOPHORE OCCASIONALLY BRANCHED. •

DACTYLINA ARNAUD EX SUBRAMANIAN 1964. J. INDIAN. BOT. SOC. 42:297 (DATED 1963) /D. TYLOPAGA (DRECHSLER) SUBRAMANIAN 1964
=NOMEN ILLEGITIMUM, A SEGREGATE FROM DACTYLELLA WITH A TERMINAL PROLONGATION OF THE CONIDIA. NON DACTYLINA NYLANDER 1860 (LICHEN), SEE LACTYDINA SUBRAMANIAN NOM. NOV. SEE ALSO DRECHSLERELLA SUBRAMANIAN 1964. •

DACTYLIUM NEES 1817. DAS SYSTEM ... P 58 /D. CANDIDUM
=NOMEN DUBIUM FIDE BARRON 1968, BUT SEE DACTYLIUM SACCARDO. •

DACTYLIUM SACCARDO 1886. SYLLOGE 4:188-190 (AS DACTYLIUM NEES, BUT EXCLUDING D. CANDIDUM) /MUCOR DENDROIDES BULLIARD 1791 (HIST. CHAMP. 1:105) =D. DENDROIDES (BULLIARD) FRIES 1832 (SYST. MYCOL. 3:414) (LECTOTYPE SEE LENTZ 1965).
=NOMEN ILLEGITIMUM, MUCOR DENDROIDES IS NOW DISPOSED AS CLADOBOTRYUM DENDROIDES (BULL. PER MERAT) GAMS & HOOZEMANS 1970 (PERSOONIA 6:103-105).
/LENTZ, REGN. VEGET. 40:10-13,1965 (NOM. CONS. PROP.,REJECTED TAXON 23:420,1974) /LENTZ, MYCOLOGIA 58:965-966,1966 (DISCUSSION) /BARRON, HYPHOMYCETES P 145-147,1968 (ILLUS,DISCUSSION) •

DACTYLOSPORIUM HARZ 1871. BULL. SOC. IMPER. MAT. MOSCOW 44:131 /D. MACROPUS (CORDA)
DICTYO,CETERI,PHAEO,RADUL,STROMA, 99D, ON LITTER. NON DACTYLOSPORIUM MECHTIEVA 1967.
/HUGHES, THE NATURALIST 1952 (APRIL-JUNE):63-64,1952 (ILLUS) /ELLIS, DH P 262-263,1971 (ILLUS) •

DACTYLOSPORIUM MECHTIEVA 1967. MIKOL. I FITOPATOL. 1:278 /DACTYLELLA LEPTOSPORA DRECHSLER D. LEPTOSPORA (DRECHSLER)
=NOMEN ILLEGITIMUM, NON DACTYLOSPORIUM HARZ 1871, REPLACED BY DACTYLELLINA MORELET 1968. •

DAMNOSPORIUM CORDA =BACTRIDIUM FIDE AINSWORTH 1971 •

DEIGHTONIELLA HUGHES 1952. CMI PAPERS 48:27-29 (ILLUS) /D. AFRICANA
DIDYMO OR PHRAGMO,CETERI,PHAEO,ANNEL, 59D, ON LEAVES & STEMS (MANY PLANTS).
/ELLIS, CMI PAPERS 66:1-12,1957 (KEY,ILLUS) /ELLIS, CMI DESCRIPTIONS 165,1968 (ILLUS) /ELLIS, DH P 163-165,1971 (ILLUS,KEY) /MATSUSHIMA, ICONES P 56 & PL 69,1975 (ILLUS) /ELLIS, MDH P 146-147,1976 (ILLUS,NEW SP) /COLE & SAMSON, CONIDIAL FUNGI P 51 & 53,1979 (ILLUS,CONIDIUM ONTOGENY) •

DELORTIA PATOUILLARD & GAILL. 1888. BULL. SOC. MYC. FRANCE 4:43-44 & PL XIII, FIG 5A TOP /D. PALMICOLA PATOUILLARD 1888
HELICO,GLOIO,HYALO,NONSP,SPORODOCHIA, 103D, ON WOOD & LITTER.
/LINDER, ANN. MISSOURI BOT. GARD. 16:338-339,1929 (ILLUS)
?=*HYALOTROCHOPHORA* FINLEY & MORRIS 1967 •

DEMATIUM PERSOON 1801. SYN. METH. FUNG. /?, SEE HUGHES 1958 (CAN. J. BOT. 36:760)
=NOMEN CONFUSUM SEE HUGHES 1958. •

DEMATOIDIUM STAUTZ 1931. PHYTOPATH. Z. 3:204-208 (ILLUS) /D. NIGRESCENS
=AUREOBASIDIUM FIDE HERMANIDES-NIJHOF 1977 (CBS STUDIES 15:143 & 158). •

DEMATOPHORA HARTIG 1883. UNTERSUCH FORSTBOT. INST. MUNCH. 3:95-140 & TAF 7 (AS A SUBGENUS) /D. NECATRIX (HARTIG) VIALA 1891 =D. STATE OF ROSELLINIA NECATRIX FIDE ELLIS 1971 (DH P 230-231,ILLUS)
AMERO,CETERI,PHAEO,RACHI OR MISC,ALEURI?,SYNNEMATA, 29G, ON ROOTS (TREES & OTHER PLANTS). SYNNEMATOUS COUNTERPART OF GENICULOSPORIUM. ASCO STATE =ROSELLINIA.
/MARTIN, S. AFRICAN J. BOT. 34:187-192,1968 (ILLUS) /ELLIS, DH P 230-231,1971 (ILLUS) /SIVANESAN & HOLLIDAY, CMI DESCRIPTIONS 351-354,1972 (ILLUS SUB ROSELLINIA)
=*PLEUROGRAPHIUM* GOIDANICH 1935 FIDE AINSWORTH 1971
?=*VRIKSHOPAMA* RAO & RAO 1964 •

DENDRINA FRIES 1825. SYST. ORB. VEG. 1:186 /?D. FLAVA FRIES 1832 (SYST. MYCOL. 3:435-454, LECTOTYPE)
?=NOMEN DUBIUM, SEE LINDAU 1907 (IN RABENHORST KRYPT.-FL. 2 AUFL. 1 BD. 8:203 & 659). •

DENDRODOCHIUM BONORDEN 1851. HANDB. MYKOL. P 135 & FIG 228 /D. AURANTIACUM (LECTOTYPE)
AMERO,GLOIO,HYALO,PHIAL,SPORODOCHIA, 12A, ON BARK & LITTER. COMPARE WITH VOLUTELLA. ASCO STATE =NECTRIA FIDE SAMUELS 1976 (MEM. N.Y. BOT. GARD. 26:22). OTHER STATE =VERTICILLIUM FIDE GAMS & DOMSCH 1969.
/BARRON, HYPHOMYCETES P 148-149,1968 (ILLUS) /GAMS & DOMSCH, NOVA HEDWIGIA

18:9-10,1969 (ILLUS) /TULLOCH, CMI PAPERS 130:5-6,1972 (DISCUSSION OF GENERIC CHARACTERS) /MATSUSHIMA, ICONES P 56-57 & PL,1975 (ILLUS)
=*OICHITONIUM* BERKELEY & CURTIS 1875 FIDE SACCARDO 1886 AS DICHITONIUM
=*PATOUILLARDIA* ROUMEGUERE 1885 FIDE HOHNEL FROM CLEMENTS & SHEAR 1931
=*VERTICILLIODOCHIUM* BUBAK 1914 FIDE TULLOCH 1972 •

DENDROGRAPHIELLA AGNIHOTHRUDU 1972. PROC. INDIAN ACAD. SCI., SECT. B 75:155-159 (ILLUS) /D. PODOSPOROIOIDES
PHRAGMO,BLASTO,PHAEO,NONSP,SYNNEMATA, 87E, ON LATEX (SAP OF HEVEA). •

DENDROGRAPHIUM MASSEE 1892. GREVILLEA 21:5 /D. ATRUM
PHRAGMO,BLASTO OR CETERI,PHAEO,RADUL,SYNNEMATA,STROMA, 87C, ON LITTER AND ON PORTULACAE.
/ELLIS, DH P 398-400,1971 (ILLUS) /GHOSH, PATHAK & SINGH, NORW. J. BOT. 24:79-81,1977 (NEW SP,ILLUS) •

DENDROSPORA INGOLD 1944. TBMS 26:104-107, DATED 1943 (ILLUS) /D. ERECTA
STAURO,CETERI,HYALO,NONSP,AQUATIC, 123D, COMPARE WITH DENDROSPOROMYCES.
/IQBAL, TBMS 59:301-307,1972 (ILLUS,NEW SP) /INGOLD, GUIDE P 32-35,1975 (ILLUS) •

DENDROSPORIUM PLAKIDAS & EDGERTON EX CRANE 1972. TBMS 58:423-426 (ILLUS,VALIDATED) /D. LOBATUM PLAKIDAS & EDGERTON 1936 (MYCOLOGIA 28:83-84, ILLUS)
DIDYMO OR PHRAGMO OR STAURO,CETERI,HYALO,RADUL OR RACHI,AQUATIC, 113D, COMPARE WITH TRICELLULA.
/MATSUSHIMA, ICONES P 57 & PL,1975 (ILLUS) /COLE & SAMSON, CONIDIAL FUNGI P 38-39,1979 (ILLUS) •

DENDROSPOROMYCES NAWAWI, WEBSTER & DAVEY 1977. TBMS 68:59-63 (ILLUS) /D. PROLIFER
STAURO,CETERI,HYALO,NONSP,AQUATIC, 123B, CLOSE TO DENDROSPORA, BUT HAS DOLIPORE SEPTA. •

DENDROSTILBELLA HOHNEL 1905. OEST. BOT. ZEITSCHR. 55:22 (NO ILLUS) /D. PRASINULA =D. STATE OF CORYNELLA PRASINULA
AMERO,GLOIO,HYALO,PHIAL,SYNNEMATA, 9G, ON FUNGI (POLYPORUS, FOMES).
?=STILBUM PERSOON 1801, SEE SUTTON 1973 (CMI PAPERS 132:17 & 46-48, ILLUS). ASCO STATE =CORYNELLA.
/BARRON, HYPHOMYCETES P 149-150,1968 (ILLUS) •

DENDRYPHIELLA BUBAK & RANOJEVIC 1914. IN RANOJEVIC, ANN. MYCOL. 12:417 /D. INTERSEMINATA (BERK. & RAV.)
=DENDRYPHION WALLROTH 1833 FIDE HUGHES 1958, BUT SEE ELLIS 1971. •

DENDRYPHION WALLROTH 1833. FLORA CRYPT. GERMAN. 2:300 /D. COMOSUM
PHRAGMO OR DIDYMO OR CHEIRO,BLASTO OR CETERI,PHAEO,RADUL OR RACHI,STROMA, 84E, ON ROOTS, STEMS & LITTER. ASCO STATE =PLEOSPORA FIDE ELLIS 1971.
/HUGHES, CAN. J. BOT. 31:636-638,1953 (ILLUS) /ELLIS, DH P 501-505,1971 (ILLUS,KEY) /MATSUSHIMA, ICONES P 57-58 & PL,1975 (ILLUS) /ELLIS, MDH P 436-437,1976 (ILLUS) /HUGHES, NEW ZEALAND J. BOT. 16:327,1978 (ILLUS)
=*DENDRYPHIELLA* BUBAK & RANOJEVIC 1914 FIDE HUGHES 1958
=*ENTOMYCLIUM* WALLROTH 1833 FIDE HUGHES 1958
=*BRACHYCLADIUM* CORDA 1838 FIDE HUGHES 1958
?=*DWAYAMALA* SUBRAMANIAN 1956
?=*PSEUDOTORULA* SUBRAMANIAN 1958 •

DENDRYPHIOPSIS HUGHES 1953. CAN. J. BOT. 31:655 /D. ATRA (CORDA) =D. STATE OF MICROTHELIA INCRUSTANS (ELLIS & EVERHART) CORLETT & HUGHES 1978
PHRAGMO,CETERI,PHAEO,NONSP OR RACHI, 92B, ON WOOD & BRANCHES. ASCO STATE =MICROTHELIA FIDE HUGHES 1978.
/HUGHES, CAN. J. BOT. 31:638-639,1953 (ILLUS) /HUGHES, CAN. J. BOT. 36:762,1958 (LIST) /ELLIS, DH P 380-381,1971 (ILLUS) /MATSUSHIMA, ICONES P 58 & PL,1975 (ILLUS) /ELLIS, MDH P 384-385,1976 (ILLUS) /HUGHES, NEW ZEALAND J. BOT. 16:327 & 361-362,1978 (ILLUS,NEW COMB SUB MICROTHELIA) •

DENDRYPHIOSPHAERA LUNGHINI & RAMBELLI 1978. GIORN. BOT. ITAL. 112:175-196 (ILLUS) /D. TAIENSIS
PHRAGMO,CETERI,PHAEO,MISC, ON LITTER. COMPARE WITH EDMUNDMASONIA. •

DENNISOGRAPHIUM RIFAI 1977. KEW BULL. 31(3):726-728 (ILLUS) /D. EPISPHAERIAE
AMERO,BASO,HYALO,ANNEL,SYNNEMATA, 11B, ON FUNGI (XYLARIACEAE). •

DENTICULARIA DEIGHTON 1972. TBMS 59:421-423 (ILLUS) /CLADOSPORIUM MODESTUM SYDOW 1939 =D. MODESTA (SYDOW)
AMERO OR DIDYMO,BLASTO,HYALO OR PHAEO,RADUL,SPORODOCHIA, 36B, ON LEAVES (ANTHOSTEMA, FICUS). COMPARE WITH RAMULARIA.
/ELLIS, MDH P 182-183,1976 (ILLUS) /DE HOOG, PERSOONIA 10:51-53,1978 (ILLUS,NEW SP,KEY) •

DEPHILIPPIA RAMBELLI 1959. MYCOPATHOLOGIA 11:136-138 (ILLUS) /D. HELICOTRICHOIDES
=CIRCINOTRICHUM NEES 1817 FIDE PIROZYNSKI 1962. •

DERMOSPORIUM LINK 1815. MAG. GES. NATURF. FREUNDE, BERLIN 7:32, TAB. 1, FIG 4A /D. FLAVICANS
?=AEGERITA PERSOON FIDE FRIES 1829 (SYST. MYCOL. 3:220). •

DESERTELLA MOUCHACCA 1979. REV. MYCOL. TOME 43:71-79 (ILLUS) /D. GLOBULIFERA
AMERO,CETERI,HYALO OR PHAEO,MISC, 49I, ON SOIL. •

DESMIDIOSPORA THAXTER 1891. BOT. GAZ. 16:203 & PL XX, FIGS 1-9 /D. MYRMECOPHILA
STAURO,CETERI,PHAEO,NONSP, 113B, ON INSECTS (ANTS). OTHER STATE ="PHIALOCONIDIA". •

DESMOTRICHUM LEVILLE 1843. ANN. SCI. NAT. 2 SER., 19:217, TAB. VII, FIG 8 /D. SIMPLEX
=GONATOBOTRYS CORDA FIDE LINDAU IN RABENHORST 1907. •

DEXHOWARDIA TAYLOR 1970. MYCOPATHOLOGIA 40:306 (305-308,ILLUS) /D. TETRASPORA
AMERO,CETERI,HYALO,NONSP, 34D, SUBSTRATE NOT KNOWN. APPEARS TO BE CONIDIAL STATE OF TRECHISPORA, SEE HUGHES IN KENDRICK 1971 (P 20-21) OR ?VARARIA SEE STALPERS 1978 (CBS STUDIES 16:40). OTHER STATE =RIESSIA SEE GOOS & TUBAKI 1973 (CAN. J. BOT. 51:1439-1442). COMPARE WITH AMBIVINA KATZ 1974. •

DIADEMOSPORA SODERSTROM & BAATH 1979. TBMS 72:340-342 (ILLUS) /D. RAMIGERA
CHEIRO,CETERI,HYALO,NONSP, 117G, ON SOIL. OTHER STATE ="ALEURIOCONIDIA". •

DIAPHANIUM FRIES 1835. FL. SCAND. 307 /D. MAXIMUM
? •

DICELLISPORA SAWADA 1944. REP. GOVT. RES. INST. FORMOSA 87:74 /D. LELEBAE
?, JAPANESE DIAGNOSIS, PROBABLY GONATOBOTRYUM. •

DICHITONIUM BERKELEY & CURTIS =CORRECTION BY SACCARDO FOR OICHITONIUM (Q.V.) •

DICHOBOTRYS HENNEBERT 1973. PERSOONIA 7:193-194 & FIG 5, P 186 /D. ABUNDANS =D. STATE OF TRICHOPHAEA ABUNDANS (KARSTEN) BOUDIER 1907
AMERO,CETERI,HYALO,AMPUL OR MISC,ALEURI?, 33C, ON SOIL. ASCO STATE =TRICHOPHAEA BOUDIER FIDE HENNEBERT 1973 OR SPHAEROSPORELLA FIDE VON ARX 1974 (GENERA P 228), OR ELEUTHERASCUS FIDE VON ARX 1977 (CBS STUDIES 14:18 & 35), OR PEZIZA SEE O'DONNELL, FIELDS & HOOPER 1976 (CAN. J. BOT. 54:1084).
/LARSEN, MYCOLOGIA 67:1058-1061,1975 (ILLUS) •

DICHOTOMELLA SACCARDO 1914. ANN. MYCOL. 12:312-313 /D. AREOLATA =NIGROSPORA ORYZAE FIDE HUGHES 1958
=NIGROSPORA ZIMMERMAN 1902 FIDE HUGHES 1958. •

DICHOTOMOPHTHORA MEHRLICH & FITZPATRICK EX ELLIS 1971. DH P 388 (ILLUS) /D. PORTULACAE MEHRLICH & FITZPATRICK 1935
PHRAGMO,CETERI,PHAEO,AMPUL OR RADUL, 84B, ON LEAVES (PORTULACA). OTHER

STATE ="SCLEROTIA".
/BARRON, HYPHOMYCETES P 151-152,1968 (ILLUS) /COLE & SAMSON, CONIDIAL FUNGI P 51 & 53,1979 (ILLUS) •

DICHOTOMOPHTHOROPSIS ELLIS 1971. CMI PAPERS 125:20-22 (ILLUS) /HELICOSPORIUM NYMPHAEARUM RAND 1917 =D. NYMPHAEARUM (RAND)
HELICO,CETERI,PHAEO,RADUL,SCLEROTIA, 106C, ON LEAVES (NYMPHEA).
/ELLIS, MDH P 386-387,1976 (ILLUS) •

DICLADIUM CESATI 1852 =COLLETOTRICHUM CORDA FIDE VON ARX 1970 •

DICOCCUM CORDA 1829. IN STURM'S DEUTSCHL. FLÖRA, III, BD 2, HEFT 9:117 /D. MINUTISSIMUM
=NOMEN DUBIUM FIDE HUGHES 1958, SEE ALSO HUGHES & PIROZYNSKI 1972 (CAN. J. BOT. 50:2521-2534,ILLUS). •

DICRANIDION HARKNESS 1885. BULL. CALIFORNIA ACAD. SCI. 1:163 /D. FRAGILE
CHEIRO OR STAURO,CETERI,HYALO,RADUL OR RACHI,SYNNEMATA, 116F, ON WOOD & FUNGI. ASCO STATE =ORBILIA FIDE VON ARX 1974 (GENERA, P 236).
/PEEK & SOLHEIM, MYCOLOGIA 50:857-860,1958 (ILLUS,EMEND) /BUTTERFIELD, CAN. J. BOT. 51:795-799,1973 (ILLUS,REVIS) /MATSUSHIMA, ICONES P 58-59 & PL,1975 (ILLUS,NEW SP)
=*PEDILOSPORA* HOHNEL 1902 FIDE HUGHES 1951 •

DICTYOARTHRINIUM HUGHES 1952. CMI PAPERS 48:29-31 (ILLUS) /D. QUADRATUM =D. SACCHARI (STEVENSON) DAMON 1953
DICTYO,CETERI,PHAEO,MISC,ALEURI, 96B, ON SOIL & LITTER. /E
/HUGHES, CMI PAPERS 48:29-33,1952 (ILLUS) /ELLIS, DH P 580-582,1971 (ILLUS,KEY) /ELLIS, MDH P 480,1976 (ILLUS) /COLE & SAMSON, CONIDIAL FUNGI P 98-99,1979 (ILLUS) •

DICTYOARTHRINOPSIS BATISTA & CIFERRI 1957. PUBL. INST. MIC. UNIV. RECIFE 94:3-5 (ILLUS) /D. COSTARICENSIS
?=STEMPHYLIOPSIS SMITH 1901. •

DICTYOCATENULATA FINLEY & MORRIS 1967. IN MORRIS & FINLEY, AMER. MIDL. NAT. 77:200-202 (ILLUS) /D. ALBA FINLEY & MORRIS 1966
DICTYO,BASO,HYALO,MISC,SYNNEMATA, 101G, ON BARK & WOOD (FAGUS, ABIES). •

DICTYOCEPHALA MEDEIROS 1962. PUBL. INST. MICOL. UNIV. RECIFE 372:13-16 & 23-24 (ILLUS) /D. ULMIFOLIAE (OBREGON-BOTERO)
=PANTOSPORA CIFERRI 1938 FIDE DEIGHTON 1976 (CMI PAPERS 140:156). •

DICTYOCHAETA SPEGAZZINI 1923. PHYSIS 7:18-19 (ILLUS) /D. FUEGIANA
?=CODINAEA, SEE GAMUNDI, ARAMBARRI & GAIOTTI 1977 (DARWINIANA 21:95-96 & FIG 6 G-K,REVIS,ILLUS). •

DICTYODESMIUM HUGHES 1951. CMI PAPERS 36:27-29 (ILLUS) /D. ULMICOLA (ELLIS & KELLERMAN)
DICTYO,CETERI,PHAEO,NONSP,SPORODOCHIA,APPEND, 94G, ON LEAVES (ULMUS).
/ELLIS, DH P 149-151,1971 (ILLUS) /SUTTON, CMI PAPERS 138:176-177,1975 (ILLUS) •

DICTYOPHRYNELLA BATISTA & CAVALCANTI 1964. PORT. ACTA BIOL. SER. B, 7:356-358 (ILLUS) /D. BIGNONIACEARUM
?, ORIGINAL ILLUSTRATION NOT DIAGNOSTIC. •

DICTYOPOLYSCHEMA ELLIS 1976. MDH P 373-374 (ILLUS) /D. PIROZYNSKII
DICTYO,CETERI,PHAEO,MISC, 100E, ON LITTER (PICEA). •

DICTYOSPIROPES ELLIS 1976. MDH P 214-215 (ILLUS) /D. ZIZIPHI
DICTYO,CETERI,PHAEO,RADUL, 94H, ON LEAVES (ZIZIPHUS). COMPARE WITH ULOCLADIUM. •

DICTYOSPORIUM CORDA 1836. WEITENWEBERS BEITRAGE ZUR NAT. P 1836 /D. ELEGANS
DICTYO OR CHEIRO,CETERI,PHAEO,NONSP,ALEURI, 98C, ON WOOD & LITTER.
/HUGHES, CAN. J. BOT. 36:762-763,1958 (LIST) /BARRON, HYPHOMYCETES P 152-153,1968 (ILLUS) /ELLIS, DH P 55-57,1971 (ILLUS,KEY) /MATSUSHIMA, ICONES P

59-60 & PL,1975 (ILLUS,NEW SP) /HUGHES, NEW ZEALAND J. BOT. 16:328,1978 (ILLUS)
=*SPEIRA* CORDA 1837 FIDE HUGHES 1958
=*CATTANEA* GAROVAGLIO 1875 FIDE ELLIS 1971 •

DICYMA BOULANGER 1897. REV. GEN. BOT. 9:17-22 & PL 1-2 /D. AMPULLIFERA =D. STATE OF ASCOTRICHA CHARTARUM BERKELEY FIDE ELLIS 1971
AMERO,CETERI,PHAEO,RADUL,SETAE, 25B, ON LITTER & CELLULOSIC PRODUCTS. ASCO STATE =ASCOTRICHA FIDE AINSWORTH 1961.
/ELLIS, DH P 212-213,1971 (ILLUS) /HAWKSWORTH, CMI PAPERS 126:1-28,1971 (ILLUS,KEY) /KHAN & CAIN, MYCOTAXON 5:409-414,1977 (ILLUS SUB ASCOTRICHA ERINACEA)
?=*GONYTRICHELLA* EMOTO & TUBAKI 1970 •

DIDYMARIA CORDA 1842. ANLEIT. STUD. MYKOL, P 32 & 199 /D. UNGERI
=RAMULARIA UNGER FIDE VON ARX 1970. •

DIDYMARIOPSIS SPEGAZZINI 1910. ANN. MUS. NAC. BUENOS AIRES 20(SER. 3,13):424 (ILLUS) /D. CUPHAEICOLA
? •

DIDYMOBOTRYOPSIS HENNINGS 1902. HEDWIGIA 41:149 /D. PARASITICA
?, NOT ILLUSTRATED BY AUTHOR. •

DIDYMOBOTRYUM SACCARDO 1886. SYLLOGE FUNGORUM 4:626 /D. RIGIDUM (BERKELEY & BROOME) (LECTOTYPE FIDE ELLIS 1971)
DIDYMO,BLASTO,PHAEO,NONSP,SYNNEMATA,STROMA, 66B, ON WOOD (SASA) & BAMBOO.
/ELLIS, DH P 381-384,1971 (ILLUS,KEY) •

DIDYMOCLADIUM SACCARDO 1886. SYLLOGE FUNGORUM 4:186 /D. TERNATUM (BONORDEN)
=CLADOBOTRYUM NEES 1817 FIDE HUGHES 1958. •

DIDYMOPSIS SACCARDO & MARCHAL 1885. IN MARCHAL, BULL. SOC. BOT. BELG. 24:61-62 & PL IV, FIGS 10 & 11 /D. PEREXIGUA
?
/LINDAU, IN RABENHORST KRYPT.-FL. 2 AUFL. 1 BD. VIII:363-364,1907 (ILLUS) •

DIDYMOSTILBE HENNINGS 1902. HEDWIGIA 41:148 (5 AUGUST) /D. COFFEAE
DIDYMO,GLOIO,HYALO,PHIAL,SYNNEMATA, 58D, ON WOOD & DUNG. NON DIDYMOSTILBE BRESADOLA & SACCARDO 1902 (31 AUGUST). ASCO STATE =NECTRIA FIDE SAMUELS 1976 (MEM. N. Y. BOT. GARD. 26:22).
/SYDOW & SYDOW, ANN. MYCOL. 1:176,1903 (PUBLICATION DATES) /MORRIS, WESTERN ILLINOIS UNIV. SER. BIOL. SCI. 3:48-49,1963 (ILLUS) /MATSUSHIMA, ICONES P 60 & PL,1975 (ILLUS,NEW SP)
?=*KUTILAKESOPSIS* AGNIHOTHRUDU & BARUA 1957
?=*HETEROSEPTATA* MORRIS 1971 •

DIDYMOTHOZETIA RANGEL 1915. BOL. AGR. SAO PAULO, SER 16A, 4:325-327 & TAB V FIGS 13-14 /D. MIMOSENSIS
?
=*DITHOZETIA* CLEMENTS & SHEAR 1931 (ORTHOGRAPHIC VARIANT) •

DIDYMOTRICHELLA ARNAUD 1954. BULL. SOC. MYCOL. FRANCE 69:284 & FIG 16B, P 301 /D. QUERCINA
?, NO LATIN DIAGNOSIS, NO TYPE SPECIMEN -SEE NICOT & CHARPENTIE 1971 (BULL. SOC. MYCOL. FRANCE 87:30). •

DIDYMOTRICHUM BONORDEN 1851. HANDB. MYK. P 89 /D. NODULOSUM (CORDA) (LECTOTYPE FIDE HUGHES 1958)
=CLADOSPORIUM LINK 1815 FIDE HUGHES 1958. •

DIDYMOTRICHUM HOHNEL 1914. S. B. AKAD. WISS. WIEN, 123:140 /D. CHRYSOSPERMUM (SACCARDO) =DACTYLARIA CHRYSOSPERMA (SACC.) BHATT & KENDRICK 1968
=NOMEN ILLEGITIMUM, NON DIDYMOTRICHUM BONORDEN 1851. •

DIDYMOZOOPHAGA SOPRUNOV & GALIULINA 1951. MIKROBIOLOGIYA 20:489-499 (ILLUS,NO LATIN DIAGNOSIS) /NOT SPECIFIED, BUT THE TYPE SPECIES OF ARTHROBOTRYS WAS

INCLUDED (SEE INDEX OF FUNGI 2(5):76-77,1952)
=NOMEN ILLEGITIMUM FOR ARTHROBOTRYS CORDA 1839. •

DIGITATISPORA DOGUET 1962. C.R. ACAD. SCI. PARIS 254:4336-4338 (ILLUS) /D. MARINA, TYPE SPECIMEN NOT SPECIFIED
STAURO,CETERI,HYALO,MISC,CLAMPS, 119D, ON WOOD IN SEA WATER. 'CONIDIA' MAY BE BASIDIOSPORES, COMPARE WITH NIA. CONIDIAL STATE =VERTICILLIUM.
/DOGUET, BULL. SOC. MYCOL. FRANCE 78:283-290,1962 (ILLUS,CYTOL) /BROOKS, MYCOLOGIA 67:172-174,1975 (DOLIPORE SEPTA) •

DIGITELLUS PAULET 1808 =STERILE GROWTH OF LENTINUS FIDE DONK 1962 (TAXON 11:82-83) •

DIHETEROSPORA KAMYSCHKO 1962. BOTAN. MATER (NOTUL. SYST. SECT. CRYPT. INST. BOTAN. ACAD. SCI. USSR) 15:137-140 /D. HETEROSPORA (LECTOTYPE) =D. CHLAMYDOSPORIA (GODDARD) BARRON & ONIONS 1966
DICTYO,CETERI,HYALO,NONSP,ALEURI, COMPARE WITH STEMPHYLIOPSIS. OTHER STATES =VERTICILLIUM (H) & PAECILOMYCES (H). MAY BE THE SAME AS CONIODICTYUM HAR. & PAT. 1909 OR DICTYOARTHRINOPSIS BATISTA & CIFERRI 1957.
/BARRON & ONIONS, CAN. J. BOT. 44:861-869,1966 (ILLUS) /CAMPBELL & GRIFFITHS, CAN. J. MICROBIOL. 21:963-971,1975 (ILLUS) /BARRON, CAN. J. BOT., IN PRESS, 1980 (NEW SPP,NEW COMBS,ILLUS)
=*POCHONIA* BATISTA & FONSECA 1965 FIDE BARRON & ONIONS 1966 •

DIMERA FRIES 1825. SYST. ORB. VEG. P 183 /BOTRYTIS DIDYMA SCHM.
=OEDEMIUM LINK 1824 FIDE HUGHES 1958. •

DIMORPHOSPORA TUBAKI 1958. J. HATTORI BOTAN. LAB. NO. 20:156-158 (ILLUS) /D. FOLIICOLA
AMERO,GLOIO OR CETERI,HYALO,RADUL,AQUATIC, 47A, OTHER STATE ="PHIALOCONIDIA".
=*FLUMINISPORA* INGOLD 1958 FIDE KENDRICK. •

DIONYSIA ARNAUD 1952. BULL. SOC. MYCOL. FRANCE 68:211-213 (ILLUS,NO LATIN DIAGNOSIS) /D. CAPITATA
=CANDELABRUM VAN BEVERWIJK 1952 FIDE KENDRICK 1973.
=*DYONISIA* ARNAUD 1952 (LAPSUS CALAMI) SEE ARNAUD 1954 (BULL. SOC. MYCOL. FRANCE 69:303) •

DIPLOCLADIELLA ARNAUD EX MATSUSHIMA 1975. ICONES P 61 & PL (SEE ALSO ELLIS, MDH P 229-230,1976) /D. SCALAROIDES
STAURO,CETERI,PHAEO,RACHI,APPEND, 126D, ON LITTER & WOOD. COMPARE WITH IYENGARINA SUBRAMANIAN.
/ARNAUD, BULL. SOC. MYCOL. FRANCE 69:295-296 & FIG 13,I-P,1954 (NO LATIN DIAGNOSIS) /TUBAKI, J. HATTORI BOT. LAB. NO. 20:158-159,1958 (ILLUS) /NICOT & CHARPENTIE, BULL. SOC. MYCOL. FRANCE 87:30,1971 (TYPE SPECIMEN) /INGOLD, GUIDE P 86-87,1975 (ILLUS) /ELLIS, MDH P 229-230,1976 (ILLUS,VALIDATION) •

DIPLOCLADIUM BONORDEN 1851. HANDB. ALLGEM. MYCOL. P 98 /TWO ORIGINAL SPECIES
=CLADOBOTRYUM NEES 1817 FIDE HUGHES 1958. •

DIPLOCOCCIUM GROVE 1885. JOUR. BOTANY, LONDON 23:167 /D. SPICATUM
DIDYMO OR PHRAGMO,CETERI OR BLASTO,PHAEO,NONSP OR RADUL, 66C, ON LITTER, WOOD, LEAVES & FUNGI. ASCO STATE =HELMINTHOSPHAERIA FIDE ELLIS 1971.
/BARRON, HYPHOMYCETES P 155-156,1968 (ILLUS) /SUTTON, CMI PAPERS 132:49-51,1973 (ILLUS,NEW SP) /ELLIS, DH P 395-397,1971 (ILLUS,KEY) /ELLIS, MDH P 389-392,1976 (ILLUS,KEY,NEW SP) •

DIPLOIDIUM ARNAUD 1923. ANN. EPIPHYT. 9:33 /D. SWEETIAE
=SEPTOIDIUM ARNAUD 1921 FIDE ELLIS 1971. •

DIPLOOSPORA GROVE 1916. JOUR. BOT. 54:220 & TAB 543, FIG 8 /D. ROSEA
DIDYMO,BLASTO,HYALO,NONSP OR RADUL, 65A, ON LITTER. DIFFERENCE FROM RAMULARIA NOT CLEAR. SEE ALSO SEPTOCYLINDRIUM.
/HUGHES, CAN. J. BOT. 46:1159-1160,1968 (ILLUS) /MATSUSHIMA, ICONES P 61,1975 (ILLUS,NEW SP) •

DIPLORHINOTRICHUM HOHNEL 1902. S. B. AKAD. WISS. WIEN 111:1040 /D. CANDIDULUM
=DACTYLARIA SACCARDO 1880 FIDE BHATT & KENDRICK 1968. •

DIPLORHYNCHUS ARNAUD 1952. BULL. SOC. MYCOL. FRANCE 68:207 & 211 (ILLUS,NO LATIN DIAGNOSIS) /D. BILOBA
HELICO,CETERI,HYALO,NONSP, 103A, ON WOOD.
/NICOT & CHARPENTIE, BULL. SOC. MYCOL. FRANCE 87:30,1971 (TYPE SPECIMEN) •

DIPLOSPORIUM BONORDEN 1851. HANDB. ALLGEMEIN. MYKOL. P 98 /D. ALBUM
=JACOBASCHELLA KUNTZE 1891, NON DIPLOSPORIUM LINK 1824. •

DIPLOSPORIUM LINK 1824. LINNAEUS SPEC. PLANT. IV,6(1):64 /D. NIGRESCENS
=OEDEMIUM LINK 1824 FIDE HUGHES 1958. •

DISCELLISPORA SAWADA 1944 NOMEN NUDUM =HAPLOBASIDION FIDE ELLIS 1957 (CMI PAPERS 67:4) •

DISCHLORIDIUM SUTTON 1976. KAVAKA 4:43-50 /CHLORIDIUM LAEENSE MATSUSHIMA 1971 =D. LAEENSE (MATSUSHIMA)
AMERO,GLOIO,HYALO,PHIAL,STROMA, 3I, ON LITTER (BROAD LEAF TREES).
/MATSUSHIMA, MICROFUNGI P 14 & FIG 112 & PL 22:7-8,1971 (ILLUS) •

DISCOCOLLA PRILLIEUX & DELACROIX 1894. BULL. SOC. MYCOL. FRANCE 10:86 /D. PIRINA
?, ORIGINAL ILLUSTRATION NOT DIAGNOSTIC.
/LINDAU, RABENH. KRYPT. FL. 2 AUFL. 1 BD. 9:506,1910 (ILLUS) •

DISCOFUSARIUM PETCH 1921. TBMS 7:164 /D. TASMANIENSE (MCALPINE)
=FUSARIUM LINK FIDE CLEMENTS & SHEAR 1931. •

DISCOSPORIUM SACCARDO & SYDOW (NON HOHNEL 1914) ?=CONIOSPORIUM FIDE AINSWORTH 1971 •

DISSOACREMONIELLA KIRILENKO 1970. NOV. SIST. NIZ. RAST. 7:235-238 (ILLUS) /D. SILVATICA
DIDYMO,CETERI,HYALO,RADUL, CLOSE TO CHLAMYDOMYCES.
/KIRILENKO, MIKRO. ZHUR. 36:153-160,1974 (ILLUS) •

DITANGIUM KARSTEN 1870. NOT. SALLSK. FAUNA FL. FENN. FORH. 11:213 (DATED 1871) /D. INSIGNE =D. STATE OF CRATEROCOLLA CERASI FIDE DONK 1962 (TAXON 11:83-84)
AMERO,GLOIO,HYALO,RADUL?,SPORODOCHIA, 26E, ON LITTER.
/NAG RAJ, CAN. J. BOT. 56:1485-1491,1978 (ILLUS)
?=*POROIDEA* GOTTINGER EX WINTER SEE DONK 1962 •

DITHOZETIA CLEMENTS & SHEAR 1931 (ORTHOGRAPHIC VARIANT), SEE DIDYMOTHOZETIA RANGEL 1915 •

DIVINIA CIFERRI 1955. SYDOWIA 9:326-327 /D. DIATRICHA
?, ORIGINAL ILLUSTRATION INADEQUATE TO CHARACTERIZE. •

DOASSANSIELLA ZAMBETTAKIS 1970. REV. MYCOL. 35:164-165 & FIG 26 /RAMULARIA AQUATILIS PECK 1885 =D. AQUATILIS (PECK) =D. STATE OF DOASSANSIOPSIS OCCULTA DIETEL
=NOMEN NUDUM (NO LATIN DIAGNOSIS), FOR CONIDIAL STATE OF DOASSANSIOPSIS. •

DOMINGOELLA PETRAK & CIFERRI 1932. ANN. MYCOL. 30:339 /D. ASTERINARUM
AMERO,CETERI,PHAEO,ANNEL, 49A, ON FUNGI (ASTERINA & OTHERS) ON WOOD & LITTER.
/ELLIS, DH P 113-114,1971 (ILLUS) /DEIGHTON & PIROZYNSKI, CMI PAPERS 128:89-92,1972 (ILLUS) /ELLIS, CMI PAPERS 131:3-5,1972 (ILLUS,NEW SPP) /ELLIS, MDH P 72-73,1976 (ILLUS) •

DORATOMYCES CORDA 1829. IN STURM'S DEUTSCHL. FLORA III, BD 2, HEFT 7:65 /D. NEESII, WHICH IS BASED ON PERICONIA SUBULATA NEES & NEES & P. STEMONITIS PERSOON SEE HUGHES 1958 (CAN. J. BOT. 36:765)
=CEPHALOTRICHUM LINK 1809 FIDE HUGHES 1958 & VON ARX 1974 (GENERA P 208-209).
/HUGHES, CAN. J. BOT. 36:744,1958 (REVIS,REFS) /MORTON & SMITH, CMI PAPERS

86:70-85,1963 (REVIS,KEY, ILLUS, SUB DORATOMYCES)
?=*SAROPHORUM* SYDOW & SYDOW 1916
?=*STILBELLULA* BOEDIJN 1951
=*STYSANUS* CORDA 1837 FIDE HUGHES 1958
=*STYSANOPSIS* FERRARIS 1909 FIDE CARMICHAEL 1973
=*SYNPENICILLLIUM* COSTANTIN 1888 FIDE MORTON & SMITH 1963
?=*TRICHURUS* CLEMENTS & SHEAR 1896 •

DRECHSLERA ITO 1930. PROC. IMP. ACAD. TOKYO 6:355 /D. TRITICI-VULGARIS (NISIKADO) (LECTOTYPE) =D. STATE OF PYRENOPHORA TRITICI-REPENTIS
PHRAGMO,CETERI,PHAEO,RACHI OR RADUL,STROMA, 83E, ON LEAVES (GRASSES & OTHER PLANTS). PROPOSED FOR CONSERVATION AGAINST ANGIOPOMA LEVILLE 1841, SEE SUTTON 1976 (MYCOTAXON 3:377-380) & 1977 (TAXON 26:591-592). COMPARE WITH CURVULARIA & EXSEROHILUM. ASCO STATE =PYRENOPHORA, TRICHOMETASPHAERIA OR COCHLIOBOLUS FIDE ELLIS 1971 OR SETOSPHAERIA FIDE VON ARX 1974 (P 141). SCLEROTIAL STATE =SCLEROTIUM SEE ELLIS 1971.
/SHOEMAKER, CAN. J. BOT. 40:809-836,1962 (ILLUS) /ELLIS, DH P 403-452,1971 (ILLUS,KEY,REVIS) /ELLIS & HOLLIDAY, CMI DESCRIPTIONS 301-302 & 304-306,1971 & 341-343 & 349,1972 & 491-494,1976 (ILLUS) /ELLIS & WALLER, CMI DESCRIPTIONS 388-390,1973 & 434,1974 (ILLUS) /MATSUSHIMA, ICONES P 63-66 & PL,1975 (ILLUS) /DANQUAH, TBMS 64:544-546,1976 (NEW SPP) /ELLIS, MDH P 396-404,1976 (ILLUS,KEY) /ALCORN, TBMS 70:61-65,1978 (ILLUS OF COCHLIOBOLUS STATES)
=*ANGIOPOMA* LEVILLE 1841 FIDE SUTTON 1976
=*BIPOLARIS* SHOEMAKER 1959 FIDE ELLIS 1971 •

DRECHSLERELLA SUBRAMANIAN 1964. J. INDIAN BOT. SOC. 42:299 (DATED 1963) /D. ACROCHAETA (DRECHSLER)
?=DACTYLELLA GROVE 1884. •

DRECHSLEROMYCES SUBRAMANIAN 1977. KAVAKA 5:93-98 (NOT SEEN) /DACTYLELLA ATRACTOIDES DRECHSLER =D. ATRACTOIDES (DRECHSLER)
?, SEE ABSTRACTS OF MYCOLOGY 67-52149,1979. •

DREPANISPORA - SEE DREPANOSPORA •

DREPANOCONIS SCHROETER & HENNINGS 1896. IN HENNINGS, HEDWIGIA 35:211-212 (NO ILLUS) /D. BRASILIENSIS
HELICO,CETERI,HYALO,MISC,SETAE, 102E, ON FRUIT & LEAVES (LAURACEAE). THIS IS A BASIDIOMYCETE FIDE MALENCON 1953 (BULL. SOC. MYCOL. FRANCE 69:91-96, ILLUS)
/LINDER, ANN. MISSOURI BOT. GARD. 16:341-345,1929 (ILLUS,NEW COMB)
?=*CURVIDIGITUS* SAWADA 1943 •

DREPANOSPORA BERKELEY & CURTIS 1875. GREVILLEA 3:105 (AS 'DREPANISPORA') /D. PANNOSA
=HELICOSPORIUM NEES 1817 FIDE MOORE 1957. •

DRUMOPAMA SUBRAMANIAN 1957. PROC. INDIAN ACAD. SCI. 46:331-333 (ILLUS) /D. GIRISA
AMERO,CETERI,HYALO,RACHI,SYNNEMATA, 29F, ON LITTER.
/MORRIS, WEST. ILLINOIS UNIV. SER. BIOL. SCI. 3:52-53,1963 (ILLUS) •

DUBIOMYCES LLOYD 1921. MYCOLOGICAL NOTES 65:1034 /D. VIRIDIS
=USTILAGINOIDEA BREFELD FIDE DIEHL FROM AINSWORTH 1971. •

DUDDINGTONIA COOKE 1969. TBMS 53:315-318 (ILLUS) /TRICHOTHECIUM FLAGRANS DUDDINGTON =D. FLAGRANS (DUDDINGTON)
?=ARTHROBOTRYS CORDA 1839 •

DUOSPORIUM THIND & RAWLA 1961. AMER. J. BOT. 48:859-862 (ILLUS) /D. CYPERI
PHRAGMO,CETERI,PHAEO,RACHI,STROMA, 82F, ON CYPERUS. FOUNDED ON TWO KINDS OF CONIDIA, BUT MACROCONIDIA ARE DISTINCTIVE. •

DWAYAANGAM SUBRAMANIAN 1977. KAVAKA 5:93-98 (NOT SEEN) /TRIPOSPORINA QUADRIDENS DRECHSLER =D. QUADRIDENS (DRECHSLER)
?, SEE ABSTRACTS OF MYCOLOGY 67-52149,1979. •

DWAYABEEJA SUBRAMANIAN 1958. J. INDIAN BOT. SOC. 37:53-57 (ILLUS) /D. SUNDARA
SCOLECO,CETERI,PHAEO,MISC, 88G, ON LITTER (PHOENIX). GENUS WAS BASED ON TWO KINDS OF CONIDIA. THE PHRAGMOCONIDIA FIT POLYSCHEMA. FOR D. GRAMINIS SEE POLYSCHEMA AND BAHUCHASHAKA.
/ELLIS, DH P 339-340,1971 (ILLUS) •

DWAYALOMA SUBRAMANIAN 1957. J. INDIAN BOT. SOC. 36:61-64 (ILLUS) /D. TRINA
DIDYMO,CETERI,HYALO,NONSP?,APPEND,SETAE, 56D, ON LITTER (GRASSES). •

DWAYALOMELLA BRISSON, PIROZYNSKI & PAUZE 1975. CAN. J. BOT. 53:2866-2871 (ILLUS) /D. VACCINII
DIDYMO,GLOIO,HYALO,PHIAL,ACERVULI, ON STEMS (VACCINIUM). (MELANCONIALES) •

DWAYAMALA SUBRAMANIAN 1956. J. INDIAN BOT. SOC. 35:473-476 (ILLUS) /D. PRATHILOMAKA
?=DENDRYPHION WALLROTH 1833. D. PRATHILOMAKA APPEARS TO BE DENDRYPHIELLA INFUSCANS (THUMEN) ELLIS 1971. •

DYONISIA ARNAUD 1952, SEE DIONYSIA •

ECHINOBOTRYUM CORDA 1831. IN STURM'S DEUTSCHL. FLORA III (PILZE), BD. 3, HEFT 12:51 & FIG 26 /E. ATRUM =E. STATE OF CEPHALOTRICHUM STEMONITIS (PERSOON) LINK 1809
AMERO,CETERI,PHAEO,NONSP OR RADUL,ALEURI,SYNNEMATA, 48E, ON LITTER, DUNG & SOIL. OTHER CONIDIAL STATES =CEPHALOTRICHUM & SCOPULARIOPSIS.
/BARRON, HYPHOMYCETES P 161,1968 (ILLUS) /HENNEBERT, TBMS 51:749-753,1968 (ILLUS,REVIS) /ELLIS, DH P 77-78,1971 (ILLUS) •

ECHINOCATENA CAMPBELL & SUTTON 1977. TBMS 69:125-131 (ILLUS) /E. ARTHRINIOIDES
AMERO,BLASTO OR CETERI,PHAEO,MISC, 37B, ON LITTER. COMPARE WITH SADASIVANIA. •

ECHINOCHONDRIUM SAMSON & VAN DER AA 1975. REVUE MYCOL. 39:103-106 (ILLUS) /E. PULCHRUM
BULBILS,CETERI,PHAEO,NONSP,APPEND, 128B, ON LEAVES (COCOS). HAS BULBILS WITH SPINES, NO CONIDIA. COMPARE WITH DACTULIOPHORA. •

ECHINODIA PATOUILLARD 1919. BULL. SOC. MYCOL. FRANCE 34:199-201 (ILLUS, DATED 1918) /E. THEOBROMAE =E. STATE OF POLYPORUS HYDNOPHORUS FIDE DONK 1962
AMERO,BLASTO,HYALO,NONSP,SYNNEMATA,CLAMPS, 45A, ON LITTER (THEOBROMA). COMPARE WITH PTYCHOGASTER. BASIDIO STATE =POLYPORUS FIDE DONK 1962 (TAXON 11:84). •

ECHINOSPORIUM WORONICHIN 1913. MONIT. JARD. BOT. TIFLIS 28:25 & FIGS 12-14 /E. ACERIS
=PETRAKIA SYDOW 1913 FIDE CARMICHAEL. •

ECTOSTROMA FRIES 1823. SYSTEMA MYCOL. 2:601-602 /E. LIRIODENDRI (KUNZE) (LECTOTYPE)
?, STERILE MYCELIUM IN LEAF SPOTS. •

ECTOTRICHOPHYTON CASTELLANI & CHALMERS 1919. MAN. TROP. MED. ED. 3 P 1003-1008 /E. MENTAGROPHYTES (ROBIN)
=TRICHOPHYTON MALMSTEN 1845. •

EDMUNDMASONIA SUBRAMANIAN 1958. J. INDIAN BOT. SOC. 37:403 (401-404,ILLUS) /E. PULCHRA
PHRAGMO,CETERI,PHAEO,RACHI OR MISC, 84A, ON LITTER (PHYLLOSTACHYS, BAMBUSA). COMPARE WITH DENDRYPHIOSPHAERA.
/RAO & RAO, MYCOPATHOLOGIA 22:242-244,1964 (ILLUS,NEW SP) /MATSUSHIMA, ICONES P 66 & PL,1975 (ILLUS) •

EGERITA AUCT. SEE AEGERITA •

EIDAMIA LINDAU 1904. RABENHORST KRYPT.-FL. BD I, 8:123,1907 /HARZIA ACREMONIOIDES (HARZ) COSTANTIN 1888 =E. ACREMONIOIDES (HARZ)
=HARZIA COSTANTIN 1888. •

ELADIA SMITH 1961. TBMS 44:47 /PENICILLIUM SACCULUM DALE 1926 (ANN. MYCOL., BERLIN 24:137) =E. SACCULA (DALE)
AMERO,BASO,PHAEO,PHIAL, 6C, ON SOIL. ?=PENICILLIUM SEE BARRON 1968.
/BARRON, HYPHOMYCETES P 162-163,1968 (ILLUS) /MATSUSHIMA, ICONES P 66 & PL,1975 (ILLUS) •

ELAEODEMA SYDOW 1922. ANN. MYCOL., BERLIN 20:64 (NO ILLUS) /E. CINNAMOMI
? •

ELLETEVERA DEIGHTON 1969. CMI PAPERS 118:17-21 (ILLUS) /PYRICULARIA PARASITICA ELLIS & EVERHART 1894 =E. PARASITICA (ELLIS & EVERHART)
?, MUCH LIKE CERCOSPORA. •

ELLISIA BATISTA & PERES 1965. MYCOPATHOLOGIA 25:166-167 (ILLUS) /E. INGAE
?, IS ANTEDATED BY THE HIGHER PLANT GENUS ELLISIA. E. INGAE MAY BELONG IN HETEROCONIUM PETRAK. •

ELLISIELLA BATISTA 1956. ANN. SOC. BIOL. PERNUMBUCO 14:19-20 (NO ILLUS) /E. HYDRANGEAE CAMARA 1941 (AGRON. LUSITANA 3:319-320 & TAB II, FIG 6-8)
=NOMEN ILLEGITIMUM, SACCARDO'S GENUS ELLISIELLA WAS EMENDED BY SUBSTITUTING A NEW TYPE SPECIES. •

ELLISIELLA SACCARDO 1880. MICHELIA 2:26 & 147, ILLUS. IN FUNGI ITALICI FIG 781 /E. CAUDATA (PECK)
=COLLETOTRICHUM CORDA 1831 FIDE VON ARX 1970. •

ELLISIELLINA DA CAMARA 1949. AGRON. LUSIT. 11:72 /ONE OF THE TWO ORIGINAL SPECIES IS THE TYPE OF ELLISIELLA SACC. 1880
=NOMEN ILLEGITIMUM. •

ELLISIOPSIS BATISTA 1956. AN. SOC. BIOL. PERNAMBUCO 14:21 /E. GALLESIAE BATISTA & NASCIMENTO 1956
=BELTRANIELLA SUBRAMANIAN 1952 FIDE PIROZYNSKI & PATIL 1970. •

ELOSIA PERSOON 1822. MYCOL. EUROP. 1:11-12 /CONOPLEA ERYNGII =ALTERNARIA ERYNGII (PERSOON) HUGHES & SIMMONS (IN HUGHES, CAN. J. BOT. 36:735,1958)
=ALTERNARIA NEES SEE HUGHES & SIMMONS 1958. •

EMBELLISIA SIMMONS 1971. MYCOLOGIA 63:380-386 (ILLUS) /E. ALLII (CAMPANILE)
PHRAGMO OR DICTYO,CETERI,PHAEO,RACHI, 82H, ON SOIL & PLANTS (ALLIUM, HYACINTHUS, DAUCUS & OTHERS). OTHER STATE ="SCLEROTIA".
/SUTTON, CMI PAPERS 132:52,1973 (ILLUS) /DE HOOG & MULLER, NETH. J. PL PATH. 79:85-93,1973 (ILLUS,NEW SP) /MUNTANJOLA-CRETKOVIC & RISTANOVIC, MYCOLOGIA 68:47-51,1976 (ILLUS,NEW SP) /ELLIS, MDH P 433-436,1976 (ILLUS,KEY) •

EMMONSIA CIFERRI & MONTEMARTINI 1959. MYCOPATHOLOGIA 10:314 /E. PARVA (EMMONS & ASHBURN)
=MYCELIOPHTHORA FIDE CARMICHAEL 1962. •

ENDOCALYX BERKELEY & BROOME 1875. J. LINN. SOC. LONDON 15:84-85 /E. THWAITESII (LECTOTYPE)
AMERO,CETERI,PHAEO,NONSP?,SPORODOCHIA,SYNNEMATA,SETAE, 46C, ON PLANTS (CISSUS, COCOS & OTHERS).
/ELLIS, DH P 579-580,1971 (ILLUS,KEY) /HUGHES, NEW ZEALAND J. BOT. 16:328,1978 (ILLUS)
?=*NIGROCUPULA* SAWADA 1944 •

ENDOCONIDIUM PRILLIEUX & DELACROIX 1891. BULL. SOC. MYCOL. FRANCE 7:116-117 (ILLUS) /E. TEMULENTUM =E. STATE OF GLOEOTINIA TEMULENTA
AMERO,BASO,HYALO,PHIAL?,SPORODOCHIA, ON FRUIT & SEEDS. ASCO STATE =GLOEOTINIA, ENDOCONIDIUM IS BASED ON THE SPERMATIAL STATE, SEE NAG RAJ & KENDRICK 1975.
/LINDAU, RABENHORST'S KRYPT.-FL. 2 ED. 1BD. 9:479-481,1910 (ILLUS) /AINSWORTH, DICTIONARY OF THE FUNGI P 167-168,1961 (REFS) /NAG RAJ & KENDRICK, CHALARA P 45-47,1975 (REFS)
?=*MYRIOCONIUM* SYDOW 1912 •

ENDOCONOSPORA GJAERUM 1971. NORWEGIAN J. BOT. 18:109-111 (ILLUS) /E. CERASTII
AMERO,GLOIO?,HYALO,PHIAL,SPORODOCHIA, 13D, ON LEAVES (CERASTIUM).
?=TUBERCULARIA TODE EX PERSOON 1801. •

ENDODERMOPHYTON CASTELLANI 1910. MAN. TROP. MED., ED. 1, P 610 /THE AGENT OF TINEA IMBRICATA (TRICHOPHYTON CONCENTRICUM)
=TRICHOPHYTON MALMSTEN 1845. •

ENDODESMIA BERKELEY & BROOME 1874. ANN. MAG. NAT. HIST. 4, 7:432 (ILLUS) /E. GLAUCA
=BROOMEOLA KUNTZE 1891, NON ENDODESMIA BENTH. 1862 •

ENDOPHRAGMIA DUVERNOY & MAIRE 1920. BULL. SOC. MYCOL. FRANCE 36:86-88 (ILLUS) /E. MIRABILIS
PHRAGMO,CETERI,PHAEO,ANNEL,SYNNEMATA,STROMA, 77A, ON LITTER & WOOD (OF MANY PLANTS) & DUNG. E. DIMORPHOSPORA AWAO & UDAGAWA 1974 WAS DESCRIBED WITH AN ASSOCIATED SELENOSPORELLA STATE.
/ELLIS, CMI PAPERS 72:19-36,1959 (ILLUS,KEY) /ELLIS, DH P 152-156,1971 (ILLUS,KEY) /ELLIS, CMI PAPERS 131:11,1972 (ILLUS,NEW SP) /SUTTON, CMI PAPERS 132:53-58,1973 (ILLUS,NEW SPP) /SHEARER, CRANE & MILLER, MYCOLOGIA 68:184-186,1976 (ILLUS,NEW SP) /MATSUSHIMA, ICONES P 67-71 & PL,1975 (ILLUS,NEW SPP) /ELLIS, MDH P 138-142,1976 (ILLUS,KEY,NEW SPP,NEW COMB) /SUTTON, BOL. SOC. ARGENT. BOT. 18:156-158,1977 (ILLUS,NEW SP) /HUGHES, NEW ZEALAND J. BOT. 16:341-343,1978 (ILLUS,SUB PHRAGMOCEPHALA) /HUGHES, FUNGI CANADENSES 150,1979 (NEW COMB,ILLUS,DISCUSSION)
=*PHRAGMOCEPHALA* MASON & HUGHES 1951 FIDE ELLIS 1959 •

ENDOPHRAGMIELLA SUTTON 1973. CMI PAPERS 132:58-62 (ILLUS) /E. PALLESCENS
DIDYMO OR PHRAGMO,CETERI,PHAEO,ANNEL, 59A, ON WOOD, BARK, LITTER & FUNGI (CYTOSPORA, VALSA & OTHERS). HUGHES (1978 P 330) REPORTS UNNAMED TELEOMORPH. OTHER STATE =SELENOSPORELLA (H).
/MATSUSHIMA, ICONES P 68 & PL,1975 (ILLUS WITH A SELENOSPORELLA STATE SUB ENDOPHRAGMIA CANADENSIS) /ELLIS, MDH P 143-145,1976 (ILLUS,KEY,NEW SPP) /BOROWSKA, ACTA MYCOL. 13:169-174,1977 (ILLUS,NEW SP) /HUGHES, FUNGI CANADENSES 122-130,1978 (ILLUS,NEW SPP,NEW COMBS) /HUGHES, NEW ZEALAND J. BOT. 16:328-332,1978 (ILLUS,NEW SPP) /HUGHES, FUNGI CANADENSES 148,1979 (NEW COMB,ILLUS) /HUGHES, FUNGI CANADENSES 162,1980 (ILLUS) •

ENDOPHRAGMIOPSIS ELLIS 1966. CMI PAPERS 106:55-57 (ILLUS) /E. PIROZYNSKII
PHRAGMO,CETERI,PHAEO,ANNEL,PODIA, 76A, ON WOOD & PLANTS (PHOENIX, ELAEIS). COMPARE WITH MITTERIELLA.
/ELLIS, DH P 156-157,1971 (ILLUS) /DE HOOG, ACTA BOT. NEERL. 22:211-212,1973 (NEW COMB) •

ENDOSPORELLA THAXTER 1920. BOT. GAZ. 69:16 & PL IV (1-27,ILLUS) /E. DIOPSIDIS
AMERO,CETERI,HYALO,PHIAL, 55A, ON INSECTS (DIPTERA). •

ENDOSPOROSTILBE SUBRAMANIAN 1958. JOUR. INDIAN BOT. SOC. 37:48-49 (ILLUS) /E. NILAGIRICA
=BLOXAMIA BERKELEY & BROOME 1854 FIDE NAG RAJ. •

ENDOSTILBUM MALENCON 1964. BULL. SOC. MYCOL. FRANCE 80:103-112 (ILLUS) /E. CERASI (BOURD. & GALZ.) =CORYNE STATE OF ASCOCORYNE SOLITARIA (REHM) DENNIS
=CORYNE NEES 1817 FIDE KORF & CANDOUSSAU 1974 (BULL. SOC. MYCOL. FRANCE 90:209-215,ILLUS). •

ENGELHARDTIELLA FUNK 1973. CAN. J. BOT. 51:1643-1645 (ILLUS) /E. ALBA
STAURO,CETERI,HYALO,AMPUL,SPORODOCHIA, 117B, ON FUNGI (CYTOSPORA). •

ENGYODONTIUM DE HOOG 1978. PERSOONIA 10:53 (33-81) /RHINOTRICHUM PARVISPORUM PETCH =E. PARVISPORUM (PETCH)
?=APHANOCLADIUM GAMS 1971. •

ENTEROBRYUS LEIDY 1849. PROC. ACAD. NAT. SCI. PHILADELPHIA 4:225-227 /E. ELEGANS
=TRICHOMYCETES (SEE 55D).
/LICHTWARDT, MYCOLOGIA 46:564-585,1954 (ILLUS,NEW SPP) /LICHTWARDT,

MYCOLOGIA 65:1-20,1973 (ILLUS,DISCUSSION) •

ENTOMOCOSMA SPEGAZZINI SEE THAXTERIOLA FOR REFERENCE •

ENTOMYCLIUM WALLROTH 1833. FLORA CRYPT. GERMAN. 2:189-190 /E. FOLLICULATUM =DENDRYPHION NANUM (NEES) HUGHES 1958
=DENDRYPHION WALLROTH 1833 FIDE HUGHES 1958. •

ENTYLOMELLA HOHNEL 1916. OESTRR. BOT. ZEITS. 66:105 /E. FICARIAE (BERKELEY) (LECTOTYPE FIDE WEESE 1932)
=NOMEN NUDUM, FOR CONIDIAL ENTYLOMA, ?=ITERSONILIA DERX FIDE SOWELL & KORF 1960. SEE ALSO CIFERRI 1959 (OMAGIU LUI TRAIAN SAVULESCU, P 176). •

EOETVOESIA SCHULZER 1866. IN SCHULZER, KANITZ & KNAPP, VERH. ZOOL.-BOT. GES. WIEN 16(ABHANDL.):36 (NO ILLUS) /E. CUCURBITAE (SCHULZER)
? •

EPHELIS FRIES 1849. SUMMA VEG. SCAND. P 370 (NO SPECIES DESCRIBED) /E. MEXICANA FRIES EX BERKELEY 1868 (LECTOTYPE FIDE CLEMENTS & SHEAR 1931, P 379)
SCOLECO,GLOIO,HYALO,PHIAL?,SPORODOCHIA,SETAE, 107E, ON PLANTS (GRASSES & SEDGES). ASCO STATE =BALANSIA FIDE ULLASA 1969 (MYCOLOGIA 61:572-579,ILLUS). COMPARE WITH SPHACELIA.
/DIEHL, J. AGRIC. RES. 41:761-766,1931 (ILLUS) /SUTTON, CMI PAPERS 141:82,1977 (REFS)
=*SPHACELIOPSIS* SPEGAZZINI 1910 FIDE DIEHL FROM SUTTON 1977 •

EPICLINIUM FRIES 1849. SUMMA VEG. SCAND. P 475 (NO ILLUS) /E. PEZIZOIDEUM (SCHWEINITZ) SACCARDO 1886
? •

EPICOCCUM LINK 1815. MAG. GES. NATURF. FREUNDE, BERLIN 7:32 /E. NIGRUM =E. PURPURASCENS EHRENBERG 1818
DICTYO,CETERI,PHAEO,NONSP,ALEURI,SPORODOCHIA, 100B, ON LEAVES (MANY PLANTS). COELO STATE =PHOMA FIDE PUNITHALINGAM & LEACH 1972 (TBMS 59:341-345,ILLUS).
/SCHOL-SCHWARZ, TBMS 42:149-173,1959 (REVIS,ILLUS) /ELLIS, DH P 72-74,1971 (ILLUS) /MATSUSHIMA, ICONES P 24 & 71 & PL,1975 (ILLUS)
=*CEREBELLA* CESATI 1851 FIDE SCHOL-SCHWARZ 1959
=*PARATRICHAEGUM* FAUREL & SCHOTTER 1966 FIDE KENDRICK 1973
?=*CLATHROCOCCUM* HOHNEL 1911 FIDE AINSWORTH 1971 •

EPIDERMOPHYTON SABOURAUD 1907. ARCH. MED. EXP. ANAT. PATHOL. 19(5):565 /E. INGUINALE =E. FLOCCOSUM (HARZ) LANGERON & MILOCHEVITCH 1930
PHRAGMO,CETERI,HYALO,NONSP,ALEURI, 89E, ON ANIMALS (MAN). HAS BEEN PROPOSED FOR CONSERVATION AGAINST EPIDERMOPHYTON MEGNIN 1881, SEE LOEFFLER 1976 (TAXON 25:208-209). OTHER CONIDIAL STATE ="ARTHROCONIDIA".
/AJELLO, SABOURAUDIA 6:147-159,1968 (REVIS) /REBEL & TAPLIN, THE DERMATOPHYTES 124 P,1970 (ILLUS) /PROCHACKI & ENGELHARDT-ZASADA, MYCOPATHOLOGIA 54:341-345,1974 (ILLUS,NEW SP) •

EPIDOCHIOPSIS KARSTEN 1892. HEDWIGIA 31:294 (NO ILLUS) & BIDRAG KANN. FINNLANDS NAT, OCH FOLK 51:499 /E. ATROVIRENS (KARSTEN)
? •

EPIDOCHIUM FRIES 1849. SUMMA VEG. SCAND. P 471 /E. ATROVIRENS
=TREMELLA (B) FIDE DONK 1958 (TAXON 7:193). •

EPILITHIA NYLANDER 1855. BOT. ZEIT. 13:66 /E. CRISTATA
=HARPOGRAPHIUM SACCARDO FIDE SACCARDO 1886. •

EPOCHNIELLA SACCARDO 1880. MICHELIA 2:176 /?
=STEMPHYLIUM WALLROTH 1833 FIDE LINDAU IN RABENHORST 1910. •

EPOCHNIUM LINK 1809. MAG. GES. NATURF. FREUNDE, BERLIN 3:18 /E. MONILIOIDES =MONILIA FRUCTIGENA PERSOON FIDE HUGHES 1958
=MONILIA PERSOON 1801 FIDE HUGHES 1958.
/DONK, TAXON 12:270,1963 (REFS) •

ERICIANELLA BRONDEAU 1830 (RECUEIL PL. CRYPT. AGENAIS, AGEN. FASC. 3, PL 10, FIG 8) =BACTRIDIUM FIDE FRIES 1849 (SUMMA VEG. SCAND. 2:474) •

ERIOCERCOSPORA DEIGHTON 1969. CMI PAPERS 118:5 (5-17,ILLUS) /E. BALLADYNAE (HANSFORD)
PHRAGMO OR SCOLECO,CETERI,PHAEO,RADUL, 82D, ON FUNGI (ASTERINA, BALLADYNA & OTHERS). FOR HYPERPARASITIC CERCOSPORA SPECIES.
/ELLIS, DH P 248-249,1971 (ILLUS) •

ERIOMENE (SACCARDO) MAIRE EX PEYRONEL 1919. BULL. SOC. MYCOL. FRANCE 35:179 /E. CILIATA (CORDA) MAIRE EX PEYRONEL 1919
=MENISPORA PERSOON 1822 FIDE HUGHES 1958 •

ERIOMENE (SACCARDO) CLEMENTS & SHEAR 1931 =MENISPORA PERSOON 1822 FIDE HUGHES 1958 •

ERIONEMA MAIRE 1906 ?=LAPSUS PRO ERIOMENE SEE HUGHES 1958 •

ERIOMENELLA PEYRONEL 1918 =MENISPORA PERSOON 1822 FIDE HUGHES 1958 •

ERIOMYCOPSIS SPEGAZZINI 1910. AN. MUS. NAC. BUENOS AIRES 3(13):429 (NO ILLUS) /E. BONPLANDII (AS 'BOMPLANDI')
DIDYMO OR PHRAGMO OR SCOLECO,CETERI,HYALO OR PHAEO,RADUL,APPEND, 79F, ON FUNGI (APPENDICULELLA, ASTERIDIELLA, IRENOPSIS, MELIOLA). COMPARE WITH SYMPODIOPHORA & PYRICULARIA. ASCO STATE =CALLORIOPSIS OR MELIOLIPHILA SEE DE HOOG 1977 (CBS STUDIES 15:137 & 140).
/DEIGHTON & PIROZYNSKI, CMI PAPERS 128:44,1972 (ILLUS) /SUTTON & HODGES, NOVA HEDWIGIA 29:600-602,1978 (ILLUS,NEW SP)
?=*PSEUDOFUSARIUM* MATSUSHIMA 1971 •

EUASPERGILLUS LUDWIG 1892 (LEHRB. NIED. KRYPTOGAM. P 258) =ASPERGILLUS FIDE THOM & CHURCH 1926 (THE ASPERGILLI P 7) •

EURASINA ARNOLD 1970. Z. PILZK. 35:305-310 (DATED 1969) /E. BONDARZEWAE (AS 'BONDARZEWIAE')
=HELMINTHOPHORA BONORDEN FIDE DE HOOG 1978 (SEE 80D). •

EURICOA BATISTA & MAIA 1955. ANN. SOC. BIOL. PERNAMBUCO 13:151-153 (ILLUS) /E. DOMINGUESII
=CYLINDROCARPON WOLLENWEBER 1913 FIDE CARMICHAEL 1973. •

EVERHARTIA SACCARDO & ELLIS 1882. IN SACCARDO, MICHELIA 2:580 /E. HYMENULOIDES
HELICO,GLOIO,HYALO,NONSP,SPORODOCHIA, 103C, SEE ALSO HYALOTROCHOPHORA & VANBEVERWIJKIA.
/LINDER, ANN. MISSOURI BOT. GARD. 16:335-338,1929 (ILLUS,KEY,NEW SP) •

EVERSIA CRANE & SCHOKNECHT 1977. IN SCHOKNECHT & CRANE, MYCOLOGIA 69:539 (539-543,ILLUS) /HELMINTHOSPORIUM SUBOPACUM COOKE & ELLIS 1878 =E. SUBOPACA (COOKE & ELLIS)
CHEIRO OR DICTYO,CETERI,PHAEO,ANNEL, 98B, ON WOOD.
/MORGAN-JONES, MYCOTAXON 7:333-336,1978 (ILLUS) •

EXCIOCONIDIUM PLUNKETT 1925. IN STEVENS, BERNICE P BISHOP MUSEUM BULL. 19:156-158 (ILLUS) /E. CIBOTII
=CHALARA FIDE NAG RAJ & KENDRICK 1975 (P 38). •

EXCIOCONIS CLEMENTS & SHEAR 1931 P 397, ILLEGITIMATE REPLACEMENT FOR EXCIOCONIDIUM PLUNKETT •

EXCIPULARIA SACCARDO 1884. SYLLOGE FUNGORUM 3:689 /E. FUSISPORA (BERKELEY & BROOME)
PHRAGMO,CETERI,PHAEO,NONSP,SETAE,SPORODOCHIA, 90C, ON WOOD & BARK. EMENDED HOHNEL 1904 (ANN. MYCOL. 2:51-53). COMPARE WITH LISTEROMYCES.
/ELLIS, MDH P 136-137,1976 (ILLUS) •

EXOBASIDIOPSIS KARAKULIN 1922. NOT. SYST. INST. CRYPT., PETROGRAD 1:83 /E. VICIAE (LECTOTYPE)

=KABATIELLA FIDE VON ARX 1970. •

EXOCHALARA GAMS & HERMANIDES-NIJHOF 1976. CBS STUDIES 13:56 (56-59,ILLUS) /CHALARA LONGISSIMA GROVE 1885 =E. LONGISSIMA (GROVE) GAMS & HERMANIDES-NIJOHF 1976
AMERO,BASO,HYALO,PHIAL, 3C, ON WOOD & LITTER. •

EXOPHIALA CARMICHAEL 1966. SABOURAUDIA 5:120-123 (ILLUS) /E. SALMONIS
AMERO OR DIDYMO,GLOIO,PHAEO,ANNEL, 20B & 60B, ON FISH, AMPHIBIA, MAN, WOOD, LITTER & SOIL. OTHER CONIDIAL STATES =RHINOCLADIELLA, RAMICHLORIDIUM OR PHAEOCOCCUS.
/DE HOOG, CBS STUDIES 15:100-121,1977 (ILLUS,REVIS,KEY) /MCGINNIS & PADHYE, MYCOTAXON 5:341-352,1977 (NEW COMB,ILLUS) /PADHYE, PROC. FOURTH INT. CONF. ON MYCOSES (PAHO SCI. PUBL. NO. 356) P 60-65,1977 (REVIS OF E. JEANSELMEI,NO ILLUS) /MCGINNIS, SABOURAUDIA 17:145-154,1979 (ILLUS,REVIS OF E. WERNECKII)
=*FOXIA* CASTELLANI 1908 (NOMEN ILLEGIT) FIDE DE HOOG 1977 •

EXOSPORELLA HOHNEL 1912. S.B. AKAD. WISS. WIEN 121:414-415 /E. SYMPLOCI
?, NOT ILLUSTRATED BY AUTHOR. •

EXOSPORIELLA KARSTEN 1892. BIDRAG KANN, FINLANDS NAT OCH FOLK 51:502-503 /E. FUNGORUM (FRIES)
PHRAGMO,CETERI,PHAEO,ANNEL,STROMA,APPEND, 75E, ON FUNGI (CORTICIUM). COMPARE WITH TOXOSPORIUM.
/ELLIS, DH P 148-149,1971 (ILLUS)
=*PHANEROCORYNELLA* HOHNEL 1919, BASED ON SAME TYPE •

EXOSPORINA ARNAUD 1921 (NON OUDEMANS 1904) =ARNAUDINA TROTTER 931 & EXOSPORINELLA BENDER 1932 •

EXOSPORINA OUDEMANS 1904. PROC. K. AKAD. WETENSCH., AMSTERDAM 6:498-501 & FIGS A-D /E. LARICIS
?, RESEMBLES OJIBWAYA. •

EXOSPORINELLA BENDER 1932 (MYCOLOGIA 24:410-411) IS A SUPERFLUOUS REPLACEMENT FOR EXOSPORINA ARNAUD NON OUDEMANS -SEE ARNAUDINA •

EXOSPORIUM LINK 1809. MAG. GES. NATURF. FREUNDE, BERLIN 3:9-10 /E. TILIAE
PHRAGMO,CETERI,PHAEO,RACHI,SPORODOCHIA OR SYNNEMATA, 83C, ON LEAVES, TWIGS & LITTER (MANY PLANTS).
/ELLIS, CMI PAPERS 82:21-39,1961 (ILLUS,KEY,REVIS) /ELLIS, DH P 400-403,1971 (ILLUS,KEY) /PIROZYNSKI, CMI PAPERS 129:47-50,1972 (ILLUS) /ELLIS, MDH P 392-395,1976 (ILLUS,KEY,NEW SPP)
=*CUSPIDOSPORIUM* CIFERRI 1955 FIDE ELLIS 1971 •

EXOTRICHUM SYDOW & SYDOW 1914. ANN. MYCOL. 12:571 (NO ILLUS) /E. LEUCOMELAS
=MYROTHECIUM TODE EX FRIES 1821 FIDE TULLOCH 1972.
/TULLOCH, CMI PAPERS 130:36,1972 (DISCUSSION) •

EXSEROHILUM LEONARD & SUGGS 1974. MYCOLOGIA 66:289 (281-297, ILLUS) /E. TURCICUM (PASSERINI) =E. STATE OF SETOSPHAERIA TURCICA (LUTRELL) LEONARD & SUGGS (IBID. P 295)
PHRAGMO,CETERI,PHAEO,RADUL, 83G, ON LEAVES (GRASSES). COMPARE WITH DRECHSLERA -SEE ELLIS 1976 (MDH P 396). ASCO STATE =SETOSPHAERIA FIDE LEONARD & SUGGS 1974
/LEONARD, MYCOLOGIA 68:402-411,1976 (ILLUS,REVIS) /HONDA & ARAGAKI, MYCOLOGIA 70:547-555,1978 (ILLUS) •

EXSERTICLAVA HUGHES 1978. NEW ZEALAND J. BOT. 16:332-334 (311-370,ILLUS) /CORDANA VASIFORMIS MATSUSHIMA =E. VASIFORMIS (MATSUSHIMA)
PHRAGMO,CETERI,HYALO,RADUL,STROMA, ON LITTER.
/MATSUSHIMA, ICONES P 40 & PL 369:1-2,1975 (ILLUS SUB CORDANA) •

FAVOTRICHOPHYTON (CASTELLANI & CHALMERS) NEVEU-LEMAIRE 1921. IN PRECIS DE PARASITOL. HUMAINE, ED. 5, P 55. /F. OCHRACEUM (SABOURAUD) NEVEU-LEMAIRE 1921 =TRICHOPHYTON VERRUCOSUM
=TRICHOPHYTON MALMSTEN 1848. •

FELLNERIA FUCKEL 1867 =COLLETOTRICHUM CORDA 1831 FIDE VON ARX 1970 •

FIBRILLARIA SOWERBY 1803. COLOURED FIGURES OF ENGLISH FUNGI 3:TEXT TO PLATE 387 /F. STELLATA (LECTOTYPE FIDE DONK, TAXON 11:84,1962)
?, STERILE. PROPOSED FOR MYCELIUM OR IMMATURE FRUIT BODIES OF HYMENOMYCETES. •

FILOBASIDIELLA KWON-CHUNG 1975. MYCOLOGIA 67:1197-1200 (ILLUS) /F. NEOFORMANS
AMERO,BASO,HYALO,PHIAL OR MISC,CLAMPS, 35F, THIS IS A BASIDIOMYCETE. OTHER STATE =CRYPTOCOCCUS (BLASTOMYCETES).
/KWON-CHUNG, MYCOLOGIA 68:821-833 & 942-946,1976 (ILLUS,NEW SP) /MALLOCH, CAN. J. BOT. 56:1823-1826,1978 (ILLUS,NEW SP) •

FILOBASIDIUM OLIVE 1968. ?CONIDIAL STATE =BLASTOMYCETES FIDE VON ARX 1974 (GENERA P 177-178,ILLUS) •

FILOSPORELLA NAWAWI 1976. TBMS 67:175 (173-176,ILLUS) /F. AQUATICA
HELICO OR SCOLECO,CETERI,HYALO,ANNEL,AQUATIC, 109A.
/DYKO & SUTTON, MYCOTAXON 7:323-326,1978 (REVIS,REFS,NOT ILLUS)
?=*ROGERSIA* SHEARER & CRANE 1976 FIDE DYKO & SUTTON 1978
=*COELOANGUILLOSPORA* DYKO & SUTTON 1978 FIDE DYKO & SUTTON 1978 •

FLABELLOSPORA ALASOADURA 1968. NOVA HEDWIGIA 15:415-418 & TAB 44 & 45 /F. CRASSA
STAURO,CETERI,HYALO,NONSP,AQUATIC, 115C & 120F.
/NAWAWI, MALAYSIAN J. SCIENCE 2:55-58,1973 (ILLUS,NEW SP) /INGOLD, GUIDE P 36-37,1975 (ILLUS) /NAWAWI, TBMS 66:543-547,1976 (ILLUS,NEW SP) •

FLAGELLOSPORA INGOLD 1942. TBMS 25:404 /F. CURVULA
SCOLECO,CETERI,HYALO,PHIAL,AQUATIC, 109C, ASCO STATE =NECTRIA FIDE RANZONI 1956 (AMER. J. BOT. 43:13-17).
/INGOLD, GUIDE P 82-83,1975 (ILLUS) •

FLAHAULTIA ARNAUD 1951. BULL. SOC. MYCOL. FRANCE 67:195 & FIG 4D-E (NO LATIN DIAGNOSIS) /F. HYALINA
AMERO,GLOIO,HYALO,RADUL, 30B, ON FUNGI (SEBACINA, TOMENTELLA). TYPE SPECIMEN WITH JACOBIA, SEE NICOT & CHARPENTIE 1971 (BULL. SOC. MYCOL. FRANCE 87:30). BASIDIO STATE =SEBACINA SEE ARNAUD 1951 & WATLING & KENDRICK 1979 (NATURALIST 104:1).
/DONK, TAXON 11:84-85,1962 (REFS) •

FLOCCARIA GREVILLE 1828. SCOTTISH CRYPT. FL. VOL 6, PL 301 /F. GLAUCA
?, PROBABLY A COREMIAL PENICILLIUM. •

FLOSCULOMYCES SUTTON 1978. MYCOLOGIA 70:788-790 & 792 (ILLUS) /F. FLORIDAENSIS
DICTYO,CETERI,PHAEO,NONSP OR MISC,ALEURI, 96F, ON LITTER (PODOCARPUS). •

FLUMINISPORA INGOLD 1958. TBMS 41:369-371 /F. OVALIS ,
=DIMORPHOSPORA TUBAKI 1958 FIDE KENDRICK. •

FONSECAEA NEGRONI 1936. REV. INST. BAC. DEP. NAC. HIG., BUENOS AIRES 7:424 (419-426,ILLUS) /HORMODENDRUM PEDROSOI BRUMPT =F. PEDROSOI (BRUMPT)
AMERO,CETERI OR GLOIO OR BLASTO,PHAEO,PHIAL, ON MAN & ANIMALS. H. PEDROSOI WAS TRANSFERRED TO RHINOCLADIELLA BY SCHOL-SCHWARZ 1968. DE HOOG (1977) MAINTAINED THIS DISPOSITION.
/SCHOL-SCHWARZ, ANTONIE VAN LEEUWENHOEK 34:119-152,1968 (ILLUS,REVIS) /CARRION & SILVA-HUTNER, INT. J. DERMATOLOGY 10:35-43,1971 (ILLUS) /BUTTERFIELD & JONG, MYCOPATHOLOGIA 58:59-62,1976 (ILLUS) /SZANISZLO, HSIEH & MARLOW, MYCOLOGIA 68:117-130,1976 (ILLUS OF YEAST & SCLEROTIC STATES) /DE HOOG, CBS STUDIES 15:89-100,1977 (ILLUS,REVIS) /COLE & SAMSON, CONIDIAL FUNGI P 39-40,1979 (ILLUS)
=*CARRIONIA* BRICENO-IRAGORRY 1938
=*HORMODENDROIDES* MOORE & ALMEIDA 1937
=*PHIALOCONIDIOPHORA* MOORE & ALMEIDA 1936 FIDE CARMICHAEL 1973 •

FONTANOSPORA DYKO 1978. TBMS 70:411-413 (ILLUS) /TRICLADIUM ECCENTRICUM PETERSON 1962 =T. ECCENTRICA (PETERSON)
STAURO,CETERI,HYALO,NONSP OR RACHI,AQUATIC, 119C. •

FOXIA CASTELLANI 1908. JOUR. TROP. MED. HYG. 11:260 (NO GENERIC DIAGNOSIS) /MICROSPORON "MANSONI" CASTELLANI 1905 (BRIT. MED. J. 2:1271,ILLUS) =F. MANSONII (CASTELLANI) =EXOPHIALA MANSONII (CASTELLANI) DE HOOG 1977
=MALASSEZIA (Y) FIDE MCGINNIS 1979
/SCHOL-SCHWARZ, ANTONIE VAN LEEUWENHOEK 34:119-152,1958 (ILLUS OF 'TYPE STRAIN' RECEIVED FROM CASTELLANI IN 1958, BUT WHICH DOES NOT FIT WITH 1905 DESCRIPTION.) /COOKE, MYCOPATHOLOGIA 17:34-36,1962 (REFS,SUB AUREOBASIDIUM) /DE HOOG, CBS STUDIES 15:100 & 114-117,1977 (ILLUS,REVIS) /MCGINNIS, SABOURAUDIA 17:145-154,1979 (DISCUSSION) •

FRACTISPORONITES CLARKE 1965 (MOUNTAIN GEOLOGIST 2:91) =FOSSIL FUNGI IMPERFECTI •

FRASERIELLA CIFERRI & CORTE 1957. IN CORTE, ATTI IST. BOT. UNIV. PAVIA, SER. 5, 14:107-109 (ILLUS) /F. BISPORA =F. STATE OF XEROMYCES BISPORA FRASER
?, ASCO STATE =XEROMYCES FIDE CIF. & CORTE 1957. THE CONIDIAL STATE IS CLOSE TO EPIDERMOPHYTON. •

FREYNELLA KUNTZE =COCCOSPORIUM FIDE AINSWORTH 1971 •

FUCKELINA SACCARDO 1875. NUOV. GIORN. BOT. ITAL. 7:326 /F. SOCIA
=STACHYBOTRYS CORDA 1837 FIDE BOOTH 1957. •

FULVIA CIFERRI 1953. ATTI IST. BOT. UNIV. PAVIA, SER. 5, 10:246 /CLADOSPORIUM FULVUM COOKE 1883 =F. FULVA (COOKE)
PHRAGMO,BLASTO,PHAEO,RADUL,STROMA, 39A, ON LEAVES (LYCOPERSICON, BERKHEYA). COMPARE WITH STENELLA & CLADOSPORIUM.
/ELLIS, DH P 306-307,1971 (ILLUS) /ELLIS, MDH P 315,1976 (ILLUS,NEW COMB) /HOLLIDAY & MULDER, CMI DESCRIPTIONS 487,1976 (ILLUS) •

FUMAGO PERSOON 1822. MYCOL. EUROP. 1:9 /F. VAGANS (LECTOTYPE FIDE HUGHES 1958)
=NOMEN CONFUSUM FIDE FRIEND 1965 (TBMS 48:371-375).
/HUGHES, MYCOLOGIA 68:773-774,1976 (REFS) •

FUMAGOPSIS SPEGAZZINI 1910. AN. MUS. NAC. BUENOS AIRES 20(SER.3,13):464-465 (ILLUS) /F. TRIGLIFIOIDES
=TRIDENTARIA PREUSS 1852 FIDE KENDRICK. •

FURCASPORA BONAR 1965. MYCOLOGIA 57:391-392 (ILLUS) /F. PINICOLA
STAURO,GLOIO,HYALO,NONSP?,ACERVULI, ON LITTER (PINUS). PERHAPS BETTER CONSIDERED A COELOMYCETE. •

FUSARIELLA SACCARDO 1884. ATTI R. ISTIT. VEN. SCI. LETT. ARTI VI, 2:463 /F. ATROVIRENS
PHRAGMO,BASO,PHAEO,PHIAL, 73B, ON LEAVES & LITTER (MANY PLANTS).
/HUGHES, CMI PAPERS 28:1-11,1949 (ILLUS,REVIS) /ELLIS, DH P 527-529,1971 (ILLUS,KEY) /MOUCHACCA & NICOT, REV. MYCOL. 37:168-182,1972 (ILLUS,NEW SPP) /MATSUSHIMA, ICONES P 71-72 & PL,1975 (ILLUS) /ELLIS, MDH P 458-461,1976 (ILLUS,KEY,NEW COMB,NEW SP)
=*KURSSANOVIA* PIDOPLICHKO 1948 FIDE VON ARX 1974
=*KUTILAKESA* SUBRAMANIAN 1956 FIDE VON ARX 1974 •

FUSARIUM LINK 1809. MAG. GES. NATURF. FREUNDE, BERLIN 3:10 /F. ROSEUM
PHRAGMO,GLOIO,HYALO,PHIAL,SPORODOCHIA, 73H, ON PLANTS (MANY), LITTER & SOIL. OTHER CONIDIAL STATE ="MICROCONIDIA". ASCO STATE =GIBBERELLA, NECTRIA, CALONECTRIA, NECTRIOPSIS, MICRONECTRIELLA, PLECTOSPHAERELLA OR GRIPHOSPHAERIA FIDE VON ARX 1974 (GENERA P 130-131).
/BOOTH & WATERSTON, CMI DESCRIPTIONS 21-30,1964 (ILLUS) /TOUSSOUN & NELSON, IDENTIFICATION OF FUSARIUM SPECIES, PENN. STATE UNIV. PRESS, 51 P,1968 (ILLUS) /CMI DESCRIPTIONS 211-220,1970 (ILLUS) /BOOTH, CMI DESCRIPTIONS 308-310,1971 & 384-385,1973 (ILLUS) /BOOTH, THE GENUS FUSARIUM, CMI, 236 P,1971 (MONOGR) /JOFFE, MYCOPATHOLOGIA 53:201-208,1974 (ILLUS,REVIS) /MATSUSHIMA, ICONES P

72-73 & PL,1975 (ILLUS,NEW SP) /BOOTH, FUSARIUM, LABORATORY GUIDE TO THE IDENTIFICATION OF THE MAJOR SPECIES, CMI, 58 P,1977 (ILLUS,KEY)

=*CEPHALODIPLOSPORIUM* KAMYSCHKO 1961 FIDE GAMS 1971
=*DISCOFUSARIUM* PETCH 1921 FIDE CLEMENTS & SHEAR 1931
=*FUSIDOMUS* GROVE 1929 FIDE SUTTON 1977
=*LACHNIDIUM* GIARD FIDE SACCARDO FROM AINSWORTH 1971
=*PIONNOTES* FRIES 1849 FIDE WOLLENWEBER FROM AINSWORTH 1971
=*PSEUDOMICROCERA* PETCH 1921 FIDE WOLLENWEBER FROM CLEMENTS & SHEAR 1931
=*RACHISIA* LINDNER 1913 FIDE CLEMENTS & SHEAR 1931
=*SELENOSPORIUM* CORDA 1837 FIDE LINDAU IN RABENHORST 1910
=*SPOROTRICHELLA* KARSTEN 1887 FIDE HUGHES 1958
=*STAGONOSTROMA* DIEDICKE 1914 FIDE SUTTON 1977 •

FUSCOPHIALIS SUTTON 1977. BOL. SOC. ARGENT. BOT. 18:154-161 (ILLUS) /F. BRASILIENSIS
?, ON LITTER (EUCALYPTUS). •

FUSELLA SACCARDO 1886. SYLLOGE FUNGORUM 4:246 (NO ILLUS) /THREE ORIGINAL SPECIES
? •

FUSICHALARA HUGHES & NAG RAJ 1973. NEW ZEALAND J. BOT. 11:661-665 (ILLUS) /F. DIMORPHOSPORA
PHRAGMO,BASO,HYALO OR PHAEO,PHIAL, 73A, ON LITTER (WEINMANNIA, MELICYTUS & OTHERS).
/NAG RAJ & KENDRICK, CHALARA P 144-147,1975 (KEY,ILLUS) /GAMS & HERMANIDES-NIJHOF, CBS STUDIES 13:76-77,1976 (ILLUS,NEW SP) •

FUSICLADIELLA HOHNEL 1919. BER. DEUTSCH. BOT. GES. 37:155 /F. ARONICI (SACCARDO) =F. MELAENA (FUCKEL) HUGHES 1952
DIDYMO,CETERI,HYALO,MISC,STROMA, 64E, ON LEAVES. ASCO STATE =MYCOSPHAERELLA FIDE HOHNEL 1919 OR DIDYMOSPHAERIA SEE DE HOOG 1977 (CBS STUDIES 15:139).
/HUGHES, CMI PAPERS 49:20-24,1952 (ILLUS) /DEIGHTON, CMI PAPERS 101:23-28,1965 (ILLUS) /ELLIS, DH P 169-170,1971 (ILLUS) •

FUSICLADIITES BABAYAN & TASLAKHCHYAN 1973 (MIKOL. FITOPATOL. 7:181) =FOSSIL FUNGI IMPERFECTI •

FUSICLADINA ARNAUD 1952. BULL. SOC. MYCOL. FRANCE 68:200 & 215, FIG 8C /F. LASIOSPHAERIAE
?, ORIGINAL ILLUSTRATION NOT DIAGNOSTIC. NO TYPE SPECIMEN, SEE NICOT & CHARPENTIE 1971 (BULL. SOC. MYCOL. FRANCE 87:30). •

FUSICLADIOPSIS MAIRE 1907. BULL. SOC. BOT. FRANCE 53:CLXXXVII (DATED 1906) /F. CONVIVA =STEMPHYLIUM BOTRYOSUM WALLROTH
=STEMPHYLIUM WALLROTH 1833 FIDE HUGHES 1958, FOR FUSICLADIOPSIS KARAK. & VASSIL. SEE KARAKULINIA. •

FUSICLADIUM BONORDEN 1851. HANDB. ALLGEM. MYKOL. P 80 /F. VIRESCENS =F. STATE OF VENTURIA PIRINA ADERH. FIDE HUGHES 1958
AMERO OR DIDYMO,CETERI,PHAEO,RADUL,STROMA, 22B, ON LEAVES, BRANCHES & OTHER PARTS (PRUNUS, PYRUS, BETULA, HEVEA & OTHERS). ASCO STATE =VENTURIA FIDE HUGHES 1958, OR APIOSPORINA OR MICROCYCLUS FIDE ELLIS 1976.
/DEIGHTON, CMI PAPERS 112:16-27,1967 (ILLUS) /BARRON, HYPHOMYCETES P 166-167,1968 (ILLUS) /SUTTON & WATERSTON, CMI DESCRIPTIONS 224-225,1970 (ILLUS) /ELLIS, DH P 271-273,1971 (ILLUS) /HUGHES & PIROZYNSKI, CAN. J. BOT. 50:2528-2530,1972 (ILLUS, NEW COMB) /CORLETT, FUNGI CANADENSES NO. 36,1974 (ILLUS) /CMI DESCRIPTIONS 224-225,1970 & 402 & 404,1974 (ILLUS) /ELLIS, MDH P 237-240,1976 (ILLUS,KEY,NEW SP,NEW COMB)

=*MEGACLADOSPORIUM* VIENNOT-BOURGIN 1949 FIDE HUGHES 1958 •

FUSICOLLA BONORDEN 1851. HANDB. ALLGEM. MYKOL. P 150 & FIG 248 /F. BETAE (DESMAZIERES)
?

/LINDAU, RABENHORST KRYPT-FL. 2 AUFL. 1 BD. 9:455,1910 (ILLUS) •

FUSIDIUM LINK 1809. MAG. GES. NATURF. FREUNDE, BERLIN 3:8 /F. GRISEUM (LECTOTYPE FIDE HUGHES 1958)
AMERO,BLASTO,HYALO,NONSP, 38A, ON LITTER (BROAD LEAF TREES) & LEAVES?. ASCO STATE =BIOSTICTUS SEE KORF 1973 (THE FUNGI IVA P 282). PHIALIDIC SPECIES ORIGINALLY PLACED IN FUSIDIUM HAVE BEEN TRANSFERRED TO ACREMONIUM (SEE GAMS 1971, P 18).
/BARRON, HYPHOMYCETES P 167-169,1968 (ILLUS) /LECHEVALIER ET AL, MYCOLOGIA 69:81-95,1977 (ILLUS,NEW SP,NOT NAMED)
=*CYLINDRIUM* BONORDEN 1851 FIDE HUGHES 1958
?=*HORMIACTELLA* SACCARDO 1886 •

FUSIDOMUS GROVE 1920 =FUSARIUM FIDE SUTTON 1977 (CMI PAPERS 141:88 & 200-201) •

FUSISPORELLA SPEGAZZINI 1910. AN. MUS. NAC. BUENOS AIRES 20(SER.3,13):454 (ILLUS) /F. BUFONIS
DIDYMO,BASO,HYALO,PHIAL,SPORODOCHIA, 58F, ASCO STATE ?=NECTRIA FIDE SAMUELS 1976
/SAMUELS, MEM. N. Y. BOT. GARD. 26:83-85,1976 (ILLUS SUB NECTRIA PALLIDULA) •

FUSISPORIUM LINK 1809. MAG. GES. NATURF. FREUNDE, BERLIN 3:19 (NO ILLUS) /F. AURANTIACUM
?, HUGHES (1958) DID NOT SEE THE TYPE. •

FUSOMA CORDA 1837. ICONES FUNGORUM 1:7 /TWO ORIGINAL SPECIES
=NOMEN DUBIUM FIDE HUGHES 1958. •

GABARNAUDIA SAMSON & GAMS 1974. CBS STUDIES 6:88-92 (ILLUS,KEY) /OOSPORA BETAE DELACROIX 1897 (BULL. SOC. MYCOL. FRANCE 13:116) =G. BETAE (DELACROIX)
?=ACREMONIUM (SEE 15F) OR SAROCLADIUM. ASCO STATE =SPHAERONAEMELLA FIDE SAMSON & GAMS 1974. •

GAMSIA MORELET 1969. ANN. SOC. SCI. NAT. ARCHEOL. TOULON VAR. 21:105 (NO ILLUS) /G. DIMERA AS '(GAMS) MORELET' =G. STATE OF WARDOMYCES DIMERUS GAMS 1968 (TBMS 51:800-802,ILLUS)
DIDYMO,BASO,HYALO,ANNEL, 59C, ON SOIL. SAME AS GYMNODOCHIUM BUT LACKING SPORODOCHIA. OTHER CONIDIAL STATE =WARDOMYCES.
/ELLIS, MDH P 349-350,1976 (ILLUS) •

GANGLIOPHRAGMA SUBRAMANIAN 1977. KAVAKA 5:93-98 (NOT SEEN) /DRECHSLERA RHOPALOTA DRECHSLER =G. RHOPALOTA (DRECHSLER)
?, SEE ASTRACTS OF MYCOLOGY 67-52149,1979. •

GANGLIOSTILBE SUBRAMANIAN & VITTAL 1976. IN VITTAL, KAVAKA 3:69-71 (DATED 1975,ILLUS) /G. INDICA
PHRAGMO,CETERI?,PHAEO,NONSP?,SYNNEMATA, 87F, ON LITTER (BAMBUSA). •

GARNAUDIA BOROWSKA 1977. ACTA MYCOL., WARSAWA 13:169-174 (ILLUS) /G. ELEGANS
DIDYMO,CETERI,PHAEO,MISC, ON WOOD & BARK (TILIA). •

GEMMOPHORA SCHKORBATOW 1912. BER. DEUTSCHL. BOT. GES. 30:478-482 (ILLUS) /G. PURPURASCENS
?, =STERILE? •

GEMMULARIA RAFINESQUE 1819. J. PHYS. CHIM. HIST. NAT. 89:106 /G. RUGOSA (LECTOTYPE FIDE DONK 1962) =G. STATE OF PORIA COCOS
=SCLEROTIA OF POLYPORES, SEE DONK 1962 (TAXON 11:85). BASIDIO STATE =PORIA AND OTHERS.
=*TUCAHUS* RAFINESQUE 1830 FIDE DONK 1962
=*RUGOSARIA* RAFINESQUE 1833 FIDE DONK 1962
=*PACHYMA* FRIES 1822 FIDE DONK 1962 •

GENICULARIA RIFAI & COOKE 1966. TBMS 49:153 /G. CYSTOPORIA (DUDDINGTON)
=NOMEN ILLEGITIMUM, FIDE RIFAI 1975 (NON GENICULARIA DE BARY 1858). SEE

GENICULIFERA. •

GENICULIFERA RIFAI 1975. MYCOTAXON 2:214-216 /TRICHOTHECIUM CYSTOSPORIUM DUDDINGTON 1951 =G. CYSTOSPORIA (DUDDINGTON)
=ARTHROBOTRYS CORDA SEE CARMICHAEL 1973 & SCHENCK ET AL 1977.
/SCHENCK, KENDRICK & PRAMER, CAN. J. BOT. 55:982,1977 (DISCUSSION)
=*GENICULARIA* RIFAI & COOKE 1966 NON DE BARY 1858 •

GENICULODENDRON SALT 1974. TBMS 63:339 (339-351, ILLUS) /G. PYRIFORME =G. STATE OF CALOSCYPHA FULGENS FIDE PADEN, SUTHERLAND & WOODS 1978 (CAN. J. BOT. 56:2375-2379,ILLUS)
AMERO,CETERI,HYALO,RADUL, 26D, ON SEEDS & LITTER. ASCO STATE =CALOSCYPHA. •

GENICULOSPORA NILSSON EX MARVANOVA & NILSSON 1971. TBMS 57:532 /G. INFLATA (INGOLD) MARVANOVA & NILSSON 1971
=ARTICULOSPORA INGOLD 1942 FIDE INGOLD 1975 (GUIDE P 22).
/NOLAN, MYCOLOGIA 64:1169-1174,1972 (ILLUS,REVIS) •

GENICULOSPORIUM CHESTERS & GREENHALGH 1964. TBMS 47:400 (393-401) & PL 18 /G. SERPENS =NODULISPORIUM CORTICIOIDES (FERR. & SACC) HUGHES =GENICULOSPORIUM STATE OF HYPOXYLON SERPENS (PERS. EX FRIES) KICKX 1835 SEE CARMICHAEL 1973
AMERO,CETERI,HYALO,RACHI, 25G, ON WOOD & BRANCHES (DECIDUOUS TREES). ASCO STATE =HYPOXYLON SEE KENERLEY & ROGERS 1976 (MYCOLOGIA 68:688-691) OR WHALLEY 1976 (TBMS 67:515-517,ILLUS). COMPARE WITH DEMATOPHORA. SEE ALSO REFERENCES UNDER NODULISPORIUM.
/ELLIS, DH P 229-230,1971 (ILLUS) /DE HOOG & HERMANIDES-NIJHOF, CBS STUDIES 15:198,1977 (NEW COMB) /ROGERS, MYCOLOGIA 71:1-42,1979 (DISCUSSION, REFS) •

GEOMYCES TRAAEN 1914. NYT. MAG. NATURV. 52:28-31 & TAF. IV, FIGS 1 & 2 /G. AURATUS (LECTOTYPE) =G. PANNORUS (LINK) SIGLER & CARMICHAEL 1976
AMERO,ARTHRO,HYALO,NONSP,ALEURI, 41C, ON SOIL & MANY OTHER SUBSTRATES. ASCO STATE =PSEUDOGYMNOASCUS FIDE SIGLER & CARMICHAEL 1976.
/SIGLER & CARMICHAEL, MYCOTAXON 4:376-380,1976 (ILLUS,REVIS,NEW SP) •

GEOSMITHIA PITT 1979. CAN. J. BOT. 57:2021-2031 (ILLUS,KEY) /PENICILLIUM LAVENDULUM RAPER & FENNELL 1948 =G. LAVENDULA (RAPER & FENNELL)
AMERO,BASO,HYALO,PHIAL, ON MANY SUBSTRATES. ASCO STATE =TALAROMYCES FIDE PITT. •

GEOTRICHELLA ARNAUD 1954. BULL. SOC. MYCOL. FRANCE 69:273-274 (ILLUS,NO LATIN DIAGNOSIS) /G. ALTERNATA
?=COREMIELLA BUBAK & KRIEGER 1912. NO TYPE SPECIMEN -SEE NICOT & CHARPENTIE 1971 (BULL. SOC. MYCOL. FRANCE 87:31).
/TUBAKI, IN KOBAYASI ET AL, ANN. REP. INST. FERMENTATION, OSAKA NO. 3:116-117,1967 (ILLUS,NEW SP) •

GEOTRICHUM LINK 1809. MAG. GES. NATURF. FREUNDE, BERLIN 3:17 /G. CANDIDUM =G. STATE OF DIPODASCUS GEOTRICHUM
AMERO,ARTHRO OR GLOIO,HYALO,NONSP, 42G, ON SOIL, LITTER, PLANTS (CITRUS & LYCOPERSION), MILK & MAN. SEE ALSO ENTRY UNDER "ARTHROCONIDIA". ASCO STATE =DIPODASCUS FIDE VON ARX 1977 OR ?DIPLOGELASINOSPORA SEE MATSUSHIMA 1975 (P 173-174) & UDAGAWA & HORIE 1972 (J. JAPANESE BOT. 47:297-305,ILLUS) OR ZENDERA, GALACTOMYCES OR MAGNUSIOMYCES FIDE REDHEAD & MALLOCH 1977.
/CARMICHAEL, MYCOLOGIA 49:820-830,1957 (ILLUS,REVIS) /VOROS-FELKAI, ACTA MICROBIOL. ACAD. SCI. HUNGARICAE 8:95-101,1961 (NEW SP,NO ILLUS) /FENG, YEN & YUE, ACTA MICROBIOL. SINICA 12:64-73,1966 (REVIS) /BARRON, HYPHOMYCETES P 172-173,1968 (ILLUS,REFS) /BUTLER & PETERSON, MYCOLOGIA 64:365-375,1972 (SUB ENDOMYCES) /KREGER-VAN RIJ & VEENHUIS, CAN. J. BOT. 50:1691-1695,1972 (ILLUS,ASCO STATE) /VON ARX, GENERA P 183-184,1974 GUEHO & BUISSIERE, ANNAL. MICROBIOLOGIE A126:483-500,1975 (BIOCHEMICAL IDENTIFICATION,NOT SEEN) /SIGLER & CARMICHAEL, MYCOTAXON 4:380-383,1976 (REFS) /REDHEAD & MALLOCH, CAN. J. BOT. 55:1705-1707,1977 (REVIS,NEW SP, NO NAME,NO ILLUS) /VON ARX ET AL, CBS STUDIES 14:32-36,1977 (NEW COMBS) /VON ARX, ANTONIE VAN LEEUWENHOEK 43:333-340,1977 (ILLUS,NEW COMBS, NEW SP)
=*OOSPOROIDEA* SUMSTINE 1913 FIDE CARMICHAEL 1957

=*POLYMORPHOMYCES* COUPIN 1914 FIDE CARMICHAEL 1973 •

GIBELLULA CAVARA 1894. ATTI IST. BOT. PAVIA 2:347 & TAB XXVI FIGS 13-19 /G. PULCHRA (SACCARDO)
AMERO,BASO,HYALO,PHIAL,SYNNEMATA, 10D, ON SPIDERS. ASCO STATE =TORRUBIELLA FIDE MORRIS 1963 OR CORDYCEPS SEE DE HOOG 1977 (CBS STUDIES 15:138). OTHER CONIDIAL STATE =DACTYLARIA-LIKE.
/MORRIS, WESTERN ILLINOIS UNIV. SER. BIOL. SCI. 3:58-59,1963 (ILLUS) /BALAZY, ACTA MYCOL. 6:71-76,1970 (ILLUS) /SAMSON & EVANS, ACTA BOT. NEERL. 22:522-528,1973 (ILLUS) /SAMSON & EVANS, PROC. KON. NEDERL. AKAD. WETEN., AMSTERDAM, SER C, 80:130-133,1977 (ILLUS,NEW COMB)
=*SYNSTERIGMATOCYSTIS* COSTANTIN FIDE PETCH FROM AINSWORTH 1971 •

GIBELLULOPSIS BATISTA & MAIA 1959. ANN. SOC. BIOL. PERNAMBUCO 16:153 /G. PISCIS
=VERTICILLIUM NEES 1817 FIDE CARMICHAEL 1973. •

GILCHRISTIA CIFERRI & REDAELLI 1934. IN REDAELLI & CIFERRI, J. TROP. MED. HYG. 37:281 /G. DERMATITIDIS (GILCHRIST & STOKES)
=CHRYSOSPORIUM CORDA 1833 FIDE CARMICHAEL 1962. •

GILLETIA TORREND 1914. BULL. JARD. BOT. BRUX. 4(1):12-13 (NO ILLUS) /G. CINNABARINA
=TELLIGIA HENDRICKX 1948, NON GILLETIA SACCARDO & PENZIG 1882 (PERONOSPORALES). •

GILMANIELLA BARRON 1964. MYCOLOGIA 56:514 /G. HUMICOLA
AMERO,BASO OR CETERI OR GLOIO,PHAEO,MISC, 49B, ON SOIL & DUNG. COMPARE WITH HUMICOLA.
/BARRON, HYPHOMYCETES P 173-174,1968 (ILLUS) /ELLIS, DH P 61-62,1971 (ILLUS) /MOUSTAFA, PERSOONIA 8:332-333,1975 (HUMICOLA GRISEA TRAAEN 1914 IS DESCRIBED AS GILMANIELLA MACROSPORA SP. NOV.) /MORINAGA, MINOURA & UDAGAWA, TRANS. MYCOL. SOC. JAPAN 19:147-148,1978 (ILLUS,NEW SP)
=*ADHOGAMINA* SUBRAMANIAN & LODHA 1964 FIDE BARRON 1968 •

GLENOSPORA BERKELEY & DESMAZIERES 1849 =SEPTOBASIDIUM PATOUILLARD 1892 NOMEN CONSERVANDUM, SEE CARMICHAEL 1962 (CAN. J. BOT. 40:1445) •

GLENOSPORELLA NANNIZZI 1931. IN AGOSTINI, ATTI IST. BOT. UNIV. PAVIA, SER IV, 2:98 & IN ATTI ACCAD. FISEOCR. SIENA, SER. 10, 6:269 /G. ALBISCANS (NIEUWENHUIS) (LECTOTYPE)
=CHRYSOSPORIUM CORDA 1833 FIDE CARMICHAEL 1962. •

GLENOSPOROPSIS FONSECA 1943 =NOMEN NUDUM FIDE CARMICHAEL 1962 •

GLIOBOTRYS HOHNEL 1902. S.B. AKAD. WISS. WIEN, ABT 1, 111:1048-1049 /G. ALBOVIRIDIS
=STACHYBOTRYS CORDA 1837 FIDE ELLIS 1971. •

GLIOCEPHALIS MATRUCHOT 1899. BULL. SOC. MYCOL. FRANCE 15:254-255, PL XIV /G. HYALINA
AMERO,GLOIO,HYALO,PHIAL, 8B, ON ROOTS (BEETS, POTATOES).
/BARRON, HYPHOMYCETES P 174-176,1968 (ILLUS) •

GLIOCEPHALOTRICHUM ELLIS & HESSELTINE 1962. BULL. TORREY BOT. CLUB 89:21-27 (ILLUS) /G. BULBILIUM
AMERO,GLOIO,HYALO,PHIAL,SETAE, 8E, ON FRUIT (MALUS), WOOD & SOIL. OTHER STATE ="BULBILS".
/BARRON, HYPHOMYCETES P 175-177,1968 (ILLUS) /WILEY & SIMMONS, MYCOLOGIA 63:575-585,1971 (ILLUS,KEY) /ELLIS, DH P 552,1971 (ILLUS)
=*ANTENAGLIUM* ALBUQUERQUE 1969 FIDE KENDRICK •

GLIOCLADIOPSIS SAKSENA 1954. MYCOLOGIA 46:662 /G. SAGARIENSIS
=CYLINDROCLADIUM MORGAN 1892 FIDE BARRON 1968. •

GLIOCLADIUM CORDA 1840. ICONES FUNGORUM 4:30-31, TAF VII, FIG 92 /G. PENICILLIOIDES =G. STATE OF NECTRIA PALLIDULA FIDE BOOTH 1959
AMERO,GLOIO,HYALO OR PHAEO,PHIAL, 7A, ON SOIL & FUNGI (SCLEROTINIA, PHOMOPSIS, FUSARIUM & OTHERS). ?=CLONOSTACHYS CORDA 1839. ASCO STATE =NECTRIA

FIDE BOOTH 1959 (CMI PAPERS 73:35-37), OR HYPOCREA FIDE MULLER ET AL 1972 (TBMS 58:1-4, ILLUS), OR HYPOMYCES FIDE SAMUELS 1970 (MEM. N. Y. BOT. GARD. 26:9). OTHER STATE ?=VALDENSIA (H) FIDE REDHEAD & PERRIN 1972.
/BARRON, HYPHOMYCETES P 177-179,1968 (ILLUS) /UDAGAWA & HORIE, J. GEN. APPL. MICROBIOL. 17:143-146,1971 (ILLUS SUB NECTRIA GLIOCLADIOIDES) /MATSUSHIMA, ICONES P 73-75 & PL,1975 (ILLUS,NEW SP) /HUANG, CAN. J. BOT. 56:2243-2246,1978 (PARASITISM)
?=*CLONOSTACHYS* CORDA 1839
=*CORYMBOMYCES* APPEL & STRUNK 1903 FIDE CARMICHAEL 1973
?=*FLAHAULTIA* ARNAUD 1951
=*LYSIPENICILLIUM* BREFELD 1908 FIDE RAPER & THOM FROM AINSWORTH 1971 •

GLIOCLADOCHIUM HOHNEL 1926. MITT. BOT. LAB. TECH. HOCHSCH. WIEN, BD. 3, HEFT 1:4 /G. TOMENTOSUM (FRIES) =PERIOLA TOMENTOSA 1823
? •

GLIODENDRON SALONEN & RUOKOLA 1969. MYCOPATHOLOGIA 38:332 /G. BALNICOLA (AS 'BALNICOLUM')
=STERIGMATOBOTRYS OUDEMANS 1886 FIDE KENDRICK. •

GLIOMASTIX GUEGUEN 1905. BULL. SOC. MYCOL. FRANCE 21:240 /G. CHARTARUM =G. MURORUM (CORDA) HUGHES 1958
AMERO,BASO OR GLOIO,PHAEO,PHIAL, 2B, ON LITTER, WOOD, SOIL & CELLULOSIC MATERIALS. ASCO STATE =WALLROTHIELLA FIDE HUGHES & DICKINSON 1968.
/DICKINSON, CMI PAPERS 115:1-24,1968 (ILLUS,KEY,REVIS) /GAMS, CEPHALO. SCHIMMEL. P 73-99,1971 (REVIS,ILLUS,KEY) /ELLIS, DH P 517-521,1971 (ILLUS,KEY) /MATSUSHIMA, ICONES P 75-77 & PL,1975 (ILLUS,NEW SPP) /ELLIS, MDH P 448-451,1976 (ILLUS,KEY,NEW COMB) /SCHOKNECHT & CRANE, MYCOLOGIA 69:537-539,1977 (ILLUS,NEW COMB)
=*BASITORULA* ARNAUD 1954 FIDE DICKINSON 1968
?=*HAPLOTRICHELLA* ARNAUD 1954
?=*SAGRAHAMALA* SUBRAMANIAN 1972
?=*NEMATOMYCES* FAUREL & SCHOTTER 1965
?=*TORULINA* SACCARDO & SACCARDO 1906 FIDE DICKINSON 1968 •

GLIOPHRAGMA SUBRAMANIAN & LODHA 1964. CAN. J. BOT. 42:1057-1060 (ILLUS) /G. SETOSUM
=WIESNERIOMYCES KOORDERS 1907 FIDE KENDRICK. •

GLIOSTROMA CORDA 1837. ICONES FUNGORUM 1:5 & TAB 1, FIG 46 /G. PEZIZOIDES
?, =AEGERITA FIDE SACCARDO FROM AINSWORTH 1971. •

GLIOTRICHUM ESCHWEILER 1824. SYLL. PLANT. A SOC. BOT. R. RATISB. EDITA 1:166 /G. CASSELIAE
=NOMEN DUBIUM FIDE HUGHES 1958. •

GLISCHRODERMA FUCKEL 1870. JB. NASSAU VER. NATURK. 23-24:34-35 /G. CINCTUM
?=OSTRACODERMA SEE HENNEBERT 1973 (SEE 33D).
/HENNEBERT, PERSOONIA 7:201 & FIG 10, P 186,1973 (ILLUS,REVIS) •

GLOEOCERCOSPORA BAIN & EDGERTON 1943 EX DEIGHTON 1971. TBMS 57:358-359 /G. SORGHI BAIN & EDGERTON 1943
?, DIFFERENCE FROM RAMULISPORA NOT CLEAR.
/RAWLA, TBMS 60:283-292,1973 (ILLUS) /MULDER & HOLLIDAY, CMI DESCRIPTIONS 300,1971 (ILLUS) •

GLOEOSPORIELLA CAVARA 1892. ATT IST. BOT. UNIV. PAVIA, SER II, 2:207-292 & FUNGI LOMBARDI EXS. NO. 41,1891 /G. ROSICOLA
STAURO,GLOIO,HYALO,NONSP,APPEND,SPORODOCHIA, 56B, ON SPINES OF ROSA.
/SUTTON, NOVA HEDWIGIA 25:161-165,1974 (ILLUS,REVIS) •

GLOEOSPORIUM AUCT. =COLLETOTRICHUM & OTHERS, SEE VON ARX 1970 (A REVISION OF THE FUNGI CLASSIFIED AS GLOEOSPORIUM, BIBLIOTHECA MYCOLOGICA (CRAMER) BD. 24:3-4) & COLLETOCONIS DE HOOG 1978 (PERSOONIA 10:48-51). •

GLOEOSPORIOPSIS SPEGAZZINI 1910 =COLLETOTRICHUM CORDA 1831 FIDE VON ARX 1970 •

GLOIOSPHAERA HOHNEL 1902. S.B. AKAD. WISS. WIEN. 111:1038-1039 /G. GLOBULIGERA =G. CLERCIANA (BOUDIER 1901) HOHNEL 1905
AMERO,GLOIO,HYALO,PHIAL, 8D, ON WOOD (CARPINUS).
/WANG, MYCOLOGIA 63:890-893,1971 (ILLUS) /POLLACK & MCKNIGHT, MYCOLOGIA 64:415-421,1972 (ILLUS,REVIS)
=*SCOPULA* ARNAUD 1952 FIDE WANG 1971 •

GLOMERULARIA PECK 1879. REP. N. Y. MUSEUM 32:43 /G. CORNI
=GLOMOPSIS, NON GLOMERULARIA KARSTEN 1849. •

GLOMOPSIS HENDERSON 1961. NOTES ROY. BOT. GARDEN EDINBURGH 23:500 /G. CORNI (PECK)
BULBILS OR STAURO,CETERI,HYALO,NONSP,APPEND, 114C, ON LEAVES (CORNUS & LONICERA). NOM. NOV. FOR GLOMERULARIA PECK 1879 NON KARSTEN 1849. COMPARE WITH CANDELABRUM. BASIDIO STATE =HERPOBASIDIUM FIDE GOULD FROM SUTTON 1973.
/SUTTON, CMI PAPERS 132:65-68,1973 (ILLUS)
=*GLOMERULARIA* PECK 1879 •

GLOMOSPORA HENDERSON 1961. NOTES ROY. BOT. GARDEN EDINBURGH 23:497-499 (ILLUS) /G. EMPETRI
AMERO,CETERI,HYALO,AMPUL?,ALEURI?,CLAMPS, 52E, ON LEAVES (EMPETRUM). •

GLOMUS TULASNE & TULASNE (?ENDOGONE) SEE MYCOTAXON 6:29-32,1977 •

GLYCOPHILA - SEE GLYCYPHILA •

GLYCYOPHILA - SEE GLYCYPHILA •

GLYCYPHILA MONTAGNE 1851. IN PAYEN, C. R. ACAD. SCI. 33:393-397 (NO ILLUS) /TWO ORIGINAL SPECIES, G. ERYTHROSPORA & G. ELAEOSPORA.
?, THE NAME HAS ALSO BEEN CITED AS GLYCYOPHILA & GLYCOPHILA. •

GOIDANICHIA ARNAUD 1954. BULL. SOC. MYCOL. FRANCE 69:280-281 (ILLUS) /SCOPULARIA SCOPULA GOIDANICH 1933 (ANN. MYCOL. 31:137 & TAV. 3, FIGS 1-4) =G. SCOPULA (GOIDANICH)
=GOIDANICHIELLA ARNAUD EX BARRON 1968, NON GOIDANICHIA TOMASELLI & CIFERRI 1952 (LICHEN).
/NICOT & CHARPENTIE, BULL. SOC. MYCOL. SOC. FRANCE 87:31,1971 (TYPE SPECIMEN) •

GOIDANICHIELLA ARNAUD EX BARRON 1968. HYPHOMYCETES P 180-182 (ILLUS, NO LATIN DIAGNOSIS FOR GENUS) /SCOPULARIA SCOPULA GOIDANICH 1933 G. SCOPULA (GOIDANICH) BARRON 1968 AS G. SCOPULA "(PREUSS) ARNAUD"
AMERO,GLOIO,HYALO,PHIAL, 8F, ON SOIL.
/MATSUSHIMA, ICONES P 77-78 & PL 289,1975 (ILLUS,NEW SP)
=*GOIDANICHIA* ARNAUD 1954 NON TOMASELLI & CIFERRI 1952 (A LICHEN) •

GOLOVINIA MECHTIEVA 1967. MIKOL. I FITOPATOL. 1:275 /G. BEMBICOIDES (DRECHSLER) =DACTYLELLA BEMBICOIDES
?, A SEGREGATE FROM DACTYLELLA GROVE 1884. •

GOMPHINARIA PREUSS 1851. LINNAEA 24:130 (ILLUS. IN STURM'S DEUTSCHL. FL., BD. 6, HEFT 35-36, TAB 60, 1862) /G. AMOENA
=NOMEN DUBIUM SEE DE HOOG & HERMINIDES-NIJHOF 1977 (CBS STUDIES 15:183). •

GONATOBOTRYS CORDA 1839. PRACHT-FLORA .... P 9 /G. SIMPLEX
AMERO,CETERI,HYALO,AMPUL, 32A, ON FUNGI (ALTERNARIA, CLADOSPORIUM, HELMINTHOSPORIUM) ON LITTER.
/BARRON, HYPHOMYCETES P 182-183,1968 (ILLUS) /GAMS & DOMSCH, NOVA HEDWIGIA 18:12-13,1969 (ILLUS) /MATSUSHIMA, ICONES P 78 & PL,1975 (ILLUS)
=*DESMOTRICHUM* LEVEILLE 1843 FIDE LINDAU 1907 •

GONATOBOTRYUM SACCARDO 1880. MICHELIA 2:24-25 /GONATOBOTRYS FUSCA SACCARDO 1877 (MICHELIA 1:84) =G. FUSCUM (SACCARDO) SACCARDO 1886 (SYLL. FUNG. 4:278)
AMERO,BLASTO,PHAEO,AMPUL, 32E, ON FUNGI (CERATOCYSTIS & OTHERS) ON WOOD.

PERHAPS THE SAME AS GONATORRHODUM CORDA.
/BARRON, HYPHOMYCETES P 183-184,1968 (ILLUS) /KENDRICK ET AL, CAN. J. BOT. 46:591-596,1968 (ILLUS) /ELLIS, DH P 173-174,1971 (ILLUS,KEY) /COLE, CAN. J. BOT. 51:1677-1684,1973 (ILLUS) /COLE & SAMSON, CONIDIAL FUNGI P 43-46,1979 (ILLUS CONIDIUM ONTOGENY)
=*THOMIELLA* DODGE 1935 FIDE CARMICHAEL
=*CHRISTIASTER* KUNTZE FIDE AINSWORTH 1971
?=*DICELLISPORA* SAWADA 1944 •

GONATOPHRAGMIUM DEIGHTON 1969. IN CEJP & DEIGHTON, CMI PAPERS 117:13-20 (ILLUS) /G. MORI (SAWADA)
PHRAGMO,CETERI,PHAEO,RADUL, 79E, ON LEAVES (FICUS & MANY OTHERS).
/ELLIS, DH P 216,1971 (ILLUS) /ELLIS, MDH P 186-187,1976 (ILLUS)
?=*SCLEROGRAPHIOPSIS* DEIGHTON 1973 •

GONATORRHODIELLA THAXTER 1891. BOTAN. GAZ. 16:501-502 /G. PARASITICA
=NEMATOGONIUM DESMAZIERES FIDE HUGHES 1953. •

GONATORRHODUM CORDA 1839. PRACHTFLORA P 5 /G. SPECIOSUM
?, PERHAPS AN EARLIER NAME FOR GONATOBOTRYUM, BUT TYPE NOT EXAMINED BY RECENT AUTHOR.
/LINDAU, IN RABENHORST KRYPT. FL. 2 ED. 1 BD. 8:686-687,1907 (ILLUS) •

GONATOSPORIUM CORDA 1839. ICONES FUNGORUM 3:8 /G. PUCCINIOIDES (DE CANDOLLE)
=ARTHRINIUM KUNZE 1817 FIDE HUGHES 1958. •

GONGROMERIZA PREUSS 1851. LINNAEA 24:106 /G. CLAVAEFORMIS
=CHLORIDIUM LINK FIDE HUGHES 1969. •

GONGYLOCLADIUM WALLROTH 1833. FL. CRYPT. GERMAN. 2:160 /G. ATRUM (LINK)
=OEDEMIUM DIDYMIUM
=OEDEMIUM LINK 1824 FIDE HUGHES 1958. •

GONIOSPORIUM LINK 1824. LINN. SPEC. PLANT., IV,6(1):45 /G. PUCCINIOIDES (DE CANDOLLE)
=ARTHRINIUM KUNZE 1817 FIDE HUGHES 1958. •

GONYELLA SYDOW & SYDOW 1919. ANN. MYCOL. 17:44 /ARTHROBOTRYUM TYPICA ROSTRUP 1916 (DANSK BOT. ARKIV. 2:46 & TAV. 3 FIG 19) =G. TYPICA (ROSTRUP)
?, FOR ARTHROBOTRYUM ROSTRUP 1916 NON CESATI 1854, APPEARS TO BE THE SAME AS ARTHROBOTRYS CORDA. •

GONYTRICHELLA EMOTO & TUBAKI 1970. TRANS. MYCOL. SOC. JAPAN 11:95-97 (ILLUS) /G. OLIVACEA
?=DICYMA BOULANGER 1897. •

GONYTRICHUM NEES & NEES 1818. NOVA ACTA ACAD. CAES. LEOP. 9:244 /G. CAESIUM =G. STATE OF CHAETOSPHAERIA INAEQUALIS (GROVE) GAMS & HOLUBOVA-JECHOVA 1976
AMERO,GLOIO,PHAEO OR HYALO,PHIAL,SETAE, 4A, ON LITTER & SOIL. SIMILAR TO UNCIGERA. ASCO STATE =MELANOPSAMMELLA FIDE HUGHES 1958 OR TRICHOSPHAERELLA SEE DE HOOG 1977 (CBS STUDIES 15:138).
/BARRON, HYPHOMYCETES P 184-186,1968 (ILLUS) /ELLIS, DH P 530-532,1971 (ILLUS,KEY) /MATSUSHIMA, ICONES P 78 & PL,1975 (ILLUS) /GAMS & HOLUBOVA-JECHOVA, CBS STUDIES 13:77-90,1976 (REVIS,KEY,ILLUS)
=*MESOBOTRYS* SACCARDO 1880 FIDE HUGHES 1958
?=*PIMINELLA* ARNAUD 1954 FIDE GAMS & HOLUBOVA-JECHOVA 1976 •

GRALLOMYCES STEVENS 1918. BOTANICAL GAZ. 45-246:245 /G. PORTORICENSIS
STAURO,CETERI,PHAEO,ANNEL,APPEND,PODIA?, 127B, ON LEAVES (MANY PLANTS).
/DEIGHTON & PIROZYNSKI, CMI PAPERS 105:10-17,1966 (ILLUS) /ELLIS, DH P 139-140,1971 (ILLUS)
=*OPHIOPODIUM* ARNAUD 1954 FIDE DEIGHTON & PIROZYNSKI 1966
=*PHIALETEA* BATISTA & NASCIMENTO 1960 FIDE DEIGHTON & PIROZYNSKI 1966 •

GRANULARIA SACCARDO 1882. MICHELIA 2:648 (NO ILLUS) /G. EUROTIOIDES SACCARDO & ELLIS 1882
?, NON GRANULARIA ROTH 1791. •

GRANULOMANUS DE HOOG & SAMSON 1978. IN DE HOOG, PERSOONIA 10:69-70 (33-81,ILLUS) /CYLINDROPHORA ARANEARUM PETCH 1944 =G. ARANEARUM (PETCH) ?=G. STATE OF TORRUBIELLA ALBOLANATA
SCOLECO,BASO?,HYALO,PHIAL, ON SPIDERS. COMPARE WITH PSEUDOMICRODOCHIUM. ASCO STATE ?=TORRUBIELLA. •

GRAPHIDIUM LINDAU 1910. RABENH. KRYPT. FL. 2 AUFL. 1 BD. 9:748 /G. CORRENSII (AS 'CORRENSI')
?, THE ILLUSTRATION SUGGESTS PAECILOMYCES, BUT THE CONIDIA ARE REPORTED AS NOT IN CHAINS OR SLIME. •

GRAPHILBUM UPADHYAY & KENDRICK 1975. MYCOLOGIA 67:800 (798-805,ILLUS) /G. SPARSUM =G. STATE OF CERATOCYSTIS SPARSA DAVIDSON
AMERO,GLOIO,HYALO,ANNEL, 19D, ON PICEA. ASCO STATE =CERATOCYSTIS FIDE UPADHYAY & KENDRICK 1975. •

GRAPHIOLA POITEAU 1824. ANN. SCI. NAT., SER. 1, 3:473-474 /G. PHOENICIS (MOUG.)
AMERO,CETERI,HYALO,AMPUL, 49E, ON LEAVES OF PHOENIX. ALSO CONSIDERED TO BE IN USTILAGINALES.
/HUGHES, CAN. J. BOT. 31:644-645,1953 (ILLUS) /TUBAKI, MYCOPATHOLOGIA 43:49-60,1971 (ILLUS)
?=*STYLINA* SYDOW 1921 •

GRAPHIOPSIS BAINIER 1907. BULL. SOC. MYCOL. FRANCE 23:19-20 (ILLUS) /G. CORNUI =PHAEOISARIA CLEMATIDIS (FUCKEL) HUGHES 1958
=PHAEOISARIA HOHNEL 1909 FIDE MASON 1937, NON GRAPHIOPSIS TRAIL 1889. •

GRAPHIOPSIS TRAIL 1899. SCOTTISH NATURALIST 10:75 /G. CHLOROCEPHALA (FRESENIUS)
=CLADOSPORIUM LINK SEE MASON & ELLIS 1953 (CMI PAPERS 56:28-29). •

GRAPHIOTHECIUM FUCKEL 1870. JAHRB. NASSAU. VER. NATURK. 23-24:366 & TAB 1, FIG 30 /G. FRESENII FUCKEL 1869
AMERO,BASO,PHAEO,PHIAL,SYNNEMATA, ON PLANTS (AMELANCHIER, VIBURNUM).
/LINDAU, RABENHORST KRYPT.-FL., 2 AUFL., 1 BD, IX ABT:389,1910 (ILLUS) /MORRIS, WESTERN ILLINOIS UNIV. SER. BIOL. SCI. 3:60-61,1963 (ILLUS OF SPECIES NOT CONGENERIC WITH TYPE) /KENDRICK & CARMICHAEL, THE FUNGI, VOL IVA,1973 (ILLUS)
=*STROMATOSTYSANUS* HOHNEL 1919 FIDE CLEMENTS & SHEAR 1931 •

GRAPHIUM CORDA 1837. ICONES FUNGORUM 1:18 /G. PENICILLIOIDES (LECTOTYPE, FIDE HUGHES 1958)
AMERO,GLOIO,PHAEO,ANNEL,SYNNEMATA, 19B, ON WOOD & LITTER (BRASSICA) & SOIL. ASCO STATE =PETRIELLA FIDE BARRON 1968, OR PETRIELLIDIUM, OPHIOSTOMA OR KERNIA FIDE VON ARX 1974 (GENERA P 11 & 115) OR ?CONIOCHAETIDIUM SEE MALLOCH & CAIN 1971 (CAN. J. BOT. 49:880). FOR CONIDIAL CERATOCYSTIS PICEAE & C. ULMI SEE PESOTUM. OTHER CONIDIAL STATES =SCEDOSPORIUM, SPOROTHRIX, HYALODENDRON.
/BARRON, HYPHOMYCETES P 186-188,1968 (ILLUS) /ELLIS, DH P 333-335,1971 (ILLUS,KEY) /SUTTON, CMI PAPERS 132:70-71,1973 (ILLUS) /CRANE & SCHOKNECHT, AMER. J. BOT. 60:346-354,1973 (ILLUS,REVIS) /FLETCHER, TBMS 64:67-71,1975 (ILLUS) /WEIJMAN & DE HOOG, ANTONIE VAN LEEUWENHOEK 41:353-360,1975 (REVIS) /MATSUSHIMA, ICONES P 78-79 & PL,1975 (ILLUS) /DABINETT & WELLMAN, CAN. J. BOT. 56:2031-2049,1978 (DISCUSSION OF GENERA,NO ILLUS)
?=*CERATOPODIUM* CORDA 1837 FIDE SACCARDO 1886 •

GRUBYELLA OTA & LANGERON 1923. ANN. PARASIT. HUM. COMP. 1:330 /G. SCHOENLEINII (LEBERT) =TRICHOPHYTON SCHOENLEINII (LECTOTYPE)
=TRICHOPHYTON MALMSTEN 1848. •

GUEDEA RAMBELLI & BARTOLI 1978. TBMS 71:340-342 (ILLUS) /G. SACRA
PHRAGMO,CETERI,PHAEO,RADUL,SETAE, ON LITTER. •

GUEGUENIA BAINIER 1907. BULL. SOC. MYCOL. FRANCE 23:106-107 /G. CESPITOSA
=AMBLYOSPORIUM FRESENIUS FIDE PIROZYNSKI 1969. •

GUELICHIA SPEGAZZINI 1886. AN. SOC. ARG. 22:220-221 /G. PARADOXA
?, NOT ILLUSTRATED BY AUTHOR. •

GUTTUROMYCES RIVOLTA ?=ASPERGILLUS FIDE AINSWORTH 1971 •

GYMNODOCHIUM MASSEE & SALMON 1902. ANN. BOT. 16:89 & PL IV, FIGS 49-51 /G. FIMICOLA (AS 'FIMICOLUM')
DIDYMO,BASO,HYALO OR PHAEO,ANNEL,SPORODOCHIA, 60A, ON DUNG. OTHER CONIDIAL STATE =WARDOMYCES SEE MALLOCH 1970.
/MALLOCH, CAN. J. BOT. 48:883-885,1970 (ILLUS SUB WARDOMYCES AGGREGATUS)
?=*GAMSIA* MORELET 1969 •

GYMNOSPORIUM CORDA 1833. STURM'S DEUTSCHL. FLORA III (PILZE), BD. 3, HEFT 13:69 /G. OLIVACEUM
=NOMEN DUBIUM SEE HUGHES 1958. •

GYOERFFIELLA KOL 1928. FOLIA CRYPTOG. 1:618 /G. TATRICA =G. ROTULA (HOHNEL) MARVANOVA 1967
STAURO,CETERI,HYALO,NONSP,APPEND,AQUATIC, 122A & C.
/VON ARX, GENERA P 215,1970 (REFS) /MARVANOVA, TBMS 65:555-565,1975 (ILLUS,NEW SP) /INGOLD, GUIDE P 38-41,1975 (ILLUS,REVIS)
=*INGOLDIA* PETERSEN FIDE MARVANOVA 1967 (PERSOONIA 5:29-44)
?=*PLEUROPEDIUM* MARVANOVA & IQBAL 1973 FIDE INGOLD 1975 •

GYROCERUS CORDA 1837. ICONES FUNGORUM 1:9 /G. AMMONIS
=NOMEN DUBIUM SEE HUGHES 1958, ALSO AS GYROCERAS SEE AINSWORTH 1961. •

GYROTHRIX (CORDA) CORDA 1842. ICONES FUNGORUM 5:13 /G. PODOSPERMA (CORDA) RABENHORST 1844 (DEUTSCHL. KRYPT. FL. 1:72)
SCOLECO OR AMERO,CETERI,HYALO,MISC,SETAE,STROMA, 54G, ON WOOD & LITTER. DIFFERS FROM CIRCINOTRICHUM ONLY IN HAVING BRANCHED SETAE.
/PIROZYNSKI, CMI PAPERS 84:8-27,1962 (ILLUS,REVIS,KEY) /PIROZYNSKI & PATIL, CAN. J. BOT. 48:579-581,1970 (DESCR,ECOLOGY) /ELLIS, DH P 358-361,1971 (ILLUS,KEY) /CUNNINGHAM, MYCOLOGIA 66:122-129,1974 (ILLUS,REVIS,KEY,NEW SP)
=*PEGLIONIA* GOIDANICH 1934 FIDE HUGHES & PIROZYNSKI 1971 •

GYROTRICHUM SPRENGEL 1827. LINN. SYST. VEG. XVI,4(1):554 /G. MACULIFORME (NEES)
=CIRCINOTRICHUM NEES 1917 FIDE HUGHES 1958. •

HADRONEMA SYDOW & SYDOW 1909. ANN. MYCOL. 7:172 /H. ORBICULARE
DIDYMO,CETERI,PHAEO,NONSP OR RACHI,SPORODOCHIA, 68C, ON LEAVES (QUERCUS). COMPARE WITH SPILODOCHIUM.
/ELLIS, DH P 198,1971 (ILLUS) •

HADROSPORIUM SYDOW 1938. ANN. MYCOL. 36:311-312 /H. FRASERIANUM
?, NOT ILLUSTRATED BY AUTHOR. •

HADROTRICHUM FUCKEL 1865. FUNGI RHENANI NO. 1522 /H. PHRAGMITIS
AMERO,CETERI,PHAEO,RADUL,SPORODOCHIA, 30G, ON GRASSES (PHRAGMITES & OTHERS). ASCO STATE =SCIRRHIA FIDE BARR 1972 (CONTRIB. UNIV. MICHIGAN HERB. 9:565,NO ILLUS).
/HUGHES, CAN. J. BOT. 31:574-575,1953 (ILLUS) /ELLIS, DH P 270-271,1971 (ILLUS) /ELLIS, MDH P 236-237,1976 (ILLUS)
=*MICROBASIDIUM* BUBAK & RANOJEVIC FIDE HOHNEL FROM AINSWORTH 1971
?=*PSEUDOEPICOCCUM* ELLIS 1971 •

HAINESIA ELLIS & SACCARDO 1884. IN SACCARDO, SYLL. FUNG. 3:699 /H. RHOINA (SACC.) =H. LYTHRI (DESM.) HOHNEL
AMERO,GLOIO,HYALO,PHIAL,SPORODOCHIA, (COELOMYCETE) ON PLANTS (MANY KINDS) & SOIL. ASCO STATE =DISCOHAINESIA FIDE VON ARX 1970. COELO STATE =SCLEROTIOPSIS FIDE SHEAR & DODGE 1921 =PILIDIUM FIDE AINSWORTH 1971.
/SHEAR & DODGE, MYCOLOGIA 13:135-170 & PL 8-10,1921 (ILLUS,ECOL,REFS) /BARRON, HYPHOMYCETES P 188-190,1968 (ILLUS,DISCUSSION) /VON ARX, GENERA P 181 & 185,1970 (ILLUS) /NAG RAJ, ICONES GENERUM COELOMYCETUM VIII:18-19,1977

(ILLUS) •

HALOBYSSUS ZUKAL 1893. OEST. BOT. Z. 43:279-281 & TAF. XII, FIG 12 /H. MONILIFORMIS
?, =MONILIA FIDE CLEMENTS & SHEAR 1931. •

HALYSIUM CORDA 1837. ICONES FUNGORUM 1:17 /H. ATRUM
=NOMEN CONFUSUM, FIDE HUGHES 1958. •

HANSFORDIA HUGHES 1951. CMI PAPERS 43:15-25 (ILLUS) /H. OVALISPORA
AMERO,CETERI,PHAEO,RADUL,SETAE, 27B, ON LEAVES, WOOD & LITTER. ASCO STATE =GEOPYXIS, SEE PADEN 1972 (PERSOONIA 6:410 & FIG 7, SUB NODULISPORIUM), OR HYPOXYLON, SEE JONG & ROGERS 1972 (WASH. AGR. EXPER. STA. TECH. BULL. 71:20-21,ILLUS).
/MATSUSHIMA, ICONES P 79-80 & PL,1975 (ILLUS) /ELLIS, DH P 233-234,1971 (ILLUS) /ELLIS, MDH P 198-200,1976 (ILLUS,NEW COMBS)
?=*RHINOTRICHELLA* ARNAUD EX DE HOOG 1977 •

HANSFORDIELLA HUGHES 1951. CMI PAPERS 47:10-15 (ILLUS) /H. ASTERINEARUM
PHRAGMO OR DICTYO,CETERI,PHAEO,NONSP,APPEND, 94C, ON FUNGI (ASTERINA & ASTEROLIBERTIA).
/ELLIS, DH P 168-169,1971 (ILLUS) •

HANSFORDIELLOPSIS DEIGHTON 1960. CMI PAPERS 78:33-35 (ILLUS) /H. ABURIENSIS
PHRAGMO OR DICTYO,CETERI,PHAEO,NONSP,PODIA?,APPEND, 94A, ON LEAVES WITH LICHENS.
/ELLIS, DH P 126-128,1971 (ILLUS) •

HANSFORDIULA MORRIS 1963. AMER. MIDL. NATURALIST 69:103-105 (ILLUS) /H. FASCICULATA
=PHAEOISARIA HOHNEL 1909 FIDE MORRIS 1963 (WESTERN ILLINOIS UNIV. SER. BIOL. SCI. 3:92). •

HANTZSCHIA AUERSWALD 1862. IN RABENHORST, FUNGI EUROPAEI NO. 441 /H. PHYCOMYCES
=NOM. REJ. SEE KENDRICK 1964 (=PHIALOCEPHALA KENDRICK)
/KENDRICK, CAN. J. BOT. 42:1291-1295,1964 (ILLUS,DISCUSSION) •

HAPLARIA LINK 1809. MAG. GES. NATURF. FREUNDE, BERLIN 3:11 /H. GRISEA
=BOTRYTIS PERSOON 1801 FIDE HUGHES 1958. •

HAPLARIELLA SYDOW 1908. ANN. MYCOL. 6:497 /H. CORDIAE (HENNINGS)
?, FOR HAPLARIOPSIS HENNINGS 1908 NON OUDEMANS 1903, NOT ILLUSTRATED BY HENNINGS OR SYDOW. HAPLARIELLA SACCARDO 1931 WAS A LATER REPLACEMENT FOR HAPLARIOPSIS HENNINGS. •

HAPLARIOPSIS HENNINGS 1908 =HAPLARIELLA SYDOW 1908 •

HAPLARIOPSIS OUDEMANS 1903. NED. ARCH. SER. 3, 2:902-903, PL 8, FIG 1 /H. FAGICOLA
DIDYMO,CETERI,HYALO,RADUL, 63A, ON LITTER. NON HAPLARIOPSIS HENNINGS 1908.
/ELLIS, DH P 293,1971 (ILLUS) •

HAPLOBASIDION ERIKSSON 1889. BOT. CENTRALBL. 38:786-787 /H. THALICTRI
AMERO,BLASTO,PHAEO,MISC, 37A, ON LEAVES (MUSA). CLOSE TO LACELLINOPSIS.
/ELLIS, CMI PAPERS 67:1-6,1957 (ILLUS,KEY) /ELLIS, DH P 353-354,1971 (ILLUS,KEY) /ELLIS, CMI PAPERS 131:19 & PL 2,1972 (ILLUS) /ELLIS & HOLLIDAY, CMI DESCRIPTIONS 496,1976 (ILLUS) •

HAPLOCHALARA LINDER 1933. MYCOLOGIA 25:347 & FIGS 7-8 /H. ANGULOSPORA
=CATENULARIA GROVE 1886 FIDE MASON 1941. •

HAPLOGRAPHIUM BERKELEY & BROOME 1859. ANN. MAG. NAT. HIST. III 3:360 /H. DELICATUM =H. STATE OF HYALOSCYPHA DEMATIICOLA (BERK. & BR.) NANNFELDT 1936
AMERO,GLOIO,HYALO,RADUL, 27C, ON WOOD & BARK (MANY PLANTS) & LEAVES (MANGO). ASCO STATE =HYALOSCYPHA.

/HUGHES, CAN. J. BOT. 31:588-589,1953 (ILLUS) /BARRON, HYPHOMYCETES P 192-193,1968 (ILLUS) /ELLIS, DH P 365-366,1971 (ILLUS) /MATSUSHIMA, ICONES P 80 & PL 66,1975 (ILLUS) /ELLIS, MDH P 364-365,1976 (ILLUS) /HUGHES, NEW ZEALAND J. BOT. 16:359-360,1978 (ILLUS SUB HYALOSCYPHA) •

HAPLOTRICHELLA ARNAUD 1954. BULL. SOC. MYCOL. FRANCE 69:280 & 282 FIG 6R /H. LINKII ?=GLIOMASTIX MURORUM VAR. FELINA FIDE CARMICHAEL
?=GLIOMASTIX GUEGUEN FIDE CARMICHAEL. NO TYPE SPECIMEN, SEE NICOT & CHARPENTIE 1971 (BULL. SOC. MYCOL. FRANCE 87:31). •

HAPLOTRICHUM ESCHWEILER 1824. SYLL. PLANT. A SOC. BOT. R. RATISB. EDITA 1:167 /H. CANDIDUM
=NOMEN DUBIUM, SEE HUGHES 1958, NON HAPLOTRICHUM LINK 1824. •

HAPLOTRICHUM LINK 1824. LINN. SPEC. PLANT. IV, 6(1):52 /H. CAPITATUM (LINK) =ALYSIDIUM STATE OF BOTRYOBASIDIUM CANDICANS ERIKSSON 1958
=ALYSIDIUM KUNZE 1817. HUGHES REFERRED THIS STATE TO ACLADIUM WHICH IN HIS TREATMENT (1958, CAN. J. BOT. 36:731-732) INCLUDED ALYSIDIUM.
/DONK, TAXON 11:85,1962 (REFS) /HOLUBOVA-JECHOVA, CESKA MYKOL. 30:3-4,1976 (REFS) •

HAPTOCARA DRECHSLER 1975. AMER. J. BOT. 62:1073-1077 (ILLUS) /H. LATIROSTRUM
PHRAGMO,CETERI,HYALO,AMPUL?,APPEND, 69D, ON NEMATODES. •

HARPAGOMYCES WILCYZNSKI 1911. KOSMOS 36:314-316 (ILLUS) /H. LOMNICKII
PHRAGMO OR STAURO,CETERI,HYALO,NONSP,APPEND, 124E. •

HARPOCEPHALUM ATKINSON 1897. BULL. CORNELL UNIV. 3:41 /H. DEMATIOIDES
=PERICONIA TODE EX PERSOON 1801 FIDE ELLIS 1971. •

HARPOGRAPHIUM SACCARDO 1880. MICHELIA 2:33 /H. FASCICULATUM (SACCARDO)
AMERO OR SCOLECO,GLOIO,HYALO,PHIAL?,SYNNEMATA, 107C, ON WOOD. COMPARE WITH IDRIELLA & SELENOSPORELLA. BASIDIO STATE =SEPTOBASIDIUM FIDE COLES & TALBOT 1976 (KEW BULL. 31:481-488).
/ELLIS, DH P 562-563,1971 (ILLUS)
=*EPILITHIA* NYLANDER 1855 FIDE SACCARDO 1886 •

HARPOSPORIUM LOHDE 1874. TAGEBLATT DER 47. VERSAMMLUNG DEUTSCHER NATURF. UND ARZTE IN BRESLAU P 206 (NO ILLUS) /H. ANGUILLULAE
HELICO OR SCOLECO,GLOIO,HYALO,PHIAL, 107G, ON NEMATODES. TYPE NOT EXAMINED BY RECENT AUTHOR(?). OTHER STATE ="ARTHROCONIDIA" SEE BARRON 1979.
/COOKE & GODFREY, TBMS 47:61-74,1964 (ILLUS,KEY) /BARRON, ANTONIE VAN LEEUWENHOEK 38:217-222,1972 (ILLUS) /MONSON & WILLIAMS, MYCOPATHOLOGIA 49:177-183,1973 (ILLUS) /MONSON & PIKUL, MYCOLOGIA 66:178-181,1974 (ILLUS,REFS,NEW SP) /BARRON, THE NEMATODE-DESTROYING FUNGI P 73-81 & 94-95,1977 (ILLUS,ECOL) /BARRON, CAN. J. BOT. 55:892-895,1977 (ILLUS) /BARRON, CAN. J. BOT. 57:886-889,1979 (NEW SP)
=*POLYRHINA* SOROKIN 1876 FIDE SOROKIN •

HARTIGIELLA SYDOW 1900. IN LINDAU IN ENGLER UND PRANTL, NACHTRAGE 1(1):558 /H. LARICIS (HARTIG)
=MERIA VUILLEMIN 1896 FIDE AINSWORTH 1971, FOR ALLESCHERIA HARTIG NON SACCARDO & SYDOW. •

HARZIA COSTANTIN 1888. LES MUCEDINEES SIMPLES P 42 /MONOSPORIUM ACREMONIOIDES HARZ 1871 =H. ACREMONIOIDES (HARZ) =ACREMONIELLA ATRA AUCT. (?NON ACREMONIUM ATRUM CORDA)
AMERO,CETERI,PHAEO,NONSP OR RACHI, 49F, ON PLANTS (MANY KINDS), DUNG, WOOD & SOIL. ASCO STATE =MELANOSPORA FIDE MASON 1933. SEE ALSO ACREMONIELLA. OTHER CONIDIAL STATE =HARZIELLA.
/MASON, CMI PAPERS 3:29-39,1933 (REVIS,ILLUS) /ELLIS, DH P 79-81,1971 (ILLUS SUB ACREMONIELLA) /KIRILENKO, MIKRO. ZHUR. 36:153-160,1974 (ILLUS) /MATSUSHIMA, ICONES, P 80 & PL,1975 (ILLUS) /ELLIS, MDH P 44-45,1976 (ILLUS SUB ACREMONIELLA VELATA)
=*EIDAMIA* LINDAU 1904 FIDE ELLIS 1971
=*MONOPODIUM* DELACROIX 1890 FIDE ELLIS 1971
?=*OLPITRICHUM* ATKINSON 1894 •

HARZIELLA COSTANTIN & MATRUCHOT 1899. BULL. SOC. MYCOL. FRANCE 15:104-107 & PL VII, FIGS 1-9 /H. CAPITATA
AMERO,GLOIO,HYALO,PHIAL, 8A, ON PLANTS, DUNG, WOOD & SOIL. A LATER HOMONYM OF HARZIELLA KUNTZE 1891. ASCO STATE =MELANOSPORA SEE UDAGAWA 1970 (TRANS. MYCOL. SOC. JAPAN 10:107-108, ILLUS SUB MELANOSPORA FUSISPORA) & FURUYA & UDAGAWA 1973 (IBID 14:20,ILLUS SUB M. ORNATA) OR MICROTHECIUM SEE UDAGAWA & CAIN 1969 (CAN. J. BOT. 47:1915-1933,ILLUS) & HAWKSWORTH & UDAGAWA 1977 (TRANS. MYCOL. SOC. JAPAN 18:149-151, ILLUS SUB H. FOVEOLATUM). OTHER CONIDIAL STATES =HARZIA, PAPULASPORA, CHLAMYDOMYCES.
/ARNAUD, BULL. SOC. MYCOL. FRANCE 68:197-199,1952 (ILLUS) /ELLIS, DH P 80,1971 (ILLUS SUB ACREMONIELLA ATRA) /ELLIS, MDH P 46,1976 (ILLUS SUB CHLAMYDOMYCES PALMARUM)
?=*PROTEOPHIALA* CIFERRI 1957 •

HARZIELLA KUNTZE 1891 (REV. GEN. PLANT. 2:855) IS A SUPERFLUOUS NAME FOR TRICHOCLADIUM HARZ 1871 •

HASSKARLINDA KUNTZE 1891 (REV. GEN. PLANT. 1:172) IS A SUPERFLUOUS NAME FOR CORALLODENDRON JUNGHUHN 1838 •

HEIMIODORA NICOT 1960. IN NICOT & PARGUEY, ANN. SCI. NAT., BOTAN., SER 12, 1:384-385 (ILLUS) /H. VERTICILLATA
AMERO,GLOIO,HYALO,AMPUL,SETAE, 34B, ON SOIL.
/BENJAMIN, ALISO 5:278-281,1963 (ILLUS) •

HELICOCEPHALUM THAXTER 1891. BOT. GAZ. 16:201-204 & PL XIX, FIG 1-6 /H. SARCOPHILUM
AMERO OR HELICO,ARTHRO OR GLOIO,PHAEO,MISC, 105F, ON NEMATODE EGGS. =ZYGOMYCETES SEE BARRON 1975.
/BARRON, TBMS 65:309-310,1975 (ILLUS) •

HELICOCERAS LINDER 1931. ANN. MISSOURI BOT. GARDEN 18:2-3 /H. CELTIDIS (BIV.-BERNH.)
=SIROSPORIUM BUBAK & SEREBRIANIKOW FIDE ELLIS 1963. •

HELICOCORYNE CORDA 1854. ICONES FUNGORUM 6:9 /H. VIRIDE =HELICOMA VIRIDIS (CORDA) HUGHES
=HELICOMA CORDA 1837 FIDE HUGHES 1958. •

HELICODENDRON PEYRONEL 1918. NUOV. GIORN. BOT. ITAL. N. S. 25:460-464 (ILLUS) /H. PARADOXUM
HELICO,BLASTO,HYALO OR PHAEO,NONSP,AQUATIC, 105A & 106B. EMENDED LINDER 1929. OTHER CONIDIAL STATE =?CHLORIDIUM SEE MATSUSHIMA 1975 (PL 257-2) OR ?=MYRIOCONIUM SEE GOOS & BANDONI 1977 & FIELD 1979 (TBMS 72:500-502). ASCO STATE =INOPERCULATE DISCOMYCETE SEE FIELD 1979.
/LINDER, ANN. MISSOURI BOT. GARD. 16:329-333,1929 (ILLUS,KEY) /VAN BEVERWIJK, TBMS 36:113-120,1953 (ILLUS,NEW SPP) /BARRON, HYPHOMYCETES P 195-196,1968 (ILLUS) /ELLIS, DH P 85-86,1971 (ILLUS) /MATSUSHIMA, ICONES P 81 & PL 252,1975 (ILLUS) /ELLIS, MDH P 48-49,1976 (ILLUS,KEY) /GOOS & BANDONI, SYESIS 10:121-123,1977 (ILLUS,REFS) /ABDULLAH, FISHER & WEBSTER, TBMS 72:324-329,1979 (ILLUS,NEW SPP)
=*HELICODESMUS* LINDER 1925 FIDE ELLIS 1971 •

HELICODESMUS LINDER 1925. AMER. J. BOT. 12:267 /H. ALBUS
=HELICODENDRON PEYRONEL 1918 FIDE ELLIS 1971. •

HELICOMA CORDA 1837. ICONES FUNGORUM 1:15 /H. MUELLERI =H. STATE OF THAXTERIELLA PEZICULA FIDE PIROZYNSKI 1972.
HELICO,CETERI,HYALO OR PHAEO,NONSP OR RADUL, 102A, ASCO STATE =THAXTERIELLA FIDE HUGHES 1958 (SUB LASIOSPHAERIA).
/LINDER, ANN. MISSOURI BOT. GARD. 16:295-317,1929 (ILLUS,KEY,NEW SP,NEW COMB) /ELLIS, DH P 190-191,1971 (ILLUS) /PIROZYNSKI, CMI PAPERS 129:26-30,1972 (ILLUS) /GOOS, MYCOLOGIA 67:888-891,1975 (ILLUS) /CRANE & SCHOKNECHT, MYCOLOGIA 67:666-671,1975 (ILLUS,NEW SP) /ELLIS, MDH P 159-160,1976 (ILLUS)
=*HELICOCORYNE* CORDA 1854 FIDE LINDER 1929
=*LITUARIA* RIESS 1853 FIDE LINDER FROM AINSWORTH 1971

=*MOORELLA* RAO & RAO 1964 FIDE KENDRICK
?=*HELICOMINOPSIS* DEIGHTON 1960
?=*HELICOSPORELLA* ARNAUD 1954
?=*HELICOSPORINA* ARNAUD 1954
?=*HELICOPSIS* KARSTEN 1889 FIDE LINDER FROM AINSWORTH 1971 •

HELICOMINA OLIVE 1948. MYCOLOGIA 40:15-17 (ILLUS) /H. CAPERONIAE
=PSEUDOCERCOSPORA SPEGAZZINI 1910 FIDE DEIGHTON 1976 (CMI PAPERS 140:8-9).
/ELLIS, DH P 209-210,1971 (ILLUS) /ELLIS, MDH P 178-179,1976 (ILLUS,NEW COMB)
?=*HYALOHELICOMINA* YOKOYAMA 1971 •

HELICOMINOPSIS DEIGHTON 1960. CMI PAPERS 78:20-22 (ILLUS) /H. FICI
?=HELICOMA CORDA 1837 (SEE 102B). •

HELICOMYCES LINK 1809. MAG. GES. NATURF. FREUNDE, BERLIN 3:21 /H. ROSEUS
HELICO,CETERI,HYALO,NONSP OR RACHI, 103B, ON LITTER. ASCO STATE
?=OPHIONECTRIA SEE SEAVER & WATERSTON 1940 (MYCOLOGIA 32:406-407 & FIG 4).
/LINDER, ANN. MISSOURI BOT. GARD. 16:270-275,1929 (ILLUS,NEW COMB) /PIROZYNSKI, CMI PAPERS 129:37-39,1972 (ILLUS) /SUTTON, CMI PAPERS 132:72-73,1973 (ILLUS) /MATSUSHIMA, ICONES P 81 & PL 256,1975 (ILLUS)
=*XYPHASMA* REBENTISCH FIDE RABENHORST FROM AINSWORTH 1971 •

HELICOON MORGAN 1892. J. CINCINNATI SOC. NAT. HIST. 14:49-51 /H. SESSILE (LECTOTYPE)
HELICO,CETERI,HYALO OR PHAEO,NONSP, 105D, ON WOOD (CONIFERS).
/LINDER, ANN. MISSOURI BOT. GARD. 16:322-329,1929 (ILLUS,KEY,NEW SP,NEW COMB) /VAN BEVERWIJK, TBMS 36:112-113,1953 (ILLUS) /BARRON, HYPHOMYCETES P 196-197,1968 (ILLUS) /ELLIS, DH P 189,1971 (ILLUS) /MATSUSHIMA, ICONES P 81-82 & PL,1975 (ILLUS) •

HELICOPSIS KARSTEN 1889. REV. MYCOL. 11:96 /H. OLIVACEA (AS 'OLIVACEUS')
=HELICOMA OLIVACEUM (KARSTEN) LINDER 1929
=HELICOMA CORDA 1837 FIDE LINDER 1929 (ANN. MO. BOT. GARD. 16:302,ILLUS) •

HELICORHOIDION HUGHES 1958. CAN. J. BOT. 36:773 /H. BOTRYOIDEUM (COOKE)
HELICO,CETERI,PHAEO,RACHI, 105E, ON WOOD.
/ELLIS, DH P 217,1971 (ILLUS) /MULDER, TBMS 61:400-402,1973 (ILLUS,NEW SP) •

HELICOSPORANGIUM KARSTEN 1865. ANN. LANDWIRTHSCHAFT. 46:229-235, FIGS 1-11 /H. PARASITICUM
=PAPULASPORA PREUSS 1851, FIDE HARZ 1890 (BOT. ZENTRALBL. 41:405-411).
/KARSTEN, HEDWIGIA 27:132-137,1888 (ILLUS) •

HELICOSPORELLA ARNAUD 1954. BULL. SOC. MYCOL. FRANCE 69:292-293, FIG 12 M /H. LUTEO-FUSCA
?=HELICOMA CORDA. NO TYPE SPECIMEN, SEE NICOT & CHARPENTIE 1971 (BULL. SOC. MYCOL. FRANCE 87:32). •

HELICOSPORINA ARNAUD 1954. BULL. SOC. MYCOL. FRANCE 69:292 & 293 (ILLUS,NO LATIN DIAGNOSIS) /H. GLOBULIFERA
?=HELICOMA CORDA 1837. SEE ALSO HELICOSPORINA RAMBELLI. NO TYPE SPECIMEN, SEE NICOT & CHARPENTIE 1971 (BULL. SOC. MYCOL. FRANCE 87:32). •

HELICOSPORINA RAMBELLI 1960. MYCOPATHOLOGIA 13:107-111 (ILLUS) /H. VERONAE
=TROPOSPORELLA KARSTEN 1892 FIDE KENDRICK. NON HELICOSPORINA ARNAUD 1954 (NOT VALIDLY PUBLISHED). •

HELICOSPORIUM NEES 1816. DAS SYSTEM ... P 68 /H. VEGETUM NEES 1817 =H. STATE OF OPHIONECTRIA CEREA (BERK. & CURT.) ELLIS & EVERHART FIDE HUGHES 1958
HELICO,CETERI,HYALO,NONSP OR RADUL,SETAE,APPEND, 102C, ON WOOD & LITTER IN WET PLACES. ASCO STATE =OPHIONECTRIA FIDE HUGHES 1958 OR TUBEUFIA FIDE ELLIS 1971.
/LINDER, ANN. MISSOURI BOT. GARD. 16:275-295,1929 (ILLUS,NEW COMB,NEW SP,KEY) /BARRON, HYPHOMYCETES P 188-189,1968 (ILLUS) /ELLIS, DH P 191-193,1971 (ILLUS) /PIROZYNSKI, CMI PAPERS 129:30-37,1972 (ILLUS) /MATSUSHIMA, ICONES P 82 & PL,1975 (ILLUS) /ELLIS, MDH P 161-162,1976 (ILLUS) /HUGHES, NEW ZEALAND J. BOT.

16:365-367,1978 (ILLUS SUB TUBEUFIA)
=*HELICOTRICHUM* NEES & NEES 1818 FIDE HUGHES 1958 •

HELICOSTILBE HOHNEL 1902. S. B. AKAD. WISS. WIEN, ABT. 1:1028-1029 (NO ILLUS) /H. HELICINA
?, LINDER (1929) EMENDED THE GENUS TO INCLUDE ONLY H. SIMPLEX & EXCLUDE HOHNELS TYPE =TROCHOPHORA MOORE 1955.
/LINDER, ANN. MISSOURI BOT. GARD. 16:333-334,1929 (EMEND) •

HELICOTRICHUM NEES & NEES 1818. NOVA ACTA ACAD. CAES. LEOP. 9:246 /H. PULVINATUM =HELICOSPORIUM STATE OF OPHIONECTRIA CEREA
=HELICOSPORIUM NEES 1817 FIDE HUGHES 1958. •

HELISCUS SACCARDO 1880. MICHELIA 2:35 & 132 /H. LUGDUNENSIS SACCARDO & THERRY 1880
STAURO OR PHRAGMO,CETERI,HYALO,PHIAL,AQUATIC, 117D, ASCO STATE =NECTRIA FIDE WEBSTER 1959 (TBMS 42:322-327)
/VON ARX, GENERA P 188-189,1970 (ILLUS) /INGOLD, GUIDE P 70-72,1975 (ILLUS) •

HELMINTHOPHORA BONORDEN 1851. HANDB. ALLGEM. MYCOL. P 93 & TAF. VI, FIG 137 /DACTYLIUM SPHAEROCEPHALUM BERKELEY 1841 =H. SPHAEROCEPHALA (BERKELEY) DE HOOG 1978
PHRAGMO,CETERI,HYALO,RADUL?, 80D, ON FUNGI (FLAMMULINA).
/DE HOOG, PERSOONIA 10:55-57,1978 (ILLUS,REVIS)
=*EURASINA* ARNOLD 1970 FIDE DE HOOG 1978 •

HELMINTHOSPORIOPSIS SPEGAZZINI 1880. AN. SOC. CIENT. ARGENTINA 10:166 (NO ILLUS) /H. TYPICA
=PODOSPORIUM SCHWEINITZ 1832 FIDE AINSWORTH 1961. •

HELMINTHOSPORIUM LINK 1809. MAG. GES. NATURF. FREUNDE, BERLIN 3:10 /H. VELUTINUM =H. CILIARE (PERSOON) HUGHES 1958
PHRAGMO,CETERI,PHAEO,RADUL,STROMA, 83F, ON PLANTS. ASCO STATE =PSEUDOCOCHLIOBOLUS FIDE TSUDA, UEYAMA & NISHIHARA 1977 (MYCOLOGIA 69:1109-1120).
/ELLIS, CMI PAPERS 82:2-21,1961 (ILLUS,REVIS,KEY) /ELLIS, CMI DESCRIPTIONS 166,1968 (ILLUS) /ELLIS, DH P 388-393,1971 (ILLUS,KEY) /MATSUSHIMA, ICONES P 83-84 & PL,1975 (ILLUS,NEW SP) /MISRA, HELMINTHOSPORIUM SPECIES OCCURRING ON CEREALS & OTHER GRAMINEAE, DHOLI MUZAFFAPUR I TIRHUT COLLEGE OF AGRICULTURE, INDIA, 289 P, 27 PL,1976 (NOT SEEN) /HUGHES, NEW ZEALAND J. BOT. 16:334-337,1978 (ILLUS) /HUGHES, FUNGI CANADENSES 163,1980 (ILLUS)
=*HELMISPORIUM* LINK 1809 (ORTHOGRAPHIC VARIANT) FIDE HUGHES 1958
=*MACROON* CORDA 1833 FIDE SACCARDO FROM AINSWORTH 1971 •

HELMISPORIUM LINK 1809 =HELMINTHOSPORIUM LINK •

HELOSTROMA PATOUILLARD 1902. BULL. SOC. MYCOL. FRANCE 18:47 /H. ALBUM (DESMAZIERES)
=MICROSTROMA NIESSL 1861 FIDE VON ARX 1970. •

HEMIBELTRANIA PIROZYNSKI 1963. CMI PAPERS 90:30-34 (ILLUS,KEY) /HANSFORDIA CINNAMOMI DEIGHTON 1960 =H. CINNAMOMI (DEIGHTON)
AMERO,CETERI,HYALO OR PHAEO,RADUL OR MISC, 24D, ON LEAVES (CINNAMOMUM, NECTANDRA) & LITTER (SYZYGIUM).
/DEIGHTON, CMI PAPERS 78:14-17,1960 (ILLUS) /ELLIS, DH P 244-245,1971 (ILLUS,KEY) /SUTTON, KAVAKA 4:43-45,1976 (ILLUS,NEW SP) •

HEMICORYNESPORA ELLIS 1972. CMI PAPERS 131:19-22 (ILLUS,KEY) /H. DEIGHTONII
AMERO OR DIDYMO,CETERI,PHAEO,ANNEL, 18B, ON LITTER (MONOCOTYLEDONS).
/ELLIS, MDH P 382-383,1976 (ILLUS,KEY) •

HEMISPORA VUILLEMIN 1906. BULL. SOC. MYCOL. FRANCE 22:128-129 (ILLUS) /H. STELLATA
=WALLEMIA JOHAN-OLSEN FIDE CARMICHAEL 1969. •

HENNEBERTIA MORELET 1969. ANN. SOC. SCI. NAT. ARCHEOL. TOULON VAR. 21:104 /TRICHOSPORIUM COLUMBINUM DEMELIUS 1922 (VERH. ZOOL. BOT. GES. WIEN

72:105-106,ILLUS) =H. COLUMBINA (DEMELIUS)
=WARDOMYCES BROOKS & HANSFORD 1922 SEE HENNEBERT 1968 (TBMS 51:753). •

HEPTASTER CIFERRI, BATISTA & NASCIMENTO 1956. PUBL. INST. MICOL. UNIV. RECIFE 46:3 /H. HUGHESII
?=TRIPOSPERMUM SPEGAZZINI 1918 SEE HUGHES 1976 (MYCOLOGIA 68:810-811). •

HERMATOMYCES SPEGAZZINI 1910. AN. MUS. NAC. BUENOS AIRES 20(SER 3,13):445-446 /H. TUCUMANENSIS
DICTYO,CETERI,PHAEO,NONSP,SETAE, 96G, ON LITTER.
/ELLIS, DH P 110-111,1971 (ILLUS) •

HERPOSIRA SYDOW 1938. ANN. MYCOL. 36:312-313 /H. VELUTINA
AMERO,CETERI,PHAEO,MISC,STROMA,SETAE, 48D, ON LITTER (XANTHORRHOEA). COMPARE WITH LACELLINA.
/ELLIS, DH P 170-171,1971 (ILLUS) •

HETEROBOTRYS SACCARDO 1880. MICHELIA 2:21 & FUNGI ITALICI FIG 807, 1881 /TWO ORIGINAL SPECIES MENTIONED, H. PARADOXA SACCARDO DESCRIBED IN MICHELIA 2:124,1880 & ILLUSTRATED IN FUNGI ITALICI, NO. 807,1881.
=ATICHIA FLOTOW 1850 FIDE MEEKER 1975 (CAN. J. BOT. 53:2487). •

HETEROCEPHALUM THAXTER 1903. BOT. GAZ. 35:157, PL IV, FIGS 1-7 /H. AURANTIACUM
AMERO,GLOIO,HYALO,RADUL?,SYNNEMATA, 28C, ON SOIL & DUNG.
/MORRIS, WESTERN ILLINOIS UNIV. SER. BIOL. SCI. 3:68-69,1963 (ILLUS) /BARRON, HYPHOMYCETES P 202-203,1968 (ILLUS,REFS) •

HETEROCONIDIUM SAWADA 1944. REP. GOVT. RES. INST. FORMOSA 87:92-93, FIGS 52-54 /H. CASSIAE
?, NO LATIN DIAGNOSIS. •

HETEROCONIUM PETRAK 1949. SYDOWIA 3:264-266 /H. CITHAREXYLI
PHRAGMO,BLASTO,PHAEO,NONSP OR ANNEL, 76G, ON LEAVES (CITHAREXYLUM) & ON FUNGI (MANY KINDS) ON LEAVES & WOOD. DOUBTFULLY DISTINCT FROM TAENIOLELLA.
/ELLIS, DH P 96-97,1971 (ILLUS) /ELLIS, MDH P 64-67,1976 (ILLUS,KEY,NEW COMBS)
?=*ELLISIA* BATISTA & PERES 1965 •

HETEROSEPTATA MORRIS 1972. MYCOLOGIA 64:889-891 & FIGS 8-10 /H. BACILLOSPORA
PHRAGMO,GLOIO,PHAEO,PHIAL,SYNNEMATA, 74C, ON LITTER. COMPARE WITH DIDYMOSTILBE. •

HETEROSPORIOPSIS PETRAK 1950. SYDOWIA 4:521-523 /H. MONILIFERA (PATOUILLARD)
?, NOT ILLUSTRATED BY PETRAK. •

HETEROSPORIUM KLOTZSCH 1877. IN COOKE, GREVILLEA 5:122-123 /H. ORNITHOGALI =CLADOSPORIUM ORNITHOGALI (KLOTZSCH) DE VRIES (LECTOTYPE)
=CLADOSPORIUM LINK 1815 FIDE HUGHES 1958. •

HEYDENIA FRESENIUS 1852. BEITR. MYKOL., HEFT 2, P 47-48, PL V, FIGS 37-45 /H. ALPINA
?=PHLEOGENA, SEE ENTRY UNDER MARTINDALIA.
/LINDAU, RABENH. KRYPT. FL. 2 ED. 1 BD. 9:373-374,1910 (ILLUS) /NICOT, REV. MYCOL. 35:223-225,1970 (COLLECTION,NO ILLUS)
=*RICCOA* CAVARA 1903 FIDE CLEMENTS & SHEAR 1931
=*RUPINIA* ROUM. & SPEGAZ. 1880 FIDE LINDAU 1910
?=*MARTINDALIA* SACCARDO & ELLIS 1884
?=*CLATHROTRICHUM* PATOUILLARD 1921 •

HEYDENIOPSIS NAUMOV 1915. MAT. MIKOL. FITOPAT. ROSSTOK. 1:23-25 & FIGS 28-29 /H. INGRICA
? •

HIMANTIA PERSOON 1801. SYN. METH. FUNG. P 703-704 /THREE ORIGINAL SPECIES
?, FOR MYCELIAL STAGES OF CORTICIUM, ETC. FIDE LINDAU 1910.
/DONK, TAXON 11:85-86,1962 (REFS) •

HIOSPIRA MOORE 1962. TBMS 45:145 /H. HENDRICKXII (HANSFORD) =H. STATE OF BROOKSIA TROPICALIS
HELICO,CETERI OR BLASTO,PHAEO,NONSP,SETAE?, 105G, ON LEAVES (ASSOCIATED WITH OTHER FUNGI). ASCO STATE =BROOKSIA HANSFORD FIDE DEIGHTON & PIROZYNSKI 1966.
/DEIGHTON & PIROZYNSKI, CMI PAPERS 105:2-10,1966 (ILLUS) /ELLIS, DH P 193-195,1971 (ILLUS) •

HIPPOCREPIDUM SACCARDO 1875. IN THUMEN, MYCOTH. UNIV. NO. 85 /H. MESPILI SACCARDO APUD THUMEN 1875
=HIRUDINARIA CESATI 1856 FIDE HUGHES 1951. •

HIRSUTELLA PATOUILLARD 1892. REV. MYCOL. 14:69 /H. ENTOMOPHILA
AMERO,GLOIO OR CETERI,HYALO,PHIAL,SYNNEMATA, 10C, ON INSECTS. ASCO STATE =CORDYCEPS, OPHIOCORDYCEPS OR TORRUBIELLA FIDE MORRIS 1963.
/MORRIS, WESTERN ILLINOIS UNIV SER BIOL SCI 3:70-71,1963 (ILLUS) /BALAZY, ACTA MYCOL. 7:7-12,1971 (ILLUS H. LECANIICOLA) /BRADY, CMI DESCRIPTIONS 607-608,1979 (ILLUS)
=*MAHEVIA* LAGARDE 1917 FIDE CARMICHAEL 1973
=*TRICHOSTERIGMA* PETCH 1923 FIDE PETCH 1924 (FROM AINSWORTH 1971) •

HIRUDINARIA CESATI 1856. KLOTZSCH HERB. MYC. EDIT. NOVA, NO. 269 /H. MACROSPORA (LECTOTYPE)
STAURO,CETERI,PHAEO,NONSP, 127F, ON LEAVES (CRATAEGUS).
/HUGHES, CMI PAPERS 39:11-24,1951 (ILLUS,REVIS) /ELLIS, DH P 52-53,1971 (ILLUS)
=*HIPPOCREPIDUM* SACCARDO 1875 FIDE HUGHES 1951 •

HISTOPLASMA DARLING 1906. J. AM. MED. ASSOC. 46:1283-1285 (ILLUS) /H. CAPSULATUM =H. STATE OF EMMONSIELLA CAPSULATA KWON-CHUNG 1972 (SCIENCE 177:368-369, NO ILLUS) =H. STATE OF AJELLOMYCES CAPSULATUS (KWON-CHUNG) MCGINNIS & KATZ 1979 (MYCOTAXON 8:157-164)
=BLASTOMYCETES. ASCO STATE =AJELLOMYCES FIDE MCGINNIS & KATZ 1979. HISTOPLASMA WAS ORIGINALLY PROPOSED FOR THE INTRACELLULAR YEAST STATE & THUS SHOULD BE CONSIDERED A FORM-GENUS OF THE BLASTOMYCETES. THE TUBERCULATE CONIDIAL STATE CAN BE REFERRED TO THE FORM-GENUS CHRYSOSPORIUM (SEE CARMICHAEL, CAN. J. BOT. 40:1146,1962).
/HOWELL, MYCOLOGIA 31:191-216,1939 (ILLUS) /BERLINER, SABOURAUDIA 6:111-118,1968 (ILLUS) /KWON-CHUNG, MYCOLOGIA 67:980-990,1975 (ILLUS,REVIS) •

HOBSONIA BERKELEY 1891. IN MASSEE, ANN. BOT. 5:509 /H. GIGASPORA
HELICO,GLOIO,HYALO,MISC?,SPORODOCHIA, 105B, ON LITTER.
/LINDER, ANN. MISSOURI BOT. GARD. 16:339-341,1929 (ILLUS,NEW COMB) •

HOEHNELIELLA BRESADOLA & SACCARDO 1902. IN STRASSER, VERH. Z-B. GES. WIEN 52:437 (NO ILLUS) /H. PERPLEXA
?=VOLUTELLA SEE RAO & SUTTON 1975
/RAO & SUTTON, KAVAKA 3:21-23,1975 (ILLUS) •

HORMIACTELLA SACCARDO 1886. SYLLOGE FUNGORUM 4:311 /HORMIACTIS FUSCA PREUSS 1851 =H. FUSCA (PREUSS)
AMERO OR DIDYMO,BLASTO,HYALO,NONSP,SETAE, 38E & 65G, ON LITTER. CLOSE TO FUSIDIUM LINK.
/HOLUBOVA-JECHOVA, FOLIA GEOBOT. PHYTOTAX., PRAHA 13:433-437,1978 (ILLUS,KEY,NEW SP)
=*POLYSCYTALINA* ARNAUD 1953 •

HORMIACTINA BUBAK 1916. HEDWIGIA 57:336-337 (ILLUS) /H. WROBLEWSKII
?, DOES NOT APPEAR TO DIFFER FROM HORMIACTIS. •

HORMIACTIS PREUSS 1851. FUNGI HOYERSWERDA P 128 /H. ALBA (LECTOTYPE, SEE SACCARDO, SYLL. FUNG. 4:186 & 311,1886)
DIDYMO,BLASTO,HYALO,NONSP, 65F, ON LITTER & SOIL.
/BARRON, HYPHOMYCETES P 206-207,1968 (ILLUS)
?=*HORMIACTINA* BUBAK 1916
?=*POLYSCYTALUM* RIESS 1853 •

HORMIOKRYPSIS BATISTA & NASCIMENTO 1957. ANN. SOC. BIOL. PERNAMBUCO 15:345-348 /H. LIBOCEDRI =H. STATE OF OPHIOCAPNOCOMA PHLOIOPHILIA (FISHER) HUGHES FIDE HUGHES 1967
STAURO,CETERI,PHAEO,NONSP, 126G, ASCO STATE =OPHIOCAPNOCOMA FIDE HUGHES 1967. OTHER STATES =CAPNOCYBE (H), CAPNOPHIALOPHORA (H), TORULOPSIELLA (H).
/HUGHES, NEW ZEALAND J. BOT. 5:117-133,1967 (ILLUS) /HUGHES, NEW ZEALAND J. BOT. 10:225-242,1972 (ILLUS) /HUGHES, MYCOLOGIA 68:707-712,1976 (ILLUS) •

HORMISCIELLA BATISTA 1956. ANN. SOC. BIOL. PERNAMBUCO 14:100-101 (ILLUS P 106) /H. ATRA =ANTENNATULA ATRA (BATISTA) HUGHES 1970
=ANTENNATULA FRIES 1850 FIDE HUGHES 1970. •

HORMISCIOIDEUS BLACKWELL & KIMBROUGH 1978. MYCOLOGIA 70:1274-1280 (ILLUS) /H. FILAMENTOSUS
PHRAGMO,ARTHRO?,HYALO,ANNEL?, ON TERMITES (ARMITERMES).
/THAXTER, BOT. GAZ. 58:238-239,1914 (ILLUS SUB HORMISCIUM MYRMECOPHILUM) •

HORMISCIOMYCES BATISTA & NASCIMENTO 1957. ANN. SOC. BIOL. PERNAMBUCO 15:349-351 (ILLUS) /H. PREPUSUM
AMERO,GLOIO,HYALO,PHIAL, 17E, ON LEAVES (SOOTY MOLD). ASCO STATE =EUANTENNARIA FIDE HUGHES 1972 & TRICHOPELTHECA SEE HUGHES 1976.
/HUGHES, NEW ZEALAND J. BOT. 8:180,181 & 187,1970 (ILLUS) /HUGHES, NEW ZEALAND J. BOT. 10:225-242,1972 (ILLUS) /HUGHES, NEW ZEALAND J. BOT. 12:299-356,1974 (ILLUS,NEW SPP) /HUGHES, MYCOLOGIA 68:703-707 & 773,1976 (ILLUS)
?=*PLOKAMIDOMYCES* BATISTA, COSTA & CIFERRI 1951 FIDE HUGHES 1976 (P 773) •

HORMISCIOPSIS SUMSTINE 1914. MYCOLOGIA 6:32 /H. GELATINOSA
?, ORIGINAL ILLUSTRATION INADEQUATE TO CHARACTERIZE. •

HORMISCIUM KUNZE 1817. IN KUNZE & SCHMIDT, MYKOL. HEFTE 1:12-13, FIG 7 /H. EXPANSUM =TORULA HERBARUM (PERSOON) LINK
=TORULA (PERSOON) LINK 1809 FIDE HUGHES 1958. •

HORMOCEPHALUM SYDOW 1939. ANN. MYCOL. 37:424-425 /H. ECUADORENSE
PHRAGMO,CETERI,PHAEO,NONSP OR MISC, 92C, ON LEAVES (ASPILIA). COMPARE WITH POLYSCHEMA.
/ELLIS, DH P 99-100,1971 (ILLUS) •

HORMOCLADIUM HOHNEL 1919. BER. DEUTSCHL. BOT. GES. 37:156 /H. KAKI (HORI & YOSHINO)
?, NOT ILLUSTRATED BY HOHNEL. •

HORMOCONIS VON ARX & DE VRIES 1973. IN VON ARX, VERH. KON. NEDERL. AKAD. WET. AFD. NAT. 61:61-62 & FIG ID /HORMODENDRUM RESINAE LINDAU 1907 =H. RESINAE (LINDAU) =HORMOCONIS STATE OF AMORPHOTHECA RESINAE.
AMERO,BLASTO,PHAEO,RADUL, 39E, ON WOOD, RESIN, CREOSOTED WOOD, SOIL & PETROLEUM PRODUCTS. ASCO STATE =AMORPHOTHECA FIDE PARBERY 1969 (AUSTRAL. J. BOT. 17:331-357,ILLUS). COMPARE WITH CLADOSPORIUM. OTHER CONIDIAL STATE =SOROCYBE FRIES.
/ELLIS, DH P 309-310,1971 (ILLUS) /REDHEAD & MALLOCH, CAN. J. BOT. 55:1707,1977 (ASCO STATE) •

HORMODENDROIDES MOORE & ALMEIDA 1937. ANN. MISSOURI BOT. GARDEN 23:547 (DATED 1936) /H. PEDROSOI (BRUMPT)
=FONSECAEA NEGRONI 1936. •

HORMODENDRON, SEE HORMODENDRUM •

HORMODENDRUM BONORDEN 1851. HANDB. ALLEGMEIN. MYKOL. P 76-77 /AMPHITRICHUM OLIVACEUM CORDA 1837 (ICON. FUNG. 1:16) =H. OLIVACEUM (CORDA) (LECTOTYPE)
?=CLADOSPORIUM LINK FIDE HUGHES 1958 (CAN. J. BOT. 36:751) OR =PENICILLIUM FIDE VON ARX 1974 (GENERA P 224). SEE ALSO DE HOOG & HERMANIDES-NIJHOF 1977 (CBS STUDIES 15:184).
=*HORMODENDRON* ORTHOGRAPHIC VARIANT •

HORMODOCHIS CLEMENTS 1909. GENERA OF FUNGI P 163 & 176 (NO ILLUS) /H. MELANOCHLORA (DESMAZIERES)
?, FOR EPIDOCHIUM MELANOCHLORUM DESMAZIERES 1851.
=*HORMODOCHIUM* (SACCARDO) TROTTER 1931
=*HORMODOCHIUM* (SACCARDO) HOHNEL 1911 •

HORMODOCHIUM -SEE HORMODOCHIS •

HORMOMYCES BONORDEN 1851. HANDB. ALLG. MYKOL. P 150 /H. AURANTIACUS
AMERO,BLASTO,HYALO,NONSP,SPORODOCHIA, 36F, ON LITTER (TSUGA).
/BANDONI & BISALPUTRA, CAN. J. BOT. 49:27-30,1971 (ILLUS) /TUBAKI, TRANS. MYCOL. SOC. JAPAN 17:243-247,1976 (DESCRIPTION,ILLUS) •

HORMONEMA LAGERBERG & MELIN 1927. IN LAGERBERG & AL., SVENSKA SKOGSV. TIDSKR. HEFT II, OCH IV:219 & 233, FIGS 38-39 /H. DEMATIOIDES =H. STATE OF SYDOWIA POLYSPORA (BREF. & TAVEL) MULLER 1953 SEE HERMANIDES-NIJHOF 1977
AMERO,GLOIO,HYALO OR PHAEO,PHIAL?, 20D, ON PLANTS (MANY KINDS). HAS BEEN CONFUSED WITH AUREOBASIDIUM. ASCO STATE =XENOMERIS, GUIGNARDIA, DOTHIORA, SYDOWIA OR POTEBNIAMYCES, SEE FUNK & SHOEMAKER 1971 (MYCOLOGIA 63:567-574,ILLUS) & VON ARX 1970 (BIBLIOTHECA MYCOLOGICA 24:69-82,ILLUS), OR PRINGSHEIMIA FIDE HERMANIDES-NIJHOF 1977, OR CONIOCHAETA SEE UDAGAWA 1969 (IN KOBAYASI ET AL., BULL. NAT. SCI. MUS. TOKYO 12:330-332,ILLUS) OR POROCONIOCHAETA SEE UDAGAWA & FURUYA 1979 (TRANS. MYCOL. SOC. JAPAN 20:5-12). COELO STATE =DOTHICHIZA, SEE VON ARX 1974 (P 153-154) & XENOPOULOS & MILLAR 1977 (TBMS 68:127-130) OR SARCOPHOMA SEE VAN DER AA 1975 (PERSOONIA 8:283-289) OR KABATINA SEE HERMANIDES-NIJHOF 1977 (P 142).
/HERMANIDES-NIJHOF, VERH. KON. NEDERL. AKAD. WET. AFD. NAT. 67:7,1976 (DISCUSSION) /HERMANIDES-NIJHOF, CBS STUDIES 15:166-173,1977 (ILLUS,REVIS,KEY) •

HUGHESIELLA BATISTA & VITAL 1956. ANN. SOC. BIOL. PERNAMBUCO 14:141-144 /H. EURICOI =CHALARA & THIELAVIOPSIS STATES OF CERATOCYSTIS PARADOXA (DADE) MOREAU
=THIELAVIOPSIS WENT 1893 FIDE ELLIS 1971. •

HUGHESINIA LINDQUIST & GAMUNDI 1970. BOL. SOC. ARGENT. BOT. 13:53-57 (ILLUS) /H. CHUSQUEAE
STAURO,CETERI,PHAEO,NONSP,PODIA, 126C, ON LEAVES (CHUSQUEA).
/ELLIS, CMI PAPERS 125:19-20,1971 (ILLUS) /ELLIS, MDH P 379-381,1976 (ILLUS) •

HUMICOLA TRAAEN 1914. NYT. MAG. NATURVIDENSK. 52:31-34 /H. FUSCOATRA
AMERO,CETERI,PHAEO,NONSP,ALEURI, 49D, ON SOIL, WOOD, LITTER & MANY OTHER SUBSTRATES. FOR THERMOPHILIC SPECIES SEE THERMOMYCES. OTHER CONIDIAL STATES =IDRIELLA, CHALARA, ACREMONIUM, TRICHOCONIS, SCOLECOBASIDIUM, PHAEOISARIA, MICRODOCHIUM, CHLORIDIUM, CHAETASBOLISIA, VERTICILLIUM. ASCO STATE =CERATOCYSTIS FIDE MATSUSHIMA 1975 (P 169), LOPHOTRICHUS SEE FURUYA & UDAGAWA 1973 (TRANS. MYCOL. SOC. JAPAN 14:19,ILLUS), CHAETOMIUM SEE DE HOOG 1977 (CBS STUDIES 15:138 & 215,1977, ILLUS SUB MICRODOCHIUM), OR ZOPFIELLA FIDE MALLOCH & CAIN 1971 (CAN. J. BOT. 49:876).
/BARRON, HYPHOMYCETES P 207-208,1968 (ILLUS,REFS) /FASSATIOVA, CASOPIS CESKOSLOV. VED. SPOLEC. MYKOL. 21:78-89, 1967 (ILLUS,KEY) /ELLIS, DH P 59-60,1971 (ILLUS,KEY) /NICOLI & RUSSO, NOVA HEDWIGIA 25:737-798,1974 (MONOGR) /GRIFFITHS, CAN. J. MICROBIOL. 20:55-58,1974 (ILLUS) /WILLCOX ET AL, CAN. J. BOT. 52:2279-2282,1974 (ILLUS SUB BDG-58) /MOORHOUSE & DE BERTOLDI, MYCOTAXON 3:109-118,1975 (BIOCHEM,NO ILLUS) /MOUSTAFA, PERSOONIA 8:332-333,1975 (ILLUS SUB GILMANIELLA MACROSPORA) /DE BERTOLDI, CAN. J. BOT. 54:2755-2768,1976 (ILLUS,NEW SPP) /CABRAL & KIRSCHEN, BOL. SOC. ARGENTINA BOT. 18:115-120,,1977 (ILLUS SUB CHAETASBOLISIA LONGISETA)
?=*CHALAROPSIS* PEYRONEL 1916, PRO PARTE
?=*CREMASTERIA* MEYERS & MOORE 1960
?=*GILMANIELLA* BARRON 1964
?=*HUMICOLOPSIS* CABRAL & MARCHAND 1976 PRO PARTE
=*MELANOGONE* WOLLENWEBER & RICHTER 1934 FIDE MASON 1941
?=*MYCOCHLAMYS* MARCHAND & CABRAL 1976 •

HUMICOLOPSIS CABRAL & MARCHAND 1976. IN MARCHAND ET AL., BOL. SOC. ARGENTINA BOT. 17:68-71 (ILLUS) /H. CEPHALOSPORIOIDES

?, A DOUBLE FORM-GENUS BASED ON A HUMICOLA STATE & "PHIALOCONIDIA". •

HUMICOLOPSIS VERONA 1977 (G. BOT. ITAL. 111:88) =HOMONYM OF HUMICOLOPSIS CABRAL & MARCHAND •

HYALOBOTRYS PIDOPLICHKO 1948. MIKROBIOL. ZH. 9(2-3):54-57 (ILLUS) /H. ELEGANS =STACHYBOTRYS CORDA 1837 FIDE CARMICHAEL 1973. •

HYALOCERAS DURIEU & MONTAGNE 1848 (FL. ALG. 587) =MONOCHAETIA • FLORA D'ALGERIE p 587

HYALOCLADIUM MOUSTAFA 1976. TBMS 67:537 /H. MOUBASHERI
DICTYO,CETERI,HYALO,RACHI, 97C, ON AIR. •

HYALOCYLINDROPHORA CRANE & DUMONT 1978. CAN. J. BOT. 56:2613-2616 (ILLUS) /H. VENEZUELENSIS
AMERO,GLOIO,HYALO,PHIAL,SYNNEMATA, 11E, ON WOOD. •

HYALODENDRON DIDDENS 1934. ZENTRALBL. BAKT. PARASIT. INFEKTIONSKR. ABT. II, 90:316-319 /H. LIGNICOLA
AMERO,BLASTO,HYALO,RADUL, 38C, ASCO STATE =CERATOCYSTIS FIDE UPADHYAY & KENDRICK 1975 (MYCOLOGIA 67:798-805), OR OPHIOSTOMA, CEPHALOASCUS OR ENDOMYCES FIDE REDHEAD & MALLOCH 1977 (CAN. J. BOT. 55:1701-1707,ILLUS), OR ?SARCOSCYPHA SEE PADEN 1972 (PERSOONIA 6:407). BASIDIO STATE ?=CONIODICTYUM (Q.V.).
/BARRON, HYPHOMYCETES P 208-210,1968 (ILLUS) /MATSUSHIMA, ICONES P 85 & PL,1975 (ILLUS) •

HYALODICTYS SUBRAMANIAN 1962. PROC. INDIAN ACAD. SCI., SECT. B 55:7-9 (ILLUS) /CLASTEROSPORIUM DEGENERANS SYDOW & SYDOW =H. DEGENERANS (SYDOW & SYDOW)
=MIURAEA HARA 1948 FIDE DEIGHTON 1973 (CMI PAPERS 133:3). •

HYALOFLOREA BATISTA & MAIA 1955. ANN. SOC. BIOL. PERNAMBUCO 13:154-156 (ILLUS) /H. RAMOSA
=CYLINDROCARPON WOLLENWEBER 1913 FIDE CARMICHAEL. •

HYALOHELICOMINA YOKOYAMA 1974. TRANS. MYCOL. SOC. JAPAN 15:158-160 (NO ILLUS) /HELICOMINA DEUTZIAE YOKOYAMA 1959 (J. FAC. AGR. HOKKAIDO UNIV. 51:191-193 FIG 1-4) =H. DEUTZIAE (YOKOYAMA)
HELICO,CETERI,HYALO,RADUL, 102D, ON LEAVES (DEUTZIA). ?=PSEUDOCERCOSPORA SPEGAZZINI 1910. •

HYALOPESOTUM UPADHYAY & KENDRICK 1975. MYCOLOGIA 67:801-802 (798-805, ILLUS) /H. INTROCITRINUM (AS 'INTROCITRINA') =H. STATE OF CERATOCYSTIS INTROCITRINA OLCHOWECKI & REID
AMERO,GLOIO,HYALO,RACHI, SEE 29B, ON TREES (BETULA). ASCO STATE =CERATOCYSTIS FIDE UPADHYAY & KENDRICK 1975. •

HYALOPUS CORDA 1838. ICONES FUNGORUM II:16 /?, 6 ORIGINAL SPECIES TRANSFERRED FROM STILBUM
?, PROBABLY A SYNONYM OF ACREMONIUM LINK FIDE LINDAU 1907 SEE ALSO GAMS 1971 (CEPHALO. SCHIMMEL. P 13). •

HYALORHINOCLADIELLA UPADHYAY & KENDRICK 1975. MYCOLOGIA 67:800 (799-805, ILLUS) /H. MINUTA-BICOLOR =H. STATE OF CERATOCYSTIOPSIS MINUTA-BICOLOR (DAVIDSON)
AMERO,CETERI OR GLOIO,HYALO,RACHI, 23B, ON BEETLES (IPS) ON TREES (PINUS). ASCO STATE =CERATOCYSTIOPSIS OR CERATOCYSTIS FIDE UPADHYAY & KENDRICK 1975. •

HYALOSTACHYBOTRYS SRINIVASAN 1958. J. INDIAN BOT. SOC. 37:340 (ILLUS P 336) /H. BISBYI
=STACHYBOTRYS CORDA FIDE BARRON 1968. •

HYALOSYNNEMA MATSUSHIMA 1975. ICONES P 85 & PL 243:1-4, PL 244:1-2 /H. MULTISEPTATUM
PHRAGMO,CETERI,HYALO,RACHI,SYNNEMATA, 86A, ON SOIL. •

HYALOTROCHOPHORA FINLEY & MORRIS 1967. IN MORRIS & FINLEY, AMER. MIDL. NAT. 77:202-204, FIGS 4-7 /EVERHARTIA LIGNITALIS THAXTER 1891 =H. LIGNITALIS (THAXTER)
?=DELORTIA PATOUILLARD & GAILL. 1888. •

HYMENELLA FRIES 1823. SYSTEMA MYCOL. 2:233-234 (NO ILLUS) /THREE ORIGINAL SPECIES. CLEMENTS & SHEAR (1931, P 400) DIFFER FROM TULLOCH (1972) IN CHOICE OF LECTOTYPE
?
/LINDAU, RABENH. KRYPT. FL. 2 AUFL. 1 BD. 9:629-630,1910 (ILLUS) /TULLOCH, CMI PAPERS 130:6,1972 (DESCRIPTION) /HAWKSWORTH & WALLER, CMI DESCRIPTIONS 501,1976 (ILLUS)
=*HYMENULA* FRIES 1828 FIDE TULLOCH 1972 •

HYMENOBACTRON (SACCARDO) HOHNEL 1923. UNTERSUCH. UND BERICHTE 1(3):342 (NO ILLUS) /H. DESMAZIERI (CASTAGNE) HOHNEL 1923
? •

HYMENOPODIUM CORDA 1837. ICONES FUNGORUM 1:7 /H. SARCOPODIOIDES =CLASTEROSPORIUM CARICINUM SCHWEINITZ
=CLASTEROSPORIUM SCHWEINITZ 1832 FIDE HUGHES 1958 •

HYMENOPSIS SACCARDO 1886. SYLLOGE FUNGORUM 4:744-747 /SIXTEEN ORIGINAL SPECIES, FIRST MENTIONED WAS H. TROCHILOIDES (SACCARDO)
?, SEE TULLOCH 1972 (CMI PAPERS 130:40). •

HYMENOSTILBE PETCH 1931. THE NATURALIST 1931:101 /H. MUSCARIUM
AMERO,CETERI,HYALO,RADUL,SYNNEMATA, 28B, ON INSECTS & SPIDERS. ASCO STATE =CORDYCEPS FIDE MORRIS 1963.
/MORRIS, WESTERN ILLINOIS UNIV. SER. BIOL. SCI. 3:74-75,1963 (ILLUS) /SAMSON & EVANS, KON. NEDERL. AKAD. VAN WET., AMSTERDAM C,78:73-80,1975 (REVIS,ILLUS,KEY) /BRADY, CMI DESCRIPTIONS 604,1979 (ILLUS) •

HYMENULA FRIES 1828, AN UNNECESSARY REPLACEMENT FOR HYMENELLA FRIES 1823. NON HYMENELLA MOCHINO & SESSE EX DE CANDOLLE 1824 •

HYPEROMYXA CORDA 1839. ICONES FUNGORUM 3:34 /H. STILBOSPOROIDES =CHEIROSPORA BOTRYOSPORA (MONT.) HUGHES
=CHEIROSPORA MOUGEOT 1825 FIDE HUGHES 1958. •

HYPHA PERSOON 1818. TRAITE CHAMP. COMEST. P 63 /H. BOMBYCINA (ROTH) (LECTOTYPE)
=NOMEN DUBIUM, SEE DONK 1962.
/DONK, TAXON 11:86,1962 (REFS)
=*CLADOBYSSUS* RITGEN 1831 FIDE SACCARDO 1901 •

HYPHASMA REBENTISCH 1804 =NOMEN DUBIUM FIDE DONK 1962 (TAXON 11:86) •

HYPHELIA FRIES 1825. SYST. ORB. VEG. P 149 /H. ROSEA (PERSOON) FRIES 1829 (SYST. MYCOL. 3:211) =TRICHOTHECIUM ROSEUM (PERSOON) LINK
=TRICHOTHECIUM LINK 1809 FIDE HUGHES 1958. BUT SEE DONK 1962 (TAXON 11:86-88). •

HYPHOCHLAENA CIFERRI 1962. ATTI IST. BOT. UNIV. PAVIA SER. 5, 19:110 /H. CAESPITOSA
? •

HYPHODERMA FRIES 1849 =HYPHELIA FRIES 1825 SEE DONK 1962 (TAXON 11:88) •

HYPHODICTYON MILLARDET 1866 =ATICHIA FLOTOW 1850 FIDE MEEKER 1975 •

HYPHODISCOSIA LODHA & CHANDRA REDDY 1974. TBMS 62:418-421 (ILLUS) /H. JAIPURENSIS
DIDYMO,CETERI OR GLOIO,HYALO,RADUL,APPEND, 56A, ON LITTER. COMPARE WITH LEPTODISCELLA.
/MATSUSHIMA, ICONES P 85-86 & PL 235:1-4,1975 (ILLUS) /ELLIS, MDH P 169-170,1976 (ILLUS) •

'HYPHOMYCES' IS BASED ON 'H. DESTRUENS', WHICH IS A SPECIES OF PYTHIUM FIDE AUSTWICK & COPLAND 1974 (NATURE 250:84). •

HYPHOPOLYNEMA NAG RAJ 1977. CAN. J. BOT. 55:760-765 (ILLUS) /H. TROPICALE
AMERO,GLOIO?,HYALO,PHIAL,APPEND,SPORODOCHIA, 56F, ON FRUITS (INGA). COMPARE WITH MINIMIDOCHIUM & VOLUTELLA. •

HYPHOSOMA SYDOW 1924. ANN. MYCOL. 22:315-317 /H. HYPOXYLOIDES
=NOMEN CONFUSUM FIDE HUGHES 1970. •

IDRIELLA NELSON & WILHELM 1956. MYCOLOGIA 48:548-550 (ILLUS) /I. LUNATA
SCOLECO,CETERI OR GLOIO,HYALO,RADUL, 21B, ON SOIL, ROOTS (FRAGARIA) & LITTER. DIFFERS FROM NODULISPORIUM IN SHAPE OF CONIDIA. COMPARE WITH MICRODOCHIUM. OTHER CONIDIAL STATE =?HUMICOLA, MONODICTYS OR DACTYLARIA SEE MATSUSHIMA 1975 (PL 182 & 217-219). ASCO STATE =HYMENOSCYPHUS FIDE KIMBROUGH 1972 (AMER. J. BOT. 59:165-171).
/ELLIS, DH P 199,1971 (ILLUS) /MATSUSHIMA, MICROFUNGI P 30-31 & FIG 22-24,1971 (ILLUS,NEW SPP) /NICOT & CHARPENTIE, BULL. SOC. MYCOL. FR. 87:421-424,1971 (ILLUS) /NICOT & MOUCHACCA, REV. MYCOL. 36:185-193,1972 (ILLUS,NEW SPP) /MATSUSHIMA, ICONES P 86-88 & PL,1975 (ILLUS,NEW COMB) /LUNGHINI & RAMBELLI, GIORN. BOT. ITAL. 112:177-185,1978 (ILLUS,NEW SP) /MORGAN-JONES, MYCOTAXON 8:402-410,1979 (ILLUS,NEW SP)
=*CHLORIDIELLA* ARNAUD 1954 (NOMEN NUDUM) FIDE NICOT & CHARPENTIE 1971 •

ILLOSPORIUM MARTIUS 1817. FL. CRYPT. ERLANG. P 325 /I. ROSEUM
?, PROBABLY A NOMEN DUBIUM SEE HUGHES 1958. SEE STRUMELLA SACCARDO. •

INDIELLA BRUMPT 1906. ARCH. PARASIT. 10:545-549 /I. MANSONI
?, ON MAN. •

INFRAFUNGUS CIFERRI 1951. MYCOPATHOLOGIA 6:26 /FUSARIUM MICROPUS SACCARDO =I. MICROPUS (SACCARDO)
?, NOT ILLUSTRATED BY CIFERRI. •

INGOLDIA PETERSEN 1962. MYCOLOGIA 54:146-148 (ILLUS) /I. CRAGINIFORMIS
=GYOERFFIELLA KOL 1928 FIDE VON ARX 1970. •

INGOLDIELLA SHAW 1972. TBMS 59:255-259 (ILLUS) /I. HAMATA
STAURO,CETERI,HYALO,NONSP,CLAMPS,AQUATIC, 124A. HAS CLAMP CONNECTIONS ON HYPHAE & CONIDIA.
/NAWAWI, TBMS 61:521-528,1973 (ILLUS,NEW SP) /INGOLD, GUIDE P 88-89,1975 (ILLUS) •

INNATOSPORA VAN BEYMA 1929. VERH. K. AKAD. WET., SECT. 2, 26:5-7 (ILLUS) /I. ROSEA
=ARTHRINIUM KUNZE 1817 FIDE ELLIS 1971. •

INSECTICOLA MAINS 1950. MYCOLOGIA 42:577-578 & FIG 20 ON P 584 /I. CLAVATA
=AKANTHOMYCES LEBERT 1858 FIDE KENDRICK.
/MORRIS, WESTERN ILLINOIS UNIV. SER. BIOL. SCI. 3:76-77,1963 (ILLUS) •

INTRAPES HENNEN & FIGUEIREDO 1979. MYCOLOGIA 71:836-840 (ILLUS) /I. PALIFORMIS
AMERO,CETERI,HYALO,PHIAL OR ANNEL,SPORODOCHIA, ON LEAVES (COUEPIA).
?=UREDINALES. •

INVOLUTISPORONITES CLARKE 1965 (MOUNTAIN GEOLOGIST 2:90) -FOSSIL FUNGI IMPERFECTI •

IRPICOMYCES DEIGHTON 1969. CMI PAPERS 118:24-26 (ILLUS) /I. SCHIFFNERULAE
SCOLECO,CETERI,PHAEO,NONSP, 110C, ON FUNGI (SCHIFFNERULA) ON LEAVES (SOLANUM). COMPARE WITH SPERMOSPORA. •

ISARIA PERSOON 1801. SYN. METH. FUNG. P 687 /? - SEE HUGHES 1958 (CAN. J. BOT. 36:781)
=NOMEN DUBIUM - SEE BARRON 1968 (HYPHOMYCETES P 212-213). MANY FUNGI PLACED IN ISARIA BELONG IN BEAUVERIA OR PAECILOMYCES. SEE BEAUVERIA FOR

DISPOSITION OF I. FELINA. SEE DE HOOG 1972 (CBS STUDIES 1:14-19,ILLUS,REVIS) & SAMSON 1974 (CBS STUDIES 6:1-119,MONOGR OF PAECILOMYCES). •

ISARIELLA HENNINGS 1909. HEDWIGIA 48:19-20 /I. AUERSWALDIAE
?, NOT ILLUSTRATED BY AUTHOR. •

ISARIOPSELLA HOHNEL 1917. BER. DEUTSCH. BOT. GES. 35:357 (NO ILLUS) /RAMULARIA VOSSIANA THUMEN =I. VOSSIANA (THUMEN)
?, NOT ILLUSTRATED BY HOHNEL.
/HOHNEL, MITTEIL. BOT. INST. TECHN. HOCHSCH. WIEN VI, 2 HEFT, P 68, 1929 (NO ILLUS) •

ISARIOPSIS FRESENIUS 1863. BEITRAGE ZUR MYKOLOGIE 3:87-88 & FIGS 18-28 /I. PUSILIS
?=CERCOSPORA SEE HODGES & HAASIS 1962 (MYCOLOGIA 54:448-454). •

ISTHMOLONGISPORA MATSUSHIMA 1971. MICROFUNGI P 32-33 & FIG 50 & PL 12 /I. INTERMEDIA
DIDYMO OR PHRAGMO OR SCOLECO,CETERI,HYALO,RADUL, 30C, ON LITTER. COMPARE WITH WIESNERIOMYCES.
/MATSUSHIMA, ICONES P 89-91 & PL,1975 (ILLUS,NEW SPP) •

ISTHMOSPORA STEVENS 1918. BOT. GAZETTE 45:244-245 /I. SPINOSA =I. STATE OF TRICHOTHYRIUM ASTEROPHORUM FIDE HUGHES 1953
DICTYO OR STAURO,CETERI,PHAEO,MISC, 96H, ON FUNGI (MELIOLA & OTHERS). ASCO STATE =TRICHOTHYRIUM FIDE HUGHES 1953.
/HUGHES, CMI PAPERS 50:77-97,1953 (ILLUS) /ELLIS, DH P 38-39,1971 (ILLUS)
=*CARLOSIA* ARNAUD 1954 FIDE KENDRICK
?=*TETRACHIA* SACCARDO 1918 •

ISTHMOTRICLADIA MATSUSHIMA 1971. BULL. NAT. SCI. MUS. TOKYO 14:478-480, FIG 51 /I. LAEENSIS
STAURO,CETERI,HYALO,NONSP,AQUATIC, 119F.
/NAWAWI, TBMS 64:243-246,1975 (ILLUS,NEW SP) /INGOLD, GUIDE P 42-43,1975 (ILLUS) /MATSUSHIMA, MICROFUNGI P 33 & FIG 48-2 & PL 15-27,1971 (ILLUS) /MATSUSHIMA, ICONES P 168 & PL,1975 (ILLUS SUB MFC-4216) •

ITERSONILIA DERX 1948. BULL. JARD. BOT. BUITENZORG SER 3,17:465-469 & 471-472 (ILLUS) /I. PERPLEXANS
AMERO,CETERI,HYALO,MISC,CLAMPS, 52B, ON FUNGI (KUNKELIA, PUCCINIA) & PLANTS.
/OLIVE, BULL. TORREY BOT. CLUB 79:126-138,1952 (ILLUS,REFS) /BARNETT, IMPERFECT FUNGI NO. 33,1960 (ILLUS) /SOWELL & KORF, MYCOLOGIA 52:934-945,1960 (EMEND,REFS)
=*ENTYLOMELLA* HOHNEL 1916 (NOMEN NUDUM) FIDE SOWELL & KORF 1960 •

IYENGARINA SUBRAMANIAN 1958. J. INDIAN BOT. SOC. 37:404-407 (ILLUS) /I. ELEGANS
STAURO,CETERI,PHAEO,ANNEL,APPEND, 112G, ON LITTER. COMPARE WITH DIPLOCLADIELLA & BIFLAGELLOSPORA.
/ELLIS, DH P 132-133,1971 (ILLUS) •

JACOBASCHELLA KUNTZE 1891. REV. GEN. PL 1:280 /J. ALBA (BONORDEN)
?, FOR DIPLOSPORIUM BONORDEN 1851 NON LINK 1824. •

JACOBIA ARNAUD 1951. BULL. SOC. MYCOL. FRANCE 67:195 & FIG 4 A-C (NO LATIN DIAGNOSIS) /J. CONSPICUA
AMERO,GLOIO,HYALO,RADUL?,AMPUL?,CLAMPS, 35H, ON FUNGI (LASIOSPHAERIA) ON LITTER.
/NICOT & CHARPENTIE, BULL. SOC. MYCOL. FRANCE 87:32,1971 (TYPE SPECIMEN) •

JACULISPORA HUDSON & INGOLD 1960. TBMS 43:475-478 (ILLUS) /J. SUBMERSA
AMERO OR STAURO,CETERI,HYALO,NONSP,APPEND,AQUATIC, 120D.
/INGOLD, GUIDE P 44-45,1975 (ILLUS) •

JACZEWSKIELLA MURASHKINSKI 1926. MAT. MIK. FIT. 5:5-6 (ILLUS) /J. ALTAJENSIS
?, ORIGINAL ILLUSTRATION DIFFICULT TO INTERPRET. •

JAINESIA FRAGOSO & CIFERRI 1925. BOL. ESPAN. HIST. NAT. 25:514 (NO ILLUS) /J. MELIOLICOLA

?=ASCOSPORES & HYPHOPODIATE MYCELIUM OF MELIOLA. •

JANETIA ELLIS 1976. MDH P 33-34 (ILLUS) /J. EUPHORBIAE
PHRAGMO,CETERI,PHAEO,RADUL, 79D, ON LITTER. •

JOHNSTONIA ELLIS 1971. CMI PAPERS 125:13-15 (ILLUS) /J. COLOCASIAE
DIDYMO,CETERI,PHAEO,NONSP, 64F.
/ELLIS MDH P 196-197,1976 (ILLUS) •

KABATIELLA BUBAK 1907. IN BUBAK & KAMAT, HEDWIGIA 46:297 /K. MICROSTICTA
AMERO,GLOIO,HYALO,AMPUL,ACERVULI, ON LEAVES OF MANY PLANTS. OTHER CONIDIAL STATE =AUREOBASIDIUM SEE HERMANIDES-NIJHOF 1977.
/VON ARX, BIBLIOTHECA MYCOLOGIA (CRAMER) BD. 24:69-82,1970 (ILLUS,REVIS) /VON ARX, GENERA P 171-172,1974 (ILLUS) /HERMANIDES-NIJHOF, CBS STUDIES 15:143-166,1977 (ILLUS,KEY)
=*EXOBASIDIOPSIS* KARAKULIN 1922 FIDE VON ARX 1970
=*PACHYBASIDIELLA* BUBAK & SYDOW 1915 FIDE VON ARX 1970
=*POLYSPORA* LAFFERTY 1921 FIDE VON ARX 1970
=*PROTOCORONOSPORA* ATKINSON & EDGERTON 1907
=*PROTOCORONIS* CLEMENTS & SHEAR 1931 FIDE VON ARX 1970
=*RHABDOGLOEOPSIS* PETRAK 1925 FIDE VON ARX 1970 •

KABATINA SCHNEIDER & VON ARX 1966. PHYTOPATH. ZEITSCHR. 57:176-182 (ILLUS) /K. THUJAE
AMERO,GLOIO,HYALO,PHIAL,SPORODOCHIA, ON BRANCHES (THUJA, JUNIPERUS). OTHER CONIDIAL STATE =HORMONEMA FIDE HERMANIDES-NIJHOF 1977 (CBS STUDIES 15:142,ILLUS). •

KAFIADDINIA MEKHTIYEVA 1978. MIKOL. FITOPATOL. 12:8-10 /K. FUSARISPORA
=DACTYLARIA SEE BARRON 1977 (THE NEMATODE-DESTROYING FUNGI P 38). •

KAMATIA RAO & SUBHEDAR 1976. TBMS 66:539-541 (ILLUS) /K. INDICA
=PSEUDODICTYOSPORIUM MATSUSHIMA 1971 FIDE KENDRICK •

KARAKULINIA GOLOVINA 1964. NOV. SIST. NIZ. RAST. P 212-213 /FUSICLADIOPSIS CERASI (RABENHORST) KARAKULIN & VASSILJEOSKY 1937 (LECTOTYPE)
=CLADOSPORIUM FIDE VON ARX 1973 (VERH. KON. NEDERL. AKAD. WET. AFT. NAT. 61:62). PROPOSED AS A NOM. NOV. FOR FUSICLADIOPSIS KARAKULIN & VASSILJEOSKY 1937 NON MAIRE 1906. •

KAUFMANNWOLFIA GALGOCZY & NOVAK 1962. ON THE WORK OF THE NATIONAL INST. FOR PUBLIC HEALTH FOR 1961, P 224-225 /K. INTERDIGITALIS (PRIESTLY) =TRICHOPHYTON MENTAGROPHYTES (ROBIN) BLANCHARD 1896
=TRICHOPHYTON MALMSTEN 1845. •

KAZULIA NAG RAJ 1977. CAN. J. BOT. 55:1621 (1619-1622,ILLUS) /YPSILONIA VAGANS SPEGAZZINI 1908 =K. VAGANS (SPEGAZZINI)
STAURO,CETERI,RADUL,HYALO,SPORODOCHIA,SETAE, 118E, ON FUNGI (ZUKALIA) ON LEAVES (SPIREA). ASCO STATE ?=ZUKALIA SEE NAG RAJ 1977 (P 1621). •

KERATINOMYCES VANBREUSEGHEM 1952. BULL. ACAD. BELG. CL. SCI. 38:1075-1076 (ILLUS) /K. AJELLOI =K. STATE OF ARTHRODERMA UNCINATUM FIDE DAWSON & GENTLES 1961
=TRICHOPHYTON FIDE AJELLO 1968.
/TUBAKI, J. HATTORI BOT. LAB. 20:184,1958 (ILLUS) •

KMETIA BRESADOLA & SACCARDO 1902. IN SACCARDO, SYLL. FUNG. 16:1158 /K. EXIGUA
SCOLECO,GLOIO,HYALO,PHIAL, 107D, ON BARK (ABIES).
/COOKE, MYCOLOGIA 44:808-809,1952 (DESCR) •

KMETIOPSIS BATISTA & PERES 1960. PUBL. INST. MIC. UNIV. RECIFE 245:4 (1-9,ILLUS) /K. HYMENAEAE
AMERO OR SCOLECO,GLOIO,HYALO,PHIAL, 107F, ON LEAVES (HYMENAEA). •

KNYARIA KUNTZE 1891 (REV. GEN. PLANT. 2:855-856) =TUBERCULARIA FIDE AINSWORTH 1971 •

KOORCHALOMA SUBRAMANIAN 1953. J. INDIAN BOT. SOC. 32:124 /K. MADREEYA
AMERO,GLOIO,HYALO,PHIAL,SPORODOCHIA,SETAE,APPEND, 12B, ON LITTER (ORYZA).
/TULLOCH, CMI PAPERS 130:7,1972 (DESCR)
?=*KOORCHALOMELLA* CHONA, MUNJAL & KAPOOR 1958
?=*AMPHITIAROSPORA* AGNIHOTHRUDU 1963 •

KOORCHALOMELLA CHONA, MUNJAL & KAPOOR 1958. INDIAN PHYTOPATH. 11:130-132 /K. ORYZAE
?=KOORCHALOMA SUBRAMANIAN 1953. •

KORDYANELLA HOHNEL 1904. ANN. MYCOL. 2:274 /K. AUSTRIACA
?, NOT ILLUSTRATED BY AUTHOR. •

KOSTERMANSINDA RIFAI 1968. REINWARDTIA 7:375-381 (ILLUS) /SCLEROGRAPHIUM MAGNUM BOEDIJN =K. MAGNA (BOEDIJN)
DICTYO,CETERI,PHAEO,NONSP,SYNNEMATA, 101D, ON LITTER (PANICUM & SACCHARUM).
/ELLIS, DH P 112,1971 (ILLUS) •

KRAMASAMUHA SUBRAMANIAN & VITTAL 1973. CAN. J. BOT. 51:1127-1132 (ILLUS) /K. SIBIKA
PHRAGMO,CETERI,PHAEO,MISC, 92A, ON LITTER.
/MATSUSHIMA, ICONES P 91 & PL,1975 (ILLUS) •

KUMANASAMUHA RAO & RAO 1964. MYCOPATHOLOGIA 22:330-334 (ILLUS) /K. SUNDARA
AMERO,CETERI,PHAEO,RADUL, 30F, ON WOOD.
/ELLIS, DH P 341-343,1971 (ILLUS) /MATSUSHIMA, ICONES P 91 & PL,1975 (ILLUS)
/HUNTER & KENDRICK, NEW ZEALAND J. BOT. 15:585,1977 (NEW SP,ILLUS) •

KURDSAWAIA HARA 1954 =NOMEN DUBIUM SEE DE HOOG & HERMANIDES-NIJHOF 1977 (CBS STUDIES 15:185) •

KURSSANOVIA PIDOPLICHKO 1948. MIKROBIOL. ZH., KIEV 9(1-2):57-58 (ILLUS) /K. TRISEPTATA
=FUSARIELLA SACCARDO 1884 FIDE VON ARX 1974 (GENERA P 205). •

KUTILAKESA SUBRAMANIAN 1956. J. INDIAN BOT. SOC. 35:478-481 (ILLUS) /K. MADREEYA
=MYROTHECIUM TODE EX LINK FIDE NAG RAJ. •

KUTILAKESOPSIS AGNIHOTHRUDU & BARUA 1957. J. INDIAN BOT. SOC. 36:306-311 (ILLUS) /K. MACALPINEI
DIDYMO,GLOIO,HYALO,PHIAL,SPORODOCHIA,SETAE, 58A, ON STEMS (ALBIZZIA). COMPARE WITH DIDYMOSTILBE. •

LACELLINA SACCARDO 1913. IN TROTTER & SACCARDO, ANN. MYCOL. BERL. 11:418-419 /L. LIBYCA SACCARDO & TROTTER 1913 =L. GRAMINICOLA (BERKELEY & BROOME) PETCH 1924
AMERO,BLASTO,PHAEO,MISC,STROMA,SETAE, 37D, ON PLANTS (GRASSES & PALMS). COMPARE WITH HERPOSIRA.
/ELLIS, CMI PAPERS 67:10-15,1957 (ILLUS,KEY) /ELLIS, DH P 171-173,1971 (ILLUS,KEY) •

LACELLINOPSIS SUBRAMANIAN 1953. PROC. INDIAN ACAD. SCI. 37:100-105 /L. SACCHARI
AMERO,BLASTO,PHAEO,MISC,STROMA,SETAE, 37C, ON LEAVES (MANY GRASSES) & LITTER.
/ELLIS, CMI PAPERS 67:6-10,1957 (ILLUS,KEY) /ELLIS, DH P 355-356,1971 (ILLUS,KEY) •

LACHNIDIUM GIARD 1891 (C. R. ACAD. SCI. 112:1520) =FUSARIUM FIDE SACCARDO FROM AINSWORTH 1971 •

LACHNODOCHIUM MARCHAL 1895. BULL. SOC. ROY. BOT. BELG. 34:144-145 & PL 1, FIG 2 /L. CANDIDUM
?
/LINDAU, RABENHORST KRYPT-FL. 2 AUFL. 1 BD. 9:472-473,1910 (ILLUS) •

LACINIOCLADIUM PETRI 1917. ANN. IST. SUP. AGR. FOR. NAZ. 2:31-32 /L. CASTANEAE
?, ORIGINAL ILLUSTRATION NOT DIAGNOSTIC. •

LACRIMASPORONITES CLARKE 1965 (MOUNTAIN GEOLOGIST 2:87) -FOSSIL FUNGI IMPERFECTI •

LACTYDINA SUBRAMANIAN 1977 (KAVAKA 5:93-98,NOT SEEN) =NOM. NOV. FOR DACTYLINA ARNAUD EX SUBRAMANIAN (SEE ABSTRACTS OF MYCOLOGY 67-52149,1979). •

LAMBDASPORIUM MATSUSHIMA 1971. IN KOBAYASI & AL., BULL. NAT. SCI. MUS. TOKYO 14:467-468 /L. WAUENSE
STAURO,CETERI,HYALO OR PHAEO,NONSP, 122B, ON LITTER.
/MATSUSHIMA, MICROFUNGI P 34 & FIG 46 & PL 15-5,1971 (ILLUS) •

LANGERONIA VANBREUSEGHEM 1950. ANN. SOC. BELGE MED. TROP. 30:885-890 (NO LATIN DIAGNOSIS) /L. SOUDANENSIS (JOYEUX)
=TRICHOPHYTON MALMSTEN 1845 FIDE AJELLO 1968. •

LANGLOISULA ELLIS & EVERHART 1889. JOUR. MYCOL. 5:68-69 /L. SPINOSA
=NOMEN CONFUSUM FIDE AINSWORTH 1971. •

LAPPODOCHIUM MATSUSHIMA 1975. ICONES P 91-92 & PL 242:1-5 /L. LAGENIFORME
AMERO,CETERI OR GLOIO?,PHAEO,RADUL,SPORODOCHIA,SETAE,APPEND, 30H, ON SOIL. •

LARIDOSPORA NAWAWI 1976. TBMS 66:344-347 (ILLUS) /DACTYLELLA APPENDICULATA ANASTASIOU 1964 =L. APPENDICULATA (ANASTASIOU)
PHRAGMO OR STAURO,CETERI,HYALO,NONSP,APPEND, 112E, ON LITTER IN WATER. •

LATERIRAMULOSA MATSUSHIMA 1971. MICROFUNGI P 34-35 & FIG 49,1-2 & PL 14-4 /L. UNIINFLATA
STAURO,CETERI,HYALO,NONSP,AQUATIC, 121D, COMPARE WITH TRICELLULA.
/MATSUSHIMA, ICONES P 92-93 & PL,1975 (ILLUS,NEW SPP) /INGOLD, GUIDE P 44-45,1975 (ILLUS) •

LEANDRIA RANGEL 1915. BOL. AGR. SAO PAULO 16:324-325 & FIGS 4-15 /L. MOMORDICAE
?, ORIGINAL ILLUSTRATION DIFFICULT TO INTERPRET. •

LECYTHOPHORA NANNFELDT 1934. IN MELIN & NANNFELDT, SVENSKA SKOGSV. TIDSKR. 2-4:433-436 (ILLUS) /L. LIGNICOLA
=PHIALOPHORA MEDLAR FIDE GOIDANICH SEE SCHOL-SCHWARZ 1970 (PERSOONIA 6:61). •

LEIGHTONIOMYCES HAWKSWORTH & SUTTON 1977. BOT. J. LINN. SOC. 75:195-209 (ILLUS) /PERICONIA PHILLIPSII BERKELEY & LEIGHTON 1875 =L. PHILLIPSII (BERKELEY & LEIGHTON)
AMERO,BASO,PHAEO,ANNEL,SYNNEMATA, 19C, ON LICHENS (THROMBIUM, STEINIA). COMPARE WITH CEPHALOTRICHUM.
/ELLIS, DH P 330-331,1971 (ILLUS SUB DORATOMYCES) •

LEIOSEPIUM SACCARDO 1900. IN SACCARDO & FAUTREY, BULL. SOC. MYCOL. FRANCE 16:24 /L. AUREUM SACCARDO & FAUTREY 1900 =SEPEDONIUM CHLORINUM (TULASNE) DAMON 1952
=SEPEDONIUM LINK 1809 FIDE DAMON 1952 (MYCOLOGIA 44:95). •

LEMONNIERA DE WILDEMAN 1894. ANN. SOC. BELGE MICR. 18:143-148 (ILLUS) /L. AQUATICA
STAURO,CETERI,HYALO,PHIAL,AQUATIC, 120A, OTHER STATE ="SCLEROTIA".
/TUBAKI, BULL. NAT. SCI. MUS. TOKYO 3:259-260,1957 (ILLUS) /VON ARX, GENERA P 188-189,1970 (ILLUS) /INGOLD, GUIDE P 72-74,1975 (ILLUS) /DESCALS, WEBSTER & DYKO, TBMS 69:89-109,1977 (EMEND,NEW SPP,ILLUS) /SINCLAIR & MORGAN-JONES, MYCOTAXON 9:469-472,1979 (ILLUS,NEW SP) •

LEPIDOPHYTON TRIBONDEAU 1899. ARCH. MED. NAVALE 72:5-52 /NO SPECIES DESCRIBED
=NOMEN ILLEGITIMUM. •

LEPTODISCELLA PAPENDORF 1969. TBMS 53:145-146 (NO ILLUS) /LEPTODISCUS AFRICANA PAPENDORF 1967 (TBMS 50:687,ILLUS) =L. AFRICANA (PAPENDORF)

DIDYMO,GLOIO,HYALO,RADUL,APPEND,SPORODOCHIA, 56C, ON SOIL & LITTER (ACACIA). COMPARE WITH HYPHODISCOSIA & NETA.
/MATSUSHIMA, ICONES P 93 & PL,1975 (ILLUS,NEW SP) •

LEPTODISCUS GERDEMANN 1953. MYCOLOGIA 45:552 /L. TERRESTRIS
=MYCOLEPTODISCUS OSTAZESKI, NON LEPTODISCUS HERTWIG (A GENUS OF ALGAE). •

LEPTODONTIDIUM DE HOOG 1979. TAXON 28:347-348 (NO ILLUS) /RHINOCLADIELLA ELATIOR MANGENOT 1952 =L. ELATIUS (MANGENOT) DE HOOG 1979 VAR. ELATIUS
AMERO,CETERI,HYALO OR PHAEO,RACHI, 23A, ON SOIL & LITTER (MANY PLANTS). CLOSE TO RHINOCLADIELLA.
/DE HOOG, CBS STUDIES 15:44 (44-59),1977 (KEY,ILLUS SUB LEPTODONTIUM)
=*LEPTODONTIUM* DE HOOG 1977 •

LEPTODONTIUM DE HOOG 1977 (CBS STUDIES 15:44) A LATER HOMONYM OF LEPTODONTIUM (MUELLER) HAMP. EX. LINDB. (A GENUS OF MOSSES). SEE LEPTODONTIDIUM. •

LEPTOGRAPHIUM LAGERBERG & MELIN 1928. IN LAGERBERG ET AL., SVENSKA SKOGSV. TIDSKR. 25:257 (DATED 1927) /L. LUNDBERGII
AMERO,GLOIO,HYALO,ANNEL, 18F, ON WOOD (CONIFERS) & INSECTS (DENDROCTONUS, IPS). ASCO STATE =CERATOCYSTIS OR EUROPHIUM FIDE BARRON 1968.
/BARRON, HYPHOMYCETES P 215-217,1968 (ILLUS) /SPENCER & GORIN, MYCOLOGIA 63:387-402,1971 (COMPARATIVE CHEMISTRY) /ELLIS, DH P 332-333,1971 (ILLUS) /MATSUSHIMA, ICONES P 93 & PL 103,1975 (ILLUS) /DALPE & NEUMANN, CAN. J. BOT. 55:2159-2167,1977 (ILLUS SUB CERATOCYSTIS PENICILLATA) /DAVIDSON, MYCOLOGIA 70:35-40,1978 (ILLUS,NEW SP) /JOOSTE, TBMS 70:152-155,1978 (ILLUS,NEW SP) •

LEPTOSPORIUM (SACCARDO) HOHNEL 1923. UNTERSUCH. UND BERICHTE 3(1):436 /FUSARIUM SALMONICOLOR BERKELEY & CURTIS
=NOMEN ILLEGITIMUM, NON LEPTOSPORIUM BONORDEN 1851. •

LEPTOTRICHUM CORDA 1842. ICONES FUNGORUM 5:10 & 51 & TAF. II, FIG 16 /L. GLAUCUM
?
/LINDAU, RABENHORST KRYPT. FL. 2 AUFL., 1 BD., 9:502,1910 (ILLUS) •

LEPTOXYPHIUM SPEGAZZINI 1918. PHYSIS (BUENOS AIRES) 4:281-295 /CAPNODIUM GRAMINUM PATOUILLARD 1897 =L. GRAMINUM (PATOUILLARD)
AMERO OR DIDYMO,GLOIO,HYALO OR PHAEO,PHIAL,SYNNEMATA, 58E, ON LEAVES (SOOTY MOLD).
/HUGHES, MYCOLOGIA 68:777-782,1976 (ILLUS,NEW SP)
=*ASTRAGOXYPHIUM* BATISTA, NASCIMENTO & CIFERRI 1963 FIDE HUGHES 1976
=*CALDARIOMYCES* WORONICHIN 1926 FIDE HUGHES 1976
=*MEGALOXYPHIUM* CIFERRI, BATISTA & NASCIMENTO 1956 FIDE HUGHES 1976 •

LEUCODOCHIUM SYDOW & SYDOW 1917. ANN. MYCOL. 15:266-267 (NO ILLUS) /L. PIPTURI
? •

LEUCOPENICILLIFER ARNOLD 1972. NOVIT. SYSTEM PLANT. NON VASCULARUM, ACAD. SCI. URSS, INST. BOT. KOMAROVII 8:134-136 (ILLUS) /L. GRACILIS ?=PAECILOMYCES PENICILLATUS (HOHNEL 1904) SAMSON 1974
?=PAECILOMYCES BAINIER 1907. •

LEUCOPHLEBS =ORTHOGRAPHIC VARIANT OF LEUCOPHLEPS •

LEUCOPHLEPS HARKNESS 1899. CALIF. ACAD. SCI. PROC. BOT. III, 1:257-259 & PL 42, FIG 7A-C /L. MAGNATA
=BASIDIOMYCETE FIDE FOGEL 1979 (CAN. J. BOT. 57:1718-1728, ILLUS,KEY,DISCUSSION). •

LEUCOSPORIUM CORDA 1833 (IN STURM'S DEUTSCHL. FL. BD 3, 13:67-68, TAB 34)
=ACHITONIUM KUNZE FIDE LINDAU 1910 (IN RABENHORST, KRYPT.-FL. 1(9):456-457,SUB PACTILIA) •

LIBERTELLA DESMAZIERES 1830. ANN. SCI. NAT. BOT. SER. 1, 19:275-278 /L. BETULINA (LECTOTYPE)
SCOLECO,GLOIO,HYALO,PHIAL,ACERVULI, ON LITTER & SOIL. (MELANCONIALES) ASCO STATE =DIATRYPE, DIATRYPELLA, QUATERNARIA OR DIAPORTHE FIDE VON ARX 1970, OR EUTYPELLA SEE JOHNSON & KUNTZ 1978 (CAN. J. BOT. 56:1518-1525,ILLUS).
/BARRON, HYPHOMYCETES P 216-218,1968 (ILLUS) •

LINDAUOMYCES KOORDERS 1907. VERH. K. AKAD. WET. AMSTERDAM, SECT 2, 13:240-245 (ILLUS) /L. JAVANICUS
=ARTHROBOTRYUM FIDE CLEMENTS & SHEAR 1931. •

LINDAUOPSIS ZAHLBRUCKNER 1906. BER. DEUTSCHL. BOT. GES. 24:144-146 & TAF. 10 /L. CALOPLACAE
?, ORIGINAL ILLUSTRATION DIFFICULT TO INTERPRET. •

LINDAVIA NIEUWLAND 1916 =SCOPULARIA PREUSS •

LINDQUISTIA SUBRAMANIAN & CHANDRASHEKARA 1977. BOL. SOC. ARGENT. BOT. 18:145-151 (ILLUS) /L. INDICA =L. STATE OF PODOSORDARIA LEPORINA
AMERO,CETERI OR ARTHRO,HYALO,AMPUL,SYNNEMATA, ON DUNG (MONKEY). ASCO STATE =PODOSORDARIA.
/FURUYA & UDAGAWA, TRANS. MYCOL. SOC. JAPAN 17:254-256,1976 (ILLUS) •

LINODOCHIUM HOHNEL 1909. S. B. AKAD. WISS. WIEN 118:1237-1239 (NO ILLUS) /L. HYALINUM (LIBERT)
SCOLECO,GLOIO,HYALO,RADUL,SPORODOCHIA, 109F & 111C, ON LITTER (PINUS).
/SUTTON, CMI PAPERS 132:73-74,1973 (ILLUS) /DYKO & SUTTON, CAN. J. BOT. 57:382-385,1979 (ILLUS) •

LISTEROMYCES PENZIG & SACCARDO 1901. MALPIGHIA 15:259 /L. INSIGNIS
PHRAGMO,CETERI,PHAEO,NONSP,ALEURI,SPORODOCHIA, 90F, ON WOOD. COMPARE WITH EXCIPULARIA & BACTRODESMIUM.
/GOOS, MYCOLOGIA 63:213-218,1971 (ILLUS) /MATSUSHIMA, ICONES P 93-94 & PL,1975 (ILLUS) /ELLIS, MDH P 134-135,1976 (ILLUS)
=*CYLOMYCES* CLEMENTS 1931 BASED ON L. INSIGNIS
?=*NYCTALOSPORA* MORRIS 1972 •

LITUARIA RIESS 1853. BOT. ZEITUNG 11:136 & TAF. III, FIGS 8-10 /L. STIGMATEA
=HELICOMA CORDA 1837 FIDE LINDER FROM AINSWORTH 1971. •

LOBATOPEDIS KIRK 1979. TBMS 73:75-79 (ILLUS) /L. FILIICOLA
PHRAGMO,BLASTO OR CETERI,PHAEO,NONSP,ALEURI?, ON LITTER (QUERCUS, FAGUS). •

LOBOA CIFERRI, AZEVEDO, CAMPOS & CARNEIRO 1956. PUBL. INST. MIC. UNIV. RECIFE 53:12 /GLENOSPORA LOBOI FONSECA & AREA LEAO 1940 =L. LOBOI (FONSECA & AREA LEAO)
=BLASTOMYCETES. LOBOA IS THE FIRST VALIDLY PUBLISHED GENUS FOR THE AGENT OF LOBOMYCOSIS. THE CULTURE DESCRIBED IN THE PROTOCOL IS A DIFFERENT FUNGUS, BUT THE NAME CAN BE RETAINED FOR THE YEAST FORM SEEN IN TISSUE, WHICH WAS ALSO BRIEFLY DESCRIBED.
/CALDWELL ET AL, AMER. J. TROP. MED. HYG. 24:105-114,1975 (ILLUS,REFS)
/BORELLI, MEDICINA CUTANEA 3:151-156,1968 (ILLUS,REFS)
=*LOBOMYCES* BORELLI 1958 NOMEN NON RITE PUBLICANDUM •

LOBOMYCES BORELLI 1958 (REVISTA DERMATOL. VENEZULANA 1:290) =PROVISIONAL NAME FOR AGENT OF LOBOMYCOSIS (SEE LOBOA). •

LOLIOMYCES MAIRE 1937. MEM. SOC. SCI. NAT. MAROC 45:140 (NO ILLUS) /L. TEMULENTIS
? •

LOMAANTHA SUBRAMANIAN 1954. J. INDIAN BOT. SOC. 33:29-33 (ILLUS) /L. POOGA
PHRAGMO,CETERI,PHAEO,NONSP,APPEND, 69B, ON LITTER (ARECA). COMPARE WITH SPORIDESMIUM BRACHYPUS, SEE ELLIS 1971 (DH P 117, FIG 77B). •

LOMACHASHAKA SUBRAMANIAN 1956. J. INDIAN BOT. SOC. 35:66-68 (ILLUS) /L. KERA
=MYROTHECIUM TODE FIDE TULLOCH 1972 (CMI PAPERS 130:36,SUB LOMAKASHAKA). •

LOMENTOSPORA HENNEBERT & DESAI 1974. MYCOTAXON 1:45 (45-50, ILLUS) /L. PROLIFICANS
AMERO,CETERI,PHAEO,RACHI, 30D, ON LITTER IN SOIL. •

LOPHOPHYTON MATRUCHOT & DASSONVILLE 1889. REV. GEN. BOT. 11:432-444 & PL 19 /L. GALLINAE (MEGNIN) =TRICHOPHYTON GALLINAE
=TRICHOPHYTON MALMSTEN 1845. •

LUNULOSPORA INGOLD 1942. TBMS 25:404-409 (ILLUS) /L. CURVULA
SCOLECO,CETERI,HYALO,NONSP?,AQUATIC, 107H.
/INGOLD, GUIDE P 84-85,1975 (ILLUS) •

LUTZIOMYCES FONSECA 1939. ANN. BRAS. DERM. SIF. 14:105-111 (ILLUS) /L. HISTOSPOROCELLULARIS (HABERFELD) =PARACOCCIDIOIDES BRASILIENSIS
=PARACOCCIDIOIDES ALMEIDA 1930 SEE DODGE 1935, P 179. •

LYCOPERDELLON TORREND 1913. BROTERIA (BOT.) 11:92 /L. TORRENDII (BRESADOLA)
=OSTRACODERMA FRIES 1825 FIDE HENNEBERT 1973 •

LYLEA MORGAN-JONES 1975. MYCOTAXON 3:129-132 (ILLUS) /L. CATENULATA
PHRAGMO,BLASTO,PHAEO,NONSP, 88C, ON LITTER (DECIDUOUS TREES) AND ON FUNGI (LOPADOSTOMA, EUTYPA, THAXTERIA & OTHERS). COMPARE WITH SEPTONEMA.
/HOLUBOVA-JECHOVA, FOLIA GEOBOT. PHYTOTAX., PRAHA 13:437-442,1978 (ILLUS,KEY,NEW COMB) •

LYSIPENICILLIUM BREFELD 1908. UNTERSUCH. GESAMMTGEB. MYKOL. 14:209-210 (NO ILLUS) /L. INSIGNE
=GLIOCLADIUM FIDE RAPER & THOM FROM AINSWORTH 1971. •

MACRAEA SUBRAMANIAN 1952 (PROC. INDIAN ACAD. SCI., SECT. B, 36:164-165) NON LINDLEY 1830 =PRATHIGADA •

MACROON CORDA 1833. IN STURM'S DEUTSCHL. FL., BD. 3, 13:93-94 & TAB 47 /M. OBSCURUM
=HELMINTHOSPORIUM FIDE SACCARDO FROM AINSWORTH 1971. •

MACROSPORIUM FRIES 1821. SYST. MYCOL. 1:XLIV /M. CHEIRANTHI (LIBERT) FRIES 1832 (SYST. MYCOL. 3:374) =ALTERNARIA CHEIRANTHI (LIBERT) BOLLE
=ALTERNARIA NEES 1817 FIDE HUGHES 1958. •

MACROSTILBUM PATOUILLARD 1898. BULL. SOC. MYCOL. FRANCE 14:197 (NO ILLUS) /M. RADICOSUM
? •

MACROTRICHUM GREVILLE 1825. EDINB. PHIL. J. 25:64-65 & PL 1 /TWO ORIGINAL SPECIES
?, ORIGINAL ILLUSTRATIONS & DESCRIPTION DIFFICULT TO INTERPRET. •

MADURELLA BRUMPT 1905. COMP. REND. SOC. BIOL. 57:997-999 /STREPTOTHRIX MYCETOMI LAVERIN 1902 =M. MYCETOMI (LAVERIN)
STERILE, ON MAN. PROPOSED FOR HYPHAE IN GRAINS IN TISSUE. SOME ISOLATES OF MADURELLA MYCETOMI HAVE A MYRIOCONIUM-LIKE STATE. OTHER STATE ?=PYRENOCHAETA (C).
/DODGE, MEDICAL MYCOLOGY P 680-685,1935 (ILLUS,KEY) /BORELLI, BOL. VENEZ. LAB. CLIN., CARACAS 2:1-15,1957 (ILLUS) •

MAGDALAENAEA ARNAUD 1952. BULL. SOC. MYCOL. FRANCE 68:209 & 211 (ILLUS,NO LATIN DIAGNOSIS) /M. MONOGRAMMA
?, NO TYPE SPECIMEN, SEE NICOT & CHARPENTIE 1971 (BULL. SOC. MYCOL. FRANCE 87:32). •

MAHABALELLA SUTTON & PATIL 1966. NOVA HEDWIGIA 11:203-207 & TAB 21 /M. ACUTISETOSA
AMERO,GLOIO,HYALO,PHIAL,SETAE,APPEND, 1B, ON LITTER (BAMBUSA). COMPARE WITH

MYCOLEPTODISCUS.
/ELLIS, DH P 517,1971 (ILLUS) •

MAHEVIA LAGARDE 1917. ARCH. ZOOL. EXPER. GEN. 56:292-294 & PL IX FIG 9-11 /M. GUIGNARDII (MAHEU)
=HIRSUTELLA PATOUILLARD 1892 FIDE CARMICHAEL 1973. •

MALBRANCHEA SACCARDO 1882. MICHELIA 2:638-639 (NO ILLUS) /M. PULCHELLA SACCARDO & PENZIG 1882
AMERO,ARTHRO,HYALO,NONSP,ALEURI, 41B, ON SOIL, DUNG, KERATIN, CELLULOSE, MAN & OTHER SUBSTRATES. ASCO STATE =GYMNOASCUS, MYXOTRICHUM, AUXARTHRON OR UNCINOCARPUS SEE SIGLER & CARMICHAEL 1976, OR PLUNKETTOMYCES FIDE ORR 1977 (MYCOTAXON 6:33-35,ILLUS) OR GYMNOASCOIDEUS FIDE ORR 1977 (MYCOTAXON 5:459-469,ILLUS) OR GYMNASCELLA FIDE ORR 1977 (MYCOLOGIA 69:130-139,ILLUS) OR MACRONODUS FIDE ORR 1977 (MYCOTAXON 5:283-290,NO ILLUS OF CONIDIA). OTHER STATE =COCCIDIOIDES FIDE SIGLER & CARMICHAEL 1976.
/COONEY & EMERSON, THERMOPHILIC FUNGI, P 93-102,1964 (ILLUS) /SIGLER & CARMICHAEL, MYCOTAXON 4:349-488 & PL,1976 (MONOGR) /COLE & SAMSON, CONIDIAL FUNGI P 120-121,1979 (ILLUS)
=*THERMOIDIUM* MIEHE 1907 FIDE SACCARDO 1908 •

MALLEOMYCES RIVOLTA ?=HYPHOMYCETE FROM AINSWORTH 1971 •

MALUSTELA BATISTA & LIMA 1960. IN BATISTA ET AL., PUBL. INST. MICOL. UNIV. RECIFE 263:5-13 /M. AERIA BATISTA ET AL. 1960 =CURVULARIA LUNATA FIDE VON ARX 1970
=CURVULARIA BOEDIJN 1933 FIDE VON ARX 1970 •

MAMMARIA CESATI 1854. IN RABENHORST, BOT. ZEITUNG 12:190 /M. ECHINOBOTRYOIDES
AMERO,CETERI,PHAEO,RADUL OR NONSP,ALEURI, 48B, ON WOOD & SOIL. OTHER CONIDIAL STATE =PHIALOPHORA FIDE HENNEBERT 1968. COMPARE WITH WARDOMYCES.
/BARRON, GENERA P 218-219,1968 (ILLUS) /HENNEBERT, TBMS 51:749-762,1968 (ILLUS,REVIS) /ELLIS, DH P 74-75,1971 (ILLUS) /MATSUSHIMA, ICONES P 94 & PL 39:1,1975 (ILLUS) •

MANGINIELLA, ORTHOGRAPHIC VARIANT OF MAUGINIELLA •

MARCOSIA SYDOW & SYDOW 1916. ANN. MYCOL. 14:96-97 (NO ILLUS) /M. ULEI
?, SEE MORGAN-JONES 1971 (CAN. J. BOT. 49:1007) FOR ILLUS OF TYPE. •

MARGARINOMYCES LAXA 1930. ZENTRLB. BAKT. PARASITKDE, ABT. 2,81:392 /M. BUBAKII =P. BUBAKII (LAXA) SCHOL-SCHWARZ
=PHIALOPHORA MEDLAR 1915 FIDE VON ARX 1970, BUT SEE COLE & KENDRICK 1973 (MYCOLOGIA 65:682-684). SEE ALSO DE HOOG AND HERMANIDES-NIJHOF 1977 (CBS STUDIES 15:185). •

MARGARITISPORA INGOLD 1942. TBMS 25:351-353 (ILLUS) /M. AQUATICA
AMERO OR PHRAGMO OR STAURO,CETERI,HYALO,PHIAL,APPEND,AQUATIC, 89B.
/TUBAKI, BULL. NAT. SCI. MUS., TOKYO 3:259-261,1957 (ILLUS) /INGOLD, GUIDE P 74-75,1975 (ILLUS) /DYKO, TBMS 70:413-414,1978 (ILLUS,NEW SP) •

MARIANNAEA ARNAUD 1952 EX SAMSON 1974. CBS STUDIES 6:74-80 (ILLUS) /PENICILLIUM ELEGANS CORDA 1838 =M. ELEGANS (CORDA) SAMSON 1974
AMERO,BASO OR GLOIO,HYALO,PHIAL, 6B, ON WOOD (PICEA, PSEUDOTSUGA, ACER) & SOIL. COMPARE WITH CLONOSTACHYS CORDA.
/ARNAUD, BULL. SOC. MYCOL. FRANCE 68:196 & FIG 2A,1952 (ILLUS) /NICOT & CHARPENTIE, BULL. SOC. MYCOL. FRANCE 87:32,1971 (TYPE SPEC) /HUGHES, NEW ZEALAND J. BOT. 16:337,1978 (ILLUS) /BISSETT, FUNGI CANADENSES 160,1979 (ILLUS) •

MARSSONINA MAGNUS 1906. HEDWIGIA 45:89-91 /M. POTENTILLAE (DESMAZIERES)
DIDYMO,GLOIO,HYALO,PHIAL OR ANNEL,SPORODOCHIA, ON LEAVES (POPULUS). ASCO STATE =DIPLOCARPON OR DREPANOPEZIZA FIDE VON ARX 1974 (GENERA P 169).
/VON ARX, GENERA P 169,1974 (ILLUS,REFS) /PIROZYNSKI, FUNGI CANADENSES 13-15,1974 (ILLUS) /MOLINE & POLLACK, PHYTOPATHOL. 66:669-674,1976 (ILLUS,REVIS)

=*MARSSONIA* FISCHER 1874 NON KARSTEN 1861 •

MARTINDALIA SACCARDO & ELLIS 1884. IN SACCARDO & BERLESE, ATTI IST. VEN. SCI. SER 6, 3:726-727 & TAB XI, FIG 31 /M. SPIRONEMA =PHLEOGENA FAGINEA (FRIES) LINK (BASIDIOMYCETE) FIDE BARR & BIGELOW 1968
AMERO,CETERI,HYALO,NONSP,CLAMPS,SETAE,SYNNEMATA, 51F, ON WOOD (ULMUS, ACER) & FUNGI (ARMILLARIA). P. FAGINEA IS A BASIDIOMYCETE THAT IS USUALLY MISTAKEN FOR A HYPHOMYCETE AND THEN CALLED M. SPIRONEMA. HEYDENIA FRESENIUS 1852 (AND ITS SYNONYMS) MAY ALSO BE BASED ON P. FAGINEA.
/MORRIS, WEST. ILLINOIS UNIV. SER. BIOL. SCI. 3:82-83,1963 (ILLUS) /BARR & BIGELOW, MYCOLOGIA 60:456-457,1968 (NO ILLUS)
?=*CLATHROTRICHUM* PATOUILLARD 1921 •

MARTINELLISIA RAO & VARGHESE 1977. NORWEGIAN J. BOT. 24:279 (NOT SEEN) /M. INDICA
DIDYMO,CETERI,PHAEO,RADUL, 61D, ON LITTER (MUSSAENDA). COMPARE WITH PSEUDOSPIROPES. •

MASONIA SMITH 1952. TBMS 35:149 /M. GRISEA
=SCOPULARIOPSIS BAINIER 1907 FIDE SMITH 1963 (NON MASONIA HANSFORD 1944). •

MASONIELLA SMITH 1952. TBMS 35:237 /M. GRISEA (SMITH) =SCOPULARIOPSIS BRUMPTII
=SCOPULARIOPSIS BAINIER 1907 FIDE SMITH 1963. •

MASONIOMYCES CRANE & DUMONT 1975. CAN. J. BOT. 53:847-849 (ILLUS) /M. CLAVIFORMIS
?=XYLOCLADIUM SYDOW 1900 (SEE 28D). •

MASTIGOCLADIUM MATRUCHOT 1911 (C. R. ACAD. SCI. 152:325-327) =ACREMONIUM FIDE GAMS 1971 (CEPHALO. SCHIMMEL. P 77-78) •

MASTIGOMYCES IMSENECKIJ & KRISS 1934. C. R. ACAD. SCI. URSS 7:111 & TAB I-II /M. PHILIPPOVII
?=BLASTOMYCETES. •

MASTIGONEMA SPEGAZZINI 1926. BOL. ACAD. CORDOBA 29:177 (NO ILLUS) /M. BRUCHIANUM
=CHAETOSPERMUM (COELOMYCETES) FIDE AINSWORTH 1971. •

MASTIGOSPORIUM RIESS 1852. IN FRESENIUS, BEITR. MYKOL. P 56-57 /M. ALBUM
PHRAGMO,GLOIO OR CETERI,HYALO,NONSP,APPEND,SPORODOCHIA, 69C, ON GRASSES (ALOPECURUS, DESCHAMPSIA & OTHERS).
/HUGHES, CMI PAPERS 36:1-3 & 40-41,1951 (ILLUS) /VON ARX, GENERA P 239,1974 (REFS) /NAG RAJ, CAN. J. BOT. 57:2492-2494,1979 (ILLUS,NEW COMB)
=*AMASTIGOSPORIUM* BONDARZEVA-MONTEVERDE 1903 FIDE HUGHES 1951 •

MATRUCHOTIA BOULANGER 1893. REV. GEN. BOT. 5:401-406 & PL 12-14, & ALSO REV. MYCOL. 16:68-72 & PL CXLII-CXLIV, 1894 /M. VARIANS
?, BASIDIOMYCETE?, HYPHOMYCETE?, FOR REFERENCES SEE DONK 1954 (REINWARDTIA 2:466,NO ILLUS). •

MATRUCHOTIELLA GRIGORAKIS 1924. C.R. ACAD. SCI., PARIS 179:1424 /M. CURRII (CHALMERS & MARSHALL)
=NOMEN DUBIUM, (PROBABLY, SEE DODGE, MEDICAL MYCOLOGY, P 431,1935). •

MATSUSHIMAEA SUBRAMANIAN 1977. KAVAKA 5:93-98 (NOT SEEN) /TORULA FASCICULATA MATSUSHIMA 1975 =M. FASCICULATA (MATSUSHIMA)
CHEIRO,CETERI,PHAEO,NONSP, ON LITTER (CINNAMOMUM)
/MATSUSHIMA, ICONES P 153-154 & PL, 1975 (ILLUS) • (61:2 & 63:1-2)

MATSUSHIMOMYCES RAO & MANI VARGHESE 1979. BOT. NOT. 132:313-314 (NOT SEEN) /M. INDICUS
?, SEE ABSTRACTS OF MYCOLOGY 18-001071,1980. •

MATULA MASSEE 1888. J. ROY. MICRO. SOC. II,8:173-176 (ILLUS) /ARTOCREAS PORONIAEFORME BERKELEY & BROOME =M. PORONIAEFORMIS (BERKELEY & BROOME) =M. STATE OF CORTICIUM HAKGALLAE FIDE DONK 1962 (TAXON 11:88)

AMERO,GLOIO,HYALO,NONSP?,SPORODOCHIA, 51D, ON LITTER. COMPARE WITH MICHENERA. BASIDIO STATE =CORTICIUM (?ALEURODISCUS).
/MARTIN, J. WASHINGTON ACAD. SCI. 30:178-182,1940 (ILLUS) /CUNNINGHAM, TRANS. ROY. SOC. NEW ZEALAND 84:235-237,1956 (ILLUS) •

MAUGINIELLA CAVARA 1925. ATTI R. ACCAD. NAZ. LINCEI, ROMA, SER 6 1-2:65-67 /M. SCAETTAE
AMERO,ARTHRO,HYALO,NONSP, 43A, ON PLANTS (INFLORESCENCE ROT OF DATE PALM).
/NICOT, FICHES DE PHYTOPATHOLOGIE TROPICALE, NO. 24:1-6, IN REV. MYCOL. 37(1972), FASC. 3, JULY 1973 (ILLUS) /SIGLER & CARMICHAEL, MYCOTAXON 4:383-384,1976 (DESCR,REFS)
?=*MONILIOPHTHORA* EVANS ET AL 1978 •

MAXILLOSPORA HOHNEL 1914. S. B. AKAD. WISS. WIEN. ABT. 1, 123:138 (NO ILLUS) /M. MAXILLIFORMIS (ROSTRUP)
=TETRACLADIUM DE WILDEMAN 1894 FIDE INGOLD FROM AINSWORTH 1971. •

MEDUSULA CORDA 1837. ICONES FUNGORUM 1:18, FIG 241 /M. OCHRACEA (LECTOTYPE)
=VOLUTELLA FIDE LINDAU IN RABENHORST 1910, NON MEDUSULA ESCHWEILER 1824 (LICHEN). •

MEGACLADOSPORIUM VIENNOT-BOURGIN 1949. CHAMPIGNONS PARASITES 1:489 /M. PIRINUM (LIBERT)
=FUSICLADIUM BONORDEN 1851 FIDE HUGHES 1958. •

MEGALODOCHIUM DEIGHTON 1960. CMI PAPERS 78:17-20 (ILLUS) /M. PALMICOLA =M. ELAEIDIS (BEELI) DEIGHTON 1969
AMERO,CETERI,PHAEO,NONSP,ALEURI?,SPORODOCHIA, 50B, ON LITTER (COCOS, ELAEIS, RAPHIA).
/CJEP & DEIGHTON, CMI PAPERS 117:8,1969 (REVIS) /ELLIS, MDH P 70-71,1976 (ILLUS) •

MEGALOXYPHIUM CIFERRI BATISTA & NASCIMENTO 1956 =LEPTOXYPHIUM SPEGAZZINI 1918 FIDE HUGHES 1976 (MYCOLOGIA 68:777 & 787) •

MEGASTER CIFERRI ET AL. 1956. PUBL. INST. MIC. UNIV. RECIFE 48:2 /M. LONGICORNIS
?=TRIPOSPERMUM SPEGAZZINI 1918 SEE HUGHES 1976 (MYCOLOGIA 68:810-811). •

MEGATRICHOPHYTON NEVEU-LEMAIRE 1921. PRECIS PARASITOL. HUMAINE, ED 5, P 46 & 54-55 /FOUR SPECIES INCLUDED
=TRICHOPHYTON MALMSTEN 1845. •

MELANCHLENUS CALANDRON 1953 =NOT VALIDLY PUBLISHED, SEE DE HOOG 1977 (CBS STUDIES 15:93) •

MELANCONIUM LINK 1809. MAG. GES. NATURF. FREUNDE, BERLIN 3:9 /M. ATRUM
AMERO,GLOIO,PHAEO,ANNEL,ACERVULI, ON PLANTS. ASCO STATE =MELANCONIS FIDE WEHMEYER 1941.
/VON ARX, GENERA P 155 & 157,1970 (ILLUS)
=*TRIMMATOSTROMA* CORDA 1837 FIDE HUGHES 1958 •

MELANODISCUS HOHNEL 1918. BERL. DEUTSCHL. BOT. GES. 36:309 (NO ILLUS) /M. NERVISEQUIUS
? •

MELANODOCHIUM SYDOW 1938. ANN. MYCOL. 36:310-311 (NO ILLUS) /M. AUSTRALIENSE
? •

MELANOGONE WOLLENWEBER & RICHTER 1934. ZENTRALBL. BAKT., ABT. 2, 90:74-76 (ILLUS) /M. PUCCINIOIDES
=HUMICOLA TRAAEN FIDE MASON 1941. •

MELANOGRAPHIUM SACCARDO 1913. ANN. MYCOL. 11:557-558 /M. SPLENISPORIUM =M. SPINULOSUM (SPEG.) HUGHES 1958
AMERO,CETERI,PHAEO,RADUL,SYNNEMATA,SPORODOCHIA, 29A, ON WOOD & BARK (MANY TREES). ASCO STATE =LASIOBERTIA FIDE SIVANESAN 1978 (TBMS

70:383-385,ILLUS).
/ELLIS, CMI PAPERS 93:14-23,1963 (KEY,ILLUS) /ELLIS, DH P 196-198,1971 (ILLUS,KEY) /MATSUSHIMA, ICONES P 94 & PL,1975 (ILLUS)
=*SPOROSTACHYS* SACCARDO 1917 FIDE HUGHES 1958
=*PSEUDOCAMPTOUM* FRAGOSO & CIFERRI 1925 FIDE HUGHES 1958 •

MELANOTRICHUM CORDA 1833. IN STURM'S DEUTSCHL. FL., III(PILZE) BD. 3, HEFT 13:89-90 & TAB 45 /M. MICROSPERMUM
?, ORIGINAL ILLUSTRATION NOT DIAGNOSTIC. •

MEMNONIELLA HOHNEL 1923. ZENTRALBL. BAKT. PARASITENKUNDE ABT. 2, 60:16-17 /M. ATERRIMA =M. ECHINATA (RIVOLTA) GALLOWAY
=STACHYBOTRYS CORDA 1837 FIDE SMITH 1962, BUT SEE JONG & DAVIS 1976 (MYCOTAXON 3:409-485) & ELLIS 1976 (MDH P 464). •

MEMNONIUM CORDA 1833. IN STURM'S DEUTSCHL. FL. III (PILZE), BD 3, HEFT 13:91-92 & TAB 46 /M. EFFUSUM
=NOMEN DUBIUM SEE HUGHES 1958 & LINDAU 1907. •

MENISPORA PERSOON 1822. MYCOLOGIA EUROPAEA 1:32 /M. GLAUCA
AMERO OR DIDYMO OR PHRAGMO,GLOIO,HYALO,PHIAL,SETAE,APPEND, 1C, ON WOOD & BARK (MANY TREES). ASCO STATE =CHAETOSPHAERIA FIDE MULLER IN KENDRICK 1971.
/HUGHES & KENDRICK, NEW ZEALAND J. BOT. 6:323-375,1968 (ILLUS,KEY) /HUGHES & KENDRICK, CAN. J. BOT. 41:693-718,1963 (ILLUS,KEY) /ELLIS, DH P 529-530,1971 (ILLUS,KEY) /HOLUBOVA-JECHOVA, FOLIA GEOBOT. PHYTOTAX., PRAHA 8:317-336,1973 (ILLUS,KEY) /MATSUSHIMA, ICONES P 95 & PL,1975 (ILLUS) /ELLIS, MDH P 461-462,1976 (ILLUS)
=*ERIOMENE* (SACCARDO) MAIRE EX PEYRONEL 1919 FIDE HUGHES 1958
=*ERIOMENE* (SACCARDO) CLEMENTS & SHEAR 1931 FIDE HUGHES 1958
=*ERIONEMA* MAIRE 1906 FIDE HUGHES 1958
=*CILIOFUSARIUM* ROSTRUP 1892 FIDE HUGHES 1958
=*ERIOMENELLA* PEYRONEL 1919 FIDE HUGHES 1958 •

MENISPORELLA AGNIHOTHRUDU 1962. PROC. INDIAN ACAD. SCI., SECT. B, 56:98-100 (ILLUS) /M. ASSAMICA
=CODINAEA MAIRE 1937 FIDE ELLIS IN HUGHES & KENDRICK 1968. •

MENISPOROPSIS HUGHES 1952. CMI PAPERS 48:59 /M. THEOBROMAE
AMERO OR DIDYMO,GLOIO,HYALO,PHIAL,SETAE,SYNNEMATA,APPEND, 56G, ON LITTER (THEOBROMA & PERSEA).
/HUGHES & KENDRICK, NEW ZEALAND J. BOT. 6:323-375,1968 (ILLUS) /ELLIS, DH P 522-523,1971 (ILLUS) /PIROZYNSKI & HODGES, CAN. J. BOT. 51:163-166,1973 (ILLUS) /MATSUSHIMA, ICONES P 96 & PL,1975 (ILLUS) /ELLIS, MDH P 451-452,1976 (ILLUS) •

MENOIDEA MANGIN & HARIOT 1907. BULL. SOC. MYC. FRANCE 23:65-68 (ILLUS) /M. ABIETIS
?, ORIGINAL ILLUSTRATION NOT DIAGNOSTIC. •

MERIA VUILLEMIN 1896. BULL. SOC. SCI. NANCY SER 2, 14:32-33 & PL 1 /M. LARICIS
AMERO,BASO OR GLOIO,HYALO,PHIAL?, 16F, ON NEMATODES.
/BARRON, THE NEMATODE-DESTROYING FUNGI P 71-74,1977 (ILLUS)
=*HARTIGIELLA* SYDOW 1900 FIDE AINSWORTH 1971 •

MERIMBLA PITT 1979. CAN. J. BOT. 57:2394-2398 (ILLUS) /PENICILLIUM INGELHEIMENSE VAN BEYMA 1942 =M. INGELHEIMENSE (VAN BEYMA) =M. STATE OF TALAROMYCES AVELLANEUS
AMERO,BASO,HYALO,PHIAL, ON MANY SUBSTRATES. ASCO STATE =TALAROMYCES. •

MERISTACRUM DRECHSLER 1940 =CONIDIAL ZYGOMYCETE SEE BARRON 1977 (THE NEMATODE-DESTROYING FUNGI P 69-71,ILLUS,REFS) •

MEROSPORIUM CORDA 1832. IN STURM'S DEUTSCHL. FL. III (PILZE), BD. 3, HEFT 12:45 /M. MINUTUM
=NOMEN CONFUSUM FIDE HUGHES 1958. •

MESOBOTRYS SACCARDO 1880. MICHELIA 2:27 /M. MACROCLADA (SACCARDO 1877) (LECTOTYPE FIDE HUGHES) =GONYTRICHUM MACROCLADUM (SACCARDO) HUGHES 1951
=GONYTRICHUM NEES & NEES 1818 FIDE HUGHES 1958 •

METARHIZIUM SOROKIN 1883. PLANT PARASITES OF MAN & ANIMALS ETC., (IN RUSSIAN) P 268-291 /M. ANISOPLIAE (METSCHNIKOV)
AMERO,BASO,HYALO,PHIAL,SPORODOCHIA, 13B, ON INSECTS & SOIL.
/BARRON, HYPHOMYCETES P 223-225,1968 (ILLUS) /TULLOCH, TBMS 66:407-411,1976 (REVIS,KEY,NO ILLUS) /VON ARX, GENERA P 195 & 197,1974 (ILLUS,REFS) /BRADY, CMI DESCRIPTIONS 609,1979 (ILLUS)
=*METARRHIZIUM* -ORTHOGRAPHIC VARIANT FIDE TULLOCH 1976 •

METARRHIZIUM -SEE METARHIZIUM •

MIAINOMYCES CORDA 1833. IN STURM'S DEUTSCHL. FL. III (PILZE), BD. 3, HEFT 13:83-84 & TAB 42 /M. FUNGICOLUS
?, ORIGINAL ILLUSTRATION NOT DIAGNOSTIC. •

MICHENERA BERKELEY & CURTIS 1868. J. LINN. SOC., LONDON (BOT.) 10:333 /M. ARTOCREAS =M. STATE OF ALEURODISCUS SUBGIGANTEUS FIDE DONK 1962 (TAXON 11:89)
AMERO,GLOIO,HYALO,NONSP?,SPORODOCHIA, 51C, ON LITTER (ACER, MAGNOLIA, CARYA). BASIDIO STATE =ALEURODISCUS (?LYCROSTROMA). COMPARE WITH MATULA & CHAETOCONIDIUM.
/LYMAN, PROC. BOSTON SOC. NAT. HIST. 33:151-160 & PL 18 & 26,1907 (ILLUS) /CUNNINGHAM, TRANS. ROY. SOC. NEW ZEALAND 84:235-237,1956 (ILLUS) /DONK, TAXON 11:89,1962 (REFS)
=*ARTOCREAS* BERKELEY & BROOME 1873 FIDE DONK 1962 •

MICROBASIDIUM BUBAK & RANOJEVIC 1914. ANN. MYCOL. 12:415-416 (ILLUS) /M. SORGHI (PASSER.)
=HADROTRICHUM FIDE HOHNEL FROM AINSWORTH 1971. •

MICROCERA DESMAZIERES 1848. ANN. SCI. NAT., SER. 3, 10:359-360 /M. COCCOPHILA =SPHAEROSTILBE COCCOPHILA (DESMAZIERES) TULASNE 1865
?=ATRACTIUM LINK 1809 SEE SEAVER 1909 (MYCOLOGIA 1:180). •

MICROCLAVA STEVENS 1917. TRANS. ILLINOIS ACAD. SCI. 10:205-206 /M. MICONIAE
AMERO OR DICTYO,CETERI,PHAEO,NONSP, 96A, ON FUNGI (CALOTHYRIUM) ON LEAVES (MICONIA). COMPARE WITH MONODICTYS.
/BONAR, MYCOLOGIA 59:596-597,1967 (ILLUS) /DEIGHTON, TBMS 52:315-321,1969 (ILLUS,NEW COMBS,KEY) /ELLIS, DH P 109-110,1971 (ILLUS)
=*OMMATOSPORA* BATISTA & CAVALCANTI 1964 FIDE DEIGHTON 1969
?=*OMMATOSPORELLA* BAT., BEZ. & POROCA 1967 •

MICRODOCHIUM SYDOW 1924. ANN. MYCOL. 22:267-268 (ILLUS) /M. PHRAGMITIS
AMERO OR DIDYMO OR SCOLECO,CETERI,HYALO,RADUL OR RACHI,SPORODOCHIA, 14E & 108B, ON LITTER. CLOSE TO IDRIELLA & SELENOSPORELLA. OTHER CONIDIAL STATE =HUMICOLA.
/SUTTON, PIROZYNSKI & DEIGHTON, CAN. J. BOT. 50:1899-1907,1972 (ILLUS,REVIS) /MOUCHACCA & SAMSON, REV. MYCOL. 37:267-275,1973 (ILLUS,NEW SPP) /DE HOOG & HERMANIDES-NIJHOF, CBS STUDIES 15:194-195 & 205 & 215,1977 (ILLUS,NEW COMBS)
?=*ACICULARIELLA* ARNAUD 1954 SEE MOUCHACCA & SAMSON 1973 •

MICROIDES DE VROEY 1970. ANN. SOC. BELGE MED. 50:24-26 (1-174) /FOUR NEW COMBINATIONS MENTIONED
=NOMEN INVALIDUM, NO LATIN DIAGNOSIS, TYPE NOT INDICATED.
/VANBREUSEGHEM, DE VROEY & TAKASHIO, PRACTICAL GUIDE TO MEDICAL & VETERINARY MYCOLOGY, MASSON PUBL., INC., P 94-103,1978 ("NEW" COMBS) •

MICROLYCHNUS FUNK 1973. CAN. J. BOT. 51:1249-1250 (ILLUS) /M. EPICORTICIS
SCOLECO,GLOIO,HYALO,PHIAL,SYNNEMATA, 111A, ON ALGAE (LICHENIZED) ON BARK (TSUGA, ABIES, PICEA). •

MICROSPATHA KARSTEN 1889. REV. MYC. 11:207 /M. GLAUCA
?, NOT ILLUSTRATED BY AUTHOR. •

MICROSPORON =ORTHOGRAPHIC VARIANT OF MICROSPORUM •

MICROSPORUM GRUBY 1843. C. R. SEANC. HEBD. ACAD. SCI. PARIS 17:301 /M. AUDOUINII
PHRAGMO,CETERI,HYALO,NONSP,ALEURI, 89F, ON SOIL, KERATIN & ANIMALS (CAT, DOG, MAN). ASCO STATE =NANNIZZIA FIDE STOCKDALE 1961, OTHER CONIDIAL STATES =CHRYSOSPORIUM ("MICROCONIDIA") & "ARTHROCONIDIA".
/STOCKDALE, SABOURAUDIA 1:41-48,1961 (ILLUS,ASCO STATE) /STOCKDALE, CMI DESCRIPTIONS 61-70,1965 (ILLUS) /BARRON, HYPHOMYCETES P 225-227,1968 (ILLUS,REFS) /REBELL & TAPLIN, DERMATOPHYTES 124 PP,1970 (ILLUS,KEY) /MORAES, PADHYE & AJELLO, MYCOLOGIA 67:1109-1113,1975 (NEW SP OF NANNIZZIA) /AJELLO, IN REC. ADV. MED. VET. MYCOL. 289-297,1977 (REFS,LIST OF SPECIES) /CARETTA & PIONTELLI, SABOURAUDIA 15:3-6,1977 (ILLUS,NEW SP)
=*MICROSPORON* GRUBY (ORTHOGRAPHIC VARIANT)
=*MICROIDES* DE VROEY 1970
=*SABOURAUDITES* OTA & LANGERON 1923
=*THALLOMICROSPORON* BENEDEK 1964 •

MICROSTROMA NIESSL 1861. OEST. BOT. ZEIT. 11:250-253 (NO ILLUS) /M. PALLIDUM =M. JUGLANDIS (BERENG.) SACCARDO 1886 (SYLL. FUNG. 4:9-10)
AMERO,CETERI,HYALO,AMPUL,SYNNEMATA, 32D, ON LEAVES (JUGLANS, CARYA). GROWS AS A YEAST IN CULTURE, COMPARE WITH ENDOSTILBUM & AMEROBOTRYUM.
/VON ARX, BIBLIOTHECA MYCOLOGICA (CRAMER) BD. 24:70-80,1970 (ILLUS)
=*HELOSTROMA* PATOUILLARD 1902 FIDE VON ARX 1970
=*ARTICULARIELLA* HOHNEL 1909 FIDE VON ARX 1970 •

MICROTRICHOPHYTON (CAST. & CHALM.) NEVEU-LEMAIRE 1921. PRECIS PARASITOL. HUM., ED. 5, P 46 & 53-54 /?
=TRICHOPHYTON MALMSTEN 1848. •

MICROTYPHA SPEGAZZINI 1910. ANN. MUS. NAC. HIST. NAT. BUENOS AIRES 20:432-433 /M. SACCHARICOLA
=ARTHRINIUM KUNZE 1817 FIDE ELLIS 1971. •

MICROXIPHIUM (HARVEY EX BERKELEY & DESMAZIERES) THUMEN 1879. MYCOTHECA UNIVERSALIS NO. 1352 /M. FAGI (PERSOON) HUGHES 1976 (MYCOLOGIA 68:790) =M. STATE OF DENNISIELLA BABINGTONII FIDE HUGHES 1976
AMERO,GLOIO,HYALO,PHIAL,SYNNEMATA,SETAE, 17F, ON LEAVES (SOOTY MOLD). ASCO STATE =DENNISIELLA FIDE HUGHES 1976.
/HUGHES, MYCOLOGIA 68:768-772 & 789-791,1976 (ILLUS,REVIS) •

MICROXYPHIUM - SEE MICROXIPHIUM •

MILOSPIUM HAWKSWORTH 1975. TBMS 65:227-232 (ILLUS) /SPILOMIUM GRAPHIDEORUM NYLANDER 1856 =M. GRAPHIDEORUM (NYLANDER)
AMERO OR STAURO OR BULBILS,CETERI,PHAEO,NONSP, 113A, ON LICHENS (OPEGRAPHA). •

MILOWIA MASSEE 1884. J. R. MICROSCOP. SOC., SER. 2, 4:841-845 & PL XII /M. NIVEA
=NOMEN DUBIUM FIDE NAG RAJ & KENDRICK 1975 (CHALARA, P 43-45). •

MINIMEDUSA WERESUB & LECLAIR 1971. CAN. J. BOT. 49:2210 (2203-2213,ILLUS) /PAPULASPORA POLYSPORA HOTSON 1912 =M. POLYSPORA (HOTSON)
BULBILS,CETERI,PHAEO,NONSP,CLAMPS, 128G, ON SOIL. COMPARE WITH PAPULASPORA. OTHER CONIDIAL STATE ?=TRICHOCONIELLA. •

MINIMIDOCHIUM SUTTON 1969. CAN. J. BOT. 47:2095-2100 /M. SETOSUM
AMERO,GLOIO,HYALO,PHIAL,SPORODOCHIA,SETAE,APPEND, 14F, COMPARE WITH NEOTTIOSPORELLA, VOLUTELLA & MAHABALELLA.
/PIROZYNSKI, CMI PAPERS 129:57-58,1972 (REFS) /ELLIS, MDH P 466-467,1976 (ILLUS) /GAMUNDI, ARAMBARRI & GIAIOTTI, DARWINIANA 21:89-90 & FIG 2,1977 (ILLUS) •

MIRANDINA ARNAUD EX MATSUSHIMA 1975. MICROFUNGI P 37,1971 (GENERIC DIAGNOSIS) AND ICONES P 96-97 (LATIN DIAGNOSIS OF TYPE SP) /M. CORTICOLA

SCOLECO,CETERI,HYALO,RADUL, 109B, ON LITTER.
/ARNAUD, BULL. SOC. MYCOL. FRANCE 68:205-206, FIG 8D-E,1952 (ILLUS) /MATSUSHIMA, MICROFUNGI P 37-38 & FIG 53:4, 54 & 55,1971 (ILLUS,NEW SPP) /SUTTON, CMI PAPERS 132:13,1973 (COMMENT) •

MITTERIELLA SYDOW 1933. IN SYDOW & MITTER, ANN. MYCOL. 31:95-96 /M. ZIZIPHINA (AS 'ZIZYPHINA')
PHRAGMO,CETERI,PHAEO,RADUL,PODIA, 82E, ON LEAVES (ZIZIPHUS). ASCO STATE =CLYPEOLELLA OR YAMAMOTOA FIDE VON ARX 1975 (CBS STUDIES 9:44), OR SCHIFFNERULA SEE DE HOOG 1977 (CBS STUDIES 15:140).
/ELLIS, DH P 228-229,1971 (ILLUS) •

MIURAEA HARA 1948. BYOGAICHU-HOTEN (MANUAL OF PESTS & DISEASES) P 260 & 779 /CLASTEROSPORIUM DEGENERANS SYDOW & SYDOW =M. DEGENERANS (SYDOW & SYDOW)
DICTYO,CETERI,HYALO,NONSP, 97F, ON LEAVES. ASCO STATE =MYCOSPHAERELLA FIDE DEIGHTON 1973 (CMI PAPERS 133:3). COMPARE WITH STEMPHYLIOPSIS SMITH.
=*HYALODICTYS* SUBRAMANIAN 1962 FIDE DEIGHTON 1973 •

MOESZIA BUBAK 1914. BOT. KOEZLEM. 13:94-96 /M. CYLINDROIDES =C. MAGNUSIANUM WOLLENWEBER FIDE VON ARX
=CYLINDROCARPON WOLLENWEBER 1913 FIDE VON ARX 1970. •

MONACROSPORIELLA SUBRAMANIAN 1977. KAVAKA 5:93-98 (NOT SEEN) /DACTYLELLA MEGALOSPORA DRECHSLER =M. MEGALOSPORA
?, SEE ABSTRACTS OF MYCOLOGY 67-52149,1979. •

MONACROSPORIUM OUDEMANS 1885. NED. KRUIDK. ARCH. SER 2, 4:250-251 & PL V FIG 9 /M. ELEGANS
?=DACTYLELLA GROVE 1884 SEE BARRON 1968. •

MONILIA BONORDON 1851. HANDB. MYKOL. P 76 FIG 78 & TAXON 23:419 /M. CINEREA BONORDEN 1851 (LECTOTYPE SEE DONK)
=MISCITATION, BONORDEN CITED "MONILIA HILL-FRIES", HE DID NOT INTEND TO PUBLISH A NEW GENUS. FOR CONSERVATION AGAINST MONILIA FRIES 1821 AND OTHER POST-1821 NAMES, SEE DONK 1963 (TAXON 12:266-271) & COMMITTEE REPORT 1974 (TAXON 23:419). •

MONILIA PERSOON 1801. SYN. METH. FUNG. P 693 /M. FRUCTIGENA (LECTOTYPE FIDE HUGHES 1958) =M. STATE OF MONILINIA FRUCTIGENA HONEY
AMERO,ARTHRO OR BLASTO,HYALO OR PHAEO,NONSP, 38B, ON PLANTS, BREAD & SOIL. ASCO STATE =MONILINIA OR NEUROSPORA FIDE VON ARX 1970 OR ASCOBOLUS SEE BREFELD 1891. OTHER CONIDIAL STATE =MYRIOCONIUM. SEE ALSO MONILIA BONORDEN.
/BARRON, HYPHOMYCETES P 227,1968 (ILLUS) /HOLLIDAY, CMI DESCRIPTIONS 226,1970 (ILLUS) /ELLIOTT, FUNGI CANADENSES NO. 38,1974 (ILLUS) /GAMS & DOMSCH, NOVA HEDWIGIA 18:12-13,1969 (ILLUS,EXCLUSIONS) /KORF & KOHN, MYCOTAXON 9:521-522,1979 (DISCUSSION) /MORDUE, CMI DESCRIPTIONS 616-619,1979 (ILLUS SUB SCLEROTINIA)
=*EPOCHNIUM* LINK 1809 FIDE HUGHES 1958
?=*HALOBYSSUS* ZUKAL 1893 FIDE CLEMENTS & SHEAR 1931 •

MONILIELLA STOLK & DAKIN 1966. ANTONIE VAN LEEUWENHOEK 32:399-405 /M. ACETOABUTANS
AMERO,ARTHRO OR BLASTO,HYALO,NONSP, 43G, ON FRUIT JUICES & SYRUPS. COMPARE WITH HORMOCONIS, HYALODENDRON & SOROCYBE. ASCO STATE =CEPHALOASCUS -SEE WELLS 1954 (MYCOLOGIA 46:37-50, SUB ASCOCYBE CONIDIAL STATE). OTHER CONIDIAL STATE ?=ARTHROGRAPHIS SEE DE HOOG 1978 (IN VERH. KON. NED. AKAD. WETEN. NATUUR., TWEEDE REEKS P 9).
/VON ARX, GENERA P 166-167,1970 (ILLUS) /RAO & DE HOOG, PERSOONIA 8:199-206,1975 (ILLUS,NEW COMB) /VON ARX ET AL., CBS STUDIES 14:32-36,1977 (NEW COMBS) •

MONILIGER LETELLIER ?=PENICILLIUM FIDE AINSWORTH 1971 •

MONILIOPHTHORA EVANS, STALPERS, SAMSON & BENNY 1978. CAN. J. BOT. 56:2528-2532 (ILLUS) /MONILIA RORERI CIFERRI 1933 =M. RORERI (CIFERRI)

AMERO,ARTHRO OR BLASTO,HYALO,NONSP OR MISC?, 40A, ON PODS & BRANCHES (THEOBROMA). COMPARE WITH MAUGINIELLA. •

MONILIOPSIS RUHLAND 1908. ARB. BIOL. ANST., BERLIN 6:71-76 (ILLUS) /M. ADERHOLDII
=RHIZOCTONIA DE CANDOLLE 1815 FIDE VON ARX 1974 (GENERA, P 270).
/DONK, TAXON 11:89,1962 (REFS) •

MONILOCHAETES HALSTEAD EX HARTER 1916. J. AGRIC. RES. 5:791 & PL LVII /M. INFUSCANS
?, ORIGINAL ILLUSTRATION NOT DIAGNOSTIC.
/BARNETT, IMPERFECT FUNGI NO. 164,1960 (ILLUS) •

MONOCERAS GUBA 1961. MONOGRAPH OF MONOCHAETIA & PESTALOTIA P 289-290 (ILLUS) /PESTALOTIA KRIEGERIANA BRESADOLA 1894 (HEDWIGIA 33:209) =M. KRIEGERIANUM (BRESADOLA)
=SEIMATOSPORIUM CORDA FIDE SUTTON FROM AINSWORTH 1971. •

MONOCHAETIA (SACCARDO) ALLESCHER 1902. IN RABENHORST, KRYPT. FL. 1, 7:665 /M. MONOCHAETA (DESMAZIERES) ALLESCHER 1902
PHRAGMO,GLOIO,PHAEO,ANNEL,APPEND,ACERVULI, (COELOMYCETE), ON LEAVES & TWIGS & BRANCHES (CUPRESSUS). SIMILAR TO PESTALOTIA. OTHER CONIDIAL STATE =SCYTALIDIUM. ASCO STATE =RHYNCHOSPHAERIA FIDE CBS LIST OF CULTURES 1972 (P 156) OR HYMENOPLEELLA SEE BARR 1975 (MYCOLOGIA 67:188 & 193).
/GUBA, MONOGRAPH OF MONOCHAETIA & PESTALOTIA, 342 PP,1961 (MONOGR) /GUBA & STEVENSON, IN TAXON 10:244-246,1961 (NOM. CONS. PROP.) /SUTTON & GIBSON, CMI DESCRIPTIONS 326,1972 (ILLUS SUB SEIRIDIUM) /MORGAN-JONES, NAG RAJ & KENDRICK, ICON. GEN. COELO. I:21-22,1972 (ILLUS) /YOKOYAMA, TBMS 65:499-503,1975 (ILLUS,NEW SP) /SIGLER & CARMICHAEL, MYCOTAXON 4:400-403,1976 (ILLUS) /SUTTON, CMI PAPERS 141:128,1977 (REFS)
=*SEIRIDIUM* NEES EX LINK 1825 FIDE GUBA & STEVENSON 1961
=*HYALOCERAS* DURIEU & MONTAGNE 1846 FIDE GUBA & STEVENSON 1961 •

MONOCHAETITES BABAYAN & TASLAKHCHYAN 1973 (MIKOL. FITOPATOL. 7:180) -FOSSIL FUNGI IMPERFECTI. NOT VALIDLY PUBLISHED, SEE SUTTON 1977 (CMI PAPERS 141:128). •

MONOCHAETOPSIS PATOUILLARD, LISTED IN AINSWORTH 1971 •

MONOCILLIUM SAKSENA 1955. INDIAN PHYTOPATH. 8:9-12 /M. INDICUM
AMERO,BASO OR GLOIO,HYALO,PHIAL, 3E, ON LITTER, LEAVES, SOIL, FUNGI, FRUITS (ARCTIUM). ASCO STATE =NIESSLIA FIDE GAMS 1971.
/GAMS, CEPHALO. SCHIMMEL. P 151-166,1971 (ILLUS,KEY,REVIS) /HASHMI, KENDRICK & MORGAN-JONES, CAN. J. BOT. 50:1461-1463,1972 (ILLUS) •

MONOCONIDIA ROZE 1897. BULL. SOC. MYCOL. FRANCE 13:82-83 & PL VI, FIG 27 /M. MINUTISSIMA
?, ORIGINAL ILLUSTRATION DIFFICULT TO INTERPRET. •

MONODICTYS HUGHES 1958. CAN. J. BOT. 36:785 (NO ILLUS) /M. PUTREDINIS (WALLROTH)
DICTYO,CETERI,PHAEO,NONSP,ALEURI, 100D, ON WOOD, LITTER, SOIL & DUNG. COMPARE WITH ACROSPEIRA & CHUPPIA. OTHER CONIDIAL STATE =DACTYLARIA (Q.V.).
/ELLIS, DH P 68-70,1971 (ILLUS,KEY) /MATSUSHIMA, ICONES P 97-98 & PL,1975 (ILLUS,NEW SPP) /ELLIS, MDH P 41-44,1976 (ILLUS,KEY,NEW COMBS) •

MONODISMA ALCORN 1975. TBMS 65:139-141 (ILLUS) /M. FRAGILIS
PHRAGMO,CETERI,HYALO,RADUL, 78B, ON GRASSES (IMPERATA, SACCHARUM & OTHERS). •

MONOGRAMMIA STEVENS 1917. TRANS. ILLINOIS ACAD. SCI. 10:202-203 /M. MINCONIAE
=TITAEA SACCARDO 1876 FIDE DAMON 1952 (J. WASH. ACAD. SCI. 42:365-367). •

MONOPODIUM DELACROIX 1890. BULL. SOC. MYCOL. FRANCE 6:99 /M. UREDOPSIS
=HARZIA COSTANTIN, SEE ELLIS 1971. •

MONOSPORELLA HUGHES 1953. CAN. J. BOT. 31:654 /M. SETOSA (BERK. & CURTIS)
=MONOTOSPORELLA HUGHES 1958 =BRACHYSPORIELLA BATISTA 1952, NON MONOSPORELLA KEILIN 1920. •

MONOSPORIELLA SPEGAZZINI 1918. PHYSIS 4:293 /M. MELIOLICOLA (SPEGAZZINI)
=NOMEN DUBIUM, SEE DEIGHTON & PIROZYNSKI 1972
/DEIGHTON & PIROZYNSKI, CMI PAPERS 128:109-110 & PL 2,1972 (DISCUSSION) •

MONOSPORIUM BONORDEN 1851. HANDB. MYKOLOGIE P 95 /?
=NOMEN ILLEGITIMUM FIDE HUGHES 1958. M. APIOSPERMUM IS THE SCEDOSPORIUM STATE OF PETRIELLIDIUM BOYDII FIDE CARMICHAEL. •

MONOSTACHYS ARNAUD 1954. BULL. SOC. MYCOL. FRANCE 69:277-278, FIG 5E-G (NO LATIN DIAGNOSIS) /M. TRANSVERSALIS
AMERO,BASO,HYALO,PHIAL, COULD BE PAECILOMYCES OR GLIOCLADIUM. SUBSTRATE NOT GIVEN.
/NICOT & CHARPENTIE, BULL. SOC. MYCOL. FRANCE 87:33,1971 (TYPE SPECIMEN) •

MONOTOSPORA CORDA 1837. ICONES FUNGORUM 1:11 /M. TORULOIDES
=NOMEN DUBIUM FIDE CARMICHAEL 1962. •

MONOTOSPORA SACCARDO 1880. MICHELIA 2:25 AS 'MONOTOSPORA CORDA EM.' /M. SPHAEROCEPHALA BERKELEY & BROOME 1859 (TYPICAL SPECIES FIDE MASON 1941)
=NOMEN ILLEGITIMUM (NON MONOTOSPORA CORDA). SEE ACROGENOSPORA. •

MONOTOSPORELLA HUGHES 1958. CAN. J. BOT. 36:786 /M. SETOSA (BERKELEY & CURTIS)
=BRACHYSPORIELLA BATISTA 1952 FIDE ELLIS 1959.
/HUGHES, NEW ZEALAND J. BOT. 16:337-338,1978 (ILLUS) •

MONOTRICHUM GAEUMANN 1922. ANN. MYCOL. 20:261 (ILLUS) /M. COMMELINAE
DIDYMO,CETERI,HYALO,NONSP,APPEND, 56E, ORIGINAL ILLUSTRATION SUGGESTS CONIDIAL BROOMELLA VITALBAE, SEE MULLER 1971 (IN KENDRICK, TAXONOMY OF FUNGI IMPERFECTI, FIG 13.2, P 194). COMPARE WITH TRICHOCONIS. FOR A DIFFERENT OPINION SEE DONK 1962 (TAXON 11:89-90). •

MONOTROPOMYCES COSTANTIN & DUFOUR 1921 =NOMEN DUBIUM SEE DE HOOG & HERMANIDES-NIJHOF 1977 (CBS STUDIES 15:185) •

MONTOYELLA CASTELLANI 1907 =NOMEN DUBIUM SEE DE HOOG & HERMANIDES-NIJHOF 977 (CBS STUDIES 15:186) •

MOORELLA RAGHUVEER RAO & DEV RAO 1964. MYCOPATHOLOGIA 22:50-52 (ILLUS) /M. SPECIOSA
=HELICOMA CORDA 1837 FIDE KENDRICK.
/ELLIS, MDH P 160-162,1976 (ILLUS) •

MORONOPSIS DELACROIX 1891. BULL. SOC. MYCOL. FRANCE 7:109 (NO ILLUS) /M. INQUINANS
=THYRSIDIUM FIDE SACCARDO FROM AINSWORTH 1971. •

MORRISOGRAPHIUM MORELET 1968. TAXON 17:528 /ISARIOPSIS PILOSA EARLE 1897 =M. PILOSUM (EARLE) =PHRAGMOGRAPHIUM ULMI MORRIS 1966
SCOLECO OR PHRAGMO,GLOIO,HYALO,RACHI OR PHIAL?,SYNNEMATA, 111D, ON BARK (ULMUS). ?=PSEUDOGRAPHIUM JACZEWSKI 1898.
=*PHRAGMOGRAPHIUM* MORRIS 1966 NON HENNINGS 1905 •

MORULARIA NANNIZZI 1925 =ATCHIA FLOTOW 1850 SEE MEEKER 1975 (CAN. J. BOT. 53:2486) •

MUCHMORIA SACCARDO 1906. ANN. MYCOL. 4:277-278 /M. PORTORICENSIS
=NOMEN CONFUSUM FIDE HUGHES 1958. •

MUCROSPORIUM PREUSS 1851. LINNAEA 24:128-129 (NO ILLUS) /TWO ORIGINAL SPECIES
=CLADOBOTRYUM FIDE DE HOOG 1978 (PERSOONIA 10:34). •

MUIARIA THAXTER 1914. BOT. GAZ. 58:241 (235-253,ILLUS) /M. GRACILIS (LECTOTYPE)
DICTYO,CETERI,PHAEO,NONSP,APPEND, 95B, ON INSECTS (FLIES). •

MUIOGONE THAXTER 1914. BOT. GAZ. 58:239 (235-253,ILLUS) /M. CHROMOPTERI
DICTYO,CETERI,PHAEO,NONSP,APPEND, 95C, ON INSECTS (FLIES). •

MUIRELLA SPRAGUE 1958. MYCOLOGIA 50:827 (NO ILLUS) /GLOEOCERCOSPORA ALASCENSIS SPRAGUE 1954 =M. ALASCENSIS (SPRAGUE)
?
/SPRAGUE, MYCOLOGIA 46:77-78,1954 (DESCR SUB GLOEOCERCOSPORA, NO ILLUS) •

MULTICLADIUM DESHPANDE & DESHPANDE 1966. MYCOPATHOLOGIA 30:184-186 (ILLUS) /M. DIGITATUM
?, PROBABLY THE SAME AS SPEIROPSIS TUBAKI 1958. •

MULTIPATINA SAWADA 1928. REP. AGR. RES. INST., FORMOSA 35:121-122 & FIGS 24-25 /M. CITRICOLA
?, ORIGINAL ILLUSTRATION NOT DIAGNOSTIC. •

MUROGENELLA GOOS & MORRIS 1965. MYCOLOGIA 57:776-781 /M. TERROPHILA
PHRAGMO,CETERI,PHAEO,NONSP,ALEURI, 90G, ON SOIL.
/BARRON, HYPHOMYCETES P 231-232,1968 (ILLUS) /ELLIS, DH P 83-84,1971 (ILLUS) •

MYCELIOPHTHORA COSTANTIN 1894. IN COSTANTIN & MATRUCHOT, REV. GEN. BOT. 6:289-293 /M. LUTEA
AMERO,BLASTO OR CETERI,HYALO,MISC,ALEURI, 48C & 53B, ON FUNGI (AGARICUS), SOIL & OTHER SUBSTRATES. COMPARE WITH CHRYSOSPORIUM & SPOROTRICHUM. ASCO STATE =THIELAVIA OR CORYNASCUS SEE VAN OORSCHOT 1977.
/CARMICHAEL, CAN. J. BOT. 40:1147-1148 & 1158-1159 & FIGS 28-29,1962 (REFS,ILLUS SUB CHRYSOSPORIUM LUTEUM) /VON KLOPOTEK, ARCH. MICROBIOL. 98:365-369,1974 (ILLUS SUB CHRYSOSPORIUM) /VON KLOPOTEK, ARCH. MICROBIOL. 107:223-224,1976 (ILLUS SUB THIELAVIA) /VAN OORSCHOT, PERSOONIA 9:401-408,1977 (ILLUS,NEW COMBS)
?=*BOTRYODERMA* PAPENDORF & UPADHYAY 1969
=*EMMONSIA* CIFERRI & MONTEMARTINI 1959 FIDE CARMICHAEL
?=*MYCOSYLVA* TULLOCH 1973 •

MYCELODERMA DUCOMET 1907. RECH. DEV. CHAMP., THESE, RENNES, P 119-120 & PL 111-112 /M. CUTICULARE
?, ORIGINAL ILLUSTRATION DIFFICULT TO INTERPRET. •

MYCOBACILLARIA NAUMOV 1915. MAT. MIK. FIT. 1:26 & FIG 30 /M. SIMPLEX
?, LOOKS LIKE OSCILLATORIA IN ILLUSTRATION. •

MYCOBANCHE PERSOON 1818. TRAITE SUR CHAMPIGN. P 133-134 /M. CHRYSOSPERMA
=SEPEDONIUM LINK 1809 FIDE HUGHES 1958. •

MYCOCENTROSPORA DEIGHTON 1972. TAXON 21:716 /CERCOSPORA ACERINA HARTIG 1880 (UNTERS. FORSTBOT. INST. MUNCHEN 1:58) =M. ACERINA (HARTIG)
SCOLECO OR PHRAGMO,CETERI,HYALO OR PHAEO,RACHI,APPEND,SPORODOCHIA,AQUATIC, 70C, ON LEAVES.
/HUGHES, CMI PAPERS 36:39,1951 (DESCR,REFS) /DEIGHTON, CMI PAPERS 124:1-13,1971 (ILLUS,SUB CENTROSPORA) /ELLIS, DH P 264-266,1971 (ILLUS SUB CENTROSPORA) /SUTTON, PIROZYNSKI & DEIGHTON, CAN. J. BOT. 50:1906-1907,1972 (ILLUS SUB CENTROSPORA) /IQBAL, BIOLOGIA, LAHORE 20:1-10,1974 (ILLUS,NEW SP) /MATSUSHIMA, ICONES P 21-22 & PL 226-228,1975 (ILLUS,NEW SP SUB CENTROSPORA) /INGOLD, GUIDE P 80-81,1975 (ILLUS SUB CENTROSPORA) /ELLIS, MDH P 225-227,1976 (ILLUS) /SINCLAIR & MORGAN-JONES, MYCOTAXON 9:472-474,1979 (ILLUS,NEW SP)
=*CENTROSPORA* NEERGARD 1942, NON CENTROSPORA TREVISAN 1845
=*ANSATOSPORA* NEWHALL 1944 (NOMEN NUDUM) FIDE DEIGHTON 1971 •

MYCOCHAETOPHORA HARA & OGAWA 1931. IN HARA, FUNGI, THE NIPPON FUNGOLOG. SOC. NR. 3-4, P 112 /M. JAPONICA
?=HYPHOMYCETE. •

MYCOCHLAMYS MARCHAND & CABRAL 1976. IN MARCHAND ET AL., BOL. SOC. ARGENTINA BOT. 17:66-69 /M. MACROSPORUM
?, A DOUBLE FORM-GENUS BASED ON HUMICOLA & SCYTALIDIUM STATES. •

MYCOENTEROLOBIUM GOOS 1970. MYCOLOGIA 62:171-175 (ILLUS) /M. PLATYSPORUM
DICTYO,CETERI,PHAEO,NONSP, 128D, ON WOOD (ARAUCARIA).
/ELLIS, DH P 50-51,1971 (ILLUS) •

MYCOGONE LINK 1809. MAG. GES. NATURF. FREUNDE, BERLIN 3:18 /M. ROSEA
DIDYMO,CETERI,HYALO,NONSP,ALEURI, 68D, ON FUNGI (AGARICS & OTHERS). ASCO STATE =HYPOMYCES FIDE BARRON 1968. OTHER CONIDIAL STATE =VERTICILLIUM.
/BARRON, HYPHOMYCETES P 232-233,1968 (ILLUS) /MATSUSHIMA, ICONES P 98-99 & PL,1975 (ILLUS) /BRADY & GIBSON, CMI DESCRIPTIONS 499-500,1976 (ILLUS) /MORGAN-JONES & GRAY, MYCOTAXON 8:144-148,1979 (NEW SP,ILLUS)
=*COCCOSPORELLA* KARSTEN 1893 FIDE HUGHES 1958 •

MYCOLEPTODISCUS OSTAZESKI 1968. MYCOLOGIA 59:970-975 (ILLUS,DATED 1967) /LEPTODISCUS TERRESTRIS GERDEMANN 1953 =M. TERRESTRIS (GERDEMANN)
AMERO OR DIDYMO,GLOIO,HYALO,PHIAL,SPORODOCHIA,APPEND, 14C, ON LEAVES (MONOCOTYLEDONS). COMPARE WITH MAHABALELLA.
/ELLIS, MDH P 447-448,1976 (ILLUS)
=*LEPTODISCUS* GERDEMANN 1953 NON HERTWIG 1877 •

MYCOMYCES WYSS-CHODAT 1928. BULL. SOC. BOT. GENEVE 19(1927):209-210 (ILLUS) /M. FUNGOIDES
?, ORIGINAL ILLUSTRATION DIFFICULT TO INTERPRET, PERHAPS NOT A FUNGUS. •

MYCOSYLVA TULLOCH 1973. TBMS 60:155-157, PL 23 /M. CLARKII
AMERO,CETERI OR BLASTO,HYALO,NONSP,SYNNEMATA, 53E, ON DUNG & PEAT (LARIX-PICEA BOG). COMPARE WITH MYCELIOPHTHORA LUTEA.
/TULLOCH, TBMS 61:198,1973 (NO ILLUS,REFS) /SAMSON & HINTIKKA, KARSTENIA 14:133-137,1974 (ILLUS,NEW SP) /CHRISTENSEN & WORLEY, MYCOLOGIA 70:456-460,1978 (ILLUS) •

MYCOVELLOSIELLA RANGEL 1918. ARCH. JARD. BOT. RIO DE JAN. 2:69-71, TAB 28-30 /CERCOSPORA CAJANI HENNINGS 1902 =M. CAJANI (HENNINGS) RANGEL EX TROTTER 1931
PHRAGMO OR SCOLECO,BLASTO,HYALO OR PHAEO,RADUL,STROMA, 72D, 82B, ON LEAVES. FOR PHRAGMOSPOROUS, RAMULARIA-LIKE FUNGI, WITH CONIDIOPHORES BORNE ON A STROMA. DIFFERENTIATED FROM CERCOSPORA BY CATENATE CONIDIA.
/ELLIS, DH P 303-305,1971 (ILLUS) /DEIGHTON, CMI PAPERS 137:1-75 & PL 1-2,1974 (ILLUS,REVIS,NEW SPP,NO KEY) /DEIGHTON, TBMS 65:509-512,1975 (ILLUS) /ELLIS, MDH P 305-307,1976 (ILLUS,KEY) /DEIGHTON, CMI PAPERS 144:14-26,1979 (ILLUS,NEW SPP,NEW COMBS)
=*CERCODEUTEROSPORA* CURZI 1932 FIDE CASTELLANI & CIFERRI
=*RAGNHILDIANA* SOLHEIM 1931 FIDE DEIGHTON 1974
?=*STENELLA* SYDOW 1930
=*VELLOSIELLA* RANGEL 1915 NON VELLOZIELLA BAILL. 1886
=*WALKEROMYCES* THAUNG 1976 FIDE DEIGHTON 1979 •

MYDONOSPORIUM CORDA 1833. IN STURM'S DEUTSCHL. FLORA III (PILZE), P 95-96, TAB 48 /M. OLIVACEUM
=CLADOSPORIUM LINK 1815 FIDE LINDAU IN RABENHORST 1907. •

MYDONOTRICHUM CORDA 1833. IN STURM'S DEUTSCHL. FLORA III (PILZE), P 37-38, TAB 19 /M. ATRUM
?, SEE LINDAU 1910 (IN RABENHORST, KRYPT.-FL. 2 AUFL., 1 BD., 9:49) •

MYLITTA FRIES =SCLEROTIA OF POLYPORUS MYLITTAE SEE MACFARLANE, KUO & HILTON 1978 (TBMS 71:359-365). •

MYRIOCEPHALUM DE NOTARIS 1842 =CHEIROSPORA MOUGEOT & FRIES 1825 FIDE HUGHES 1958 •

MYRIOCOCCUM FRIES 1823. SYST. MYCOL. 2:304 /M. PRAECOX (NO ILLUS)
?, SEE VON ARX 1974 (GENERA P 268). •

MYRIOCONIUM SYDOW 1912. ANN. MYCOL. 10:448-450, FIGS 4-7 /M. SCIRPI =M. SCIRPICOLA (FERDINANDSEN & WINGE) SYDOW 1913 (IN FERDINANDSEN & WINGE, ANN. MYCOL. 11:22)

AMERO,GLOIO OR BASO,HYALO,PHIAL,SPORODOCHIA, 5D, ON PLANTS, LITTER, SOIL & OTHER SUBSTRATES. FOR SPERMATIAL STATES SIMILAR TO PHIALOPHORA. ASCO STATE =SCLEROTINIA, CIBORIA, RUTSTROEMIA, MONILINIA, NEOBULGARIA, MYRIOSCLEROTINIA, DIBELONIELLA, WHETZELINIA & OTHERS. OTHER CONIDIAL STATES =BOTRYTIS, MONILIA, CRISTULARIELLA, HELICODENDRON & OTHERS.

/MULLER & DEFAGO, SYDOWIA 20:161-164,1967 (ILLUS SUB DIBELONIELLA RAINERI) /VON ARX, GENERA P 180,1970 (REFS) /MORDUE, CMI DESCRIPTIONS 616-619,1979 (ILLUS SUB SCLEROTINIA)

=*BOTRYOPHIALOPHORA* LINDER 1944 FIDE VON ARX 1970
?=*ENDOCONIDIUM* PRIL. & DELACR. 1891
=*RHACODIELLA* PEYRONEL 1919 FIDE CARMICHAEL 1973 •

MYRIODONTIUM SAMSON & POLONELLI 1978. PERSOONIA 9:505-509 (ILLUS) /M. KERATINOPHILUM

AMERO,CETERI,HYALO,NONSP,ALEURI, ON SOIL.

/COLE & SAMSON, CONIDIAL FUNGI P 34-35,1979 (ILLUS) •

MYRIOPHYSA FRIES 1849. SUMMA VEG. SCAND. 481 /M. ATRA

?=ATICHIA FIDE HOHNEL FROM AINSWORTH 1971. •

MYRIOPHYSELLA SPEGAZZINI 1910 =ATICHIA FLOTOW 1850 FIDE SANTESSON FROM AINSWORTH 1971. SEE ALSO MEEKER 1975 (CAN. J. BOT. 53:2489) •

MYROPYXIS CESATI 1851. FLORA 34:73 /M. CARICICOLA

?, PROBABLY A NOMEN DUBIUM. •

MYROTHECIELLA SPEGAZZINI 1911. AN. MUS. NAC. BUENOS AIRES 3(13):460-461 (ILLUS) /M. CATENULIGERA

=MYROTHECIUM TODE FIDE TULLOCH 1972. •

MYROTHECIUM TODE EX LINK 1809. MAG. GES. NATURF. FREUNDE, BERLIN 3:23 /M. INUNDATUM

AMERO,GLOIO,HYALO OR PHAEO,PHIAL,SPORODOCHIA,SETAE,SYNNEMATA, 12F, ON LEAVES, LITTER, SOIL, FUNGI (RUSSULA) & OTHER SUBSTRATES. ASCO STATE =NECTRIA FIDE BARRON 1968. COMPARE WITH VOLUTELLA & COLLETOCONIS.

/ELLIS, DH P 552-556,1971 (ILLUS,KEY) /TULLOCH, CMI PAPERS 130:1-42,1972 (ILLUS,MONOGR) /MATSUSHIMA, ICONES P 99-100 & PL,1975 (ILLUS) /ELLIS, MDH P 465-466,1976 (ILLUS) /ALFIERI, MYCOTAXON 10:217-218,1979 (ILLUS,NEW SP SUB KUTILAKESA)

=*EXOTRICHUM* SYDOW 1914 FIDE HOHNEL FROM AINSWORTH 1971
=*KUTILAKESA* SUBRAMANIAN 1956 FIDE NAG RAJ
=*LOMACHASHAKA* SUBRAMANIAN 1956 FIDE TULLOCH 1972
=*MYROTHECIELLA* SPEGAZZINI 1910 FIDE TULLOCH 1972
=*MYXORMIA* BERKELEY & BROOME 1850 FIDE TULLOCH 1972
=*PHAEOSTILBELLA* HOHNEL 1919 FIDE TULLOCH 1972
?=*PIROBASIDIUM* HOHNEL 1902
=*STARKEYOMYCES* AGNIHOTHRUDU 1956 FIDE TULLOCH 1972 •

MYSTROSPORIELLA MUNJAL & KULSHRESTHA 1969. MYCOPATHOLOGIA 39:356-357 (ILLUS) /M. LITSEAE

DICTYO OR PHRAGMO,CETERI,PHAEO,RADUL, 99A, ON LEAVES (LITSEA).

/ELLIS, DH P 301-302,1971 (ILLUS) /MATSUSHIMA, ICONES P 100 & PL 151,1975 (ILLUS,NEW SP) •

MYSTROSPORIUM CORDA 1837. ICONES FUNGORUM 1:12 /M. DUBIUM

=NOMEN DUBIUM SEE HUGHES 1958. •

MYXOCLADIUM CORDA 1837. ICONES FUNGORUM 1:12 /M. ARUNDINIS =CLADOSPORIUM HERBARUM (PERSOON) LINK 1815

=CLADOSPORIUM LINK 1815 FIDE HUGHES 1958. •

MYXOCYCLUS RIESS APUD FRESENIUS 1852. BEITRAGE ZUR MYKOLOGIE 2:63-64 /M. CONFLUENS =M. POLYCISTIS (BERKELEY & BROOME) SACCARDO 1908 FIDE SUTTON 1975 (CMI PAPERS 138:120-123, ILLUS)

DICTYO,CETERI,PHAEO,NONSP,ACERVULI OR SPORODOCHIA, ON BRANCHES (BETULA). (MELANCONIACEAE). •

MYXODOCHIUM ARNAUD 1951. BULL. SOC. MYCOL. FRANCE 67:196 & FIG 4H-I, P 189 (NO LATIN DIAGNOSIS) /M. HYALINUM
? •

MYXONEMA CORDA 1837. ICON. FUNG. 1:10 & TAB. 2, FIG 150 /M. ASSIMILE
=NOMEN DUBIUM FIDE FRIES FROM AINSWORTH 1971. •

MYXORMIA BERKELEY & BROOME 1850. ANN. MAG. NAT. HIST., SER. 2, 5:457 & PL 12, FIG 9 /M. ATROVIRIDIS =MYROTHECIUM ATROVIRIDE (BERKELEY & BROOME) TULLOCH 1972
=MYROTHECIUM TODE FIDE TULLOCH 1972 (CMI PAPERS 130:36). •

MYXOTRICHELLA (SACCARDO) SACCARDO 1892. SYLLOGE FUNGORUM 10:593 (NO ILLUS) /M. SPELAEA (LECTOTYPE)
?, FOR EMPTY ASCOCARPS OF MYXOTRICHUM. •

NAKATAEA HARA 1939. THE DISEASES OF THE RICE PLANT, 2 ED., P 185 /HELMINTHOSPORIUM SIGMOIDEUM CAVARA 1880 =N. SIGMOIDEA (CAVARA) =N. STATE OF LEPTOSPHAERIA SALVINII FIDE ELLIS 1971
PHRAGMO,CETERI,PHAEO,RADUL,PODIA, 79B, ON STEMS (ORYZA). ASCO STATE =LEPTOSPHAERIA FIDE SHOEMAKER 1959 =MAGNAPORTHE KRAUSE & WEBSTER 1972 (MYCOLOGIA 64:103-114,ILLUS). COMPARE WITH PYRICULARIA. OTHER STATE =SCLEROTIUM (H) SEE ELLIS & HOLLIDAY 1972 (CMI DESCRIPTIONS 344) & BARR 1977 (MYCOLOGIA 69:952-953).
/ELLIS, DH P 219-221,1971 (ILLUS) /MATSUSHIMA, ICONES P 100 & PL 193,1975 (ILLUS,NEW COMB)
=*VAKRABEEJA* SUBRAMANIAN 1956 FIDE ELLIS 1971 •

NALANTHAMALA SUBRAMANIAN 1956. J. INDIAN BOT. SOC. 35:476-478 (ILLUS) /N. MADREEYA
AMERO,BASO,HYALO,PHIAL,SPORODOCHIA, 14A, ON APHIDS & FUNGI (SOOTY MOLD). COMPARE WITH PAECILOMYCES & AKANTHOMYCES. •

NAPICLADIUM THUMEN 1875. MYCOTH. UNIV. NO 91, ANNO 1875 /N. SORAUERI =SPILOCAEA POMI FRIES
=SPILOCAEA FRIES 1825 FIDE HUGHES 1958. •

NASCIMENTOA CIFERRI & BATISTA 1956. PUBLCOES INST. MICOL. RECIFE 44:4 /N. PSEUDOENDOGENA =SPIROPES CAPENSIS (THUMEN) ELLIS 1968
=SPIROPES CIFERRI 1955 FIDE ELLIS 1968 (CMI PAPERS 114:2 & 6). •

NECATOR MASSEE 1898. KEW BULLETIN 138:119 (NO ILLUS) /N. DECRETUS =N. STATE OF CORTICIUM SALMONICOLOR
AMERO,ARTHRO OR BASO?,HYALO,NONSP,SPORODOCHIA, 45C, ON TROPICAL PLANTS (HEVEA). BASIDIO STATE =CORTICIUM FIDE DONK 1962 (TAXON 11:90) & MORDUE & GIBSON 1976 (CMI DESCRIPTIONS 511,ILLUS). •

NECRAPHIDIUM CIFERRI 1951. MYCOPATHOLOGIA 6:25 (NO ILLUS) /N. ZOOPHILUM (SACCARDO & TROTTER) =RAMULARIA ZOOPHILA
? •

NEGERIELLA HENNINGS 1897. HEDWIGIA 36:245 (NO ILLUS) /N. CHILENSIS
PHRAGMO,CETERI,PHAEO,NONSP?,SYNNEMATA, 87H, ON LITTER & WOOD. MAY BE A SYNONYM OF PODOSPORIUM SCHWEINITZ 1832.
/HOHNEL, S. B. AKAD. WISS. WIEN, ABT. 1, 120:401-402,1911 (EMEND, NO ILLUS) /MORRIS, MYCOPATHOLOGIA 33:181-184,1967 (ILLUS) •

NEMATOCTONUS DRECHSLER 1941. PHYTOPATHOLOGY 31:779-780 (773-780,ILLUS) /N. TYLOSPORUS (LECTOTYPE FIDE DONK 1962, TAXON 11:90)
AMERO,CETERI,HYALO,NONSP,CLAMPS, 52C, ON NEMATODES & SOIL. BASIDIO STATE =HOHENBUEHELIA FIDE BARRON 1977.
/COOKE & GODFREY, TBMS 47:61-74,1964 (ILLUS,KEY) /VON ARX, GENERA P 163-164,1970 (ILLUS) /MATSUSHIMA, ICONES P 100 & PL 375,1975 (ILLUS) /BARRON, THE NEMATODE-DESTROYING FUNGI P 40-46, 81-84 & 97,1977 (ILLUS) /BARRON, CAN.

J. BOT. 55:3054-3062,1977 (ILLUS) •

NEMATOGONUM DESMAZIERES 1834. ANN. SCI. NAT. II,2:70 /N. AURANTIACUM =N. FERRUGINEUM (PERSOON) HUGHES 1958
AMERO,BLASTO,HYALO,AMPUL, 32C, ON SOIL & LITTER. OTHER CONIDIAL STATE ="PHIALOCONIDIA" FIDE MATSUSHIMA 1975.
/BARRON, HYPHOMYCETES P 235-236,1968 (ILLUS) /GAMS, REV. MYCOL. 39:273-278,1974-1975 (ILLUS,NEW SP) /MATSUSHIMA, ICONES P 101 & PL 6,1975 (ILLUS,NEW COMB)
=*BOTRYOCLADIUM* PREUSS 1851 FROM AINSWORTH 1971
=*GONATORRHODIELLA* THAXTER 1891 FIDE HUGHES 1953 •

NEMATOGRAPHIUM GOIDANICH 1935. ANNALI DE BOT. 21:46 (NO ILLUS) /FIVE ORIGINAL NEW COMBINATIONS
?, FOR GRAPHIUM WITH SIMPLE SYNNEMATA. •

NEMATOMYCES FAUREL & SCHOTTER 1965. REV. MYCOL. 30:344-348 (ILLUS) /N. COPROPHILA
=GLIOMASTIX GUEGUEN 1905 FIDE VON ARX 1970. NO TYPE MATERIAL FIDE GAMS 1971 (CEPHALO. SCHIMMEL. P 18). •

NEMATOPHAGUS MEKHTIEVA 1975. MIKOL. FITOPATOL. 9:250-252 (ILLUS) /N. AZERBAIDZHANICUS
?=ARTHROBOTRYS CORDA 1839. •

NEOMICHELIA PENZIG & SACCARDO 1901. MALPIGHIA 15:246 /N. MELAXANTHA
=PITHOMYCES BERKELEY & BROOME 1873 FIDE CLEMENTS & SHEAR 1931. •

NEOTRICHOPHYTON CASTELLANI & CHALMERS 1919. MAN. TROP. MED. ED 3 P 1003-1004 /N. FLAVUM (BODIN)
=TRICHOPHYTON MALMSTEN 1845. •

NEOTTIOSPORELLA HOHNEL EX GRANITI 1951. NUOVO G. BOT. ITAL. 58:148-151 /N. TRISETI GRANITI 1951
AMERO,GLOIO,HYALO,PHIAL,SPORODOCHIA,SETAE,APPEND, 9H, ON LITTER.
/PIROZYNSKI, CMI PAPERS 129:57-58,1972 (REFS)
?=*MINIMIDOCHIUM* SUTTON 1969
?=*SCHIZOTRICHELLA* MORRIS 1956 •

NEOTTIOSPORIS CLEMENTS & SHEAR 1931 IS A SUPERFLUOUS NAME FOR NEOTTIOSPORELLA HOHNEL •

NETA SHEARER & CRANE 1971. MYCOLOGIA 63:239-242 (ILLUS) /N. PATUXENTICA SHEARER & CRANE 1970
DIDYMO,CETERI OR GLOIO,HYALO,RACHI OR RADUL,SETAE, 62H, ON WOOD IN WATER.
/SUTTON, CMI PAPERS 132:81-82,1973 (ILLUS) /ELLIS, MDH P 166-167,1976 (ILLUS) •

NIA MOORE & MEYERS 1959. MYCOLOGIA 51:874 (871-876,ILLUS) /N. VIBRISSA
AMERO OR STAURO,CETERI,HYALO,NONSP OR AMPUL,APPEND,SPORODOCHIA,CLAMPS, 121B, ON WOOD IN SEA WATER. HAS CLAMP CONNECTIONS, THE 'CONIDIA' ARE BASIDIOSPORES, SEE DOGUET 1967 (C.R. ACAD. SCI. PARIS, SERIES D, 265:1780-1783,ILLUS), & DOGUET 1969 (BULL. SOC. MYCOL. FRANCE 85:93-104,ILLUS).
/BROOKS, MYCOLOGIA 67:172-174,1975 (DOLIPORE SEPTA) •

NIGROCUPULA SAWADA 1944. REP. GOVT. RES. INST. FORMOSA 87:91 & PL 5, FIGS 46-49 (NO LATIN DIAGNOSIS) /N. FORMOSANA
?, PERHAPS ENDOCALYX. •

NIGROSPORA ZIMMERMAN 1902. ZENTRALBL. BAKT. ABT. 2,8:220 /N. PANICI
AMERO,CETERI,PHAEO,MISC, 49C, ON PLANTS (ORYZA, SACCHARUM & OTHERS) & SOIL. ASCO STATE =APIOSPORA FIDE VON ARX 1974 (GENERA P 129 & 255).
/HUDSON, TBMS 46:355-360,1963 (ILLUS,REFS) /ELLIS, DH P 319-320,1971 (ILLUS,KEY)
/SIVANESAN & HOLLIDAY, CMI DESCRIPTIONS 311,1971 (ILLUS SUB KHUSKIA)
/MATSUSHIMA, ICONES P 101 & PL,1975 (ILLUS)
=*BASISPORIUM* MOLLIARD 1902 FIDE MASON 1933
=*DICHOTOMELLA* SACCARDO 1914 FIDE HUGHES 1958

=*PHAEOCONIS* CLEMENTS 1909 FIDE HUGHES 1958 •

NODULISPORIUM PREUSS 1849. KLOTZSCHII HERB. VIV. MYCOL. NO 1272 /N. OCHRACEUM
AMERO,CETERI,HYALO OR PHAEO,RADUL OR AMPUL,SYNNEMATA, 27D, ON WOOD. SEE ALSO GENICULOSPORIUM, VIRGARIA, PHAEOISARIA, IDRIELLA & SPOROTHRIX. ASCO STATE =HYPOXYLON, XYLARIA, ROSELLINIA, NUMMULARIA, DALDINIA & RELATED GENERA FIDE MARTIN 1967 & ELEUTHERASCUS SEE SAMSON & LUITEN 1975 (TBMS 64:338-340 & PL 28,ILLUS) OR GRAPHOSTROMA SEE DE HOOG 1977 (CBS STUDIES 15:139) OR CONIOCHAETA FIDE HAWKSWORTH 1978 (NORW. J. BOT. 25:15-18,ILLUS,NEW SP). OTHER CONIDIAL STATE =?BEAUVERIOPHORA.
/MARTIN, J. S. AFRICAN BOT. 33:205-240,1967 (ILLUS) /GREENHALGH & CHESTERS, TBMS 51:57-82,1968 (ILLUS) /ELLIS, DH P 232,1971 (ILLUS) /SIVANESAN & HOLLIDAY, CMI DESCRIPTIONS 355-357 & 359,1972 (ILLUS) /JONG & ROGERS, WASHINGTON AGRIC. EXPER. STA. TECH. BULL. 71:1-51,1972 (ILLUS,REVIS) /WHALLEY & GREENHALGH, TBMS 64:222-233,1975 (REFS) /ROGERS, MYCOLOGIA 67:1061-1065,1975 (ILLUS) /DE HOOG & HERMANIDES-NIJHOF, CBS STUDIES 15:191 (ILLUS OF TYPE SP OF N. OCHRACEUM) & 15:203,1977 (NEW COMB) /ROGERS, MYCOLOGIA 71:1-42,1979 (DISCUSSION,REFS)
=*ACROSTAPHYLUS* ARNAUD 1954 FIDE VON ARX 1970
?=*THAROOPAMA* SUBRAMANIAN 1956 •

NOMURAEA MAUBLANC 1903. BULL. SOC. MYCOL. FRANCE 19:295-296 & PL 14, FIG 7 /N. PRASINA =N. RILEYI (FARLOW 1883) SAMSON 1974
AMERO,BASO,HYALO,PHIAL,SYNNEMATA, 4D, ON SPIDERS.
/SAMSON, CBS STUDIES 6:80-85,1974 (REVIS,ILLUS) /KISH & ALLEN, MYCOLOGIA 68:436-439,1976 (ECOLOGY) /HOCKING, TBMS 69:511-513,1977 (ILLUS,NEW SP) /SAMSON & EVANS, PROC. K. NEDERL. AKAD. WETEN., AMSTERDAM, SER C, 80:133,1977 (DISTRIBUTION,REFS) /ONIONS, CMI DESCRIPTIONS 612,1979 (ILLUS) •

NOTHOCLAVULINA SINGER 1970. FLORA NEOTROPICA MONOGR. 3:18 (NO ILLUS) /N. DITOPA =N. STATE OF ARTHROSPORELLA DITOPA SINGER 1970 (IBID. P 17)
AMERO,ARTHRO,HYALO,NONSP,SYNNEMATA, ON FUNGI (AGARICS). BASIDIO STATE =ARTHROSPORELLA. COMPARE WITH SCLEROSTILBUM.
/SINGER, AGARICALES IN MODERN TAXONOMY P 22,1975 (DISCUSSION) /WATLING & KENDRICK, MICHIGAN BOT. 16:71,1977 (REFS) •

NOTHOSPORA PEYRONEL 1913 =DUBIUS FIDE CLEMENTS & SHEAR 1931 •

NYCTALINA ARNAUD 1952. BULL. SOC. MYCOL. FRANCE 68:189-190 /N. LIGNICOLA
?, NO TYPE SPECIMEN, SEE NICOT & CHARPENTIE 1971 (BULL. SOC. MYCOL. FRANCE 87:33). SEE ALSO OSTEOMORPHA & ASTEROPHORA.
/ARNAUD, BULL. SOC. MYCOL. FRANCE 67:197, FIG 1Q,1951 (ILLUS) •

NYCTALOSPORA MORRIS 1972. MYCOLOGIA 64:890-892 (FIGS 4-7) /N. COMPACTA
?=LISTEROMYCES PENZIG & SACCARDO 1901. •

OBSTIPISPORA SINCLAIR & MORGAN-JONES 1979. MYCOTAXON 8:152-156 (ILLUS) /O. CHEWACLENSIS
?=CONDYLOSPORA NAWAWI 1976. •

OCHROCONIS DE HOOG & VON ARX 1973. KAVAKA 1:57 (55-60,ILLUS) /SCOLECOBASIDIUM CONSTRICTUM ABBOTT 1937 =O. CONSTRICTA (ABBOTT)
DIDYMO OR PHRAGMO,CETERI,PHAEO,RADUL, 62E, ON SOIL & LITTER. DIFFERENTIATION FROM DACTYLARIA NOT CLEAR.
/WELLMAN, CAN. J. BOT. 53:1630-1633,1975 (ILLUS,NEW SP) •

OEDEMIUM LINK 1824. LINN. SPEC. PLANT. IV,6(1):42 /O. ATRUM =O. STATE OF THAXTERIA FUSCA (FUCKEL) BOOTH
DIDYMO,BLASTO OR CETERI,PHAEO,AMPUL,SETAE, 66D, ON LITTER (BRANCHES OF DECIDUOUS TREES). ASCO STATE =THAXTERIA FIDE ELLIS 1971 OR CHAETOSPHAERELLA FIDE MULLER IN KENDRICK 1971.
/HUGHES & HENNEBERT, CAN. J. BOT. 41:773-809,1963 (ILLUS,REVIS) /ELLIS, DH P 175-176,1971 (ILLUS,KEY)
=*DIPLOSPORIUM* LINK 1824 FIDE HUGHES 1958
=*CLADOTRICHUM* CORDA 1831/32 FIDE HUGHES 1958
=*GONGYLOCLADIUM* WALLROTH 1833 FIDE HUGHES 1958

=*DIMERA* FRIES 1825 FIDE HUGHES 1958
?=*SPHINCTROSPORIUM* KUNZE EX FRIES 1849 •

OEDOCEPHALUM PREUSS 1851. LINNAEA 24:131-132 /O. ELEGANS (LECTOTYPE)
AMERO,CETERI,HYALO,AMPUL, 35G, ON DUNG, LITTER, WOOD & SOIL. ASCO STATE =ASCOPHANUS SEE TUBAKI 1958 (J. HATTORI BOT. LAB. 20:181), OR PEZIZA OR IODOPHANUS FIDE STALPERS 1974 OR CLEISTOIODOPHANUS FIDE BEZERRA & KIMBROUGH 1976 (AMER. J. BOT. 63:838-844,ILLUS) OR PYRONEMA FIDE DE HOOG 1977 (CBS STUDIES 15:137).
/VON ARX, GENERA P 208-209,1970 (ILLUS,REFS) /STALPERS, PROC. KON. NED. AKAD. WET., SER. C 77:383-401,1974 (MONOGR) /MATSUSHIMA, ICONES P 102 & PL,1975 (ILLUS)
?=*SPHONDYLOCEPHALUM* STALPERS 1974 •

OEDOTHEA SYDOW 1930. ANN. MYCOL. 28:202-203 /O. VISMIAE
DIDYMO,CETERI,PHAEO,ANNEL,SPORODOCHIA, 59B, ON LEAVES (VISMIA).
/ELLIS, DH P 143-144,1971 (ILLUS) •

OICHITONIUM BERKELEY & CURTIS 1875. IN BERKELEY, GREVILLEA 3:146 /O. MELLEUM
?=DENDRODOCHIUM FIDE SACCARDO 1886 (SYLL. FUNG. 4:653). •

"OIDIA" SEE "ARTHROCONIDIA" •

OIDIODENDRON ROBAK 1932. NYT. MAG. NATURVIDENSK. 71:251 (243-255) /O. FUSCUM (LECTOTYPE)
AMERO,ARTHRO,HYALO OR PHAEO,NONSP,ALEURI, 43B, ON LITTER, WOOD, SOIL & OTHER SUBSTRATES. ASCO STATE =MYXOTRICHUM FIDE SIGLER & CARMICHAEL 1976 OR BYSSOASCUS FIDE VON ARX 1974 (GENERA P 99,REFS).
/BARRON, CAN. J. BOT. 40:589-607,1962 (ILLUS) /ELLIS, DH P 34-35,1971 (ILLUS) /SIGLER & CARMICHAEL, MYCOTAXON 4:384-391,1976 (ILLUS,NEW SP, NEW COMB) /ELLIS, MDH P 20-24,1976 (ILLUS,KEY) /TERRACINA, CAN. J. BOT. 55:580-584,1977 (ILLUS,CONIDIUM ONTOGENY) •

OIDIOPSIS SCALIA 1902. AGRICOLT. CALABRO-SICULO 27:393-397 (NO ILLUS) /O. SICULA
?=OIDIUM LINK 1824 NOM. CONS. 1975.
/RAYCHAUDHURI, TBMS 32:288-290,1949 (POOR ILLUS) /THIRUMALACHAR & MUNDKUR, CMI PAPERS 40:14-15,1951 (NEW SPP,NO ILLUS) •

OIDIUM LINK 1809. MAG. GES. NATURF. FREUNDE, BERLIN 3:18 /O. AUREUM (PERSOON) =SPOROTRICHUM AUREUM LINK 1809
=SPOROTRICHUM LINK 1809 FIDE HUGHES 1958. NON OIDIUM LINK 1824 NOM. CONS. 1975. OIDIUM SENSU LINDER 1942 WAS USED FOR ACLADIUM, ALYSIDIUM & OLPITRICHUM.
/HUGHES, CAN. J. BOT. 36:732-733,1958 (REFS) •

OIDIUM LINK 1824 NOM. CONS. 1975. IN WILLD., LINNE SP. PL ED. 4, VI(1):121 /ACROSPORIUM MONILIOIDES NEES 1816/17 (DAS SYSTEM ...P 53) =O. MONILIOIDES (NEES) LINK 1824 =O. STATE OF ERYSIPHE GRAMINIS DE CANDOLLE 1815 EX MERAT 1821 (NUOV. FL. ENV. PARIS, ED II, 1:133)
AMERO,BASO,HYALO,MISC, 40C, ON LEAVES (POWDERY MILDEW). NOMEN CONSERVANDUM, SEE WERESUB, TAXON 22:696-701,1973 & TAXON 24:534,1975. ASCO STATE =ERYSIPHE, SPHAEROTHECA, PHYLLACTINIA, MICROSPHAERA, PODOSPHAERA, LEVEILLULA, UNCINULA, SEE YARWOOD 1973 & PARMELEE 1977, & PLEOCHAETA SEE KIMBROUGH & KORF 1963 (MYCOLOGIA 55:608).
/KAPOOR, CMI DESCRIPTIONS 151-156 & 158-160,1967 (ILLUS) /MUKERJI, CMI DESCRIPTIONS 182, 187-190,1968 (ILLUS) /PURNELL & SIVANESAN, CMI DESCRIPTIONS 251-252 & 254,1970 (ILLUS) /YARWOOD, THE FUNGI, IVA:71-81,1973 (ILLUS,KEY) /PARMELEE, FUNGI CANADENSES 63 & 71,1975 (ILLUS) /SIVANESAN & HOLLIDAY, CMI DESCRIPTIONS 508-509,1976 (ILLUS) /SIGLER & CARMICHAEL, MYCOTAXON 4:390-391,1976 (EXCLUSION) /PARMELEE, CAN. J. BOT. 55:1940-1987,1977 (ILLUS) /HAMMETT, NEW ZEALAND J. BOT. 15:687-711,1977 (ILLUS)
=*ACROSPORIUM* NEES 1816/17 NOM. REJ. 1975
?=*OVULARIOPSIS* PATOUILLARD & HARIOT 1900
?=*OIDIOPSIS* SCALIA 1902 •

OJIBWAYA SUTTON 1973. CMI PAPERS 132:82-86 (ILLUS) /O. PERPULCHRA
AMERO,ARTHRO OR BASO,PHAEO,MISC,SPORODOCHIA, 40E, ON LITTER (JUNIPERUS).

SIMILAR TO BRIOSIA.
/ELLIS, MDH P 25-26,1976 (ILLUS)
?=*EXOSPORINA* OUDEMANS 1904 •

OLPITRICHUM ATKINSON 1894. BOT. GAZ. 19:244-245 & PL 23 /O. CARPOPHILUM =O. MACROSPORUM (FARLOW) SUMSTINE 1911
AMERO,CETERI,HYALO,NONSP OR RADUL, 31F, ON LITTER. OTHER CONIDIAL STATE =?HARZIELLA. COMPARE WITH HARZIA.
/BARRON, HYPHOMYCETES P 242-243,1968 (ILLUS) /HOLUBOVA-JECHOVA, FOLIA GEOBOT. PHYTOTAX., PRAHA 9:425-432,1974 (REVIS,ILLUS) /MATSUSHIMA, ICONES P 102-103 & PL,1975 (ILLUS,NEW SP)
=*OIDIUM* LINDER 1942 PRO PARTE, NON OIDIUM LINK 1809 OR 1824
=*PLECTOTHRIX* SHEAR 1902 FIDE LINDER 1942 •

OMMATOSPORA BATISTA & CAVALCANTI 1964. RIV. PATOL. VEG. PAVIA SER. 3, 4:565-566 (ILLUS) /O. RORAIMENSIS
=MICROCLAVA STEVENS 1917 FIDE DEIGHTON 1969 (TBMS 52:315-321). •

OMMATOSPORELLA BATISTA, BEZERRA & POROCA 1967. ATAS INST. MICOL. UNIV. RECIFE 5:424-428 & FIG 1 /O. BURSERACEARUM
?, PROBABLY MICROCLAVA. •

ONCOCLADIUM WALLROTH 1833. FLOR. CRYPT. GERMAN. P 289-290 (NO ILLUS) /O. FLAVUM
AMERO,ARTHRO,HYALO,NONSP,ALEURI, 43E, ON KERATIN. SEE ALSO MALBRANCHEA.
/HUGHES, CAN. J. BOT. 46:941-942 & PL II,1968 (ILLUS) /ORR & KUEHN, MYCOLOGIA 63:198-199,1971 (REVIS) /SIGLER & CARMICHAEL, MYCOTAXON 4:448-453,1976 (ILLUS,REVIS)
=*ACTINODENDRON* ORR & KUEHN 1963 PRO PARTE •

ONCOPODIELLA ARNAUD EX RIFAI 1965. PERSOONIA 3:407-411 (ILLUS) /O. TETRAEDRICA ARNAUD 1954 =O. TRIGONELLA (SACCARDO) RIFAI 1965
DICTYO OR STAURO,CETERI,PHAEO,NONSP,APPEND, 95D, ON WOOD & BARK (ULMUS & OTHER TREES).
/ARNAUD, BULL. SOC. MYCOL. FRANCE 69:296 & FIG 13Q & R,1954 (ILLUS) /ELLIS, DH P 210-211,1971 (ILLUS) /HAWKSWORTH & PUNITHALINGAM, TBMS 64:93-94,1975 (ILLUS,NEW SP) •

ONCOPODIUM SACCARDO & SACCARDO 1904. IN SACCARDO, ANN. MYCOL., BERLIN 2:19, TAB III FIG V /O. ANTONIAE
DICTYO,CETERI,PHAEO,NONSP OR MISC,APPEND,SPORODOCHIA, 95F, ON LITTER.
/ELLIS, CMI PAPERS 87:17-20,1963 (ILLUS) /ELLIS, DH P 111-112,1971 (ILLUS,KEY) /ELLIS, MDH P 72,1976 (ILLUS,NEW SP) /SUTTON, MYCOLOGIA 70:790-796,1978 (ILLUS,NEW SP) •

OOSPORA WALLROTH 1833. FL. CRYPT. GERM. 2:182 /?, MANY ORIGINAL SPECIES INCLUDING THE TYPES OF FOUR OTHER GENERA
=NOMEN ILLEGITIMUM FIDE HUGHES 1958. HAS BEEN USED FOR SPECIES OF OIDIUM (=ACROSPORIUM) & GEOTRICHUM AMONG OTHERS. FOR O. SULPHUREO-OCHRACEA, SEE OVADENDRON. •

OOSPOROIDEA SUMSTINE 1913. MYCOLOGIA 5:52-53 /O. LACTIS (FRESENIUS) =GEOTRICHUM CANDIDUM LINK
=GEOTRICHUM LINK 1809 FIDE CARMICHAEL 1957. •

OPHIOCLADIUM CAVARA 1893. ZEITS. PFLANZENKR. 3:26 /O. HORDEI
=OVULARIA SACCARDO 1880 FIDE SPRAGUE FROM AINSWORTH 1971. •

OPHIODENDRON ARNAUD 1952. BULL. SOC. MYCOL. FRANCE 69:208 & 211 (ILLUS) /O. LAOCOONI
=NOMEN DUBIUM FIDE HENNEBERT 1968. NO TYPE SPECIMEN, SEE NICOT & CHARPENTIE 1971 (BULL. SOC. MYCOL. FRANCE 87:33). SEE ALSO SPIROSPHAERA. •

OPHIOPODIUM ARNAUD 1954. BULL. SOC. MYCOL. FRANCE 69:300 /O. HUMIRIAE
=GRALLOMYCES STEVENS 1918 FIDE DEIGHTON & PIROZYNSKI 1966. NO TYPE SPECIMEN, SEE NICOT & CHARPENTIE 1971 (BULL. SOC. MYCOL. FRANCE 87:33). •

OPHIOTRICHUM KUNZE APUD FRIES 1849. SUMMA VEG. SCAND. 2:503 (NO ILLUS) /O. PHLOMIDIS FRIES 1849
=NOMEN DUBIUM, NO TYPE SPECIMEN FIDE NANNFELDT 1960 (IN LIT.) •

ORAMASIA URRIES 1956. AN. INST. BOT. A. J. CAVANILLES 14:168 (NOT SEEN) /O. HIRSUTA
AMERO,GLOIO,HYALO,PHIAL,SPORODOCHIA,SETAE, ON LITTER.
/SUTTON, MYCOLOGIA 70:796-798,1978 (ILLUS,NEW VAR)
=*SINGERA* BATISTA & BEZERRA 1960 FIDE SUTTON 1977 (CMI PAPERS 141:191) •

ORBIMYCES LINDER 1944. FARLOWIA 1:404-405 & PL I, FIG 5-6 /O. SPECTABILIS
STAURO,CETERI,PHAEO,NONSP?,APPEND, 115F, ON WOOD IN SEA WATER.
/BARNETT, IMPERFECT FUNGI NO. 272,1960 (ILLUS) •

OREOPHYLLA CIFERRI 1954. SYDOWIA 8:253-255 (ILLUS) /O. ANGELAEMARIAE
?, ORIGINAL ILLUSTRATION NOT DIAGNOSTIC. •

ORMATHODIUM SYDOW 1928. ANN. MYCOL. 26:138-139 (NO ILLUS) /O. STYRACIS
=NOMEN DUBIUM, SEE DEIGHTON 1974 (CMI PAPERS 137:4). •

OSTEOMORPHA ARNAUD EX WATLING & KENDRICK 1979. NATURALIST 104:1-4 (ILLUS) /O. FRAGILIS
AMERO,ARTHRO,HYALO,NONSP,CLAMPS, 45E, ON FUNGI (TRECHISPORA) ON WOOD. COMPARE WITH SCLEROSTILBUM. BASIDIO STATE ?=TRECHISPORA SEE WATLING & KENDRICK.
/ARNAUD, BULL. SOC. MYCOL. FRANCE 67:197 & FIG 1P,1951 (ILLUS) /ARNAUD, BULL. SOC. MYCOL. FRANCE 68:192,1952 (NO LATIN DIAGNOSIS) /NICOT & CHARPENTIE, BULL. SOC. MYCOL. FRANCE 87:33,1971 (TYPE SPECIMEN) /WATLING & KENDRICK, MICHIGAN BOT. 16:69-71,1977 (ILLUS) •

OSTRACODERMA FRIES 1825. SYST. ORB. VEG., P 150 /O. PULVINATUM FRIES 1829 (SYST. MYCOL. 3:213-214)
AMERO,CETERI,HYALO,AMPUL,PYCNIDIA, ON SOIL (COELOMYCETE). OSTRACODERMA HAS BEEN USED FOR CHROMELOSPORIUM SPECIES.
/HENNEBERT, PERSOONIA 7:199-201 & FIG 9, P 186,1973 (ILLUS,REVIS) /BRONCHART & DEMOULIN, CAN. J. BOT. 53:1549-1553,1975 (CYTOLOGY)
=*LYCOPERDELLON* TORREND 1913 FIDE HENNEBERT 1973
?=*GLISCHRODERMA* SEE HENNEBERT 1973 •

OSTRACODERMIDIUM MUKERJI 1973. TRANS. MYCOL. SOC. JAPAN 14:175-176 (NO ILLUS) /ACREMONIELLA SERPENTINA RAI & TEWARI 1963 (CAN. J. BOT. 41:331-334,ILLUS) =O. SERPENTINUM (RAI & TEWARI)
AMERO,CETERI,HYALO,AMPUL,SETAE, 32G, ON SOIL. CLOSE TO PHYMATOTRICHOPSIS HENNEBERT 1973. •

OVADENDRON SIGLER & CARMICHAEL 1976. MYCOTAXON 4:391-394 (ILLUS) /OOSPORA SULPHUREO-OCHRACEA VAN BEYMA 1933 =O. SULPHUREO-OCHRACEUM (VAN BEYMA)
AMERO,ARTHRO,HYALO,NONSP,ALEURI, 41A, ON MAN. •

OVEREEMIA ARNAUD 1954. BULL. SOC. MYCOL. FRANCE 69:300 & FIG 3N & O (NO LATIN DIAGNOSIS) /O. GLOMERELAE (OVEREEM)
STERILE, ON LEAVES OF TROPICAL PLANTS. STERILE MYCELIUM (OF BROOKSIA TROPICALIS).
/DEIGHTON & PIROZYNSKI, CMI PAPERS 105:3,1966 NICOT & CHARPENTIE, BULL. SOC. MYCOL. FRANCE 87:33,1971 (NO TYPE SPECIMEN) •

OVULARIA SACCARDO 1880. MICHELIA 2:17 /O. OBOVATA (FUCKEL) (LECTOTYPE) =O. OBLIQUA (COOKE) OUDEMANS 1883 (HEDWIGIA 22:85)
AMERO OR DIDYMO,CETERI,HYALO,RADUL, 24E, ON LEAVES. ASCO STATE =MYCOSPHAERELLA FIDE VON ARX 1970. COMPARE WITH RAMULARIA.
/VON ARX, GENERA P 202-203,1970 (ILLUS)
=*OPHIOCLADIUM* CAVARA 1893 FIDE SPRAGUE FROM AINSWORTH 1971
=*PSEUDOVULARIA* SPEGAZZINI 1911 FIDE CLEMENTS & SHEAR 1931 •

OVULARIOPSIS PATOUILLARD & HARIOT 1900. JOUR. DE BOTANIQUE 14:245 /O. ERYSIPHOIDES
?=OIDIUM LINK 1824 NOM. CONS. 1975 (SEE 40B). •

OVULITIS BUCHWALD 1949 EX BUCHWALD 1970. FRIESIA 9:326-329 (NO ILLUS) /O. AZALEAE BUCHWALD 1970 =O. STATE OF OVULINIA AZALEAE WEISS 1940
AMERO,CETERI,HYALO,NONSP,ALEURI, 53D, ON FLOWERS (AZALEAE). ASCO STATE =OVULINIA WEISS 1940.
/WEISS, PHYTOPATHOLOGY 30:236-244,1940 (ILLUS) •

OXYSPORIUM LEVILLE 1863 ='HELMINTHOSPORIUM', SEE SACCARDO SYLL. FUNG. 10:613,1892 •

OZONIUM LINK 1809. MAG. GES. NATURF. FREUNDE, BERLIN 3:21 /O. AURICOMUM =O. STATE OF COPRINUS DOMESTICUS FIDE WATLING 1979 (IN KENDRICK 1979)
?, FOR STERILE MYCELIUM. HUGHES 1958 DID NOT FIND TYPE. BASIDIO STATE =COPRINUS SEE DONK 1962 (TAXON 11:94) OR PSATHYRELLA FIDE BULLER 1924 (RESEARCHES ON FUNGI 3:29-59,SEE FIG 21). •

PAATHRAMAYA SUBRAMANIAN 1956. J. INDIAN BOT. SOC. 35:68-70 (ILLUS) /P. SUNDARA
AMERO,CETERI,PHAEO,RADUL OR MISC,SYNNEMATA, 9C, ON WOOD.
/ELLIS, DH P 321-322,1971 (ILLUS) •

PACHNOCYBE BERKELEY 1836. SMITHS ENGLISH FLORA 5(2):333-335 /P. FERRUGINEA (SOWERBY) (LECTOTYPE)
AMERO,CETERI,HYALO,AMPUL OR RADUL?,SYNNEMATA, 32F, ON WOOD. ASCO STATE ?=PORONIA SEE PADEN 1977 (CAN. J. BOT. 56:1774-1776, ILLUS SUB PORONIA).
/ELLIS, DH P 184-185,1971 (ILLUS) •

PACHNODIUM UPADHYAY & KENDRICK 1975. MYCOLOGIA 67:802 (798-865,ILLUS) /P. CANUM =P. STATE OF CERATOCYSTIS CANA (MUNCH) MOREAU
AMERO,BLASTO OR GLOIO,HYALO,NONSP OR MISC, 53F, ON TREES (PINUS). ASCO STATE =CERATOCYSTIS FIDE UPADHYAY & KENDRICK 1975. OTHER STATE =SPOROTHRIX. •

PACHYBASIDIELLA BUBAK & SYDOW 1915. ANN. MYCOL. 13:9-10 /P. POLYSPORA =K. POLYSPORA (BUBAK & SYDOW) KARAKULIN
=KABATIELLA BUBAK 1907 FIDE VON ARX 1970. •

PACHYBASIUM SACCARDO 1885. REV. MYCOLOGIQUE 7:160-161 PL LV FIG 6 /P. HAMATUM (BONORDEN) =TRICHODERMA SPORULOSUM (LINK) HUGHES
=TRICHODERMA PERSOON 1801 FIDE HUGHES 1958. •

PACHYMA FRIES 1822 =GEMMULARIA RAFINESQUE 1819, SEE DONK 1962 (TAXON 11:85 & 94) •

PACHYTRICHUM SYDOW 1925. ANN. MYCOL. 23:420-421 (ILLUS) /P. GAUZUMAE
=PERICONIA PERSOON 1801 FIDE LINDER FROM AINSWORTH 1971. •

PACTILIA FRIES 1835. FL. SCAND. P. 365 (NO ILLUS) /P. MYCOPHILA MONTAGNE & FRIES 1849
=ACHITONIUM KUNZE 1819 FIDE FRIES FROM SACCARDO 1886 (SYLL. FUNG. 4:673). •

PADIXONIA SUBRAMANIAN 1972. CURRENT SCIENCE 41:282-283 (ILLUS) /P. BISPORA =P. STATE OF XYLOSPHAERIA FURCATA (FRIES) DENNIS
AMERO,CETERI,HYALO,AMPUL OR MISC, 34F, ON NESTS (MACROTERMES). ASCO STATE =XYLOSPHAERIA. CONIDIA FORCIBLY DISCHARGED.
/DIXON, TBMS 48:211-217,1965 (ILLUS) •

PAECILOMYCES BAINIER 1907. BULL. SOC. MYCOL. FRANCE 23:26-27 PL 7 /P. VARIOTI
AMERO,BASO,HYALO OR PHAEO,PHIAL,SYNNEMATA, 3B & 5A, ON INSECTS, SOIL, FRUIT, WOOD & MANY OTHER SUBSTRATES. ASCO STATE =BYSSOCHLAMYS, TALAROMYCES OR THERMOASCUS, SEE SAMSON 1974, OR APHANOASCUS FIDE JONG & DAVIS 1975 (MYCOLOGIA 67:1143-1147,ILLUS,NEW SP) OR DACTYLOMYCES FIDE MALLOCH & CAIN 1972 OR CORDYCEPS FIDE PACIONI & FRIZZI 1978 (CAN. J. BOT. 56:391-394,ILLUS). OTHER STATE =CYLINDROCARPON FIDE BOOTH 1966 (CMI PAPERS 104:11-12). SEE ALSO MARIANNAEA ARNAUD 1952 EX SAMSON 1974.

/BROWN & SMITH, TBMS 40:17-89,1957 (MONOGR) /BARRON, HYPHOMYCETES P 244-246,1968 (ILLUS) /MALLOCH & CAIN, CAN. J. BOT. 50:2613-2628,1972 (ILLUS) /STOLK & SAMSON, CBS STUDIES 2:1-64,1972 (ILLUS) /SAMSON, CBS STUDIES 6:1-119,1974 (MONOGR) /SAMSON & TANSEY, TBMS 65:512-514,1975 (ILLUS,NEW SP) /MATSUSHIMA, ICONES P 88-89 & P 104-106 & PL,1975 (ILLUS SUB ISARIA,NEW SPP) /UDAGAWA, FURUYA & HORIE, BULL. NAT. MUS., TOKYO 16:508-510, 1973 (ILLUS SUB THERMOASCUS CRUSTACEUS) /SAMSON & EVANS, PROC. KON. NEDERL. AKAD. WETEN., AMSTERDAM, SER C, 80:128-134,1977 (ENTOMOGENOUS SPECIES,ILLUS OF SYNNEMATA) /BISSETT, FUNGI CANADENSES 151-159,1979 (ILLUS,NEW SP) /ONIONS, CMI DESCRIPTIONS 613-615,1979 (ILLUS ENTOMOGENOUS SPECIES)

?=*ALPHITOMYCES* REISSEK 1856
?=*COREMIOPSIS* SIZOVA & SUPRUN 1957
=*COROLLIUM* SOPP 1912 FIDE CARMICHAEL 1973
?=*CULICINOMYCES* COUCH, ROMNEY & RAO 1974
?=*GABARNAUDIA* SAMSON & GAMS 1974
?=*GRAPHIDIUM* LINDAU 1910
=*ISARA* AUCT., PRO PARTE SEE SAMSON 1974
?=*LEUCOPENICILLIFER* ARNOLD 1972
?=*PSEUDOFUSIDIUM* DEIGHTON 1969
=*SPICARIA* AUCT. NON HARTING
?=*SPICARIOPSIS* HEIM 1939 •

PAEPALOPSIS KUHN, SEE PAIPALOPSIS, ORTHOGRAPHIC VARIANT •

PAGIDOSPORA DRECHSLER 1960. SYDOWIA 14:246-247 & PL 34-35 /P. AMOEBOPHILA
AMERO,CETERI,HYALO,NONSP,CLAMPS, 52A, ON AMOEBAE. •

PAIPALOPSIS KUHN 1882. SEE HEDWIGIA 22:11 & 28-31,1883 (NO ILLUS,NO FORMAL DESCRIPTON) /P. IRMISCHIAE KUHN 1883
?, BASIDIO STATE =TUBERCINIA FIDE ZAMBETTAKIS 1970
/ZAMBETTAKIS, REV. MYCOL. 35:158-175,1970 (REFS) •

PALMOMYCES MAIRE 1926 (BULL. SOC. MYCOL. FRANCE 41:42) SEE ANDREAEANA •

PANCHANANIA SUBRAMANIAN & NAIR 1966. ANTONIE VAN LEEUWENHOEK 32:381-384 (ILLUS) /P. JAIPURENSIS
AMERO OR DIDYMO,CETERI,PHAEO,RADUL,APPEND,SYNNEMATA, 63F, ON LITTER (GREWIA). •

PANTOSPORA CIFERRI 1938. ANN. MYCOL. 36:240-242 (NO ILLUS) /P. GUAZUMAE
DICTYO,CETERI,PHAEO,ANNEL,SYNNEMATA, 101H, ON LEAVES (GUAZUMA).
/DEIGHTON, CMI PAPERS 140:156-159 & PL 5,1976 (ILLUS,REVIS)
=*DICTYOCEPHALA* MEDEIROS 1962 FIDE DEIGHTON 1976 •

PAPILIONOSPORA RAO & SUTTON 1976. KAVAKA 3:23-25 (DATED 1975,ILLUS) /GRAPHIUM ASPERGILLOIDES SPEGAZZINI 1879 =P. ASPERGILLOIDES (SPEGAZZINI)
AMERO OR DIDYMO,CETERI,PHAEO,RADUL,SYNNEMATA, 67C, ON LEAVES (SORBUS, PYRUS). •

PAPPIMYCES SUTTON & HODGES 1975. NOVA HEDWIGIA 26:527-530 (ILLUS) /P. HASTATUS
AMERO OR DIDYMO OR SCOLECO,CETERI,HYALO,RADUL, 110F, ON LITTER (EUCALYPTUS) & FUNGI (STENELLA ON CANAVALIA). •

PAPULARIA FRIES 1825. SYST. ORB. VEG. P. 195 /P. FAGI
=ARTHRINIUM KUNZE 1817 FIDE ELLIS 1965. •

PAPULASPORA PREUSS 1851. LINNAEA 24:112 /P. SEPEDONIOIDES
BULBILS,CETERI,HYALO OR PHAEO,NONSP, 128E, ON WOOD, SOIL, LITTER & OTHER SUBSTRATES. OTHER CONIDIAL STATES =PHIALOPHORA OR CLADORRHINUM OR "ARTHROCONIDIA" SEE GAMS & DOMSCH 1969 (NOVA HEDWIGIA 18:7-8, 10-11, 13 & 16,ILLUS). FOR P. VIRIDIS SEE AEGERITA. ASCO STATE =ASCOBOLUS FIDE DODGE 1920 (MYCOLOGIA 12:115-134 & PL 7-8), OR MELANOSPORA SEE UDAGAWA 1970 (TRANS. MYCOL. SOC. JAPAN 10:107-108,ILLUS), OR MICROTHECIUM SEE UDAGAWA & TAKADA 1974 (TRANS. MYCOL. SOC. JAPAN 15:25, BULBILS NOT ILLUS), OR LOPHIUM SEE TUBAKI 1958 (J. HATTORI BOT. LAB. 20:217), OR CUBONIA, CHAETOMIUM, NEUROSPORA OR SORDARIA FIDE FERGUS 1971. BASIDIO STATE

=CORTICIUM, GRANDINIA OR TRECHISPORA FIDE FERGUS 1971. SEE ALSO ENTRY UNDER "BULBILS." COMPARE WITH MINIMEDUSA.
/BARRON, HYPHOMYCETES P 246-247,1968 (ILLUS) /WERESUB & LECLAIR, CAN. J. BOT. 49:2203-2213,1971 (ILLUS,REVIS) /FERGUS, MYCOLOGIA 63:426-431,1971 (ILLUS,NEW SP,REFS) /WERESUB, FUNGI CANADENSES NO. 27,1974 (ILLUS) /CRANE & SCHOKNECHT, MYCOLOGIA 67:666-671,1975 (ILLUS)
=*BARYEIDAMIA* KARSTEN 1888 FIDE SACCARDO
=*HELICOSPORANGIUM* KARSTEN 1865 FIDE HARZ 1890
=*PAPULOSPORA* -ORTHOGRAPHIC VARIANT FIDE WERESUB & LECLAIR 1971 •

PAPULOSPORA, SEE PAPULASPORA •

PARACERCOSPORA DEIGHTON 1979. CMI PAPERS 144:47-52 (ILLUS) /CERCOSEPTORIA EGENULA SYDOW 1935 =P. EGENULA (SYDOW)
PHRAGMO,CETERI,PHAEO,RADUL,STROMA, ON LEAVES (MANY PLANTS). ASCO STATE =MYCOSPHAERELLA SEE DEIGHTON. COMPARE WITH CERCOSPORA. •

PARACHINOMYCES THAUNG 1979. TBMS 72:333-335 (ILLUS) /P. ACROCONIDIELLINAE
PHRAGMO,CETERI,HYALO,RADUL, ON FUNGI (ACROCONIDIELLINA) ON LITTER. COMPARE WITH CHIONOMYCES. •

PARACOCCIDIOIDES ALMEIDA 1930. C. R. SOC. BIOL. PARIS 106:315-316 (NO ILLUS) /ZYMONEMA BRASILENSE SPLENDORE 1912 (BULL. SOC. PATH. EXOT., PARIS 5:313-319, NO ILLUS) =P. BRASILIENSIS (SPLENDORE)
?=BLASTOMYCETES. THE ORIGINAL DESCRIPTIONS OF THE TYPE SPECIES AND THE GENUS MENTION ONLY BUDDING CELLS AND STERILE MYCELIUM. CAUSE OF SOUTH AMERICAN BLASTOMYCOSIS. AN OCCASIONALLY SEEN CONIDIAL STATE IS NOT DIFFERENT FROM CHRYSOSPORIUM. SEE BORELLI 1955 (CONGR. VENEZ. DE CIENCEAS MED. IV:2241-2253) FOR ILLUSTRATION OF CONIDIAL STATE.
=*LUTZIOMYCES* FONSECA 1939 SEE DODGE 1935, P 179 •

PARADENDRYPHIOPSIS ELLIS 1976. MDH P 385-386 (ILLUS) /P. CAMBRENSIS
PHRAGMO,BLASTO,PHAEO,NONSP, 92F, ON WOOD (QUERCUS). •

PARAPERICONIA ELLIS 1976. MDH P 359-360 (ILLUS) /P. ANGUSII
AMERO,BLASTO,PHAEO,MISC,SPORODOCHIA,SETAE, 50F, ON LEAVES (GLYCINE). •

PARAPHAEOISARIA DE HOOG & MORGAN-JONES 1978. MYCOTAXON 7:133-138 (ILLUS) /P. ALABAMENSIS
AMERO,CETERI,HYALO,RADUL, 23D, ON FUNGI (CRONARTIUM). •

PARAPITHOMYCES THAUNG 1976. TBMS 66:211-215 (ILLUS) /P. BRIDELIAE
DICTYO,BLASTO OR CETERI,PHAEO,RADUL, 99F, ON LEAVES (BRIDELIA). •

PARAPYRICULARIA ELLIS 1972. CMI PAPERS 131:12-14 /P. MUSAE ELLIS & PEREGRINE 1972
PHRAGMO,CETERI,PHAEO,RADUL, 84F, ON LEAVES (MUSA).
/ELLIS, MDH P 187-189,1976 (ILLUS) •

PARASPORA GROVE 1884. JOUR. BOT. 22:196 & TAB. 246, FIG 9 /P. TRISEPTATA
?, ORIGINAL ILLUSTRATION DIFFICULT TO INTERPRET. •

PARASYMPODIELLA PONNAPPA 1975. TBMS 64:344-345 (NO ILLUS) /SYMPODIELLA LAXA SUBRAMANIAN & VITTAL 1973 =P. LAXA (SUBRAMANIAN & VITTAL)
AMERO OR DIDYMO OR PHRAGMO,ARTHRO,HYALO,RACHI, ON LITTER (PINUS, CARYOTA, EUCALYPTUS & OTHERS). COMPARE WITH SYMPODIELLA (43D).
/SUBRAMANIAN & VITTAL, CAN. J. BOT. 51:1127-1132,1973 (ILLUS) /SUTTON, TBMS 71:167-168,1978 (ILLUS,NEW SP) •

PARATOMENTICOLA ELLIS 1976. MDH P 174-175 (ILLUS) /HELMINTHOSPORIUM LANCEOLATUM COOKE 1883 =P. LANCEOLATUS (COOKE) ELLIS 1976
PHRAGMO,CETERI,PHAEO,RACHI, 72E, ON BARK (BERCHEMIA). •

PARATRICHAEGUM FAUREL & SCHOTTER 1966. REV. MYCOL. 30:348-349 & FIG 9 (DATED 1965) /P. STERCORARIUM
=EPICOCCUM LINK 1815 FIDE KENDRICK 1973. •

PARATRICHOCONIS DEIGHTON & PIROZYNSKI 1972. CMI PAPERS 128:32-33 (ILLUS) /P. CHINENSIS (HANSFORD)
PHRAGMO,CETERI,HYALO,MISC OR RADUL, 80E, ON FUNGI (ASTERINA, LINDERA). COMPARE WITH PYRICULARIA & ERIOMYCOPSIS.
/MATSUSHIMA, ICONES P 106 & PL,1975 (ILLUS,NEW SP) /ELLIS, MDH P 189-190,1976 (ILLUS) •

PASPALOMYCES LINDER 1933. MYCOLOGIA 25:345-346 & PL 42, FIGS 4-6 /P. AUREUS
DIDYMO,BLASTO OR CETERI,HYALO,RACHI,SETAE, 63C, ON BARK (ACER).
?=*SCOLECOBASIDIELLA* ELLIS 1971 •

PASSALORA FRIES 1849. SUMMA VEG. SCAND. 2:500 /P. BACILLIGERA (MONTAGNE & FRIES) MONTAGNE & FRIES 1856
DIDYMO OR PHRAGMO,CETERI,PHAEO,RADUL, 61C, ON LEAVES (ALNUS). COMPARE WITH CERCOSPORA & ASPERISPORIUM. ASCO STATE =MYCOSPHAERELLA FIDE VON ARX 1970 OR MICROCYCLUS SEE DE HOOG 1977 (CBS STUDIES 15:139).
/HUGHES, CAN. J. BOT. 31:570-572,1953 (ILLUS) /ELLIS, DH P 266-267,1971 (ILLUS) /BROWN & MORGAN-JONES, MYCOTAXON 4:303-304,1976 (ILLUS,NEW COMB) •

PATOUILLARDIA ROUMEGUERE 1885. REV. MYC. 7:177-178 /P. LICHENOIDES
=DENDRODOCHIUM FIDE HOHNEL FROM CLEMENTS & SHEAR 1931. •

PATOUILLARDIELLA SPEGAZZINI 1889. BOL. ACAD. CIENC. CORDOBA 11:620 /GYMNOSPORANGIUM GUARANITICA SPEGAZZINI 1884 =P. GUARANITICA (SPEGAZZINI)
?, NOT ILLUSTRATED BY AUTHOR. •

PEDILOSPORA HOHNEL 1902. S. B. AKAD. WISS. WIEN 111:1047 (NO ILLUS) /P. PARASITANS
=DICRANIDION HARKNESS 1885 FIDE PEEK & SOLHEIM 1958 •

PEGLIONIA GOIDANICH 1934. MALPIGHIA 33:5-9 (ILLUS) /P. VERTICICLADA
=GYROTHRIX (CORDA) CORDA 1842 FIDE HUGHES & PIROZYNSKI 1971 •

PENDULISPORA ELLIS 1961. CMI PAPERS 82:41-43 (ILLUS) /P. VENEZUELANICA
DICTYO,CETERI,PHAEO,RACHI, 94D, ON WOOD (FAGUS & OTHERS). ASCO STATE =TUBEUFIA FIDE HUGHES 1978 (NEW ZEALAND J. BOT. 16:339 & 367-368).
/ELLIS, DH P 161,1971 (ILLUS) /MATSUSHIMA, ICONES P 107 & PL 204,1975 (ILLUS) •

PENICILLARIA CHEVALLIER 1826 =STERILE MYCELIUM, SEE DONK 1962 (TAXON 11:95) •

PENICILLIFER VAN EMDEN 1968. ACTA BOT. NEERL. 17:54-57 /P. PULCHER
DIDYMO OR PHRAGMO,BASO,HYALO,PHIAL, 57E, ON SOIL & LITTER (QUERCUS, PASANIA & OTHERS).
/VON ARX, GENERA P 187,1970 (ILLUS) /MATSUSHIMA, MICROFUNGI P 43 & FIG 118-2,1971 (ILLUS) /SAMSON, CBS STUDIES 6:85-88,1974 (ILLUS,SUB SEPTOFUSIDIUM) /MATSUSHIMA, ICONES P 107 & PL,1975 (ILLUS,NEW SP)
?=*SEPTOFUSIDIUM* GAMS 1971 •

PENICILLIUM LINK 1809. MAG. GES. NATURF. FREUNDE BERLIN 3:16 /P. EXPANSUM (LECTOTYPE SEE HAWKSWORTH ET AL. 1976)
AMERO,BASO,HYALO OR PHAEO,PHIAL,SYNNEMATA, 7B, ON MANY SUBSTRATES. ASCO STATE =EUPENICILLIUM, TALAROMYCES, HAMIGERA, PENICILLIOPSIS, TRICHOCOMA, TRICHOSKYTALE OR DICHLAENA, SEE VON ARX 1970 & 1974, MALLOCH & CAIN 1972, STOLK & SAMSON 1972 (CBS STUDIES 2:1-64,ILLUS), SAMSON ET AL 1976 (CBS STUDIES 11:1-47,ILLUS,REVIS) & SAMSON & MAHONEY 1977 (TBMS 69:158-160,ILLUS). COMPARE WITH GEOSMITHIA.
/RAPER & THOM, A MANUAL OF THE PENICILLIA, WILLIAMS & WILKINS, BALTIMORE, 875 P, 1949 (MONOGR,ILLUS) /KULIK, USDA AGR. HANDBOOK NO. 351, 80 P,1968 (COMPILATION, KEY,DESCR) /MALLOCH & CAIN, CAN. J. BOT. 50:2613-2628,1972 (ILLUS,REVIS) /UDAGAWA & TAKADA, TRANS. MYCOL. SOC. JAPAN 9:12-15,1968 (ILLUS) /SAMSON, STOLK & HADLOK, CBS STUDIES 11:1-47,1976 (ILLUS,REVIS, KEY,SYNOS) /UDAGAWA & HORIE, J. JAPANESE BOT. 47:338-346 & PL VIII,1972 (ILLUS,ASCO STATES) /HAWKSWORTH, PITT & SUTTON, TAXON 25:665-670,1976 (ILLUS, TYPIFICATION, REFS) /PITT, THE GENUS PENICILLIUM AND ITS TELEOMORPHIC STATES EUPENICILLIUM AND TALAROMYCES, ACADEMIC PRESS, IN PRESS,1979 (MONOGR)
=*COREMIUM* LINK 1809 FIDE AINSWORTH 1960

=*CITROMYCES* WEHMER 1893 FIDE THOM FROM AINSWORTH 1971
?=*ELADIA* SMITH 1961
?=*FLOCCARIA* GREVILLE 1828
?=*MONILIGER* LETEL FIDE AINSWORTH 1971
=*PRITZELIELLA* HENNINGS 1903 FIDE CLEMENTS & SHEAR 1931
=*RHODOCEPHALUS* CORDA 1837 FIDE LINDAU 1907 •

PENOMYCES GIARD 1891. COMP. REND. ACAD. SCI. PARIS 112:1518-1520 /ENTOMOPHTHORA TELARIA GIARD 1888 (BULL. SCIENT. FRANCE & BELGIQUE 19:301) =P. TELARIUS (GIARD) (LECTOTYPE)
?, NOT ILLUSTRATED BY AUTHOR. •

PENTAPOSPORIUM BATISTA 1957. REV. BIOL. LISBOA 1:106 & FIGS 4-5 /P. FOURCROYAE
?=TRIPOSPERMUM SPEGAZZINI 1918, SEE HUGHES 1976 (MYCOLOGIA 68:810-811). •

PERIBOTRYON FRIES 1832. SYSTEMA MYCOL. 3:287-288 (NO ILLUS) /P. PAVONI
=LICHEN FIDE AINSWORTH 1971. •

PERICONIA TODE EX PERSOON 1801. SYN. METH. FUNGORUM P. 686 /P. BYSSOIDES PERSOON 1801 (LECTOTYPE FIDE HUGHES 1958)
AMERO,BLASTO,PHAEO,AMPUL OR RADUL OR NONSP,STROMA,SETAE, 37E, ON LEAVES & LITTER (MANY PLANTS). ASCO STATE =DIDYMOSPHAERIA FIDE ELLIS 1971 OR HALOSPHAERIA FIDE KOHLMEYER 1972 (CAN. J. BOT. 50:1962) OR REMISPORA SEE DE HOOG 1977 (CBS STUDIES 15:138).
/MASON & ELLIS, CMI PAPERS 56:1-127,1953 (MONOGR,ILLUS) /ELLIS, DH P 344-353,1971 (ILLUS,KEY) /DE HOOG & RAO, PERSOONIA 8:210-212,1975 (ILLUS,NEW SP) /ELLIS, MDH P 352-359,1976 (ILLUS,KEY,NEW SP) /KOHLMEYER, REV. MYCOL. 41:202-203,1977 (ILLUS,NEW SP) /COLE & SAMSON, CONIDIAL FUNGI P 47-50,1979 (ILLUS, CONIDIUM ONTOGENY)
=*BERKLEYNA* KUNZE 1898 FIDE ELLIS 1971
=*HARPOCEPHALUM* ATKINSON 1897 FIDE ELLIS 1971
=*PACHYTRICHUM* SYDOW 1925 FIDE LINDER FROM AINSWORTH 1971
=*SPOROCYBE* FRIES 1825 FIDE HUGHES 1958
=*SPORODUM* CORDA 1837 FIDE HUGHES 1958
?=*SUBRAMANIA* RAO & RAO 1964
=*TRICHOCEPHALUM* COSTANTIN 1887 FIDE ELLIS 1971 •

PERICONIELLA SACCARDO 1885. IN SACCARDO & BERLESE, ATTI IST. VENETO SCI., SER. 6, 3:727 /P. VELUTINA (WINTER)
AMERO OR DIDYMO OR PHRAGMO,CETERI OR BLASTO,PHAEO,RADUL,SYNNEMATA, 84G, ON LEAVES (MANY PLANTS). ASCO STATE ?=HYPOXYLON SEE JONG & ROGERS 1972 (WASH. AGRIC. EXPER. STA. TECH. BULL. 71:5).
/ELLIS, CMI PAPERS 111:2-35,1967 (ILLUS,KEY) /BARRON, HYPHOMYCETES P 250,1968 (ILLUS) /ELLIS, DH P 295-301,1971 (ILLUS,KEY) /MATSUSHIMA, ICONES P 108 & PL,1975 (ILLUS) /ELLIS, MDH P 304-305,1976 (ILLUS) /DE HOOG & HERMANIDES-NIJHOF, CBS STUDIES 15:204,1977 (NEW COMB) /STEWART & CORDEN, MYCOTAXON 8:302-306,1979 (NEW SP,ILLUS)
=*ACRODESMIS* SYDOW 1926 FIDE ELLIS 1967
=*RAMICHLORIDIUM* STAHEL 1937 FIDE ELLIS 1967 •

PERIDIOMYCES KARSTEN ?=HYPHOMYCETE, FROM AINSWORTH 1971 •

PERIOLA FRIES 1822. SYSTEMA MYCOL. 2:266-267 (NO ILLUS) /P. TOMENTOSA (LECTOTYPE FIDE AINSWORTH 1971)
?, SEE ALSO GLIOCLADOCHIUM. •

PERIOLOPSIS MAIRE 1913. ANN. MYCOL. 11:357 (ILLUS) /P. HELICOCHAETA
=SARCOPODIUM EHRENBERG 1818 FIDE PIROZYNSKI •

PESOTUM CRANE & SCHOKNECHT 1973. AMER. J. BOT. 60:346-354 (ILLUS) /GRAPHIUM ULMI SCHWARZ =P. ULMI (SCHWARZ) =PESOTUM STATE OF CERATOCYSTIS ULMI (BUISMAN) MOREAU
AMERO,GLOIO,HYALO,RACHI OR RADUL,SYNNEMATA, 29B, ON TREES (ULMUS) & INSECTS (BEETLES). HAS BEEN CONFUSED WITH GRAPHIUM. OTHER CONIDIAL STATES =SPOROTHRIX & HYALODENDRON SEE MATSUSHIMA 1975 (PL 384:7). ASCO STATE =CERATOCYSTIS FIDE CRANE & SCHOKNECHT 1973 OR OPHIOSTOMA FIDE

VON ARX 1974 (GENERA P 111 & 230).
/SPENCER & GORIN, MYCOLOGIA 63:387-402,1971 (COMPARATIVE CHEMISTRY) /HARRIS & TABER, CAN. J. BOT. 51:1565-1571,1973 (ILLUS SUB CERATOCYSTIS ULMI) /MATSUSHIMA, ICONES P 170 & PL 384,1975 (ILLUS SUB GRAPHIUM STATE OF CERATOCYSTIS PICEAE) /ELLIS, MDH P 184-186,1976 (ILLUS) /DALPE & NEUMANN, CAN. J. BOT. 55:2159-2167,1977 (ILLUS SUB CERATOCYSTIS ULMI & C. PICEAE) /HIRATSUKA & TAKAI, CAN. J. BOT. 56:1909-1914,1978 (ILLUS) /DABINETT & WELLMAN, CAN. J. BOT. 56:2031-2049,1978 (DISCUSSION OF GENERA,NO ILLUS) •

PESTALOTIA DE NOTARIS 1839. MEM. R. ACCAD. SCI. TORINO 3:80-81 & ICON IX /P. PEZIZOIDES
PHRAGMO,GLOIO,PHAEO,PHIAL,ACERVULI,APPEND, ON LEAVES & STEMS (THEA & OTHERS). (MELANCONIALES) ASCO STATE =BROOMELLA SACCARDO FIDE VON ARX 1970 OR LEPTOSPHAERIA FIDE MATSUSHIMA 1971 (MICROFUNGI P 73-74,ILLUS), OR PESTALOPEZIZA OR PESTALOSPHAERIA SEE BARR 1975 (MYCOLOGIA 67:187-194,ILLUS,REFS).
/GUBA, MONOGRAPH OF MONOCHAETIA & PESTALOTIA, 342 P,1961 (MONOGR) /VON ARX, GENERA P 156-159,1970 (ILLUS,REFS) /MORDUE & HOLLIDAY, CMI DESCRIPTIONS 318-320,1971 (ILLUS SUB PESTALOTIOPSIS) /MATSUSHIMA, ICONES P 108 & PL,1975 (ILLUS) /YOKOYAMA, MYCOLOGIA 67:1032-1035,1975 (ILLUS) /MORDUE, CMI DESCRIPTIONS 514-515,1976 (ILLUS SUB PESTALOTIOPSIS,NEW COMB)
=*PESTALOTIOPSIS* STEYAERT 1949 FIDE VON ARX 1970
=*PESTALOZZIA*, ORTHOGRAPHIC VARIANT
=*TRUNCATELLA* STEYAERT 1949 FIDE VON ARX 1970 •

PESTALOTIOPSIS STEYAERT 1949. BULL. JARD. BOT. ETAT, BRUXELLES 19:300 /P. GUEPINI (DESMAZIERES)
=PESTALOTIA DE NOTARIS 1839 FIDE VON ARX 1970 BUT SEE SUTTON 1975 (CMI PAPERS 138:171-172,ILLUS) AND BARR 1975 (MYCOLOGIA 67:192).
/JONES, CAN. J. BOT. 55:766-771,1977 (ILLUS, CONIDIUM ONTOGENY) •

PESTALOZZIA -SEE PESTALOTIA •

PESTALOZZIELLA SACCARDO & ELLIS 1882. IN SACCARDO, MICHELIA 2:575 /P. SUBSESSILIS
AMERO,GLOIO,HYALO,RADUL,ACERVULI,APPEND, ON LEAVES (LUDWIGIA, GERANIUM, APOCYNUM) & LITTER (ARTOCARPUS). (MELANCONIALES). OTHER STATE ="SCLEROTIA".
/NAG RAJ & KENDRICK, CAN. J. BOT. 50:607-617,1972 (ILLUS,KEY, NEW SP, EXCLUSIONS) /NAG RAJ, TBMS 52:205-211,1969 (ILLUS,NEW SP) •

PESTALOZZINA (SACCARDO) SACCARDO 1894. IN SORAUER, Z. PFLKRANKH. 4:213-215 /P. UNICOLOR (BERKELEY & BROOME) SACCARDO 1895 (SYLL. FUNG. 11:580, LECTOTYPE FIDE GUBA 1961)
?, SEE SUTTON 1977 (CMI PAPERS 141:143-144)
/HUGHES, CMI PAPERS 36:2-3,1951 (ILLUS) •

PETRAKIA SYDOW & SYDOW 1913. ANN. MYCOL. 11:406-407 (ILLUS) /P. ECHINATA (PEGLION)
DICTYO OR STAURO,CETERI,PHAEO,NONSP,APPEND,SPORODOCHIA, 115E, ON LEAVES (ACER).•
/VON ARX, GENERA P 159-160,1970 (ILLUS) /VAN DER AA, ACTA BOT. NEERL. 17:221-225,1968 (ILLUS) /ELLIS, DH P 151-152,1971 (ILLUS)
=*ECHINOSPORIUM* WORONICHIN 1913 FIDE CARMICHAEL •

PETRAKIOPSIS SUBRAMANIAN & REDDY 1968. SYDOWIA 20:339-341 (ILLUS,DATED 1966) /P. ELEGANS
SCOLECO OR STAURO OR CHEIRO,CETERI,HYALO,NONSP, 116E, ON LITTER (QUERCUS). CLOSE TO OR THE SAME AS TETRACRIUM.
/MATSUSHIMA, ICONES P 110 & PL,1975 (ILLUS) •

PEYLIA OPIZ 1857 (IN PEYL, LOTOS 7:26) =BOTRYOSPORIUM CORDA 1831 FIDE AINSWORTH 1971 •

PEYRONELIA CIFERRI & FRAGOSO 1927. IN FRAGOSO & CIFERRI, BOL. REAL SOC. ESPAN. HIST. NAT. 27:333-334 (ILLUS) /P. SIRODESMIOIDES
DICTYO,BLASTO,PHAEO,NONSP, 101B, ON LEAVES (VITIS, MALUS, CITRUS, PRUNUS,

SOLANUM & OTHERS) & WOOD (POPULUS, SALIX & OTHERS). ASCO STATE =GLYPHIUM FIDE SUTTON 1970. OTHER STATE =PHOMA SEE MATSUSHIMA 1975 (SUB PEYRONELLAEA).
/MORGAN-JONES, CMI DESCRIPTIONS 134-135,1967 (ILLUS SUB PHOMA) /SUTTON, TBMS 54:255-264,1970 (ILLUS) /MATSUSHIMA, ICONES P 110 & PL 372,1975 (ILLUS SUB PEYRONELLAEA)
=*BILGRAMIA* PANWAR, PUROHIT & CHOUHAN 1974 FIDE CARMICHAEL
=*PEYRONELLAEA* (GOID.) EX GOIDANICH 1952 PRO PARTE SEE SUTTON 1977 •

PEYRONELINA ARNAUD EX FISHER, WEBSTER & KANE 1976. TBMS 67:351-354 (ILLUS) /P. GLOMERULATA
BULBILS,CETERI,HYALO,NONSP,APPEND,AQUATIC, 128C.
/ARNAUD, BULL. SOC. MYCOL. FRANCE 68:213 & 215,1952 (ILLUS) •

PEYRONELLAEA GOIDANICH EX GOIDANICH 1952. IN TOGLIANI, ANN. SPER. AGR. N.S. 6:92 /CONIOTHYRIUM GLOMERATUM CORDA =P. GLOMERATA (CORDA) GOIDANICH 1952
=PHOMA, PRO PARTE & PEYRONELIA, PRO PARTE, SEE SUTTON 1977 (CMI PAPERS 141:144) •

PEZIOTRICHUM (SACCARDO) LINDAU 1900. NAT. PFLANZENF. 1(1**):467 /P. LACHNELLA
=NOMEN DUBIUM SEE DOWNING 1953 (MYCOLOGIA 45:938). •

PHACELLULA SYDOW 1927. ANN. MYCOL. 25:139-140 /P. GOUANIAE
?, NOT ILLUSTRATED BY AUTHOR. •

PHAEOCOCCOMYCES DE HOOG 1979. TAXON 28:349 (NO ILLUS) /CRYPTOCOCCUS NIGRICANS RICH & STERN 1958 =P. NIGRICANS (RICH & STERN)
AMERO,GLOIO,PHAEO,NONSP, 50A, SUBSTRATE NOT KNOWN. =BLASTOMYCETES. OTHER CONIDIAL STATES =EXOPHIALA & RHINOCLADIELLA.
/DE HOOG, CBS STUDIES 15:122-128,1977 (ILLUS,KEY)
=*PHAEOCOCCUS* DE HOOG NON BORZI •

PHAEOCOCCUS DE HOOG 1977 (CBS STUDIES 15:122-128) A LATER HOMONYM OF PHAEOCOCCUS BORZI (ALGAE). SEE PHAEOCOCCOMYCES. •

PHAEOCONIS CLEMENTS 1909. GENERA P 148 /P. PANICI (ZIMMERMANN)
=NIGROSPORA ZIMMERMANN 1902 FIDE CLEMENTS & SHEAR 1931 •

PHAEODACTYLIUM AGNIHOTHRUDU 1968. PROC. INDIAN ACAD. SCI. B 68:206-209 (ILLUS) /P. VENKATESANUM =P. ALPINIAE (SAWADA) ELLIS 1971
PHRAGMO,CETERI,HYALO,RADUL, 81E, ON LEAVES (ALPINIA & ELETTARIA).
/ELLIS, DH P 214-216,1971 (ILLUS) •

PHAEODAEDALEA FIDALGO 1961. MYCOLOGIA 53:203-206 (201-210,ILLUS) /"P. SPRUCEI (BERK) FIDALGO" 1961 =P. STATE OF DAEDALEA SPRUCEI BERKELEY
AMERO,CETERI,PHAEO,RADUL,CLAMPS, 24G, ON WOOD & BARK. BASIDIO STATE =DAEDALEA FIDE RYVARDEN 1974 (SVENSK. BOT. TIDSKR. 68:273-284). •

PHAEODOCHIUM FARR 1968. IN FARR & HORNER, NOVA HEDWIGIA 15:268-269 & FIGS 52,70,87 /P. MYROTHECIOIDES
? •

PHAEOHARZIELLA LOUBIERE 1924. RECHERCHES SUR QUELQUE MUCEDINEES CASEICOLES, THESE, PARIS, P 52 (48-52 & PL V) /P. HETEROSPORA
=ARTHRINIUM KUNZE 1817 FIDE ELLIS 1965 •

PHAEOHYMENULA PETRAK 1954. SYDOWIA 8:77-79 /P. FUSISPORA
?, NOT ILLUSTRATED BY AUTHOR. •

PHAEOISARIA HOHNEL 1909. S. B. AKAD. WISS. WIEN 118:329-330 /P. BAMBUSAE =P. CLEMATIDIS (FUCKEL) HUGHES 1958
AMERO OR DIDYMO,CETERI,HYALO,RADUL,SYNNEMATA, 28A, ON LITTER (MANY PLANTS). COMPARE WITH NODULISPORIUM. OTHER CONIDIAL STATE =HUMICOLA FIDE DE HOOG & PAPENDORF 1976 (P 411) OR ?SPOROTHRIX SEE MATSUSHIMA 1975 (PL 247:1). ASCO STATE =PERONEUTYPELLA FIDE DEIGHTON 1974 (TBMS 62:243-252).
/ELLIS, DH P 213-214,1971 (ILLUS,KEY) /MATSUSHIMA, ICONES P 110-111 & PL,1975

(ILLUS) /DE HOOG & PAPENDORF, PERSOONIA 8:407-414,1976 (ILLUS,KEY,REVIS) /ELLIS, DH P 183-184,1976 (ILLUS) /HUGHES, NEW ZEALAND J. BOT. 16:339-340,1978 (ILLUS)
=*GRAPHIOPSIS* BAINIER 1907 FIDE MASON 1937 (CMI PAPERS 4:94)
=*HANSFORDIULA* MORRIS 1963 FIDE MORRIS 1963
?=*SPICULOSTILBELLA* MORRIS 1963 SEE DE HOOG & PAPENDORF 1976
?=*THAROOPAMA* SUBRAMANIAN 1956 •

PHAEOISARIOPSIS FERRARIS 1909. ANN. MYCOL. 7:280 /P. GRISEOLA (SACCARDO) (LECTOTYPE FIDE ELLIS 1971)
DIDYMO OR PHRAGMO,CETERI,PHAEO,RADUL,SYNNEMATA,STROMA, 86D, ON LEAVES. CLOSE TO CERCOSPORA. ASCO STATE =MYCOSPHAERELLA FIDE CBS LIST OF CULTURES 1972 (P 167).
/ELLIS, DH P 268-269,1971 (ILLUS) /ELLIS, MDH P 230-233,1976 (ILLUS,NEW COMBS) /BROWN & MORGAN-JONES, MYCOTAXON 4:303-306,1976 (ILLUS,NEW COMB) /MORGAN-JONES, MYCOTAXON 7:130-132,1978 (ILLUS,NEW SP) •

PHAEORAMULARIA MUNTANOLA 1960. LILLOA 30:182-183 & 209-213 (ILLUS) /CERCOSPORA GOMPHRENICOLA SPEGAZZINI 1882 =P. GOMPHRENICOLA (SPEGAZZINI)
=CLADOSPORIUM LINK 1815 FIDE VON ARX 1973 (VERH. KON. NEDERL. AKAD. WET. AFD. NAT. 61:62). SEE ALSO STENELLA & CERCOSPORA.
/ELLIS, DH P 307-308,1971 (ILLUS) /ELLIS, MDH P 316-325,1976 (ILLUS,KEY,NEW COMBS) /DEIGHTON, TBMS 67:140-142 & 223-232,1976 (ILLUS,NEW COMBS) /DEIGHTON, CMI PAPERS 144:26-47,1979 (ILLUS,NEW SPP,NEW COMBS) •

PHAEOSCOPULARIOPSIS OTA 1928. JAPANESE Z. DERM. UROL. 28:405-407 (AS 'GENUS PROVISUM') /THREE ORIGINAL SPECIES
=SCOPULARIOPSIS BAINIER 1907 FIDE MORTON & SMITH 1963. •

PHAEOSTALAGMUS GAMS 1976. IN GAMS & HOLUBOVA-JECHOVA, CBS STUDIES 13:90 (90-93,ILLUS) /STACHYLIDIUM CYCLOSPORUM GROVE 1884 =P. CYCLOSPORUS (GROVE)
AMERO,GLOIO,HYALO,PHIAL, 6D, ON WOOD (ALNUS, BETULA, PINUS & OTHERS) & LITTER (QUERCUS).
/HUGHES, NEW ZEALAND J. BOT. 16:341,1978 (ILLUS,NEW SP) •

PHAEOSTILBELLA HOHNEL 1919. BER. DEUTSCH. BOT. GES. 27:153 /P. ATRA (DESMAZIERES) HOHNEL 1925 (MITT. BOT. TECHN. HOCHSCH. WIEN 2:72, LECTOTYPE) =SACCARDAEA ATRA (DESM.) MASON & ELLIS
=MYROTHECIUM TODE FIDE TULLOCH 1972 •

PHAEOTRICHOCONIS SUBRAMANIAN 1956. PROC. INDIAN ACAD. SCI., SECT. B,44:2 /P. CROTALARIAE (SALAM & RAO)
PHRAGMO,CETERI,PHAEO,RACHI,APPEND, 70A, ON PLANTS & SOIL. OTHER STATE ="SCLEROTIA" SEE ELLIS 1971.
/BARRON, HYPHOMYCETES P 251-254,1968 (ILLUS SUB PHAEOTRICHONIS) /ELLIS, DH P 462-464,1971 (ILLUS) •

PHANEROCORYNELLA HOHNEL 1919. BERL. DEUTSCHL. BOT. GES. 37:157 /P. FUNGORUM (FRIES)
=EXOSPORIELLA KARSTEN 1892, BASED ON SAME TYPE. •

PHIALETEA BATISTA & NASCIMENTO 1960. ATAS INST. MICOL. UNIV. RECIFE 1:264-265 /P. AEROSPORA
=GRALLOMYCES STEVENS 1918 FIDE DEIGHTON & PIROZYNSKI 1966 •

PHIALOARTHROBOTRYUM MATSUSHIMA 1975. ICONES P 111-112 & PL 332:4-7 /P. TRISEPTATUM
PHRAGMO,GLOIO,PHAEO,PHIAL,SYNNEMATA, 74H, ON LITTER (BAMBUSA). COMPARE WITH ALBOSYNNEMA. •

PHIALOCEPHALA KENDRICK 1961. CAN. J. BOT. 39:1079-1081 (ILLUS) /P. DIMORPHOSPORA
AMERO,BASO OR GLOIO,HYALO,PHIAL, 7D, ON WOOD, LITTER & SOIL. ASCO STATE =LILLIPUTIA FIDE CARMICHAEL 1973 OR CHAETOSPHAERELLA FIDE MULLER IN KENDRICK 1971, OR CERATOCYSTIS FIDE UPADHYAY & KENDRICK 1975 (MYCOLOGIA 67:798-805) OR ROUMEGUERIELLA SEE DE HOOG 1977 (CBS STUDIES 15:138).

OTHER STATE =ANAVIRGA (H).
/KENDRICK, CAN. J. BOT. 41:1015-1023,1963 (ILLUS,KEY) /ELLIS, DH P 514,1971 (ILLUS) /JONG & DAVIS, MYCOLOGIA 64:1351-1356,1972 (ILLUS,KEY) /SIVASITHAMPARAM, TBMS 64:335-338,1975 (ILLUS,NEW SP) /SUTTON, TBMS 64:405-426,1975 (ILLUS,NEW SP,NEW COMB) /SHEARER, CRANE & MILLER, MYCOLOGIA 68:186-189,1976 (ILLUS,NEW SP) /MATSUSHIMA, ICONES P 112 & PL,1975 (ILLUS) /ELLIS, MDH P 445-447,1976 (ILLUS,KEY)
=*SPORENDOCLADIA* ARNAUD EX NAG RAJ & KENDRICK 1975 FIDE SUTTON 1975 •

PHIALOCLADUS KREISEL 1972. Z. ALLG. MIKROBIOL. 12:649 (649-654,ILLUS) /P. ZSOLTII (NO TYPE MATERIAL SPECIFIED)
AMERO,BASO,HYALO OR PHAEO,PHIAL, 16C, ON ANT NESTS. ALSO HAS A BLASTOCONIDIAL STATE.
/SAMSON, CBS STUDIES 6:5,1974 (DESCR) /COLE & SAMSON, CONIDIAL FUNGI P 56-57,1979 (ILLUS) •

"PHIALOCONIDIA" REPORTED AS CONIDIAL STATES OF ANGUILLOSPORA (H), CASARESIA (H), SCLEROTIUM (H), DIMORPHOSPORA (H), Q.V. OR ?HYPHODONTIA (B) SEE STALPERS 1978 (CBS STUDIES 16:41), OR APODOSPORA (A) SEE CAIN & MIZRA 1970 (CAN. J. BOT. 48:891), OR ZOPFIELLA (A) SEE MALLOCH & CAIN 1971 (CAN. J. BOT. 49:178), OR CHAETOMIUM (A) FIDE GAMS & DOMSCH 1969 (NOVA HEDWIGIA 18:5-6), OR ASTERIDIELLA (A) SEE HUGHES 1978 (NEW ZEALAND J. BOT. 16:358) OR GYMNOSPORANGIUM (B) SEE MIMS, SEABURY & THURSTON 1976 (AMER. J. BOT. 13:997-1002,ILLUS). •

PHIALOCONIDIOPHORA MOORE & ALMEIDA 1937. ANN. MISSOURI BOT. GARD. 23:547-550 & PL 26 (DATED 1936) /TWO ORIGINAL SPECIES, HORMODENDRUM COMPACTUM CARRION 1935 & P. GUGGENHEIMIA =RHINOCLADIELLA PEDROSOI FIDE DE HOOG & HERMANIDES-NIJHOF 1977 (CBS STUDIES 15:96, 98 & 186)
=FONSECAEA NEGRONI 1936 FIDE CARMICHAEL 1973 •

PHIALOGENICULATA MATSUSHIMA 1971. IN KOBAYASI ET AL., BULL. NAT. SCI. MUS. TOKYO 14:471-473, FIG 44 /P. GUADALCANALENSIS
AMERO OR DIDYMO,GLOIO,HYALO,PHIAL, 57G, ON WOOD (DECIDUOUS TREES). COMPARE WITH CYPHELLOPHORA.
/MATSUSHIMA, MICROFUNGI P 44-45 & FIG 128 & PL 22-1,1971 (ILLUS) •

PHIALOGRAPHIUM UPADHYAY & KENDRICK 1974. MYCOLOGIA 66:182-183 (NO ILLUS) /P. SAGMATOSPORAE =P. STATE OF CERATOCYSTIS SAGMATOSPORA WRIGHT & CAIN 1961
AMERO,GLOIO,HYALO,PHIAL,SYNNEMATA, 11A, ON TREES (PINUS) & LITTER (CATHORMIONUS & CHRYSALIDOCARPUS). ASCO STATE =CERATOCYSTIS FIDE UPADHYAY & KENDRICK.
/WRIGHT & CAIN, CAN. J. BOT. 39:1226-1229,1961 (ILLUS) /RAO & SUTTON, KAVAKA 3:25-28,1975 (ILLUS,NEW SPP) •

PHIALOMYCES MISRA & TALBOT 1964. CAN. J. BOT. 42:1287-1290 /P. MACROSPORUS
AMERO,BASO,PHAEO,PHIAL, 7E, ON SOIL.
/BARRON, HYPHOMYCETES P 255,1968 (ILLUS) /ELLIS, DH P 546-547,1971 (ILLUS) /MATSUSHIMA, ICONES P 112-113 & PL,1975 (ILLUS) •

PHIALOPHORA MEDLAR 1915. MYCOLOGIA 7:201-203 (ILLUS) /P. VERRUCOSA
AMERO,BASO OR GLOIO,HYALO OR PHAEO,PHIAL, 15B, ON WOOD, SOIL, MAN & MANY OTHER SUBSTRATES. ASCO STATE =CONIOCHAETA, GAEUMANNOMYCES, PYRENOPEZIZA, LASIOSPHAERIA, PODOSPORA, MOLLISIA, ASCOCORYNE, ETC. FIDE VON ARX 1970 OR MICROTHECIUM (A) SEE HAWKSWORTH & UDAGAWA 1977 (TRANS. MYCOL. SOC. JAPAN 18:149-150,ILLUS) OR PTERIDIOSPERMA (A) SEE KRUG & JENG 1979 (MYCOTAXON 10:44-45). SEE ALSO FONSECAEA & MYRIOCONIUM. OTHER CONIDIAL STATES =CLADOSPORIUM, RHINOCLADIELLA, MAMMARIA, PAPULASPORA, ANTHOPSIS, MADURELLA, SPADICOIDES.
/SCHOL-SCHWARZ, PERSOONIA 6:59-94,1970 (ILLUS,KEY,REVIS) /ELLIS, DH P 524,1971 (ILLUS,REFS) /FURUYA & UDAGAWA, J. GEN. APPL. MICROBIOL. 18:433-454,1972 (ILLUS SUB PODOSPORA SPP FIGS 5 & 14) /WALKER, CMI DESCRIPTIONS 381-383,1973 (ILLUS OF GAEUMANNOMYCES) /COLE & KENDRICK, MYCOLOGIA 65:661-688,1973 (ILLUS,KEY,REVIS) /UDAGAWA & FURUYA, TRANS. MYCOL. SOC. JAPAN 15:211,1974 (ILLUS SUB TRIANGULARIA ANGULOSPORA) /UDAGAWA & TAKADA,

TRANS. MYCOL. SOC. JAPAN 15:23-26,1974 (ILLUS SUB MICROTHECIUM CILIATUM) /SIVASITHAMPARAM, AUSTRAL. J. BOT. 23:193-212,1975 (ILLUS,REVIS) /UDAGAWA & FURUYA TRANS. MYC. SOC. JAPAN 16:215-216,1975 (ILLUS SUB CONIDIAL APIOSORDARIA VERRUCULOSA) /SZANISZLO, HSIEH & MARLOW, MYCOLOGIA 68:117-130,1976 (ILLUS YEAST & SCLEROTIC STATES) /HAWKSWORTH, GIBSON & GAMS, TBMS 66:427-431,1976 (ILLUS, P. PARASITICA) /GAMS & HERMANIDES-NIJHOF, CBS STUDIES 13:59-75,1976 (REVIS,KEY,ILLUS) /ELLIS, MDH P 452-455,1976 (ILLUS,KEY) /HAWKSWORTH & GIBSON, CMI DESCRIPTIONS 503-505,1976 (ILLUS) /HORNBY ET AL, TBMS 69:21-25,1977 (ILLUS SUB GAEUMANNOMYCES CYLINDROSPORUS) /MORGAN-JONES, MYCOTAXON 7:327-332,1978 (ILLUS,NEW SP)

?=*ANTHOPSIS* MARCHISIO, FONTANA & MOSCA 1977
=*CADOPHORA* LAGERB. & MELIN 1928 FIDE HUGHES 1958
?=*CYSTODENDRON* BUBAK 1914
=*FONSECAEA* NEGRONI 1936 PRO PARTE
=*LECYTHOPHORA* NANNFELDT 1934 FIDE GOIDANICH FROM AINSWORTH 1971
=*MARGARINOMYCES* LAXA 1930 FIDE VON ARX 1970 •

PHIALOPHOROPSIS BATRA 1968. MYCOLOGIA 59:1008-1009 (ILLUS P 995,DATED 1967) /P. TRYPODENDRI
AMERO,BASO,HYALO,PHIAL, 17B, ON WOOD, INSECTS (BEETLES) & LEAVES (EMBROTHRIUM).
/BRADY & SUTTON, TBMS 72:337-339,1979 (NEW SP,ILLUS) •

PHIALOSTELE DEIGHTON 1969. IN CEJP & DEIGHTON, CMI PAPERS 117:11-13 (ILLUS) /P. SCYTOPETALI
AMERO,BASO,HYALO OR PHAEO,ANNEL? OR PHIAL,SYNNEMATA, 9D, ON FRUIT (SCYTOPETALUM). PERHAPS NOT DIFFERENT FROM STILBELLA OR CEPHALOTRICHUM. •

PHIALOTUBUS ROY & LEELAVATHY 1966. TBMS 49:495-498 /P. MICROSPORUS
AMERO,BASO,HYALO OR PHAEO,PHIAL, 5C, ON SOIL.
/BARRON, HYPHOMYCETES P 256-258,1968 (REFS) •

PHLOEOCONIS FRIES 1825. SYST. ORB. VEG. P 199-200 /P. VIOLACEA (CESATI) FRIES 1849 (SUMMA VEG. SCAND. 2:520)
?, STERILE.
=*PSEUDOPROTOMYCES* GIBELLI FIDE SACCARDO 1899 (SYLL. FUNG. 14:1197)
?=*RHIZOSPORIUM* RABENHORST FIDE AINSWORTH 1971 •

PHRAGMOCEPHALA MASON & HUGHES 1951. NATURALIST, LONDON NO. 838:97-105 (ILLUS) /P. COOKEI =ENDOPHRAGMIA ELLIPTICA (BERKELEY & CURTIS)
=ENDOPHRAGMIA DUVERNOY & MAIRE 1920 FIDE ELLIS 1959 (CMI PAPERS 72:19-20). •

PHRAGMODOCHIUM HOHNEL 1924. BULL. JARD. BOT. BUITENZ. SER 3, 6:6-7 (NO ILLUS) /P. MODESTUM
=FUSARIUM LINK 1809 FIDE CLEMENTS & SHEAR 1931. •

PHRAGMOGRAPHIUM MORRIS 1966. MYCOPATHOLOGIA 28:99-101 (ILLUS) /P. ULMI
=NOMEN ILLEGITIMUM, NON PHRAGMOGRAPHIUM HENNINGS 1905, NOW REPLACED BY MORRISOGRAPHIUM MORELET 1968. •

PHRAGMOSPATHULA SUBRAMANIAN & NAIR 1966. ANTONIE VAN LEEUWENHOEK 32:384-386 /P. PHOENICIS
DIDYMO OR PHRAGMO,CETERI,PHAEO,ANNEL,STROMA, 76D, ON LITTER (PHOENIX).
/ELLIS, DH P 162-163,1971 (ILLUS) /SUTTON, CMI PAPERS 138:144-146,1975 (ILLUS) •

PHRAGMOSTACHYS COSTANTIN 1888. LES MUCEDINEES SIMPLES P 97 /P. ELATA (SACCARDO)
=STERIGMATOBOTRYS OUDEMANS 1886 FIDE HUGHES 1958 •

PHRAGMOSTILBE SUBRAMANIAN 1959. MYCOPATHOLOGIA 10:351-354 (ILLUS) /P. LINDERI =A. CANDIDUM (SCHWEINITZ) HUGHES 1958 FIDE WANG 1972
=ARTHROSPORIUM SACCARDO 1880 FIDE WANG 1972 (MYCOLOGIA 64:1177). •

PHRAGMOXYPHIUM BATISTA & CIFERRI 1963 =CIFERRIOXYPHIUM BATISTA & MAIA 1963 FIDE HUGHES 1976 (MYCOLOGIA 68:762 & 800) •

PHYMATOSTROMA CORDA 1837. ICONES FUNGORUM 1:5 /?
=ACHITONIUM KUNZE FIDE LINDAU IN RABENHORST 1910 SUB PACTILIA. •

PHYMATOTRICHOPSIS HENNEBERT 1973. PERSOONIA 7:199 & FIG 8 /PHYMATOTRICHUM OMNIVORUM DUGGAR 1916 =P. OMNIVORA (DUGGAR)
AMERO,CETERI,HYALO,AMPUL,ALEURI, 33A, ON PLANTS. CLOSE TO BOTRYTIS. •

PHYMATOTRICHUM BONORDEN 1851. HANDB. ALLGEM. MYKOL. P 116 & FIG 138,181,197 /?, ONE OF THE THREE ORIGINAL SPECIES APPEARS TO BE A BOTRYOSPORIUM, ONE A BOTRYTIS & ONE AN OSTRACODERMA.
=BOTRYTIS PERSOON 1801, IF P. GEMELLUM IS SELECTED AS LECTOTYPE. SEE HENNEBERT 1973 (PERSOONIA 7:190-191) FOR A REDISPOSITION OF SPECIES ASSIGNED TO PHYMATOTRICHUM. FOR P. FUNGICOLA SEE AEGERITA. •

PHYSALIDIUM MOSCA 1965. ALLIONIA 11:78-79 /P. ELEGANS
AMERO OR STAURO,CETERI,PHAEO,NONSP OR ANNEL?,APPEND, 114D, ON LITTER (TRITICUM).
/ELLIS, DH P 161-162,1971 (ILLUS) /MATSUSHIMA, MICROFUNGI P 45 & FIG 90-2 & PL 15-4,1971 (NEW SP) •

PHYSOSPORA FRIES 1849. SUMMA VEG. SCAND. 2:495 /P. RUBIGINOSA (FRIES 1832) (LECTOTYPE)
?=NOMEN DUBIUM, SEE DONK 1962 (TAXON 11:95). •

PIETRAIA =THE PSEUDOSCLEROTIUM OF POLYPORUS TUBERASTER FIDE AINSWORTH 1971. SEE ALSO VANTERPOOL & MACRAE 1951 (CAN. J. BOT. 29:147). •

PILULINA ARNAUD 1954. BULL. SOC. MYCOL. FRANCE 69:268-269 (ILLUS,NO LATIN DIAGNOSIS) /P. NIGROSPORA
?, NO TYPE SPECIMEN -SEE NICOT & CHARPENTIE 1971 (BULL. SOC. MYCOL. FRANCE 87:34). •

PIMINA GROVE 1888. J. BOT., LONDON 26:206 /P. PARASITICA =Z. GIBBUM (SACC., ROUSS. & BOMM.) HUGHES 1958
=ZYGOSPORIUM MONTAGNE 1842 FIDE HUGHES 1958. •

PIMINELLA ARNAUD 1954. BULL. SOC. MYCOL. FRANCE 69:280-281 (ILLUS,NO LATIN DIAGNOSIS) /P. CASTANEAE
?=GONYTRICHUM FIDE GAMS & HOLUBOVA-JECHOVA 1976 (CBS STUDIES 13:77 & 83). •

PINOYELLA CASTELLANI & CHALMERS 1919. MAN. TROP. MED., ED. 3, P 1023 /P. SIMII (PINOY)
=TRICHOPHYTON MALMSTEN FIDE STOCKDALE, MACKENZIE & AUSTWICK 1965. •

PIONNOTES FRIES 1849. SUMMA VEG. SCAND. 2:481 /P. CAPITATA (SCHWEINITZ)
=FUSARIUM LINK FIDE WOLLENWEBER FROM AINSWORTH 1971 •

PIRICAUDA BUBAK 1914. ANN. MYCOL. 12:218 /P. ULEANA (SACCARDO & SYDOW) =P. PARAGUAYENSIS (SPEG.) MOORE 1959
DICTYO,CETERI,PHAEO,NONSP,APPEND, 94B, ON LEAVES (BIGNONIA & OTHERS).
/HUGHES, CAN. J. BOT. 38:921-924,1960 (ILLUS,REVIS) /ELLIS, DH P 369-370,1971 (ILLUS) /ELLIS, MDH P 367-369,1976 (ILLUS,NEW COMBS) •

PIRICULARIA -SEE PYRICULARIA (ORTHOGRAPHIC VARIANT) •

PIROBASIDIUM HOHNEL 1902. S. B. AKAD WISS. WIEN 111:1001-1002 /P. SARCOIDES =CORYNE STATE OF ASCOCORYNE SARCOIDES
=CORYNE NEES 1817 FIDE GROVES & WILSON 1967. •

PIROSTOMELLA SACCARDO 1914. ANN. MYCOL. 12:308 (NO ILLUS) /P. RAIMUNDII (AS 'RAIMUNDI')
? •

PIROZYNSKIA SUBRAMANIAN 1972. CURRENT SCIENCE, INDIA 41:711-712 (NO ILLUS) /P. FARRIAE =P. STATE OF MAURODOTHINA FARRAE PIROZYNSKI & SHOEMAKER 1970 (CAN. J. BOT. 48:1325-1328, FIG 3A, 9 & 10)
DIDYMO,CETERI,PHAEO,NONSP,PODIA, 64G, ON LEAVES (ABIES). ASCO STATE

=MAURODOTHINA FIDE PIROZYNSKI & SHOEMAKER, OR EUPELTE SEE DE HOOG 1977 (CBS STUDIES 15:140).
/ELLIS, MDH P 365-366,1976 (ILLUS) •

PITHOMYCES BERKELEY & BROOME 1873. J. LINN. SOC. LONDON 14:100 /P. FLAVUS
PHRAGMO OR DICTYO,CETERI,PHAEO,NONSP,ALEURI, 90A & 100C, ON LITTER (MANY PLANTS) & SOIL. COMPARE WITH MONODICTYS.
/ELLIS, CMI PAPERS 76:7-19,1960 (ILLUS,KEY) /ELLIS, DH P 43-46,1971 (ILLUS,KEY) /MATSUSHIMA, ICONES P 113-114 & PL,1975 (ILLUS) /ELLIS, MDH P 30-32,1976 (ILLUS,KEY) /PALM ET AL, MYCOTAXON 10:246-250,1979 (ILLUS,NEW SP)
=*BIOCONIOSPORIUM* BATISTA & BEZERRA 1964 FIDE VON ARX 1970
=*NEOMICHELIA* PENZIG & SACCARDO 1901 FIDE CLEMENTS & SHEAR 1931
=*SCHELEOBRACHEA* HUGHES 1958 FIDE VON ARX 1970 •

PITHOSIRA PETRAK 1949. SYDOWIA 3:259-261 /P. SYDOWII
?, NOT ILLUSTRATED BY AUTHOR. •

PLACENTARIA AUERSWALD & RABENHORST '?=PERIOLA' FIDE AINSWORTH 1971 •

PLANCTOMYCES GIMESI 1924. HYDROBIOL. TANULMANYOK, BUDAPEST, P 5-8 /P. BEKEFII
?=CAULOBACTER OR OTHER STALKED BACTERIUM FIDE CARMICHAEL.
/SCHEIBBS, SYDOWIA 6:443-452,1952 (ILLUS) •

PLECOTRICHUM CORDA 1833. STURM'S DEUTSCHL. FL. III (PILZE) BD 3, HEFT 13:87-88 & TAB 44 /P. FUSCUM
?, STERILE FIDE HUGHES 1958. •

PLECTOTHRIX SHEAR 1902. BULL. TORREY BOT. CLUB 29:457 (NO ILLUS) /P. GLOBOSA
=OLPITRICHUM MACROSPORUM ATKINSON FIDE LINDER 1942
=OLPITRICHUM ATKINSON 1849 FIDE LINDER 1942 (LLOYDIA 5:173). •

PLEIOCHAETA (SACCARDO) HUGHES 1951. CMI PAPERS 36:32-39 (ILLUS) /P. SETOSA (KIRCHNER) HUGHES 1951
PHRAGMO,CETERI,HYALO OR PHAEO,RACHI,APPEND, 69A, ON PLANTS (CROTALARIA, LUPINUS & OTHERS).
/ELLIS, DH P 263,1971 (ILLUS) /PIROZYNSKI, FUNGI CANADENSES NO. 12,1974 (ILLUS) /INGOLD, GUIDE P 86-87,1975 (ILLUS) /ELLIS & HOLLIDAY, CMI DESCRIPTIONS 495,1976 (ILLUS) •

PLESIOSPORA DRECHSLER 1970. SYDOWIA 24:173-176 (ILLUS) /P. GLOBOSA
AMERO,BASO,HYALO,PHIAL?, 2F, ON NEMATODES. •

PLEUROCATENA ARNAUD 1952. BULL. SOC. MYCOL. FRANCE 68:193 & 195 (ILLUS,NO LATIN DIAGNOSIS) /P. ACICULARIS
?, (SEE FIG 3A), ON NEMATODES. NO TYPE SPECIMEN, SEE NICOT & CHARPENTIE 1971 (BULL. SOC. MYCOL. FRANCE 87:34). •

PLEUROCOLLA PETRAK 1924. ANN. MYCOL. 22:15-16 (NO ILLUS) /P. TILIAE
?, FOR ILLUSTRATION OF P. COMPRESSA SEE HUGHES 1978 (NEW ZEALAND J. BOT. 16:342 & 344). •

PLEUROGRAPHIUM GOIDANICH 1935. ANNALI DI BOT. 21(1):48-49 /P. DESMAZIERI (SACCARDO) (LECTOTYPE)
=DEMATOPHORA FIDE AINSWORTH 1971. •

PLEUROPEDIUM MARVANOVA & IQBAL 1973. ANTONIE VAN LEEUWENHOEK 39:401-408 (ILLUS) /P. TRICLADIOIDES
?=GYOERFFIELLA KOL 1928 FIDE INGOLD 1975 (GUIDE P 40). (SEE FIG 122D). •

PLEUROPHRAGMIUM COSTANTIN 1888. LES MUCEDINEES SIMPLES P 100 /P. BICOLOR =P. SIMPLEX (BERKELEY & BROOME) HUGHES 1958 =P. PARVISPORUM (PREUSS) HOLUBOVA-JECHOVA 1972
PHRAGMO,CETERI,PHAEO,RADUL, 81A, ON LITTER & FUNGI (LEPTOSPHAERIA). COMPARE WITH DACTYLARIA, SEE DE HOOG & VON ARX 1973 (KAVAKA 1:55-60) & DE HOOG 1977 (CBS STUDIES 14:60).
/ELLIS, CMI PAPERS 114:42-44,1968 (ILLUS) /ELLIS, DH P 202-204,1971 (ILLUS)

/HOLUBOVA-JECHOVA, CESKA MYKOL. 26:223-225,1972 (ILLUS,NEW COMB) /MATSUSHIMA, ICONES P 114-117 & PL,1975 (ILLUS,NEW SPP) /ELLIS, MDH P 163-166,1976 (ILLUS,NEW SP,NEW COMBS) •

PLEUROPYXIS CORDA 1837. ICONES FUNGORUM 1:23 /P. MICROSPERMA
=NOMEN CONFUSUM FIDE HUGHES 1958. •

PLEUROTHECIOPSIS SUTTON 1973. TBMS 61:417-422 (ILLUS) /P. PUSILLA
DIDYMO OR PHRAGMO,BLASTO,HYALO,RADUL, 65D, ON LITTER & BARK. COMPARE WITH CACUMISPORIUM.
/ELLIS, MDH P 168-169,1976 (ILLUS) •

PLEUROTHECIUM HOHNEL 1919. BERL. DEUTSCHL. BOT. GES. 37:154 (NO ILLUS) /ACROTHECIUM RECURVATUM MORGAN 1895 =P. RECURVATUM (MORGAN)
PHRAGMO,CETERI OR GLOIO,HYALO OR PHAEO,RACHI, 80C, ON BARK (FAGUS). HAS BEEN CONFUSED WITH CACUMISPORIUM.
/GOOS, MYCOLOGIA 61:1048-1053,1969 (ILLUS) /MATSUSHIMA, ICONES P 118 & PL 195:1,1975 (ILLUS) /KIFFER, MANGENOT & REISINGER, REV. ECOL. BIOL. SOL 6:181-194,1969 (ILLUS)
?=*"ACROTHECIUM ?MULTISPORUM"* ARNAUD 1954 (BULL. SOC. MYCOL. FRANCE 69:288 & FIG 3,I-M) •

PLOKAMIDOMYCES BATISTA, COSTA & CIFERRI 1957. PUBL. INST. MICOL. RECIFE 90:15-17 (ILLUS) /P. COLENSOI =P. STATE OF TRICHOPELTHECA ASIATICA FIDE HUGHES 1965
?=HORMISCIOMYCES BATISTA & NASCIMENTO 1957 FIDE HUGHES 1976 (MYCOLOGIA 68:773).
/HUGHES, NEW ZEALAND J. BOT. 3:320-332,1965 (ILLUS,REVIS) /HUGHES, NEW ZEALAND J. BOT. 10:225-242,1972 (ILLUS) •

POCHONIA BATISTA & FONSECA 1965. PUBL. INST. MICOL. RECIFE NO. 462:4-10 (ILLUS) /P. HUMICOLA =D. HETEROSPORA KAMYSCHKO 1962
=DIHETEROSPORA KAMYSCHKO 1962 FIDE BARRON & ONIONS 1966. •

PODOBACTRIDIUM PETCH 1916. ANN. R. BOT. GARD. PERADENIYA 6:180 /BACTRIDIUM CLAVATUM BERKELEY & BROOME 1873 =P. CLAVATUM (BERKELEY & BROOME)
=BACTRIDIUM KUNZE 1817 FIDE HUGHES 1966 (NEW ZEALAND J. BOT. 4:523). •

PODOCONIS BOEDIJN 1933. BULL. JARD. BOT. BUITENZ. SER 3, 13:133-134 & FIG 3-4 ON P 125-126 /P. THEAE (BERN.) =SPORIDESMIUM THEAE (BERN.) HUGHES
=SPORIDESMIUM LINK 1809 FIDE ELLIS 1958. •

PODOSPORIELLA ELLIS & EVERHART 1895. PROC. ACAD. NAT. SCI. PHILA. 1894:385 /P. HUMILIS =P. GLOMERATA (HARKNESS) BONAR 1965
PHRAGMO,CETERI,PHAEO,RADUL,SYNNEMATA,SPORODOCHIA, 87A, ON LEAVES (GARRYA, FAUREA).
/ELLIS, DH P 291-292,1971 (ILLUS) /ELLIS, MDH P 303-304,1976 (ILLUS,NEW COMB) •

PODOSPORIUM SCHWEINITZ 1832. TRANS. AMER. PHIL. SOC. II,4:278 /P. RIGIDUM (LECTOTYPE) =DEMATIUM CRUCIGERAE SCHWEINITZ 1822 PRO PARTE, FIDE SCHWEINITZ 1832.
PHRAGMO,CETERI,PHAEO,NONSP OR ANNEL,SYNNEMATA,STROMA, 77B, ON LITTER (AMPELOPSIS & RHUS).
/ELLIS, DH P 378-380,1971 (ILLUS) /ELLIS, MDH P 383-384,1976 (ILLUS)
=*HELMINTHOSPORIOPSIS* SPEGAZZINI 1880 FIDE MORRIS 1963
?=***NEGERIELLA*** **HENNINGS 1897 FIDE KENDRICK 1973** •

POLLACCIA BALDACCI & CIFERRI 1937. ATTI IST. BOT. UNIV. PAVIA SER 4, 10:61 & 71 (NO ILLUS) /OIDIUM RADIOSUM LIBERT =P. RADIOSA (LIBERT) =P. STATE OF VENTURIA MACULARIS FIDE ELLIS 1976
?=SPILOCAEA. ASCO STATE =VENTURIA FIDE VON ARX 1975 (CBS STUDIES 8:99).
/HUGHES, CAN. J. BOT. 31:572-573,1953 (ILLUS) /ELLIS, DH P 141-142,1971 (ILLUS) /SIVANESAN, CMI DESCIPTIONS 403,1974 & 482-483,1976 (ILLUS) /ELLIS, MDH P 109-110,1976 (ILLUS) •

POLYACTIS LINK 1809. MAG. GES. NATURF. FREUNDE, BERLIN 3:16 /P. VULGARIS =BOTRYTIS CINEREA PERSOON 1801
=BOTRYTIS MICHELI EX PERSOON 1801 FIDE HUGHES 1958. •

POLYCEPHALOMYCES KOBAYASI 1941. SCI. REP. TOKYO BUN. DAIG., SECT B, 5:245-247 (ILLUS) /P. FORMOSUS
?=STILBUM PERSOON 1801. •

POLYCLADIUM INGOLD 1959. TBMS 42:112-114 (ILLUS) /P. EQUISETI
STAURO,CETERI,HYALO,NONSP OR RADUL,AQUATIC, 122E.
/INGOLD, GUIDE P 46-47,1975 (ILLUS) •

POLYDESMUS MONTAGNE 1845. ANN. SCI. NAT. III,4:365 /P. ELEGANS DURIEU & MONTAGNE IN MONT. 1845 =S. DENSUM (SACC. & ROUM.) MASON & HUGHES IN HUGHES 1953
=SPORIDESMIUM LINK 1809, SEE HUGHES 1958 (CAN. J. BOT. 36:799) & ELLIS 1958 (CMI PAPERS 70:61-62). •

POLYMORPHOMYCES COUPIN 1914. REV. GEN. BOT. 26:245-248 & PL 5 /P. BONNIERI
=GEOTRICHUM LINK 1809 FIDE CARMICHAEL 1973. •

POLYPAECILUM SMITH 1961. TBMS 44:437-440 (ILLUS) /P. INSOLITUM
AMERO,BASO,HYALO,ANNEL?, 20A, ON MAN, LITTER (CAPSICUM), SOIL, WOOD CHIPS & BIRD NESTS (THERMOPHILIC). ASCO STATE =DICHOTOMOMYCES OR THERMOASCUS FIDE VON ARX 1970, OR DACTYLOMYCES FIDE MALLOCH & CAIN 1972.
/UDAGAWA, TRANS. MYCOL. SOC. JAPAN 10:105-106,1970 (ILLUS) /MALLOCH & CAIN, CAN. J. BOT. 50:2613-2628,1972 (ILLUS) /RAO & DE HOOG, PERSOONIA 8:202,1975 (NEW COMB) /COLE & SAMSÓN, CONIDIAL FUNGI P 69-71,1979 (ILLUS) •

POLYRHINA SOROKIN 1876. ANN. SCI. NAT. SER. 6, 4:65 (ILLUS) /P. MULTIFORMIS
=HARPOSPORIUM LOHDE 1874 FIDE SOROKIN.
=*POLYRRHINA* - ORTHOGRAPHIC VARIANT •

POLYRHIZIUM GIARD 1889. BULL. SCIENT. FRANCE & BELGIQUE 20:217-222 & PL V /P. LEPTOPHYEI (GIARD)
?=CLADOSPORIUM FIDE PETCH FROM AINSWORTH 1971. •

POLYRRHINA -SEE POLYRHINA. •

POLYSCHEMA UPADHYAY 1966. MYCOPATHOLOGIA 30:276-282 (ILLUS) /P. TERRICOLA
PHRAGMO,CETERI,PHAEO,MISC, 90H, ON SOIL, WOOD & LITTER. COMPARE WITH HORMOCEPHALUM, DWAYABEEJA, BAHUCHASHAKA & TORULA. THE TYPE SPECIES IS NOT AS 'TORULA-LIKE' AS MOST OF THE SPECIES ADDED SINCE.
/MATSUSHIMA, ICONES P 66 & PL,1975 (ILLUS,NEW SP SUB DWAYABEEJA) /ELLIS, MDH P 370-373,1976 (ILLUS,KEY, NEW COMBS) /TIWARI, AGRAWAL & SUTTON, TBMS 69:514-516,1977 (ILLUS,NEW SP) /PANDEYA & SAKSENA, MYCOLOGIA 70:876-879,1978 (ILLUS,NEW SP) •

POLYSCYTALINA ARNAUD 1954. BULL. SOC. MYCOL. FRANCE 69:283-284 (ILLUS,NO LATIN DIAGNOSIS) /P. GRISEA
=HORMIACTELLA SACCARDO FIDE HOLUBOVA-JECHOVA 1978. •

POLYSCYTALUM RIESS 1853. BOT. ZEIT. 11:138-139 (ILLUS) /P. FECUNDISSIMUM
AMERO OR DIDYMO OR PHRAGMO,BLASTO,HYALO,RADUL,SPORODOCHIA, 65C, ON LITTER (BETULA, FAGUS, QUERCUS, SALIX, URTICA, PINUS, & OTHERS). TYPE NOT EXAMINED BY MODERN AUTHOR, PERHAPS THE SAME AS HORMIACTIS PREUSS 1851.
/BARRON, HYPHOMYCETES P 261-263,1968 (ILLUS) /ELLIS, DH P 188-189,1971 (ILLUS) /MATSUSHIMA, ICONES P 118 & PL 50,1975 (ILLUS,NEW SP,SUB POLYSCYTALINA) /ELLIS, MDH P 158-159,1976 (ILLUS,NEW NAME,NEW COMB) /SUTTON & HODGES, NOVA HEDWIGIA 28:487-490,1977 (NEW SP,NEW COMB,ILLUS) /GAMUNDI, ARAMBARRI & GIAIOTTI, DARWINIANA 21:87-88 & FIG 1,1977 (ILLUS,NEW COMB) /SUTTON, TBMS 71:167-171,1978 (ILLUS,NEW SP)
?=*SCOPULARIELLA* GJAERUM 1971 •

POLYSPORA LAFFERTY 1921 =KABATIELLA BUBAK 1907 FIDE VON ARX 1970 •

POLYTHRINCIELLA BATISTA & MAIA 1960. PUBL INST. MICOL. UNIV. RECIFE 283:20-23 /P. BOMBACIFOLIA (AS 'BOMBACISFOLIA')
?, ORIGINAL ILLUSTRATION NOT DIAGNOSTIC. •

POLYTHRINCIOPSIS WALKER 1966. AUSTRALIAN J. BOT. 14:195-200 & PL 1 /P. PHRAGMITIS
=POLYTHRINCIUM KUNZE & SCHMIDT 1817 FIDE KENDRICK 1973. •

POLYTHRINCIUM KUNZE 1817. IN KUNZE & SCHMIDT, MYK. HEFT 1:13-15 /P. TRIFOLII KUNZE & SCHMIDT 1817 =P. STATE OF MYCOSPHAERELLA KILLANII PETRAK (=CYMADOTHEA TRIFOLII WOLF) FIDE VON ARX 1970
DIDYMO,CETERI,PHAEO,RACHI,STROMA, 61F, ON LEAVES (TRIFOLIUM). ASCO STATE =MYCOSPHAERELLA FIDE VON ARX 1970.
/HUGHES, CMI PAPERS 49:19-20,1952 (ILLUS) /ELLIS, DH P 284-285,1971 (ILLUS) /BOOTH & O'ROURKE, CMI DESCRIPTIONS 393,1973 (ILLUS)
=*POLYTHRINCIOPSIS* WALKER 1966 FIDE KENDRICK 1973 •

POROCLADIUM DESCALS 1976. IN DESCALS, NAWAWI & WEBSTER, TBMS 67:211 (207-222,ILLUS) /P. AQUATICUM
STAURO,CETERI,HYALO,RACHI,AQUATIC, 120G. •

POROIDEA GOETTINGER EX WINTER 1882 ?=DITANGIUM KARSTEN, SEE DONK 1962 (TAXON 11:95) •

POROISARIOPSIS MORELET 1971. BULL. SOC. SCI. NAT. ARCHEOL. TOULON VAR 195:7 /PHAEOISARIOPSIS ARMILLATA JONG & MORRIS 1968 =P. ARMILLATA (JONG & MORRIS)
?=SPIROPES CIFERRI 1955
/JONG & MORRIS, MYCOPATHOLOGIA 34:264-267,1968 (ILLUS) •

POSADASIA CANTON 1898. IN POSADAS, COMMUN. A LA ACAD. DE BUENOS AIRES, J. PEUSER, BUENOS AIRES (NOT SEEN) /P. ESFERIFORMIS
=COCCIDIOIDES STILES 1896 FIDE DODGE 1935 (MED. MYCOL. P 148-149). •

PRATHIGADA SUBRAMANIAN 1956. IN SUBRAMANIAN & RAMAKRISHNAN, J. MADRAS UNIV. 26:366-367 /NAPICLADIUM CRATAEVAE SYDOW 1913 (ANN. MYCOL. 11:329) =P. CRATAEVAE (SYDOW)
PHRAGMO,CETERI,PHAEO,RADUL,SPORODOCHIA,APPEND, 72C, ON LEAVES (CRATAEVA).
/ELLIS, DH P 278-279,1971 (ILLUS)
=*MACRAEA* SUBRAMANIAN 1952, NON LINDLEY 1830 •

PRATHODA SUBRAMANIAN 1956. J. INDIAN BOT. SOC. 35:71-74 (ILLUS) /P. SAPARVA =A. SAPARVA (SUBRAMANIAN) DEIGHTON 1969
=ALTERNARIA NEES 1817 SEE CEJP & DEIGHTON 1969 (CMI PAPERS 117:30-31). •

PREUSSIASTER KUNTZE 1891. REV. GEN. PLANT. 2:850 /P. PAUCISEPTATUS (PREUSS)
=CORDANA PREUSS 1851 FIDE HUGHES 1958. •

PRISMARIA PREUSS 1851. LINNAEA 24:125 & FUNGI HOYERSW. NO. 86 /P. ALBA
?, THE ILLUSTRATION (IN STURM'S DEUTSCHL. FL., VI, 35-6:TAB. 55) IS DIFFICULT TO INTERPRET.
/LINDAU, RABENHORST KRYPT. FL. 2 AUFL. 1 BD. 8:537-539,1907 (ILLUS) •

PRITZELIELLA HENNINGS 1903. HEDWIGIA BEIBLATT 42 P 88 /P. CAERULEA
=PENICILLIUM LINK 1809 FIDE CLEMENTS & SHEAR 1931 (AS COREMIUM LINK). •

PROPHYTROMA SOROKIN 1877. HEDWIGIA 16:87-88 & FIGS 1-12 /P. TUBULARIS
?, ORIGINAL ILLUSTRATION DIFFICULT TO INTERPRET. •

PROSTHEMIELLA SACCARDO 1881. MICHELIA 2:356 (NO ILLUS) /P. FORMOSA SACCARDO & MALBRANCHE 1881
? •

PROTEOPHIALA CIFERRI 1957. SYDOWIA 11:286-289 (NO ILLUS) /P. MATTIROLIANA
?=HARZIELLA. ASCO STATE =MELANOSPORA FIDE AINSWORTH 1971. •

PROTOCORONIS CLEMENTS & SHEAR 1931 A SUPERFLUOUS NAME FOR PROTOCORONOSPORA •

PROTOCORONOSPORA ATKINSON & EDGERTON 1907. JOUR. MYCOL. 13:185-186 /P. NIGRICANS

=KABATIELLA BUBAK 1907 FIDE VON ARX 1970.
=*PROTOCORONIS* CLEMENTS & SHEAR 1931 •

PSAMMINA ROUSSEL & SACCARDO 1901. BULL. SOC. BOT. BELG. 29:295 (NO ILLUS) /P. BOMMERIAE
STAURO,CETERI,HYALO OR PHAEO,NONSP,ACERVULI?, 115A, ON LEAVES (AMMOPHILA). •

PSEUDOANGUILLOSPORA IQBAL 1974. BIOLOGIA, LAHORE 20:11-14 (ILLUS) /P. STRICTA
SCOLECO,CETERI,HYALO,RACHI,AQUATIC, 108E, COMPARE WITH ANGUILLOSPORA.
/SINCLAIR & MORGAN-JONES, MYCOTAXON 9:474-477,1979 (ILLUS,NEW SP) •

PSEUDOBASIDIUM TENGWALL 1924. MEDED. PHYTOPATH. LAB. WILLIE COMMELIN SCHOLTEN 6:38 /P. BICOLOR
=ARTHRINIUM KUNZE 1817 FIDE ELLIS 1971. •

PSEUDOBELTRANIA HENNINGS 1902. HEDWIGIA 41:310 /P. CEDRELAE
AMERO,CETERI,PHAEO,RADUL, 22A, ON LEAVES (CEDRELA) & LITTER.
/PIROZYNSKI, CMI PAPERS 90:28-30,1963 (ILLUS) /ELLIS, DH P 243-244,1971 (ILLUS) /ELLIS, MDH P 206-207,1976 (ILLUS) •

PSEUDOBOTRYTIS KRZEMIENIEWSKA & BADURA 1954. ACTA SOC. BOT. POLON. 23:761-762 /P. FUSCA =P. TERRESTRIS (TIMONIN) SUBRAMANIAN 1956
AMERO OR DIDYMO,CETERI,PHAEO,RADUL, 62C, ON LITTER, WOOD & SOIL.
/BARRON, HYPHOMYCETES P 263-264,1968 (ILLUS) /ELLIS, DH P 361-362,1971 (ILLUS) /MATSUSHIMA, ICONES P 118-119 & PL,1975 (ILLUS)
=*UMBELLULA* MORRIS 1955 FIDE SUBRAMANIAN 1956 •

PSEUDOCAMPTOUM FRAGOSO & CIFERRI 1925. BOL. SOC. ESPANA. HIST. NAT. 25:454 /P. CITRI =MELANOGRAPHIUM CITRI (FRAG. & CIF.) ELLIS
=MELANOGRAPHIUM SACCARDO 1913 FIDE ELLIS 1963. •

PSEUDOCERCOSPORA SPEGAZZINI 1910. AN. MUS. NAC. HIST. NAT. BUENOS AIRES 20(SER 3,13):437-438 (NO ILLUS) /P. VITIS (LEVEILLE)
SCOLECO OR PHRAGMO OR HELICO,CETERI,HYALO OR PHAEO,RACHI OR RADUL OR ANNEL,STROMA, 85B, ON LEAVES (VITIS, SPHENOCLEA & PTEROCAULON). COMPARE WITH CERCOSPORA. ASCO STATE =MYCOSPHAERELLA FIDE DEIGHTON 1976 (P 10).
/ELLIS, DH P 208-210,1971 (ILLUS) /DEIGHTON, CMI PAPERS 140:1-168,1976 (MONOGR) /ELLIS, MDH P 175-177,1976 (ILLUS,NEW COMBS) /SIVANESAN, TBMS 72:157-161,1979 (NEW SP,ILLUS) /DEIGHTON, CMI PAPERS 144:4-13,1979 (ILLUS,NEW SPP)
=*ANCYLOSPORA* SAWADA 1944 FIDE DEIGHTON 1976
=*CERCOCLADOSPORA* AGARWAL & SINGH 1974 FIDE DEIGHTON 1976
=*CERCOSPORIOPSIS* MIURA 1928 FIDE DEIGHTON 1976
=*HELICOMINA* OLIVE 1948 FIDE DEIGHTON 1976
?=*HYALOHELICOMINA* YOKOYAMA 1974
?=*SPIRALUM* MULDER 1975 •

PSEUDOCERCOSPORELLA DEIGHTON 1973. CMI PAPERS 133:38-41 (ILLUS) /CERCOSPORELLA IPOMOEAE SAWADA 1973 =P. IPOMOEAE (SAWADA EX)
PHRAGMO OR SCOLECO,CETERI,HYALO,RACHI, 81D, ON LEAVES (PHASEOLUS, PUNICA & OTHERS). FOR CERCOSPORA SPECIES WITH INCONSPICUOUS SCARS.
/BOOTH & WALLER, CMI DESCRIPTIONS 386,1973 (ILLUS) /SUTTON, CMI PAPERS 138:97-99,1975 (ILLUS) /DEIGHTON, TBMS 66:547-548,1976 (ILLUS,NEW SP,NEW COMBS) •

PSEUDOCERCOSPORIDIUM DEIGHTON 1973. CMI PAPERS 133:55-58 (ILLUS) /P. VENEZUELANUM (SYDOW)
?, ON LEAVES (CORDIA). NOT VERY DIFFERENT FROM CERCOSPORIDIUM EARLE 1901.
/ELLIS, MDH P 297-299,1971 (ILLUS) •

PSEUDOCOCCIDIOIDES FONSECA 1928. BOL. INST. CLIN. QUIR., BUENOS AIRES 4:495-501 /P. MAZZAE =COCCIDIOIDES IMMITIS
=COCCIDIOIDES STILES 1896 FIDE DODGE 1935. •

PSEUDOCORDYCEPS HAUMAN 1936. BULL. SOC. BOT. BELG. 69:116-118 & FIG 5 /P. SEMINICOLA
?, ORIGINAL ILLUSTRATION DIFFICULT TO INTERPRET. •

PSEUDODICTYOSPORIUM MATSUSHIMA 1971. IN KOBAYASI ET AL., BULL. NAT. SCI. MUS. TOKYO 14:473-474 (ILLUS) /P. WAUENSE
DICTYO OR CHEIRO,CETERI,PHAEO OR HYALO,NONSP, 98E, ON LITTER (BROAD LEAF TREES). COMPARE WITH DICTYOSPORIUM.
/MATSUSHIMA, MICROFUNGI P 46 & FIGS 92 & 93 & PL 5-1,1971 (ILLUS)
/MATSUSHIMA, ICONES P 119 & PL,1975 (ILLUS)
=*KAMATIA* RAO & SUBHEDAR FIDE KENDRICK •

PSEUDOEPICOCCUM ELLIS 1971. DH P 270 (ILLUS) /P. COCOS (STEVENS)
AMERO,CETERI,PHAEO,RADUL,SPORODOCHIA, 30E, ON LEAVES (ARECA & COCOS). CLOSE TO HADROTRICHUM. •

PSEUDOFUMAGO BRIOSI & FARNETI 1906. ATT. IST. BOT. UNIV. PAVIA SER 2, 10:31 & TAV X, FIGS 18-24 /P. CITRI
=NOMEN DUBIUM SEE DE HOOG & HERMANIDES-NIJHOF 1977 (CBS STUDIES 15:186-187). •

PSEUDOFUSARIUM MATSUSHIMA 1971. MICROFUNGI P 46-47 FIG 62 /P. FUSARIOIDEUM
=P. SEMITECTUM (BERKELEY & RAVENEL) MATSUSHIMA 1975
?=ERIOMYCOPSIS OR PSEUDOHANSFORDIA (SEE 78C).
/GAMS, PERSOONIA 7:165,1973 (REFS) /MATSUSHIMA, ICONES P 119 & PL,1975 (ILLUS,NEW COMB) •

PSEUDOFUSIDIUM DEIGHTON 1969. CMI PAPERS 118:26-28 (ILLUS) /P. HANSFORDII
AMERO,BASO? OR BLASTO?,HYALO,PHIAL?, 36D, ON FUNGI (MYCOVELLOSIELLA ON ABUTILON LEAVES). ?=ACREMONIUM. •

PSEUDOGASTER HOHNEL 1907. DENK. AKAD. WIEN 83:38 /P. SINGULARIS
?, NOT ILLUSTRATED BY AUTHOR. •

PSEUDOGIBELLULA SAMSON & EVANS 1973. ACTA BOT. NEERL. 22:524 (522-528,ILLUS) /GIBELLULA FORMICARUM MAINS 1949 =P. FORMICARUM (MAINS)
AMERO,CETERI,HYALO,RADUL,SYNNEMATA, 28F, ON INSECTS & ARACHNIDS. •

PSEUDOGRAPHIELLA MORRIS 1966. MYCOPATHOLOGIA 28:97-99 (ILLUS) /P. VARIASEPTATA
DIDYMO OR PHRAGMO,GLOIO,HYALO,PHIAL?,SYNNEMATA, 74A, ON FUNGI (GEOGLOSSUM). •

PSEUDOGRAPHIUM JACZEWSKI 1898. NOUV. MEM. SOC. IMP. NAT. MOSCOU XV(XX):370 (NOT SEEN) /P. PERSICAE (SCHW.)
?, MAY BE AN EARLIER NAME FOR MORRISOGRAPHIUM.
/HOHNEL, SITZ. AKAD. WIEN 124:117,1915 (LECTOTYPE CHOSEN?)
=*SUBULARIELLA* HOHNEL 1915 NOMEN ILLEGITIMUM •

PSEUDOHANSFORDIA ARNOLD 1970. Z. PILZK. 35:307-310 (ILLUS,DATED 1969) /P. IRREGULARIS
AMERO OR DIDYMO OR PHRAGMO,CETERI,HYALO,RACHI, 63B, ON FUNGI (TRAMETES, STEREUM, DOTHIDELLA, GLOMERELLA & OTHERS ON LEAVES). COMPARE WITH SIBIRINA, CLADOBOTRYUM & ERIOMYCOPSIS. ASCO STATE =HYPOMYCES.
/DE HOOG, PERSOONIA 10:57-62,1978 (ILLUS,NEW COMBS,KEY)
=*SYMPODIOPHORA* ARNOLD 1970 FIDE DE HOOG 1978
?=*PSEUDOFUSARIUM* MATSUSHIMA 1971 •

PSEUDOHANSFORDIA REDDY & BILGRAMI 1975. NOVA HEDWIGIA 26:151-153 (ILLUS) /P. HUGHII
=NOMEN ILLEGITIMUM, NON PSEUDOHANSFORDIA ARNOLD 1970. SEE DE HOOG 1978 (PERSOONIA 10:58) FOR NOTES ON THE TYPE SPECIMEN. •

PSEUDOMERIA BARRON 1980. CAN. J. BOT. (IN PRESS) /P. MUCOSA
AMERO,CETERI,HYALO,RACHI, ON ROTIFERS (ADENITA). •

PSEUDOMICROCERA PETCH 1921. TBMS 7:164 /P. HENNINGSII (KOORDERS)
=FUSARIUM FIDE WOLLENWEBER FROM CLEMENTS & SHEAR 1931 •

PSEUDOMICRODOCHIUM SUTTON 1975. TBMS 64:417-421 (ILLUS) /P. ACICULARE
DIDYMO OR SCOLECO,GLOIO,HYALO,PHIAL,SPORODOCHIA, 108C, ON LITTER (CASTANEA CUPULES). FOR P. CANDIDUM SEE CYLINDRODENDRUM. •

PSEUDOPAPULASPORA SHARMA 1977. J. INDIAN BOT. SOC. 56:100 (NO MICROSCOPIC ILLUS) /P. KENDRICKII
?, ON SOIL. TYPE COLLECTION IN CMI IS NOW DISPOSED AS CURVULARIA SP. (A. ONIONS,IN LITT,1979). •

PSEUDOPETRAKIA ELLIS 1971. CMI PAPERS 125:3-4 (ILLUS) /PETRAKIA KAMBAKKAMENSIS SUBRAMANIAN 1957 =P. KAMBAKKAMENSIS (SUBRAMANIAN)
DICTYO OR STAURO,CETERI,PHAEO,NONSP,APPEND, 94F, ON WOOD.
/ELLIS, MDH P 35-36,1976 (ILLUS) •

PSEUDOPOLYSTIGMINA MURASHKINSKY 1928. IN MURASHKINSKI & ZILING, TRANS. SIBER. INST. 9:234-235 (ILLUS) /P. SPIRAEICOLA
?, ORIGINAL DESCRIPTION & ILLUSTRATION DOES NOT INDICATE ARRANGEMENT OR GENESIS OF CONIDIA. •

PSEUDOPROTOMYCES GIBELLI 1874 =PHLOEOCONIS FIDE SACCARDO 1899 (SYLL. FUNG. 14:1197) •

PSEUDOPUCCINIA HOHNEL 1919. BER. DEUTSCH. BOT. GES. 37:157 (NO ILLUS) /STIGMINA THERMOPSIDIS HARKNESS 1884 (BULL. CALIF. ACAD. SCI. 1:37, NO ILLUS) =P. THERMOPSIDIS (HARKNESS) HOHNEL 1925 (MITT. BOT. TECHN. HOCHSCHULE, WIEN 2:41-42, NO ILLUS)
=STIGMINA FIDE LAUNDON FROM AINSWORTH 1971. •

PSEUDOSPIROPES ELLIS 1971. DH P 258-260 (ILLUS,KEY) /P. NODOSUS (WALLROTH)
PHRAGMO,CETERI,PHAEO,RADUL,STROMA, 83A, ON WOOD, BARK & LITTER. DIFFERS FROM SPIROPES MAINLY IN HABITAT. COMPARE WITH MARTINELLISIA. ASCO STATE =MELANOMMA SEE ELLIS 1976.
/MATSUSHIMA, ICONES P 119-120 & PL,1975 (ILLUS,NEW SP) /ELLIS, MDH P 217-225,1976 (ILLUS,KEY, NEW SPP,NEW COMBS) /MORGAN-JONES, MYCOTAXON 5:481-483,1977 (ILLUS,NEW SP) /HUGHES, NEW ZEALAND J. BOT. 16:343-345,1978 (ILLUS) •

PSEUDOSTEMPHYLIUM (WILTSHIRE) SUBRAMANIAN 1961. CURR. SCI. 30:423 /STEMPHYLIUM LANUGINOSUM HARZ =P. LANUGINOSUM (HARZ) SUBRAMANIAN 1961 =U. CONSORTIALE (THUM.) SIMMONS
=ULOCLADIUM PREUSS 1851 FIDE SIMMONS 1967 (MYCOLOGIA 59:80-82). •

PSEUDOTORULA SUBRAMANIAN 1958. J. INDIAN BOT. SOC. 37:57-61 (ILLUS) /P. HETEROSPORA
?=DENDRYPHION WALLROTH 1833. •

PSEUDOVULARIA SPEGAZZINI 1911. ANN. MUS. NAC. BUENOS AIRES 20(SER 3,13):418-419 (ILLUS) /P. TRIFOLII
=OVULARIA SACCARDO 1880 FIDE CLEMENTS & SHEAR 1931. •

PSILOBOTRYS SACCARDO 1879. MICHELIA 1:538 /RHINOTRICHUM MINUTUM SACCARDO 1877 (MICHELIA 1:87) =P. MINUTA (SACCARDO)
=CHLORIDIUM LINK 1809 FIDE HUGHES 1958. •

PSILONIA FRIES 1825. SYST. ORB. VEG. 1:187 /P. GILVA (PERSOON) FRIES 1832 (SYST. MYCOL. 3:451,LECTOTYPE)
?=VOLUTELLA FIDE AINSWORTH 1971. •

PSILONIELLA COSTANTIN 1888. LES MUCEDINEES SIMPLES P 190-191 (ILLUS) /P. CUNEIFORMIS (RICHON)
=CATENULARIA GROVE 1886 FIDE ELLIS 1971. •

PTEROCONIUM SACCARDO EX GROVE 1914. HEDWIGIA 55:146 /P. PTEROSPERMUM (COOKE & MASSEE) GROVE 1914

AMERO,CETERI,PHAEO,MISC,ALEURI,SPORODOCHIA, 46D, ON LITTER (GRASSES). ASCO STATE =APIOSPORA FIDE ELLIS 1971. COMPARE WITH ARTHRINIUM & CORDELLA. /ELLIS, DH P 577-579,1971 (ILLUS,KEY) /ELLIS, MDH P 479-480,1976 (ILLUS,NEW SP) •

PTERODINIA CHEVALLIER 1826 =BOTRYTIS FIDE LEVEILLE 1849 (IN ORB. DICT. 8:494) •

PTERULOPSIS WAKEFIELD & HANSFORD 1943. IN HANSFORD, PROC. LINN. SOC. LONDON 1942-3, P 64 & FIG 15, P 63 /P. DUMMERI
SCOLECO,CETERI,HYALO,RACHI,SYNNEMATA, 111B, ON WOOD.
/MORRIS, WEST. ILLINOIS UNIV. SER. BIOL. SCI. 3:104-105,1963 (ILLUS) •

PTYCHOGASTER CORDA 1838. ICONES FUNGORUM 2:23-24 & PL 12 FIG 90 /P. ALBUS =P. FULIGINOIDES (PERS. EX STEUDEL) DONK 1972 =P. STATE OF TYROMYCES PTYCHOGASTER
AMERO,BLASTO,HYALO,NONSP,ALEURI,CLAMPS, 51A, ON WOOD. BASIDIO STATE =TYROMYCES FIDE VON ARX 1973, OR INONOTUS SEE ROY 1971 (INDIAN PHYTOPATH. 24:590-592) OR PUNCTULARIA OR OLIGOPORUS FIDE STALPERS 1978 (CBS STUDIES 16:39 & 152). COMPARE WITH ECHINODIA. SEE ALSO CERIOMYCES.
/DONK, PROC. K. NEDERL. ACAD. WET. (C), 75:165-177,1972 (REFS) /VON ARX, PERSOONIA 7:127-130,1973 (ILLUS) /SIGLER & CARMICHAEL, MYCOTAXON 4:394-397,1976 (ILLUS,REFS) •

PUCCINIOPSIS SPEGAZZINI 1888. AN. SOC. ARG. 26:74 /P. GUARANITICA
?, NOT ILLUSTRATED BY AUTHOR.
/BARNETT, IMPERFECT FUNGI NO. 310,1960 (ILLUS) •

PUCIOLA DE BERTOLDI 1976. MYCOTAXON 3:553-557 (ILLUS) /P. SPINOSA
AMERO,CETERI,PHAEO,RACHI,SETAE, 25D, ON SOIL. COMPARE WITH ACRODONTIUM. •

PULCHROMYCES HENNEBERT 1973. PERSOONIA 7:198-199 & FIG 7 /PHYMATOTRICHUM FIMICOLA DRING 1959 (TBMS 42:406,ILLUS) =P. FIMICOLA (DRING)
AMERO,CETERI,HYALO,AMPUL OR MISC, 33E, ON DUNG.
/PFISTER, WINN & CACAVIO, MYCOTAXON 1:137-141,1974 (ILLUS) •

PULLULARIA BERKHOUT 1923. DIE SCHIMMELGESCHLACTEN..., THESIS, UTRECHT, P 54-55 & PL 4 FIG 32 /P. PULLULANS (DE BARY)
=AUREOBASIDIUM VIALA & BOYER 1891 FIDE COOKE 1962. •

PUTAGRAIVAM SUBRAMANIAN & BHAT 1978. PROC. INDIAN ACAD. SCI. SECT. B 87:99-104 (ILLUS) /P. SUNDARAM =P. STATE OF PEETHAMBARA SUNDARA (REV. MYCOL. 42:49-55, ILLUS)
DIDYMO,GLOIO,HYALO,PHIAL,SYNNEMATA, ON LITTER (MACARANGA). ASCO STATE =PEETHAMBARA. COMPARE WITH DIDYMOSTILBE. •

PYCNOSTYSANUS LINDAU 1904. VERH. BOT. VER. BRANDENB. 45:160-161 (ILLUS) /P. RESINAE =SOROCYBE RESINAE (FRIES) FRIES 1849
=SOROCYBE FRIES 1849 (SEE 39F). HUGHES (1958, P 759) SAYS P. RESINAE =PYCNOSTYSANUS STATE OF S. RESINAE, BUT BOTH THESE NAMES APPEAR TO BE BASED ON THE SYNNEMATOUS STATE. THE DIFFUSE CONIDIAL STATE WAS NAMED HORMOCONIS BY VON ARX & DE VRIES IN 1973 & HORMODENDRUM RESINAE BY LINDAU IN 1902.
/ELLIS, MDH P 154-155,1976 (ILLUS) •

PYRAMIDOSPORA NILSSON 1962. SVENSK BOT. TIDSKR. 56:358-360 (ILLUS) /P. CASUARINAE
STAURO,CETERI,HYALO,NONSP,AQUATIC, 113E, COMPARE WITH UVARISPORA.
/INGOLD, GUIDE P 48-49,1975 (ILLUS) /SINCLAIR & MORGAN-JONES, MYCOTAXON 9:477-479,1979 (ILLUS,NEW SP) •

PYRENIOPSIS KUNTZE 1898. REVIS. GEN. PL. 3(2):508 /NEW NAME FOR TRICHODERMA
=TRICHODERMA PERSOON 1801 FIDE RIFAI 1969. •

PYRENIUM PERSOON 1801. SYN. METH. FUNG. P 236 /P. TERRESTRE
=NOMEN DUBIUM SEE HUGHES 1958. •

PYRICULARIA SACCARDO 1880. MICHELIA 2:20 /P. GRISEA (COOKE) =P. STATE OF MAGNAPORTHE GRISEA FIDE BARR 1977 (MYCOLOGIA 69:954)

PHRAGMO OR DIDYMO,CETERI,HYALO OR PHAEO,RACHI OR RADUL, 80A, ON GRASSES (ORYZA & OTHERS). ASCO STATE =MASSARINA FIDE VON ARX 1970 OR MAGNAPORTHE FIDE BARR 1977.
/BARRON, HYPHOMYCETES P 265-266,1968 (ILLUS) /ELLIS, DH P 218-219,1971 (ILLUS,KEY) /ELLIS, CMI PAPERS 131:11-12,1972 (ILLUS,NEW SP) /MATSUSHIMA, ICONES P 120-121 & PL,1975 (ILLUS) /ELLIS, MDH P 186-188,1976 (ILLUS) /YAEGASHI & UDAGAWA, CAN. J. BOT. 56:180-183 & 2184,1978 (ILLUS,REVIS)
?=*ERIOMYCOPSIS* SPEGAZZINI 1910
?=*NAKATEA* HARA 1939
?=*PARATRICHOCONIS* DEIGHTON & PIROZYNSKI 1972
=*PIRICULARIA* SACCARDO 1880 (ORTHOGRAPHIC VARIANT) •

PYRICULARIOPSIS ELLIS 1971. DH P 206-207 (ILLUS) /P. PARASITICA (SACCARDO & BERLESE)
PHRAGMO,CETERI,PHAEO,RADUL, 72A, ON LITTER (MUSA & OTHERS).
/MATSUSHIMA, ICONES P 121 & PL,1975 (ILLUS,NEW SPP) •

QUESTIERIELLA ARNAUD 1954. BULL. SOC. MYCOL. FRANCE 69:284 & 287, FIG 8A (NO LATIN DIAGNOSIS) /Q. CORNI (LECTOTYPE)
?, NO TYPE SPECIMEN, SEE NICOT & CHARPENTIE 1971 (BULL. SOC. MYCOL. FRANCE 87:34). ASCO STATE =QUESTIERIA FIDE ARNAUD (=SCHIFFNERULA FIDE PETRAK FROM AINSWORTH 1971). •

RACEMOSPORIUM MOREAU & MOREAU 1941. REV. MYCOLOGIE N.S. 6:80-82 (NO LATIN DIAGNOSIS) /R. SATURNUS
=ARTHRINIUM KUNZE 1817 FIDE ELLIS 1965. •

RACHISIA LINDNER 1913. DEUT. ESSIGIND. 17:467 (ILLUS) /R. SPIRALIS
=FUSARIUM LINK FIDE CLEMENTS & SHEAR 1931. •

RACODIUM PERSOON 1801. SYN. METH. FUNG. P 701 /R. CELLARE (LECTOTYPE FIDE FRIES 1821, SYST. MYCOL. I:XLVI OR GRAY 1821, NAT. ARR. BRIT. PL. 1:557)
AMERO,CETERI,PHAEO,RADUL, 24C, ON WOOD (IN CELLARS) & SOIL. SEE ALSO RACÓDIUM FRIES 1829, RHACODIOPSIS DONK 1975 & ZASMIDIUM FRIES 1849. RACODIUM AS TYPIFIED BY R. CELLARE HAS BEEN PROPOSED FOR REJECTION BY HAWKSWORTH & RIEDL.
/GUEGEN, BULL. SOC. MYCOL. FRANCE 22:77-95 & 146-163,1906 (ILLUS OF RHINOCLADIELLA STATE SUB RHACODIUM, PL. III & V FOLLOWING P 108) /DONK, TAXON 11:96,1962 (REFS,DISCUSSION) /RIEDL, TAXON 17:34-37,1968 (DESCR,REFS) /HAWKSWORTH, TBMS 54:323-325,1970 (REFS) /ELLIS, DH P 248,1971 (NEW COMB IN RHINOCLADIELLA,ILLUS) /DE HOOG, CBS STUDIES 15:42-44,1977 (EMENDED TO INCLUDE RHINOCLADIELLA STATE) /HAWKSWORTH & RIEDL, TAXON 26:208,1977 (PROPOSAL TO CONSERVE RACODIUM FRIES 1829) /DE HOOG, TAXON 28:347-348,1979 (DISCUSSION)
=*RHACODIUM* (ORTHOGRAPHIC VARIANT) •

RACODIUM FRIES 1829. SYST. MYCOL. 3:229 /R. RUPESTRE PERSOON 1801
?, A LATER HOMONYM OF RACODIUM PERSOON PROPOSED FOR CONSERVATION FOR LICHENIZED STERILE MYCELIUM BY HAWKSWORTH & RIEDL 1977 (TAXON 26:208). SEE ALSO RHACODIOPSIS DONK 1975 (PERSOONIA 8:273-276). •

RADAISIELLA BAINIER 1910. BULL. SOC. MYCOL. FRANCE 26:382-384 & PL XX /R. ELEGANS
=BOTRYOSPORIUM FIDE MAIRE FROM AINSWORTH 1971. •

RADICISETA SAWADA & KATSUKI 1959. PUBL. COLL. AGRIC. NAT. TAIWAN UNIV. 8:205 /R. BLECHNI
?, NOT ILLUSTRATED BY AUTHORS. •

RAFFAELEA VON ARX & HENNEBERT 1965. MYCOPATHOLOGIA 25:309-312 (ILLUS) /R. AMBROSIAE
AMERO,CETERI,HYALO,RADUL,SPORODOCHIA, 23F, ON WOOD & INSECTS (BEETLES). ASCO STATE =BOTRYOASCUS OR OPHIOSTOMA FIDE VON ARX 1977 (CBS STUDIES 14:32).
/BATRA, MYCOLOGIA 59:976-1017,1967 (REVIS,KEY,ILLUS) /WIEJMAN & DE HOOG, ANTONIE VAN LEEUWENHOEK 41:353-360,1975 (ILLUS,REVIS)
?=*SPICELLUM* NICOT & ROQUEBERT 1976 •

RAGNHILDIANA SOLHEIM 1931. IN SOLHEIM & STEVENS, MYCOLOGIA 23:402-403 (ILLUS) /R. AGERATI (STEVENS) STEVENS & SOLHEIM
=MYCOVELLOSIELLA RANGEL FIDE DEIGHTON 1974. •

RAMALIA BATISTA 1957. REV. BIOL. LISBOA 1:110-111 (ILLUS ON P 107-108) /R. VERONICAE
?, ORIGINAL ILLUSTRATION NOT DIAGNOSTIC. •

RAMICHLORIDIUM STAHEL EX DE HOOG 1977. CBS STUDIES 15:59 (59-86,ILLUS,KEY) /CHLORIDIUM APICULATUM MILLER, GIDDENS & FOSTER 1957 =R. APICULATUM (MILLER, GIDDENS & FOSTER) DE HOOG 1977
AMERO OR DIDYMO,CETERI,HYALO OR PHAEO,RADUL, 25F, ON LITTER. CLOSE TO RHINOCLADIELLA, VERONAEA & DACTYLARIA.
/STAHEL, TROP. AGRIC. TRIN. 14:44,1937 (NO LATIN DIAGNOSIS) /SUTTON, MYCOLOGIA 70:798,1978 (REVIS,NEW COMB,NO ILLUS) •

RAMULARIA UNGER 1833. DIE EXANTH. DER PFL. P 169 & TAB 2 FIG 12 /R. DIDYMA (LECTOTYPE FIDE VON ARX 1970 BUT TYPE SPECIMEN IS NOT KNOWN)
AMERO OR DIDYMO,BLASTO,HYALO,RADUL, 61B, ON LEAVES (MANY PLANTS). COMPARE WITH OVULARIA, CERCOSPORA & CLADOSPORIUM. SEE DISCUSSION UNDER CERCOSPORA & CLADOSPORIUM. ASCO STATE =MYCOSPHAERELLA FIDE MULLER IN KENDRICK 1971, OR MELANODOTHIS SEE ARNOLD 1974 (FUNGI CANADENSES NO. 30).
/VON ARX, GENERA P 202-203,1970 (ILLUS) /PIROZYNSKI, FUNGI CANADENSES NO. 23,1974 (ILLUS)
=*ACROTHECA* FUCKEL 1860 FIDE HUGHES 1958
=*DIDYMARIA* CORDA 1842 FIDE VON ARX 1970
?=*DIPLOOSPORA* GROVE 1916 •

RAMULARIITES BABAYAN & TASLAKHYCHYAN 1973 (MIKOL. FITOPATOL. 7:181) -FOSSIL FUNGI IMPERFECTI •

RAMULARIOPSIS SPEGAZZINI 1910. AN. MUS. NAC. BUENOS AIRES 20(SER 3,13):421-422 /R. CNIDOSCOLI
PHRAGMO OR SCOLECO,BLASTO,HYALO,RADUL, 81G, ON LEAVES. CLOSE TO RAMULARIA & MYCOVELLOSIELLA.
/DEIGHTON, TBMS 59:419-421,1972 (ILLUS) /DEIGHTON, CMI PAPERS 137:71-73,1974 (ILLUS,NEW SP) •

RAMULASPERA LINDROTH 1902. ACT. SOC. FENN. 22:5-6 /R. SALICINA (VESTGREN)
?, NOT ILLUSTRATED BY LINDROTH, VESTGREN'S ILLUSTRATION NOT DIAGNOSTIC. •

RAMULISPORA MIURA 1920. S. MANCHURIAN RLY. CO. AGRIC. EXP. STA. BULL. 11:43 /R. ANDROPOGONIS
SCOLECO OR STAURO,GLOIO,HYALO,RADUL,SPORODOCHIA, 110A, ON LEAVES (SORGHUM, ANDROPOGON).
/RAWLA, TBMS 60:283-292,1973 (ILLUS) /ANAHOSUR, CMI DESCRIPTIONS 585-586,1978 (ILLUS)
?=*GLOEOCERCOSPORA* BAIN & EDGERTON 1943 •

RANOJEVICIA BUBAK 1910. IN RANOJEVIC, ANN. MYCOL. 8:400-401 (ILLUS) /R. VAGANS RANOJEVIC & BUBAK
?, ORIGINAL ILLUSTRATION & DESCRIPTION NOT DIAGNOSTIC. •

RAPERIA SUBRAMANIAN & RAJENDRAN 1976. KAVAKA 3:129-133 (ILLUS,DATED 1975) /ASPERGILLUS SPINULOSUS WARCUP =R. SPINULOSA (WARCUP)
=ASPERGILLUS FIDE SAMSON 1979 (CBS STUDIES 18:1). •

RECTICHARELLA SCHEER 1944. Z. PARASITENK. 13:275-282 (ILLUS,NO LATIN DIAGNOSIS) /R. ASELII
?, ORIGINAL ILLUSTRATION DIFFICULT TO INTERPRET. •

REDAELLIA CIFERRI 1930. ARCH. PROTISTENK. 71:424-428 (ILLUS) /R. ELEGANS =ASPERGILLUS PENICILLOIDES (TYPE STRAIN FROM CBS DET. D. FENNELL)
=ASPERGILLLUS LINK 1809 FIDE ALECRIM 1958 (ANN. FAC. MED. UNIV. RECIFE 18:81-85). •

REDBIA DEIGHTON & PIROZYNSKI 1972. CMI PAPERS 128:83-85 (ILLUS) /R. PUCCINIICOLA
PHRAGMO,CETERI,HYALO,RADUL, 82C, ON FUNGI (PUCCINIA) & LITTER (PERSEA).
/PIROZYNSKI & HODGES, CAN. J. BOT. 51:165-166,1973 (ILLUS) /ELLIS, MDH P 180-181,1976 (ILLUS) •

REFRACTOHILUM HAWKSWORTH 1977. BOT. J. LINN. SOC. 75:195-209 /R. GALLIGENUM
AMERO OR PHRAGMO,BASO,HYALO,ANNEL, 77G, ON LICHENS (NEPHROMA, PELTIGERA, PARMELIA) & ON WOOD OR BARK (FRAXINUS, ABIES, CORNUS, LARIX & OTHERS).
/SUTTON, CMI PAPERS 132:105-106,1973 (ILLUS SUB SPORIDESMIUM) •

RETIARIUS OLIVIER 1978. TBMS 71:193-201 (ILLUS) /R. SUPERFICIARIS
STAURO,CETERI,HYALO,RADUL, ON POLLEN GRAINS. •

RHABDOGLOEOPSIS PETRAK 1925. ANN. MYCOL. 23:51-52 /R. BALSAMEAE (DAVIS)
=COELOMYCETES, SEE NAG RAJ & MORGAN-JONES 1973 (CAN. J. BOT. 51:565-569,ILLUS). •

RHABDOSPORIUM CHEVALLIER 1826. FLORE GEN. ENV. PARIS 1:428-429 /R. DIFFUSUM
=CHEIROSPORA MOUGEOT & FRIES 1825 FIDE HUGHES (AS '?=') •

RHACODIELLA PEYRONEL 1919. STAZ. SPER. AGR. ITAL. 52:39-41 (ILLUS) /R. CASTANEAE (BAINIER)
=MYRIOCONIUM SYDOW 1912 FIDE CARMICHAEL 1973. •

RHACODIOPSIS DONK 1975. PERSOONIA 8:273-276 /RACODIUM RUPESTRE PERSOON 1801 =RHACODIOPSIS RUPESTRIS (PERSOON)
=LICHEN. PROPOSED TO REPLACE RACODIUM FRIES 1829 NON PERSOON 1801. •

RHACODIUM, ORTHOGRAPHIC VARIANT OF RACODIUM •

RHACOPHYLLUS BERKELEY & BROOME 1871. J. LINN. SOC. LONDON (BOT.) 11:559 /R. LILACINUS =R. STATE OF COPRINUS CLASTOPHYLLUS FIDE WATLING 1979 (IN KENDRICK 1979, P 465 & 468)
BULBILS,SYNNEMATA, ON WOOD (QUERCUS). SEE DONK 1962 (TAXON 11:96-97).
/REYNDERS & MALENCON, BULL. SOC. MYCOL. FRANCE 85:295-304 & PL I-III,1969 (ILLUS) •

RHINOCEPHALUM KAMYSCHKO 1961. BOT. MATER. (NOT. SYST. SECT. CRYPT. INST. BOT. ACAD. SCI. USSR) 14:224-225 (ILLUS) /R. CHOCHRAJAKOVII
?=ARTHRINIUM KUNZE 1817. •

RHINOCLADIELLA KAMYSCHKO 1960. NOT. SYST. SECT. CRYPT. INST. BOT. ACAD. SCI. USSR 13:165 & FIG 5 /R. SPOROTRICHOIDES
=NOMEN ILLEGITIMUM, NON RHINOCLADIELLA NANNFELDT 1934, REPLACED BY RHINOCLADIOPSIS KAMYSCHKO 1961. •

RHINOCLADIELLA NANNFELDT 1934. IN MELIN & NANNFELDT, SVENSKA SKOGSVFOREN. TIDSKR. HEFTE 3-4:461-462 /R. ATROVIRENS
AMERO,CETERI,HYALO OR PHAEO,RADUL, ~~49G~~, ON TREES, WOOD, LITTER & SOIL. ASCO STATE ?=THECOTHEUS SEE CONWAY 1975 (MYCOLOGIA 67:241-252,ILLUS), OR DICTYOTRICHELLA FIDE SCHOL-SCHWARZ 1968, OR BIOSTICTIS SEE SHERWOOD 1978 (PAPERS OF THE FARLOW HERBARIUM 13:39-41,ILLUS). OTHER CONIDIAL STATES =EXOPHIALA, SPOROTHRIX, PHAEOCOCCUS, OR RACODIUM. SEE ALSO FONSECAEA & RAMICHLORIDIUM. L25
/BARRON, HYPHOMYCETES P 267-268,1968 (ILLUS) /SCHOL-SCHWARZ, ANTONIE VAN LEEUWENHOEK 34:119-152,1968 (ILLUS,REVIS) /ELLIS, DH P 246-248,1971 (ILLUS) /MATSUSHIMA, ICONES P 122-125 & PL,1975 (ILLUS,NEW SPP) /SIVASITHAMPARAM, AUSTRAL. J. BOT. 23:193-212,1975 (ILLUS) /DE HOOG, CBS STUDIES 15:1-140 & 211,1977 (MONOGR)
=*BOTRYTOIDES* MOORE & ALMEIDA 1936 FIDE DE HOOG 1977 •

RHINOCLADIOPSIS KAMYSCHKO 1961. BOT. MATER. (NOT. SYST. SECT. CRYPT. INST. BOT. ACAD. SCI. USSR) 14:243 /NOM. NOV. FOR RHINOCLADIELLA KAMYSCHKO NON NANNFELDT
=CHRYSOSPORIUM CORDA 1833 FIDE CARMICHAEL 1973. •

RHINOCLADIUM SACCARDO & MARCHAL 1885. IN MARCHAL, BULL. SOC. BOT. BELG. 24:65-66 & FIGS 10-12 /R. COPROGENUM
AMERO,CETERI,PHAEO,NONSP, 49G, ON DUNG & SOIL.
/BARRON, HYPHOMYCETES P 268-269,1968 (ILLUS) •

RHINOTRICHELLA ARNAUD EX DE HOOG 1977. CBS STUDIES 15:86 (86-89,ILLUS) /R. GLOBULIFERA
AMERO,CETERI,HYALO OR PHAEO,RADUL, 25A, ON LITTER. ?=HANSFORDIA HUGHES 1951.
/ARNAUD, BULL. SOC. MYCOL. FRANCE 69:269 & 272,1954 (ILLUS,NO LATIN DIAGNOSIS)
/UDAGAWA & HORIE, J. GEN. APPL. MICROBIOL. 17:153-158,1971 (ILLUS) •

RHINOTRICHUM CORDA 1837. ICONES FUNGORUM 1:17 /R. SIMPLEX
=NOMEN DUBIUM FIDE HUGHES 1958. FUNGI PLACED IN RHINOTRICHUM MAY BELONG IN ACLADIUM, ALYSIDIUM OR OLPITRICHUM. FOR R. SUBALUTACEUM SEE BASIFIMBRIA. •

RHIPIDOCEPHALUM TRAIL 1888. SCOTTISH NAT., SER. 2, 3:41 (NO ILLUS) /R. ABIETIS TRAIL 1886
=NOMEN NUDUM FIDE KENDRICK 1961 (CAN. J. BOT. 39:818-820). •

RHIZOCTONIA DE CANDOLLE 1815. FLOR. FRANCE 6:110-111 /R. CROCORUM (PERSOON) DE CANDOLLE (LECTOTYPE SEE DONK 1962, TAXON 11:97) =R. STATE OF HELICOBASIDIUM BREBISSONII (DESM.) DONK 1962
STERILE OR BULBILS, BASIDIO STATE =HELICOBASIDIUM FIDE BUDDIN & WAKEFIELD 1927 OR THANATEPHORUS, CERATOBASIDIUM, WAITEA, BOTRYOBASIDIUM, UTHATOBASIDIUM OR ATHELIA FIDE TU ET AL., 1977 OR TULASNELLA SEE WARCUP & TALBOT 1967 (NEW PHYTOL. 66:631-641) OR PELLICULARIA FIDE VON ARX 1974 OR LAETISARIA SEE BURDSALL 1979 (TBMS 72:419-422) OR AQUATHANATEPHORUS FIDE TU & KIMBROUGH 1978 (BOT. GAZ. 139:454-466). OTHER CONIDIAL STATE =TUBERCULINA OR SCLEROTIUM.
/BUDDIN & WAKEFIELD, TBMS 12:116-140,1927 (ILLUS) /BARRON, HYPHOMYCETES P 269-270,1968 (ILLUS,REFS) /PARMETER, RHIZOCTONIA SOLANI, U. CALIFORNIA PRESS, P 255,1970 (ILLUS, REVIS) /MORDUE, CMI DESCRIPTIONS 406-409,1974 (ILLUS) /TU & KIMBROUGH, CAN. J. BOT. 53:2282-2296,1975 (ILLUS,REVIS) /VON ARX, GENERA P 269-270,1974 (ILLUS,REFS) /OGOSHI, REV. PLANT PROTECT. RES. 8:93-103,1975 (TELEOMORPHS, REFS,NOT SEEN) /TU, KIMBROUGH & ALDRICH, CAN. J. BOT. 55:2419-2436,1977 (ILLUS,CYTOL)
=*ALLOSPHAERIUM* LINK FIDE SACCARDO FROM AINSWORTH 1971
=*BRYOCHYSIUM* FIDE RABENHORST FROM AINSWORTH 1971
=*MONILIOPSIS* RUHLAND 1908 FIDE VON ARX 1974
=*THANATOPHYTUM* NEES 1817 FIDE LINDAU IN RABENHORST 1910 •

RHIZOHYPHA CHODAT & SIGRIANSKY 1911. BULL. SOC. BOT. GENEVE 2(3):350 /R. RADICIS
?, STERILE. •

RHIZOMORPHA ROTH EX PERSOON 1801. SYN. METH. FUNGORUM P 704 /R. SUBCORTICALIS PERSOON 1801 (LECTOTYPE)
?, STERILE, FOR RHIZOMORPHS OF ARMILLARIA MELLEA FIDE DONK 1962 (TAXON 11:97-98) & ?LAETISARIA SEE BURDSALL 1979 (TBMS 72:419).
/SNIDER, MYCOLOGIA 51:693-707,1959 (ILLUS)
?=*APHOTISTUS* HUMB. SEE AINSWORTH 1971 •

RHIZOSPORIUM RABENHORST 1844 (DEUTSCHL. KRYPT.-FL. 1:2) ?=PHLOEOCONIS FIDE SACCARDO 1901 (SYLL. FUNG. 15:322) •

RHIZOSTILBELLA VAN DER WOLK 1914. MYCOL. CENTRALB. 4:236-241 & PL /R. RUBRA
?, ORIGINAL ILLUSTRATION NOT DIAGNOSTIC. •

RHIZOSTROMA FRIES 1819. NOVIT. FL. SUEC., PART. V, CONT. P 79 /RHIZOMORPHA XYLOSTROMA ACHARIUS (LECTOTYPE FIDE DONK 1962)
?, STERILE MYCELIUM FIDE DONK 1962 (TAXON 11:98). THE WHITE MYCELIAL CORDS ARE OFTEN ASSOCIATED WITH PHALLUS IMPUDICUS AND TRICHOLOMOPSIS PLATYPHYLLA FIDE WATLING 1979 (IN KENDRICK 1979). •

RHODOCEPHALUS CORDA 1837. ICONES FUNGORUM 1:21 & TAB 6 FIG 282 /R. CANDIDUS
=PENICILLIUM LINK 1809 FIDE LINDAU 1907. •

RHODOTHALLUS BATISTA & CIFERRI 1959. MYCOPATHOLOGIA 11:96-97 (ILLUS) /R. CASEARIAEFOLIUM BATISTA, CIFERRI & COSTA 1959
?, STERILE. •

RHOMBOSTILBELLA ZIMMERMANN 1902. ZENTRALBL. BAKT. PARASIT. ABT.2, 8:221 (NO ILLUS) /R. ROSEA
AMERO,CETERI,HYALO,RACHI,SYNNEMATA, 29E, ON FUNGI (PHAEOSACCARDINULA & OTHERS ON LEAVES).
/PIROZYNSKI, CMI PAPERS 90:34-37,1963 (ILLUS,KEY) •

RHOPALIDIUM MONTAGNE 1836. ANN. SCI. NAT. SER. 2, 6:30 (ILLUS IN ANN. SCI. NAT. SER. 6, 5:PL 12, FIG 4) /R. BRASSICAE (MONTAGNE)
=ALTERNARIA NEES 1817 FIDE HOHNEL 1910 (S.B. AKAD. WIS. 119:670-671)
/MONTAGNE & FRIES, SYLL. CRYPT. P 297-298,1858 (VALIDATED?) •

RHOPALOCONIDIUM PETRAK 1952. SYDOWIA 6:299-301 (NO ILLUS) /PHLEOSPORA ASIMINAE ELLIS & MORGAN 1887 (J. MYCOL. 3:88,NO ILLUS) =R. ASIMINAE (ELLIS & MORGAN)
? •

RHOPALOCYSTIS GROVE 1911. J. ECONOM. BIOL. 6:40-41 /R. NIGRA (VAN TIEGHEM) =ASPERGILLUS NIGER VAN TIEGHEM
=ASPERGILLUS LINK 1809 FIDE RAPER & FENNELL 1965 •

RHOPALOMYCES CORDA 1839 ?=ZYGOMYCETE SEE BARRON 1977 (THE NEMATODE-DESTROYING FUNGI P 84-87,ILLUS,REFS) •

RHYNCHOMYCES WILLKOMM 1866. DIE MIKR. FEINDES DES WALDES, DRESDEN P 87-88, TAF IV, FIGS 34-38 /R. VIOLACEUS
?, POSSIBLY THE SAME AS BRACHYSPORIUM, NON RHYNCHOMYCES (A) SACCARDO & MARCHAL 1885. •

RHYNCHOSPORINA VON ARX 1957. VERH. AKAD. WET. AMST. 51:19 & 103 /GLOEOSPORIUM MINERSII SPRAGUE =R. MEINERSII (SPRAGUE)
AMERO,GLOIO,HYALO,PHIAL, 2E, ON LEAVES (GRAMINEAE). ASCO STATE ?=BUERGENERULA SEE BARR 1976 (MYCOLOGIA 68:612-614,ILLUS).
/VON ARX, BIBLIO. MYCOL. (CRAMER) BD. 24:22-23,1970 (ILLUS) •

RHYNCHOSPORIUM HEINSEN 1901. JB. HAMB. WISSEN. VEREIN. 18:43 /R. GRAMINICOLA =R. SECALIS (OUDEMANS) DAVIS 1919 FIDE VON ARX 1974 (GENERA P 169)
DIDYMO,GLOIO,HYALO,ANNEL,SPORODOCHIA, 59E, ON LEAVES & SEED (GRASSES). SCARCELY DIFFERENT FROM MARSSONINA MAGNUS 1906, SEE VON ARX 1970. OTHER STATE ="SCLEROTIA".
/OWEN, TBMS 41:99-108,1958 (ILLUS) /OWEN, CMI DESCRIPTIONS 387,1973 (ILLUS) /MATSUSHIMA, ICONES P 125-126 & PL 100:4,1975 (ILLUS) •

RICCOA CAVARA 1903. ANN. MYCOL. 1:41-45 (ILLUS) /R. AETENSIS
=HEYDENIA FIDE HOHNEL FROM CLEMENTS & SHEAR 1931 •

RICLARETIA PEYRONEL 1916. MEM. R. ACC. SCI. TORINO, II, 66:39-40 (ILLUS) /R. URTICAE
SCOLECO,BASO,HYALO,PHIAL,SPORODOCHIA, 17C, ON LITTER (URTICA).
?=*STACHYBOTRYNA* TUBAKI & YOKOYAMA 1971 •

RIESSIA FRESENIUS 1852. BEITR. Z. MYCOL. 2:74-75 /R. SEMIOPHORA
DICTYO OR STAURO,CETERI,HYALO,NONSP OR MISC,CLAMPS, 113G, ON WOOD & LITTER (THEA). OTHER STATE =DEXHOWARDIA.
/GOOS & TUBAKI, CAN. J. BOT. 51:1439-1442,1973 (ILLUS,REVIS) •

ROGERSIA SHEARER & CRANE 1976. MYCOLOGIA 68:946-950 (ILLUS) /R. ANNELIDICA
?=FILOSPORELLA NAWAWI 1976 (SEE FIG 109E) SEE CRANE & SHEARER 1977 (MYCOTAXON 6:27-28) & DYKO & SUTTON 1978 (MYCOTAXON 7:323-326). •

ROSULOMYCES MARCHAND & CABRAL 1976. IN MARCHAND ET AL., BOL. SOC. ARGENTINA BOT. 17:64-66 (ILLUS) /R. ARTHROSPORIOIDES
AMERO,ARTHRO,HYALO,NONSP OR RADUL, 42H, ON SOIL. ?=SCYTALIDIUM.
/DE HOOG, CBS STUDIES 15:128-130,1977 (ILLUS) •

ROTAEA CESATI 1851. BOT. ZEIT. 9:180 OR FLORA, P 75 /R. FLAVA
?, NOT ILLUSTRATED IN FLORA 1851.
/LINDAU, RABENH. KRYPT. FL. 2 AUFL. 1 BD. 8:401,1907 (ILLUS) •

RUBROMADURELLA TALICE 1935. ANN. PARASIT. HUM. COMP. 13:584-590 (ILLUS) /R. LANGERONI
?=BACTERIA. •

RUGOSARIA RAFINESQUE 1833 SEE GEMMULARIA •

RUPINIA SPEGAZZINI & ROUMEGUERE 1879. IN ROUMEGUERE, REV. MYCOL. 1:171-173 (ILLUS) /R. PYRENAICA =R. BAYLACII ROUMEGUERE 1880 (REV. MYCOL. 2:2-4)
=HEYDENIA FRESENIUS 1852 FIDE SACCARDO 1886. •

RUTOLA CRANE & SCHOKNECHT 1977. CAN. J. BOT. 55:3013-3019 (ILLUS) /TORULA GRAMINIS DESMAZIERES 1826 =R. GRAMINIS (DESM.)
AMERO OR DIDYMO OR PHRAGMO,BLASTO,PHAEO,NONSP, 88H, ON PLANTS (GRASSES). •

SABOURAUDIELLA BOEDIJN 1951. MYCOPATHOLOGIA 6:123-125 /S. PURPUREA (BANG) =T. RUBRUM (CASTELLANI) SABOURAUD 1911
=TRICHOPHYTON MALMSTEN 1845 •

SABOURAUDITES OTA & LANGERON 1923. ANN. PARASIT. HUM. COMP. 1:326-329 /THREE SUBGENERA, EACH WITH ITS OWN TYPE
=MICROSPORUM PRO PARTE & TRICHOPHYTON PRO PARTE •

SACCARDAEA CAVARA 1894. ATTI. IST. BOT. PAVIA 3:346-347 & TAB 26 FIG 1-5 /S. ECHINOCEPHALA
=NOMEN DUBIUM FIDE TULLOCH 1972. S. ATRA IS A MYROTHECIUM FIDE TULLOCH 1972 (CMI PAPERS 130:36). •

SADASIVANIA SUBRAMANIAN 1957. J. INDIAN BOT. SOC. 36:64-67 (ILLUS) /S. GIRISA
AMERO,CETERI,PHAEO,MISC, 37F, ON LITTER (GRASSES).
/ELLIS, DH P 343-344,1971 (ILLUS) /SUNDBERG & WICKLOW, MYCOLOGIA 65:925-929,1973 (ILLUS) /WICKLOW & SUNDBERG, MYCOLOGIA 68:891-901,1976 (ILLUS) •

SAGENOMELLA GAMS 1978. PERSOONIA 10:97-112 (ILLUS) /SCOPULARIOPSIS DIVERSISPORA VAN BEYMA 1937 =S. DIVERSISPORA (VAN BEYMA)
?, ON SOIL. FOR PAECILOMYES SPECIES FIRST SEGREGATED BY GAMS (1971) IN ACREMONIUM SECTION GLIOMASTIX. FOR S. VERTICILLATA SEE SAROCLADIUM. SEE ALSO SAGRAHAMALA. •

SAGRAHAMALA SUBRAMANIAN 1972. CURRENT SCIENCE 41:48-49 (NO ILLUS) /TORULA LUZULAE FUCKEL 1870 =GLIOMASTIX LUZULAE (FUCKEL) MASON EX HUGHES 1958 (CAN. J. BOT. 36:769,NO ILLUS) =S. LUZULAE (FUCKEL)
?, FOR DRY-SPORED, MONOPHIALIDIC SPECIES OF GLIOMASTIX & PAECILOMYCES. SEE ALSO SAGENOMELLA.
/SUBRAMANIAN & PUSHKARAN, KAVAKA 3:82-83,1975 (ILLUS) •

SAPROCHAETE WAGNER & DAWES 1970. MYCOLOGIA 62:792-796 (ILLUS) /S. SACCHAROPHILA (COKER & SHANOR)
STERILE,AQUATIC, ORIGINALLY DESCRIBED AS AN ALGA. •

SAPROPHRAGMA DESHPANDE & DESHPANDE 1966. MYCOPATHOLOGIA 30:200-202 (ILLUS) /S. ACEROSUM (AS 'ACEROSIAE')
?, ORIGINAL ILLUSTRATION NOT DIAGNOSTIC. •

SARCINELLA SACCARDO 1877. FUNGI ITALICI TAB 126 (DIAGNOSIS IN MICHELIA 2:31,1880) /S. HETEROSPORA =S. STATE OF SCHIFFNERULA PULCHRA (SACC.) PETRAK 1928 FIDE ELLIS 1971

DICTYO,CETERI,PHAEO,NONSP,PODIA, 96C, ON LEAVES (FRAXINUS & LIGUSTRUM). ASCO STATE =SCHIFFNERULA FIDE ELLIS 1971 OR CLYPEOLELLA, YAMAMOTOA OR NEOPARODIA FIDE VON ARX 1975 (CBS STUDIES 8:44 & 52).
/ELLIS, DH P 49-50,1971 (ILLUS) •

SARCINODOCHIUM HOHNEL 1905. OEST. BOT. ZEITS. 55:16-17 (NO ILLUS) /S. HETEROSPORUM
? •

SARCINOMYCES LINDNER 1898. MIKROSCOPISCHE BETRIEBSKONTROLLE, BERLIN, ED. 2:228-230 (ILLUS) /S. CRUSTACEUS
AMERO,GLOIO,PHAEO,MISC, REDESCRIPTION TO BE PUBLISHED IN MYCOTAXON.
/JONG & KING, MYCOTAXON 3:397,1976 (REFS) /HERMANIDES-NIJHOF, CBS STUDIES 15:173-175,1977 (REVIS,DOUBTFUL SYNONYMS) /MCGINNIS, SABOURAUDIA 17:145-154,1979 (DISCUSSION SUB EXOPHIALA WERNECKII) •

SARCOPODIUM EHRENBERG 1818. SYLVAE MYC. BEROL. P 12 & 23 /S. CIRCINATUM
AMERO,GLOIO,HYALO,PHIAL,SPORODOCHIA,SETAE, 13C, ON LITTER (MANY HERBACEOUS PLANTS), WOOD (FAGUS, POPULUS), QUERCUS & RHAMNUS) & LEAVES (COFFEA & ERYTHRINA). DIFFERS FROM MYROTHECIUM IN HAVING BROWN SETAE.
/HUGHES, CAN. J. BOT. 36:801-802,1958 (REFS) /SUTTON, CMI PAPERS 132:97-99,1973 (ILLUS) /ELLIS, MDH P 467-469,1976 (ILLUS,KEY)
=*TRICHOLECONIUM* CORDA 1837 FIDE HUGHES 1958
=*PERIOLOPSIS* MAIRE 1913 FIDE PIROZYNSKI •

SAROCLADIUM GAMS & HAWKSWORTH 1975. KAVAKA 3:57-61 (ILLUS) /ACROCYLINDRIUM ORYZAE SAWADA 1922 (REP. GOVT. RES. INST. DEPT. AGRIC. FORMOSA 2:135) =S. ORYZAE (SAWADA)
AMERO,GLOIO,HYALO,PHIAL, 5F, ON LEAVES (ORYZA). COMPARE WITH GABARNAUDIA, GLIOCLADIUM, ACREMONIUM & SAGENOMELLA.
/GAMS, PERSOONIA 10:97-112,1979 (NEW SP SUB SAGENOMELLA VERTICILLATA) •

SAROPHORUM SYDOW & SYDOW 1916. BOT JB. 54:260-261 (ILLUS) /S. LEDERMANNII SYD. (SIC) 1916
?=CEPHALOTRICHUM LINK 1809 •

SATCHMOPSIS SUTTON & HODGES 1975. IN SUTTON, NOVA HEDWIGIA 26:1-5 (ILLUS) /S. BRASILIENSIS
AMERO,BASO,HYALO,PHIAL,SPORODOCHIA, ON LITTER (EUCALYPTUS). ACERVULOID CONIDIOMATA, SEE KENDRICK & NAG RAJ IN KENDRICK 1979. •

SATWALEKERA RAO, RAO & RAO 1969. NOVA HEDWIGIA 18:637-641 (ILLUS) /S. SUNDARA ?=TORULA HERBARUM (PERSOON) LINK 1815
=TORULA (PERSOON) LINK 1809 FIDE CARMICHAEL 1973 •

SAVULESCUELLA CIFERRI 1959. OMAGIU LUI TRAIAN SAVULESCU P 179 (NO ILLUS) /CYLINDROSPORIUM ALISMACEARUM SACCARDO =S. ALISMACEARUM (SACCARDO)
?, BASIDIO STATE =DOASSANSIA & TRACYA (TILLETIACEAE)
/ZAMBETTAKIS, REV. MYCOL. 35:158-175,1970 (REVIS) •

SCEDOSPORIUM SACCARDO EX CASTELLANI & CHALMERS 1919. MANUAL OF TROPICAL MEDICINE ED. 3, P 1122 (ILLUS) /S. APIOSPERMUM (SACCARDO) CASTELLANI & CHALMERS 1919 =SCEDOSPORIUM STATE OF PETRIELLIDIUM BOYDII (SHEAR) MALLOCH 1970
AMERO,GLOIO,PHAEO,ANNEL, ON MAN, DUNG & SOIL. ASCO STATE =PETRIELLIDIUM. OTHER CONIDIAL STATE =GRAPHIUM. •

SCENOMYCES STEVENS 1927. ILL. BIOL. MONOGR. 11:212-213 & FIGS 89-91, 114 & 123-125 /S. PERPLEXANS
?, STERILE. •

SCEPTRIFERA DEIGHTON 1965. IN DEIGHTON & PIROZYNSKI, CMI PAPERS 101:37-39 (ILLUS) /S. PULCHRA
AMERO OR DIDYMO,CETERI,PHAEO,RADUL, 61E, ON LEAVES (MALLOTUS). •

SCEPTROMYCES CORDA 1831. STURM'S, DEUTSCHL. FL. III (PILZE) 3:7-8 & TAB. 4 /S. OPIZII

?, LINDAU 1907 LISTS THIS AS A MONSTROUS FORM OF ASPERGILLUS NIGER, BASED ON A REPORT OF ENGELKE 1902 (IN HEDWIGIA 41:219). •

SCHELEOBRACHEA HUGHES 1958. CAN. J. BOT. 36:802 /S. ECHINULATA (SPEGAZZINI) =PITHOMYCES CHARTARUM (BERK. & CURT.) ELLIS 1960
=PITHOMYCES BERKELEY & BROOME FIDE ELLIS 1960. •

SCHIZOCEPHALUM PREUSS 1852. LINNAEA 25:77 /S. ATROFUSCUM
?, SEE LINDAU IN RABENHORST 1907 (P 696) FOR DESCRIPTION. •

SCHIZOTRICHELLA MORRIS 1956. MYCOLOGIA 48:733-735 /S. LUNATA
=COLLETOTRICHUM CORDA FIDE TULLOCH 1972 (CMI PAPERS 130:36) BUT SEE ALSO PIROZYNSKI 1972 (CMI PAPERS 129:57-58, RE NEOTTIOSPORELLA). •

SCHIZOTRICHUM MCALPINE 1903. PROC. LINN. SOC. N. S. WALES 28:562 (NO ILLUS) /S. LOBELIAE
SCOLECO,CETERI,HYALO,RADUL,STROMA,SETAE, 110E, ON STEMS & LEAVES (LOBELIA).
/ELLIS, DH P 278,1971 (ILLUS) •

SCHOENLEINIUM JOHAN-OLSEN 1897. ZENTRALBL. BAKT. ABT 2, 3:276-284 & TAF. IV /TWO ORIGINAL SPECIES
=TRICHOPHYTON MALMSTEN 1845 FIDE CARMICHAEL 1973 •

SCINIATOSPORIUM "(REICHB.) CORDA" 1844. IN RABENHORST, DEUTSCHL. KRYPT.-FL. 1:47,1844 /SEIMATOSPORIUM ROSAE CORDA 1833, IN STURM'S DEUTSCHL. FL. III, 3, PL 40
=SEIMATOSPORIUM CORDA (COELOMYCETES). SCINIATOSPORIUM IS AN ORTHOGRAPHIC VARIANT BASED ON IMPERFECTIONS IN THE PRINT FOR THE NAME UNDER CORDA'S PLATE 40. BUT SEE MORGAN-JONES 1971 (CAN. J. BOT. 49:993-1009). •

SCLEROCOCCUM (ACHARIUS) FRIES EX FRIES 1825. SYST. ORB. VEG. P 172-173 /SPILOMA SPHAERALE ACHARIUS 1814 =S. SPHAERALE (ACHARIUS) FRIES EX FRIES 1825
DICTYO,BLASTO,PHAEO,NONSP,SPORODOCHIA, 50E, ON LICHENS (PERTUSARIA & OTHERS).
/FRIES, NOV. FL. SUECIAE, LUNDAE, 5:79,1819 (NO ILLUS) /ELLIS, MDH P 62,1976 (ILLUS) /HAWKSWORTH, BULL. BRITISH MUSEUM (NATURAL HISTORY) /6:249-250,1979 (ILLUS) •

SCLERODISCUS PATOUILLARD 1890. JOUR. DE BOT. 4:66-67 (ILLUS) /S. NITENS
?, ORIGINAL ILLUSTRATION NOT DIAGNOSTIC. •

SCLEROGRAPHIOPSIS DEIGHTON 1973. TBMS 61:193-195 (ILLUS) /S. DALBERGIAE
PHRAGMO,BLASTO,PHAEO,RADUL,SYNNEMATA, 62A, ON LEAVES (DALBERGIA). COMPARE WITH GONATOPHRAGMIUM.
/ELLIS, MDH P 216-217,1976 (ILLUS) •

SCLEROGRAPHIUM BERKELEY 1854. HOOKER J. BOT. 6:209 & PL 8 FIG 4 /S. ATERRIMUM
DICTYO,CETERI,PHAEO,RADUL,SYNNEMATA, 101A, ON LEAVES (INDIGOFERA & PHYLLANTHUS).
/HUGHES, INDIAN PHYTOPATH. 4:5-6,1951 (ILLUS) /ELLIS, DH P 256-258,1971 (ILLUS,KEY) •

SCLEROMA FRIES =PSEUDO-SCLEROTIUM OF PANUS BADIUS FIDE SINGER 1975 (THE AGARICALES IN MODERN TAXONOMY, 3RD ED.) •

SCLEROSTILBUM POVAH 1932. MYCOLOGIA 24:242-243 (ILLUS,MACROSCOPIC ONLY) /S. SEPTENTRIONALE =S. STATE OF COLLYBIA RACEMOSA FIDE DONK 1962 (TAXON 11:98-99) =S. STATE OF MICROCOLLYBIA RACEMOSA SEE LENNOX 1979 (MYCOTAXON 9:191-193,ILLUS)
AMERO,ARTHRO OR GLOIO,HYALO,NONSP,SYNNEMATA,CLAMPS, 45D, ON LITTER. BASIDIO STATE =COLLYBIA.
/MORRIS, WESTERN ILLINOIS UNIV. SER. BIOL. SCI. 3:116-117,1963 (ILLUS) /WATLING & KENDRICK, MICHIGAN BOT. 16:65-72,1977 (ILLUS) •

"SCLEROTIA" ARE REPORTED AS OCCURRING IN MANY FUNGI. WHETHER THESE SHOULD MOSTLY BE REFERRED TO THE FORM-GENUS SCLEROTIUM OR TO SEVERAL FORM-GENERA IS NOT CLEAR. SEE FOR EXAMPLE, BOTRYTIS, DICHOTOMOPHTHORA, PHAEOTRICHOCONIS. •

SCLEROTIELLA SARBHOY & SARBHOY 1975. MAHARASHTRA V. M. PATRIKA 9:91-92 (DATED 1974, NOT SEEN) /S. INDICA
?, FROM SOIL. •

SCLEROTIOMYCES WORONICHIN 1926 (ANN. MYCOL. 24:233-235) ?=SCLEROTIUM FIDE AINSWORTH 1971 •

SCLEROTIUM TODE EX PERSOON 1801. SYN. METH. FUNG. P 119-122 /S. COMPLANATUM TODE 1790 (LECTOTYPE) =S. STATE OF TYPHULA GYRANS FIDE DONK 1962 (TAXON 11:99)
SCLEROTIA, ON PLANTS, LITTER & SOIL. BASIDIO STATE =PELLICULARIA, CORTICIUM, TYPHULA OR LEUCOGYROPHANA SEE STALPERS 1978 (CBS STUDIES 16:157). ASCO STATE =SCLEROTINIA SEE KOSASIH & WILLETTS 1975 (MYCOLOGIA 67:89-97) OR MORDUE 1976 (CMI DESCRIPTIONS 512-513,ILLUS) OR STROMATINIA FIDE DRAYTON & GROVES 1952 (MYCOLOGIA 44:119) OR MAGNAPORTHE SEE BARR 1977 (MYCOLOGIA 69:952-953). COELO STATE =MACROPHOMINA OR COLLETOTRICHUM FIDE VON ARX 1970 & BARRON 1968. CONIDIAL STATES ="PHIALOCONIDIA" SEE MORDUE 1976, OR BOTRYOSPORIUM, BOTRYTIS, DRECHSLERA, PHAEOTRICHOCONIS, THYSANOPHORA, NAKATAEA; VALDENSIA, ACROCONIDIELLINA, ALTERNARIA, STREPTOBOTRYS, AMPHOBOTRYS.
/BARRON, HYPHOMYCETES P 272-274,1968 (ILLUS) /VON ARX, GENERA P 268-270,1974 (DICUSSION) /MORDUE, CMI DESCRIPTIONS 410,1974 (ILLUS SUB CORTICIUM) /KOSKE, MYCOLOGIA 67:128-146,1975 (SUB TYPHULA ERYTHROPUS) /CHAN & SAKSTON, CAN. J. BOT. 55:999-1002,1977 (ILLUS OF S. BATATICOLA)
=*APIOSPORIUM* KUNZE 1817 FIDE HOHNEL FROM AINSWORTH 1971
=*ASTOMA* GRAY FIDE SACCARDO FROM AINSWORTH 1971
=*BROMICOLLA* EICHWALD 1843 FIDE SACCARDO FROM AINSWORTH 1971
=*COCCOPLEUM* EHRENBERG FIDE LINDAU IN RABENHORST 1910
?=*COCCOSPORA* WALLROTH 1833
=*PACHYMA* FRIES, SEE AINSWORTH 1971
?=*SCLEROTIOMYCES* WORONICHIN FIDE AINSWORTH 1971
=*SPERMOEDIA* FRIES 1822, SEE AINSWORTH 1971
=*XYLOCHOERAS* FRIES 1849 FIDE SACCARDO FROM AINSWORTH 1971 •

SCOLECOBASIDIELLA ELLIS 1971. CMI PAPERS 125:12-13 (ILLUS) /DIPLOCOCCIUM AVELLANEUM SAPPA & MOSCA 1965 =S. AVELLANEA (SAPPA & MOSCA)
DIDYMO,CETERI,HYALO OR PHAEO,RADUL, 62D, ON SOIL. ?=PASPALOMYCES LINDER 1933. VERONAEA SIMPLEX PAPENDORF 1969 (TBMS 52:485-487 & PL 35) MAY BELONG IN SCOLECOBASIDIELLA.
/ELLIS, MDH P 195-196,1976 (ILLUS) •

SCOLECOBASIDIUM ABBOTT 1927. MYCOLOGIA 19:29-31 (ILLUS) /S. TERREUM
DIDYMO OR PHRAGMO OR STAURO,CETERI,PHAEO,RADUL, 117F, ON SOIL, LITTER & OTHER SUBSTRATES. COMPARE WITH DACTYLARIA. MOST SPECIES WERE TRANSFERRED TO OCHROCONIS BY DE HOOG & VON ARX. OTHER CONIDIAL STATE =?HUMICOLA SEE MATSUSHIMA 1975 (ICONES PL 178:3).
/BARRON & BUSCH, CAN. J. BOT. 40:77-84,1962 (ILLUS) /MATSUSHIMA, MICROFUNGI, P 50-53 & FIGS 25-30, PL 10:2 & 14:6-7,1971 (ILLUS,NEW SPP) /ELLIS, DH P 224-226,1971 (ILLUS,KEY) /PIROZYNSKI & HODGES, CAN. J. BOT. 51:161-164,1973 (ILLUS) /DE HOOG & VON ARX, KAVAKA 1:55-60,1973 (ILLUS,REVIS) /SUTTON, TBMS 61:424-426,1973 (ILLUS,NEW SP) /MATSUSHIMA, ICONES P 126-128 & PL,1975 (ILLUS,NEW SPP) /ELLIS, MDH P 190-195,1976 (ILLUS,KEY,NEW SPP,NEW COMBS)
=*OCHROCONIS* DE HOOG & VON ARX 1974 FIDE ELLIS 1976 •

SCOLECOSPORITES LANGE & SMITH 1971 -FOSSIL FUNGI IMPERFECTI (SEE INDEX OF FUNGI 4:455,1977) •

SCOLICOTRICHUM KUNZE 1817. IN KUNZE & SCHMIDT, MYK. HEFTE 1:10-11 /S. VIRESCENS
=NOMEN CONFUSUM FIDE HUGHES 1958
=*SCOLECOTRICHUM* KUNZE EX AUCT. (ORTHOGRAPHIC VARIANT) •

SCOPULA ARNAUD 1952. BULL. SOC. MYCOL. FR. 68:197-198 (ILLUS,NO LATIN DIAGNOSIS) /S. HYALINA
=GLOIOSPHAERA HOHNEL 1902 FIDE WANG 1971. •

SCOPULARIA PREUSS 1851. LINNAEA 24:133-134 /S. VENUSTA (=NOMEN DUBIUM FIDE KENDRICK 1964)
=NOMEN ILLEGITIMUM, NON SCOPULARIA LINDLEY 1835, SEE KENDRICK 1964 (CAN. J. BOT. 42:1119-1122). LINDAVIA NIEUWLAND 1916 (AMER. MIDL. NAT. 4:385) WAS PROPOSED TO REPLACE SCOPULARIA PREUSS. •

SCOPULARIELLA GJAERUM 1971. FRIESIA 9:415-417 (ILLUS) /S. VACCINII
?=POLYSCYTALUM RIESS 1853 •

SCOPULARIOPSIS BAINIER 1907. BULL. SOC. MYCOL. FR. 23:98-99 & 104 & PL XI /S. BREVICAULIS (SACCARDO) (LECTOTYPE)
AMERO,BASO,HYALO,ANNEL, 18E, ON MANY SUBSTRATES. ASCO STATE =MICROASCUS OR KERNIA FIDE VON ARX 1974 (GENERA P 115) OR CHAETOMIUM OR ?=DICHOTOMOMYCES SEE UDAGAWA 1970 (TRANS. MYCOL. SOC. JAPAN 10:105-107,ILLUS). OTHER CONIDIAL STATES =WARDOMYCES & CEPHALOTRICHUM. COMPARE WITH GYMNODOCHIUM, GAMSIA, POLYPAECILUM & WARDOMYCOPSIS.
/MORTON & SMITH, CMI PAPERS 86:1-96,1963 (ILLUS,MONOGR) /BARRON, ANTONIE VAN LEEUWENHOEK 32:293-298,1966 (ILLUS,NEW SP) /UDAGAWA & AWAO, TRANS. MYCOL. SOC. JAPAN 10:4,1969 (ILLUS SUB MICROASCUS ALBONIGRESCENS) /MULLER & PACHA-AUE, NOVA HEDWIGIA 18:159-162,1969 (ILLUS,NEW SPP) /ELLIS, DH P 326-328,1971 (ILLUS) /MATSUSHIMA, ICONES P 128-129 & PL,1975 (ILLUS)
=*ACAULIUM* SOPP 1912 FIDE ELLIS 1971
?=*BAINIERIA* ARNAUD 1952 SEE GAMS 1969
=*MASONIA* SMITH 1952 FIDE SMITH 1963
=*MASONIELLA* SMITH 1952 FIDE SMITH 1963
=*PHAEOSCOPULARIOPSIS* OTA 1928 FIDE MORTON & SMITH 1963
?=*TORULOIDEA* SUMSTINE 1913
?=*TYMPANOSPORIUM* GAMS 1974 •

SCORIOMYCES ELLIS & SACCARDO 1885. IN SACCARDO & BERLESE, ATTI VENET. SCI. LETT. ARTI, SER 6, 3:728-729 /S. CRAGINII (AS 'CRAGINI')
?, NOT ILLUSTRATED BY AUTHORS. •

SCORPIOSPORIUM IQBAL 1974. BIOLOGIA, LAHORE 20:17-20 (ILLUS) /S. MINUTUM
STAURO,CETERI,HYALO,NONSP,AQUATIC, 123C, COMPARE WITH TRICLADIUM. •

SCUTISPORIUM PREUSS 1851. LINNAEA 24:112 /TWO ORIGINAL SPECIES
=STEMPHYLIUM WALLROTH FIDE LINDAU IN RABENHORST 1910 •

SCYTALIDIUM PESANTE 1957. ANN. SPER. AGR. N. S. 11:264-265 (CCLXIV-CCLXV) /S. LIGNICOLA
AMERO OR DIDYMO OR PHRAGMO,ARTHRO,PHAEO,NONSP, 42E, ON WOOD, SOIL, FUNGI (CRONARTIUM) & OTHER SUBSTRATES. OTHER CONIDIAL STATES =MONOCHAETIA OR HENDERSONULA, SEE YOKOYAMA 1975 (TBMS 65:499-503,ILLUS) & SIGLER & CARMICHAEL 1976. ASCO STATE ?=ANTHRACOBIA SEE LARSEN 1975 (MYCOLOGIA 67:1058-1061,ILLUS).
/VON ARX, GENERA P 168,1970 (ILLUS) /PUNITHALINGAM & WATERSTON, CMI DESCRIPTIONS 274,1970 (ILLUS SUB HENDERSONULA) /ELLIS, DH P 28,1971 (ILLUS) /RAO & DE HOOG, PERSOONIA 8:199-206,1975 (ILLUS,NEW SP) /SIGLER & CARMICHAEL, MYCOTAXON 4:395-404,1976 (ILLUS,KEY, NEW COMB) /KUHLMAN, CARMICHAEL & MILLER, MYCOLOGIA 68:1188-1194,1976 (ILLUS,NEW SP) /ELLIS, MDH P 16-17,1976 (ILLUS) /AUSTWICK, NEW ZEALAND J. AGRIC. RES. 19:29,1976 (NEW COMB) /CAMPBELL & MULDER, SABOURAUDIA 15:161-166,1977 (ILLUS,NEW SP) /NICOT, BULL. SOC. MYCOL. MEDICALE 6:245-249,1977 (ILLUS) /HIRATSUKA, TSUNEDA & SIGLER, PLANT DIS. REPTR. 63:512-513,1979 (ILLUS,ECOL)
?=*ROSULOMYCES* MARCHAND & CABRAL 1976 •

SEARCHOMYCES MEHROTRA & MEHROTRA 1963. SYDOWIA 16:212-214 (ILLUS,DATED 1962) /M. COPROPHILOIDES
=AMBLYOSPORIUM FRESENIUS 1863 FIDE PIROZYNSKI 1969 •

SEIMATOSPORIUM CORDA 1833. STURM'S DEUTSCHL. FLORA III (PILZE), BD. 3, HEFT 13:79-80 & PL 40 /S. ROSAE

PHRAGMO,GLOIO,PHAEO,ANNEL,ACERVULI, ON PLANTS. (MELANCONIALES) ASCO STATE =GRIPHOSPHAERIA, DISCOSTROMA (=CLATHRIDIUM BERLESE NON SACCARDO) OR PARADIDYMELLA SEE BARR 1975 (MYCOLOGIA 67:187-194,REFS).
/VON ARX, GENERA P 156-158,1970 (ILLUS) /SUTTON, CMI PAPERS 138:2-4,1975 (REFS) /VON ARX, GENERA P 130, 173-174,1974 (ILLUS,SYNOS) /MATSUSHIMA, ICONES P 129-130 & PL,1975 (ILLUS)
=*BASIPILUS* SUBRAMANIAN 1961 FIDE VON ARX 1970
=*MONOCERAS* GUBA 1961 FIDE VON ARX 1970
=*SCINIATOSPORIUM* (ORTHOGRAPHIC VARIANT) •

SEIRIDIUM NEES EX FRIES 1821 (SYST. MYCOL. 1:15) =MONOCHAETIA •

SELENOSPORELLA ARNAUD EX MACGARVIE 1968. SCIENT. PROC. R. DUBL. SOC. SER. B, 2:153-155 /S. CURVISPORA
AMERO OR SCOLECO,GLOIO,HYALO,RADUL, 21A, ON FUNGI (MANY KINDS). OTHER CONIDIAL STATES =SPORIDESMIUM, CERATOSPORIUM, ENDOPHRAGMIA, ENDOPHRAGMIELLA, TERATOSPERMA & ?SOPAGRAHA. COMPARE WITH HARPOGRAPHIUM, IDRIELLA, UMBELLIDION & MICRODOCHIUM. ASCO STATE =DIATRYPE SEE VON ARX 1976 (CBS REPORT P 6).
/ARNAUD, BULL. SOC. MYCOL. FR. 69:292 & 293,1954 (ILLUS) /ELLIS, DH P 564-565,1971 (ILLUS) /SUTTON, CMI PAPERS 132:100-102,1973 (ILLUS,NEW SPP) /AWAO & UDAGAWA, TRANS. MYCOL. SOC. JAPAN 15:99-104,1974 (ILLUS) /MATSUSHIMA, ICONES P 130 & PL 176,1975 (ILLUS,SEE ALSO PL 135 & TOP OF P 69 & PL 349) /SUTTON & HODGES, NOVA HEDWIGIA 28:490-493,1977 (NEW SP,ILLUS,KEY) /UECKER, AYERS & ADAMS, MYCOTAXON 7:275-282,1978 (ILLUS SUB SPORIDESMIUM SCLEROTIVORUM) /SUTTON & HODGES, NOVA HEDWIGIA 29:602-604,1978 (ILLUS,NEW SP) •

SELENOSPORIUM CORDA 1837. ICONES FUNGORUM 1:7 /S. PALLENS (NEES)
=FUSARIUM FIDE LINDAU IN RABENHORST 1910 •

SELENOTILA LAGERHEIM 1892. BERL. DEUTSCHL. BOT. GES. 10:53 & FIGS 24-28 /S. NIVALIS
?, PERHAPS AN ALGA OR A SMUT. •

SEPEDONIUM LINK 1809. MAG. GES. NATURF. FREUNDE, BERLIN 3:18 /UREDO MYCOPHILA PERSOON =S. MYCOPHILUM (PERSOON) NEES 1816/17
AMERO,CETERI,HYALO,NONSP,ALEURI, 47D, ON FUNGI (AGARICS, BOLETES, RHIZOPOGON). ASCO STATE =HYPOMYCES, APIOCREA,THIELAVIA, CORYNASCUS OR PECKIELLA FIDE VON ARX 1974 (GENERA P 117, 127 & 250). OTHER CONIDIAL STATES =?("PHIALOCONIDIA") & ? FOR S. AMPULLOSPORUM.
/BARRON, HYPHOMYCETES P 278-279,1968 (ILLUS) /TUBAKI, REPT. TOTTORI MYCOL. INST. (JAPAN) 12:161-169,1975 (ILLUS) /MATSUSHIMA, ICONES P 130-131 & PL,1975 (ILLUS,NEW SP) /CHU-CHOU & GRACE, TBMS 68:473-475,1977 (ILLUS)
=*LEIOSEPIUM* SACCARDO 1900 FIDE DAMON 1952
=*MYCOBANCHE* PERSOON 1818 FIDE HUGHES 1958 •

SEPTOCYLINDRIUM BONORDEN EX SACCARDO 1880. MICHELIA 2:15 /S. SEPTATUM (BONORDEN) LINDAU 1907 (IN RABENHORST, KRYPT. FL. ED. 2, 1(8):405) =S. BONORDENII SACCARDO EX SACCARDO 1880 (MICHELIA 2:122)
PHRAGMO,BLASTO,HYALO,NONSP, 88B, ON LITTER (GALANTHUS). S. SEPTATUM WAS DESCRIBED AS A HYALINE COUNTERPART OF SEPTONEMA SECEDENS, SEE ALSO DIPLOOSPORA.
/BONORDEN, HANDB. MYKOL. P 35 & FIG 16,1851 (ILLUS) /GAMS & DOMSCH, NOVA HEDWIGIA 18:18-19,1969 (ILLUS OF SP NOT CONGENERIC WITH S. SEPTATUM)
=*TAPEINOSPORIUM* BONORDEN 1853 FIDE LINDAU IN RABENHORST 1907 •

SEPTODOCHIUM MATSUSHIMA 1971. MICROFUNGI P 54 & FIG 151:4-6 & PL 27:6-8 /S. TRISEPTATUM
PHRAGMO,GLOIO,HYALO,PHIAL,SPORODOCHIA, 74E, ON SOIL. •

SEPTOFUSIDIUM GAMS 1971. CEPHALO. SCHIMMEL. P 147-148 (ILLUS) /FUSIDIUM ELEGANTULUM PIDOPLICHKO 1953 =S. ELEGANTULUM (PIDOPLICHKO)
?, DISTINCTION FROM PENICILLIFER NOT CLEAR.
/SAMSON, CBS STUDIES 6:85-88,1974 (ILLUS,NEW SPP,KEY) •

SEPTOIDIUM ARNAUD 1921. ANN. EPIPHYT. 7:106 (NO ILLUS) /S. CLUSIACEAE
PHRAGMO,CETERI,HYALO,ANNEL,PODIA, 75H, ON LEAVES (HURA, ANACARDIUM & OTHERS). ASCO STATE =PARODIOPSIS FIDE ARNAUD 1921 OR EUPELTE, PARODIELLINA, PILGERIELLA OR OTHERS FIDE VON ARX 1975 (CBS STUDIES 8:44 & 50-51) OR ALINA, PERISPORIOPSIS OR OPHIOPARODIA SEE DE HOOG 1977 (CBS STUDIES 15:140).
/ELLIS, DH P 115-116,1971 (ILLUS) /HUGHES & PIROZYNSKI, CAN. J. BOT. 50:2530-2532,1972 (ILLUS,NEW COMB) /ELLS, MDH P 74-76,1976 (ILLUS,KEY,NEW COMBS)
?=*ARNAUDINA* TROTTER 1931
=*DIPLOIDIUM* ARNAUD 1923 FIDE ELLIS 1971 •

SEPTOMYROTHECIUM MATSUSHIMA 1971. IN KOBAYASI ET AL., BULL. NAT. SCI. MUS. TOKYO 14:469-471 (ILLUS) /S. UNISEPTATUM
DIDYMO,GLOIO,HYALO,PHIAL,SPORODOCHIA, 58C, ON LITTER (BROAD-LEAF TREES).
/MATSUSHIMA, MICROFUNGI P 54-55 & FIG 152-3,1971 •

SEPTONEMA CORDA 1837. ICONES FUNGORUM 1:9 /S. SECEDENS
PHRAGMO,BLASTO,PHAEO,NONSP, 88A, ON WOOD, BARK (BETULA, PINUS & OTHERS). SEE ALSO TAENIOLELLA. ASCO STATE =AKAROPELTELLA FIDE FARR 1972 (MYCOLOGIA 64:252-264) OR MYTILIDION SEE DE HOOG 1977 (CBS STUDIES 15:140).
/HUGHES, THE NATURALIST P 173-176,1951 (ILLUS) /HUGHES, THE NATURALIST P 7-12,1952 (ILLUS) /ELLIS, DH P 97-98,1971 (ILLUS) /MATSUSHIMA, ICONES P 131-132 & PL,1975 (ILLUS) /ELLIS, MDH P 67,1976 (ILLUS) /HOLUBOVA-JECHOVA, FOLIA GEOBOT. PHYTOTAX., PRAHA 13:421-433,1978 (ILLUS,KEY,NEW SP) •

SEPTORIOPSIS STEVENS & DALBY 1918 (MYCOLOGIA 11:4-5,ILLUS) =CERCOSEPTORIA, NON SEPTORIOPSIS FRAGOSO & PAUL 1915 •

SEPTOSPORIUM CORDA 1832. IN STURM'S DEUTSCHL. FLORA III, BD. 3, HEFT 12:33-34 /S. ATRUM
DICTYO,CETERI,PHAEO,NONSP OR ANNEL,SETAE,APPEND, 95E, ON WOOD & LITTER. OTHER CONIDIAL STATE ="MICROPYCNIDIA".
/ELLIS, CMI PAPERS 79:1-5,1961 (ILLUS) /ELLIS, DH P 105-108,1971 (ILLUS) /MATSUSHIMA, ICONES P 133 & PL,1975 (ILLUS) /HUGHES, NEW ZEALAND J. BOT. 16:345-346,1978 (ILLUS) •

SEPTOTIS BUCHWALD 1949 EX VON ARX 1970. A REVISION OF THE FUNGI CLASSIFIED IN GLOEOSPORIUM, ED. 2, P 158 /S. PODOPHYLLINA (ELLIS & EVERHART 1888) VON ARX 1957 =S. STATE OF SEPTOTINIA PODOPHYLLINA WHETZEL 1937
PHRAGMO,GLOIO,HYALO,RADUL,SPORODOCHIA, ON LEAVES & LITTER (PODOPHYLLUM). ASCO STATE =SEPTOTINIA WHETZEL. OTHER STATE =MYRIOCONIUM (H) SEE WHETZEL 1937 (FIGS 17-18).
/WHETZEL, MYCOLOGIA 29:128-146,1937 (ILLUS) /BUCHWALD, FRIESIA 9:326-329,1970 (NO ILLUS) •

SEPTOTRICHUM CORDA 1840. ICONES FUNGORUM 4:4-6 /?
=LEAF HAIRS FIDE HUGHES 1958 •

SEPTOTRULLULA HOHNEL 1902. S. B. AKAD. WISS. WIEN, ABT. 1, 111:1025-1027 /S. BACILLIGERA (LECTOTYPE)
PHRAGMO,GLOIO OR ARTHRO,HYALO OR PHAEO,NONSP,SPORODOCHIA, ON BARK (BETULA, FAGUS & QUERCUS).
/ELLIS, DH P 98-99,1971 (ILLUS) /MORGAN-JONES, NAG RAJ & KENDRICK, ICONES GENERUM COELOMYCETUM V:31-32,1972 (ILLUS) •

SESQUICILLIUM GAMS 1968. ACTA BOT. NEERL. 17:455-460 (ILLUS) /S. BUXI (SCHMIDT)
AMERO,BASO OR GLOIO,HYALO,PHIAL, 15A, ON PLANTS (BUXUS, PINUS), LITTER & SOIL. ASCO STATE =PSEUDONECTRIA FIDE VON ARX 1974 (GENERA P 121) & GNOMONIA FIDE SIVANESAN & SHAW 1977 (TBMS 68:85-90,ILLUS).
/VON ARX, GENERA P 179-181,1970 (ILLUS) /VITTAL, KAVAKA 2:43-46,1974 (ILLUS,NEW SP) •

SESSILIOSPORA HAWKSWORTH 1979. BULL. BRITISH MUSEUM (NAT. HIST.) 6:250-253 (ILLUS) /S. BICOLOR
PHRAGMO,NONSP,PHAEO,MISC, 91H, ON LICHENS (LASIOLOMA). •

SETODOCHIUM BATISTA & CIFERRI 1957. PUBL. INST. MIC. UNIV. RECIFE 94:5 /S. CASEARIAE BATISTA, CIFERRI & NASCIMENTO 1957
?=VOLUTINA PENZIG & SACCARDO 1904 •

SIBIRINA ARNOLD 1970. NOVA HEDWIGIA 19:299-301 (ILLUS) /S. FUNGICOLA
DIDYMO,GLOIO,HYALO,MISC, 58B, ON FUNGI (POLYPORUS, LENTINUS) & WOOD. COMPARE WITH CLADOBOTRYUM. ASCO STATE =HYPOMYCES FIDE VON ARX 1974 (GENERA P 127).
/GAMS, PERSOONIA 7:163-165,1973 (REVIS,ILLUS) •

SIGMATOMYCES SACCARDO & SYDOW 1913. ANN. MYCOL. 11:319 /S. BAKERI
?, NOT ILLUSTRATED BY AUTHORS. •

SIGMOIDEA CRANE 1968. AMER. J. BOT. 55:996-999 (ILLUS) /FLAGELLOSPORA PROLIFERA PETERSON =S. PROLIFERA (PETERSEN)
SCOLECO,CETERI,HYALO,NONSP?,AQUATIC, 108D. •

SIGMOIDEOMYCES THAXTER 1891. BOT. GAZ. 16:22-23 & PL IV, FIGS 15-18 /S. DISPIROIDES
AMERO,CETERI,HYALO,AMPUL,SETAE, 31D, ON WOOD. •

SINGERA BATISTA & BEZERRA 1960. PUBL. INST. MIC. UNIV. RECIFE 298:5 /S. ESENBECKIAE
=ORAMASIA URRIES 1956 FIDE SUTTON 1977 (CMI PAPERS 141:191). •

SIRODESMIUM DE NOTARIS 1849. MEM. ACCAD. TORINO, II, 10:348-350 & ICON. X /S. GRANULOSUM =C. STATE OF HYSTERIUM INSIDENS SCHWEINITZ FIDE HUGHES 1958
=CONIOSPORIUM LINK 1809 FIDE HUGHES 1958 •

SIRODOCHIELLA HOHNEL 1925. MITT. BOT. TECHN. HOCHSCH. WIEN 2:67-68 (NO ILLUS) /S. RHODELLA
?, NOTE ALSO SIRODOCHIELLA HOHNEL 1919 (BER. DEUTSCH. BOT. GES. 37:153, NO ILLUS,NOMEN NUDUM) •

SIROSPORIUM BUBAK & SEREBRIANIKOW 1912. IN BUBAK, HEDWIGIA 52:272 /S. ANTENNIFORME (BERK. & CURTIS)
DICTYO OR PHRAGMO,CETERI,PHAEO,RADUL,STROMA, 99E, ON LEAVES (CELTIS, MORUS, CARYA & OTHERS).
/ELLIS, CMI PAPERS 87:2-11,1961 (ILLUS) /ELLIS, DH P 288-290,1971 (ILLUS,KEY)
/ELLIS, MDH P 299-303,1976 (ILLUS,KEY,NEW COMBS)
=*HELICOCERAS* LINDER 1931 FIDE ELLIS 1963 •

SOLHEIMIA MORRIS 1967. MYCOPATHOLOGIA 33:181-182 (ILLUS) /S. COSTASPORA
AMERO,GLOIO,PHAEO,PHIAL,SYNNEMATA, 9F, ON LEAVES (PALM). •

SOLOSYMPODIELLA MATSUSHIMA 1971. MICROFUNGI P 55 & FIG 64:1-3 & PL 33-6 /S. CLAVATA
AMERO OR DIDYMO,CETERI,HYALO,RACHI OR RADUL, 23C, ON SOIL & LITTER. COMPARE WITH RHINOCLADIELLA.
/MATSUSHIMA, ICONES P 133-134,1975 (ILLUS,NEW SPP) •

SOPAGRAHA SUBRAMANIAN & SUDHA 1979. CAN. J. BOT. 57:34-39 (ILLUS) /S. SIBIKA
BULBILS OR STAURO,CETERI,PHAEO,NONSP, 114F, ON LITTER (GLYCOSMIS). OTHER CONIDIAL STATE ?=SELENOSPORELLA. •

SOREDOSPORA CORDA 1837. ICONES FUNGORUM 1:12 TAB 3 FIG 173 /S. GRAMINIS =FUMAGO GRAMINIS (CORDA) HUGHES 1958
?, AFTER EXAMINING THE TYPE SPECIMEN, HUGHES TRANSFERRED S. GRAMINIS TO FUMAGO, BUT FUMAGO IS NOW ACCEPTED AS A NOMEN CONFUSUM. •

SOREYMATOSPORIUM CAMARA 1930 (ACAD. SCI. LISBON LA CLASS. P 18-20)
=STEMPHYLIUM FIDE WILTSHIRE 1938 (TBMS 21:223) •

SOROCYBE FRIES 1849. SUMMA VEGET. SCAND. 2:468 /RACODIUM RESINAE FRIES 1815, NON MYXOTRICHUM RESINAE FRIES 1832 FIDE HUGHES 1958 (CAN. J. BOT. 36:731) =S. RESINAE (FRIES)
AMERO,BLASTO,PHAEO,NONSP,SYNNEMATA, 39G, ON SOIL, CREOSOTED WOOD, RESIN

& PETROLEUM DERIVATIVES & ON PLANTS (AZALEA, RHODODENDRONS). ASCO STATE =AMORPHOTHECA PARBERY SEE PARBERY 1969. COMPARE WITH ALYSIDIUM. OTHER CONIDIAL STATE =HORMOCONIS VON ARX & DE VRIES 1973. /LINDAU, IN RABENHORST KRYPT. FLORA 8:699-700,1907 & 9:379-380,1910 (ILLUS) /PARBERY, AUSTR. J. BOT. 17:331-357,1969 (ILLUS,ECOL) /HUGHES, CAN. J. BOT. 36:805,1958 (REVIS,REFS) /ELLIS, DH P 308-310,1971 (ILLUS SUB CLADOSPORIUM STATE OF AMORPHOTHECA RESINAE) /VON ARX, VERH. KON. NEDERL. AKAD. WET. AFD. NAT. 61:61-62 & FIG 1D,1973 (ILLUS) /ELLIS, MDH P 154-156,1976 (ILLUS SUB PYCNOSTYSANUS AZALEAE)
=*PYCNOSTYSANUS* LINDAU 1904 SEE HUGHES 1958 •

SOROSPORELLA SOROKIN 1888. ZENTRALBL. BAKT. ABT. 2, 4:644-647 & TAF. 4, FIGS 6-18 /S. AGROTIDIS
?, YEAST-LIKE(?) INSECT PARASITE •

SOROSPORIELLA ZAMBETTAKIS 1970 (REV. MYCOL. 35:173) IS A NOMEN NUDUM FOR SOROSPORELLA LAVROV 1936 NON SOROKIN 1888. •

SPADICOIDES HUGHES 1958. CAN. J. BOT. 36:805-806 /S. BINA (CORDA)
AMERO OR DIDYMO OR PHRAGMO,CETERI,PHAEO,RADUL, 64C, ON WOOD & BARK (MANY TREES). COMPARE WITH HELMINTHOSPORIUM. OTHER STATE =PHIALOPHORA (H) SEE WANG IN KENDRICK 1979.
/ELLIS, CMI PAPERS 93:6-14,1963 (ILLUS,KEY) /ELLIS, DH P 393-395,1971 (ILLUS,KEY) /HUGHES, FUNGI CANADENSES NOS. 4-9,1973 (ILLUS) /MATSUSHIMA, ICONES P 134-135 & PL,1975 (ILLUS) /ELLIS, MDH P 388-389,1976 (ILLUS) /WANG, MEM. N.Y. BOT. GARD. 28:218-224,1976 (ILLUS,KEY,NEW SP,NOT SEEN) /HUGHES, NEW ZEALAND J. BOT. 16:346-347,1978 (ILLUS) /SUTTON & HODGES, NOVA HEDWIGIA 129:604-607,1978 (ILLUS,NEW SP) •

SPEGAZZINIA SACCARDO 1880. MICHELIA 2:37 & 172 /S. ORNATA =S. TESSARTHRA (BERKELEY & CURTIS) SACCARDO 1886 FIDE HUGHES 1953
STAURO OR DICTYO,CETERI,PHAEO,MISC,APPEND, 114B, ON PLANTS (MANY KINDS) & SOIL.
/HUGHES, CMI PAPERS 50:62-66,1953 (ILLUS) /ELLIS, DH P 565-567,1971 (ILLUS,KEY) /SIVASITHAMPARAM, TBMS 62:427-429,1974 (ILLUS,NEW SP) /ELLIS, MDH P 474-477,1976 (ILLUS,KEY,NEW SP) /COLE & SAMSON, CONIDIAL FUNGI P 96-97,1979 (ILLUS,CONIDIUM ONTOGENY) •

SPEIRA CORDA 1837. ICONES FUNGORUM 1:9 TAB 2 FIG 140 /S. TORULOIDES
=DICTYOSPORIUM CORDA 1836 FIDE HUGHES 1958 •

SPEIROPSIS TUBAKI 1958. J. HATTORI BOT. LAB. 20:171-173 (ILLUS) /S. PEDATOSPORA
STAURO,CETERI,HYALO OR PHAEO,RADUL,AQUATIC, 116D, ON LITTER & DUNG.
/INGOLD, GUIDE P 48-49,1975 (ILLUS) /MATSUSHIMA, ICONES P 135 & PL 231,1975 (ILLUS) /ELLIS, MDH P 363-364,1976 (ILLUS)
?=*MULTICLADIUM* DESHPANDE & DESHPANDE 1966 •

SPELAEOMYCES FRESENIUS 1863 (BEITR. MYKOL. P 106-108) ?=XYLOSTROMA FIDE AINSWORTH 1971 •

SPERMATOLONCHA SPEGAZZINI 1909. AN. MUS. NAC. HIST. NAT. BUENOS AIRES 17(SER 3,10):139 (ILLUS) /S. MATICOLA
STAURO,CETERI,HYALO,NONSP,STROMA, 125D, ON FUNGI (MELIOLA).
/DEIGHTON & PIROZYNSKI, CMI PAPERS 128:98-100,1972 (ILLUS) •

SPERMODERMIA TODE 1790. FUNGI MECKLENB. 1:1 /S. CLANDESTINA
=A YOUNG HYPOXYLON FIDE HOHNEL FROM AINSWORTH 1971 •

SPERMOEDIA FRIES 1822 (SYST. MYCOL. 2:268-269) =SCLEROTIUM SEE AINSWORTH 1971 •

SPERMOSPORA SPRAGUE 1948. MYCOLOGIA 40:177-180 (ILLUS) /S. SUBULATA (SPRAGUE)
PHRAGMO,GLOIO OR CETERI,HYALO,NONSP?,APPEND, 70G, ON PLANTS (FESTUCA, AVENA, AMMOPHILA).
/DEIGHTON, TBMS 51:41-49,1968 (ILLUS,REVIS)
?=*SPERMOSPORELLA* DEIGHTON 1969

?=*IRPICOMYCES* DEIGHTON 1969 •

SPERMOSPORELLA DEIGHTON 1969. CMI PAPERS 118:21-22 (ILLUS) /S. AGGREGATA
PHRAGMO,CETERI,HYALO,RADUL,APPEND, 71D, ON FUNGI (MELIOLA). COMPARE WITH SPERMOSPORA. •

SPHACELIA LEVEILLE 1827. MEM. SOC. LINN., PARIS 5:578-579 (NO ILLUS) /S. SEGETUM
AMERO,GLOIO,HYALO,PHIAL,SPORODOCHIA, 5B, ON PLANTS (GRASSES). ASCO STATE =CLAVICEPS OR EPICHLOE FIDE VON ARX 1974 (GENERA P 125). COMPARE WITH EPHELIS.
/LOVELESS, TBMS 47:205-213,1964 (ILLUS,ECOL)
?=*CERATONEMA* PERSOON 1822 •

SPHACELIOPSIS SPEGAZZINI 1910. AN. MUS. NAC. BUENOS AIRES 20:457 /S. CYPERICOLA
=EPHELIS FRIES 1849 FIDE DIEHL FROM SUTTON 1977 (CMI PAPERS 141:195). •

SPHAERIDIUM FRESENIUS 1852. MYKOL. HEFT II P 46-47 & TAF 5 FIG 31-36 /S. VITELLINUM
AMERO,BLASTO,HYALO,MISC,SYNNEMATA, 38D, ON LITTER. NOT VERY DIFFERENT FROM FUSIDIUM LINK.
/BARRON, HYPHOMYCETES P 282-283,1968 (ILLUS) •

SPHAEROCOLLA KARSTEN 1892. HEDWIGIA 31:294 /S. AURANTIACA
?=HORMOMYCES BONORDEN 1851 FIDE DONK 1962 (TAXON 11:99). •

SPHAEROCYBE MAGROU 1945. C. R. ACAD. SCI. PARIS 220:220-222 (NO LATIN DIAGNOSIS) /S. CONCENTRICA
=NOMEN NUDUM, NOT ILLUSTRATED BY AUTHOR. •

SPHAEROMYCES ARNAUD 1954. BULL. SOC. MYCOL. FR. 69:279-280 (ILLUS,NO LATIN DIAGNOSIS) /S. CLAVISPORUS
AMERO,GLOIO,HYALO,PHIAL, 3G, ON LITTER (BROAD-LEAF TREES). NICOT & CHARPENTIE 1971 (BULL. SOC. MYCOL. FRANCE 87:35) LIST TYPE SPECIMEN. NON SPHAEROMYCES MONTAGNE 1845.
/MATSUSHIMA, ICONES P 28 & PL 273:5-7,1975 (ILLUS SUB CHLORIDIUM RENIFORME) •

SPHAEROMYCES MONTAGNE 1845. ANN. SCI. NAT., SER 3, 4:365 (NO ILLUS) /S. ALGERIENSIS DURIEU & MONT. 1845
?=ASPERGILLUS FIDE ? •

SPHAEROMYCETELLA ARNAUD 1954. BULL. SOC. MYCOL. FR. 69:279 & 280 (ILLUS,NO LATIN DIAGNOSIS) /S. LEUCOCEPHALA
?, DOES NOT APPEAR TO DIFFER FROM CHLORIDIUM LINK 1809. SEE ALSO PIMINELLA. NICOT & CHARPENTIE 1971 (BULL. SOC. MYCOL. FRANCE 87:35) LIST TYPE SPECIMEN. •

SPHAEROSPORIUM SCHWEINITZ 1832. TRANS. AMER. PHIL. SOC. II, 4:303 /S. LIGNATILE
AMERO,ARTHRO OR BLASTO,HYALO,NONSP,SPORODOCHIA, 40D, ON WOOD.
/DAMON & DOWNING, MYCOLOGIA 46:214-216,1954 (ILLUS) •

SPHAERULOMYCES MARVANOVA 1977. TBMS 68:485-488 (ILLUS) /S. CORALLOIDES
AMERO,BLASTO? OR CETERI,HYALO,MISC?,AQUATIC, 36G, ON FUNGI (AQUATIC HYPHOMYCETES). •

SPHINCTROSPORIUM KUNZE IN REICHENBACH 1828 (CONSPECT. REGNI VEG. P 10)
=CLADOTRICHUM FIDE SACCARDO 1886 (SYLL. FUNG. 4:374) •

SPHONDYLOCEPHALUM STALPERS 1974. PROC. KON. NED. AKAD. WET., SER. C, 77:399-400 (ILLUS) /OEDOCEPHALUM VERTICILLATUM THAXTER 1891 =S. VERTICILLATUM (THAXTER)
AMERO,CETERI,HYALO,AMPUL, 31A, ON DUNG (NEWT). COMPARE WITH OEDOCEPHALUM. •

SPICARIA HARTING 1846. NIEUWE VERH. KON. INST. WETENSCH. AMSTERDAM 12:203-297 /S. SOLANI

=NOMEN CONFUSUM SEE BROWN & SMITH 1957 (TBMS 40:22-24). SPICARIA HARZ 1871 WAS AN ILLEGITIMATE REVISION WHICH EXCLUDED HARTING'S TYPE SPECIES. MOST FUNGI IDENTIFIED AS SPICARIA ARE SPECIES OF PAECILOMYCES. •

SPICARIOPSIS HEIM 1939. IN HEIM & BOURIQUET, REV. PATH. VEG. 26:25-29 (ILLUS) /S. TROPICALE
?=PAECILOMYCES BAINIER 1907, SEE SAMSON 1974 (CBS STUDIES 6:10). •

SPICELLUM NICOT & ROQUEBERT 1976. REV. MYCOL. (PARIS) 39:269-272 (DATED 1975,ILLUS) /S. ROSEUM
?=RAFFAELEA VON ARX & HENNEBERT 1965 •

SPICULARIA PERSOON 1822. MYC. EUR. 1:39 /?, SEE HUGHES 1958 (CAN. J. BOT. 36:806)
=NOMEN DUBIUM SEE HUGHES 1958. •

SPICULOSTILBELLA MORRIS 1963. AMER. MIDL. NATURALIST 69:101-103 (ILLUS) /S. DENDRITICA
?=PHAEOISARIA HOHNEL 1909. NO TYPE MATERIAL AVAILABLE, FIDE DE HOOG & PAPENDORF 1976 (PERSOONIA 8:408). •

SPILOCAEA FRIES EX FRIES 1821. SYST. MYCOL. 1:XXXIX (NOV. FL. SUECIAE 5:79,1819) /S. POMI FRIES 1825 (SYST. ORB. VEG. P 198) =S. STATE OF VENTURIA INAEQUALIS (COOKE) WINTER APUD THUMEN 1875
AMERO OR DIDYMO,CETERI,PHAEO,ANNEL,SPORODOCHIA, 18C, ON LEAVES (MALUS, PYRUS, OLEA & OTHERS). ASCO STATE =VENTURIA. SEE ALSO POLLACCIA.
/HUGHES, CAN. J. BOT. 31:560-565,1953 (ILLUS) /DANCE, CAN. J. BOT. 39:875-890,1971 (ILLUS,REVIS,SUB POLLACCIA) /ELLIS, DH P 142-143,1971 (ILLUS) /HUGHES & PIROZYNSKI, CAN. J. BOT. 50:2529-2531,1972 (ILLUS, NEW COMB) /CORLETT, FUNGI CANADENSES NO. 35,1974 (ILLUS) /CORLETT, CHONG & KOKKO, CAN. J. BOT. 22:1144-1152,1976 (ILLUS,CONIDIUM ONTOGENY) /ELLIS, MDH P 111-114,1976 (ILLUS,KEY,NEW COMBS,NEW SP)
=*CYCLOCONIUM* CASTAGNE 1845 FIDE HUGHES 1958
=*NAPICLADIUM* THUMEN 1875 FIDE HUGHES 1958
=*BASIASCUM* CAVARA 1888 FIDE HUGHES 1958
?=*POLLACCIA* BALDACCI & CIFERRI 1937 •

SPILODOCHIUM SYDOW 1927. ANN. MYCOL. 25:158-159 /S. VERNONIAE
AMERO OR DIDYMO,BLASTO,PHAEO,NONSP,SPORODOCHIA, 66F, ON LEAVES. COMPARE WITH HADRONEMA.
/ELLIS, DH P 94-95,1971 (ILLUS) /ELLIS, MDH P 63,1976 (ILLUS) •

SPILOMIUM NYLANDER 1856. PROD. LICH. GALL. P 91 /S. SILICEUM (FEE)
?, NOT ILLUSTRATED BY NYLANDER •

SPINIGER STALPERS 1974. PROC. KON. NED. AKAD. WET., SER. C, 77:402-407 (ILLUS,KEY) /S. MEINECKELLUS (OLSON) =S. STATE OF HETEROBASIDION ANNOSUM (FRIES) BREFELD FIDE STALPERS
AMERO,CETERI,HYALO,AMPUL,CLAMPS, 35B-C, ON WOOD. COMPARE WITH OEDOCEPHALUM & OLPITRICHUM. BASIDIO STATE =VARARIA, LAURILIA, BONDARZEWIA, CORTICIUM, ASTEROSTROMELLA, TRAMETES & HETEROBASIDION FIDE STALPERS 1973, & PENIOPHORA SEE MCKEEN 1952 (CAN. J. BOT. 30:764-787) & RESINICIUM FIDE PARMASTO 1968 (CONSPECT. SYST. CORTICIACEARUM, TARTU) & STEREUM SEE MAXWELL 1954 (CAN. J. BOT. 32:259-280, ILLUS) & HYPHODERMA FIDE VON ARX 1974 (GENERA P 179), & DICHOSTEREUM SEE BOIDIN & LANQUETIN 1977 (MYCOTAXON 6:287,ILLUS) & FOMITOPSIS SEE TUBAKI 1958 (BULL. HATTORI BOT. LAB. 20:217-218,ILLUS).
/LYMAN, PROC. BOSTON SOC. NAT. HIST. 33:125-209, PL 21-22,1907 (ILLUS SUB CORTICIUM) /STALPERS, PROC. KON. NED. AKAD. WET., SER. C, 77:383-401,1974 (ILLUS) /STALPERS, CBS STUDIES 16:40, 108-111 & 179-180,1978 (KEY) •

SPINULOSPORA DEIGHTON 1973. TBMS 61:195-196 (ILLUS) /S. PUCCINIIPHILA
BULBILS,CETERI,HYALO,NONSP,CLAMPS?,SPORODOCHIA, 129F, ON FUNGI (PUCCINIA & SCOLECOBASIDIUM ON SMILACIS). COMPARE WITH AEGERITA. •

SPIRALIA GRIGORAKIS 1924. COMP. REND. 179:1424 /NO SPECIES MENTIONED IN 1924. TWO SPECIES INCLUDED IN ANN. SCI. NAT. BOT. 7:415,1925.

=NOMEN ILLEGITIMUM (FOR TRICHOPHYTON PRO PARTE). •

SPIRALOTRICHUM YATES 1918. PHILIPP. J. SCI. (BOT.) 13:383-384 /S. PIPERIS
=SPIRALUM MULDER PRO PARTE, TYPE SPECIMEN CONTAINED TWO ELEMENTS, FIDE MULDER 1975 (TBMS 65:517-521). •

SPIRALUM MULDER 1975. TBMS 65:517-521 (ILLUS) /S. HELICOSPORUM
PHRAGMO OR HELICO,CETERI,HYALO OR PHAEO,RADUL, 106F, ON LEAVES (PIPER, FICUS). ?=PSEUDOCERCOSPORA SPEGAZZINI 1910.
=*SPIRALOTRICHUM* YATES PRO PARTE, TYPE SPECIMEN CONTAINED TWO ELEMENTS •

SPIROPES CIFERRI 1955. SYDOWIA 9:301-303 /S. GUAREICOLA (STEVENS)
PHRAGMO,CETERI,PHAEO,RACHI OR RADUL,SYNNEMATA, 83D, ON FUNGI (ASTERIDIELLA, IRENOPSIS, MELIOLA & OTHERS) ON LEAVES.
/ELLIS, CMI PAPERS 114:1-42,1968 (ILLUS,MONOGR) /ELLIS, DH P 249-256,1971 (ILLUS,KEY) /ELLIS, MDH P 212-214,1976 (ILLUS) /HUGHES & COOK, FUNGI CANADENSES 161,1980 (ILLUS)
?= *ANNELLOPHRAGMIA* SUBRAMANIAN 1963
= *NASCIMENTOA* CIFERRI & BATISTA 1956 FIDE ELLIS 1968
?= *POROISIARIOPSIS* MORELET 1971
?= *PSEUDOSPIROPES* ELLIS 1971 •

SPIROSPHAERA VAN BEVERWIJK 1953. TBMS 36:120-122 (ILLUS) /S. FLORIFORMIS (AS 'FLORIFORME')
HELICO OR BULBILS,CETERI,HYALO OR PHAEO,NONSP,AQUATIC, 106A, COMPARE WITH STRUMELLA SACCARDO.
/HENNEBERT, TBMS 51:13-24,1968 (ILLUS,REVIS,NEW SP) •

SPIROSPORA MANGIN & VINCENS 1920. BULL. SOC. MYCOL. FR. 36:96 /S. CASTANEAE
=A. MIRABILIS BERKELEY & BROOME
=ACROSPEIRA BERKELEY & BROOME 1857 FIDE HUGHES 1958 •

SPIROTRICHUM SAITO EX VAN BEYMA 1940 =TRITIRACHIUM FIDE DE HOOG 1972 (CBS STUDIES 1:19). •

SPONDYLOCLADIELLA LINDER 1934. MYCOLOGIA 26:437 & PL 50 FIG 1-3 /S. BOTRYTIOIDES
PHRAGMO,CETERI,PHAEO,MISC, 84C, ON FUNGI (CORTICIUM).
/ELLIS, DH P 386-387,1971 (ILLUS) •

SPONDYLOCLADIOPSIS ELLIS 1963. CMI PAPERS 87:15-17 (ILLUS) /S. CUPULICOLA
PHRAGMO,CETERI,HYALO,RADUL,SETAE, 80F, ON LITTER (FAGUS).
/ELLIS, DH P 293-295,1971 (ILLUS) •

SPONDYLOCLADIUM MARTIUS 1817. FLORA CRYPT. ERLANG. P 354-355 /S. FUMOSUM
=STACHYLIDIUM VERTICILLATUM (PERS.) HUGHES
=STACHYLIDIUM LINK 1809 FIDE HUGHES 1958 •

SPORENDOCLADIA ARNAUD EX NAG RAJ & KENDRICK 1975. CHALARA P 39 & 162-163 (ILLUS,LATIN DIAGNOSIS) /S. CASTANEAE ARNAUD 1954 =PHIALOCEPHALA FUMOSA (ELL. & EV.) SUTTON 1975
=PHIALOCEPHALA KENDRICK FIDE SUTTON 1975 (TBMS 64:411-414)
/ARNAUD, BULL. SOC. MYCOL. FR. 69:277 & 279,1954 (ILLUS) •

SPORENDONEMA DESMAZIERES 1827. ANN. SCI. NAT. 11:246-249 & PL XXI /S. CASEI
AMERO,ARTHRO,HYALO,NONSP,ALEURI, 41E, ON CHEESE & FUNGI (AGARICUS). ASCO STATE ?=ONYGENA SEE TUBAKI 1960 (BULL. NAT. SCI. MUS. TOKYO 46:36-43) OR SAMSON & VAN DER AA 1973 (COOLIA 16:30-35) OR SIGLER & CARMICHAEL 1976 (MYCOTAXON 4:378 & 413). BASIDIO STATE ?=COLLYBIA SEE SIGLER & CARMICHAEL 1976 (P 368-372) OR ?CONIOPHORA OR ?PUNCTULARIA SEE STALPERS 1978 (CBS STUDIES 16:39).
/VON ARX, GENERA P 166-167,1970 (ILLUS) /SIGLER & CARMICHAEL, MYCOTAXON 4:404-410,1976 (ILLUS) /COLE & SAMSON, CONIDIAL FUNGI P 117-119,1979 (ILLUS,CONIDIUM ONTOGENY)
=*COPROTRICHUM* BONORDEN 1851, BASED ON MASON & HUGHES 1953 •

SPORHELMINTHIUM SPEGAZZINI 1918. PHYSIS 4:292 /S. ANOMALUM
=CLASTEROSPORIUM ANOMALUM (SPEG.) HUGHES 1958
=CLASTEROSPORIUM SCHWEINITZ 1832 FIDE HUGHES 1958 •

SPORIDESMIUM LINK 1809. MAG. GES. NATURF. FREUNDE, BERLIN 3:41 /S. ATRUM
PHRAGMO OR SCOLECO,CETERI,PHAEO,ANNEL,STROMA, 75C-D, ON WOOD & FUNGI.
ASCO STATE =EUPELTE FIDE ELLIS 1958 OR AKAROPELTELLA FIDE FARR 1972 OR ?PLACOSOMA SEE HUGHES 1978. OTHER STATE =SELENOSPORELLA (H).
/ELLIS, CMI PAPERS 70:16-84,1958 (ILLUS,MONOGR) /ELLIS, DH P 116-122,1971 (ILLUS) /HUGHES & ILLMAN, FUNGI CANADENSES NOS. 57-60,1974 (ILLUS) /MATSUSHIMA, ICONES P 135-142 & PL,1975 (ILLUS,NEW SPP) /ELLIS, MDH P 76-98,1976 (ILLUS,KEY,NEW COMBS,NEW SP) /MORGAN-JONES, MYCOTAXON 6:356-358,1977 (ILLUS,NEW SP) /HUGHES, NEW ZEALAND J. BOT. 16:347-350 & 363-365,1978 (ILLUS,DISCUSSION)
?=*CERCOSPERMA* ARNAUD 1954
?=*LOMAANTHA* SUBRAMANIAN 1954
=*PODOCONIS* BOEDIJN 1933 FIDE HUGHES 1958
=*POLYDESMUS* DURIEU & MONTAGNE 1845 SEE ELLIS 1958
=*SPORODESMIUM* -ORTHOGRAPHIC VARIANT •

SPORIDIOBOLUS NYLAND 1949. MYCOLOGIA 41:686-701 (ILLUS) /S. JOHNSONII
?, HAS CLAMP CONNECTIONS. NOT VERY DIFFERENT FROM TILLETIOPSIS & ITERSONILIA.
/SOWELL & KORF, MYCOLOGIA 52:934-945,1960 (REVIS) •

SPOROCEPHALIUM CHEVALLIER 1826. FLORE GEN. ENV. PARIS 1:59-60 /S. CAPITATUM (LINK) (LECTOTYPE) =ACLADIUM STATE OF BOTRYOBASIDIUM CANDICANS
=ACLADIUM LINK 1809 FIDE HUGHES 1958 •

SPOROCEPHALUM ARNAUD 1952. BULL. SOC. MYCOL. FR. 68:187-189 (ILLUS,NO LATIN DIAGNOSIS) /S. PENIOPHORAE
AMERO,CETERI,HYALO,AMPUL?, 31E, ON FUNGI (PENIOPHORA). NICOT & CHARPENTIE 1971 (BULL. SOC. MYCOL. FRANCE 87:36) LIST TYPE SPECIMEN. •

SPOROCLADIUM CHEVALLIER 1826. FLORE GEN. ENV. PARIS 1:36 & P 647 /CLADOSPORIUM HERBARUM (PERSOON) LINK (LECTOTYPE FIDE HUGHES 1958)
=CLADOSPORIUM LINK 1815 FIDE HUGHES 1958 •

SPOROCLEMA TIESSENHAUSEN 1912. ARCH. HYDR. PLANKT. 7:302-303 (ILLUS) /S. PIRIFORME
?, PERHAPS AN IMMATURE OOMYCETE. •

SPOROCYBE FRIES 1825. SYST. ORB. VEG. 1:170 /PERICONIA BYSSOIDES PERSOON 1801 (LECTOTYPE)
=PERICONIA PERSOON 1801 FIDE HUGHES 1958 •

SPOROCYSTIS MORGAN 1902. JOUR. MYCOL. 8:169 /S. CONDITA
?, NOT ILLUSTRATED BY AUTHOR. •

SPORODERMA MONTAGNE 1856. SYLLOGE CRYPT. P 291, 1856 & IN SACCARDO, SYLLOGE FUNG. 4:676,1886 /S. CHLOROGENUM
=TRICHODERMA PERSOON FIDE HOHNEL 1910 FROM RIFAI 1969 •

SPORODESMIUM -ORTHOGRAPHIC VARIANT OF SPORIDESMIUM •

SPORODINIOPSIS HOHNEL 1903. ANN. MYCOL. 1:528-529 /S. DICHOTOMA (AS 'DICHOTOMUS')
=NOMEN DUBIUM FIDE HAWKSWORTH 1971 (CMI PAPERS 126:26) •

SPORODUM CORDA 1837. ICONES FUNGORUM 1:18 TAB 4 FIG 247 /S. CONOPLEOIDES
=PERICONIA HISPIDULA (PERSOON) MASON & ELLIS
=PERICONIA PERSOON 1801 FIDE HUGHES 1958 •

SPOROGLENA SACCARDO 1894. IN HENNINGS, ENGLER'S BOT. JB. 18 BAND, 3:40 (NO ILLUS) /S. VELUTINA
? •

SPOROPHIALA RAO 1970. MYCOPATHOLOGIA 41:315-320 (ILLUS) /S. PROLIFICA
?=TAENIOLELLA HUGHES 1958. ON LITTER. •

SPOROPHLEUM NEES EX LINK 1824. IN LINN. SPEC. PLANT. IV, 6(1):45 /S. GRAMINEUM
=ARTHRINIUM SPOROPHLEUM KUNZE 1817
=ARTHRINIUM KUNZE 1817 FIDE HUGHES 1958 •

SPOROPHORA LUTERAAN 1952. REV. BIOL. GEN. THEOR. APPL. 2:1-12 & PLATE /S. TORULIFORMIS
?, NO LATIN DIAGNOSIS, PERHAPS YEAST STAGE OF USTILAGO. •

SPOROSCHISMA BERKELEY & BROOME 1847. IN BERKELEY, GARD. CHRON. P 540 /S. MIRABILE
PHRAGMO,BASO,PHAEO OR HYALO,PHIAL,SETAE,STROMA, 73C, ON WOOD, BARK & LITTER. ASCO STATE =MELANOCHAETA FIDE MULLER ET AL 1969 (REV. MYCOL. 33:369-378,ILLUS) OR CHAETOSPHAERIA SEE DE HOOG 1977 (CBS STUDIES 15:138).
/HUGHES, CMI PAPERS 31:1-22,1949 (ILLUS,KEY) /ELLIS, DH P 505-507,1971 (ILLUS,KEY) /NAG RAJ & KENDRICK, CHALARA P 39 & 157-161,1975 (KEY,ILLUS) /MATSUSHIMA, ICONES P 142 & PL 261,1975 (ILLUS) •

SPOROSCHISMOPSIS HOLUBOVA-JECHOVA & HENNEBERT 1972. BULL. JARD. BOT. NAT. BELGIUM 42:385-391 (ILLUS) /S. MORAVICA
PHRAGMO,BASO OR GLOIO,PHAEO,PHIAL,STROMA, 73D, ON WOOD.
/ELLIS, MDH P 437-438,1976 (ILLUS) /HUGHES, NEW ZEALAND J. BOT. 16:350-351,1978 (ILLUS,NEW SP) •

SPOROSTACHYS SACCARDO 1917. ATTI ACCAD. SCI. VEN.-TRENT.-ISTR. III, 10:92 /S. MAXIMA (LECTOTYPE) =MELANOGRAPHIUM SPINULOSUM (SPEGAZZINI) HUGHES
=MELANOGRAPHIUM SACCARDO 1913 FIDE HUGHES 1958 •

SPOROTHRIX HEKTOEN & PERKINS 1900. J. EXPER. MEDICINE 5:77 /S. SCHENCKII
AMERO,CETERI,HYALO OR PHAEO,RACHI, 23E, ON WOOD, INSECTS (BEETLES), SOIL, LITTER, MAN & OTHER SUBSTRATES. ASCO STATE =CERATOCYSTIS OR CERATOCYSTIOPSIS FIDE UPADHYAY & KENDRICK 1975 (MYCOLOGIA 67:798-805) OR VALSONECTRIA SEE LOEFFLER & MULLER 1961 (BERICHTE SCHWEIZ. BOT. GESELL. 71:409) OR OPHIOSTOMA FIDE DE HOOG 1974 OR PSEUDEUROTIUM SEE BOOTH 1961 (MYCOL PAPERS 83:8-11) OR PETRIELLA FIDE CARMICHAEL, OR FRAGOSPHAERIA SEE DE HOOG 1977 (CBS STUDIES 15:137) OR LEUCOSPHAERIA SEE VON ARX ET AL 1978 (PERSOONIA 10:141-143). OTHER CONIDIAL STATES =GRAPHIUM, PESOTUM, ?BEAUVERIPHORA, PACHNODIUM OR ZYGOSPORIUM.
/BARRON, HYPHOMYCETES P 284-285,1968 (ILLUS) /NICOT & MARIAT, MYCOPATHOLOGIA 49:53-65,1973 (ILLUS,REVIS) /DE HOOG, CBS STUDIES 7:1-84,1974 (MONOGR) /WEIJMAN & DE HOOG, ANTONIE VAN LEEUWENHOEK 41:353-360,1975 (PERFECT STATE) /MATSUSHIMA, ICONES P 143 & PL,1975 (ILLUS,NEW COMB & SUB CERATOCYSTIS PICEAE) /TAYLOR, MYCOLOGIA 69:651-653,1977 (ILLUS,NEW SP) /DE HOOG, PERSOONIA 10:62-67,1978 (ILLUS,NEW SPP,NEW COMBS) /COLE & SAMSON, CONIDIAL FUNGI P 39-40,1979 (ILLUS)
=*SPOROTRICHUM* AUCT. NON LINK FIDE CARMICHAEL 1962
=*SPOROTRICHOPSIS* DE BEURMANN & GOUGEROT 1911 FIDE CARMICHAEL 1973 •

SPOROTRICHELLA KARSTEN 1887. MEDDEL. SOC. F. F. FENNICA 14:96 /S. ROSEA
=FUSARIUM LINK 1809 FIDE HUGHES 1958 •

SPOROTRICHOPSIS GUEGUEN 1911. IN DE BEURMANN & GOUGEROT, ARCH. PARASIT. 15:103-105 (ILLUS) /"SPOROTRICHOPSIS? BEURMANNI"
=SPOROTHRIX HEKTOEN & PERKINS 1900 FIDE CARMICHAEL 1973 •

SPOROTRICHUM LINK 1809. MAG. GES. NATURF. FREUNDE, BERLIN 3:13 /S. AUREUM (LECTOTYPE, FIDE HUGHES 1958) =S. STATE OF POLYPORUS METAMORPHOSUS FIDE STALPERS 1978 (CBS STUDIES 16:181)
AMERO,CETERI,HYALO OR PHAEO,NONSP,ALEURI,CLAMPS, 52D, ON WOOD. BASIDIO STATE ?=CORTICIUM, COMPARE S. DIMORPHOSPORUM WITH BIGGS 1937 (MYCOLOGIA 29:686-706) OR POLYPORUS, LAETIPORUS, PHANEROCHAETE OR FISTULINA SEE STALPERS 1978 (CBS STUDIES 16:40 & PL 1). OTHER CONIDIAL STATE =? (SEE S. DIMORPHOSPORUM).
/VON ARX, GENERA P 163-164,1970 (ILLUS) /VON ARX, PERSOONIA 6:179-184,1971

(ILLUS,REVIS) /VON ARX, PERSOONIA 7:127-130,1973 (ILLUS,REVIS) /MATSUSHIMA, ICONES P 29 & PL 20,1975 (ILLUS SUB CHRYSOSPORIUM DIMORPHUM =SPOROTRICHUM DIMORPHOSPORUM VON ARX 1971 FIDE CARMICHAEL) /STALPERS, CBS STUDIES 16:169,1978 (DESCR S. DIMORPHOSPORUM)
=*OIDIUM* LINK 1809 FIDE HUGHES 1958, NON OIDIUM LINK 1824 NOM. CONS. •

STACHYBOTRYELLA ELLIS & BARTHOLOMEW 1902. JOUR. MYCOL. 8:177 /S. REPENS
?, NOT ILLUSTRATED BY AUTHORS. •

STACHYBOTRYNA TUBAKI & YOKOYAMA 1971. TRANS. MYCOL. SOC. JAPAN 12:18-20 (ILLUS) /S. COLUMARE
?, LOOKS LIKE RICLARETIA PEYRONEL 1916. •

STACHYBOTRYS CORDA 1837. ICONES FUNGORUM 1:21 /S. ATRA =S. CHARTARUM (EHRENBERG) HUGHES 1958
AMERO,BASO OR GLOIO,HYALO OR PHAEO,PHIAL,SYNNEMATA, 7C, ON LITTER, SOIL & PAPER. ASCO STATE =MELANOPSAMMA FIDE HUGHES 1958 OR CHAETOSPHAERIA FIDE MULLER IN KENDRICK 1971 (SUB GLIOBOTRYS).
/ELLIS, DH P 540-546,1971 (ILLUS,KEY) /CAMPBELL, MYCOLOGIA 67:760-769,1975 (ILLUS) /JONG & DAVIS, MYCOTAXON 3:409-485,1976 (MONOGRAPH) /MATSUSHIMA, ICONES P 94-95 & P 144-146 & PL, 1975 (ILLUS SUB MEMNONIELLA,NEW SP) /ELLIS, MDH P 463-464,1976 (ILLUS,PARTLY SUB MEMNONIELLA) /HUGHES, NEW ZEALAND J. BOT. 16:351-352,1978 (ILLUS)
=*MEMNONIELLA* HOHNEL 1923 FIDE SMITH 1962
=*GLIOBOTRYS* HOHNEL 1902 FIDE ELLIS 1971
=*HYALOBOTRYS* PIDOPLICHKO 1948 FIDE CARMICHAEL 1973
=*HYALOSTACHYBOTRYS* SRINIVASAN 1958 FIDE BARRON 1964
=*FUCKELINA* SACCARDO 1875 FIDE BOOTH 1957
=*SYNSPORIUM* PREUSS 1849 FIDE HUGHES 1958 •

STACHYLIDIUM LINK 1809. MAG. GES. NATURF. FREUNDE, BERLIN 3:15 /S. BICOLOR (LECTOTYPE FIDE HUGHES 1958) =S. VERTICILLATUM (PERSOON) HUGHES 1958
AMERO,GLOIO,PHAEO OR HYALO,PHIAL,STROMA, 4G, ON LITTER (MANY PLANTS). ASCO STATE =NECTRIOPSIS SEE DOI 1978 (MEM. NATN. SCI. MUSEUM, TOKYO 11:17-20).
/BARRON, HYPHOMYCETES P 286-289,1968 (ILLUS) /MATSUSHIMA, MICROFUNGI P 61 & FIG 134:1,1971 (ILLUS) /ELLIS, DH P 537-539,1971 (ILLUS) /HUGHES, NEW ZEALAND J. BOT. 16:353,1978 (ILLUS)
=*SPONDYLOCLADIUM* MARTIUS 1817 FIDE HUGHES 1958 •

STAGONOSTROMA DIEDICKE 1914. KRYPT. FL. MARK BRANDENB. 9:561-562 (NO ILLUS) /S. DULCAMARAE (PASSERINI) =STAGONOSPORA DULCAMARAE PASSERINI
=FUSARIUM FIDE SUTTON 1977 (CMI PAPERS 141:200-201) •

STAHELIELLA VAN EMDEN 1974. ACTA BOT. NEERL. 23:251-256 (ILLUS) /S. NODOSA
AMERO,ARTHRO,PHAEO,NONSP, 44A, ON SOIL. COMPARE WITH BOTRYOMONILIA.
/COLE & SAMSON, CONIDIAL FUNGI P 111-112,1979 (ILLUS) •

STAPHYLOTRICHUM MEYER & NICOT 1957. BULL. SOC. MYCOL. FR. 72:318-323 (ILLUS,DATED 1956) /S. COCCOSPORUM
AMERO,CETERI,HYALO,NONSP,ALEURI, 47F, ON SOIL. NOT VERY DIFFERENT FROM BOTRYOTRICHUM.
/BARRON, HYPHOMYCETES P 289-291,1968 (ILLUS) /MATSUSHIMA, MICROFUNGI P 61 & FIG & PL,1971 (ILLUS)
=*BOTRYDIELLA* BADURA 1963 SEE BARRON 1968 •

STARKEYOMYCES AGNIHOTHRUDU 1956. J. INDIAN BOT. SOC. 35:40 /S. KOORCHALOMOIDES
=MYROTHECIUM TODE EX LINK 1809 SEE TULLOCH 1972 •

STEIROCHAETE BRAUN & CASPARY 1890 (KRANKH. PFL. P 28) =COLLETOTRICHUM CORDA 1831 FIDE VON ARX 1970 (P 153) •

STELLIFERA LEMAN 1824 (DICT. SCI. NAT. 33:489) =ASTEROPHORA DITMAR 1809 FIDE DONK 1962 (TAXON 11:100-101) •

STEMMARIA PREUSS 1851. LINNAEA 24:137, ILLUS IN STURM'S DEUTSCHL. FL. 6, 35-6 & TAB 67,1862 /S. GLOBOSA

?, ORIGINAL ILLUSTRATION NOT DIAGNOSTIC. •

STEMPHYLIITES BABAYAN & TASLAKHYCHYAN 1973 (MIKOL. FITOPATOL.) -FOSSIL FUNGI IMPERFECTI •

STEMPHYLIOMMA SACCARDO & TRAVERSO 1911. IN SACCARDO, SYLL. FUNG. 20:886 /S. VALPARADISIACUM (SPEGAZZINI)
PHRAGMO,CETERI,PHAEO,NONSP,ALEURI?, 76H, ON LITTER (PUYA) & SOIL.
/ELLIS, DH P 82-83,1971 (ILLUS) /ELLIS, MDH P 46-47,1976 (ILLUS)
=*STEMPHYLIOPSIS* SPEGAZZINI 1910, NON SMITH 1901 •

STEMPHYLIOPSIS SMITH 1901. J. ROYAL MICROSCOP. SOC. 13:617 (614-618 & PL XIII, FIG 4) /S. HETEROSPORA
DICTYO,CETERI,HYALO,NONSP,ALEURI, 97A, ON SOIL, PLANTS, EGGS, DUNG & OTHER SUBSTRATES. OTHER STATES =CYLINDROCARPON (H), SEE BARRON & ONIONS 1966, VERTICILLIUM (H), ?PAECILOMYCES (H) & NEOCOSMOSPORA (A) SEE UDAGAWA 1963 (TRANS. MYCOL. SOC. JAPAN 4:121-125, FIGS 1,2 & 4).
/BARRON & ONIONS, CAN. J. BOT. 44:861-862,1966 (DESCR OF TYPE SPECIMEN)
?=*CONIODICTYUM* HARIOT & PATOUILLARD 1909
?=*DICTYOARTHRINOPSIS* BATISTA & CIFERRI 1957
?=*DIHETEROSPORA* KAMYSCHKO 1962
?=*MIURAEA* HARA 1948
?=*POCHONIA* BATISTA & FONSECA 1965 •

STEMPHYLIOPSIS SPEGAZZINI 1910. REV. FAC. AGRON. 6:193 /S. VALPARADISIACUM
=STEMPHYLIOMMA SACC. & TRAV. 1911, NON STEMPHYLIOPSIS SMITH 1901. •

STEMPHYLIUM WALLROTH 1833. FL. CRYPT. GERMANIAE P 300 /S. BOTRYOSUM =S. STATE OF PLEOSPORA HERBARUM (PERSOON) RABENHORST FIDE SIMMONS 1952
DICTYO,CETERI,PHAEO,ANNEL, 99C, ON LEAVES, LITTER, WOOD & OTHER SUBSTRATES. ASCO STATE =PLEOSPORA FIDE SIMMONS 1952 (MYCOLOGIA 44:332).
/SIMMONS, MYCOLOGIA 59:67-92,1967 (ILLUS) /CMI DESCRIPTIONS 150,1967 & 471-472,1975 (ILLUS) /ELLIS, DH P 165-168,1971 (ILLUS,KEY) /MATSUSHIMA, ICONES P 146-147 & PL,1975 (ILLUS) /ELLIS, MDH P 147-148,1976 (ILLUS) /REIFSCHNEIDER, CAN. J. BOT. 57:1484-1485,1979 (ILLUS,NEW SP)
=*FUSICLADIOPSIS* MAIRE 1906 FIDE HUGHES 1958
=*EPOCHNIELLA* SACCARDO 1880 FIDE LINDAU IN RABENHORST 1910
=*SCUTISPORIUM* PREUSS 1851 FIDE LINDAU IN RABENHORST 1910
=*SOREYMATOSPORIUM* CAMARA 1930 FIDE AINSWORTH 1971
=*THYRODOCHIUM* WERDERMANN 1924 FIDE WILTSHIRE FROM AINSWORTH 1971
=*THYROSPORA* TEHON & DANIELS 1925 FIDE SMITH FROM AINSWORTH 1971 •

STENELLA SYDOW 1930. ANN. MYCOL. 28:205-206 /S. ARAGUATA
PHRAGMO,BLASTO,PHAEO,RADUL,STROMA, 82A, ON LEAVES. CLOSE TO MYCOVELLOSIELLA & CLADOSPORIUM SEE VON ARX 1973 (VERH. KON. NEDERL. AKAD. WET. AFD. NAT. 61:62).
/DEIGHTON, TBMS 56:411-418,1971 (ILLUS) /ELLIS, DH P 305-306,1971 (ILLUS) /MULDER, TBMS 65:514-517,1975 (ILLUS,NEW SP,NEW COMBS) /MATSUSHIMA, ICONES P 147 & PL,1975 (ILLUS) /ELLIS, MDH P 307-314,1976 (ILLUS,NEW COMB) /MCGINNIS & PADHYE, MYCOTAXON 7:415-418,1978 (ILLUS) /DEIGHTON, CMI PAPERS 144:52-54,1979 (NEW COMBS)
=*BIHARIA* THIRUMALACHAR & MISHRA 1953 FIDE ELLIS 1971 & DEIGHTON 1979 •

STENELLOPSIS HUGUENIN 1966. BULL. SOC. MYCOL. FR. 81:693-696 (ILLUS,DATED 1965) /S. FAGRAEAE
PHRAGMO,CETERI,PHAEO,RACHI,STROMA, 85C, ON LEAVES (FAGRAEA).
/ELLIS, DH P 267-268,1971 (ILLUS) •

STENOSPORA DEIGHTON 1969. CMI PAPERS 118:22-24 (ILLUS) /S. UREDINICOLA
?, NOT VERY DIFFERENT FROM CERCOSPORA. •

STEPHANOMA WALLROTH 1833. FLORA CRYPT. GERMAN. 2:269 /S. STRIGOSUM
AMERO OR STAURO,CETERI,HYALO,NONSP,ALEURI,APPEND, 113C, ON FUNGI (DISCOMYCETES, FUSARIUM & OTHERS). ASCO STATE = HYPOMYCES FIDE BARRON 1968. OTHER CONIDIAL STATE ="PHIALOCONIDIA" FIDE BARRON 1968 (HYPHOMYCETES P 292-293,ILLUS).

/BUTLER & MCCAIN, MYCOLOGIA 60:955-959,1968 (ILLUS)
=*ASTEROTHECIUM* WALLROTH 1836 FIDE LINDAU 1907
=*SYNTHETOSPORA* MORGAN 1892 FIDE CLEMENTS & SHEAR 1931 •

STEPHANOSPORIUM DAL VESCO 1961. ALLIONIA 7:182 /S. ATRUM =S. CEREALE (THUMEN) SWART 1965
AMERO,ARTHRO,PHAEO,NONSP,ALEURI, 43B, ON LITTER (BAMBUSA, SECALE), SOIL & PAPER.
/ELLIS, DH P 35-36,1971 (ILLUS) •

STERIGMATOBOTRYS OUDEMANS 1886. NEDERL. KRUIDK. ARCH. II, 4:548 /S. ELATA (SACCARDO) (LECTOTYPE) =S. MACROCARPA (CORDA) HUGHES 1958
PHRAGMO,GLOIO,PHAEO OR HYALO,RACHI,STROMA, 84D, ON WOOD (ABIES, PICEA, TAXUS & OTHERS).
/HUGHES, CAN. J. BOT. 36:814,1958 (REFS) /JONG & DAVIS, NORWEGIAN J. BOT. 18:177-181,1971 (ILLUS) /ELLIS, DH P 368-369,1971 (ILLUS) /MATSUSHIMA, ICONES P 148-149 & PL,1975 (ILLUS) /HUGHES, NEW ZEALAND J. BOT. 16:352-353,1978 (ILLUS)
=*ATRACTINA* HOHNEL 1904 FIDE HUGHES 1958
=*PHRAGMOSTACHYS* COSTANTIN 1888 FIDE HUGHES 1958
?=*GLIODENDRON* SALONEN & RUOKOLA 1969 FIDE JONG & DAVIS 1971 •

STERIGMATOCYSTIS CRAMER 1859. VIERT. NAT. GES., ZURICH 4:323 /S. NIGRA (VAN TIEGHEM) =A. NIGER VAN TIEGHEM
=ASPERGILLUS LINK 1809 FIDE CLEMENTS & SHEAR 1931 •

STEVENSOMYCES MORRIS & FINLEY 1965. MYCOLOGIA 57:483-485 (ILLUS) /S. PALMAE (STEVENS & KING)
AMERO,CETERI,HYALO,ANNEL? OR MISC,SYNNEMATA, 19F, ON PLANTS (PALM INFLORESCENCES). •

STEVENSONULA PETRAK 1952. SYDOWIA 6:1-4 & FIG 3 /S. CILIATA
DIDYMO,CETERI,PHAEO,NONSP,SPORODOCHIA,APPEND,SETAE, ON PLANTS (BACCHARIS & GRASS CULMS).
/SUTTON, NOVA HEDWIGIA 26:6-9,1975 (ILLUS,REVIS) •

STIGMINA SACCARDO 1880. MICHELIA 2:22 /S. PLATANI (FUCKEL) (LECTOTYPE, FIDE HUGHES 1958)
PHRAGMO OR DICTYO,CETERI,PHAEO,ANNEL,SPORODOCHIA, 76B & 98A, ON LEAVES & TWIGS (MANY PLANTS). ASCO STATE = MYCOSPHAERELLA FIDE BARR 1972 (CONTRIB. UNIV. MICHIGAN HERB. 9:601) OR OTTHIA FIDE VON ARX 1975 (CBS STUDIES 9:79).
/ELLIS, CMI PAPERS 72:36-71,1959 (ILLUS,KEY) /ELLIS, DH P 144-148,1971 (ILLUS) /SUTTON, CMI PAPERS 132:113-121,1973 (ILLUS,NEW SP) /MATSUSHIMA, ICONES P 149 & PL 147,1975 (ILLUS, CONIDIOGENOUS CELLS SHOWN AS SYMPODIAL RACHI) /ELLIS, MDH P 114-134,1976 (ILLUS,KEYS, NEW SPP,NEW COMB) /THUANG, TBMS 66:211-215,1976 (ILLUS,NEW SP) /HUGHES, NEW ZEALAND J. BOT. 16:353,1978 (NEW COMB,ILLUS)
=*COLUMNOPHORA* BUBAK & VLEUGEL 1916 FIDE NAG RAJ & KENDRICK 1975
=*PSEUDOPUCCINIA* HOHNEL 1919 FIDE LAUNDON FROM AINSWORTH 1971
=*THYROSTROMA* HOHNEL 1911 FIDE ELLIS 1971
=*THYROSTROMELLA* SYDOW 1924 FIDE ELLIS 1971 •

STIGMOPSIS BUBAK 1914. ANN. MYCOL. 12:217-218 /S. MONTELLICA (SACCARDO) (LECTOTYPE) =CHEIROMYCES MONTELLICA (SACC.) MOORE 1958
=CHEIROMYCES BERKELEY & BROOME 1857 FIDE MOORE 1958 •

STILBELLA LINDAU 1900. IN ENGLER UND PRANTL, NAT. PFLANZENFAM. TEIL I, ABT. 1**, P 489 & 491 (ILLUS) /STILBUM ERYTHROCEPHALUM DITMAR 1816 =S. ERYTHROCEPHALA (DITMAR 1816) (LECTOTYPE, SEE TAXON 17:525,1968)
AMERO,GLOIO,HYALO,?PHIAL,SYNNEMATA, 11D, ON LITTER, DUNG, FUNGI, SOIL, WOOD & OTHER SUBSTRATES. DITMAR'S "ROTHKOPFIGES STILBUM" FROM EXCREMENT MIXED WITH BEETLE PARTS HAS APPARENTLY NOT BEEN RE-DISCOVERED. HE DESCRIBED IT AS HAVING A WHITE, HAIRY STALK, & HEAD WHICH CHANGED FROM GREEN TO RED AS IT MATURED. CORDA'S CILICIOPODIUM VIOLACEUM HAS A SIMILAR DESCRIPTION & ILLUSTRATIONS. COMPARE WITH CONIDIAL NECTRIA AURANTIACA, SEE BOOTH 1959 (CMI PAPERS 73:29-32,ILLUS). FOR CONSERVATION AGAINST BOTRYONIPHA PREUSS, SEE TULLOCH 1972 (CMI PAPERS 130:7,

COMMENTS ON SPECIMENS), OR BENJAMIN 1968 (TAXON 17:521-522) & COMMITTEE REPORT 1974 (TAXON 23:420). FOR ILLUSTRATION LABELED AS S. ERYTHROCEPHALA, SEE SUTTON 1973 (CMI PAPERS 132:18). ASCO STATE =NECTRIA FIDE SAMUELS 1976 (MEM. N. Y. BOT. GARDEN 26:22) OR BYSSOSTILBE FIDE ROSSMAN 1977 (MYCOLOGIA 69:372).
/SINGH & WEBSTER, TBMS 61:487-495,1973 (ILLUS OF COLONY) /SAMUELS, MYCOLOGIA 65:409,1973 (NEW COMB) /MATSUSHIMA, ICONES P 149-150 & PL,1975 (ILLUS,NEW SPP) /ESTEY, TBMS 68:120-122,1977 (ILLUS,NEW SP, S. FLAVESCENS APPEARS TO BE THE SAME AS S. BULBICOLA HENNINGS FIDE CARMICHAEL) •

STILBELLULA BOEDIJN 1951. SYDOWIA 5:227-228 (ILLUS) /S. PALLIDA
?=CEPHALOTRICHUM PHILLIPSII (BERKELY & LEIGHTON) HUGHES 1958
?=CEPHALOTRICHUM LINK 1809 •

STILBOCHALARA FERDINANDSEN & WINGE 1910. BOT. TIDSSKR. 30:220-222 (ILLUS) /S. DIMORPHA
=THIELAVIOPSIS WENT 1893 FIDE ELLIS 1971. SEE ALSO NAG RAJ & KENDRICK 1975 (CHALARA P 35). •

STILBODENDRON SYDOW & SYDOW 1916. ANN. MYCOL. 14:260-262 (ILLUS) /S. CAMERUNENSE
=CORALLODENDRON JUNGHUHN 1838 FIDE MORRIS 1963 •

STILBOMYCES ELLIS & EVERHART 1896. PROC. ACAD. PHIL. 1895:441 /S. BERENICE
?, NOT ILLUSTRATED BY AUTHORS. DESCRIPTION SUGGESTS ARTHROSPORIUM SACC. •

STILBOSPORA PERSOON 1801. SYN. METH. FUNG. P 96 /S. MACROSPERMA (LECTOTYPE SEE SUTTON 1977, CMI PAPERS 141:204)
PHRAGMO,CETERI,PHAEO,ANNEL,ACERVULI, ON LEAVES & STEMS (CARPINUS & OTHER PLANTS). ASCO STATE =MELANCONIS OR PROSTHECIUM FIDE MULLER & VON ARX, IN AINSWORTH ET AL 1973 (IVA:113).
/SUTTON, CMI PAPERS 138:6-7,1975 (DESCR) •

STILBOTHAMNIUM HENNINGS 1897. ENGLER'S BOT. JHB. 23, 4:542 & TAF 14, FIG 69A TO C /S. TOGOENSE
?=ASPERGILLUS, SEE ROQUEBERT & ABADIE 1973 (C. R. ACAD. SCI. PARIS 276:2883-2885) & ROQUEBERT 1975 (REV. MYCOL. 38:3-8,ILLUS) & SAMSON 1979 (CBS STUDIES 18:25). •

STILBUM PERSOON 1801. SYN. METH. FUNG. P 680 /S. RIGIDUM (LECTOTYPE, FIDE HUGHES 1958)
AMERO,GLOIO,HYALO,PHIAL,SYNNEMATA, ON WOOD & SOIL. DIFFERENTIATED FROM STILBELLA BY BROWN STIPE. FOR A DIFFERENT APPLICATION, SEE DONK 1966 (PERSOONIA 4:161). ASCO STATE ?=EMERICELLOPSIS SEE MATHUR & THIRUMALACHAR 1960 (MYCOLOGIA 52:695) & HYPOCREOPSIS SEE WEHMEYER 1975 (PYRENOMYCETOUS FUNGI P 200).
/BARRON, HYPHOMYCETES P 294-295,1968 (ILLUS) /MATSUSHIMA, MICROFUNGI P 62-63, FIGS 153-154 & PL 24,1971 (ILLUS,NEW SP) /SUTTON, CMI PAPERS 132:16-20,1973 (ILLUS,REVIS)
?=*BLISTUM* SUTTON 1973
?=*CRINULA* FRIES 1821, SEE SUTTON 1973
?=*DENDROSTILBELLA* HOHNEL 1905, SEE SUTTON 1973
?=*PHIALOSTELE* DEIGHTON 1969
?=*POLYCEPHALOMYCES* KOBAYASI 1941
?=*STILBELLA* LINDAU 1900 •

STREBLOCAULIUM CHEVALLIER 1837. FUNG. & BYSS. ILLUS., FASC. 1, NO. 19 /S. ATRO-VIRENS
?, LOOKS LIKE CONOPLEA PERSOON 1801. •

STREPTOBOTRYS HENNEBERT 1973. PERSOONIA 7:191-192 & FIG 2, P 186 /S. STREPTOTHRIX (COOKE & ELLIS 1878)
AMERO,CETERI,PHAEO,AMPUL OR NONSP,ALEURI, 34A, ON PLANTS. CLOSE TO BOTRYTIS. ASCO STATE =STREPTOTINIA WHETZEL FIDE HENNEBERT 1973 (=SCLEROTINIA). SCLEROTIAL STATE =SCLEROTIUM FIDE HENNEBERT. SPERMATIAL STATE =MYRIOCONIUM FIDE HENNEBERT. •

STREPTOPODIUM ZHENG & CHEN 1978. ACTA MICROBIOL. SIN. 18:181-188 (ILLUS) /OVULARIOPSIS BONARIENSIS SPEGAZZINI =S. BONARIENSE (SPEGAZZINI)
?, NOT VALIDLY PUBLISHED, NO REF TO BASIONYM. A SEGREGATE FROM OIDIUM (SUB OVULARIOPSIS). •

STREPTOTHRIX CORDA 1839. PRACHTFLORA, P 27 /S. FUSCA =CONOPLEA FUSCA PERSOON 1822
=CONOPLEA PERSOON 1801 FIDE HUGHES 1958 •

STROMATERIA CORDA 1837. ICONES FUNGORUM 1:5, TAB 1 FIG 81 /S. CARNEA
=TUBERCULARIA FIDE SACCARDO FROM AINSWORTH 1971 •

STROMATOCREA COOKE 1952. MYCOLOGIA 44:248-250 (ILLUS) /S. CEREBRIFORME
AMERO,BASO,PHAEO,PHIAL,SPORODOCHIA, 13A, ON LITTER (ACER). ASCO STATE =HYPOCREOPSIS FIDE AINSWORTH 1971. •

STROMATOGRAPHIUM HOHNEL 1907. DENK. AKAD. WIEN 83:37 /STILBUM STROMATICUM BERKELEY 1843 (LONDON J. BOT. 2:642) =S. STROMATICUM (BERK.)
?, NOT ILLUSTRATED BY HOHNEL. •

STROMATOSTYSANUS HOHNEL 1919. BER. DEUTSCHL. BOT. GES. 37:153 /S. CAPRIFOLIORUM (DESMAZIERES)
=GRAPHIOTHECIUM FUCKEL FIDE CLEMENTS & SHEAR 1931 •

STRONGYLOTHALLUS BATISTA & CIFERRI 1959. MYCOPATHOLOGIA 11:97 /S. SCUTATUS BATISTA, CIFERRI & COSTA 1959
?, STERILE. •

STRUMELLA FRIES 1825. SYST. ORB. VEG., P 199 /S. CUCURBITACEARUM FRIES 1849 (SUMMA VEG. SCAND. 2:482,LECTOTYPE FIDE HUGHES 1958)
=NOMEN DUBIUM FIDE HUGHES 1958 •

STRUMELLA SACCARDO 1880. MICHELIA 2:36 (NO ILLUS) /ILLOSPORIUM OLIVATRUM SACCARDO 1877 (MICHELIA 1:90) (LECTOTYPE) =S. OLIVATRA (SACCARDO)
HELICO OR BULBILS,CETERI,PHAEO,NONSP, 97E, ON WOOD. ILLEGITIMATE BUT IN USE (NON STRUMELLA FRIES 1825). PROBABLY THE SAME AS SPIROSPHAERA (SEE HENNEBERT, TBMS 51:23,1968). COMPARE WITH CHEIROSPORA. ASCO STATE =URNULA SEE DAVIDSON 1950 (MYCOLOGIA 32:735-742).
/MATSUSHIMA, ICONES P 150 & PL,1975 (ILLUS,NEW SPP,S. MACROSPORA MATS. IS NOT CONGENERIC WITH S. OLIVATRA)
=*DACRINA* FRIES 1832 (NON FRIES 1825) FIDE SACCARDO 1886 •

STRUMELLOPSIS HOHNEL 1909. S. B. AKAD. WISS. WIEN 118:896-897 /S. ANNULARIS (RACIBORSKI)
?, NOT ILLUSTRATED BY HOHNEL. •

STYLINA SYDOW 1921. IN FISCHER, ANN. MYCOL. 81:192 (NO ILLUS,DATED 1920) /S. DISTICHA (EHRENBERG) =SPHAERIA DISTICHA EHRENBERG EX FRIES 1823 (SYST. MYCOL. 2:434)
?, A SEGREGATE FROM GRAPHIOLA. •

STYSANOPSIS FERRARIS 1909. ANN. MYCOL. 7:281 /S. MEDIA (SACCARDO) (LECTOTYPE FIDE CARMICHAEL 1973) =CEPHALOTRICHUM MEDIUM (SACC.) HUGHES 1958
=CEPHALOTRICHUM LINK 1809 •

STYSANUS CORDA 1837. ICONES FUNGORUM 1:21-22 /S. STEMONITIS (PERSOON) (LECTOTYPE FIDE HUGHES 1958) =C. STEMONITIS (PERSOON) LINK 1809
=CEPHALOTRICHUM LINK 1809 FIDE HUGHES 1958 •

SUBRAMANIA RAO & RAO 1964. TRANS. AM. MICROSC. SOC. 83:399-406 (ILLUS) /S. SUNDARA
?=PERICONIA TODE EX PERSOON 1801 •

SUBULARIELLA HOHNEL 1915. SITZB. AKAD. WIEN 124:118 /S. MACROSPORA (BERKELEY & CURTIS)
=A SUPERFLUOUS NAME FOR PSEUDOGRAPHIUM JACZEWSKI FIDE DI COSMO. •

SUBULISPORA TUBAKI 1971. IN TUBAKI & YOKOYAMA, TRANS. MYCOL. SOC. JAPAN 12:20-22 (ILLUS) /S. PROCURVATA
SCOLECO,CETERI,HYALO,RACHI,APPEND, 71E, ON LITTER (CASTANOPSIS, QUERCUS, EUCALYPTUS).
/PIROZYNSKI, CMI PAPERS 129:44-47,1972 (ILLUS) /SUTTON, TBMS 61:422-424,1973 (ILLUS,NEW SP) /MATSUSHIMA, ICONES P 151 & PL,1975 (ILLUS) /ELLIS, MDH P 227-228,1976 (ILLUS) /SUTTON, TBMS 71:170-171,1978 (ILLUS,NEW SP) •

SUTRAVARANA SUBRAMANIAN & CHANDRASHEKARA 1977. CAN. J. BOT. 55:248-251 (245-253,ILLUS) /S. SAMALA ?=BEAUVERIA CRETACEA (VAN BEYMA) MATSUSHIMA
?=BEAUVERIA VUILLEMIN
/MATSUSHIMA, MICROFUNGI P 7,1971 (ILLUS,NEW COMB SUB BEAUVERIA CRETACEA) /DE HOOG, CBS STUDIES 1:14-17,1972 (ILLUS) •

SYAMITHABEEJA SUBRAMANIAN & NATARAJAN 1976. MYCOLOGIA 67:1213-1217 (ILLUS,DATED 1975) /S. GIRISA
DICTYO OR BULBILS,CETERI,PHAEO,RACHI OR MISC, 99G, ON LITTER. •

SYMPHYOSIRA PREUSS 1853. LINNAEA 25:742 (DATED 1852, NO ILLUS) /S. LUTEA
?, ASCO STATE =SYMPHYOSIRINIA FIDE AINSWORTH. •

SYMPODIELLA KENDRICK 1958. TBMS 41:519-521 (ILLUS) /S. ACICOLA
AMERO OR DIDYMO OR PHRAGMO,ARTHRO,HYALO,RADUL OR RACHI, 43D, ON LITTER. COMPARE WITH PARASYMPODIELLA.
/BARRON, HYPHOMYCETES P 296-297,1968 (ILLUS) /ELLIS, DH P 36-37,1971 (ILLUS) /TUBAKI & YOKOYAMA, TRANS. MYCOL. SOC. JAPAN 12:24,1971 (ILLUS,NEW SP) /MATSUSHIMA, ICONES P 151-152 & PL,1925 (ILLUS,NEW SP, ASSOCIATED CONIDIAL STATE REPORTED FOR S. LAXA MAY BE A PARASITE) •

SYMPODINA SUBRAMANIAN & LODHA 1964. ANTONIE VAN LEEUWENHOEK 30:317-328 (317-330,ILLUS) /S. COPROPHILA =VERONAEA BOTRYOSA FIDE DE HOOG & HERMANIDES-NIJHOF 1977 (CBS STUDIES 15:198)
=VERONAEA CIFERRI & MONTEMARTINI 1957 FIDE VON ARX 1970 •

SYMPODIOPHORA ARNOLD 1970. NOVA HEDWIGIA 19:301-304 (ILLUS) /S. STEREICOLA =S. STATE OF HYPOMYCES SEMITRANSLUCENS ARNOLD
=PSEUDOHANSFORDIA ARNOLD 1970 FIDE DE HOOG 1978 (PERSOONIA 10:57-62).
/DEIGHTON & PIROZYNSKI, CMI PAPERS 128:61-74,1972 (ILLUS,NEW SPP) /GAMS, PERSOONIA 7:167-168,1973 (REVIS) /MATSUSHIMA, ICONES P 152 & PL,1975 (ILLUS) /ROQUEBERT & CHARPENTIE, REV. MYCOL. 38:124-128,1975 (ILLUS,NOT SEEN) /COLE & SAMSON, CONIDIAL FUNGI P 36-39,1979 (ILLUS) •

SYNCOLLESIA AGARDH 1824. SYST. ALG., P 32 /S. FOLIORUM
?, PROBABLY A NOMEN DUBIUM. •

SYNGLIOCLADIUM PETCH 1932. TBMS 17:177 /S. ARANEARUM
?, NOT ILLUSTRATED BY AUTHOR.
/MORRIS, WESTERN ILLINOIS UNIV. SER. BIOL. SCI. 3126-127,1963 (ILLUS) •

SYNNEMATIUM SPEARE 1920. MYCOLOGIA 12:74-75 & PL 4, FIGS 1-15 /S. JONESII
AMERO,GLOIO,HYALO,PHIAL,SYNNEMATA, 10A, ON INSECTS (BEETLES & OTHERS). A SYNNEMATOUS ACREMONIUM (CEPHALOSPORIUM AUCT.). SEE ALSO CORALLINOPSIS.
/MORRIS, WESTERN ILLINOIS UNIV. SER. BIOL. SCI. 3:128-129,1963 (ILLUS) •

SYNPENICILLIUM COSTANTIN 1888. BULL. SOC. MYCOL. FR. 4:67 /S. ALBUM =DORATOMYCES PUTREDINIS (CORDA) MORTON & SMITH 1963
=CEPHALOTRICHUM LINK 1809 FIDE MORTON & SMITH 1963 SUB DORATOMYCES •

SYNPHRAGMIDIUM STRAUSS 1853. STURM'S DEUTSCHL. FLORA III (PILZE) BD. 7, HEFT 33/34:41 /S. KUMMERI
?, HUGHES 1958 COULD NOT FIND THE TYPE, ?NOMEN DUBIUM. •

SYNSPORIUM PREUSS 1849. KLOTZSCHII HERB. VIV. MYCOL. NO 1285, ANNO 1849 /S. BIGUTTATUM =STACHYBOTRYS CHARTARUM (EHRENBERG) HUGHES 1958
=STACHYBOTRYS CORDA 1837 FIDE HUGHES 1958. •

SYNSTERIGMATOCYSTIS COSTANTIN =GIBELLULA FIDE PETCH FROM AINSWORTH 1971 •

SYNTHETOSPORA MORGAN 1892. BOT. GAZ. 46:192 /S. ELECTA =STEPHANOMA WALLROTH 1833 FIDE CLEMENTS & SHEAR 1931. •

SYZYGOSPORA MARTIN 1937. J. WASHINGTON ACAD. SCI. 27:112 /S. ALBA =S. STATE OF CHRISTIANSENIA ALBA BOIDIN 1970
AMERO OR DIDYMO,CETERI OR GLOIO,HYALO,MISC,CLAMPS, 51G, ON FUNGI (PHANEROCHAETE). BASIDIO STATE =CHRISTIANSENIA FIDE BOIDIN 1970.
/KAO, MYCOLOGIA 48:677-684,1956 (ILLUS,REVIS) /BOIDIN, BULL. MENS. SOC. LINN. LYON 39:132-137,1970 (ILLUS,REVIS,KEY) /ERIKSSON & RYVARDEN, THE CORTICIACEAE OF NORTH EUROPE 2:242-243, 1973 (ILLUS SUB CHRISTIANSENIA) •

TAENIOLA BONORDEN 1851. HANDB. ALLGEM. MYKOL. P36 /SIX ORIGINAL SPECIES =NOMEN ILLEGITIMUM FIDE HUGHES 1958 •

TAENIOLELLA HUGHES 1958. CAN. J. BOT. 36:816-817 (REFS) /T. EXILIS (KARSTEN)
PHRAGMO,BLASTO,PHAEO,NONSP, 91D, ON WOOD, BARK & LITTER. DIFFERENCE FROM HETEROCONIUM NOT CLEAR.
/ELLIS, DH P 91-94,1971 (ILLUS,KEY) /ELLIS, MDH P 55-61,1976 (ILLUS,KEY,NEW COMB) /REDDY & REDDY, MYCOTAXON 6:508-510,1978 (ILLUS,NEW SP)
?=*SPOROPHIALA* RAO 1970 •

TAENIOLINA ELLIS 1976. MDH P 61 (61-62,ILLUS) /TORULA CENTAURII FUCKEL =T. CENTAURII (FUCKEL)
CHEIRO OR STAURO,CETERI OR BLASTO,PHAEO,NONSP, 127C, ON LEAVES (CENTAURIUM). COMPARE WITH CHEIROMYCES. •

TAENIOSPORA MARVANOVA 1977. TBMS 69:146-148 (ILLUS) /T. GRACILIS
STAURO,CETERI,HYALO,NONSP,CLAMPS,AQUATIC, 124B, BASIDIO STATE =LEPTOSPOROMYCES SEE MARVANOVA 1977 & NAWAWI ET AL 1977. COMPARE WITH INGOLDIELLA.
/NAWAWI, DESCALS & WEBSTER, TBMS 68:31-36,1977 (ILLUS) •

TANDONELLA PRASAD & VERMA 1970. INDIAN PHYTOPATH. 23:111-113 (ILLUS) /T. ZIZIPHI (AS 'ZIZYPHI')
DIDYMO OR PHRAGMO,BLASTO,PHAEO,RADUL,SYNNEMATA, 86E, ON LEAVES (ZIZIPHUS). SIMILAR TO CLADOSPORIUM.
/ELLIS, MDH P 216-218,1976 (ILLUS) •

TAPEINOSPORIUM BONORDEN 1853. BOT. ZEIT. 11:285, TAB. VII, FIG 6 /T. VIRIDE =SEPTOCYLINDRIUM BONORDEN 1851 FIDE LINDAU 1907. •

TAWDIELLA DESHPANDE & DESHPANDE 1966. MYCOPATHOLOGIA 28:206-208 (ILLUS) /T. DOLIIFORMIS
?, ORIGINAL ILLUSTRATION NOT DIAGNOSTIC. •

TELLIGIA HENDRICKX 1948. PUBL. INST. NAT. ETUDE AGRON. CONGO BELGE, SER. SCI. 35:8 /GILLETIA CINNABARINA TORREND 1914 =T. CINNABARINA (TORREND)
?, NOT ILLUS. BY HENDRICKX. FOR GILLETIA TORREND NON SACCARDO. •

TERATOSPERMA SYDOW & SYDOW 1909. ANN. MYCOL. 7:172-173 (ILLUS) /T. SINGULARE
PHRAGMO OR STAURO,CETERI,PHAEO,ANNEL,APPEND, 75F, ON FUNGI, ALGAE & LICHENS ON LEAVES & BARK. OTHER CONIDIAL STATE ?=SELENOSPORELLA (Q.V.) SEE MATSUSHIMA 1975 (ICONES P 152).
/ELLIS, CMI PAPERS 69:1-7,1957 (ILLUS,KEY) /ELLIS, DH P 133-135,1971 (ILLUS) /MATSUSHIMA, ICONES P 152-153 & PL,1975 (ILLUS,NEW SP) •

TERMITARIA THAXTER 1920. BOT. GAZ. 69:3-8 (ILLUS) /T. SNYDERI (LECTOTYPE)
AMERO,BASO,HYALO,PHIAL,SPORODOCHIA, ON INSECTS (TERMITES).
/KHAN & ALDRICH, CAN. J. BOT. 51:2307-2314,1973 (ILLUS) /KHAN & KIMBROUGH, MYCOLOGIA 66:446-462,1974 (ILLUS,REFS) /COLE & SAMSON, CONIDIAL FUNGI P 61-63,1979 (ILLUS,CONIDIUM ONTOGENY) •

TERMITOSPHAERA CIFERRI 1935. ATTI IST. BOT. UNIV. PAVIA, SER 4, 6:242 /AEGERITA DUTHIEI BERKELEY =T. DUTHIEI (BERKELEY)
?, COMPARE WITH AEGERITA. BASIDIO STATE =TERMITOMYCES HEIM 1942 FIDE DONK 1962 (TAXON 11:101).
/HEIM, TERMITES ET CHAMPIGNONS, PARIS, 208 P,1977 (NOT SEEN) •

TETRABRUNNEOSPORA DYKO 1978. TBMS 70:414-416 (ILLUS) /T. ELLISII
STAURO,CETERI,PHAEO,ANNEL,AQUATIC, 119G. •

TETRACHAETUM INGOLD 1942. TBMS 25:377-381 & FIG 27-29 /T. ELEGANS
STAURO,CETERI,HYALO,NONSP,AQUATIC, 118A.
/TUBAKI, BULL. NAT. SCI. MUSEUM, TOKYO 41:261-262,1957 (ILLUS) /INGOLD, GUIDE P 50-51,1975 (ILLUS) •

TETRACHIA SACCARDO 1918. BULL. ORT. BOT. NAPOLI 6:65 (NO ILLUS) /T. SINGULARIS
?=ISTHMOSPORA STEVENS 1918 •

TETRACIUM HENNINGS -IN ERROR FOR TETRACRIUM •

TETRACLADIUM DE WILDEMAN 1893. ANN. SOC. BELG. MICR. 17:35 /T. MARCHALIANUM
STAURO,CETERI,HYALO,NONSP,AQUATIC, 112H.
/TUBAKI, BULL. NAT. SCI. MUSEUM, TOKYO 41:261-262,1957 (ILLUS) /INGOLD, GUIDE P 25-53,1975 (ILLUS)
=*MAXILLOSPORA* HOHNEL 1914 FIDE INGOLD FROM AINSWORTH 1971 •

TETRACOCCOSPORIS CLEMENTS 1909 =SUPERFLUOUS NAME FOR TETRACOCCOSPORIUM SZABO •

TETRACOCCOSPORIUM SZABO 1905. HEDWIGIA 44:76-77 (ILLUS) /T. PAXIANUM
DICTYO,CETERI,PHAEO,NONSP,ALEURI?, 96D, ON DUNG & SOIL.
/ELLIS, DH P 81-83,1971 (ILLUS) /MISRA & SRIVASTAVA, MYCOTAXON 4:276-278,1976 (ILLUS,NEW SP)
=*TETRACOCCOSPORIS* CLEMENTS 1909, NOMEN ILLEGITIMUM •

TETRACOLIUM KUNZE EX LINK 1824. IN LINN. SPEC. PLANT. IV, 61(1):125 /T. TUBERCULARIAE (NEES & NEES) KUNZE IN LINK 1824
=NOMEN DUBIUM FIDE HUGHES 1958 •

TETRACRIUM HENNINGS 1902. HEDWIGIA 41:116 (NO ILLUS) /T. AURANTII
STAURO,CETERI,HYALO,NONSP,SPORODOCHIA, 125C, ON INSECTS (SCALE INSECTS).
ASCO STATE =PUTTEMANSIA FIDE HOHNEL 1911 OR PODONECTRIA FIDE ROSSMAN 1977 (MYCOLOGIA 69:377).
/SHERWOOD, MYCOTAXON 1:117-120,1974 (ILLUS,NEW SP) /ROSSMAN, MYCOTAXON 7:163-182,1978 (ILLUS,KEY TO PODONECTRIA)
=*TETRACIUM* HENNINGS 1902, ORTHOGRAPHIC VARIANT
?=*PETRAKIOPSIS* SUBRAMANIAN & REDDY 1968 •

TETRACYTUM VANDERWALLE 1945. PARASITICA 1:145-150 (NO LATIN DIAGNOSIS) /T. LAURI
=CYLINDROCLADIUM MORGAN 1892 (SUB CANDELOSPORA) FIDE WORMALD FROM AINSWORTH 1971 •

TETRAPLOA BERKELEY & BROOME 1850. ANN. & MAG. NAT. HIST. SER. 2, 5:459 /T. ARISTATA
DICTYO OR STAURO,CETERI,PHAEO,NONSP,APPEND, 125A, ON LEAVES & STEMS (GRASSES & PALMS) & ON WOOD IN WATER.
/ELLIS, DH P 51-52,1971 (ILLUS,KEY) /MATSUSHIMA, ICONES P 153 & PL,1975 (ILLUS) •

TETRAPOSPORIUM HUGHES 1951. CMI PAPERS 46:25-28 (ILLUS) /T. ASTERINEARUM
STAURO,CETERI,PHAEO,NONSP, 120B, ON FUNGI (ASTERINA).
/ELLIS, DH P 54-55,1971 (ILLUS) •

THALLOMICROSPORON BENEDEK 1964. MYCOPATHOLOGIA 23:96-98 /T. KUEHNII ?=M. GYPSEUM FIDE WEITZMAN IN AJELLO 1968 (SABOURAUDIA 6:153-155)
=MICROSPORUM GRUBY 1843. •

THALLOSPORA OLIVE 1948. MYCOLOGIA 40:11-12 & FIG 3B-E ON P 18 /T. ASPERA
STAURO,CETERI,HYALO,NONSP, 122F, ON PLANTS (OVARIES OF VERONICA).
/BARNETT, IMPERFECT FUNGI NO. 127,1960 (ILLUS) •

THANATOPHYTUM NEES 1817 =RHIZOCTONIA FIDE LINDAU IN RABENHORST 1910 •

THAROOPAMA SUBRAMANIAN 1956. J. INDIAN BOT. SOC. 35:83-86 (ILLUS) /T. TRINA
?=NODULISPORIUM
/MORRIS, WESTERN ILLINOIS UNIV. SER. BIOL. SCI. 3:130-131,1963 (ILLUS) •

THAXTERIOLA SPEGAZZINI 18 . ANN. SOC. NAT. ARGENTINA 85:314 (NOT SEEN, CITATION FROM THAXTER 1920) /?
AMERO,CETERI,HYALO,PHIAL, 55C, ON INSECTS (BEETLES & MITES).
/THAXTER, BOT. GAZ. 69:14-16 & PL IV,1920 (ILLUS,NEW SP) /MAJEWSKI & WISNIEWSKI, MYCOTAXON 7:508-510,1978 (ILLUS) •

THECLOSPORA HARKNESS 1884. BULL. CALIF. ACAD. SCI. 1:41 /T. BIFIDA
=EMERICELLA (A) BERKELEY & BROOME FIDE PEEK & SOLHEIM 1958 •

THEDGONIA SUTTON 1973. TBMS 61:426-429 (ILLUS) /CERCOSPORA LIGUSTRINA BOEREMA 1962 =T. LIGUSTRINA (BOEREMA)
PHRAGMO,ARTHRO,HYALO,RACHI,SPORODOCHIA, 67D, ON LEAVES (LIGUSTRUM). •

THERMOIDIUM MIEHE 1907. BER. DEUTSCH. BOT. GES. 25:515 /T. SULFUREUM
=MALBRANCHEA SULFUREA (MIEHE) SIGLER & CARMICHAEL 1976
=MALBRANCHEA SACCARDO 1882 FIDE SACCARDO 1908
/SIGLER & CARMICHAEL, MYCOTAXON 4:441-446,1976 (REFS) •

THERMOMYCES TSIKLINSKY 1899. ANN. INST. PASTEUR 13:500 /T. LANUGINOSUS
AMERO,CETERI,PHAEO,NONSP,ALEURI, 49E, ON COMPOST & LITTER. FOR NON-THERMOPHILIC SPECIES, SEE HUMICOLA. ASCO STATE =CHAETOMIUM FIDE GAMS 1971 (P 19).
/BARRON, HYPHOMYCETES P 297-298,1968 (ILLUS) /ELLIS, DH P 58-59,1971 (ILLUS,KEY)
?=*HUMICOLA* TRAAEN 1914 •

THIELAVIOPSIS WENT 1893. MEDED. PROEFST. W. JAVA 5:4 /T. ETHACETICA =T. STATE OF CERATOCYSTIS PARADOXA FIDE ELLIS 1971
AMERO,BASO,PHAEO,MISC OR PHIAL?, 50C, ON PLANTS (ANANAS, COCOS, MUSA, SACCHARUM, THEOBROMA & OTHERS). ASCO STATE =CERATOCYSTIS. ORIGINALLY PROPOSED FOR TWO CONIDIAL STATES. THE NAME IS HERE RETAINED FOR THE DARK, THICK-WALLED CONIDIA BORNE IN SHORT CHAINS. OTHER CONIDIAL STATE =CHALARA. WE EXCLUDE A SECOND SPECIES, THIELAVIOPSIS BASICOLA (BASIONYM TORULA BASICOLA BERKELEY & BROOME 1850 ANN. MAG. NAT. HIST. SER 2, 5:455-466 =**TRICHOCLADIUM BASICOLA** (BERKELEY & BROOME) CARMICHAEL COMB. NOV.).
/ELLIS, DH P 31-32,1971 (ILLUS) /HENNEBERT, ANTONIE VAN LEEUWENHOEK 33:333-340,1967 (ILLUS) /MORGAN-JONES, CMI DESCRIPTIONS 141-143,1967 (ILLUS, SUB CERATOCYSTIS) /NAG RAJ & KENDRICK, CHALARA 200 P,1975 (MONOGR) /ELLIS, MDH P 19-20 & 440-441,1976 (ILLUS,ALSO UNDER CHALARA STATES OF CERATOCYSTIS ADIPOSA & CER. FIMBRIATA) /HAWES & BECKETT, TBMS 68:259-265 & 267-276,1977 (ILLUS,CONIDIUM ONTOGENY) /HAWES, TBMS 72:177-187,1979 (ILLUS,ONTOGENY) /COLE & SAMSON, CONIDIAL FUNGI P 66-67,1979 (ILLUS,CONIDIUM ONTOGENY SUB CHALARA)
=*HUGHESIELLA* BATISTA & VITAL 1956 FIDE ELLIS 1971
=*STILBOCHALARA* FERD. & WINGE 1910 FIDE ELLIS 1971 •

THOMIELLA DODGE 1935. MEDICAL MYCOLOGY P 834 (NO ILLUS) /T. DESSYI (SPEGAZZINI)
=GONATOBOTRYUM SACCARDO 1886 FIDE CARMICHAEL 1973 •

THOZETELLA KUNTZE 1891. REVISIO GENERUM PLANTARUM, A. FELIX, LEIPZIG 2:873 (NO ILLUS) /THOZETIA NIVEA BERKELEY & MUELLER 1881 =T. NIVEA (BERKELEY & MUELLER)
AMERO,GLOIO,HYALO,PHIAL,APPEND,SYNNEMATA,SETAE, 1D, ON LITTER (PERSEA, CAMELLIA), BARK & WOOD.
/PIROZYNSKI & HODGES, CAN. J. BOT. 51:168-173,1973 (REVIS,ILLUS) /SHERWOOD,

MYCOTAXON 1:119-120,1974 (ILLUS,NEW SP,SUB THOZETELLOPSIS) /NAG RAJ, CAN. J. BOT. 54:1374-1376,1976 (ILLUS,NEW SP)
=*THOZETIA* BERKELEY & MUELLER 1881 NON THOZETIA MUELLER
=*THOZETELLOPSIS* AGNIHOTHRUDU 1958 FIDE PIROZYNSKI & HODGES 1973 •

THOZETELLOPSIS AGNIHOTHRUDU 1958. MYCOLOGIA 50:576-577 (570-579,ILLUS) /T. TOCKLAIENSIS
=THOZETELLA KUNTZE 1891 FIDE PIROZYNSKI & HODGES 1973 (SEE 9A) •

THOZETIA BERKELEY & MUELLER 1881. JOUR. LINN. SOC. 18:388 /T. NIVEA
=THOZETELLA, NON THOZETIA MUELLER •

THYRODOCHIUM WERDERMANN 1924. IN SYDOW & WERDERMANN, ANN. MYCOL. 22:188-189 (ILLUS) /T. DRACAENAE
=STEMPHYLIUM FIDE WILTSHIRE FROM AINSWORTH 1971 •

THYROSPORA TEHON & DANIELS 1925. PHYTOPATH. 15:716-718 (ILLUS) /T. SARCINIFORME
=STEMPHYLIUM FIDE SMITH FROM AINSWORTH 1971 •

THYROSTROMA HOHNEL 1911. S. B. AKAD. WISS. WIEN., ABT. 1, 120:472 /T. COMPACTUM (SACCARDO) (LECTOTYPE)
=STIGMINA SACCARDO 1880 FIDE ELLIS 1971 •

THYROSTROMELLA HOHNEL 1919. BERL. DEUTSCHL. BOT. GES. 37:157 /T. MYRIANA (DESMAZIERES)
DICTYO,CETERI,PHAEO,RACHI,SPORODOCHIA, 99H, ON LEAVES (AMMOPHILA). NON THYROSTROMELLA SYDOW 1924.
/HUGHES, CAN. J. BOT. 33:341-343,1955 (ILLUS) /ELLIS, DH P 287-288,1971 (ILLUS) •

THYROSTROMELLA SYDOW 1924. IN SYDOW & PETRAK, ANN. MYCOL. 22:405-407 /T. TRIMERA (SACCARDO)
=STIGMINA SACCARDO 1880 FIDE ELLIS 1971 •

THYRSIDINA HOHNEL 1905. ANN. MYCOL. 3:337-338 (NO ILLUS) /T. CARNEO-MINIATA
DICTYO,GLOIO,HYALO,NONSP,CLAMPS,ACERVULI, 97D, (COELOMYCETE), ON HERBACEOUS STEMS.
/NAG RAJ & DI COSMO, ICONES GENERUM COELOMYCETUM X:40-41,1978 (ILLUS) •

THYRSIDIUM MONTAGNE 1849. IN DURIEU, EXPL. SCI. ALGERIE BOT. 1:325 /T. BOTRYOSPORUM (MONTAGNE)
=CHEIROSPORA MOUGEOT & FRIES 1825 FIDE HUGHES 1958 •

THYSANOPHORA KENDRICK 1961. CAN. J. BOT. 39:820 (817-832,ILLUS) /HAPLOGRAPHIUM PENICILLIOIDES ROUMEGUERE =T. PENICILLIOIDES (ROUMEGUERE)
AMERO,BASO,HYALO,PHIAL, 7F, ON LEAVES (TSUGA) & LITTER (CONIFERS). OTHER STATE ="SCLEROTIA" SEE ELLIS 1971.
/ELLIS, DH P 550-551,1971 (ILLUS) /MATSUSHIMA, ICONES P 153 & PL,1975 (ILLUS) •

THYSANOPYXIS CESATI 1850. IN RABENHORST IN KLOTZSCH, HERB. MYCOL. NO. 1482 /T. PULCHELLA =VOLUTELLA PULCHELLA (CESATI) SACCARDO 1886 (SYLL. FUNGI. 4:684)
=VOLUTELLA TODE EX FRIES 1832 FIDE SACCARDO 1886 •

TILACHLIDIOPSIS KEISSLER 1924. ANN. NATURH. MUS. WIEN 37:215-216 (ILLUS) /T. RACEMOSA
AMERO,GLOIO,HYALO,PHIAL,SYNNEMATA, 10E, ON INSECTS (BEETLES) & SOIL.
/MORRIS, WESTERN ILLINOIS UNIV. SER. BIOL. SCI. 3:132-133,1963 (ILLUS) •

TILACHLIDIUM PREUSS 1851. LINNAEA 24:126-127 (NO ILLUS) /T. ~~PLANATUM~~ =T. BRACHIATUM (BATSCH EX FRIES) PETCH 1937
AMERO,GLOIO,HYALO,PHIAL,SYNNEMATA, 10B, ON FUNGI (TRAMETES, HYPHOLOMA, MYCENA & OTHERS) & INSECTS. A SYNNEMATOUS ACREMONIUM. ASCO STATE =PSEUDONECTRIA FIDE GAMS 1975 (PERSOONIA 8:329-331,ILLUS) OR BYSSOSTILBE SEE DE HOOG 1977 (CBS STUDIES 15:138).
/MORRIS, WESTERN ILLINOIS UNIV. SER. BIOL. SCI. 3:134-135,1963 (ILLUS) /GAMS,

CEPHALO. SCHIMMEL. P 141-144 & TAF 1F,1971 (ILLUS,REVIS) •

TILLETIELLA ZAMBETTAKIS 1970. REV. MYCOL. 35:165 (NO ILLUS) /FOR T. STATE OF TILLETIA ALOPECURI LING
=NOMEN NUDUM, NO DIAGNOSIS •

TITAEA SACCARDO 1876. NUOV. GIORN. ITAL. 8:193 (NO ILLUS) /T. CALLISPORA
STAURO,CETERI,HYALO,NONSP, 112A, 112C & 121E, ON FUNGI (DIMEROSPORIUM) ON LEAVES.
/HANSFORD, CMI PAPERS 15:207-210,1946 (ILLUS,NEW SPP)
=*AORATE* SYDOW 1929 FIDE AINSWORTH 1961
=*ARANEOMYCES* HOHNEL 1909 FIDE DAMON 1952
=*MONOGRAMMIA* STEVENS 1917 FIDE DAMON 1952 •

TITAEELLA ARNAUD 1951. BULL. SOC. MYCOL. FRANCE 67:196 & FIG 1-I P 191 (NO LATIN DIAGNOSIS) /T. CAPNOPHILA
STAURO,CETERI,HYALO,NONSP,CLAMPS, 124D, ON FUNGI (SOOTY MOLDS). •

TOLYPOCLADIUM GAMS 1971. PERSOONIA 6:185-191 (ILLUS) /T. INFLATUM
AMERO,GLOIO,HYALO,PHIAL, 16A, ON SOIL & ROTIFERS. COMPARE WITH TRICHODERMA & HARPOSPORIUM.
/BARRON, CAN. J. BOT., IN PRESS,1980 (NEW SP,ILLUS) •

TOLYPOMYRIA PREUSS 1852. LINNAEA 25:726-727 /T. PRASINA
=TRICHODERMA PERSOON 1801 FIDE HUGHES 1969 (FRIESIA 9:64). •

TOMENTELLEOPSIS ORLOVA 1959. IN ORLOVA & GOLODNAYA, MANUAL OF THE PHYTOPATHOLOGICAL ANALYSIS OF SEEDS OF TREES & BUSHES, MOSCOW P 125-126 (NO LATIN DIAGNOSIS) /TWO NEW SPECIES DESCRIBED
=NOMEN ILLEGITIMUM, NON TOMENTELLOPSIS HJORSTAM 1970 •

TOMENTICOLA DEIGHTON 1969. IN CEJP & DEIGHTON, CMI PAPERS 117:20-25 (ILLUS) /T. TREMATIS
PHRAGMO,CETERI,PHAEO,RACHI, 83B, ON LEAVES (TREMA).
/ELLIS, DH P 207-208,1971 (ILLUS) •

TORULA (PERSOON) LINK 1809. MAG. GES. NATURF. FREUNDE, BERLIN 3:21 /MONILIA HERBARUM PERSOON 1801 =TORULA HERBARUM (PERSOON) LINK 1815
PHRAGMO,BLASTO,PHAEO,MISC, 88D, ON LITTER, SOIL & DUNG. FOR T. FASCICULATA SEE MATSUSHIMAEA. SEE ALSO RUTOLA.
/ELLIS, DH P 336-339,1971 (ILLUS,KEY) /RAO & DE HOOG, PERSOONIA 8:199-206,1975 (ILLUS,NEW SP,KEY,REVIS) /ELLIS & GRIFFITHS, CAN. J. MICROBIOL. 21:1661-1675 & 1921-1929,1975 (ILLUS) /ELLIS & GRIFFITHS, CAN. J. MICROBIOL. 22:858-866,1976 (ILLUS) /ELLIS, MDH P 351-352,1976 (ILLUS) /SCHOKNECHT & CRANE, MYCOLOGIA 69:533-546,1977 (EXCLUSIONS)
?=*BAHUSAGANDA* SUBRAMANIAN 1971
=*BAHUSANDHIKA* SUBRAMANIAN 1966 FIDE KENDRICK 1973
=*HORMISCIUM* KUNZE 1817 FIDE HUGHES 1958
=*SATWALEKERA* RAO, RAO & RAO 1969 FIDE CARMICHAEL 1973 •

TORULELLA GYELNIK 1938. LILLOA 4:64-65 /T. ASPERELLAE
?, ORIGINAL ILLUSTRATION NOT DIAGNOSTIC •

TORULINA SACCARDO & SACCARDO 1906. SYLLOGE FUNGORUM 18:566-567 /T SEROTINAE (OUDEMANS) =GLIOMASTIX MURORUM
=GLIOMASTIX GUEGUEN 1905 FIDE DICKINSON 1968 (CMI PAPERS 115:23) •

TORULOIDEA SUMSTINE 1913. MYCOLOGIA 5:53 & PL 34, FIG 1 /T. EFFUSA SUMSTINE 1912
?, ILLUSTRATION & DESCRIPTION SUGGEST SCOPULARIOPSIS. •

TORULOMYCES DELITSCH 1943. IN LEMBKE UND DELITSCH, SYSTEMATIK DER SCHIMMELPILZE, P 91-93 & TAF. 30, FIGS 232-235 /T. LAGENA
AMERO,BASO,PHAEO,PHIAL, 3D, ON SOIL.
/HASHMI, KENDRICK & MORGAN-JONES, CAN. J. BOT. 50:1461-1463,1972 (ILLUS) /GAMS, CEPHALO. SCHIMMEL. P 17, 22 & 239,1971 (REVIS) /MATSUSHIMA, ICONES P 154 & PL,1975 (ILLUS) •

TORULOPSIELLA BENDER 1932. MYCOLOGIA 24:411 (NO ILLUS, FOR TORULOPSIS SPEGAZZINI NON BERLESE) /T. FUMAGINEA (SPEGAZZINI)
PHRAGMO,ARTHRO,PHAEO,NONSP, 91A, ON LEAVES (SOOTY MOLDS). ASCO STATES =OPHIOCAPNOCOMA & METACAPNODIUM FIDE HUGHES 1972. OTHER STATES =CAPNOCYBE (H), CAPNOPHIALOPHORA (H) & HORMOKRYPSIS (H).
/HUGHES, NEW ZEALAND J. BOT. 10:225-242,1972 (ILLUS) •

TORULOPSIS OUDEMANS NON BERLESE, SEE TORULINA •

TORULOPSIS SPEGAZZINI 1918 (PHYSIS 4:292-293) NON BERLESE 1895, SEE TORULOPSIELLA •

TOXOSPORIUM VUILLEMIN 1896. BULL. SOC. MYCOL. FRANCE 12:34-35 (33-44) /T. ABIETINUM =T. CAMPTOSPERMUM (PECK) MAUBLANC 1907
PHRAGMO,CETERI,PHAEO,ANNEL,SPORODOCHIA, 77F, ON LEAVES (NEEDLES OF ABIES, CRYPTOMERIA & PINUS). COMPARE WITH EXOSPORIELLA.
/SUTTON, CMI PAPERS 138:65-68 & 102-103,1975 (ILLUS,REFS) /ELLIS, MDH P 135-136,1976 (ILLUS) •

TRACYELLA ZAMBETTAKIS 1970. REV. MYCOL. 35:165 & 170 (NO ILLUS) /SAVULESCUELLA HYDROCHARIDIS CIFERRI 1959 =T. HYDROCHARIDIS (CIFERRI)
=NOMEN NUDUM, NO DIAGNOSIS, NON TRACYELLA (SACCARDO) TASSI (COELO), FOR CONIDIAL TRACYA HYDROCHARIDIS LAGERHEIM. •

TRETOPHRAGMIA SUBRAMANIAN & NATARAJAN 1973. PROC. INDIAN NAT. SCI. ACAD., SER B, 39:549-552 (ILLUS) /ARTHROBOTRYUM NILGIRENSE SUBRAMANIAN =T. NILGIRENSE (SUBRAMANIAN)
PHRAGMO,CETERI,PHAEO,NONSP,SYNNEMATA, 87G, ON LITTER (BAMBUSA). •

TRETOPILEUS DODGE 1946. BULL. TORREY BOT. CLUB 73:223 & FIGS 1 & 2 /T. OPUNTIAE =T. SPHAEROPHORUS (BERKELEY & CURTIS) HUGHES & DEIGHTON 1960
DICTYO OR BULBILS,CETERI,PHAEO,MISC OR ANNEL,SYNNEMATA, 128A, ON LITTER (CAJANUS, CITRUS, HIBISCUS & OTHERS).
/DEIGHTON, CMI PAPERS 78:1-4,1960 (ILLUS) •

TRETOSPEIRA PIROZYNSKI 1972. CMI PAPERS 129:58-60 /T. UGANDENSIS (HANSFORD)
CHEIRO OR DICTYO,CETERI,PHAEO,NONSP, 98G, ON WOOD & LITTER (MONANTHOTAXIS).
/ELLIS, MDH P 381-382,1976 (ILLUS) •

TRETOSPORA ELLIS 1976. MDH P 374-376 (ILLUS) /T. NEGRII
PHRAGMO,CETERI,PHAEO,NONSP,APPEND,PODIA, 72B, ON LEAVES (RANDIA, GARDENIA & OTHERS). ASCO STATE =BALLADYNOPSIS FIDE ELLIS 1976. •

TRIADELPHIA SHEARER & CRANE 1971. MYCOLOGIA 63:247-249 (ILLUS) /T. HETEROSPORA
DIDYMO OR PHRAGMO,CETERI,PHAEO,NONSP OR MISC?,SPORODOCHIA, 70F, ON WOOD (IN WATER), BARK & SOIL (LOUDETIA RHIZOSPHERE). GENUS WAS BASED ON TWO KINDS OF CONIDIA, THE LARGER OF WHICH FIT BACTRODESMIUM.
/HUGHES & PIROZYNSKI, CAN. J. BOT. 50:2524-2525,1972 (ILLUS,NEW COMB) /ELLIS, MDH P 345-346,1976 (ILLUS) /MAGGI, BARTOLI & RAMBELLI, TBMS 71:148-154,1978 (ILLUS,NEW SPP) •

TRICELLULA VAN BEVERWIJK 1954. ANTONIE VAN LEEUWENHOEK 20:11-15 (ILLUS) /T. INAEQUALIS
STAURO,CETERI,HYALO,RADUL? OR NONSP,AQUATIC, 117C.
/VON ARX, GENERA P 216,1970 (ILLUS) /INGOLD, GUIDE P 76-77,1975 (ILLUS SUB VOLUCRISPORA)
?=*DENDROSPORIUM* PLAKIDAS & EDGERTON EX CRANE 1972
?=*LATERIRAMULOSA* MATSUSHIMA 1971
=*VOLUCRISPORA* HASKINS 1958 FIDE VON ARX 1970 •

TRICHAEGUM CORDA 1837. ICONES FUNGORUM 1:15 (ILLUS) /T. CLADOSPORIOIDES (LECTOTYPE)
?, ORIGINAL ILLUSTRATION NOT DIAGNOSTIC. •

TRICHOBOTRYS PENZIG & SACCARDO 1901. MALPIGHIA 15:245 /T. PANNOSA
AMERO,BLASTO,PHAEO,NONSP,SETAE, 39D, ON PLANTS (BAMBUSA, PHRAGMITES & OTHERS). COMPARE WITH PERICONIA.
/ELLIS, DH P 340-341,1971 (ILLUS) •

TRICHOCEPHALUM COSTANTIN 1887. LES MUCEDINEES SIMPLES P 106 /T. CURTUM (BERKELEY)
=PERICONIA TODE EX PERSOON 1801 FIDE ELLIS 1971 •

TRICHOCLADIUM HARZ 1871. BULL. SOC. IMP. MOSCOW 44:125-127 /T. ASPERUM
DIDYMO OR PHRAGMO,CETERI,PHAEO,NONSP,ALEURI, 68E & 92D, ON WOOD, LITTER & SOIL. ASCO STATE =CHAETOMIUM FIDE GAMS 1971 (P 19) & HALOSPHAERIA FIDE SHEARER & CRANE 1977 (MYCOLOGIA 69:1218-1223) OTHER STATES =IDRIELLA (H), CHALARA (H) & ?CLADOBOTRYUM (H). ALSO "PHIALOCONIDIA", SEE MATSUSHIMA 1975 (PL 23). SEE ALSO NEW COMBINATION UNDER THIELAVIOPSIS.
/BARRON, HYPHOMYCETES P 305-306,1968 (ILLUS) /ELLIS, DH P 66-68,1971 (ILLUS,KEY) /SHEARER & CRANE, MYCOLOGIA 63:244-245,1971 (ILLUS,NEW COMB) /DE HOOG & GRINBERGS, TBMS 64:341-343,1975 (ILLUS,NEW SP) /HUGHES & PIROZYNSKI, CAN. J. BOT. 50:2526-2528,1972 (ILLUS,NEW COMB) /NAG RAJ & KENDRICK, CHALARA P 110-113,1975 (ILLUS SUB CHALARA ELEGANS) /MATSUSHIMA, ICONES P 154-155 & PL,1975 (ILLUS,NEW SP) /ELLIS, MDH P 38,1976 (ILLUS SUB CULCITALNA) /CRANE & SHEARER, MYCOLOGIA 70:866-874,1978 (ILLUS,NEW SPP) /ELLIS, TMBS 70:67-71,1978 (ILLUS,CONIDIUM ONTOGENY)
=*CULCITALNA* MEYERS & MOORE 1940 FIDE HUGHES 1969
=*HARZIELLA* KUNTZE 1891 (Q.V.) •

TRICHOCONIELLA JAIN 1976. KAVAKA 3:37-39 (ILLUS,DATED 1975) /TRICHOCONIS PADWICKII GANGULY 1947 (J. INDIAN BOT. SOC. 26:238) =T. PADWICKII (GANGULY)
PHRAGMO,CETERI,HYALO,NONSP,APPEND, 70H, ON LEAVES (ORYZA). OTHER STATE ="BULBILS" (?MINIMEDUSA) SEE ELLIS 1971.
/ELLIS, DH P 494-495,1971 (ILLUS SUB ALTERNARIA PADWICKII) •

TRICHOCONIS CLEMENTS 1909. GENERA OF FUNGI P 145 & 176 (NO ILLUS) /=PIRICULARIA CAUDATA APPEL & STRUNK (ILLUS IN ZENTRALB. BAKT., ABT. 2,11:556-557,1903) =T. CAUDATA (APPEL & STRUNK)
PHRAGMO,CETERI,HYALO,NONSP OR RACHI,APPEND, 71F, ON FUNGI (BOTRYODIPLODIA, ASTERINA, MELIOLA & MANY OTHERS). COMPARE WITH MONOTRICHUM. OTHER STATE =HUMICOLA (H) FIDE DEIGHTON & PIROZYNSKI 1972 (P 22-24).
/DEIGHTON & PIROZYNSKI, CMI PAPERS 128:4-31,1972 (ILLUS,NEW SPP,NEW COMBS)
=*APPELIA* "(SACC.) SACCARDO" FIDE AINSWORTH 1971 •

TRICHODERMA PERSOON 1801. SYN. METH. FUNG. P 230 /T. VIRIDE (LECTOTYPE FIDE HUGHES 1958)
AMERO,GLOIO,HYALO,PHIAL, 16E, ON SOIL, WOOD, LITTER & OTHER SUBSTRATES. ASCO STATE =HYPOCREA FIDE RIFAI 1969 OR PODOSTROMA FIDE VON ARX 1974 (GENERA P 123) OR ?THUEMENELLA SEE RIFAI & WEBSTER 1965 (TBMS 48:409-413).
/RIFAI, CMI PAPERS 116:1-56,1969 (ILLUS,KEY,REVIS)
=*ALEURISMA* LINK 1809 FIDE HUGHES 1958
=*PACHYBASIUM* SACCARDO 1885 FIDE HUGHES 1958
=*PYRENIOPSIS* KUNTZE 1898 FIDE RIFAI 1969
=*SPORODERMA* MONTAGNE 1856 SEE RIFAI 1969
=*TOLYPOMYRIA* PREUSS FIDE HUGHES 1969
?=*TOLYPOCLADIUM* GAMS 1971 •

TRICHODERMIA HOFFMAN 1791 (DEUTSCHL. FL. DESCRIPT. P 7, TAB 10 FIG 1)
=TRICHOTHECIUM FIDE SACCARDO 1901 (SYLL. FUNG. 15:424) •

TRICHODOCHIUM SYDOW 1927. ANNALES MYCOLOGICI 25:159-160 /T. DISSEMINATUM
DIDYMO OR PHRAGMO,BASO,PHAEO,ANNEL,SPORODOCHIA,STROMA,SETAE, 60C, ON LEAVES (RAPANEA).
/ELLIS, CMI PAPERS 111:36-39,1967 (ILLUS) /ELLIS, DH P 149,1971 (ILLUS) •

TRICHOFUSARIUM BUBAK 1906. BULL. HERB. BOISS. SER 2, 6:488 (NO ILLUS) /FUSARIUM ROSEUM VAR. RUSCI SACCARDO
=FUSARIUM LINK 1809 FIDE AINSWORTH 1961 •

TRICHOLECONIUM CORDA 1837. ICONES FUNGORUM 1:17 /T. ROSEUM =SARCOPODIUM CIRCINATUM EHRENBERG 1818 FIDE HUGHES
=SARCOPODIUM EHRENBERG 1818 FIDE HUGHES 1958 •

TRICHOMYCES MALMSTEN 1845. TRANSLATION IN ARCH. ANAT. PHYSIOL. WISS. MED. 1848:1-19 /T. TONSURANS
=TRICHOPHYTON MALMSTEN 1845 FIDE DODGE 1935 •

TRICHOPHYTON MALMSTEN 1845. ARCH. ANAT. PHYS. WISS. MED. 1848:1 (THIS IS A TRANSLATION INTO GERMAN OF THE RARE SWEDISH ORIGINAL PUBLICATION OF 1845) /T. TONSURANS
PHRAGMO,CETERI,HYALO,NONSP,ALEURI, 89G, ON SOIL, MAN, ANIMALS & KERATINACEOUS LITTER. ASCO STATE =ARTHRODERMA (SEE DAWSON & GENTLES 1961). OTHER CONIDIAL STATES =CHRYSOSPORIUM ("MICROCONIDIA") & "ARTHROCONIDIA".
/DAWSON & GENTLES, SABOURAUDIA 1;49-57,1961 (ILLUS,ASCO STATES) /AJELLO, SABOURAUDIA 6:147-159,1968 (REVIS) /REBELL & TAPLIN, DERMATOPHYTES 124 PP,1970 (ILLUS,KEY) /PADHYE & CARMICHAEL, CAN. J. BOT. 49:1525-1540,1971 (ILLUS, KEY TO ARTHRODERMA) /AJELLO, IN RECENT ADV. MED. & VET. MYCOL. P 289-297,1977 (REFS,LIST OF SPECIES) /AJELLO & PADHYE, MYCOTAXON 8:383-384,1979 (EMEND)
=*FAVOTRICHOPHYTON* NEVEU-LEMAIRE 1921
=*ACHORION* REMAK 1845
=*CHLAMYDOALEUROSPORIA* GRIGORAKIS 1924
=*ECTOTRICHOPHYTON* CASTELLANI & CHALMERS 1919
=*ENDODERMOPHYTON* CASTELLANI 1910
=*GRUBYELLA* OTA & LANGERON 1923
=*KAUFMANNWOLFIA* GALGOCZY & NOVAK 1962 FIDE AJELLO 1968
=*KERATINOMYCES* VANBREUSEGHEM 1952 FIDE AJELLO 1968
=*LANGERONIA* VANBREUSEGHEM 1950 FIDE AJELLO 1968
=*LOPHOPHYTON* MATRUCHOT & DASSONVILLE 1889
=*MEGATRICHOPHYTON* NEVEU-LEMAIRE 1921
=*MICROIDES* DE VROEY 1970
=*MICROTRICHOPHYTON* NEVEU-LEMAIRE 1921
=*NEOTRICHOPHYTON* CASTELLANI & CHALMERS 1919
=*PINOYELLA* CASTELLANI & CHALMERS 1919 FIDE STOCKDALE, MACKENZIE & AUSTWICK 1965
=*SABOURAUDIELLA* BOEDIJN 1951
=*SCHOENLEINIUM* JOHAN-OLSEN 1897 FIDE CARMICHAEL 1973
=*TRICHOMYCES* MALMSTEN 1845 FIDE DODGE 1935 •

TRICHOSPORIELLA KAMYSCHKO EX GAMS & DOMSCH 1969. NOVA HEDWIGIA 18:19-20 (ILLUS) /T. HYALINA =T. CEREBRIFORMIS (DE VRIES & KLEINE-NATROP) GAMS 1971 (IN VON ARX, PERSOONIA 6:184)
AMERO,CETERI,HYALO,NONSP,ALEURI, 53A, ON SOIL. ASCO STATE =THIELAVIA SEE MATSUSHIMA 1975 (ICONES PL 404:1), & DIPLOGELASINOSPORA SEE UDAGAWA & HORIE 1972 (J. JAPANESE BOT. 47:297-305,ILLUS)
/VON ARX, GENERA P 232-234,1970 (ILLUS) •

TRICHOSPORIUM - SEE TRICHOSPORUM •

TRICHOSPORON BEHREND 1890 (BERL. KLIN. WSCHR. 27:464) =BLASTOMYCETES. SEE FIG 42D. •

TRICHOSPORONOIDES HASKINS & SPENCER 1967. CAN. J. BOT. 45:515-520 (ILLUS) /T. OEDOCEPHALIS
AMERO,CETERI,PHAEO,AMPUL, 31B, ON BROOD CELLS IN HONEY COMB (APIS). COMPARE WITH COLLECEPHALUS. OTHER STATE =TRICHOSPORON (BLASTOMYCETES).
/DE HOOG, CBS STUDIES 15:130-132,1977 (ILLUS) •

TRICHOSPORUM FRIES 1825. SYST. ORB. VEG. P 306-307 (WITHOUT ANY SPECIES) /?, 12 SPECIES INCLUDED IN SUMMA VEG. SCAND. P 492, 1849
=NOMEN ILLEGITIMUM, NON TRICHOSPORIUM DON 1822 FIDE HUGHES 1958. •

TRICHOSTERIGMA PETCH 1923. TBMS 8:215 /T. CLAVISPORUM
=HIRSUTELLA FIDE PETCH 1924 (TBMS 10:43) •

TRICHOSTROMA CORDA 1829. STURMS DEUTSCHL. FLORA III, (PILZE), BD. 2, HEFT 9:131 /T. PURPURASCENS (LECTOTYPE)
=NOMEN DUBIUM, SEE HUGHES 1958, NON TRICHOSTROMA LINK 1826 SINE SPECIEI NOMINE. •

TRICHOTHALLUS STEVENS 1925. BULL. BERNICE P BISHOP MUSEUM 19:85-86 /T. HAWAIENSIS
?=ANTENNATULA FRIES EX STRAUSS 1850 SEE HUGHES 1976 (MYCOLOGIA 68:773). ASCO STATE =TRICHOPELTHECA FIDE HUGHES 1972.
/HUGHES, NEW ZEALAND J. BOT. 10:225-242,1972 (ILLUS) /HUGHES, MYCOLOGIA 68:703-707 & 773,1976 (ILLUS,REVIS) •

TRICHOTHECA KARSTEN 1887. MEDDEL SOC. FAUNA FL. FENN. 14:101 /T. ALBA
?, NOT ILLUSTRATED BY AUTHOR. •

TRICHOTHECIUM LINK 1809. MAG. GES. NATURF. FREUNDE, BERLIN 3:18 /T. ROSEUM (PERSOON)
DIDYMO,BASO,HYALO,MISC, 67E, ON FUNGI (CLAVICEPS) & LITTER. ASCO STATE =HYPOMYCES FIDE TUBAKI 1960.
/RIFAI & COOKE, TBMS 49:147-168,1966 (ILLUS,REVIS) /MADELIN, TBMS 49:275-288,1966 (ILLUS,NEW COMB) /MATSUSHIMA, ICONES P 156 & PL,1975 (ILLUS)
=*CEPHALOTHECIUM* CORDA 1838 (PROBABLY), SEE HUGHES 1958
=*HYPHELIA* FRIES 1825 FIDE HUGHES 1958
=*HYPHODERMA* FRIES 1849 SEE DONK 1962
=*TRICHODERMIA* HOFFMAN FIDE SACCARDO FROM AINSWORTH 1971 •

TRICHURUS CLEMENTS & SHEAR 1896. IN CLEMENTS & POUND, BOT. SURV. NEBRASKA 4:7 /T. CYLINDRICUS
AMERO,BASO,PHAEO,ANNEL,SYNNEMATA,SETAE, 19A, ON SOIL & LITTER.
?=CEPHALOTRICHUM LINK 1809. T. CYLINDRICUS MATCHES LINK'S DESCRIPTION & ILLUSTRATION OF C. RIGESCENS (FIDE CARMICHAEL).
/SWART, ANTONIE VAN LEEUWENHOEK 30:257-260,1964 (ILLUS) /BARRON, HYPHOMYCETES P 310-311,1968 (ILLUS) /ELLIS, DH P 331-332,1971 (ILLUS) /FAHMY & YUSEF, ACTA PHYTOPATH. ACAD. SCI. HUNGARICAE 9:81-87 & 89-98,1974 (SYNNEMA,ILLUS,NOT SEEN) /MATSUSHIMA, ICONES P 156 & PL,1975 (ILLUS) /SHOEMAKER & KOKKO, FUNGI CANADENSES 100,1977 (ILLUS) /HAMMILL, CAN. J. BOT. 55:233-244,1977 (ILLUS) •

TRICLADIUM INGOLD 1942. TBMS 25:388-389 & PL 13 FIG 8-9 /T. SPLENDENS
STAURO,CETERI,HYALO,NONSP,AQUATIC, 123F.
/TUBAKI, NAGAOA 1960:18-19,1960 (ILLUS) /NAWAWI, TBMS 63:267-272,1974 (ILLUS,NEW SP) /SUTTON, TBMS 64:422-424,1975 (ILLUS,NEW SP) /INGOLD, GUIDE P 54-59,1975 (ILLUS) /MATSUSHIMA, ICONES P 156 & PL,1975 (ILLUS,NEW SP) •

TRICORNISPORA BONAR 1967. MYCOLOGIA 59:596-599 /T. BAMBUSAE
=TRIDENTARIA PREUSS 1852 FIDE KENDRICK 1973 •

TRIDENTARIA PREUSS 1852. LINNAEA 25:74 (NO ILLUS) /T. ALBA
STAURO,CETERI,HYALO,NONSP,AQUATIC, 118F.
/INGOLD, GUIDE P 42-43,1975 (ILLUS) /MATSUSHIMA, ICONES P 156 & PL,1975 (ILLUS)
=*FUMAGOPSIS* SPEGAZZINI 1910 FIDE KENDRICK 1973
=*TRICORNISPORA* BONAR 1967 FIDE KENDRICK 1973 •

TRIGLYPHIUM FRESENIUS 1852. BEITR. ZUR. MYKOL. P 44 & PL 9, FIG 6 /T. ALBUM
?
/LINDAU, RABENH. KRYPT. FL. 2 AUFL. 1 BD. 9:590-591,1910 (ILLUS) •

TRIMMATOSTROMA CORDA 1837. ICONES FUNGORUM 1:9 /T. SALICIS
DICTYO OR PHRAGMO,ARTHRO,PHAEO,MISC,SPORODOCHIA, 91B, ON BRANCHES (SALIX, BETULA), LITTER & SOIL.
/ELLIS, DH P 41-42,1971 ELLIS, MDH P 27-30,1976 (ILLUS,KEY) /MANOHARACHARY, RAGHUVEER & RAMARAO, CURR. SCI. 46:788,1977 (ILLUS,NEW SP) /SUTTON & GANAPATHI, NEW ZEALAND J. BOT. 16:529-533,1978 (ILLUS,NEW SP) •

TRINACRIUM RIESS 1852. IN FRESENIUS, BEITR. MYKOLOGIE P 42 & TAF. V, FIGS 14-17 /T. SUBTILE

STAURO,CETERI,HYALO,NONSP OR RACHI,AQUATIC, 125F, ON WOOD, LITTER & FUNGI (PYTHIUM).
/BARNETT, IMPERFECT FUNGI NO. 112,1960 (ILLUS) /MATSUSHIMA, ICONES P 157 & PL,1975 (ILLUS,NEW SP) •

TRIPLICARIA KARSTEN 1889. HEDWIGIA 28:195 (NO ILLUS) /T. HYPOXYLOIDES
?, ASCO STATE =HYPOXYLON FIDE AINSWORTH 1961. •

TRIPOCONIDIUM SUBRAMANIAN 1977. KAVAKA 5:93-98 (NOT SEEN) /TRIPOSPORINA APHANOPHAGA DRECHSLER =T. APHANOPHAGUM (DRECHSLER)
STAURO,CETERI,HYALO,NONSP, ON NEMATODES. 117A
/DRECHSLER, MYCOLOGIA 29:531-535,1937 (ILLUS) •

TRIPOSPERMUM SPEGAZZINI 1918. PHYSIS 4:295 /T. ACERINUM (SYDOW)
STAURO,CETERI,PHAEO,NONSP,AQUATIC, 126F, ASCO STATE =TRIPOSPORIOPSIS FIDE HUGHES 1976.
/HUGHES, CMI PAPERS 46:10-22,1951 (ILLUS,REVIS) /ELLIS, DH P 86-87,1971 (ILLUS) /INGOLD, GUIDE P 60-61,1975 (ILLUS) /MATSUSHIMA, ICONES P 157-158 & PL,1975 (ILLUS,NEW SP) /HUGHES, MYCOLOGIA 68:712-714 & 809-811,1976 (ILLUS) /SINCLAIR & MORGAN-JONES, MYCOTAXON 9:479-481,1979 (ILLUS,NEW SP)
?=*HEPTASTER* CIFERRI, BATISTA & NASCIMENTO 1956 SEE HUGHES 1976
?=*MEGASTER* CIFERRI ET AL. 1956 SEE HUGHES 1976
?=*PENTAPOSPORIUM* BATISTA 1957 SEE HUGHES 1976 •

TRIPOSPORINA HOHNEL 1912. S.B. AKAD. WISS. WIEN 121:410-411 (ILLUS) /T. UREDINICOLA
STAURO,CETERI,HYALO,ANNEL, ~~117A,~~ ON FUNGI (PUCCINIA). FOR T. APHANOPHAGA SEE TRIPOCONIDIUM.
/DEIGHTON & PIROZYNSKI, CMI PAPERS 128:96-99,1972 (ILLUS) /MATSUSHIMA, ICONES P 158 & PL,1975 (ILLUS,NEW SP-PROBABLY NOT CONGENERIC WITH T. UREDINICOLA) •

TRIPOSPORIUM CORDA 1837. ICONES FUNGORUM 1:16 & TAB. IV, FIG 220 /T. ELEGANS
STAURO,CETERI,PHAEO,ANNEL, 126A, ON WOOD & LITTER. ASCO STATE =BATISTINULA SEE DE HOOG 1977 (CBS STUDIES 15:140).
/HUGHES, CMI PAPERS 46:1-10,1951 (ILLUS) /ELLIS, DH P 136-137,1971 (ILLUS) /CRANE, TBMS 63:596-598,1974 (ILLUS) /CRANE & DUMONT, CAN. J. BOT. 53:843-851,1975 (ILLUS,NEW SP) /GAMUNDI, ARAMBARRI & GIAIOTTI, DARWINIANA 21:91-92 & FIG 4A-E,1977 (ILLUS,NEW SP) •

TRIRAMULISPORA MATSUSHIMA 1975. ICONES P 158-159 & PL 334:4, PL 361:1-2 /T. OBCLAVATA
STAURO,CETERI,PHAEO,NONSP, 121C, ON LITTER (BROAD-LEAF TREES). SIMILAR TO TRISCELOPHORUS. •

TRISCELOPHORUS INGOLD 1943. TBMS 26:148-152 & PL VIII /T. MONOSPORUS
STAURO,CETERI,HYALO,NONSP,AQUATIC, 120H.
/TUBAKI, BULL. NAT. SCI. MUS. TOKYO 3:263-264,1957 (ILLUS) /NAWAWI, TBMS 64:345-347,1975 (ILLUS,NEW SP) /INGOLD, GUIDE P 62-63,1975 (ILLUS) •

TRISULCOSPORIUM HUDSON & SUTTON 1964. TBMS 47:197-200 & FIG 1 /T. ACERINUM
STAURO,CETERI,HYALO,NONSP,AQUATIC, 120E.
/INGOLD, GUIDE P 76-77,1975 (ILLUS) •

TRITIRACHIUM LIMBER 1940. MYCOLOGIA 32:23-26 (ILLUS) /T. DEPENDENS
AMERO,CETERI,HYALO OR PHAEO,RACHI, 26C, ON ROOTS & BULBS (YUCCA, LILIUM) & LEAVES (MUSA) & TWIGS (SALIX). COMPARE WITH ACRODONTIUM. ASCO STATE ?=CEPHALOTHECA SEE BOOTH 1961 (CMI PAPERS 83:12-13).
/BARRON, HYPHOMYCETES P 313-314,1968 (ILLUS) /DE HOOG, CBS STUDIES 1:19-23,1972 (ILLUS) /MATSUSHIMA, ICONES P 159-160, & PL,1975 (ILLUS,NEW SPP) /COLE & SAMSON, CONIDIAL FUNGI P 36-37,1979 (ILLUS CONIDIUM ONTOGENY)
=*SPIROTRICHUM* SAITO EX VAN BEYMA 1940 FIDE DE HOOG 1972 •

TROCHOPHORA MOORE 1955. MYCOLOGIA 47:90 (NO ILLUS) /T. SIMPLEX (PETCH)
HELICO,CETERI,PHAEO,RADUL,SYNNEMATA, 104E, ON LEAVES & LITTER (DAPHNIPHYLLUM). NOM. NOV. FOR HELICOSTILBE HOHNEL EMEND. LINDER 1929.
/ELLIS, DH P 185-186,1971 (ILLUS) /MATSUSHIMA, ICONES P 161 & PL 259,1975

(ILLUS)
=*CURVIDIGITUS* SAWADA 1943 (NOMEN NUDUM) FIDE PIROZYNSKI 1972 •

TROPOSPORELLA KARSTEN 1892. HEDWIGIA 31:299 /T. FUMOSA
HELICO,CETERI,PHAEO,NONSP,SPORODOCHIA, 104F-G, ON BARK (POPULUS) & LITTER (JUNIPERUS & PINUS).
/LINDER, ANN. MISSOURI BOT. GARD. 16:334-335,1929 (ILLUS) /ELLIS, DH P 85,1971 (ILLUS) /SUTTON, CMI PAPERS 132:123-124,1973 (ILLUS) /ELLIS, MDH P 47,1976 (ILLUS,NEW COMB)
=*HELICOSPORINA* RAMBELLI 1960 FIDE KENDRICK •

TROPOSPORIUM HARKNESS 1884. BULL. CALIFORNIA ACAD. SCI. 1:39-40 (NO ILLUS) /T. ALBUM
HELICO,CETERI,HYALO,NONSP,SPORODOCHIA, 105C, ON LITTER & WOOD (CORYLUS).
/LINDER, ANN. MISSOURI BOT. GARD. 16:345,1929 (ILLUS) /PEEK & SOLHEIM, MYCOLOGIA 50:847-851,1958 (ILLUS) •

TUBAKIA SUTTON 1973. TBMS 60:164-165 (NO ILLUS) /ACTINOPELTE JAPONICA SACCARDO 1913 =T. JAPONICA (SACCARDO)
?=COELOMYCETES, REPLACES ACTINOPELTE SACCARDO NON ACTINOPELTE SITZENB. 1861. HAS A RHIZOTHYRIUM STATE & IS USUALLY CONSIDERED A COELOMYCETE.
/YOKOYAMA & TUBAKI, INST. FOR FERMENTATION, OSAKA, RESEARCH COMMUN. 5:43-77,1971 (ILLUS,REVIS) /VON ARX, GENERA P 173,1974 (REFS) •

TUBERCULARIA TODE EX PERSOON 1801. SYN. METH. FUNG. P 111 /T. VULGARIS TODE 1790 =T. STATE OF NECTRIA CINNABARINA FIDE BOOTH 1959 (CMI PAPERS 73:27)
AMERO,GLOIO,HYALO,PHIAL,SPORODOCHIA, 12C, ON STEMS & BRANCHES (MANY WOODY PLANTS). ASCO STATE =NECTRIA FIDE BARRON 1968 OR THYRONECTRIA FIDE VON ARX 1974 (GENERA P 122-123). TUBERCULARIA TODE EX FRIES 1832, WITH THE SAME TYPE SPECIES, WAS CONSERVED AGAINST TUBERCULARIA WIGGERS 1780 (LICHEN) 1961 ICBN P 219.
/BARRON, HYPHOMYCETES P 314-315,1968 (ILLUS) /SMYK, UKRAIN. BOT. ZHUR. 32:655-657,1975 (ILLUS)
=*KNYARIA* KUNTZE FIDE AINSWORTH 1971
=*STROMATERIA* CORDA 1837 FIDE SACCARDO FROM AINSWORTH 1971
?=*ENDOCONOSPORA* GJAERUM 1971 •

TUBERCULARIELLA HOEHNEL 1914. ZEIT. GAR. 5:209 /T. SANGUINEA (FUCKEL)
?=CRYPTOSPORIOPSIS SEE PETRAK 1966 (SYDOWIA 19:227) FROM AINSWORTH 1971. •

TUBERCULARIOPSIS HOHNEL 1909. S.B. AKAD. WISS. WIEN 118:421-422 (ILLUS) /T. ANOMALA
? •

TUBERCULARIS CLEMENTS & SHEAR 1931 FOR TUBERCULARIOPSIS •

TUBERCULINA SACCARDO 1880. MICHELIA 2:34 /T. PERSICINA (DITMAR)
AMERO,CETERI,PHAEO,PHIAL,ACERVULI, 14D, ON FUNGI (RUSTS). BASIDIO STATE =HELICOBASIDIUM SEE BUDDIN & WAKEFIELD 1927 (TBMS 12:116-140).
/VON ARX, GENERA P 156-157,1970 (ILLUS) /GARDNER, MILLER & KUHLMAN, MYCOLOGIA 71:848-852,1979 (ILLUS)
=*CORDALIA* GOBI 1885 FIDE LINDAU IN RABENHORST 1910
=*UREDINULA* FIDE SPEGAZZINI FROM AINSWORTH 1971 •

TUBERCULIS CLEMENTS & SHEAR 1931 FOR TUBERCULARIELLA •

TUBERCULISPORA DEIGHTON & PIROZYNSKI 1972. CMI PAPERS 128:92-94 (ILLUS) /T. JAMAICENSIS
AMERO,CETERI,HYALO,NONSP, 24H, ON FUNGI (IRENOPSIS). ASCO STATE =NECTRIA SEE DE HOOG 1977 (CBS STUDIES 15:138). •

TUBURCINIELLA ZAMBETTAKIS 1970. REV. MYCOL. 35:165 /NO SPECIES MENTIONED
=NOMEN NUDUM, REPLACEMENT FOR PAIPALOPSIS KUHN 1883 •

TUCAHUS RAFINESQUE 1830 =GEMMULARIA RAFINESQUE 1819 FIDE DONK 1962 (TAXON 11:101-102) •

TUREENIA HALL 1915. PHYTOPATHOLOGY 5:57-58 (ILLUS) /T. JUNCOIDEA
=ARTHRINIUM KUNZE 1817 FIDE ELLIS 1971 •

TYLOMYCES CORTINI 1921. RENDIC. R. ACCAD. NAZ. LINCEI ROMA 30:113-116 /T. GUMMIPARUS
?, ORIGINAL PUBLICATION NOT SEEN. •

TYMPANOSPORIUM GAMS 1974. ANTONIE VAN LEEUWENHOEK 40:478 (471-479,ILLUS) /T. PARASITICUM
AMERO,BASO,HYALO,ANNEL,SPORODOCHIA, 5E, ON FUNGI (TUBERCULARIA). OTHER STATE ="PHIALOCONIDIA". •

UBERISPORA PIROZYNSKI & HODGES 1973. CAN. J. BOT. 51:166-168 (ILLUS) /ARACHNOPHORA SIMPLEX ICHINOE =U. SIMPLEX (ICHINOE)
STAURO,CETERI,PHAEO,ANNEL,APPEND, 114E, ON LITTER (PERSEA, MYRICA).
/ELLIS, MDH P 107-108,1976 (ILLUS) •

ULOCLADIUM PREUSS 1851. IN STURM'S, DEUTSCHL. FLORA III (PILZE), HEFT 30:83-84 TAB 42 /U. BOTRYTIS
DICTYO,CETERI,PHAEO,RACHI, 99B, ON SOIL, LITTER, PLANTS & MANY OTHER SUBSTRATES. DIFFERENCE FROM ALTERNARIA NOT CLEAR. ASCO STATE =LASIOBOTRYS SEE BONAR 1928 (MYCOLOGIA 20:292-294,ILLUS). COMPARE WITH DICTYOSPIROPES.
/SIMMONS, MYCOLOGIA 59:67-92,1967 (ILLUS,KEY) /ELLIS, DH P 497-498,1971 (ILLUS) /MATSUSHIMA, ICONES P 161 & PL,1975 (ILLUS) /ELLIS, MDH P 427-433,1976 (ILLUS,KEY) /MOUBASHER & ABDEL-HAFEZ, TBMS 69:164-166,1977 (ILLUS,NEW SP)
=*PSEUDOSTEMPHYLIUM* (WILTSHIRE) SUBRAMANIAN 1961 FIDE SIMMONS 1967 •

UMBELLIDION SUTTON & HODGES 1975. NOVA HEDWIGIA 26:529-533 (ILLUS) /U. RADULANS
AMERO,CETERI,HYALO,RADUL, 21D, ON LITTER (EUCALYPTUS). •

UMBELLULA MORRIS 1955. MYCOLOGIA 47:602-605 (ILLUS) /U. TERRESTRIS (TIMONIN)
=PSEUDOBOTRYTIS KRZEMIENIEWSKA & BADURA 1954 FIDE SUBRAMANIAN 1956 •

UMBELOPSIS AMOS & BARNETT 1966. MYCOLOGIA 58:805-808 (ILLUS) /U. VERSIFORMIS
AMERO,CETERI,HYALO,NONSP, ~~9~~7E, ON ROOTS (BROAD-LEAF TREES) & SOIL. 14
/BARRON, HYPHOMYCETES P 316-318,1968 (ILLUS) /WATANABE, TRANS. MYCOL. SOC. JAPAN 18:242-244,1977 (ILLUS,NEW SP) •

UNCIGERA SACCARDO 1885. IN SACCARDO & BERLESE, ATTI R. ISTIT. VEN. SCI. LETT. ARTI VI,3:741, TAB IX, FIG 10 /U. CORDAE SACCARDO & BERLESE 1885
AMERO OR DIDYMO,GLOIO,HYALO,PHIAL, 4E, ON LITTER. SIMILAR TO GONYTRICHUM.
/LINDAU, RABENH. KRYPT. FL., 2 AUFL., 1 BD.,8:338,1907 (ILLUS) •

UNCISPORA SINCLAIR & MORGAN-JONES 1979. MYCOTAXON 8:140-143 (ILLUS) /U. HARROLDII
SCOLECO,CETERI,PHAEO,NONSP,SYNNEMATA, 70E, ON LITTER (BETULA). •

UREDINULA SPEGAZZINI 1880. ANAL. SOC. CIENTIF. ARGENT. 10:10-11 /?
=TUBERCULINA FIDE SPEGAZZINI FROM AINSWORTH 1971 •

UROBASIDIUM GIESENHAGEN 1892. FLORA, JENA, 76:139-141 /U. ROSTRATUM
=ZYGOSPORIUM MONTAGNE 1842 FIDE ELLIS 1971 •

UROPHIALA VUILLEMIN 1910. C. R. ACAD. SCI. PARIS 150:883 & BULL. SEANC. SOC. SCI. NANCY, SER. 3, 11:158 /U. MYCOPHILA
=ZYGOSPORIUM MONTAGNE 1842 FIDE HUGHES 1958 •

UROSPORIUM FINGERHUTH 1836. LINNAEA 10:231-232 /U. CURVATUM
?, NOT ILLUSTRATED BY AUTHOR. •

USTILAGINOIDEA BREFELD 1895. UNTERS. MYKOL. 12:195 & 235 & TAF 12, FIG 22-30 /U. ORYZAE (PATOUILLARD) (LECTOTYPE) =U. VIRENS (COOKE) TAKAHASHI 1896 FIDE ELLIS 1971
AMERO,CETERI,PHAEO,NONSP,ALEURI?,SPORODOCHIA, 50D, ON OVARIES (ORYZA & OTHER GRASSES).

/ELLIS, DH P 42-43,1971 (ILLUS)
=*DUBIOMYCES* LLOYD 1921 FIDE DIEHL FROM AINSWORTH 1971 •

UVARISPORA GOOS & PIROZYNSKI 1975. CAN. J. BOT. 53:2930-2932 (ILLUS) /U. LIGNICOLA
BULBILS OR STAURO,GLOIO,HYALO,NONSP,SPORODOCHIA, 113F, ON WOOD. COMPARE WITH PYRAMIDOSPORA. •

VAKRABEEJA SUBRAMANIAN 1956. J. INDIAN BOT. SOC. 35:464-466 (ILLUS) /V. SIGMOIDEA (CAVARA)
=NAKATAEA HARA 1939 FIDE ELLIS 1971 (DH P 219) •

VALDENSIA PEYRONEL 1923. STAZ. SPERIM. ITAL. AGRIC. 56:521-538 /V. HETERODOXA
STAURO,CETERI,PHAEO,NONSP, 125E, ON LEAVES (GAULTHERIA). ASCO STATE =VALDENSINIA PEYRONEL 1953. OTHER STATES ?=GLIOCLADIUM (H) & "SCLEROTIA" FIDE REDHEAD & PERRIN 1972.
/REDHEAD & PERRIN, CAN. J. BOT. 50:409-412,1972 (ILLUS SUB ASTEROBOLUS)
=*ASTEROBOLUS* REDHEAD & PERRIN 1972 •

VANBEVERWIJKIA AGNIHOTHRUDU 1961. TBMS 44:51-54 (ILLUS) /V. SPIROSPORA
HELICO,GLOIO,HYALO,PHIAL,SPORODOCHIA, 103E, ON WOOD (SOMETIMES IN WATER). EMENDED BY SHEARER & CRANE 1971, PERHAPS THE SAME AS EVERHARTIA.
/SHEARER & CRANE, MYCOLOGIA 63:249-251,1971 (ILLUS,EMEND) •

VANBREUSEGHEMIA BALABANOFF 1965. MYCOPATHOLOGIA 25:345 /?
=KERATINOMYCES VANBREUSEGHEM 1952 FIDE INDEX OF FUNGI 3:448,1968 •

VARICOSPORINA MEYERS & KOHLMEYER 1965. CAN. J. BOT. 43:915-921 (ILLUS) /V. RAMULOSA
STAURO,CETERI,HYALO,NONSP,AQUATIC, 123E. •

VARICOSPORIUM KEGEL 1906. BER. DEUTSCHL. BOT. GES. 24:213-216 (ILLUS) /V. ELODEAE
STAURO,CETERI OR BLASTO,HYALO,NONSP,AQUATIC, 123A, ASCO STATE =HYMENOSCYPHUS FIDE TUBAKI 1966.
/BARRON, HYPHOMYCETES P 318-319,1968 (ILLUS) /TUBAKI, TBMS 49:345-349,1966 (ILLUS,NEW SP) /NAWAWI, NOVA HEDWIGIA 24:39-43,1973 (ILLUS,NEW SP) /INGOLD, GUIDE P 64-67,1975 (ILLUS) /MATSUSHIMA, ICONES P 162 & PL,1975 (ILLUS) •

VASCULOMYCES ASHBY 1913. BULL. DEPT. AGR. JAMAICA 2:151 & PL 28, FIGS 22-23 /V. XANTHOSOMAE
?, STERILE. •

VELLOSIELLA RANGEL 1915. BOL. AGRIC. SAO PAULO 16A:144-151 (ILLUS) /V. CAJANI (HENNINGS)
=MYCOVELLOSIELLA RANGEL 1917, NON VELLOZIELLA BAILL. 1886 •

VERMICULARIA FRIES 1849 =COLLETOTRICHUM CORDA 1831 FIDE VON ARX 1970 •

VERMICULARIOPSIELLA BENDER 1932. MYCOLOGIA 24:412 (NO ILLUS) /EXCIPULA IMMERSA DESMAZIERES 1857 (BULL. SOC. BOT. FRANCE 4:911 & PL CRYPT. DE FRANCE, SER 2, NO. 268,1855) =V. IMMERSA (DESMAZIERES)
?, FOR VERMICULARIOPSIS HOHNEL 1918 NON TORREND 1912, NOT ILLUSTRATED BY BENDER OR HOHNEL OR DESMAZIERES.
/HOHNEL, MITT. BOT. INST. TECHN. HOCHSCH., WIEN 6:30-32,1929 (NO ILLUS) •

VERMICULARIOPSIS HOHNEL 1918 (BER. DEUTSCH. BOT. GES. 36:317) -SEE VERMICULARIOPSIELLA •

VERMISPORA DEIGHTON & PIROZYNSKI 1972. CMI PAPERS 128:87-88 (ILLUS) /V. GRANDISPORA
SCOLECO,CETERI,HYALO,NONSP, 78E, ON FUNGI (IRENOPSIS). COMPARE WITH DACTYLELLA & SPERMOSPORELLA. •

VERONAEA CIFERRI & MONTEMARTINI 1957. ATTI IST. BOT. LAB. CRITT. PAVIA 5,15:68-70 (ILLUS) /V. BOTRYOSA
AMERO OR DIDYMO,CETERI,HYALO OR PHAEO,RADUL , 61G, ON LITTER, LEAVES

(MUSA, BAMBUSA), FUNGI (HEMILEIA), WOOD, SOIL & DUNG. CLOSE TO SCOLECOBASIDIUM & DACTYLARIA. FOR V. SIMPLEX SEE ALSO SCOLECOBASIDIELLA.
/VON ARX, GENERA P 217-218,1970 (ILLUS) /ELLIS, DH P 245-246,1971 (ILLUS) /GEESON, TBMS 64:348-351,1975 (ILLUS,NEW SP) /ELLIS, MDH P 207-213,1976 (ILLUS,KEY,NEW COMBS,NEW SPP) /DE HOOG, CBS STUDIES 14:59-86,1977 (SUB RAMICHLORIDIUM) /DE HOOG & HERMANIDES-NIJHOF, CBS STUDIES 15:198,1977 (SYNO) /MORGAN-JONES, MYCOTAXON 8:149-151,1979 (ILLUS,NEW SP)
=*SYMPODINA* SUBRAMANIAN & LODHA 1964 FIDE VON ARX 1970 •

VERONAIA BENEDEK 1961. MYCOPATHOLOGIA 14:115 /V. CASTELLANII
=NOMEN CONFUSUM SEE AJELLO 1968 •

VERRUCISPORA SHAW & ALCORN 1967. PROC. LINN. SOC. N. S. WALES 92:171-173 & PL VI /V. PROTEACEARUM
PHRAGMO,CETERI,PHAEO,RADUL,STROMA, 85A, ON LEAVES (FINSCHIA, SMILACIS & HAKEA).
/ELLIS, DH P 274-275,1971 (ILLUS) /KAMAL & SINGH, CAN. J. BOT. 56:2785-2786,1978 (ILLUS,NEW SP) •

VERRUCOBOTRYS HENNEBERT 1973. PERSOONIA 7:193 & FIG 4, P 186 /V. GERANII (SEAVER 1947) =V. STATE OF SEAVERINA GERANII (SEAVER & HORNE) WHETZEL 1945
AMERO,CETERI,PHAEO,NONSP OR AMPUL,ALEURI,STROMA, 34C, ON PLANTS (GERANIUM). ASCO STATE =SEAVERINA WHETZEL FIDE HENNEBERT 1973 (=SCLEROTINIA).
/SEAVER, MYCOLOGIA 39:117,1947 (ILLUS SUB SEAVERINA) •

VERTICICLADIELLA HUGHES 1953. CAN. J. BOT. 31:653 & FIG 29 /V. ABIETINA (PECK)
AMERO,GLOIO,HYALO,RADUL, 27A, ON TRUNKS & ROOTS (PICEA, PINUS) & INSECTS (BARK BEETLES). ASCO STATE =CERATOCYSTIS FIDE KENDRICK 1962 OR CERATOCYSTIOPSIS FIDE UPADHYAY & KENDRICK 1975 (MYCOLOGIA 67:798-805) OR OPHIOSTOMA OR EUROPHIUM FIDE VON ARX 1974 (GENERA P 111-112 & 236-237).
/KENDRICK, CAN. J. BOT. 40:771-797,1962 (ILLUS) /KENDRICK & MOLNAR, CAN. J. BOT. 43:39-43,1965 (ILLUS) /ELLIS, DH P 366-367,1971 (ILLUS) /GAMBOGI & LORENZINI, TBMS 69:217-223,1977 (ILLUS) /DABINETT & WELLMAN, CAN. J. BOT. 56:2031-2049,1978 (DISCUSSION OF GENERA,NO ILLUS) /JOOSTE, TBMS 70:152-155,1978 (ILLUS)

VERTICICLADIUM PREUSS 1851. LINNAEA 24:127 /V. TRIFIDUM
AMERO,CETERI,HYALO,RADUL, 27E, ON LITTER (PINUS). ASCO STATE =DESMAZIERELLA FIDE HUGHES 1951.
/HUGHES, CMI PAPERS 43:6-15,1951 (ILLUS) /ELLIS, DH P 367-368,1971 (ILLUS) •

VERTICILLIASTRUM DASZEWSKA 1912. BULL SOC. BOT. GENEV, SER. 2, 4:302-303 & 313 (ILLUS) /V. GLAUCUM
?, ORIGINAL ILLUS. & DESCR. NOT DIAGNOSTIC.

VERTICILLIODOCHIUM BUBAK 1914. ANN. MYCOL. 12:219-220 & TAF 8 FIG 1-6 /V. TUBERCULARIOIDES (SPEGAZZINI)
=DENDRODOCHIUM BONORDEN 1851 FIDE TULLOCH 1972 (CMI PAPERS 130:36).
=*VERTICILLIS* CLEMENTS & SHEAR 1931 (ORTHOGRAPHIC VARIANT) •

VERTICILLIOPSIS COSTANTIN 1892. C. R. ACAD. SCI. PARIS 114:850 (NO ILLUS) /V. INFESTANS
=NOMEN DUBIUM FIDE GAMS 1971 (CEPHALO. SCHIMMEL. P 18). •

VERTICILLIS CLEMENTS & SHEAR 1931 FOR VERTICILLIODOCHIUM •

VERTICILLIUM NEES 1817. DAS SYSTEM .... P 57 & TAB. IV, FIG 55 /V. TENERUM =V. STATE OF NECTRIA INVENTA FIDE HUGHES 1958
AMERO,GLOIO,HYALO,PHIAL, 4C, ON PLANTS, WOOD, LITTER, SOIL, FUNGI & OTHER SUBSTRATES. ASCO STATE =NECTRIA OR CORDYCEPS OR TORRUBIELLA FIDE GAMS 1971 OR EPHEMEROASCUS, CALONECTRIA, HYPOCREA OR HYPOMYCES FIDE VON ARX 1974. OTHER STATES =MYCOGONE, DIHETEROSPORA, VOLUTELLA, DENDRODOCHIUM, HUMICOLA. SEE ALSO DIGITATISPORA DOGUET 1961.

/VON ARX, GENERA P 184-185,1970 (ILLUS,REFS) /CMI DESCRIPTIONS 255-260,1970 & 497-498,1976 (ILLUS) /ELLIS, DH P 535-536,1971 (ILLUS) /GAMS, CEPHALO. SCHIMMEL. P 167-195,1971 (REVIS,KEY,ILLUS) /SAMUELS, MYCOLOGIA 65:409-412,1973 (ILLUS WITH NECTRIA STATE) /VON ARX, GENERA P 103, 122-123 & 127,1974 (ILLUS,REFS) /BARRON, THE NEMATODE-DESTROYING FUNGI P 71-75 & 84,1977 (ILLUS,NEMATOPHAGOUS)
=*ACROSTALAGMUS* CORDA 1838 FIDE HUGHES 1958
=*GIBELLULOPSIS* BATISTA & MAIA 1959 FIDE CARMICHAEL 1973 •

VERTICIMONOSPORIUM MATSUSHIMA 1971. MICROFUNGI P 68 & FIG 91 & PL 4-8 & PL 33-3 /V. DIFFRACTUM
AMERO,CETERI,HYALO,MISC, 4F, ON LITTER (COCOS, PINUS, SEQUOIA). OTHER STATE(?) FORMED BY DISARTICULATION OF CONIDIOGENOUS CELLS & HYPHAE.
/GAMS, PERSOONIA 7:163,1973 (DISCUSSION) /MATSUSHIMA, ICONES P 162-163 & PL,1975 (ILLUS,NEW SP) •

VINCULUM ROY, DWIVEDI & KHANNA 1965. TBMS 48:113-115 & PL 9 /V. INDICUM
=BARNETTELLA INDICA (ROY, DWIVEDI & KHANNA) SATYANARAYANA & RAO 1971
=BARNETTELLA RAO & RAO 1964 FIDE SATYANARAYANA & RAO 1971 (MYCOPATHOLOGIA 45:267-268). •

VIRGARIA NEES 1816. DAS SYSTEM .....P. 54 /V. NIGRA (LINK)
AMERO,CETERI,PHAEO,RADUL, 21E, ON WOOD & BARK. SEE ALSO NODULISPORIUM COMPLEX.
/BARRON, HYPHOMYCETES P 323-324,1968 (ILLUS) /ELLIS, DH P 211-212,1971 (ILLUS) /MATSUSHIMA, ICONES P 163 & PL,1975 (ILLUS,NEW SP) /HUGHES, NEW ZEALAND J. BOT. 16:353-354,1978 (ILLUS) •

VIRGARIELLA HUGHES 1953. CAN. J. BOT. 31:653-654 /V. GLOBIGERA (SACCARDO & ELLIS)
AMERO,CETERI,PHAEO,RADUL, 24B, ON WOOD & LITTER. SEE ALSO NODULISPORIUM COMPLEX. ASCO STATE =DALDINIA SEE TUBAKI 1958 (J. HATTORI BOT. LAB. 20:211).
/ELLIS, DH P 195-196,1971 (ILLUS) /MATSUSHIMA, ICONES P 163 & PL,1975 (ILLUS,NEW SPP) /DE HOOG & HERMANIDES-NIJHOF, CBS STUDIES 15:204-205,1977 (NEW COMB) /LUNGHINI & RAMBELLI, GIORN. BOT. ITAL. 112:177-179,1978 (ILLUS) •

VIRGASPORIUM COOKE 1875. GREVILLEA 3:182 & PL 48, FIG 5 /V. MACULATUM
=CERCOSPORA FRESENIUS 1863 FIDE ELLIS 1971 •

VIRGATOSPORA FINLEY 1967. MYCOLOGIA 59:538-541 (ILLUS) /V. ECHINOFIBROSA
PHRAGMO,GLOIO,PHAEO,PHIAL,SYNNEMATA, 74G, ON LITTER (THEOBROMA).
/ELLIS, DH P 557-558,1971 (ILLUS) •

VISCOMACULA SPRAGUE 1950. MYCOLOGIA 42:758-759 /V. AENEA
?, NOT ILLUSTRATED BY AUTHOR. •

VOGLINOANA KUNTZE 1891. REV. GEN. PLANT. 2:874 (NO ILLUS) /CYSTOPHORA CRATERIOIDES RABENHORST =V. CRATERIOIDES (RABENHORST)
=NOMEN DUBIUM FIDE LINDAU 1907 (SUB CYSTOPHORA) •

VOLUCRISPORA HASKINS 1958. CAN. J. MICROBIOL. 4:278-279 (ILLUS P 275) /V. AURANTIACA
=TRICELLULA VAN BEVERWIJK 1954 FIDE VON ARX 1970 •

VOLUTELLA TODE 1790 EX FRIES 1832 NOM. CONS. 1974. SYSTEMA MYCOL. III:466-467 /V. CILIATA (ALBERTINI & SCHWEINITZ) FRIES EX FRIES (LECTOTYPE) =V. STATE OF NECTRIA CONSORS SEE SAMUELS 1977
AMERO,GLOIO,HYALO,PHIAL,SPORODOCHIA,SETAE, 12E, ON PLANTS, LITTER & SOIL. ASCO STATE =PSEUDONECTRIA FIDE VON ARX 1970 OR NECTRIA FIDE SAMUELS 1976 (MEM. N. Y. BOT. GARD. 26:22). CONSERVED AGAINST AN EARLIER NON-FUNGUS NAME (VOLUTELLA FORSK. 1775, LAURACEAE), SEE TAXON 23:647, 1974 & 24:242 & 244,1975. OTHER STATES =VERTICILLIUM & ACREMONIUM FIDE MATSUSHIMA 1975 (P 164). COMPARE WITH MYROTHECIUM.
/VON ARX, GENERA P 182-183,1970 (ILLUS) /TULLOCH, CMI PAPERS 130:8,1972 (REFS) /HAWKSWORTH & TULLOCH, TAXON 21:707-708,1972 (NOM. CONS. PROP.)

/MATSUSHIMA, ICONES P 164 & PL,1975 (ILLUS) /SAMUELS, MYCOLOGIA 69:255-262,1977 (ILLUS,NECTRIA STATE)
=*CHAETODOCHIUM* HOHNEL 1932 FIDE VON ARX 1970
?=*HOEHNELIELLA* BRESADOLA & SACCARDO 1902
=*MEDUSULA* CORDA 1837 FIDE LINDAU IN RABENHORST 1910
?=*MINIMIDOCHIUM* SUTTON 1969
?=*PSILONIA* FRIES 1825 FIDE AINSWORTH 1971
=*SCHIZOTRICHELLA* MORRIS 1956 FIDE CARMICHAEL 1973
=*THYSANOPYXIS* RABENHORST 1864 FIDE LINDAU 1910
?=*VOLUTINA* PENZIG & SACCARDO FIDE PIROZYNSKI 1972 •

VOLUTELLARIA (SACCARDO) SACCARDO 1886. SYLLOGE FUNGORUM 4:682 /ILLOSPORIUM ACAROIDES SACCARDO 1882 (MICHELIA 2:580) =V. ACAROIDES
?, NOT ILLUSTRATED BY AUTHOR. •

VOLUTELLIS CLEMENTS & SHEAR 1931. GENERA OF FUNGI P 389 /VOLUTELLOPSIS SULPHUREA TORREND 1914 (BULL. JARD. BOT. BRUX 4:12)
?, FOR VOLUTELLOPSIS TORREND NON SPEGAZZINI 1910 •

VOLUTELLOPSIS SPEGAZZINI 1910. REV. FAC. AGRON. Y VET. DE LA PLATA 6:197-198 (ILLUS) /V. CHILENSIS
?, ORIGINAL ILLUSTRATION & DESCRIPTION NOT DIAGNOSTIC. •

VOLUTELLOPSIS TORREND 1914 (IN DE WILDEMAN, BULL. JARD. BOT. BRUX. 4:12) -SEE VOLUTELLIS •

VOLUTINA PENZIG & SACCARDO 1902. MALPIGHIA 15:257 (NO ILLUS) /V. CONCENTRICA
AMERO,BASO,HYALO,PHIAL?, ON PLANTS & FUNGI (FIDE ATCC 22253 & 22371).
/PENZIG & SACCARDO, ICONES FUNGORUM JAVAN. P 113 & TAB 79, FIG 2,1904 (ILLUS) /PIROZYNSKI, CMI PAPERS 129:57,1972 (STATUS)
?=*SETODOCHIUM* BATISTA & CIFERRI 1957 •

VRIKSHOPAMA RAO & RAO 1964. MYCOPATHOLOGIA 23:287-290 (ILLUS) /V. SWETASAKHA
?=DEMATOPHORA HARTIG 1883 •

WALKEROMYCES THAUNG 1976. TBMS 66:211-215 (ILLUS) /W. GREWIAE
=MYCOVELLOSIELLA FIDE DEIGHTON 1979 (SEE 72D).
/DEIGHTON, CMI PAPERS 144:18-20,1979 (ILLUS,REVIS) •

WALLEMIA JOHAN-OLSON 1887. FORH. CHRISTIANIA VID. SELSK. 12:6 & TAB I-IV /W. ICHTHYOPHAGA (AS 'ICHTYOPHAGA') =W. SEBI (FRIES) VON ARX 1970
AMERO,BASO OR ARTHRO,HYALO,PHIAL OR MISC,SPORODOCHIA, 42A, ON SOIL & OSMOTICALLY ARID SUBSTRATES.
/BARRON, HYPHOMYCETES P 49, 134, & 326-327,1968 (ILLUS) /WILSON, PADHYE & CARMICHAEL, ANTONIE VAN LEEUWENHOEK 35:529-532 (SYNO,PHYSIOL) /ELLIS, DH P 37-38,1971 (ILLUS) /HASHMI & MORGAN-JONES, CAN. J. BOT. 51:1669-1671,1973 (ILLUS) /MADELIN & DORABJEE, TBMS 63:121-130 & PL 14-15,1974 (ILLUS) /HILL, J. STORED PRODUCTS RES. 10:209-215,1974 (ILLUS,CONIDIUM ONTOGENY) /COLE & SAMSON, CONIDIAL FUNGI P 103-104,1979 (ILLUS)
=*HEMISPORA* VUILLEMIN 1906 FIDE CARMICHAEL IN WILSON ET AL 1969
=*BARGELLINIA* BORZI 1888 FIDE VON ARX 1970 •

WALZIA SOROKIN 1873. ARBEIT. DRITT. VERSAMML. RUSS. NATURF., KIEV, BOT. ABT. P 45-47 /W. RACEMOSA
? •

WANGIELLA MCGINNIS 1977. MYCOTAXON 5:353-363 (ILLUS) & 6:367-369 (ILLUS) /HORMISCIUM DERMATITIDIS KANO 1934 =W. DERMATITIDIS (KANO)
AMERO,GLOIO,HYALO,PHIAL, 15E, ON MAN. COMPARE WITH EXOPHIALA & PHIALOPHORA.
/DE HOOG, CBS STUDIES 15:100 & 118-119,1977 (ILLUS SUB EXOPHIALA DERMATITIDIS) /MCGINNIS, SABOURAUDIA 17:145-154,1979 (ILLUS SUB "MICROSPORUM MANSONII") •

WARDOMYCES BROOKS & HANSFORD 1922. TBMS 8:135-137 (ILLUS) /W. ANOMALA
AMERO OR DIDYMO,CETERI,PHAEO,MISC,ALEURI, 68A, ON SOIL, LITTER & OTHER

SUBSTRATES. ASCO STATE =MICROASCUS FIDE VON ARX 1974 (GENERA P 115). COMPARE WITH MAMMARIA & WARDOMYCOPSIS. OTHER CONIDIAL STATES =SCOPULARIOPSIS, CEPHALOTRICHUM, GAMSIA & GYMNODOCHIUM.
/BARRON, GENERA P 327-329,1968 (ILLUS,REFS) /HENNEBERT, CAN. J. BOT. 40:1203-1216,1962 (REVIS,ILLUS) /HENNEBERT, TBMS 51:749-762,1968 (REVIS,ILLUS) /GAMS, TBMS 51:798-802,1968 (ILLUS,NEW SPP) /MALLOCH, CAN. J. BOT. 48:883-885,1970 (ILLUS,NEW SP) /ELLIS, DH P 75-77,1971 (ILLUS,KEY) /MATSUSHIMA, ICONES P 164 & PL 39:2,1975 (ILLUS)
=*HENNEBERTIA* MORELET 1969 SEE HENNEBERT 1968 •

WARDOMYCOPSIS UDAGAWA & FURUYA 1978. MYCOTAXON 7:91-96 (ILLUS) /W. INOPINATA =W. STATE OF MICROASCUS INOPINATUS
AMERO,BASO,PHAEO,ANNEL, 18D, ON SOIL. COMPARE WITH SCOPULARIOPSIS & WARDOMYCES. ASCO STATE =MICROASCUS FIDE UDAGAWA & FURUYA. •

WETTSTEINIELLA KUNTZE 1891 =ARTHROBOTRYUM CESATI 1852, SEE SUTTON 1973 (CMI PAPERS 132:117) •

WIESNERIOMYCES KOORDERS 1907. VERH. AKAD. WETENSCH. AMSTERDAM 2, 13:246-247 (ILLUS) /W. JAVANICUS
PHRAGMO OR SCOLECO,GLOIO,HYALO,RADUL,SPORODOCHIA,SETAE, 111D, ON LITTER. COMPARE WITH ISTHMOLONGISPORA & SPEIROPSIS.
/BARRON, HYPHOMYCETES P 83,1968 (ILLUS SUB ABGLIOPHRAGMA) /MANIOTIS & STRAIN, MYCOLOGIA 60:203-208,1968 (ILLUS) /ELLIS, DH P 362-364,1971 (ILLUS) /PIROZYNSKI, CMI PAPERS 129:57,1972 (SYNONYMS) /MATSUSHIMA, ICONES P 164-165 & PL,1975 (ILLUS)
=*ABGLIOPHRAGMA* ROY & GUJARATI 1966 FIDE KENDRICK 1973
=*ASCHIZOTRICHUM* RIEUF 1962 FIDE KENDRICK 1973
=*CHAETOSIRA* CLEMENTS 1931 FIDE CLEMENTS & SHEAR 1931
=*GLIOPHRAGMA* SUBRAMANIAN & LODHA 1964 FIDE KENDRICK 1973 •

XANTHORIICOLA HAWKSWORTH 1973. IN HAWKSWORTH & PUNITHALINGAM, TBMS 61:66-69 (ILLUS) /GYMNOSPORIUM PHYSCIAE KALCHBRENNER 1865 =X. PHYSCIAE (KALCHBRENNER)
AMERO,BASO OR CETERI,PHAEO,PHIAL, 6G, ON LICHENS (XANTHORIA).
/ELLIS, MDH P 456-457,1976 (ILLUS) •

XENODIELLA SYDOW 1935. ANN. MYCOL. 33:98 (NO ILLUS) /X. PETRAKII =X. STATE OF XENODIUM PETRAKII SYDOW 1935
?, ASCO STATE =XENODIUM SYDOW 1935 •

XENOPLACA PETRAK 1949. SYDOWIA 3:261-263 (NO ILLUS) /X. AEQUATORIENSIS
? •

XENOPUS PENZIG & SACCARDO 1901. MALPIGHIA 15:240 /X. FARINOSUS
=BASIDIOMYCETE FIDE HUGHES 1958
/DONK, TAXON 11:102,1962 (REFS) •

XENOSPORELLA HOHNEL 1923. ZENTRALBL. BAKT. ABT. 2, 60:17 /X. PLEUROCOCCA =X. PLEUROCOCCUM (HOHNEL) PIROZYNSKI 1966
=XENOSPORIUM PENZIG & SACCARDO 1902 FIDE PIROZYNSKI 1966 •

XENOSPORIUM PENZIG & SACCARDO 1901. MALPIGHIA 15:248 /X. MIRABILE
DICTYO OR HELICO,CETERI,PHAEO,ANNEL OR NONSP, 104C, ON WOOD & LITTER. ASCO STATE =ACANTHOSTIGMELLA SEE LINDER 1929 (ANN. MISSOURI BOT. GARD. 16:317-322,ILLUS SUB XENOSPORELLA).
/ELLIS, CMI PAPERS 87:13-15,1963 (ILLUS) /PIROZYNSKI, CMI PAPERS 105:24-35,1966 (ILLUS,KEY) /ELLIS, DH P 131-132,1971 (ILLUS) /MATSUSHIMA, ICONES P 165 & PL,1975 (ILLUS) /ELLIS, MDH P 104-105,1976 (ILLUS,KEY) /HUGHES, NEW ZEALAND J. BOT. 16:354-357,1978 (ILLUS,NEW SP)
=*XENOSPORELLA* HOHNEL 1923 FIDE PIROZYNSKI 1966 •

XENOSTILBUM PETRAK 1959. SYDOWIA 13:105-108 /X. SYDOWII
?, NOT ILLUSTRATED BY AUTHOR. •

XIPHOMYCES SYDOW & SYDOW 1916. ANN. MYCOL. 14:374 /X. SACCHARI
?, NOT ILLUSTRATED BY AUTHORS. •

XYLOCHAERAS FRIES, ORTHOGRAPHIC VARIANT OF XYLOCHOERAS •

XYLOCHOERAS FRIES 1849 (SUMMA VEG. SCAND. 2:476) =SCLEROTIUM FIDE ACCARDO FROM AINSWORTH 1971 •

XYLOCLADIUM SYDOW 1900. IN LINDAU, NAT. PFLANZENF. 1(1):494 /CERATOCLADIUM CLAUTRIAVII PATOUILLARD 1898 (BULL. SOC. MYCOL. FRANCE 14:196) =X. CLAUTRIAVII (PATOUILLARD) SYDOW 1902 (IN SACCARDO, SYLL. FUNG. 16:1090)
AMERO,CETERI,HYALO,AMPUL OR RADUL, 28E, ON TREES (BRANCHES & TWIGS), WOOD & LITTER. ASCO STATE =HYPOXYLON FIDE BARNETT 1957, OR PORONIA FIDE JONG & ROGERS 1969 (MYCOLOGIA 61:853-862 ILLUS,NOT NAMED), & SEE ALSO KRUG & CAIN 1974 (CAN. J. BOT. 52:590 & 605,REFS), OR NUMMULARIA FIDE ROGERS 1975 (AMER. J. BOT. 62:761-764,ILLUS).
/BARNETT, MYCOLOGIA 49:588-595,1957 (ILLUS,ECOL) /MARTIN, SOUTH AFRICAN J. BOT. 33:205-240,1967 (ILLUS,ECOL) /JONG & ROGERS, CONIDIAL STATES OF HYPOXYLON, WASHINGTON AGRIC. EXPER. STATION TECHN. BULL. 71:1-51,1972 (ILLUS,REVIS,KEY) /ROGERS, MYCOLOGIA 71:1-42,1979 (DISCUSSION,REFS)
=*CERATOCLADIUM* PATOUILLARD 1898 NON CORDA 1839
=*BASIDIOBOTRYS* HOHNEL 1909 BASED ON SAME TYPE SPECIES
?=*MASONIOMYCES* CRANE & DUMONT 1975 •

XYLOHYPHA (FRIES) MASON 1960. IN DEIGHTON, CMI PAPERS 78:43 (REFS) /X. NIGRESCENS (PERSOON) MASON 1960
AMERO OR DIDYMO,BLASTO,PHAEO,NONSP, 36C, ON WOOD. COMPARE WITH HORMOCONIS & ALYSIDIUM.
/ELLIS, DH P 95,1971 (ILLUS) /HUGHES & SUGIYAMA, NEW ZEALAND J. BOT. 10:447-460,1972 (REVIS,ILLUS) /SUTTON, CMI PAPERS 132:126-129,1973 (ILLUS,NEW SP) /HAWKSWORTH & PUNITHALINGAM, TBMS 64:98-99,1975 (ILLUS,NEW SP) /ELLIS, MDH P 63-64,1976 (ILLUS) •

XYLOMYCES GOOS, BROOKS & LAMORE 1977. MYCOLOGIA 69:282 (280-286,ILLUS) /X. CHLAMYDOSPORIS
PHRAGMO,BLASTO,PHAEO,NONSP, 91C, ON WOOD (IN WATER). COMPARE WITH TAENIOLELLA & PEYRONELIA. •

XYLOSTROMA TODE 1790 =SHEETS OF MYCELIUM FIDE DONK 1962 (TAXON 11:102) •

XYPHASMA REBENTISCH =MISPRINT FOR HYPHASMA IN DEUTSCHL. KRYPT. FL. 1:38,1844 •

ZAKATOSHIA SUTTON 1973. CMI PAPERS 132:129-131 (ILLUS) /Z. HIRSCHIOPORI
AMERO,GLOIO,HYALO,PHIAL, 16D, ON FUNGI (HIRSCHIOPORUS ON ABIES).
/ELLIS, MDH P 469-470,1976 (ILLUS) •

ZALERION MOORE & MEYERS 1962. CAN. J. MICROBIOL. 8:408 (407-416,ILLUS) /Z. NEPURA =Z. MARITIMA (LINDER) ANASTASIOU 1963
HELICO,CETERI,PHAEO,NONSP, 106E, ON WOOD IN WATER & TREES (LARIX, PICEA). COMPARE WITH CIRRENALIA.
/ANASTASIOU, CAN. J. BOT. 41:1135-1139,1963 (ILLUS,REVIS) /COLE, MARINE BIOL. 38:147-158,1976 (ILLUS) /ELLIS, MDH P 40-41,1976 (ILLUS) •

ZANCLOSPORA HUGHES & KENDRICK 1965. NEW ZEALAND J. BOT. 3:151-158 (ILLUS) /Z. NOVA-ZELANDIAE
AMERO,GLOIO,HYALO,PHIAL,SETAE, 4H, ON LITTER, WOOD & BARK. ASCO STATE =CHAETOSPHAERIA FIDE MULLER IN KENDRICK 1971.
/ELLIS, DH P 536-537,1971 (ILLUS) /SUTTON & HODGES, NOVA HEDWIGIA 26:522-524,1975 (ILLUS,NEW SP) /MATSUSHIMA, ICONES P 165-166 & PL,1975 (ILLUS) •

ZASMIDIUM FRIES 1849. SUMMA VEG. SCAND. 2:407 (NO ILLUS) /RACODIUM CELLARE PERSOON 1801 =Z. CELLARE (PERSOON) (LECTOTYPE FIDE RIEDL 1968, TAXON 17:34-37)
?, A SUPERFLUOUS NAME FOR RACODIUM PERSOON, Q.V. •

ZOPHEROMYCES SUTTON & HODGES 1977. NOVA HEDWIGIA 28:496-498 (ILLUS) /Z. FASCIATUS
DIDYMO,CETERI,PHAEO,AMPUL OR MISC, 64D, ON LITTER (EUCALYPTUS). •

ZYGODESMELLA GONZALEZ FRAGOSO 1917. BOLL. SOC. ESPAN. HIST. NAT. 17:260-262 (ILLUS) /Z. CASARESI
?=NOMEN DUBIUM, SEE DONK 1962 (TAXON 11:103). •

ZYGODESMUS CORDA 1837 (ICON. FUNG. 1:11, TAB 2 FIG 164) =NOMEN DUBIUM FIDE DONK 1962 (TAXON 11:103) •

ZYGOPHIALA MASON 1945. IN MARTYN, CMI PAPERS 13:3-5 (ILLUS) /Z. JAMAICENSIS
DIDYMO,CETERI,HYALO,RADUL, 63E, ON LEAVES (MUSA & OTHER PLANTS).
/HUGHES, CMI PAPERS 48:82,1952 (ILLUS) /ELLIS, DH P 292,1971 (ILLUS) •

ZYGOSPORIUM MONTAGNE 1842. ANN. SCI. NAT. BOT. SER. 2, 17:120-121 /Z. OSCHEOIDES
AMERO,CETERI,HYALO OR PHAEO,MISC,SETAE, 54C, ON LITTER. OTHER STATE =SPOROTHRIX (H) SEE WANG IN KENDRICK 1979.
/HUGHES, CMI PAPERS 44:1-18,1951 (ILLUS,REVIS) /ELLIS, DH P 322-326,1971 (ILLUS,KEY) /WANG & BAKER, CAN. J. BOT. 45:1945-1951,1967 (ILLUS) /MATSUSHIMA, ICONES P 166 & PL,1975 (ILLUS) /ELLIS, MDH P 344-345,1976 (ILLUS,NEW SP)
=*PIMINA* GROVE 1888 FIDE HUGHES 1958
=*UROPHIALA* VUILLEMIN 1910 FIDE HUGHES 1958
=*UROBASIDIUM* GIESENHAGEN 1892 FIDE ELLIS 1971 •

ZYGOTHRIX REINSCH BASED ON HYPHAE, FROM AINSWORTH 1971. •

ZYMONEMA BEURMANN & GOUGEROT 1909. TRIBUNE MEDICALE, PARIS 42:503 /Z. GILCHRISTI =ZYMONEMA DERMATITIDIS (GILCHRIST & STOKES) DODGE 1935
=BLASTOMYCETES, SEE BLASTOMYCES GILCHRIST & STOKES. CONIDIAL STATE =CHRYSOSPORIUM CORDA 1833 FIDE CARMICHAEL 1962 (CAN. J. BOT. 40:1151-1155). •

Illustrations

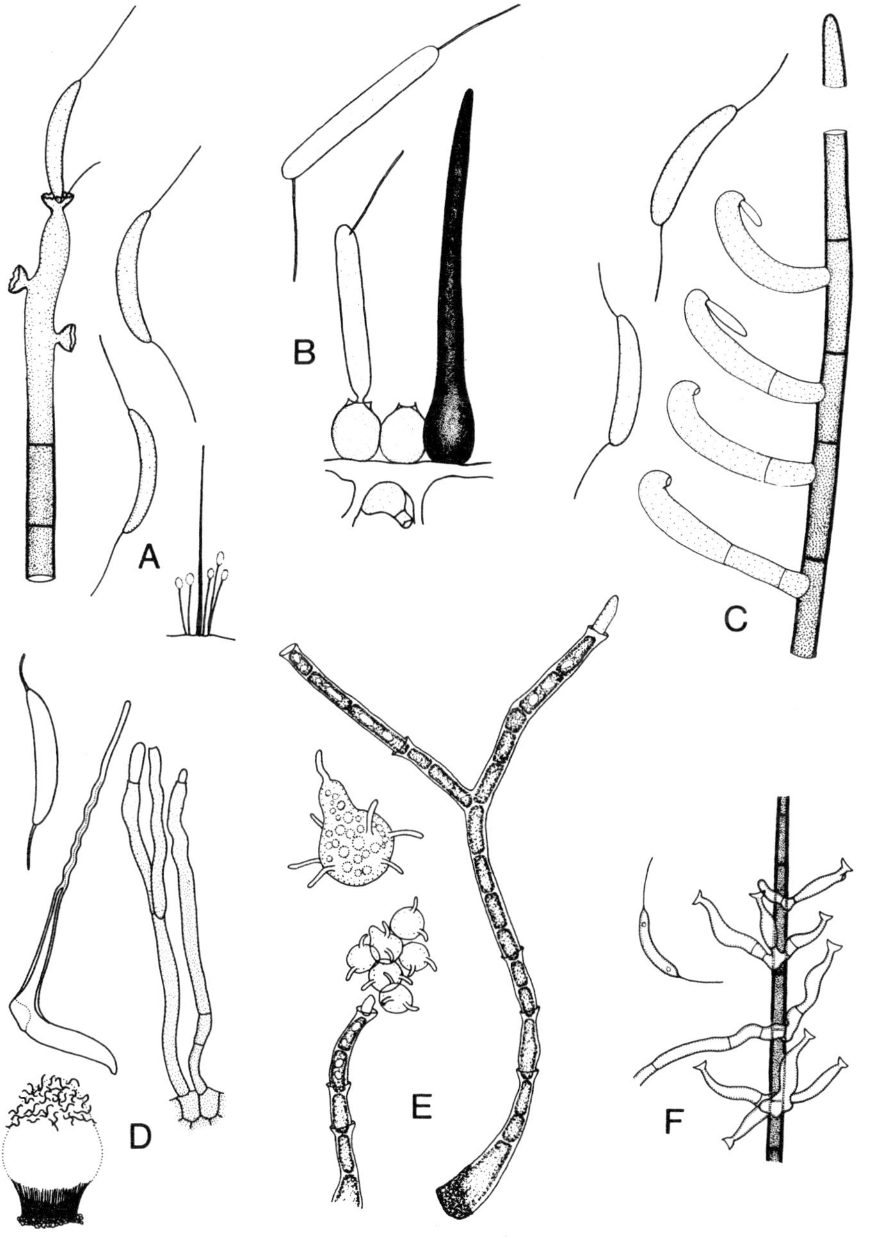

Plate 1: A *Codinaea assamica;* B *Mahabalella acutisetosa;* C *Menispora ciliata;* D *Thozetella nivea;* E *Bahusutrabeeja dwaya;* F *Codinaeopsis gonytrichoides.*

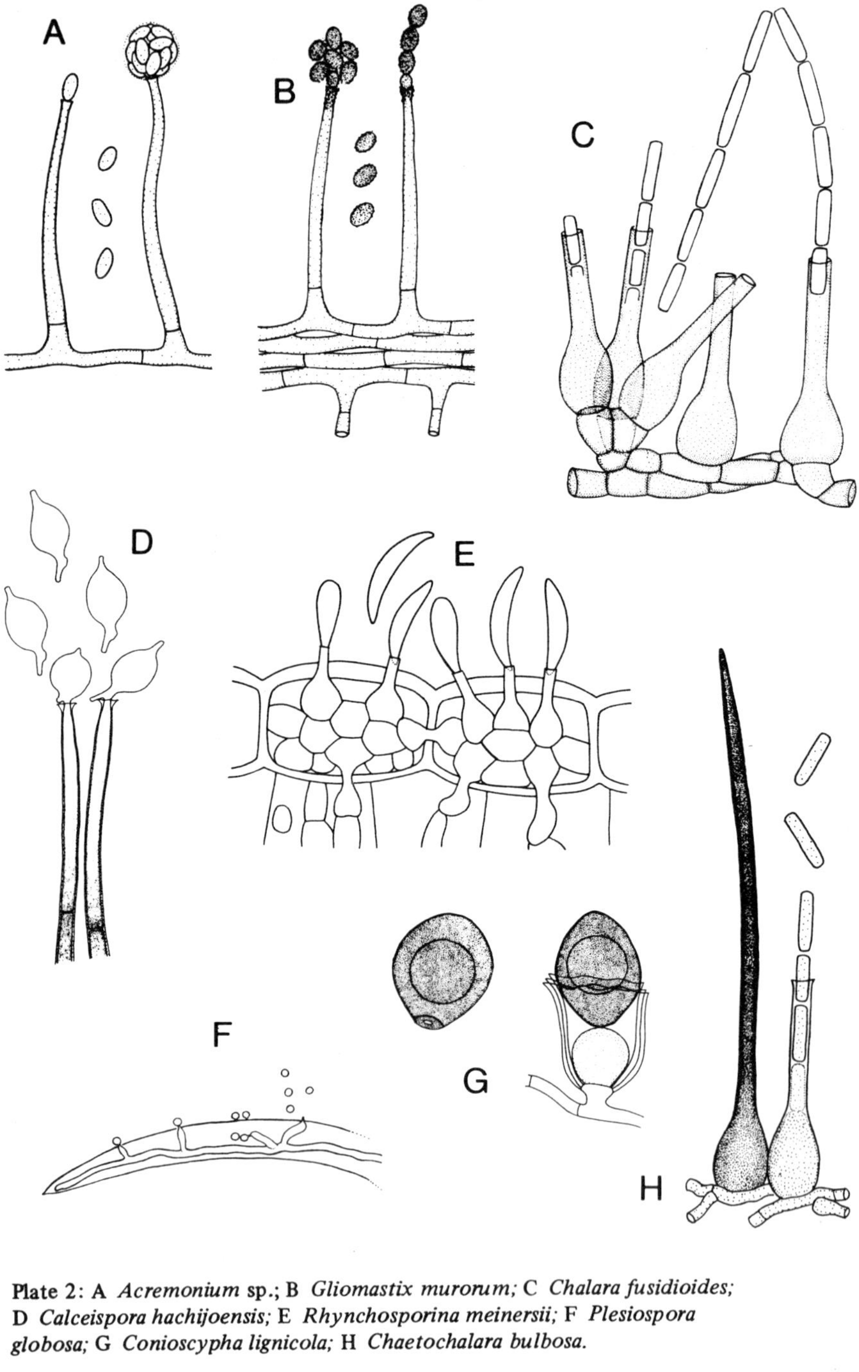

Plate 2: A *Acremonium* sp.; B *Gliomastix murorum;* C *Chalara fusidioides;* D *Calceispora hachijoensis;* E *Rhynchosporina meinersii;* F *Plesiospora globosa;* G *Conioscypha lignicola;* H *Chaetochalara bulbosa.*

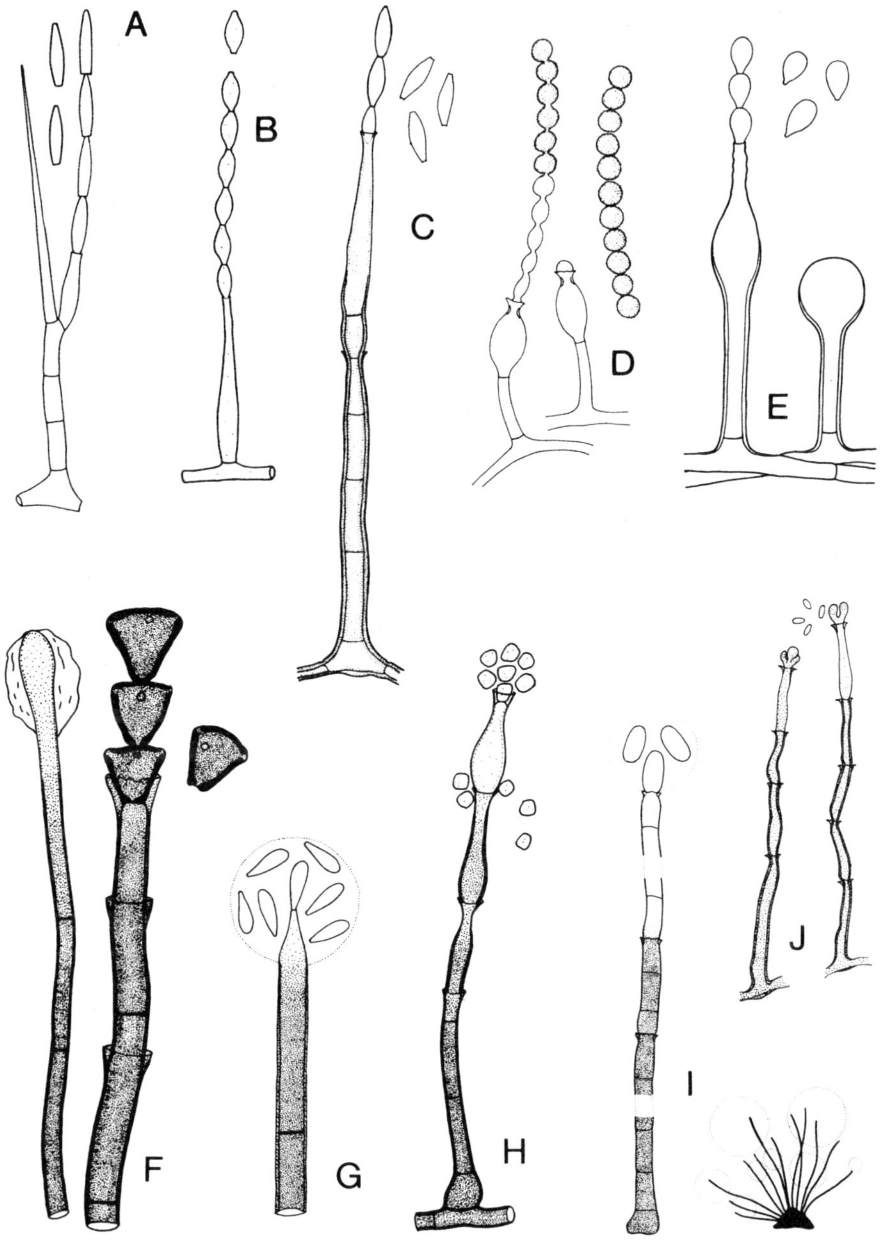

Plate 3: A *Pleurocatena acicularis;* B *Paecilomyces roseolus (=Acremonium roseolum);* C *Exochalara longissima;* D *Torulomyces lagena;* E *Monocillium indicum;* F *Catenularia* state of *Chaetosphaeria novae-zelandiae;* G *Sphaeromyces clavisporus;* H *Chloridium* state of *Chaetosphaeria myriocarpa;* I *Dischloridium laeense;* J *Bisporomyces chlamydosporis.*

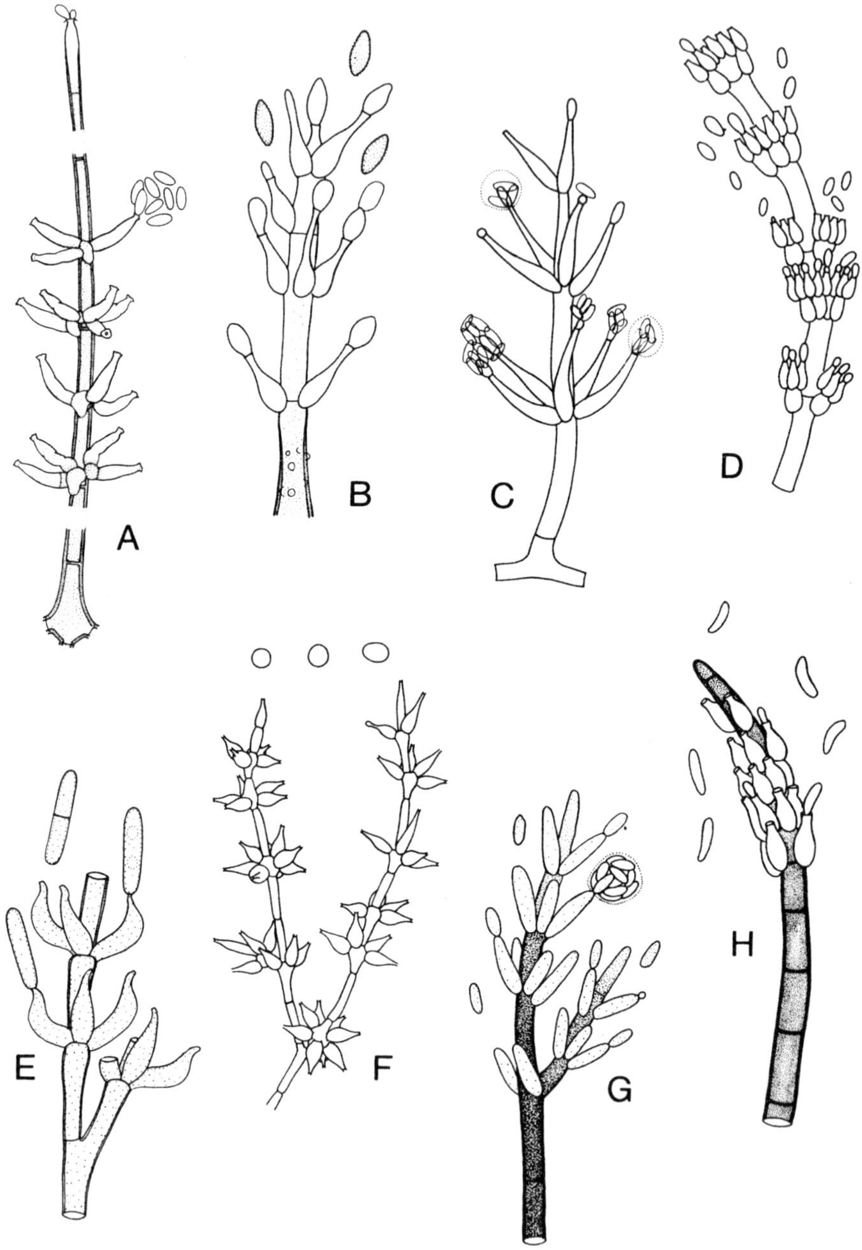

Plate 4: A *Gonytrichum macrocladum;* B *Acrophialophora fusispora;* C *Verticillium lateritium;* D *Nomuraea atypicola;* E *Uncigera cordae;* F *Verticimonosporium diffractum;* G *Stachylidium bicolor;* H *Zanclospora brevispora.*

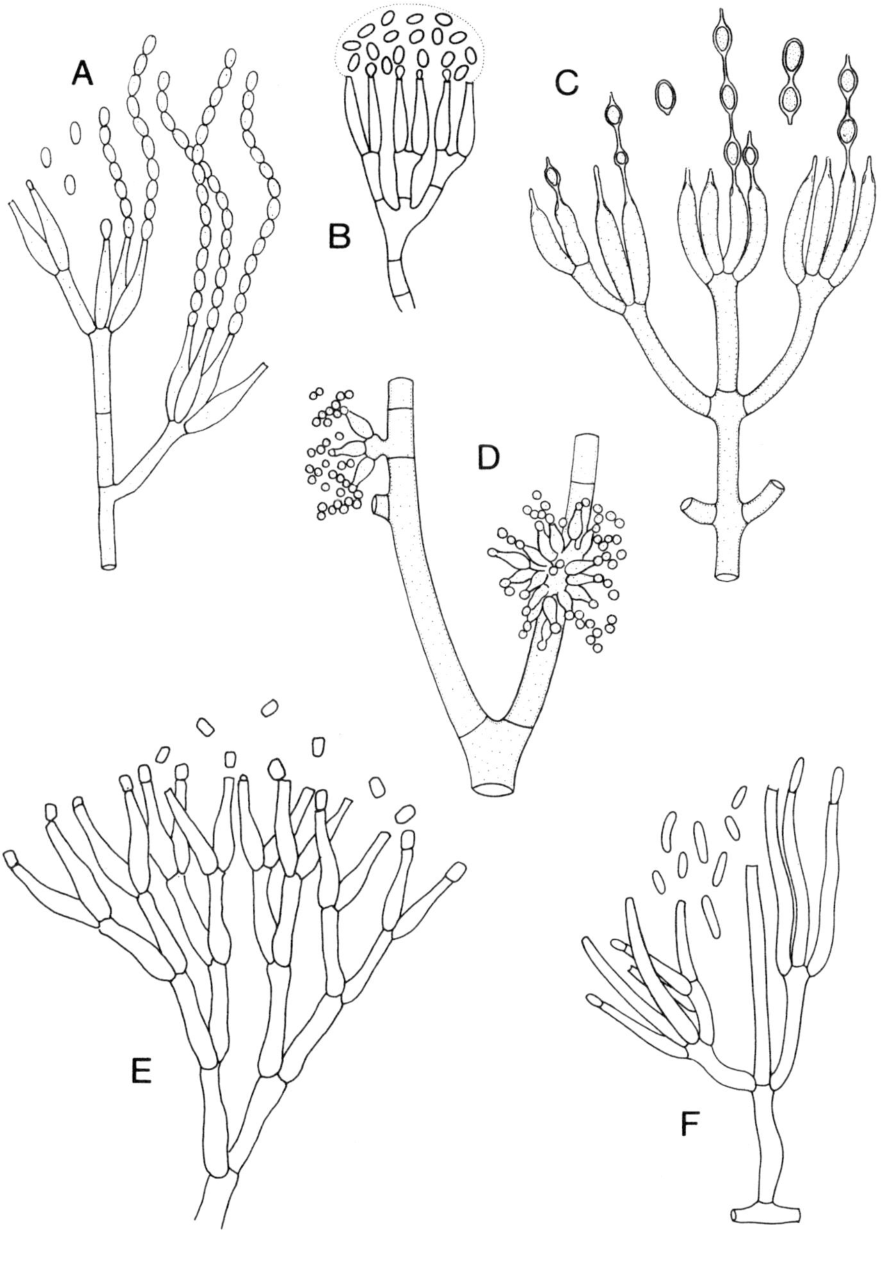

Plate 5: A *Paecilomyces varioti;* B *Sphacelia* state of *Claviceps purpurea;* C *Phialotubus microsporus;* D *Myrioconium marina;* E *Tympanosporium parasiticum;* F *Sarocladium oryzae.*

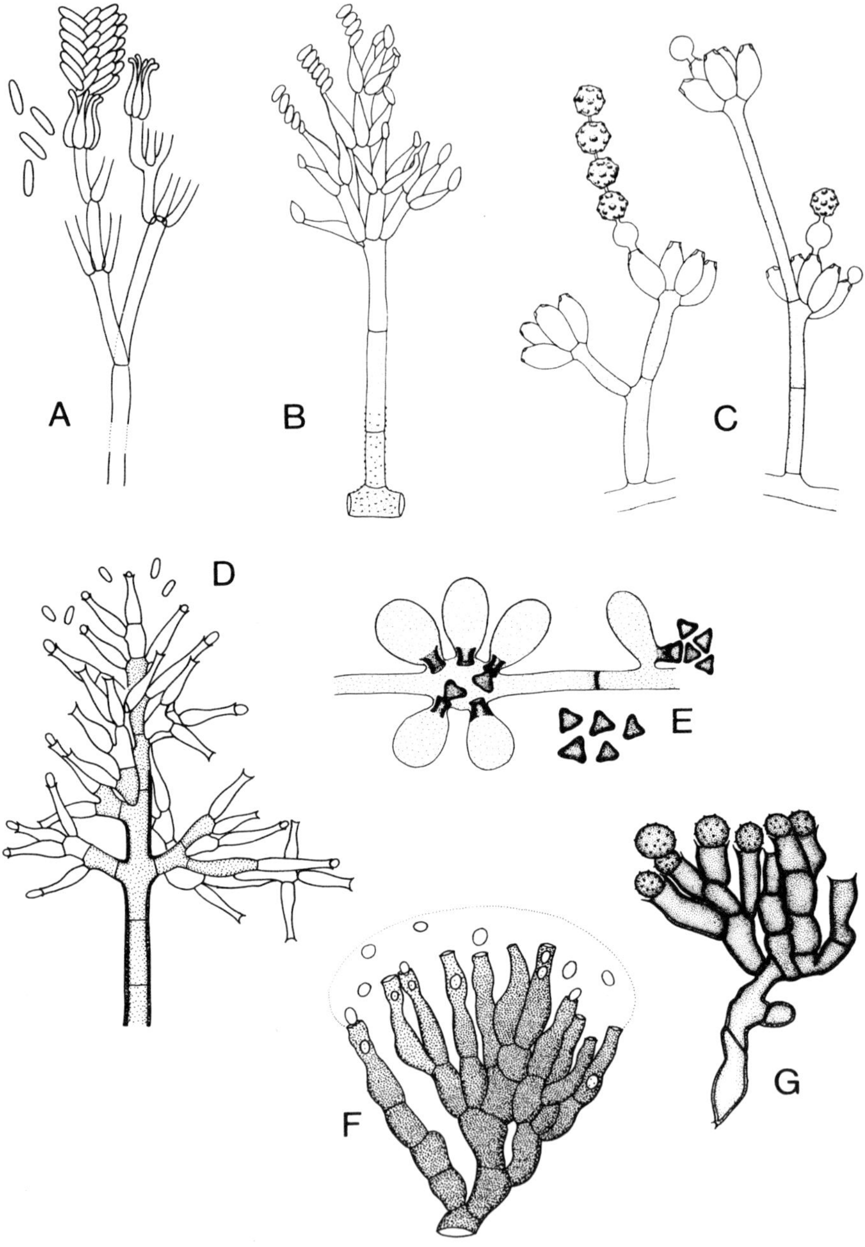

Plate 6: A *Clonostachys cylindrospora;* B *Mariannaea elegans;* C *Eladia saccula;* D *Phaeostalagmus cyclosporus;* E *Anthopsis deltoidea;* F *Cystodendron dryophilum;* G *Xanthoriicola physciae.*

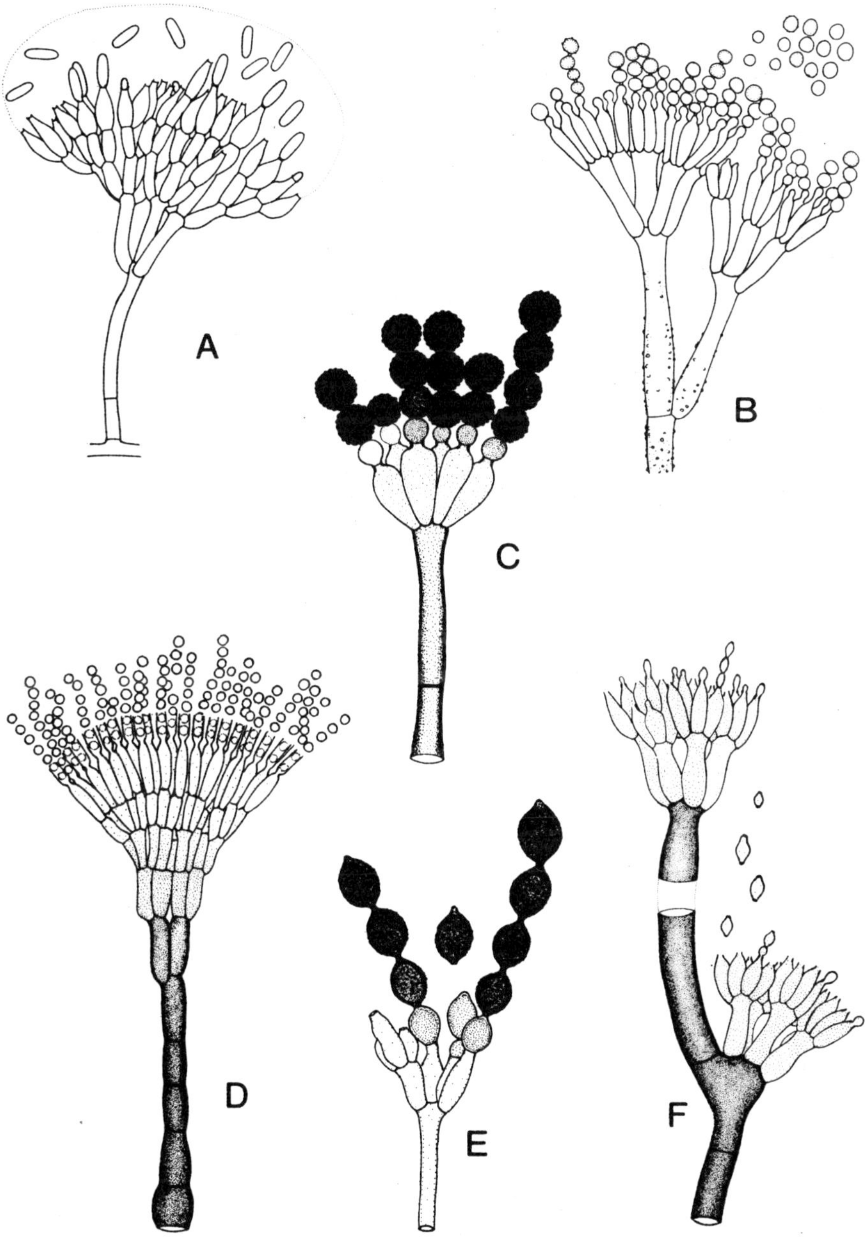

Plate 7: A *Gliocladium cylindrospermum;* B *Penicillium atrovenetum;* C *Stachybotrys* sp.; D *Phialocephala dimorphospora;* E *Phialomyces macrosporus;* F *Thysanophora penicillioides.*

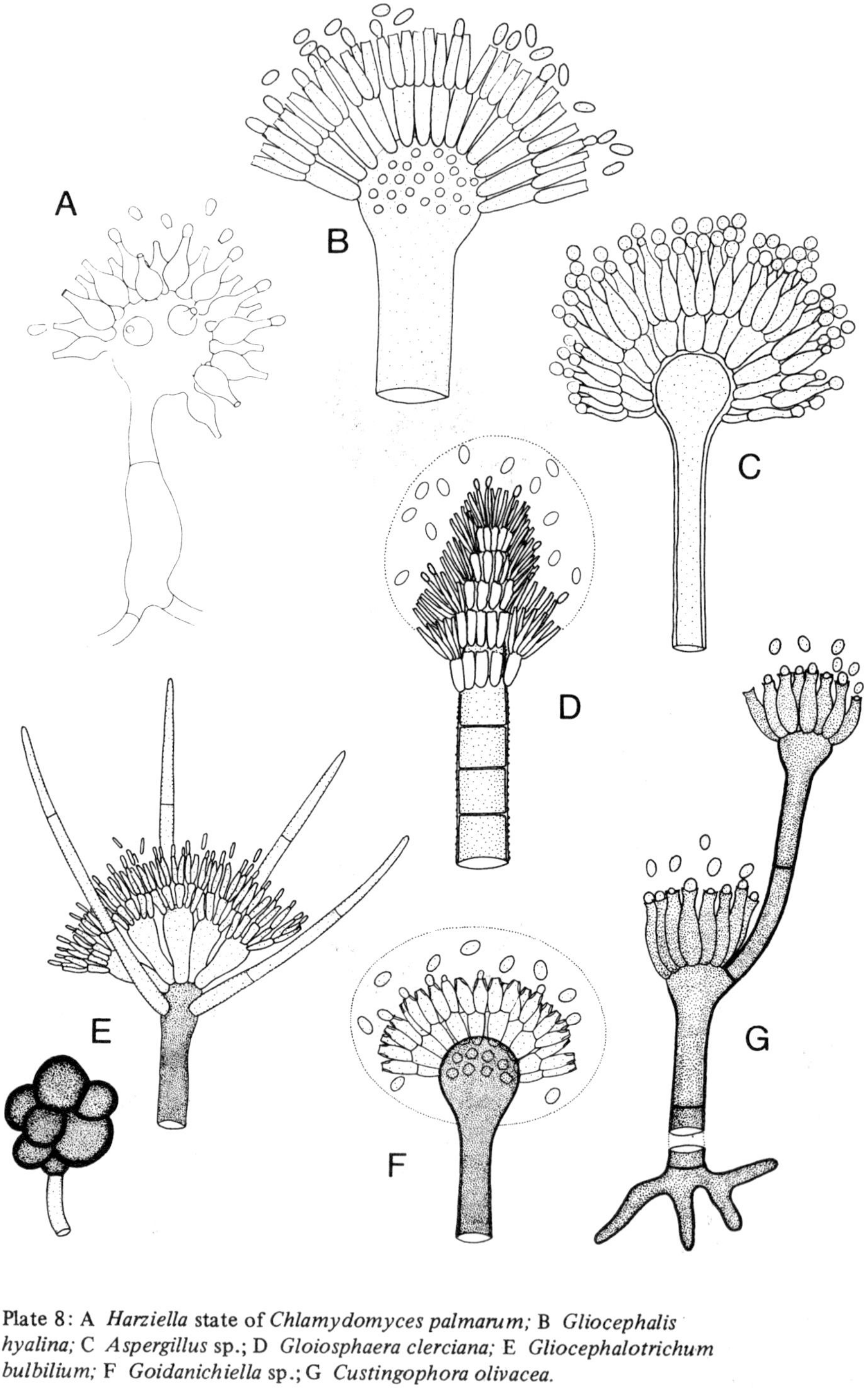

Plate 8: A *Harziella* state of *Chlamydomyces palmarum;* B *Gliocephalis hyalina;* C *Aspergillus* sp.; D *Gloiosphaera clerciana;* E *Gliocephalotrichum bulbilium;* F *Goidanichiella* sp.; G *Custingophora olivacea.*

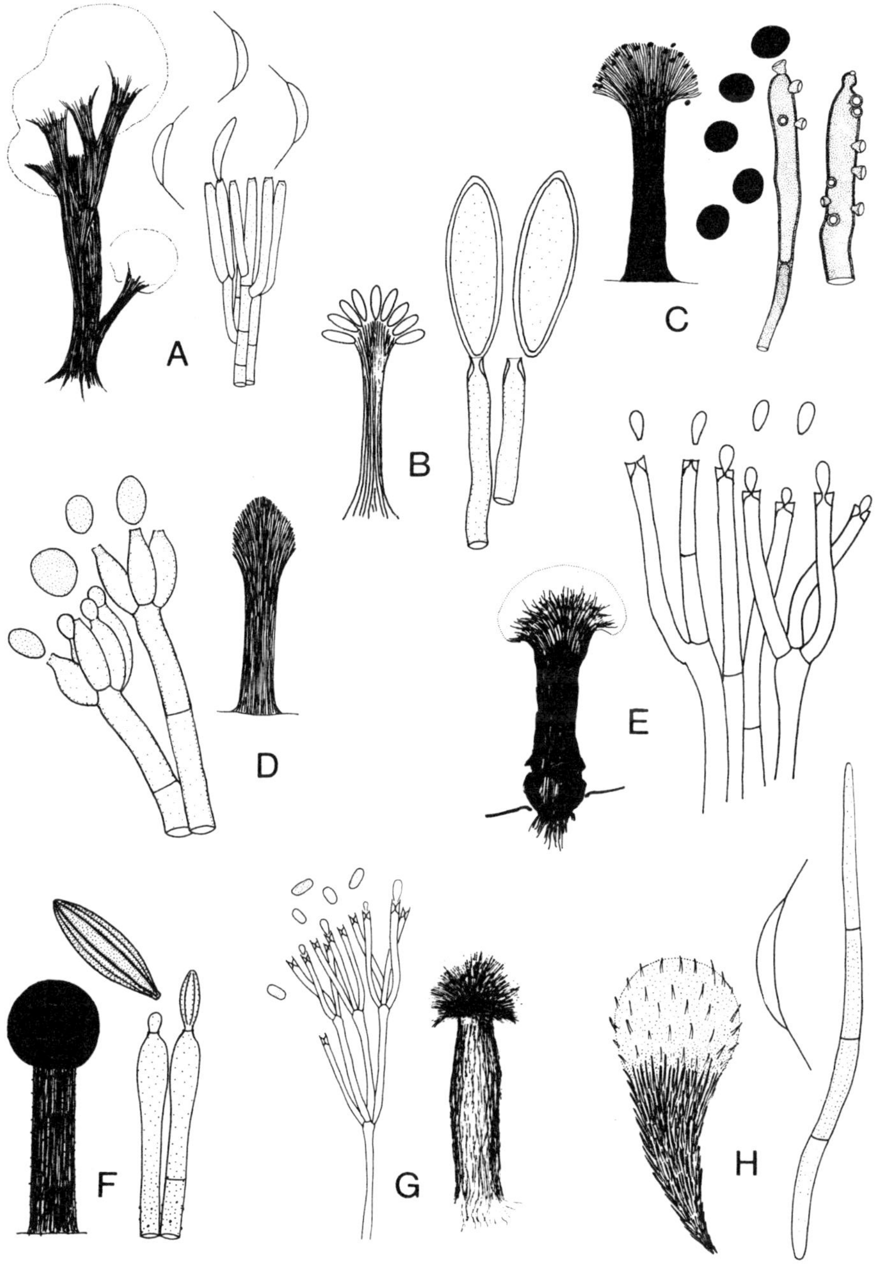

Plate 9: A *Thozetella toklaiensis;* B *Ciliciopodium macrosporum;* C *Paathramaya sundara;* D *Phialostele scytopetali;* E *Crinula caliciiformis;* F *Solheimia costaspora;* G *Dendrostilbella prasinula;* H *Neottiosporella radicata.*

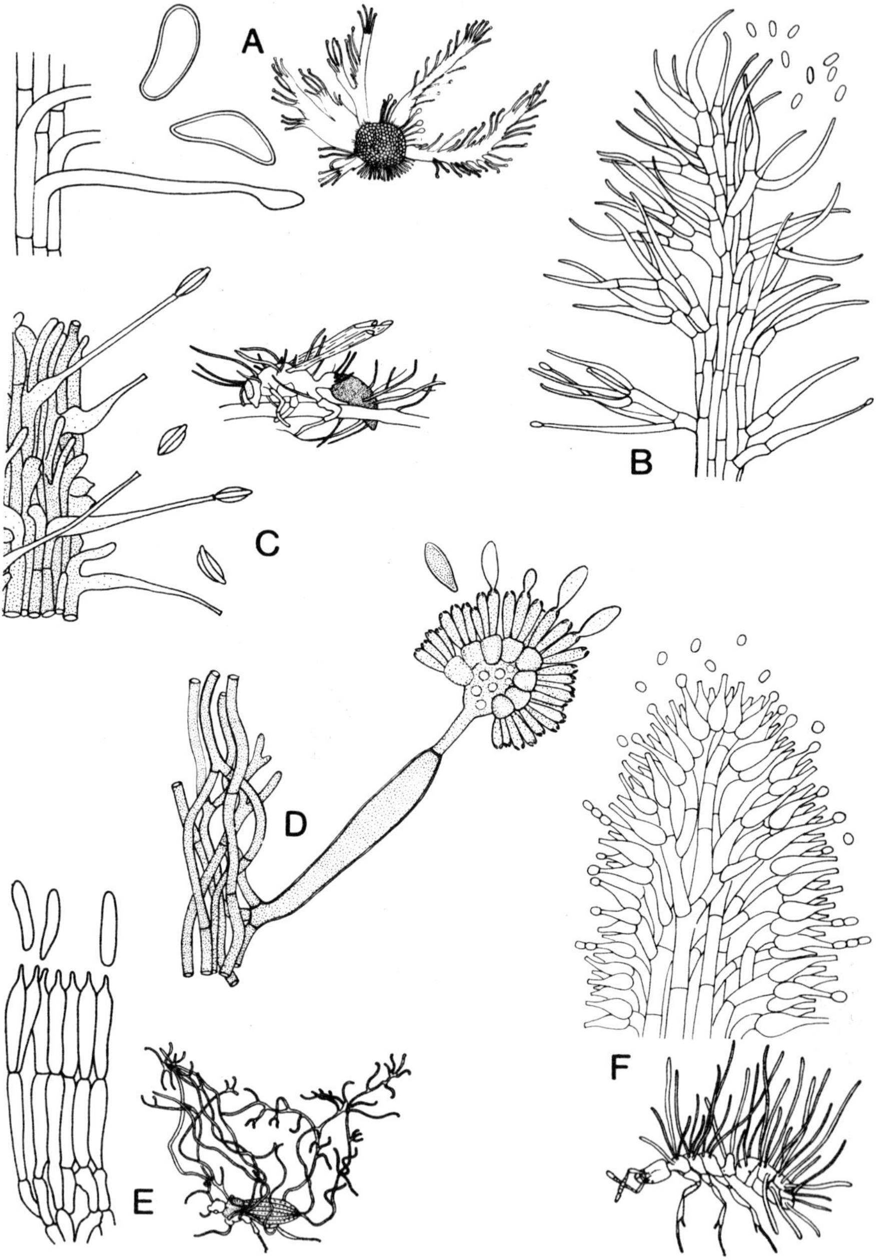

Plate 10: A *Synnematium jonesii;* B *Tilachlidium brachiatum;* C *Hirsutella saussurei;* D *Gibellula suffulta;* E *Tilachlidiopsis gracilis;* F *Akanthomyces gracilis.*

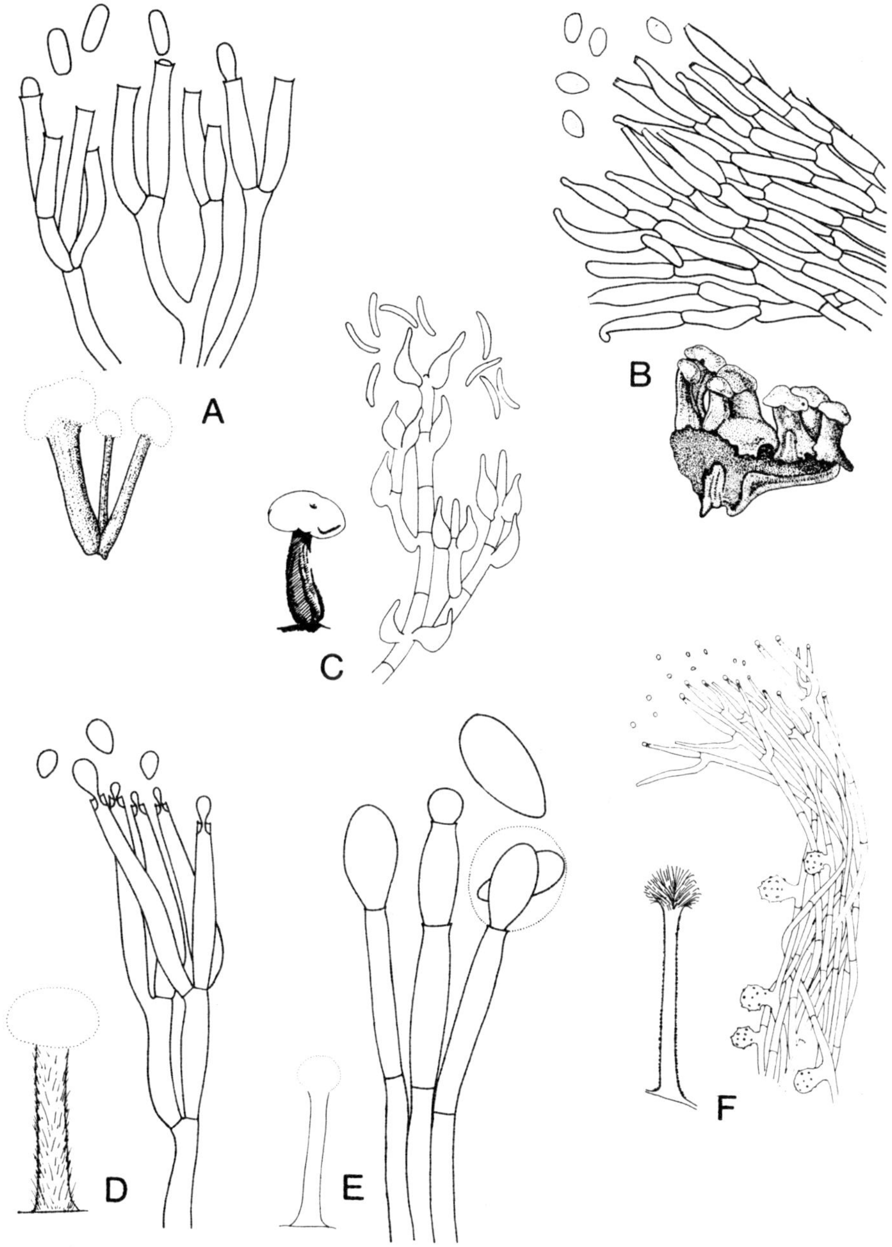

Plate 11: A *Phialographium* state of *Ceratocystis sagmatospora;*
B *Dennisographium episphaeriae;* C *Coryne* sp.; D *Stilbella erythrocephala;*
E *Hyalocylindrophora venezuelensis;* F *Blistum tomentosum.*

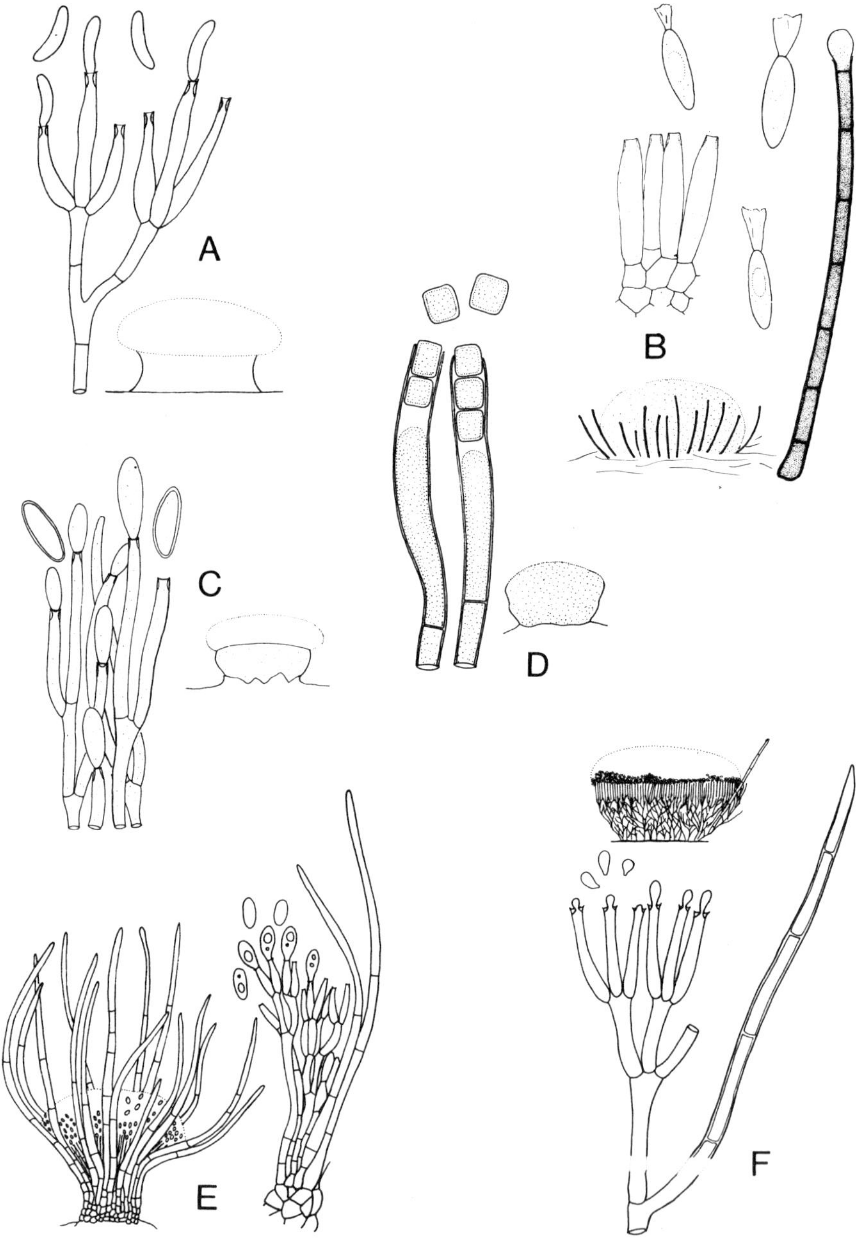

Plate 12: A *Dendrodochium* state of *Nectria magnusiana;* B *Koorchaloma madreeya;* C *Tubercularia vulgaris;* D *Bloxamia truncata;* E *Volutella* sp.; F *Myrothecium inundatum.*

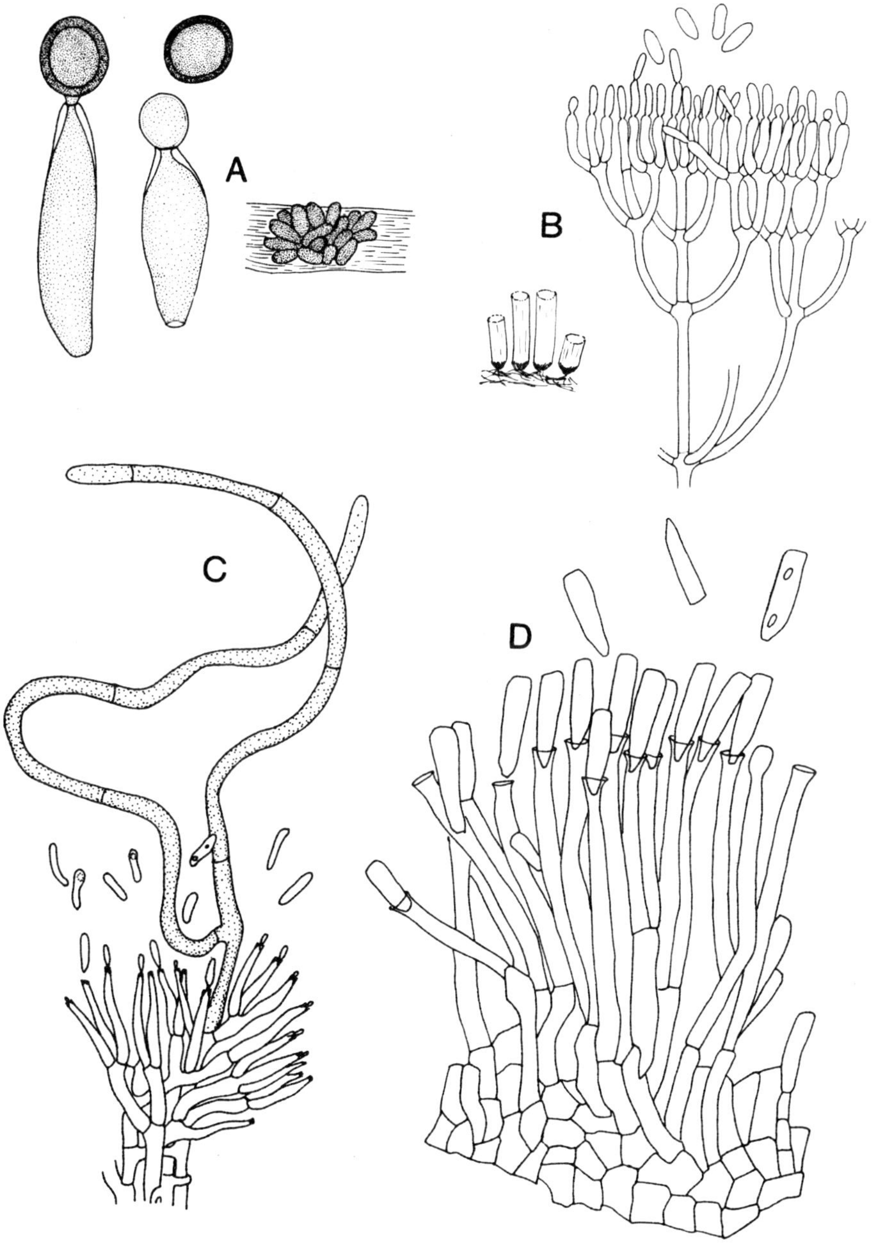

Plate 13: A *Stromatocrea cerebriforme;* B *Metarhizium anisopliae;* C *Sarcopodium tortuosum;* D *Endoconospora cerastii.*

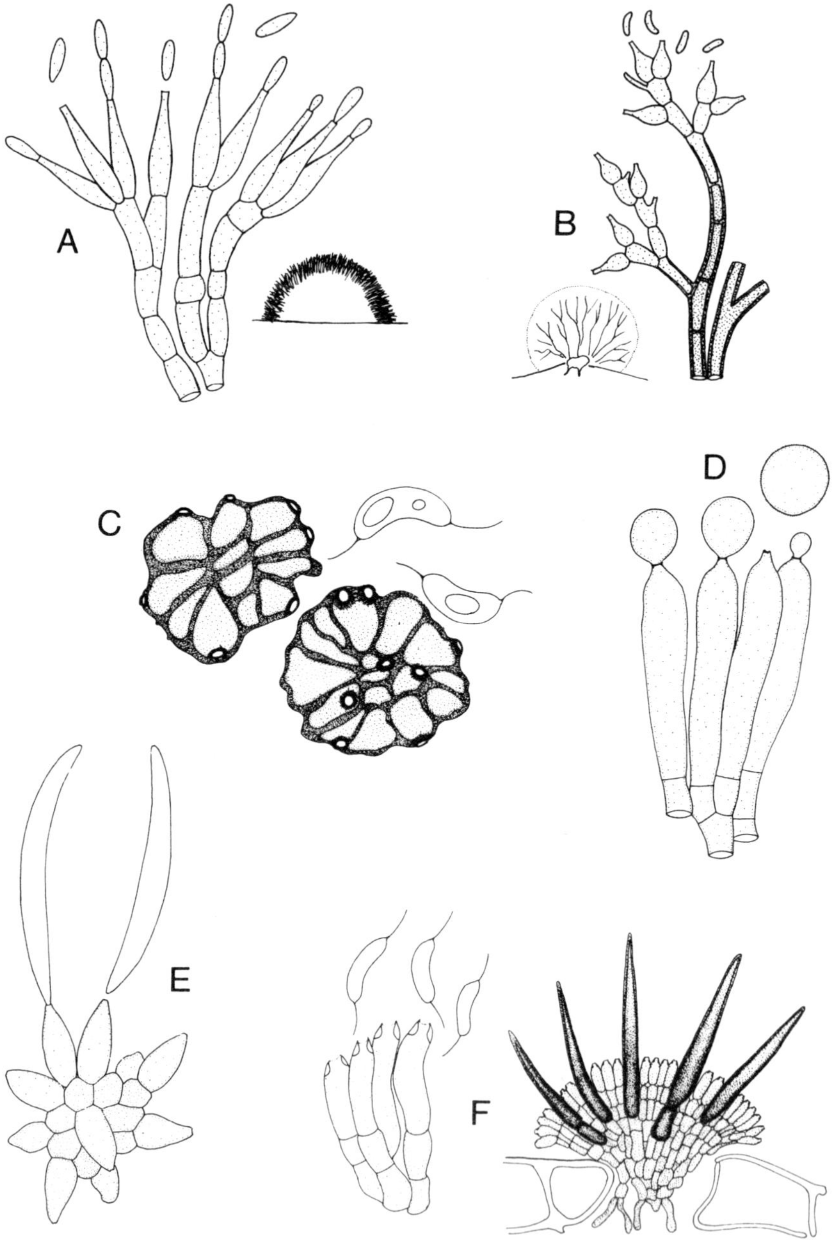

Plate 14: A *Nalanthamala madreeya;* B *Agyriella nitida;* C *Mycoleptodiscus indicus;* D *Tuberculina persicina;* E *Microdochium phragmitis;* F *Minimidochium setosum.*

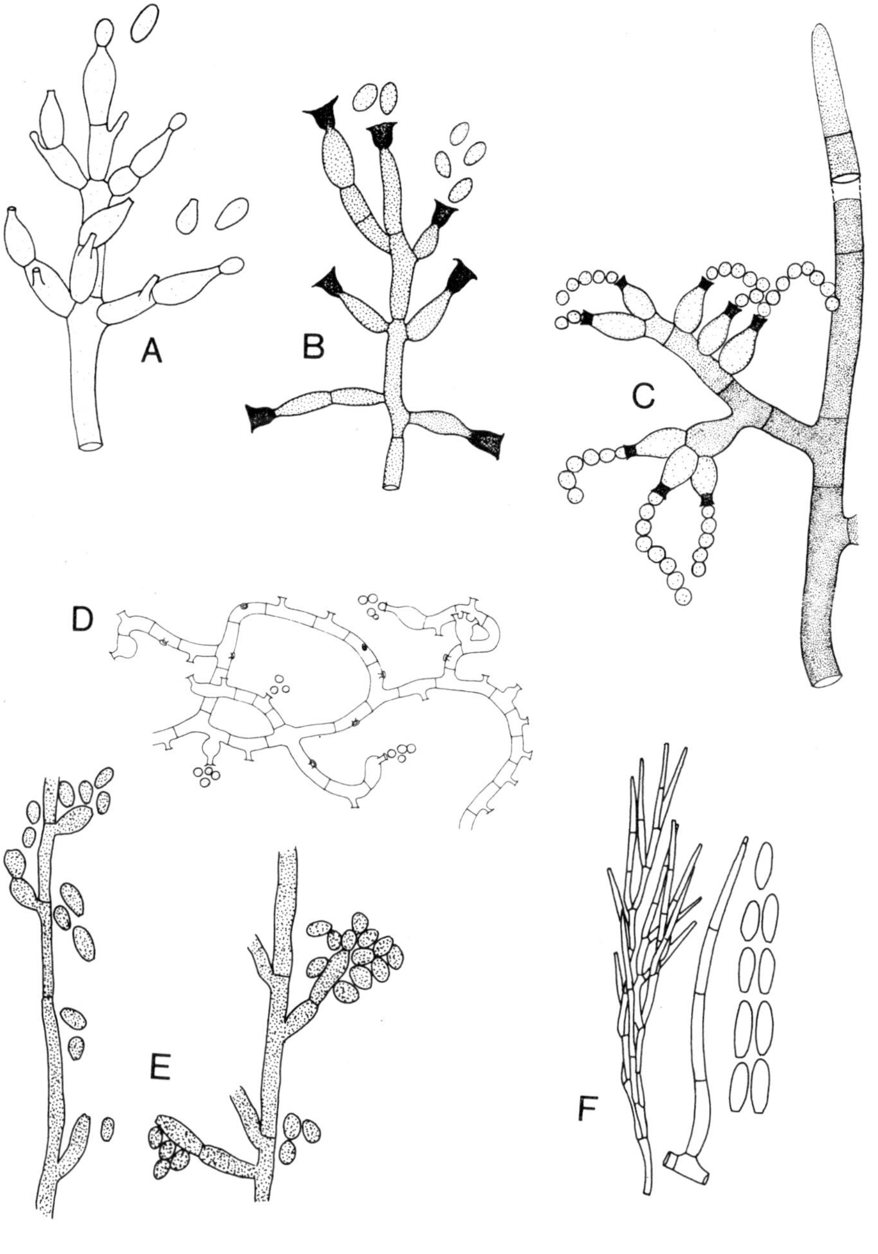

Plate 15: A *Sesquicillium candelabrum;* B *Phialophora verrucosa;* C *Angulimaya sundara;* D *Cladorrhinum foecundissimum;* E *Wangiella dermatitidis;* F *Gabarnaudia betae.*

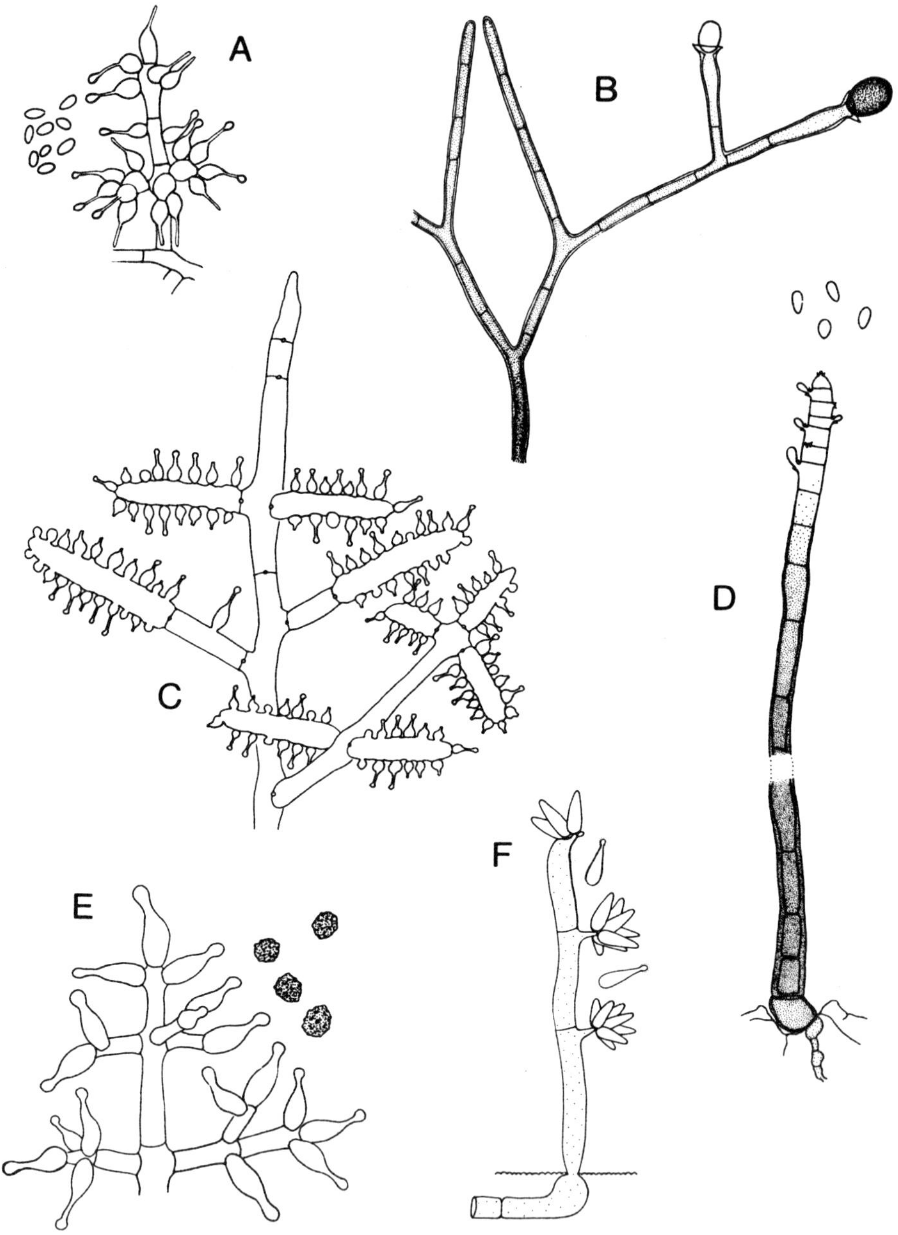

Plate 16: A *Tolypocladium inflatum;* B *Craspedodidymum elatum;* C *Phialocladus zsoltii;* D *Zakatoshia hirschiopori;* E *Trichoderma viride;* F *Meria coniospora.*

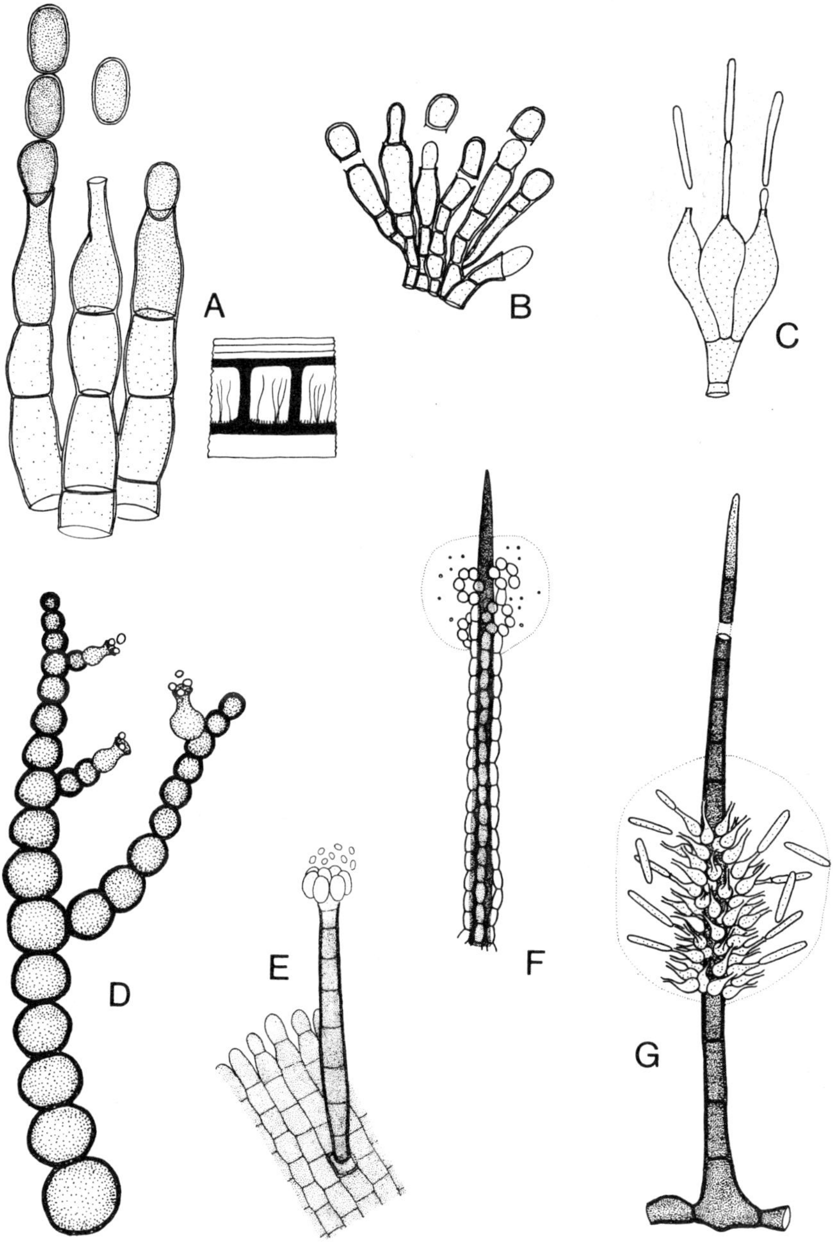

Plate 17: A *Cryptostroma corticale;* B *Phialophoropsis trypodendri;* C *Riclaretia urticae;* D *Capnophialophora* state of *Metacapnodium juniperi;* E *Hormisciomyces* state of *Trichopeltheca asiatica;* F *Microxiphium* state of *Dennisiella babingtonii;* G *Chaetopsina fulva.*

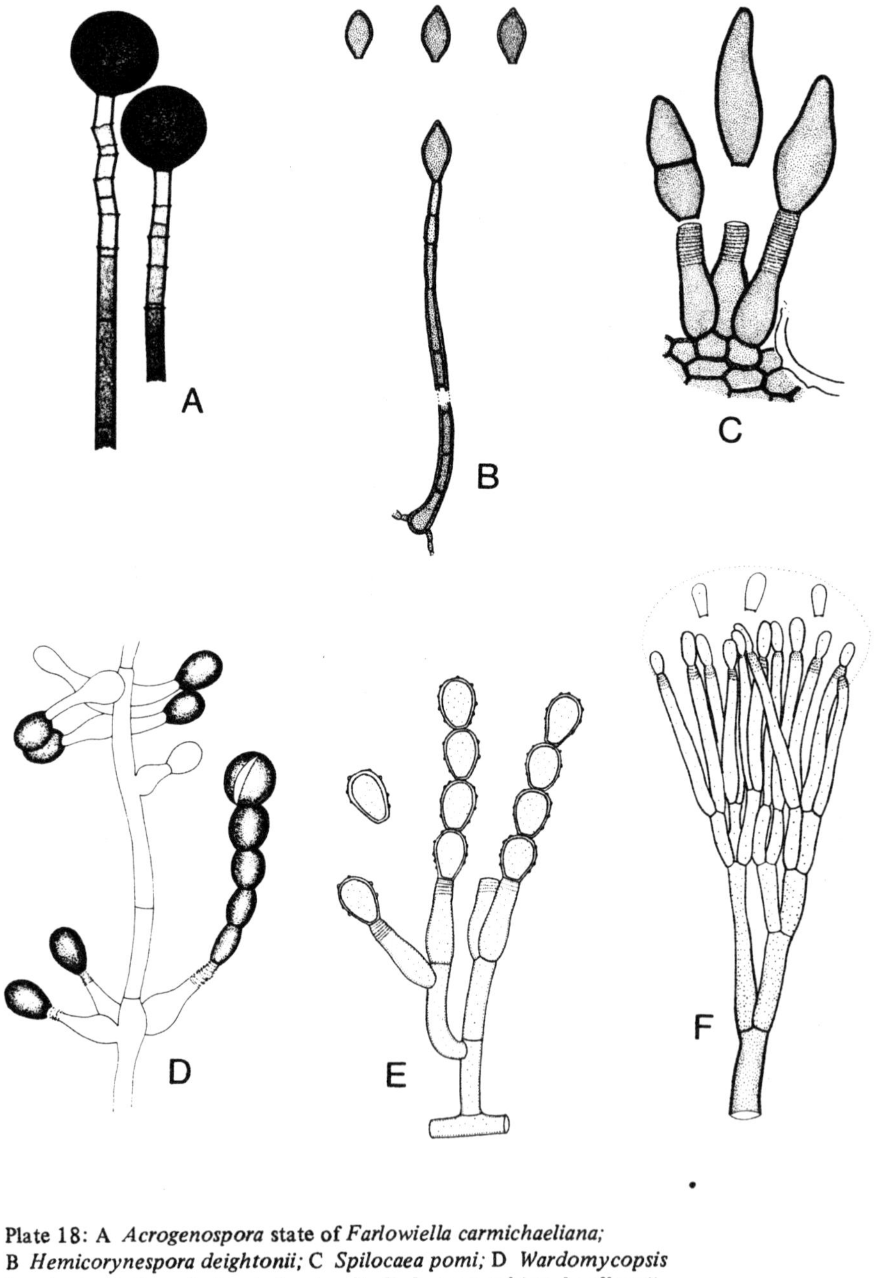

Plate 18: A *Acrogenospora* state of *Farlowiella carmichaeliana;* B *Hemicorynespora deightonii;* C *Spilocaea pomi;* D *Wardomycopsis inopinata;* E *Scopulariopsis brevicaulis;* F *Leptographium lundbergii.*

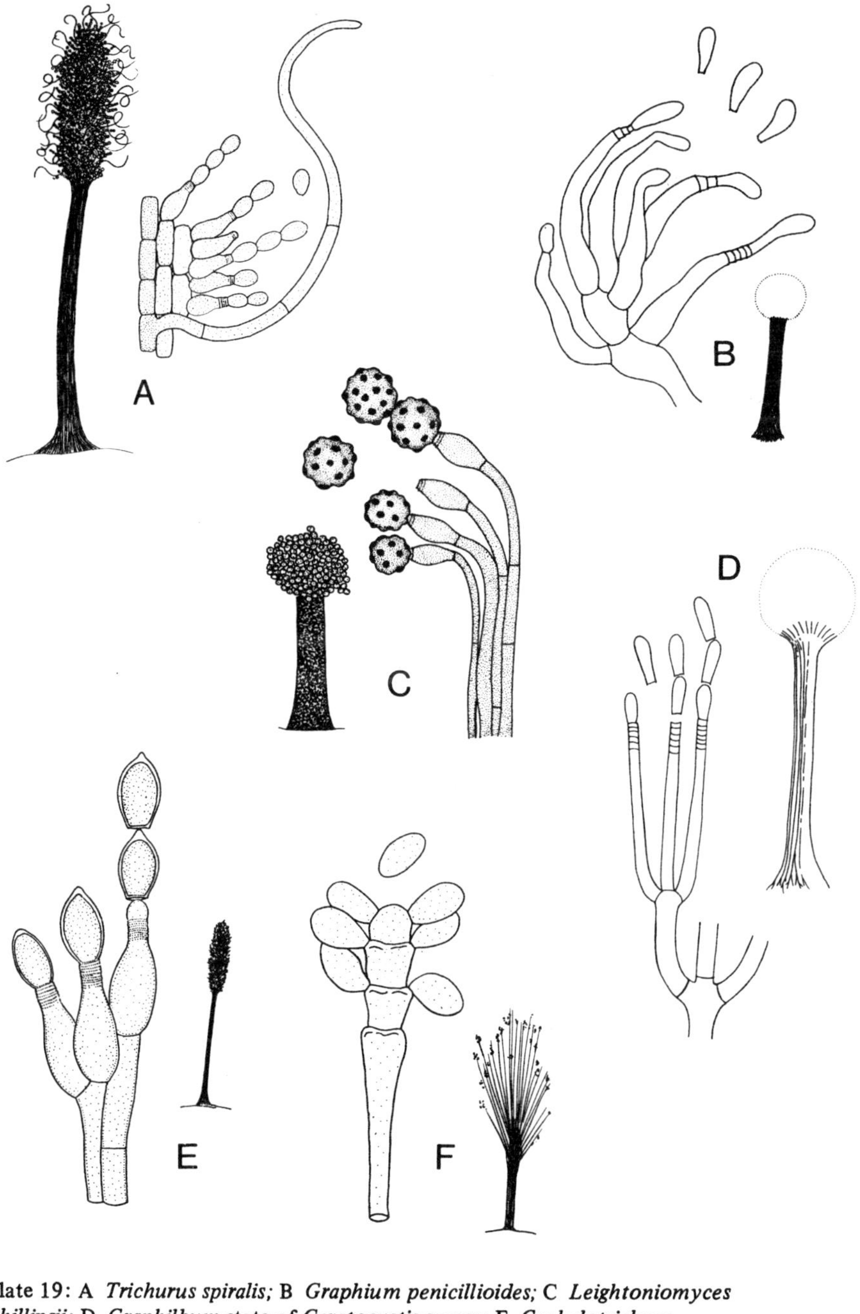

Plate 19: A *Trichurus spiralis;* B *Graphium penicillioides;* C *Leightoniomyces phillipsii;* D *Graphilbum* state of *Ceratocystis sparsa;* E *Cephalotrichum stemonitis;* F *Stevensomyces palmae.*

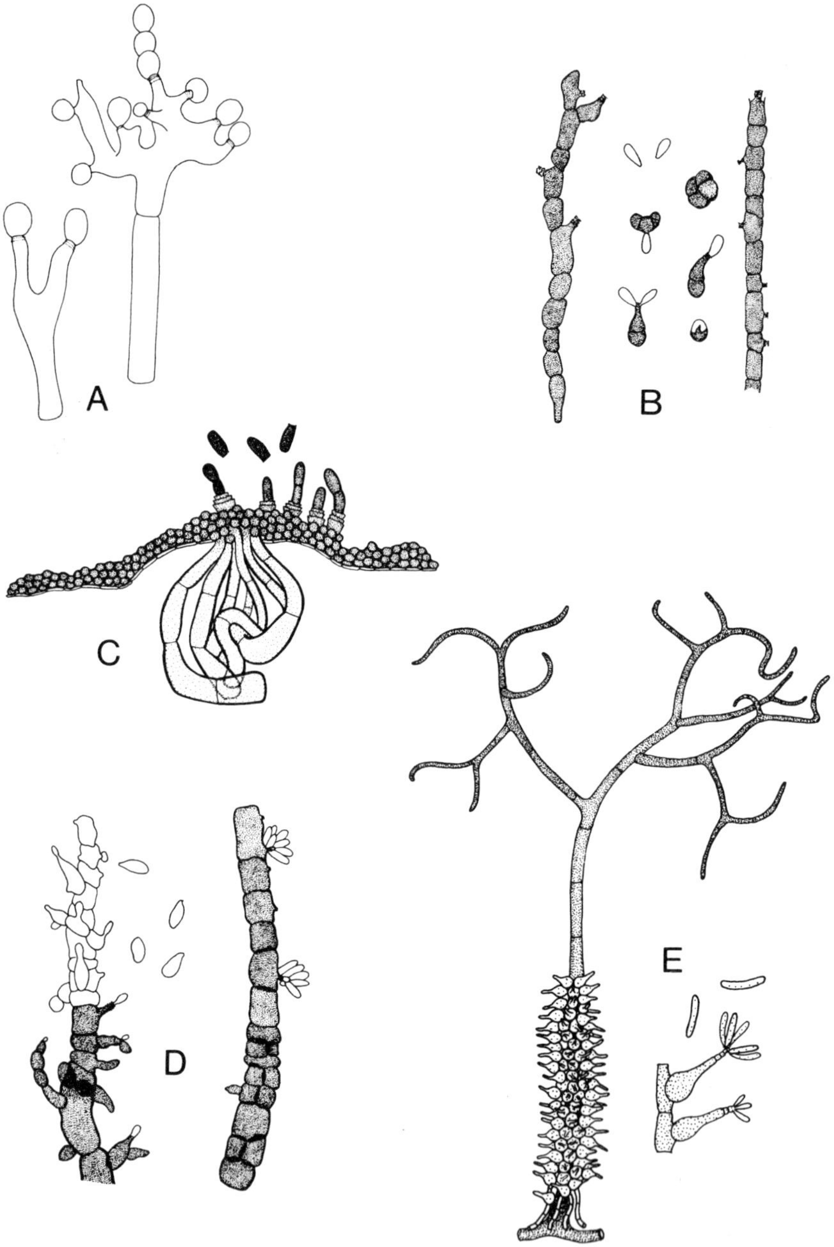

Plate 20: A *Polypaecilum insolitum;* B *Exophiala werneckii;*
C *Batcheloromyces proteae;* D *Hormonema* state of *Sydowia polyspora;*
E *Ceratocladium microspermum.*

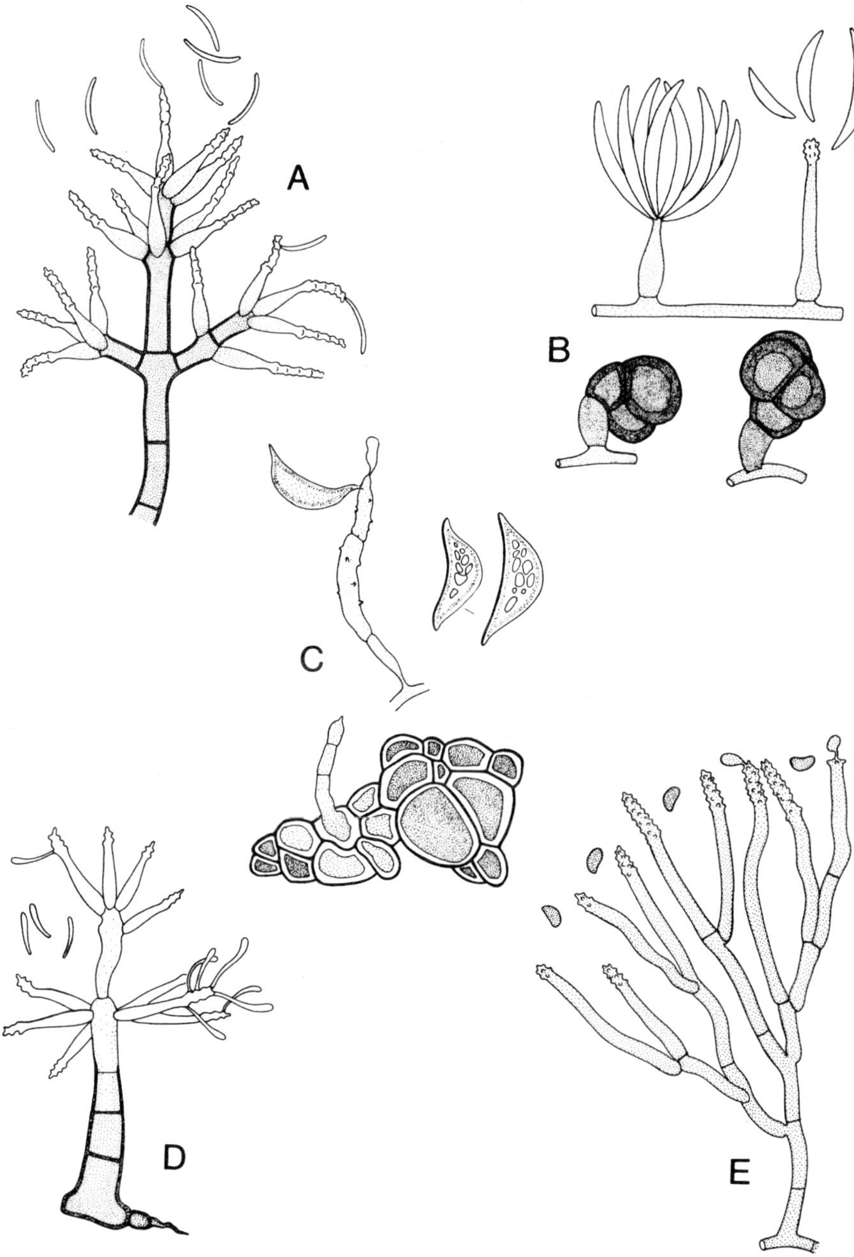

Plate 21: A *Selenosporella curvispora;* B *Idriella lunata;* C *Ardhachandra selenoides;* D *Umbellidion radulans;* E *Virgaria nigra.*

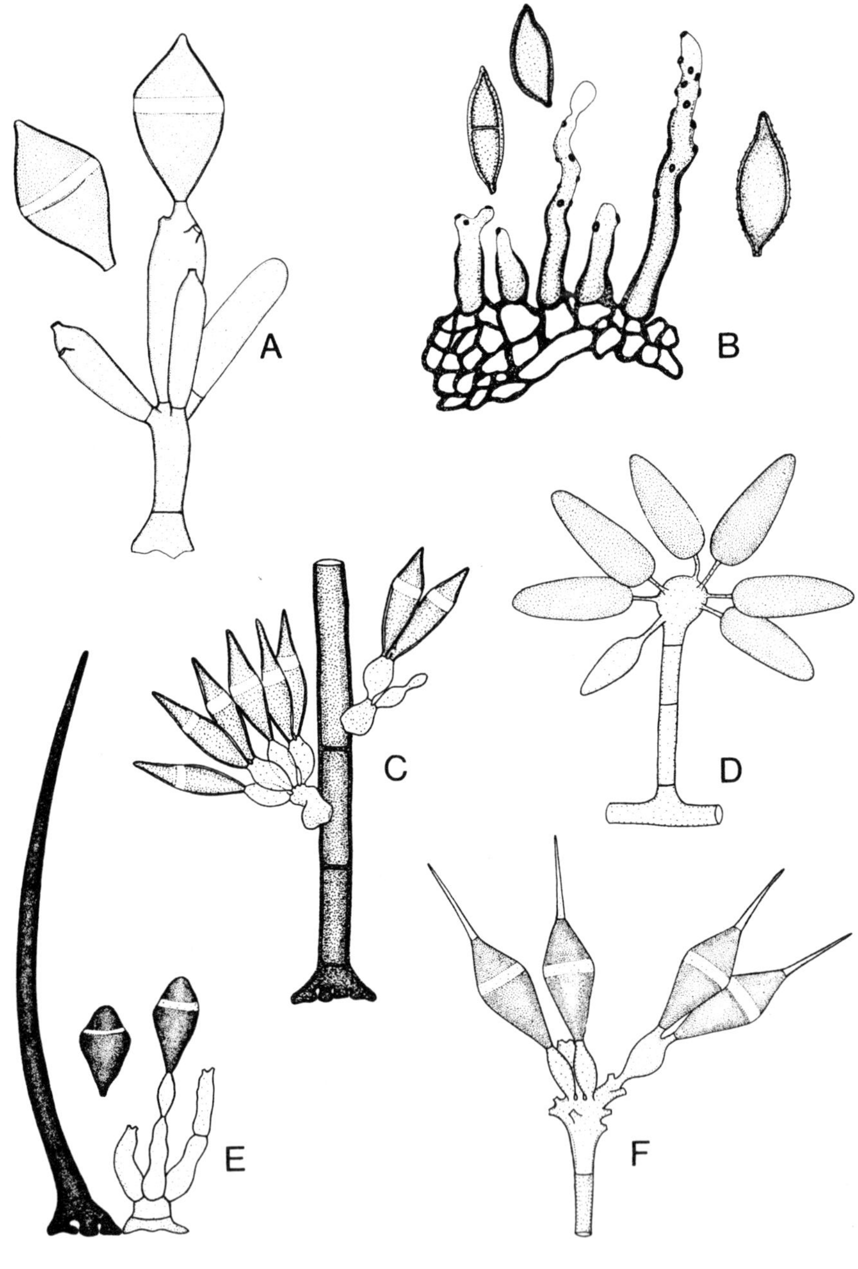

Plate 22: A *Pseudobeltrania cedrelae;* B *Fusicladium* state of *Venturia pirina;* C *Beltraniopsis esenbeckiae;* D *Asteromyces cruciatus;* E *Beltraniella humicola;* F *Beltrania rhombica.*

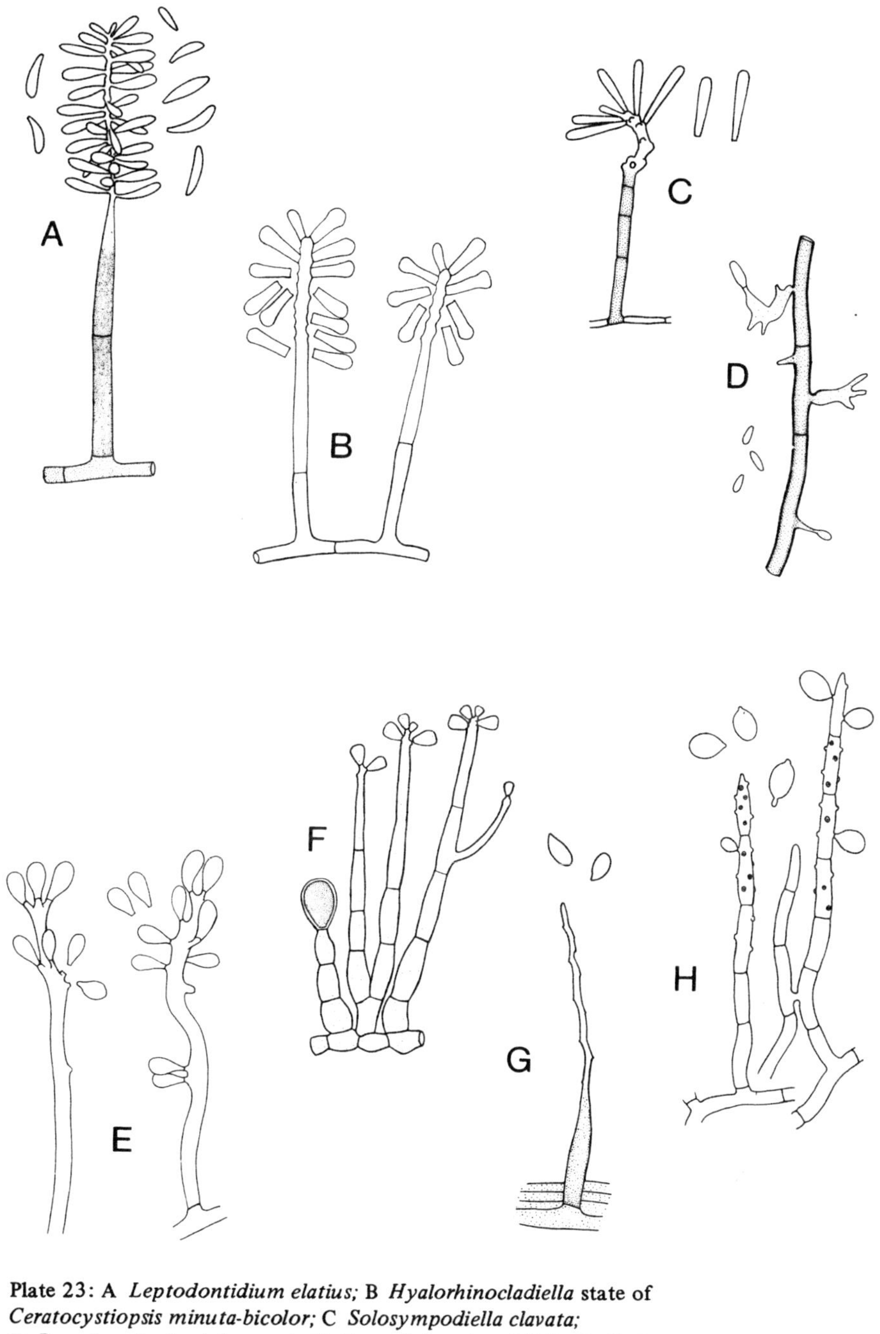

Plate 23: A *Leptodontidium elatius;* B *Hyalorhinocladiella* state of *Ceratocystiopsis minuta-bicolor;* C *Solosympodiella clavata;* D *Paraphaeoisaria alabamensis;* E *Sporothrix schenckii;* F *Raffaelea ambrosiae;* G *Acrodontium crateriforme;* H *Acladium* state of *Botryobasidium conspersum.*

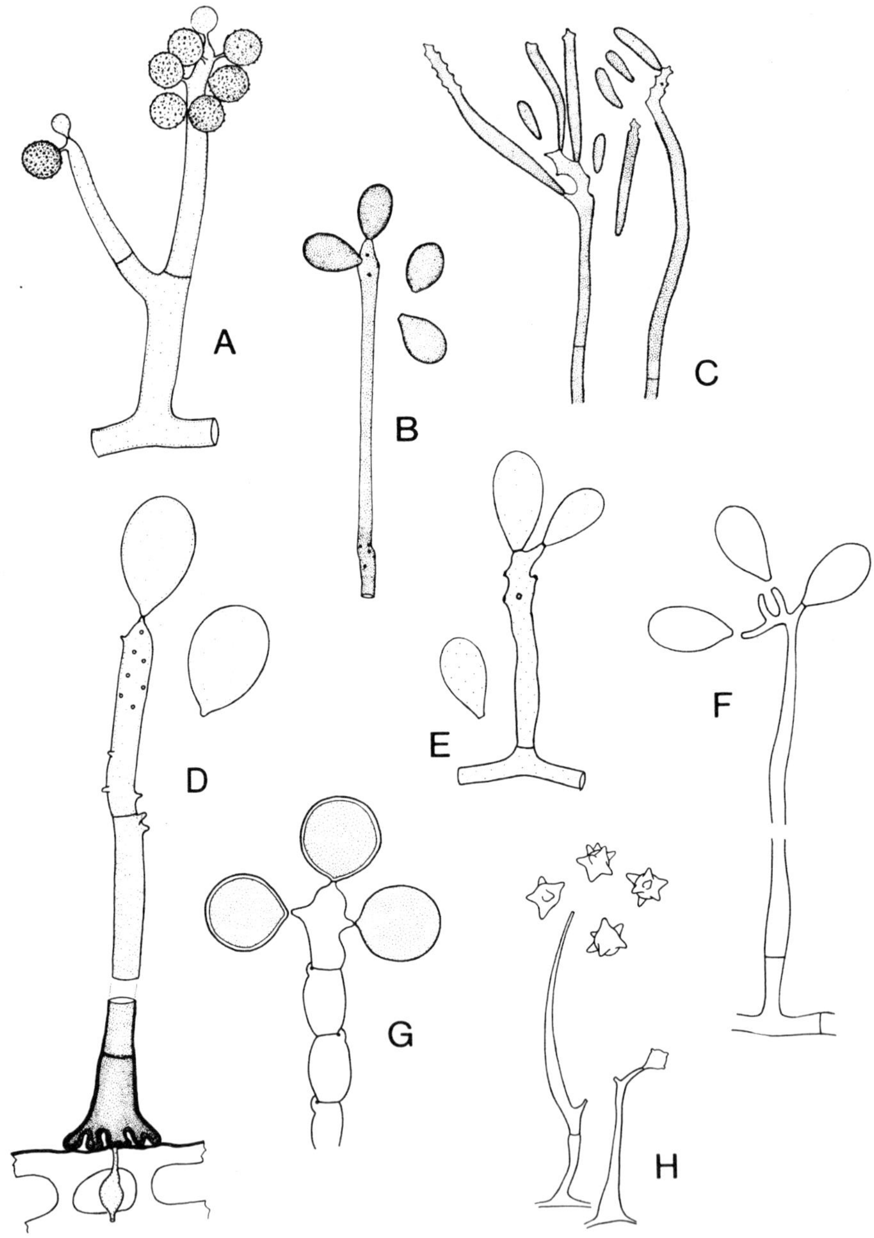

Plate 24: A *Basifimbria aurea;* B *Virgariella atra;* C *Racodium cellare;* D *Hemibeltrania cinnamomi;* E *Ovularia obliqua;* F *Calcarisporiella thermophila;* G *Phaeodaedalea* state of *Daedalea sprucei;* H *Tuberculispora jamaicensis.*

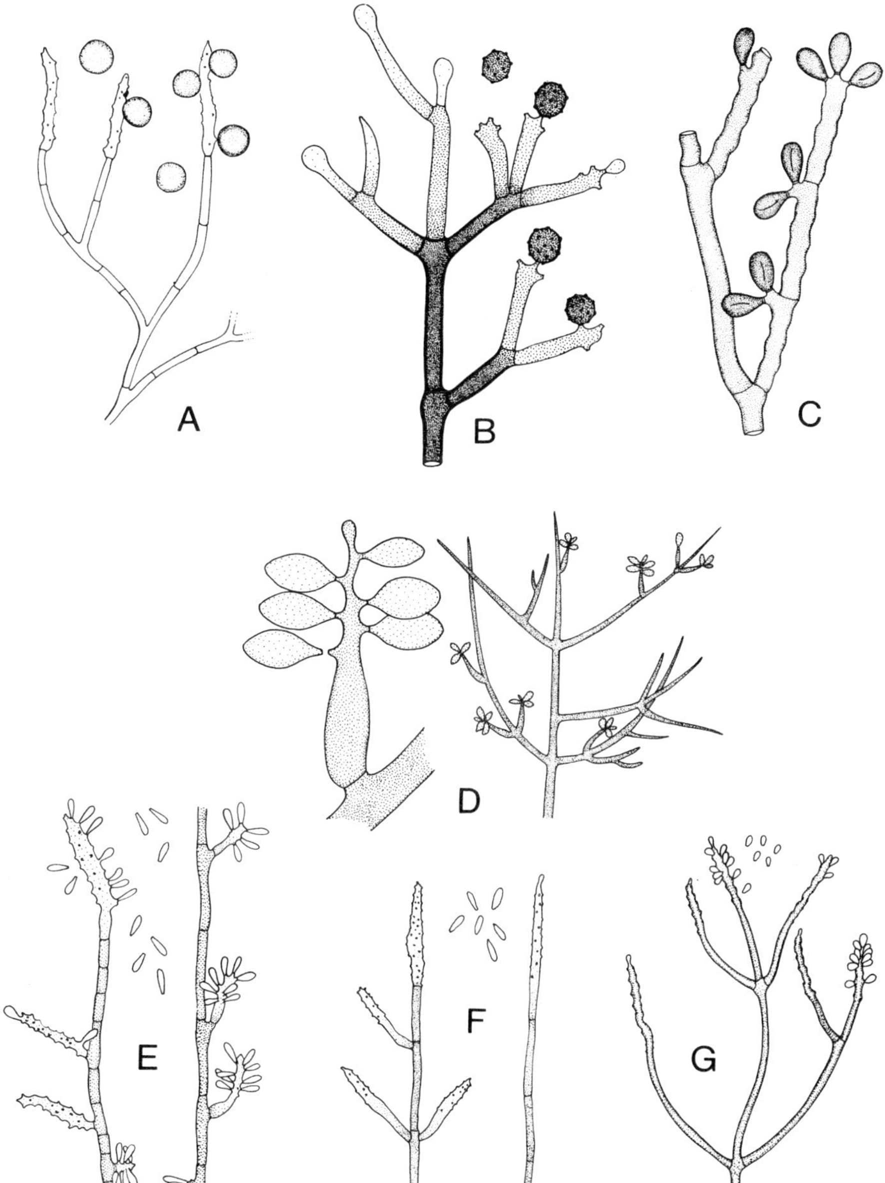

Plate 25: A *Rhinotrichella globulifera;* B *Dicyma* state of *Ascotricha chartarum;* C *Conoplea sphaerica;* D *Puciola spinosa;* E *Rhinocladiella atrovirens;* F *Ramichloridium musae;* G *Geniculosporium* state of *Hypoxylon serpens.*

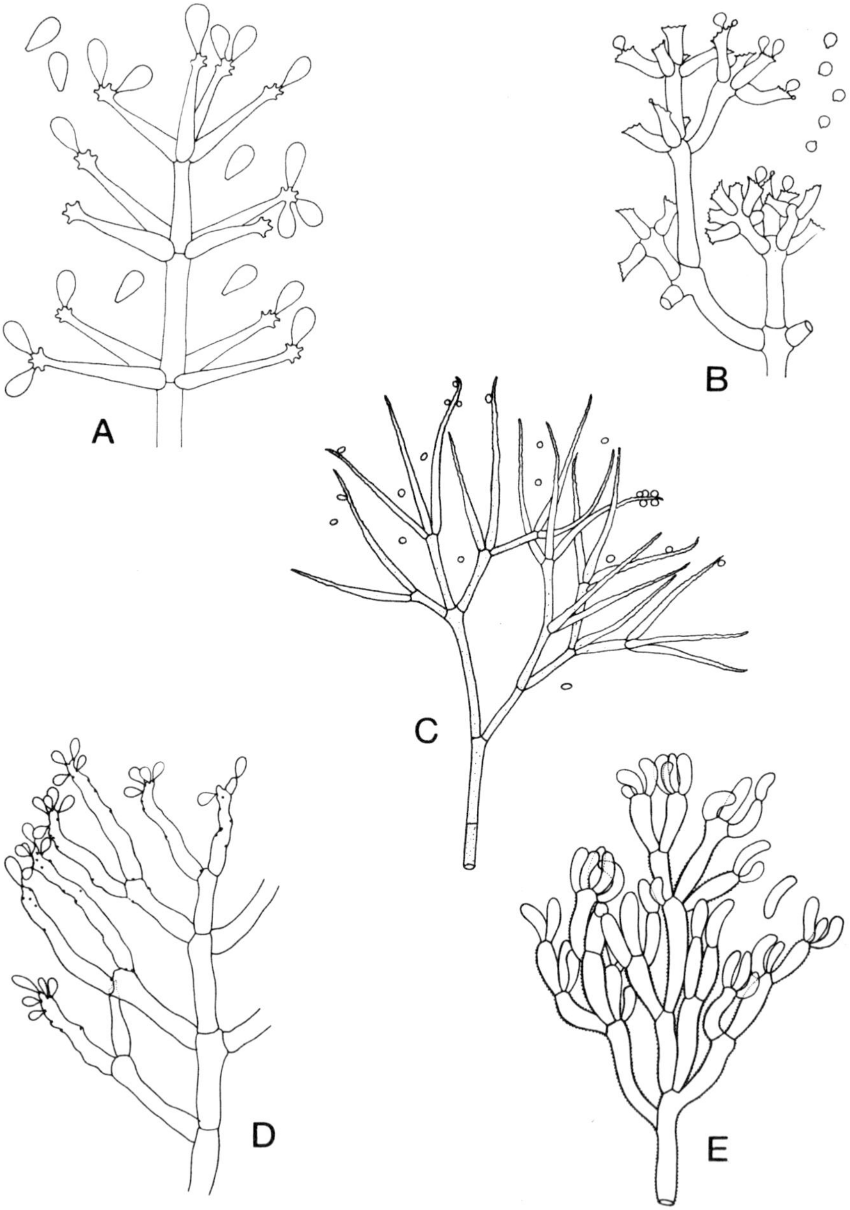

Plate 26: A *Calcarisporium arbuscula;* B *Costantinella terrestris;* C *Tritirachium hydnicola;* D *Geniculodendron pyriforme;* E *Ditangium insigne.*

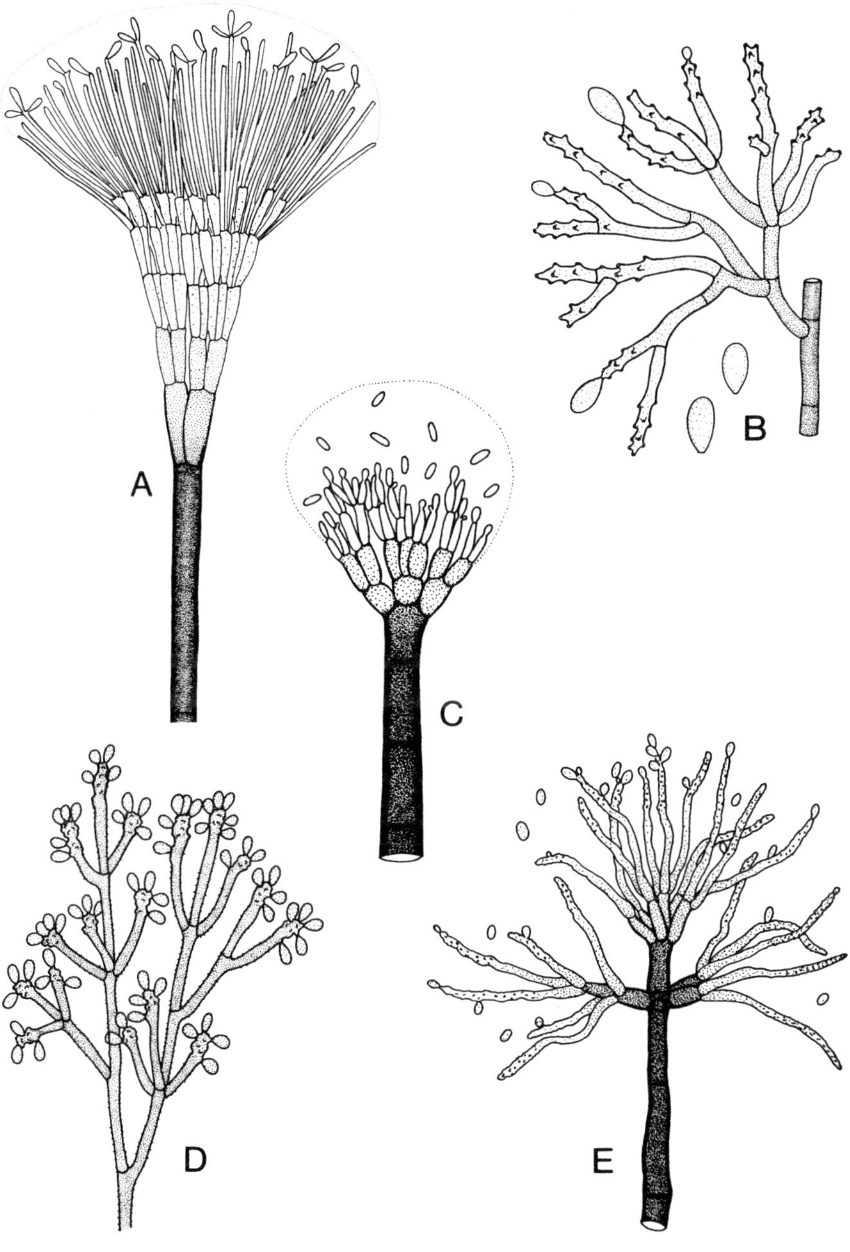

Plate 27: A *Verticicladiella abietina;* B *Hansfordia ovalispora;* C *Haplographium* state of *Hyaloscypha dematiicola;* D *Nodulisporium* state of *Hypoxylon fragiforme;* E *Verticicladium trifidum.*

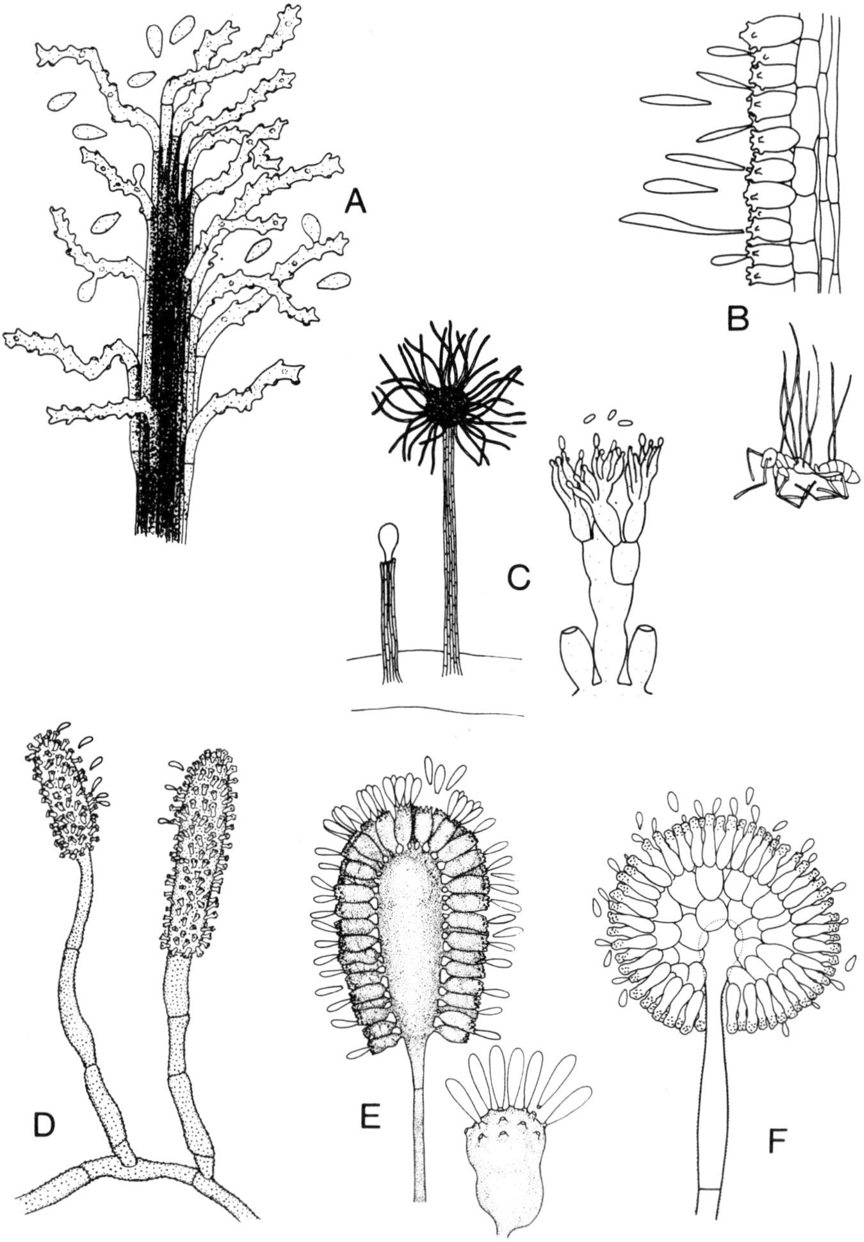

Plate 28: A *Phaeoisaria clematidis;* B *Hymenostilbe longispora;* C *Heterocephalum aurantiacum;* D *Masoniomyces claviformis;* E *Xylocladium* state of *Hypoxylon tinctor;* F *Pseudogibellula formicarum.*

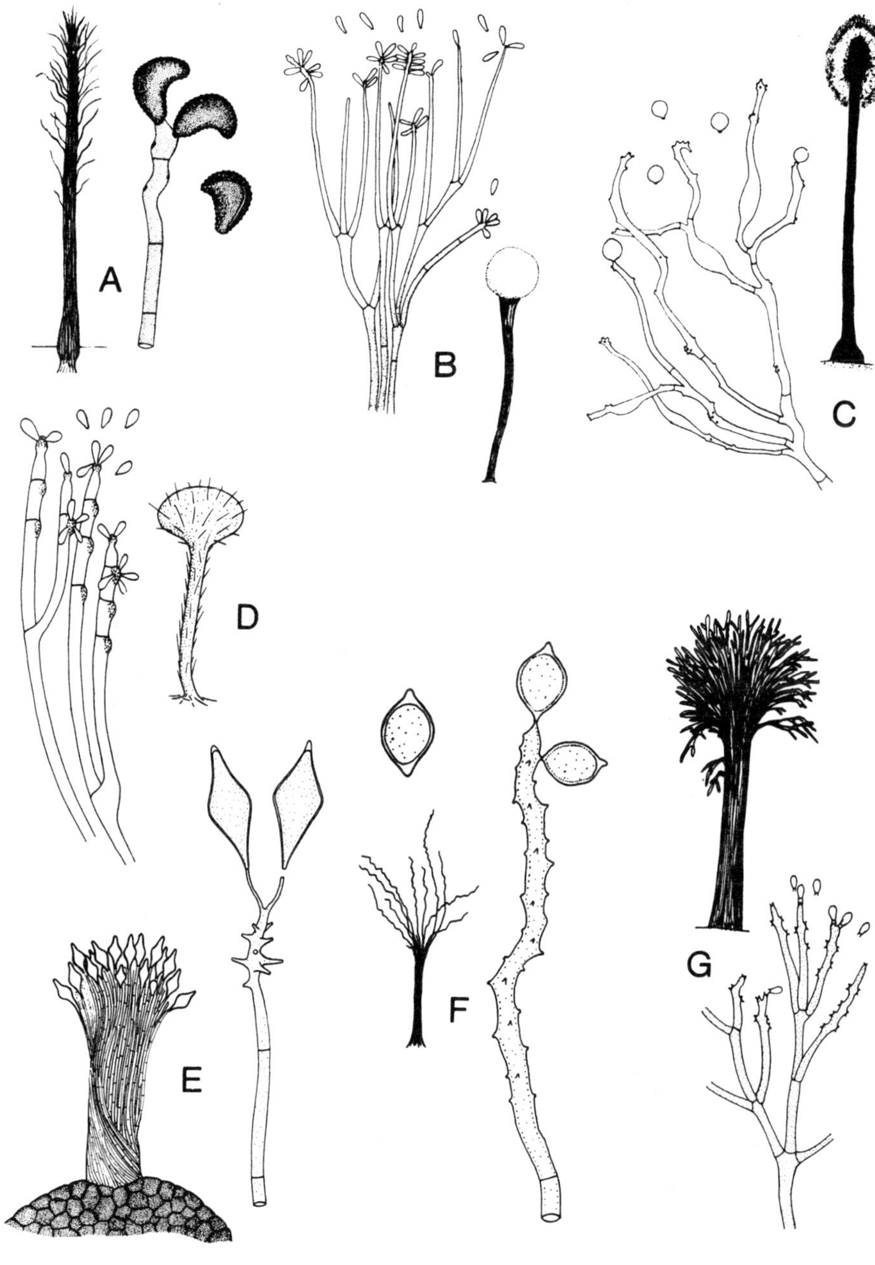

Plate 29: A *Melanographium selenioides;* B *Pesotum* state of *Ceratocystis ulmi;* C *Beauveriphora sporotrichis;* D *Agaricostilbum palmicola;* E *Rhombostilbella rosea;* F *Drumopama girisa;* G *Dematophora necatrix.*

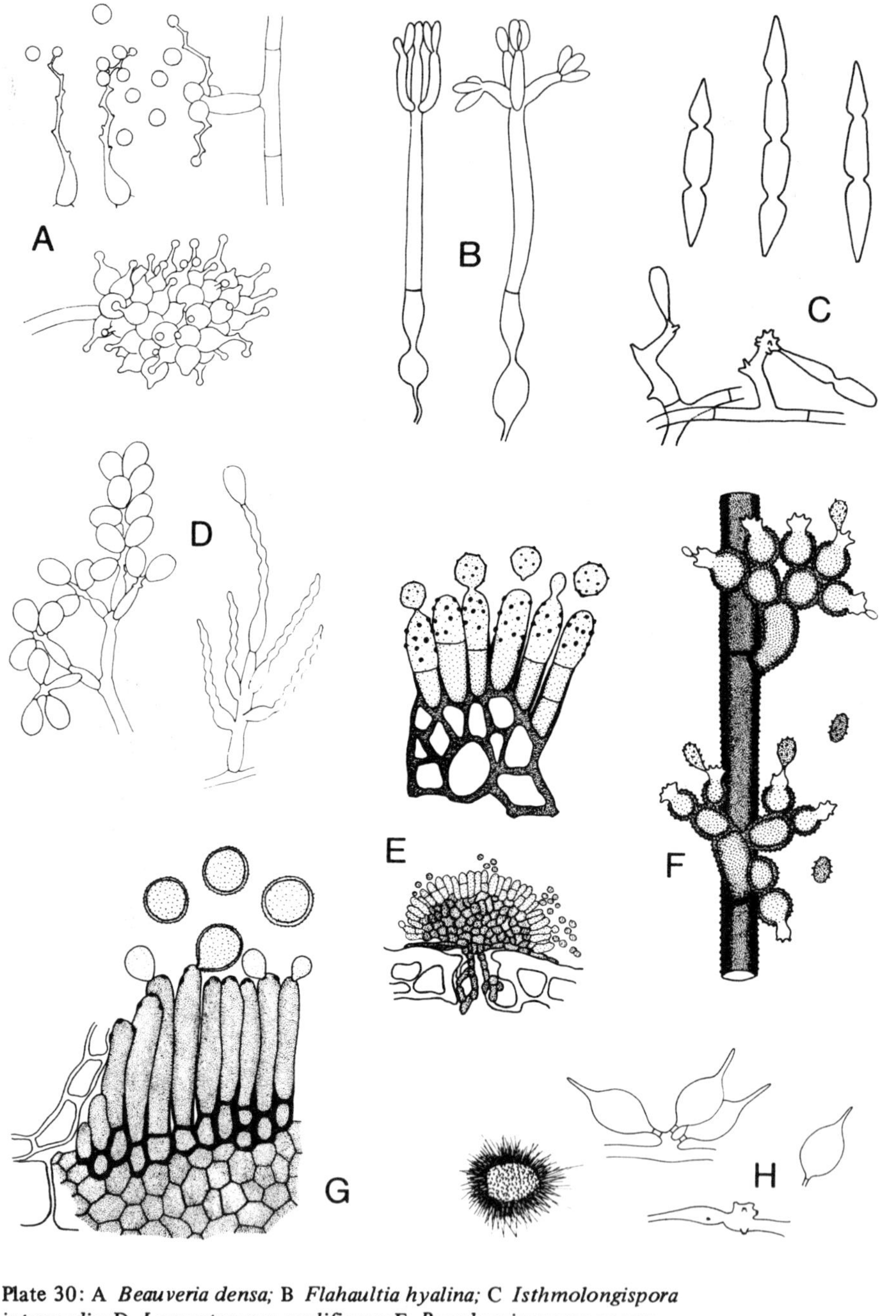

Plate 30: A *Beauveria densa;* B *Flahaultia hyalina;* C *Isthmolongispora intermedia;* D *Lomentospora prolificans;* E *Pseudoepicoccum cocos;* F *Kumanasamuha sundara;* G *Hadrotrichum phragmitis;* H *Lappodochium lageniforme.*

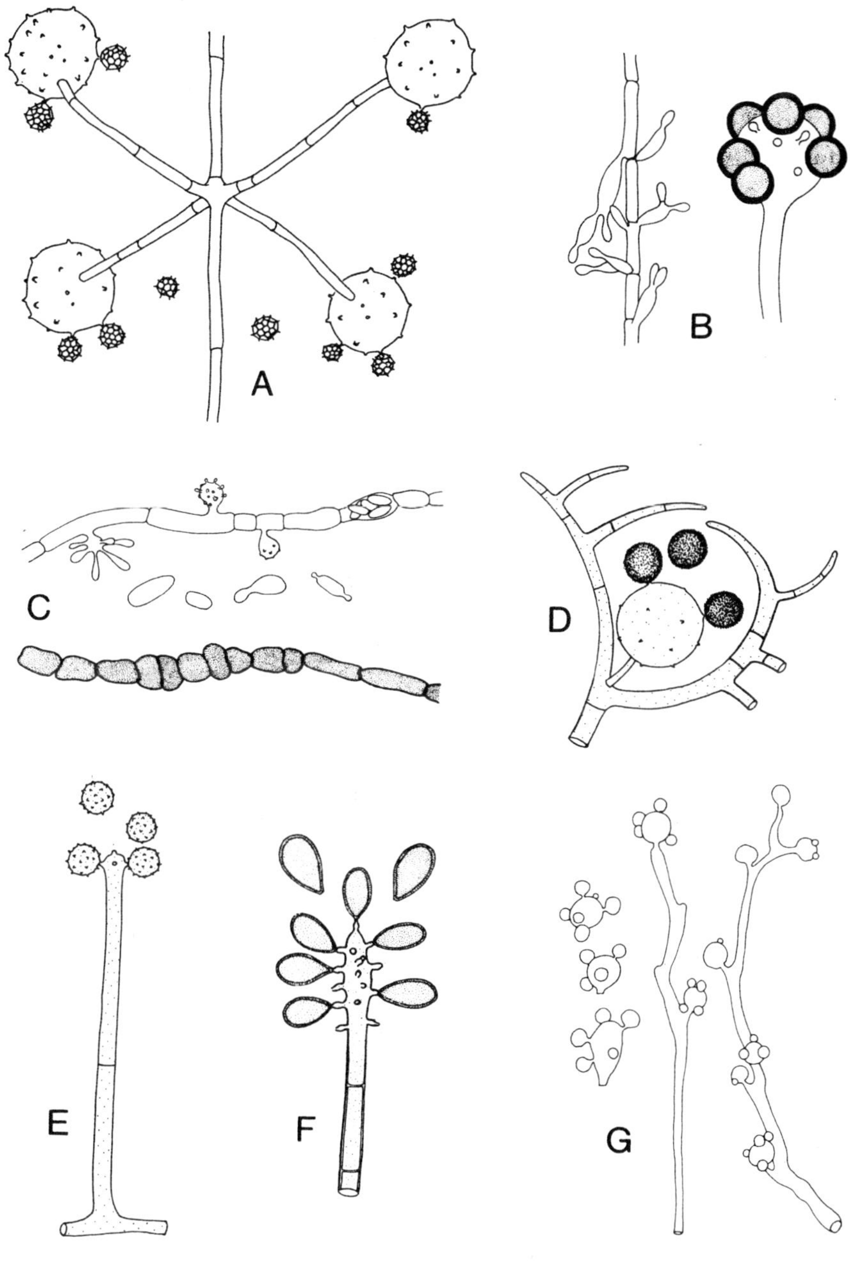

Plate 31: A *Sphondylocephalum verticillatum;* B *Trichosporonoides oedocephalis;* C *Aureobasidium pullulans;* D *Sigmoideomyces dispiroides;* E *Sporocephalum peniophorae;* F *Olpitrichum macrosporum;* G *Blastobotrys nivea.*

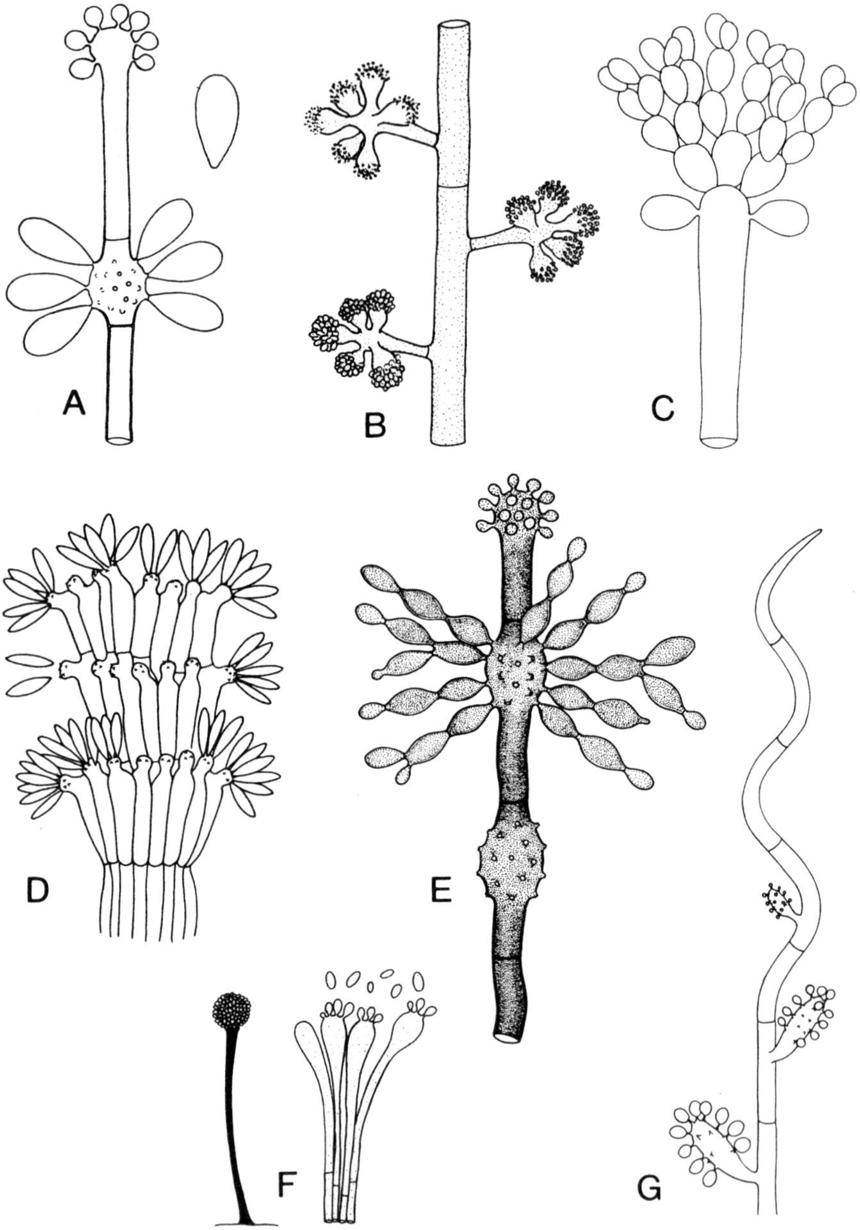

Plate 32: A *Gonatobotrys simplex;* B *Botryosporium pulchrum;* C *Nematogonum ferrugineum;* D *Microstroma album;* E *Gonatobotryum apiculatum;* F *Pachnocybe ferruginea;* G *Ostracodermidium serpentinum.*

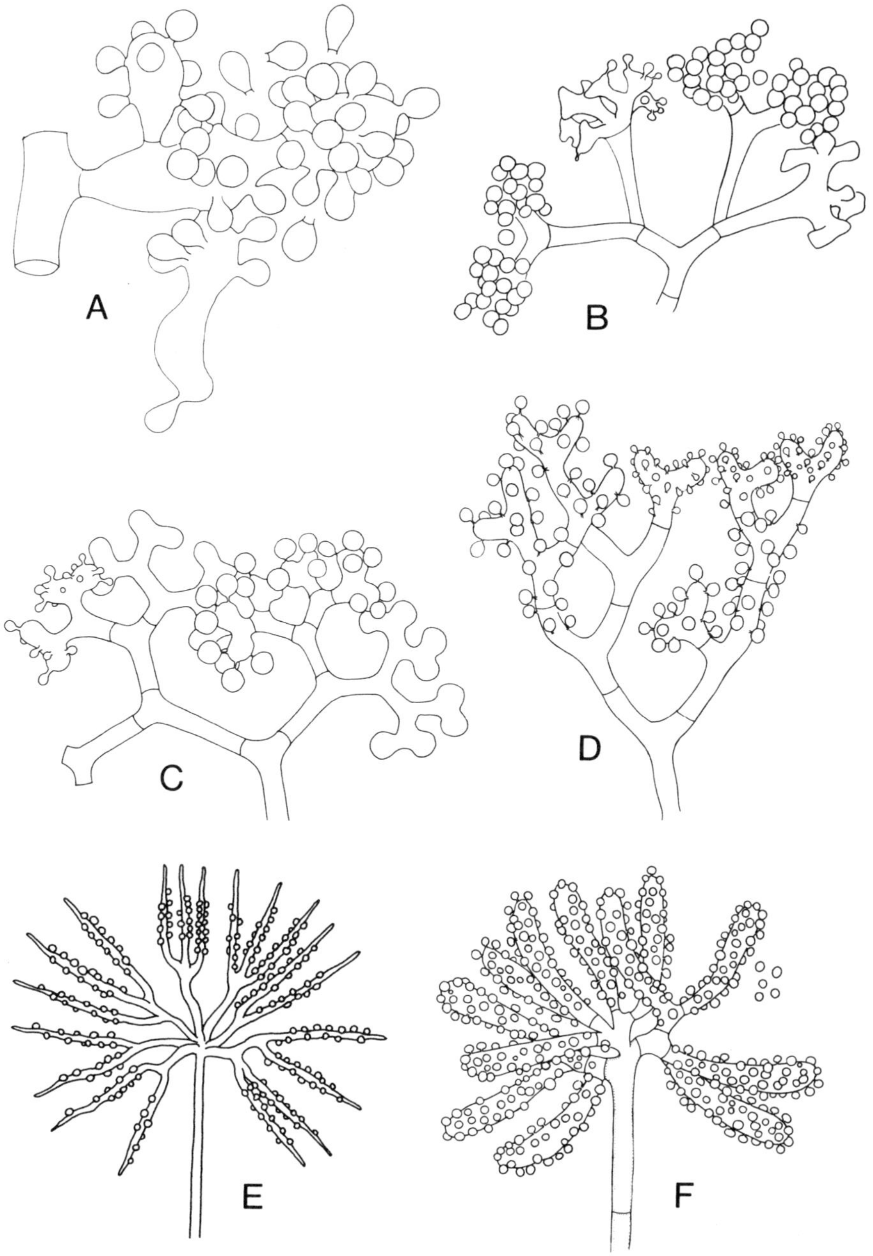

Plate 33: A *Phymatotrichopsis omnivora;* B *Amphobotrys ricini;* C *Dichobotrys* state of *Trichophaea abundans;* D *Glischroderma* sp.; E *Pulchromyces fimicola;* F *Chromelosporium ollare.*

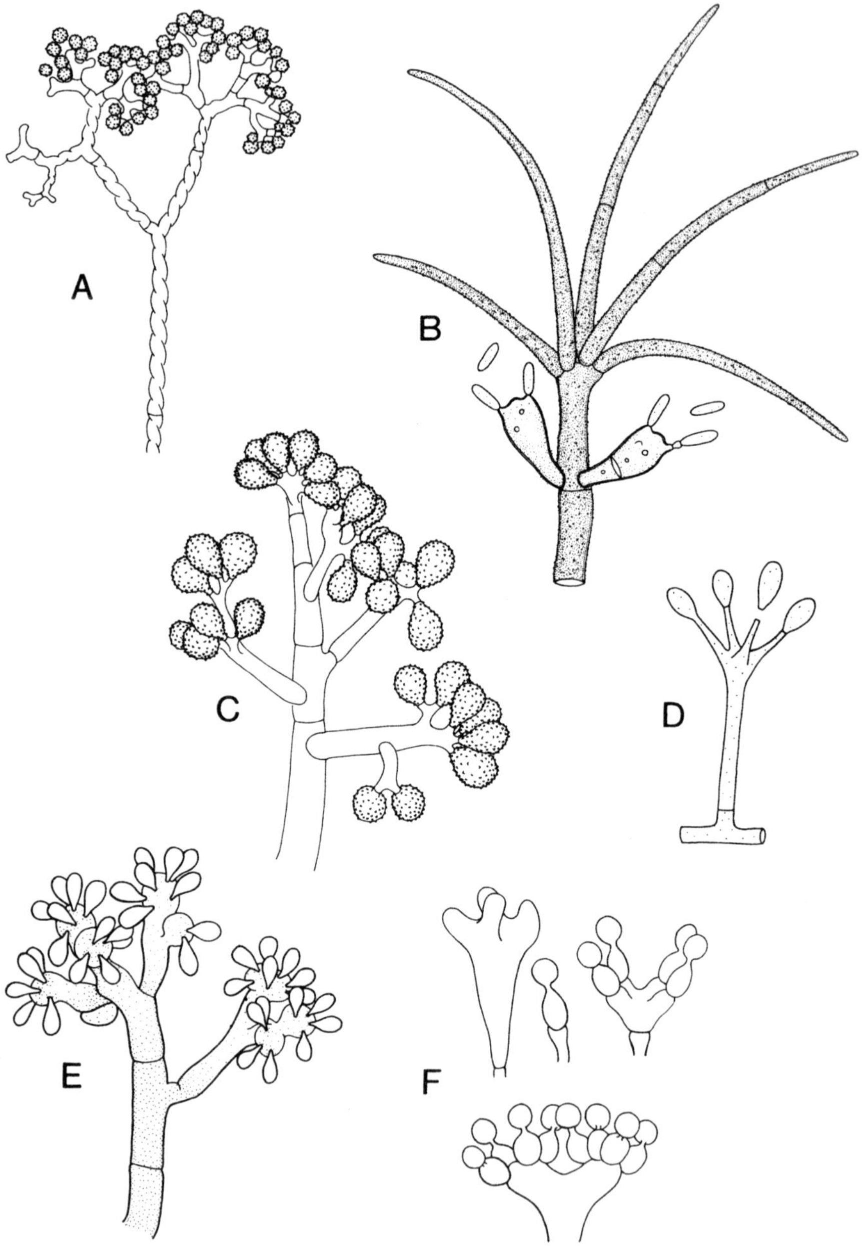

Plate 34: A *Streptobotrys* state of *Streptotinia arisaemae;* B *Heimiodora verticillata;* C *Verrucobotrys* state of *Seaverinia geranii;* D *Dexhowardia tetraspora;* E *Botrytis cinerea;* F *Padixonia bispora.*

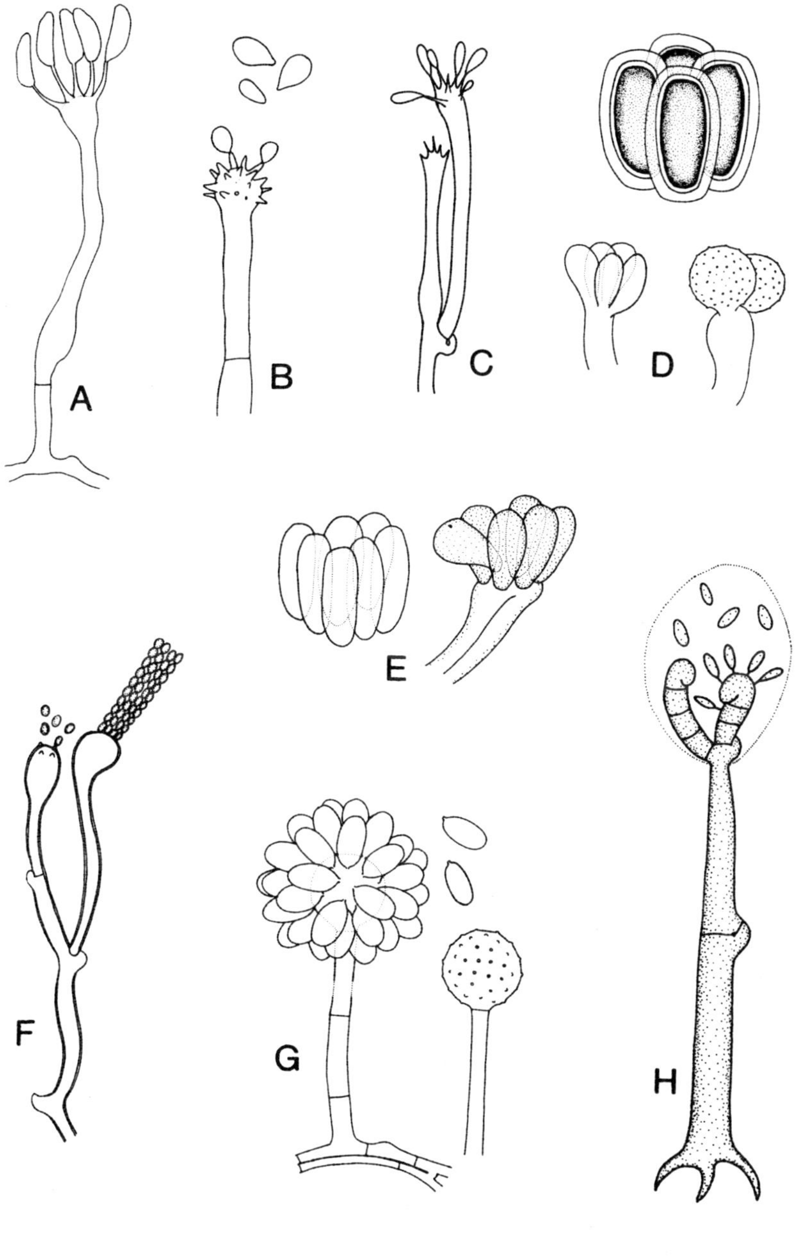

Plate 35: A *Ambivina filobasidia;* B *Spiniger meineckellus;* C *Spiniger* state of *Resinicium furfuraceum;* D *Clinoconidium bullatum;* E *Botryoconis saccardoi;* F *Filobasidiella neoformans;* G *Oedocephalum glomerulosum;* H *Jacobia conspicua.*

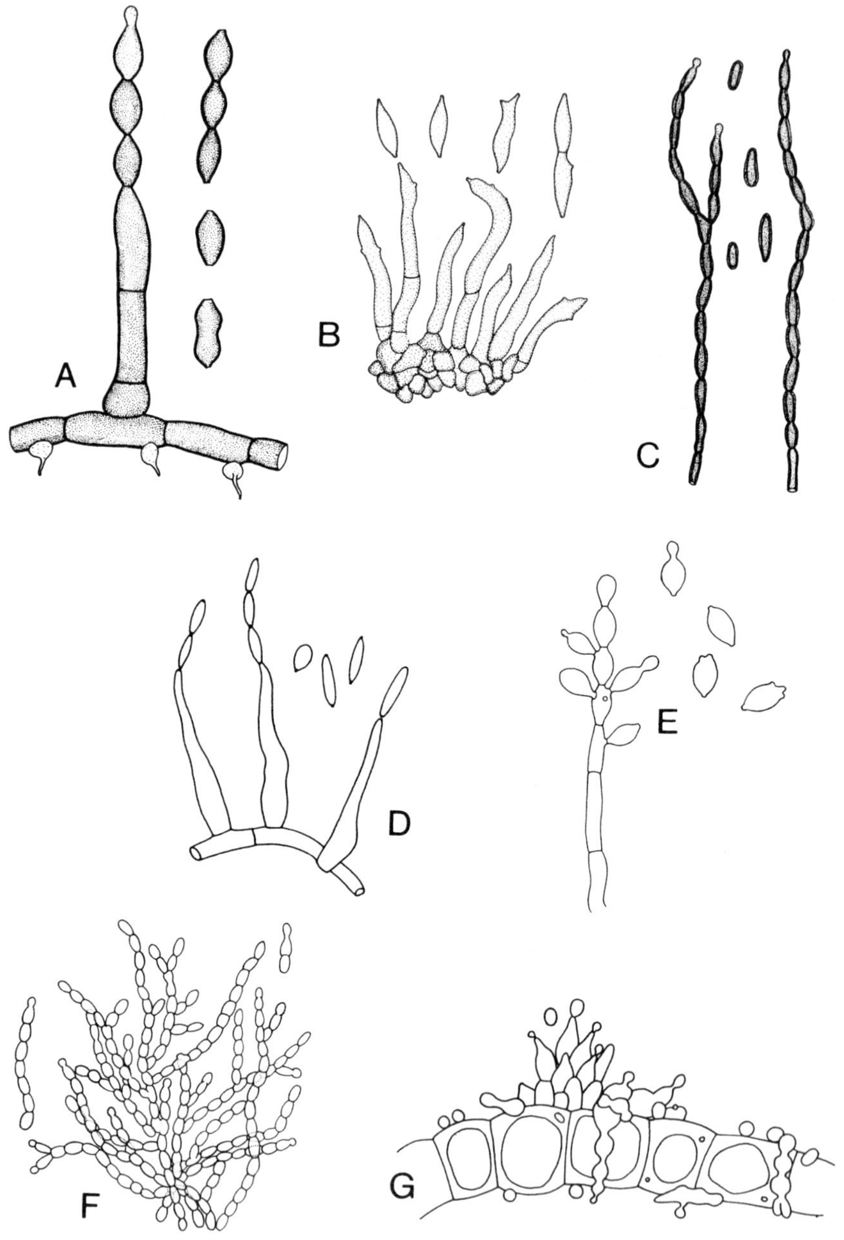

Plate 36: A *Ampullifera foliicola;* B *Denticularia modesta;* C *Xylohypha nigrescens;* D *Pseudofusidium hansfordii;* E *Alysidium* state of *Botryobasidium aureum;* F *Hormomyces aurantiacus;* G *Sphaerulomyces coralloides.*

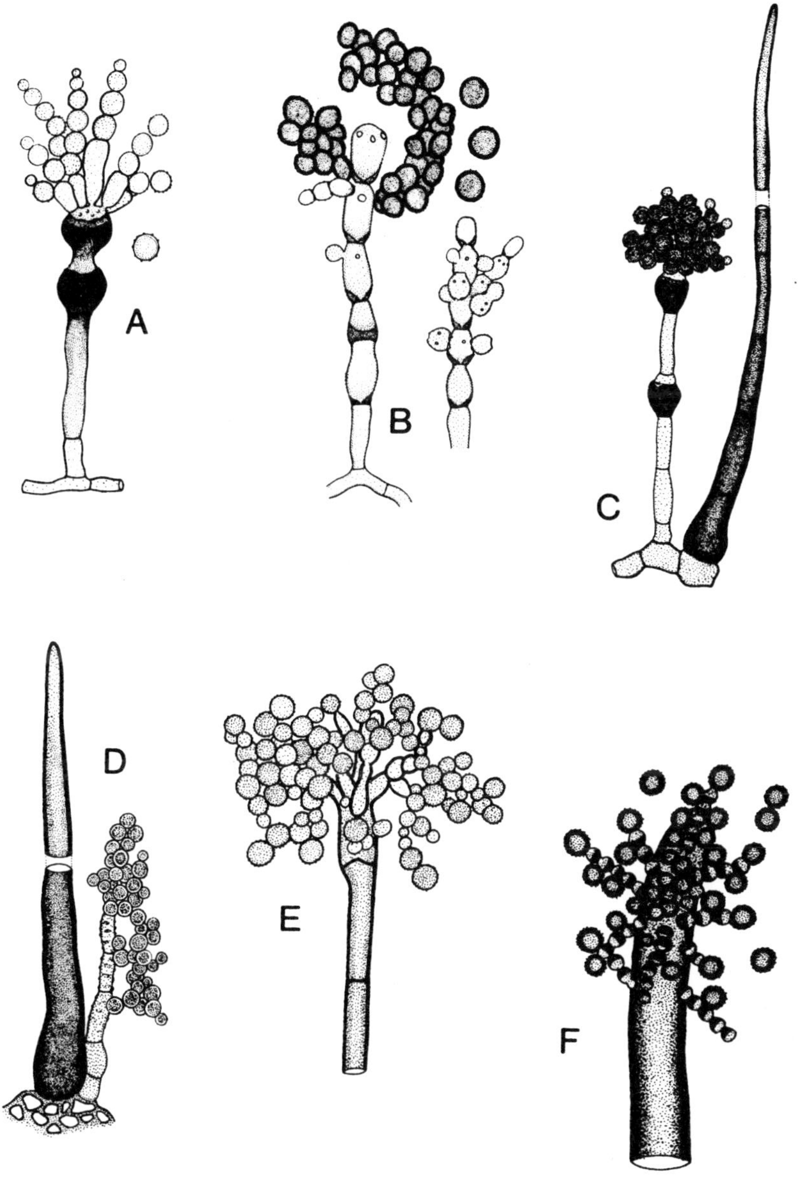

Plate 37: A *Haplobasidion thalictri;* B *Echinocatena arthrinioides;* C *Lacellinopsis sacchari;* D *Lacellina graminicola;* E *Periconia paludosa;* F *Sadasivania girisa.*

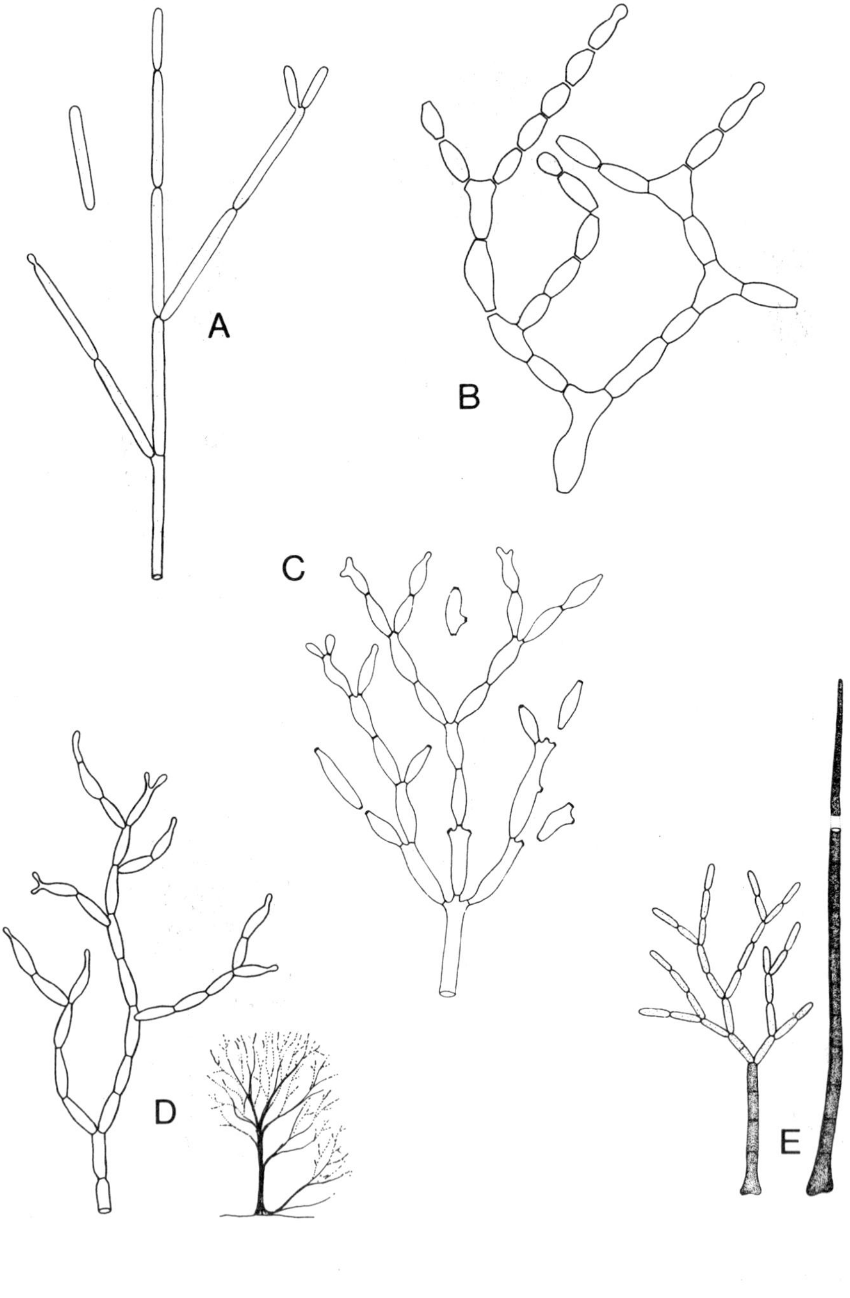

Plate 38: A *Fusidium griseum;* B *Monilia* state of *Neurospora tetrasperma;* C *Hyalodendron* sp.; D *Sphaeridium candidum;* E *Hormiactella* sp.

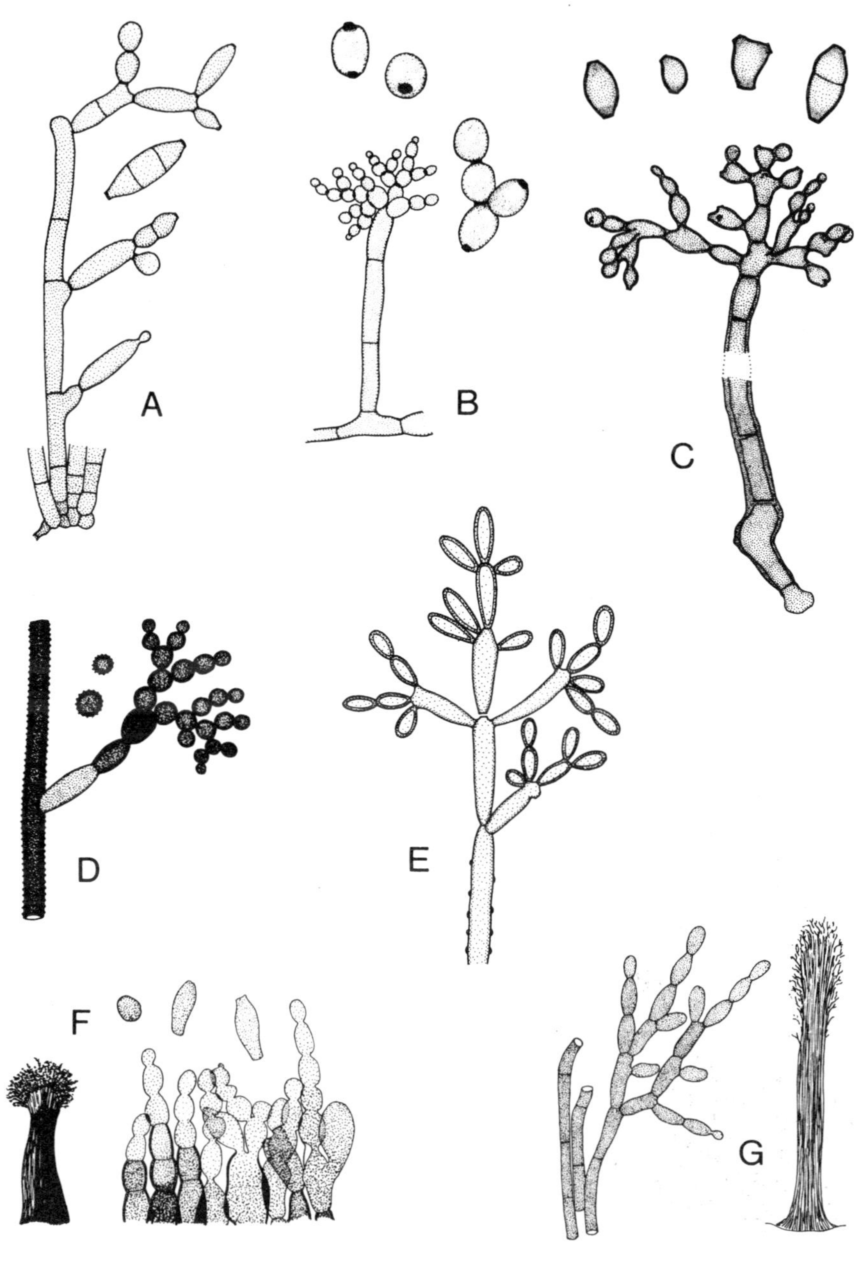

Plate 39: A *Fulvia fulva;* B *Agarwalia terricola;* C *Alysidiopsis pipsissewae;*
D *Trichobotrys effusa;* E *Hormoconis* state of *Amorphotheca resinae;*
F *Pycnostysanus azaleae;* G *Sorocybe resinae.*

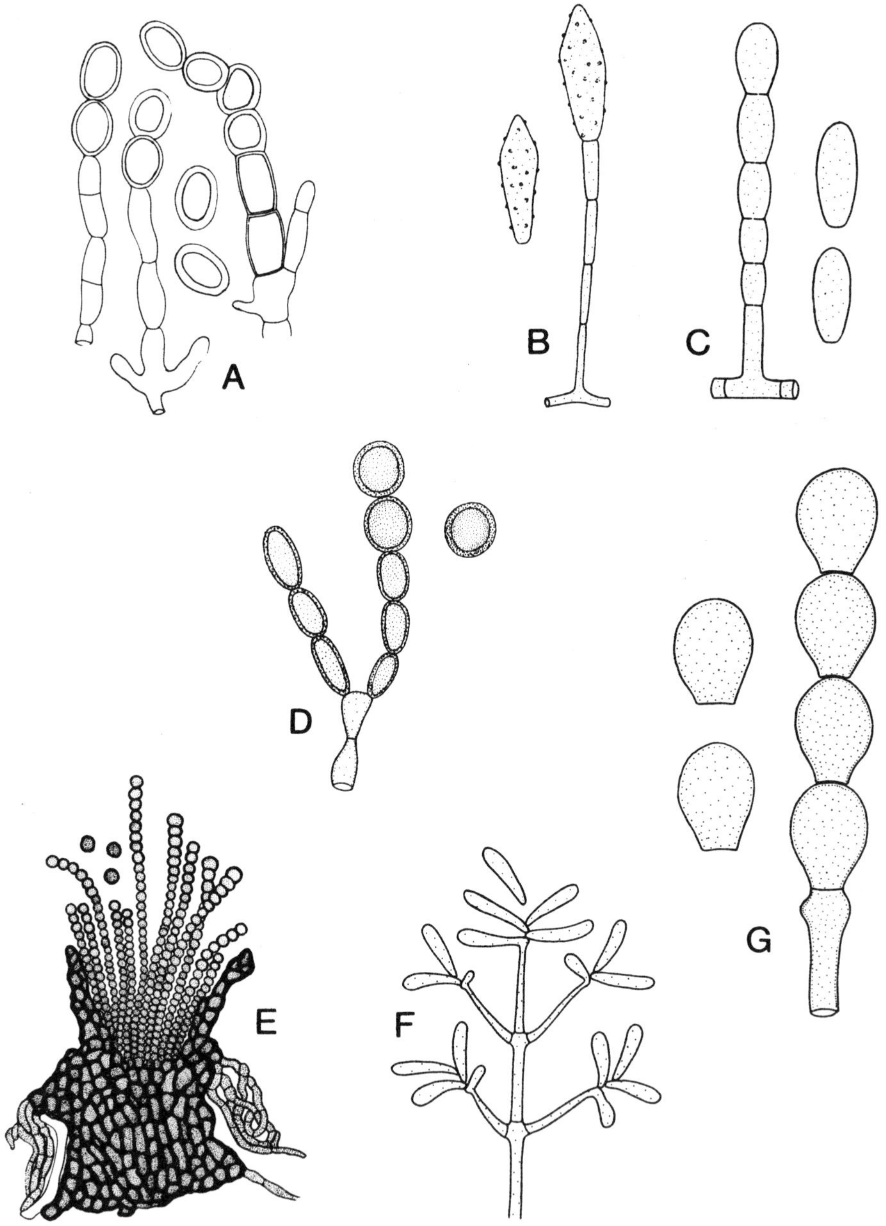

Plate 40: A *Moniliophthora roreri;* B *Ovulariopsis* state of *Phyllactinia* sp.; C *Oidium* state of *Erisyphe graminis;* D *Sphaerosporium lignatile;* E *Ojibwaya perpulchra;* F *Cladobotryum apiculatum;* G *Basipetospora rubra.*

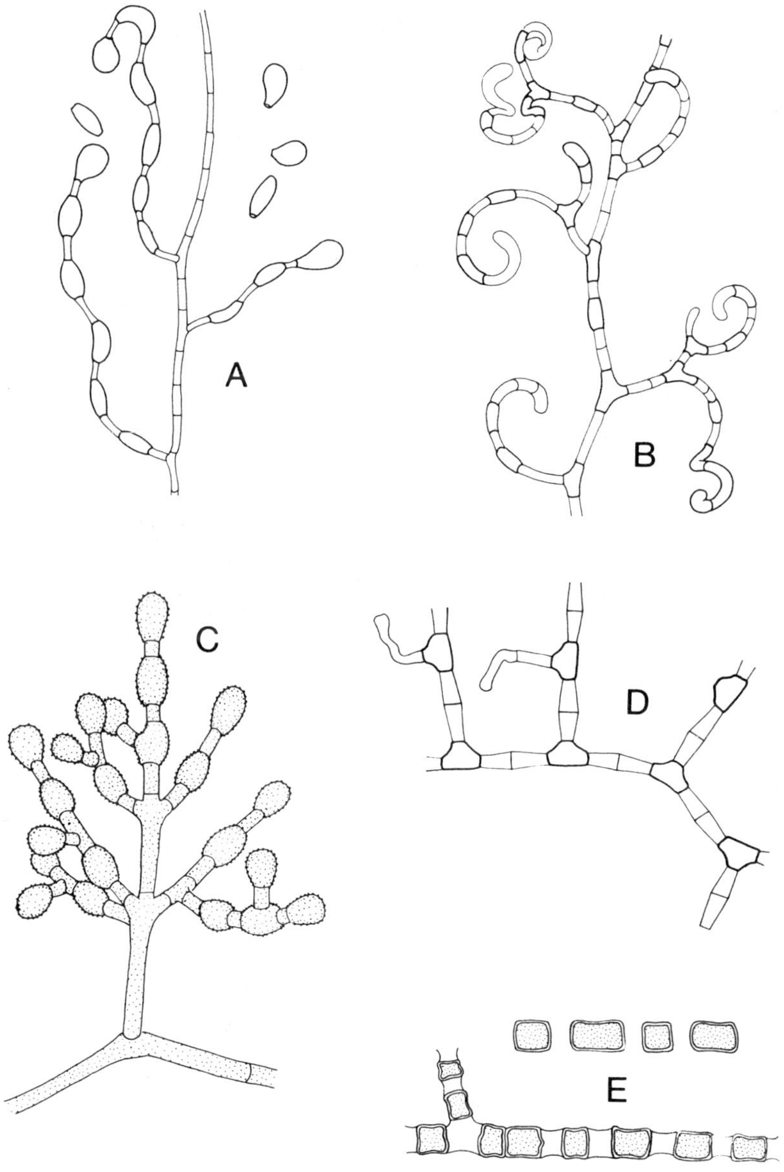

Plate 41: A *Ovadendron sulphureo-ochraceum;* B *Malbranchea pulchella;* C *Geomyces* sp.; D *Arcuadendron triangularis;* E *Sporendonema purpurascens.*

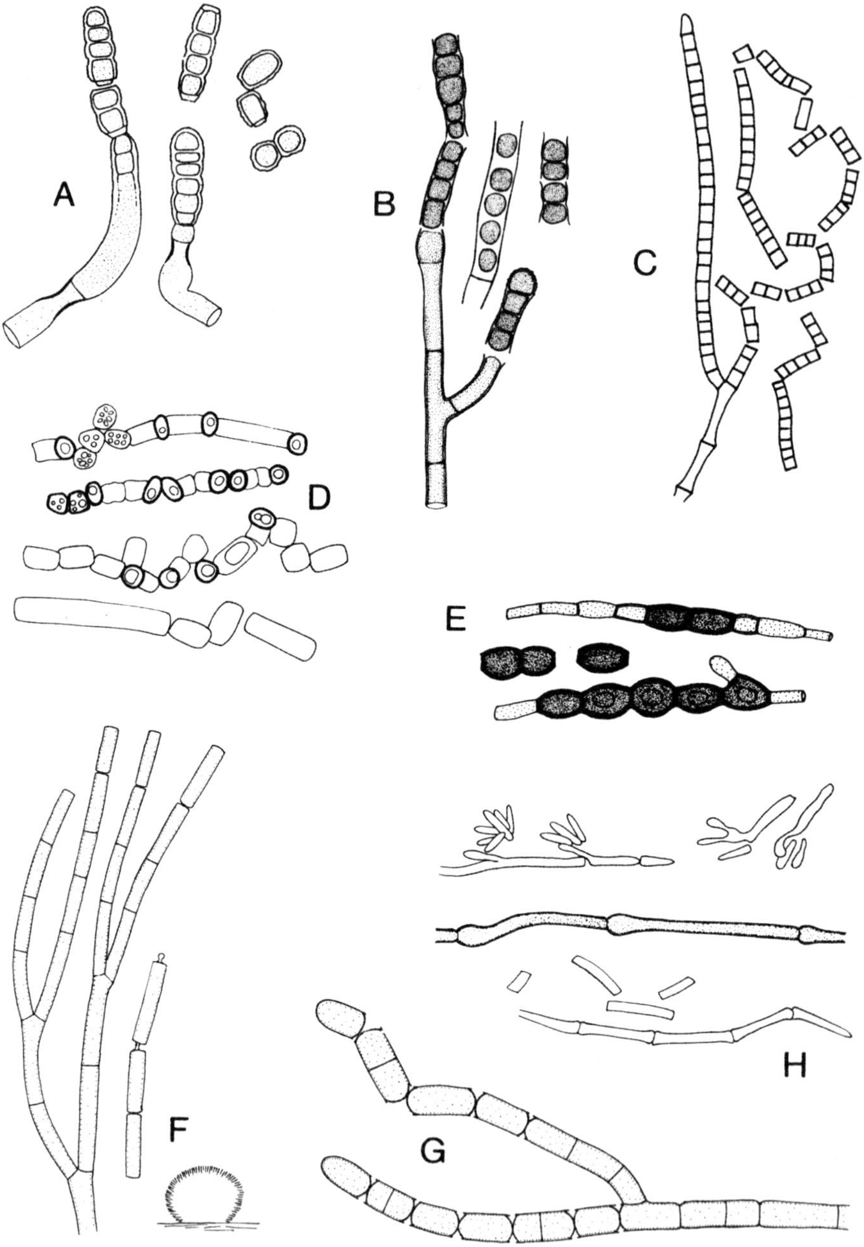

Plate 42: A *Wallemia ichthyophaga;* B *Bahusakala olivaceo-nigra;* C *Arthrographis cuboidea;* D *Trichosporon cutaneum;* E *Scytalidium lignicola;* F *Cylindrocolla urticae;* G *Geotrichum candidum;* H *Rosulomyces arthrosporioides.*

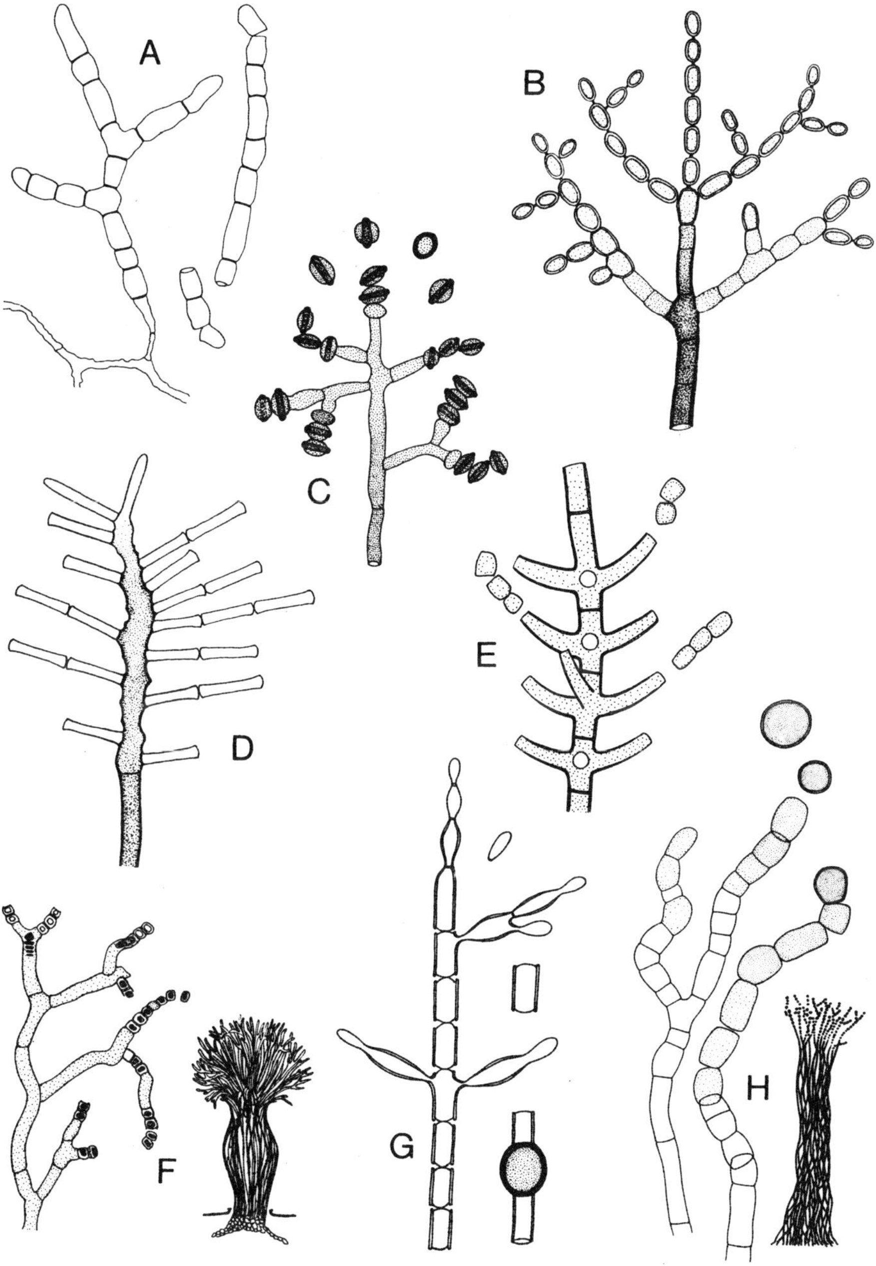

Plate 43: A *Mauginiella scaettae;* B *Oidiodendron* sp.; C *Stephanosporium cereale;* D *Sympodiella acicola;* E *Oncocladium flavum;* F *Coremiella cubispora;* G *Moniliella acetoabutans;* H *Briosia ampelophaga.*

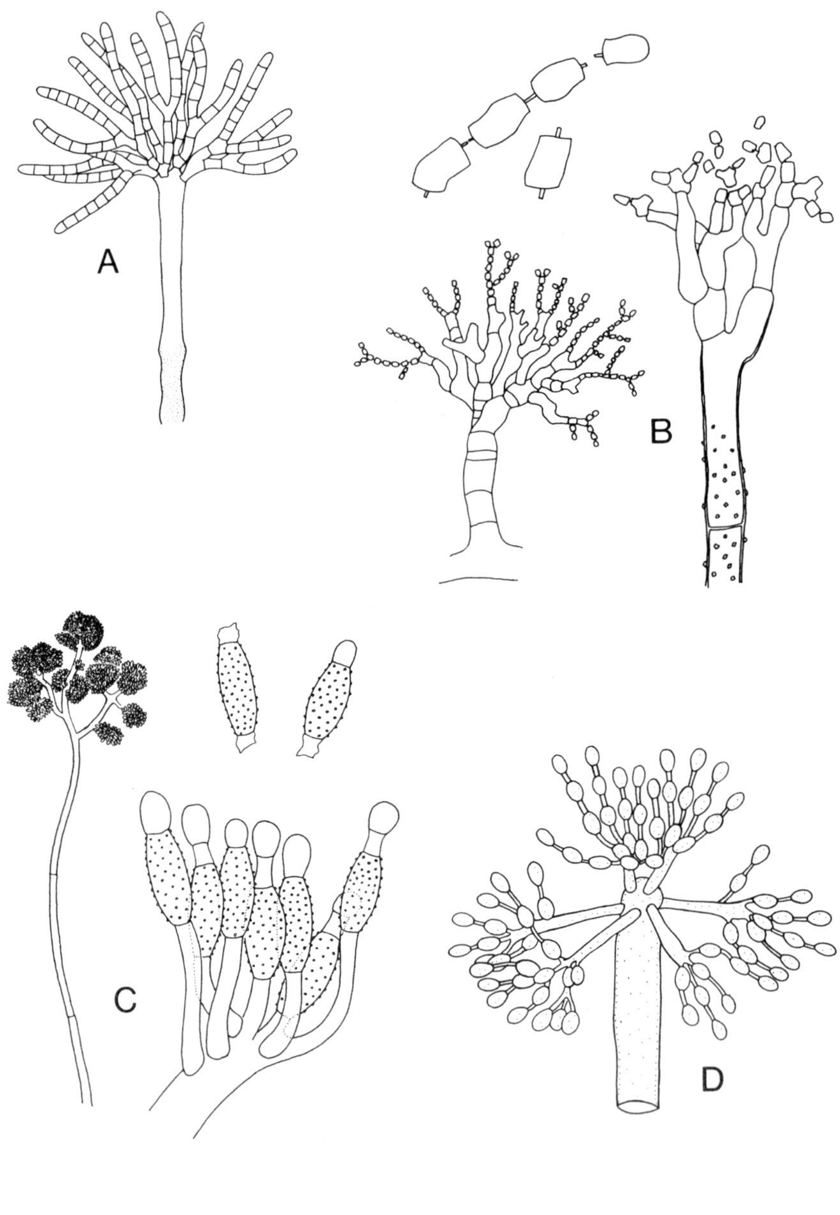

Plate 44: A *Staheliella nodosa;* B *Botryomonilia scheeleae;* C *Amblyosporium botrytis;* D *Amblyosporium spongiosum.*

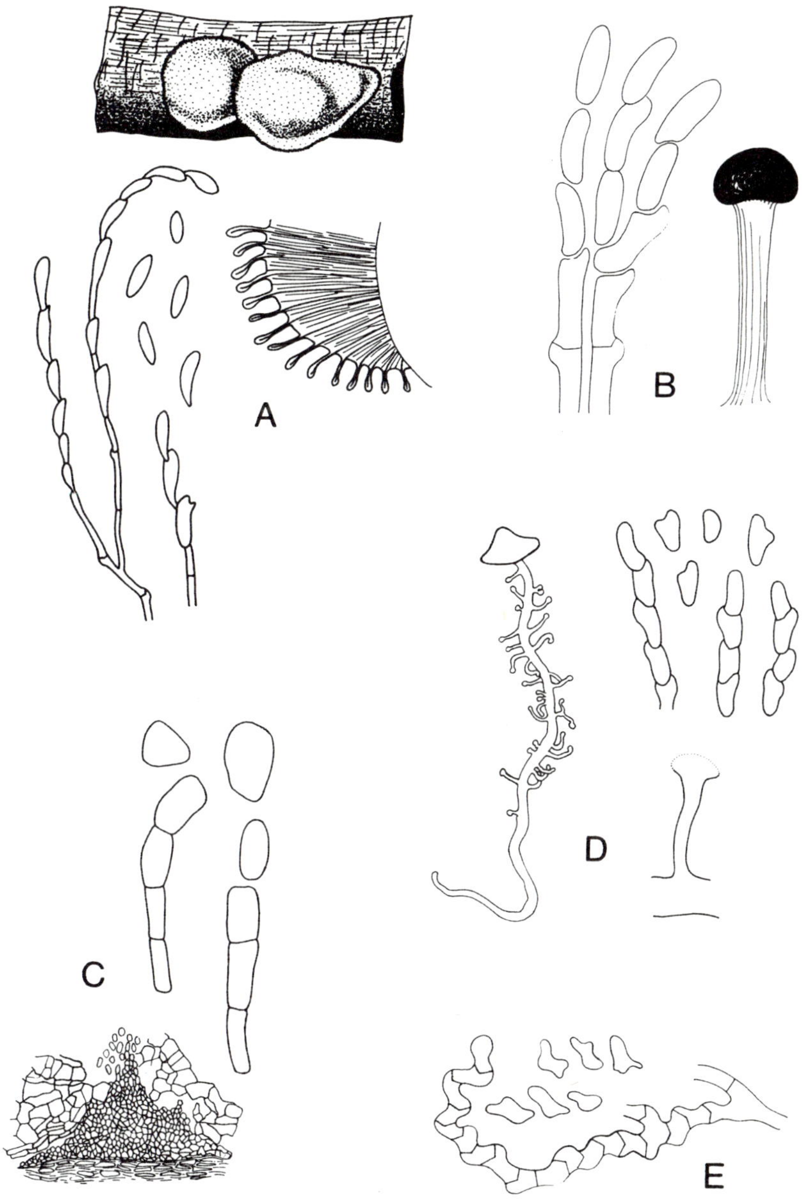

Plate 45: A *Echinodia theobromae;* B *Antromycopsis* state of *Pleuotus cystidiosus;* C *Necator* state of *Corticium salmonicolor;* D *Sclerostilbum* state of *Collybia racemosa;* E *Osteomorpha fragilis.*

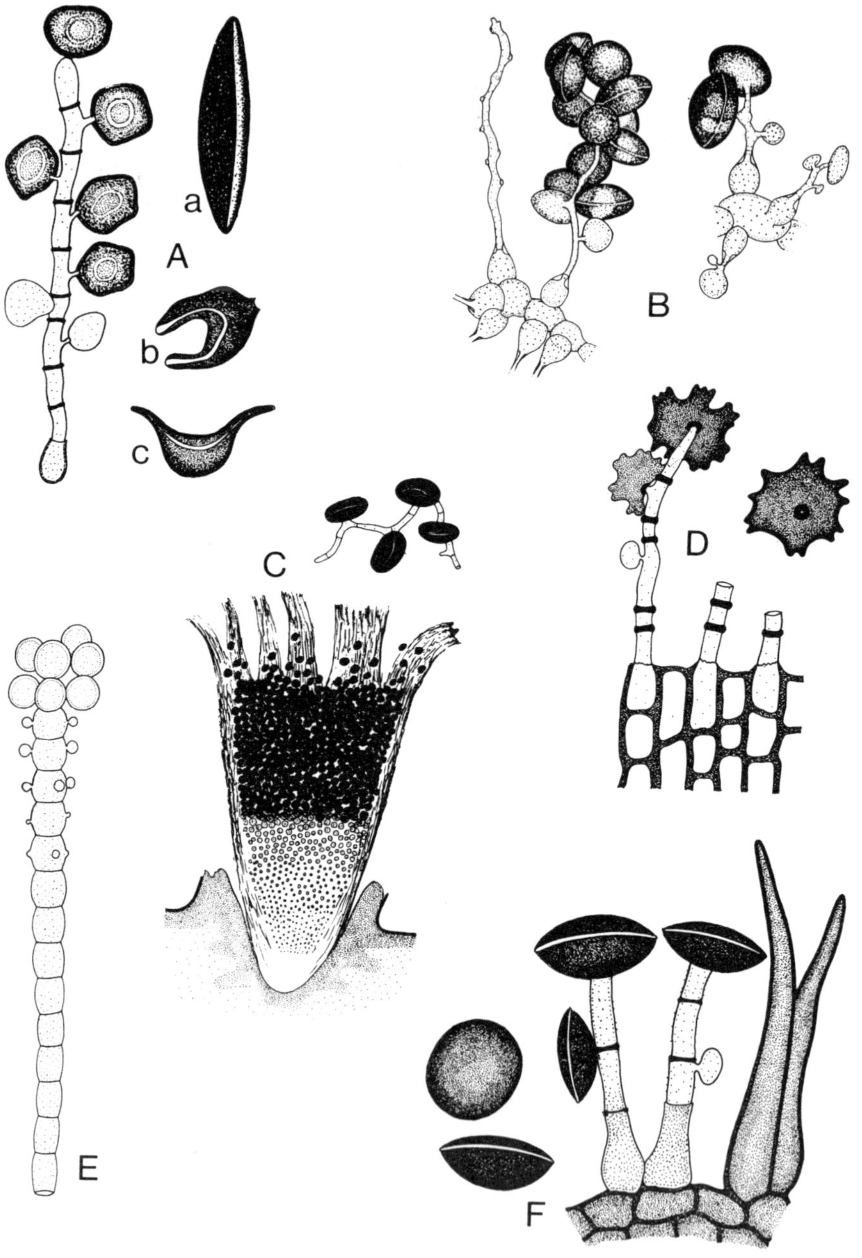

Plate 46: A *Arthrinium puccinioides* (a) *caricicola* (b) *luzulae* (c) *cuspidatum;* B *Arthrinium phaeospermum;* C *Endocalyx melanoxanthus;* D *Pteroconium pterospermum;* E *Graphiola thaxteri;* F *Cordella coniosporioides.*

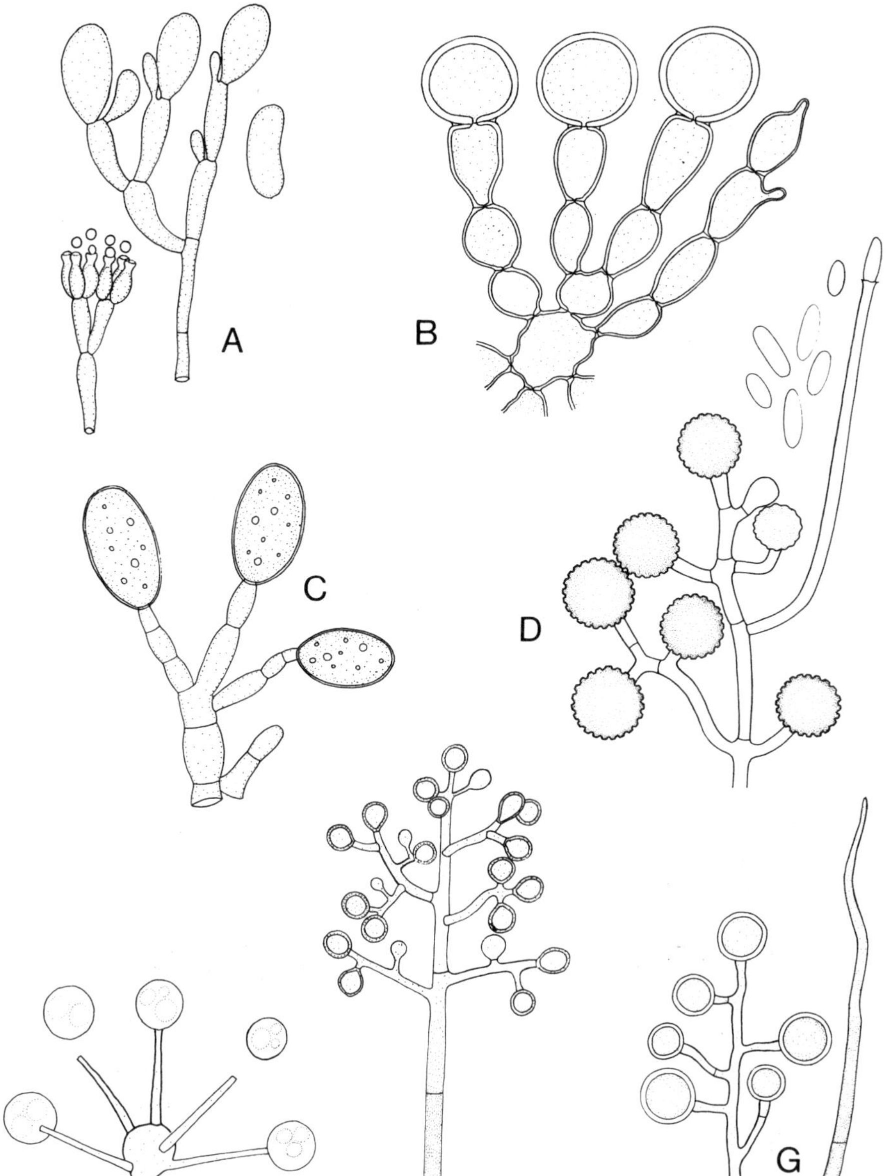

Plate 47: A *Dimorphospora foliicola;* B *Ambrosiella xylebori;* C *Bactridiopsis ulei;* D *Sepedonium chrysospermum;* E *Umbelopsis versiformis;* F *Staphylotrichum coccosporum;* G *Botryotrichum piluliferum.*

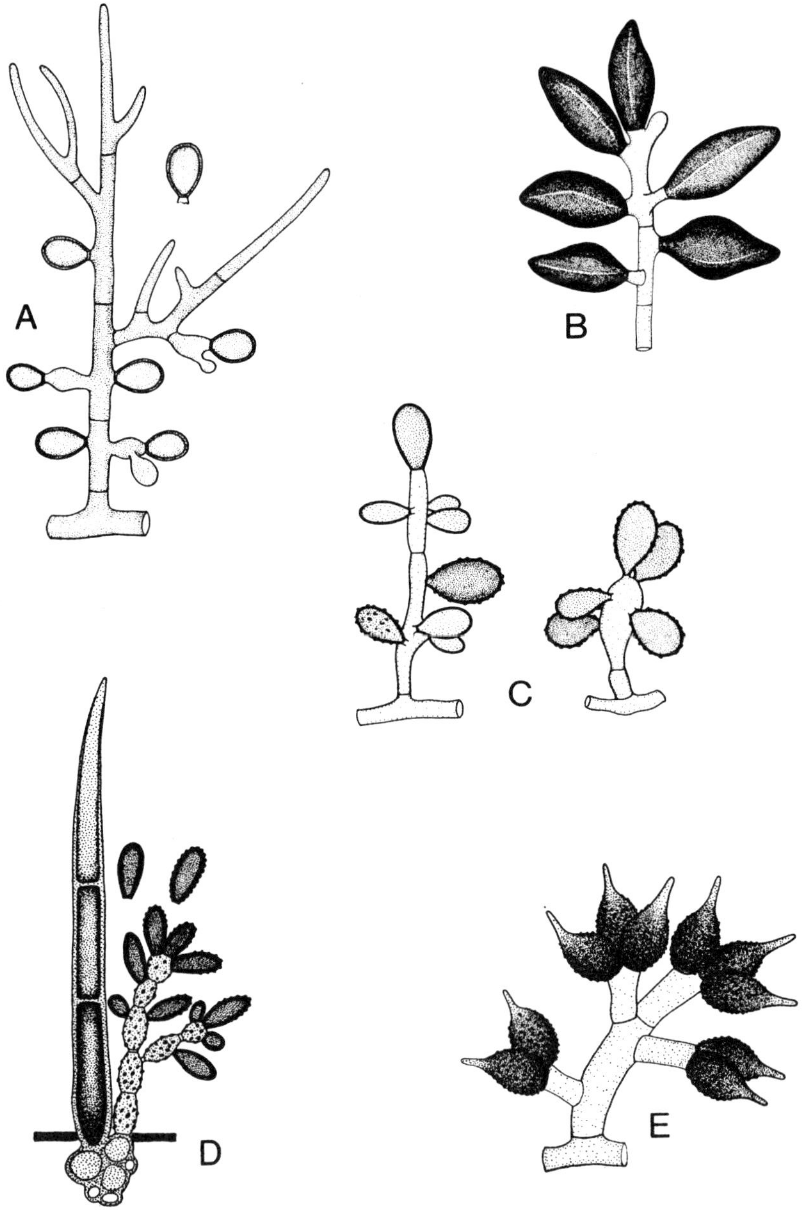

Plate 48: A *Botryoderma lateritium;* B *Mammaria echinobotryoides;* C *Myceliophthora thermophila;* D *Herposira velutina;* E *Echinobotryum atrum.*

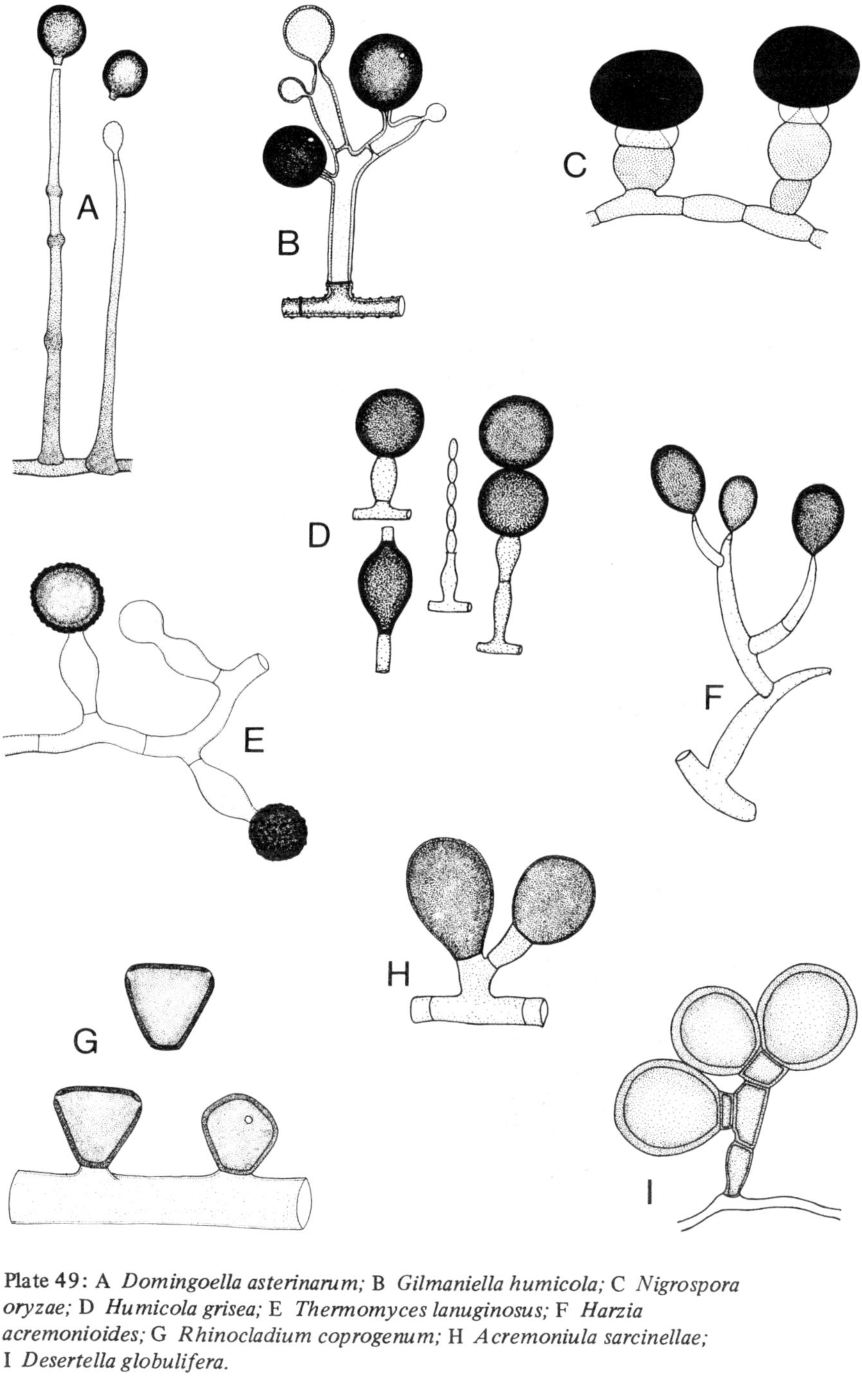

Plate 49: A *Domingoella asterinarum;* B *Gilmaniella humicola;* C *Nigrospora oryzae;* D *Humicola grisea;* E *Thermomyces lanuginosus;* F *Harzia acremonioides;* G *Rhinocladium coprogenum;* H *Acremoniula sarcinellae;* I *Desertella globulifera.*

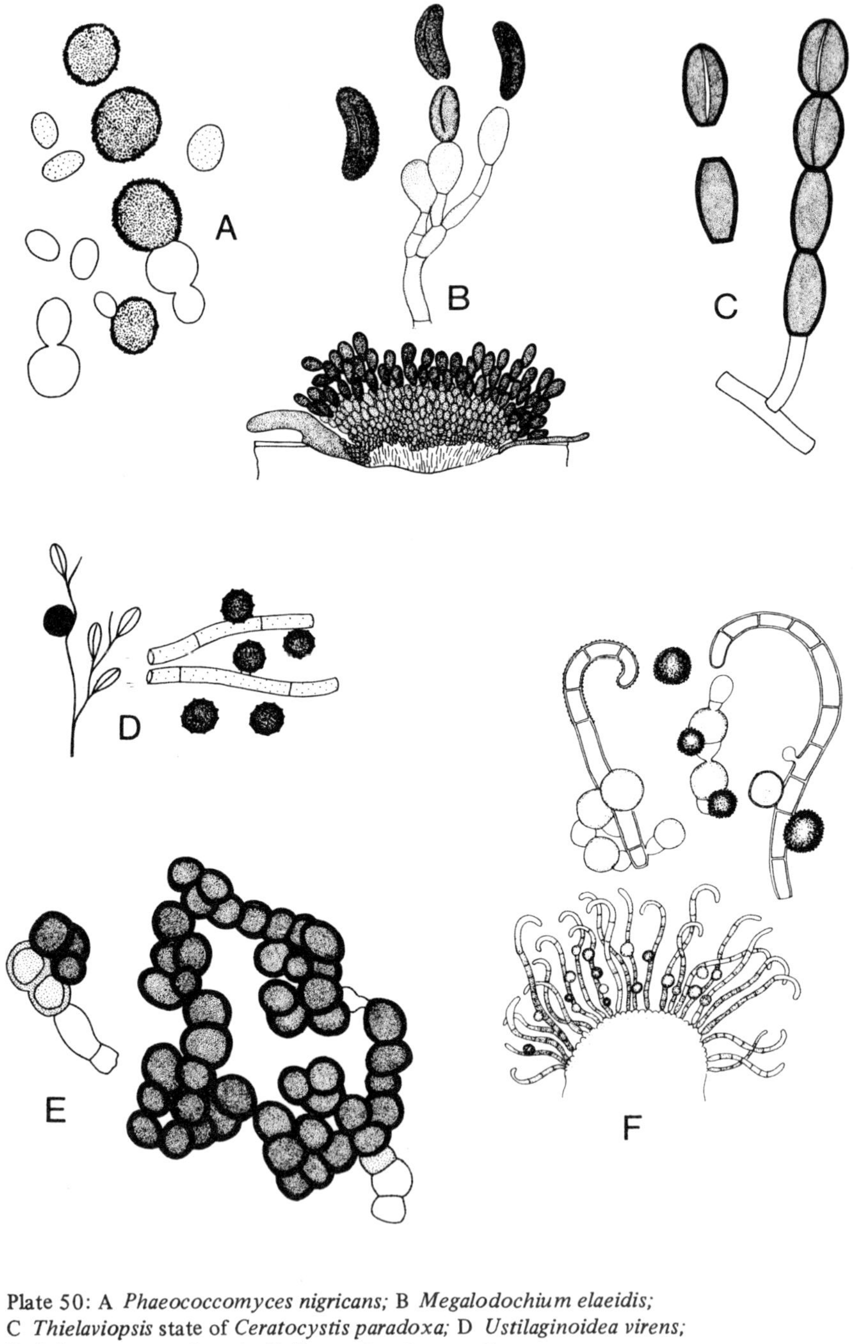

Plate 50: A *Phaeococcomyces nigricans;* B *Megalodochium elaeidis;* C *Thielaviopsis* state of *Ceratocystis paradoxa;* D *Ustilaginoidea virens;* E *Sclerococcum sphaerale;* F *Parapericonia angusii.*

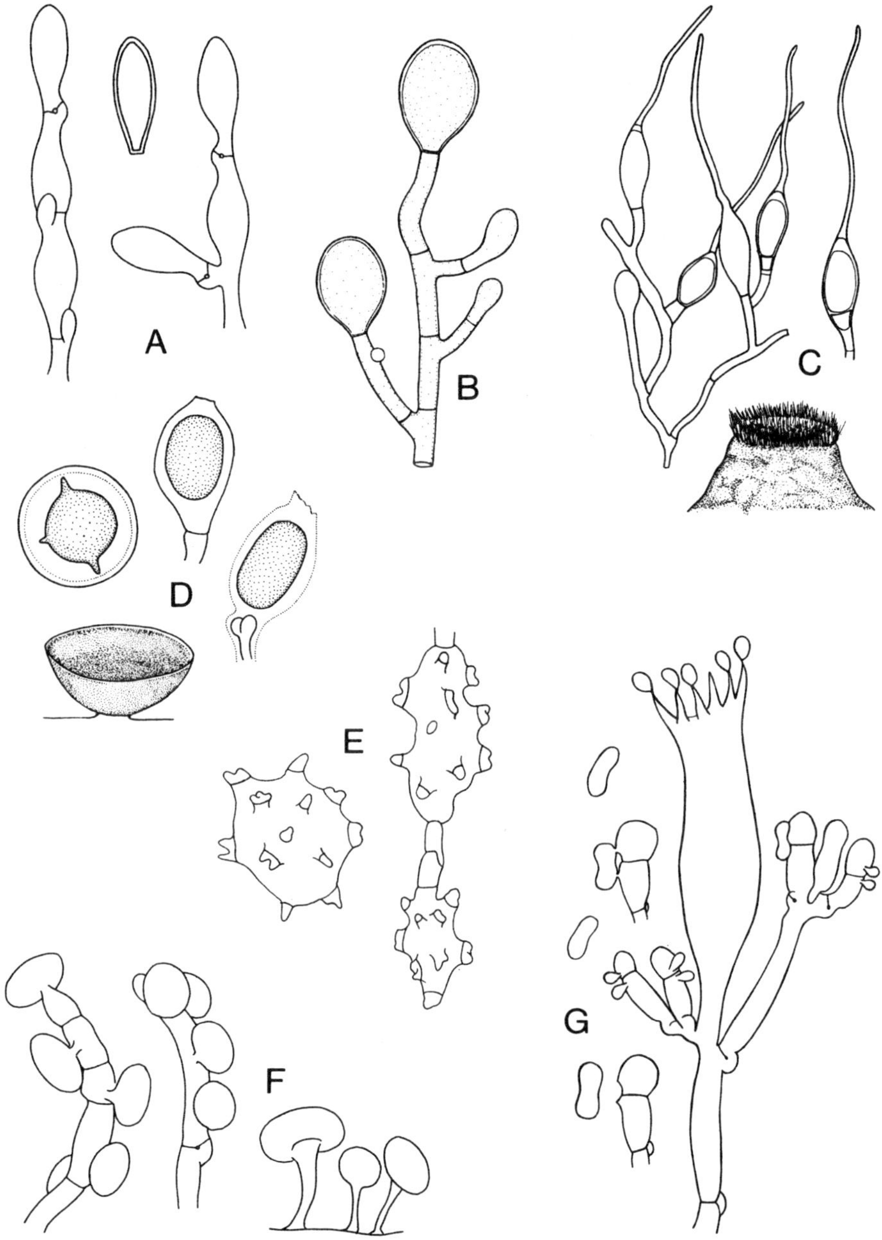

Plate 51: A *Ptychogaster rubescens;* B *Allescheriella* state of *Botryobasidium croceum;* C *Michenera artocreas;* D *Matula rompelii;* E *Asterophora lycoperdoides;* F *Martindalia spironema;* G *Syzygospora* state of *Christiansenia alba.*

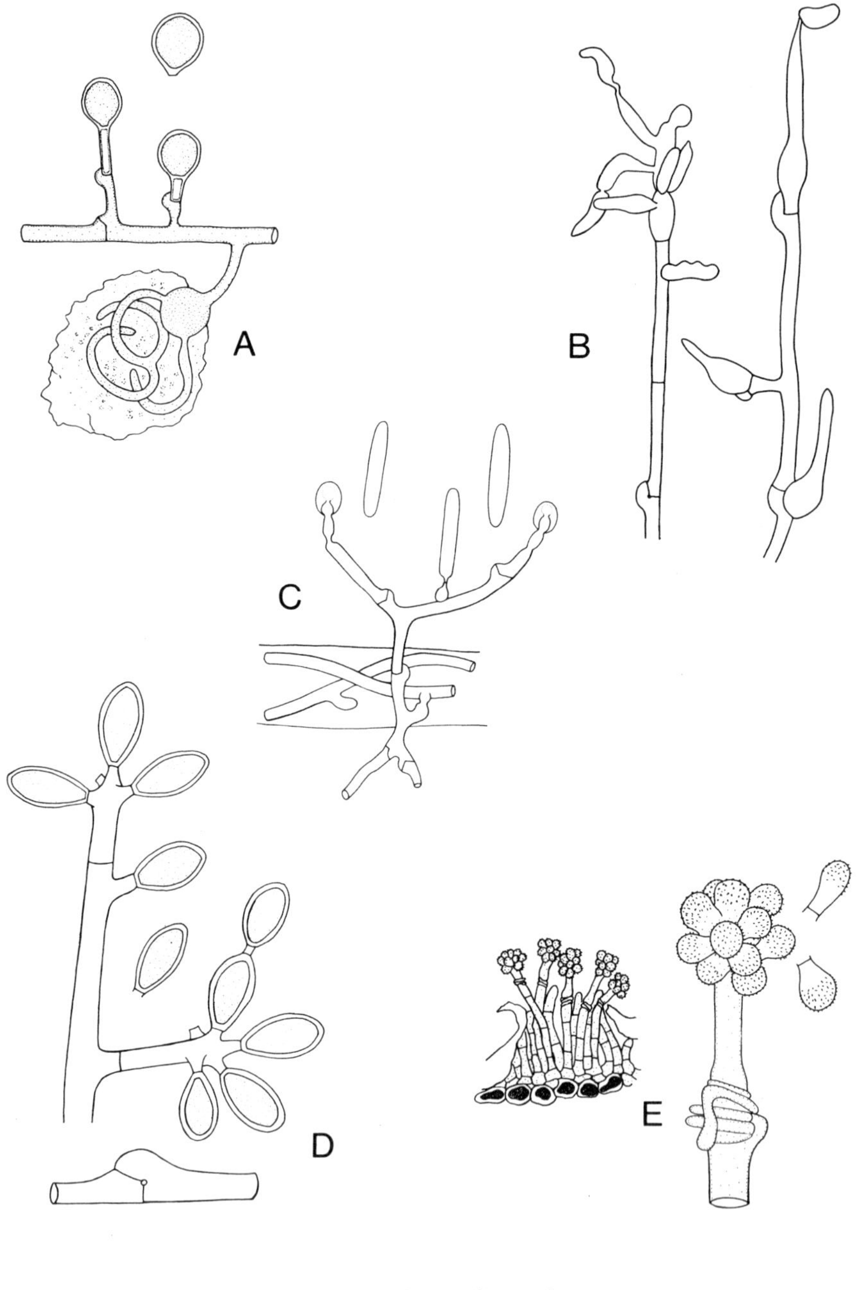

Plate 52: A *Pagidospora amoebophila;* B *Itersonilia perplexans;* C *Nematoctonus haptocladus;* D *Sporotrichum aureum;* E *Glomospora empetri.*

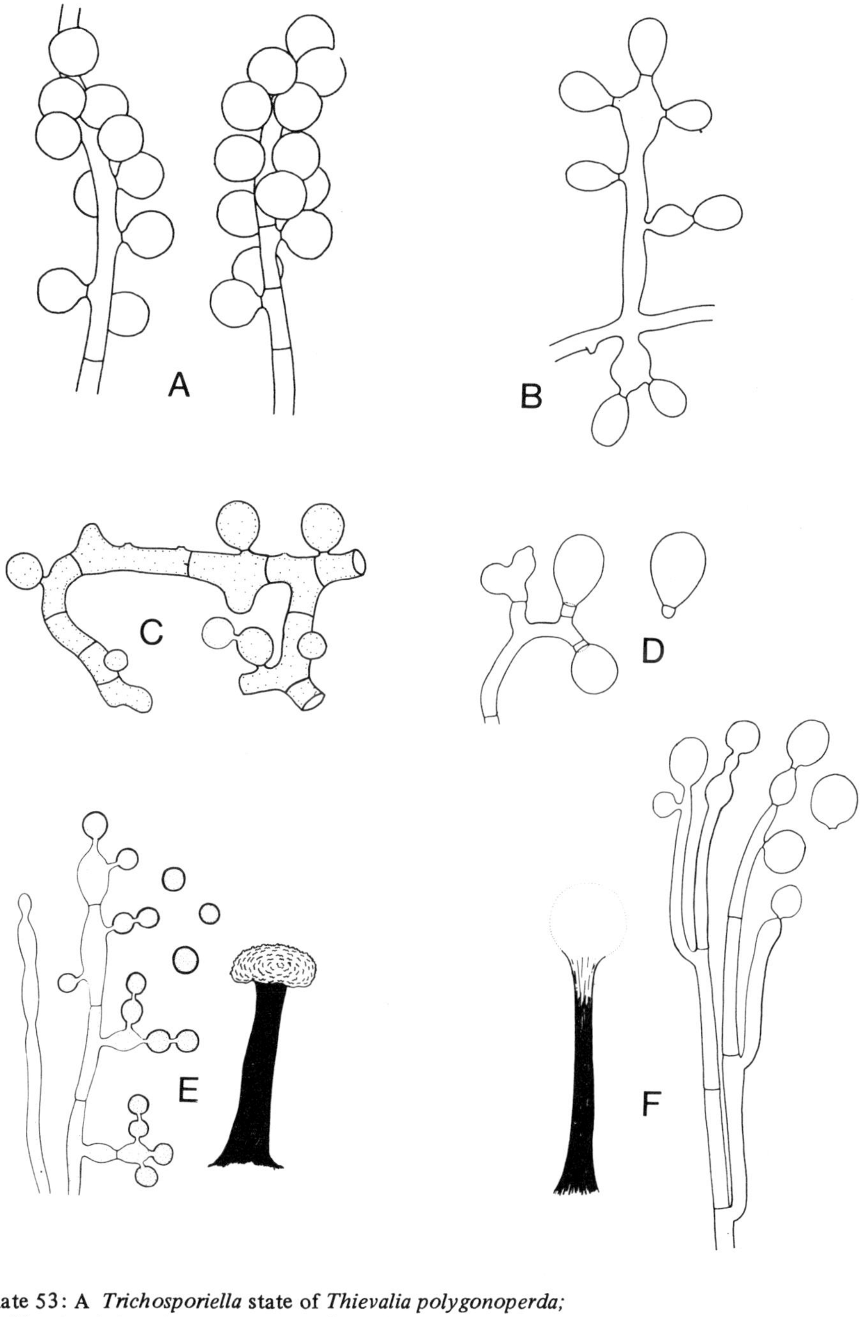

Plate 53: A *Trichosporiella* state of *Thievalia polygonoperda;*
B *Myceliophthora lutea;* C *Beniowskia sphaeroidea;* D *Ovulitis azaleae;*
E *Mycosylva clarkii;* F *Pachnodium* state of *Ceratocystis cana.*

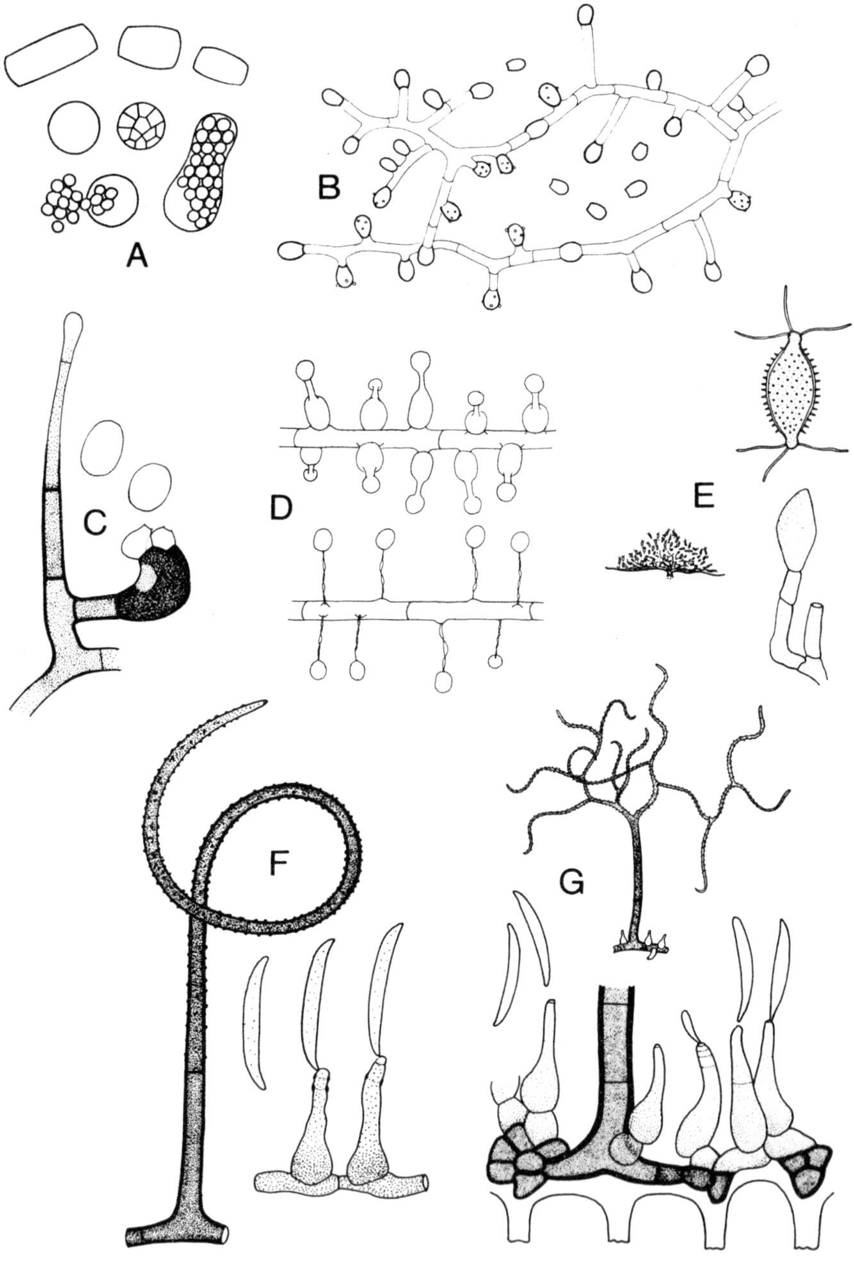

Plate 54: A *Coccidioides immitis;* B *Chrysosporium merdarium;* C *Zygosporium oscheoides;* D *Aphanocladium aranearum;* E *Amphichaetella echinata;* F *Circinotrichum maculiforme;* G *Gyrothrix podosperma.*

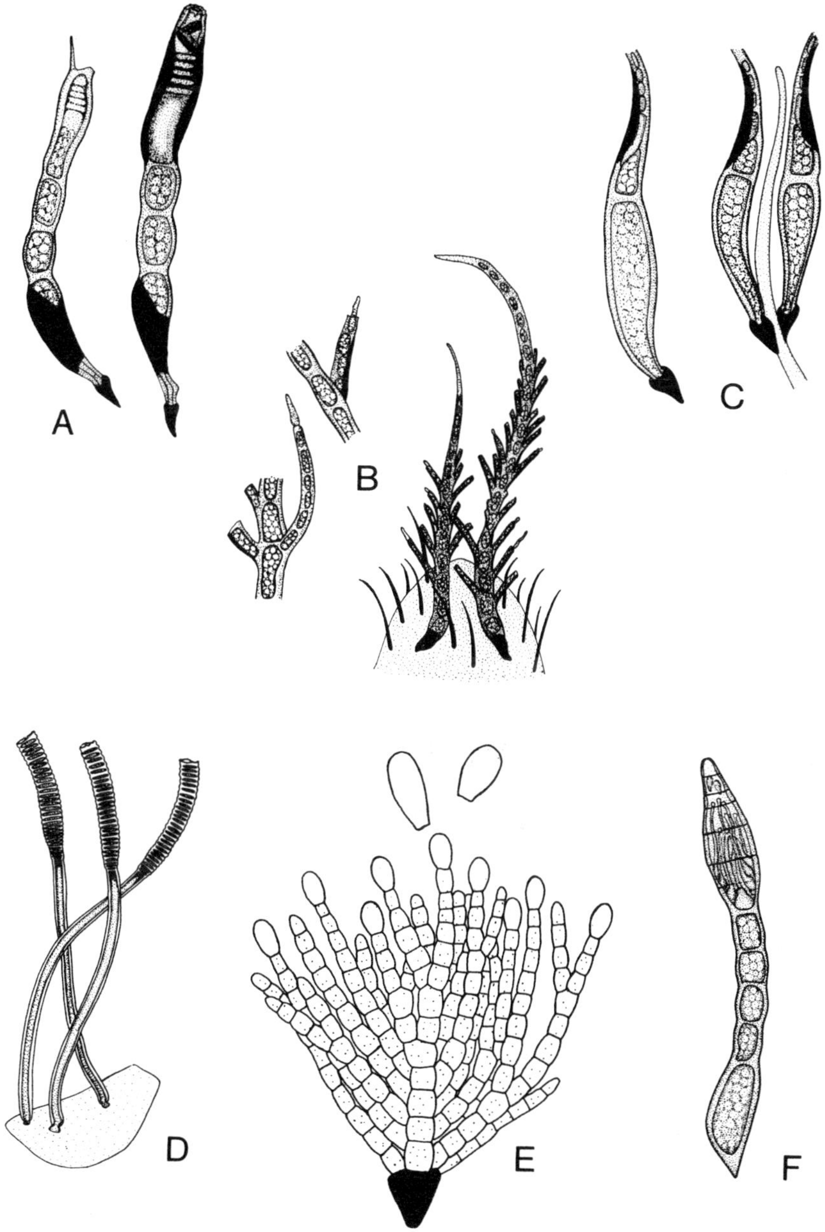

Plate 55: A *Endosporella diopsidis;* B *Aposporella elegans;* C *Thaxteriola nigromarginata;* D *Enterobryus compressus;* E *Chantransiopsis xantholini;* F *Coreomycetopsis oedipus.*

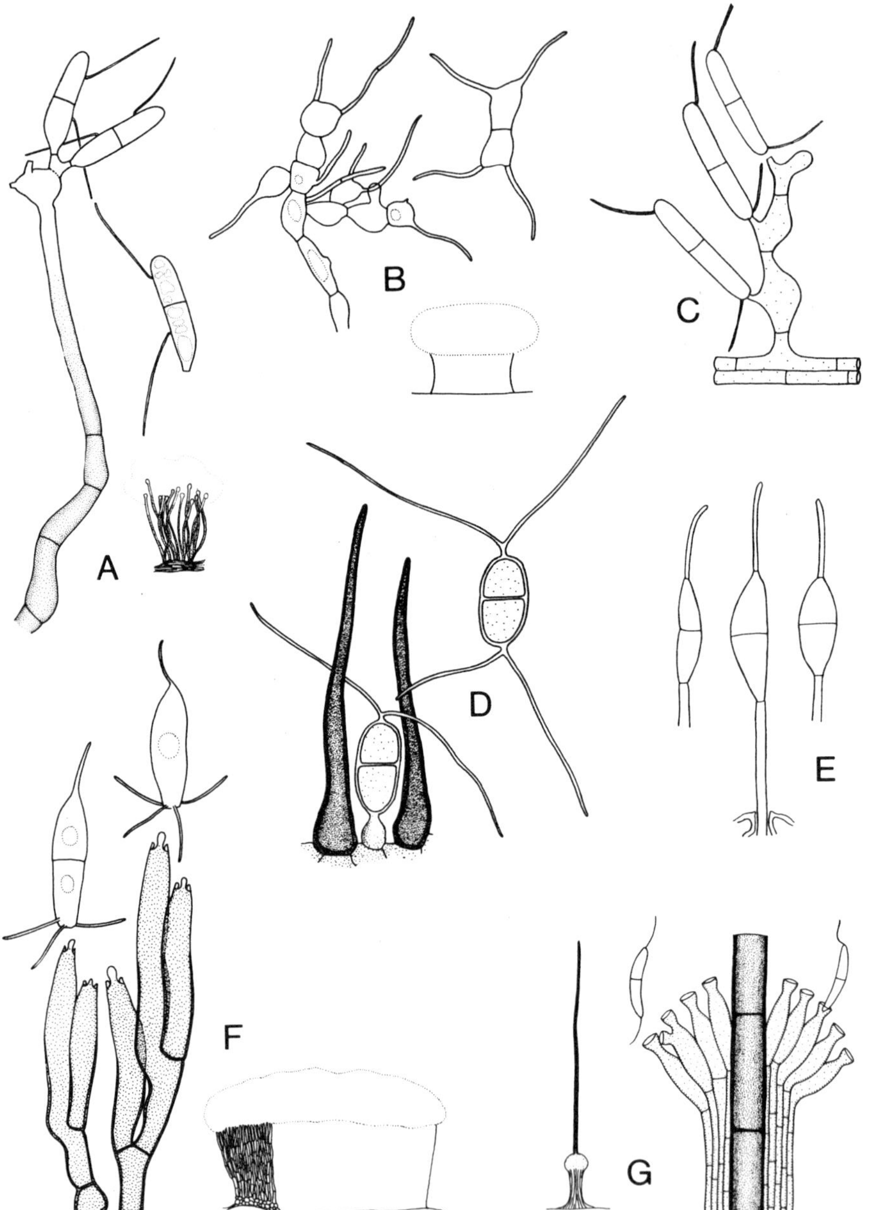

Plate 56: A *Hyphodiscosia jaipurensis;* B *Gloeosporiella rosicola;* C *Leptodiscella africana;* D *Dwayaloma trina;* E *Monotrichum commelinae;* F *Hyphopolynema tropicale;* G *Menisporopsis novae-zelandiae.*

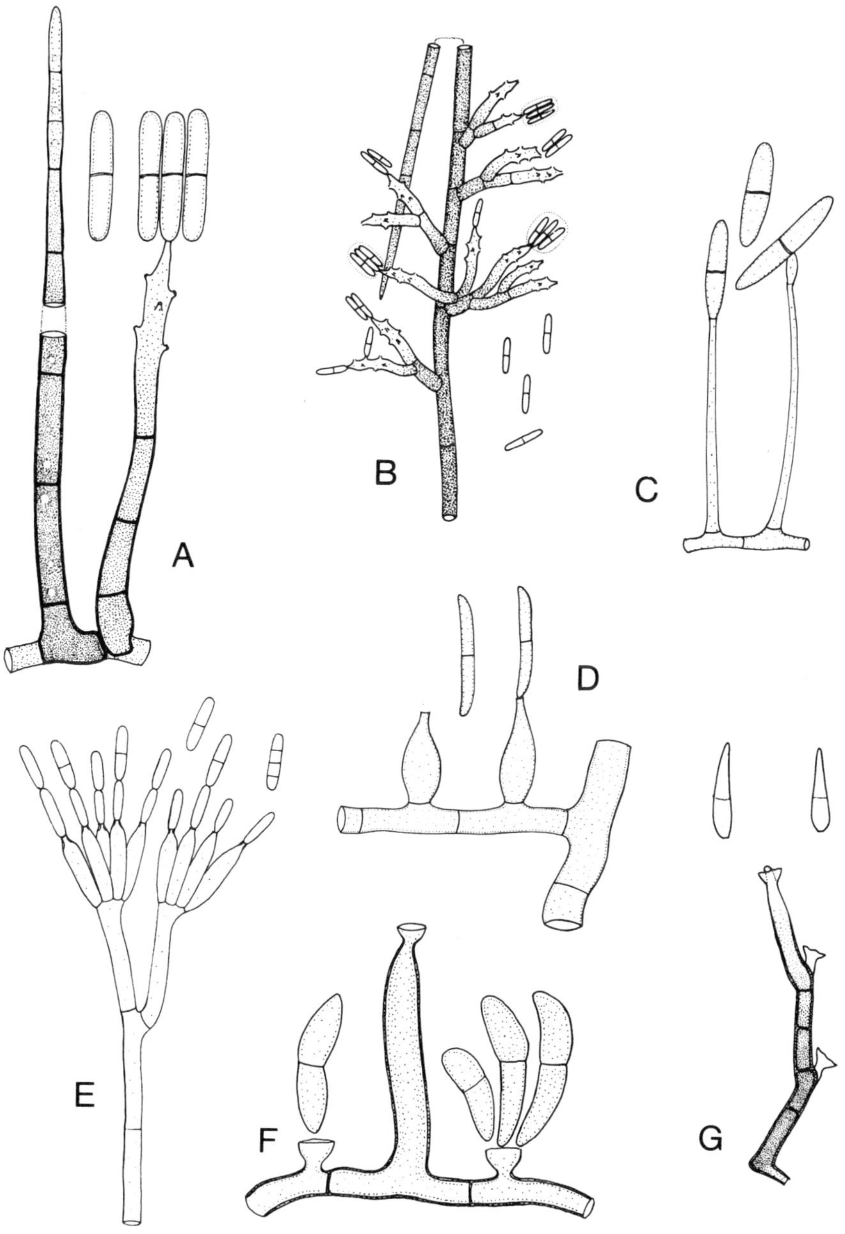

Plate 57: A *Cylindrotrichum oligospermum;* B *Chaetopsis grisea;* C *Cephalosporiopsis* sp.; D *Cylindrodendrum album;* E *Penicillifer pulcher;* F *Cyphellophora laciniata;* G *Phialogeniculata guadalcanalensis.*

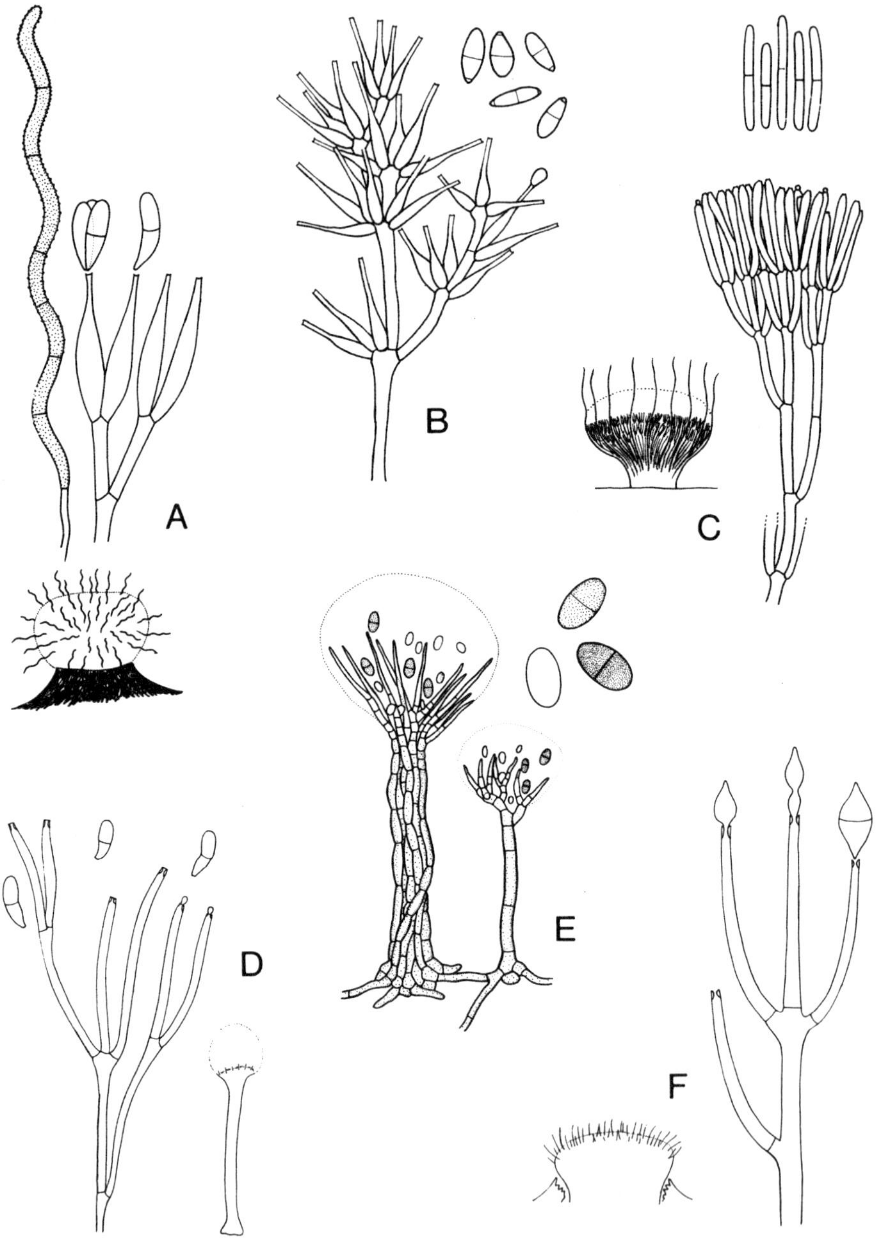

Plate 58: A *Kutilakesopsis macalpinei;* B *Sibirina fungicola;*
C *Septomyrothecium uniseptatum;* D *Didymostilbe obovoidea;*
E *Leptoxyphium* sp.; F *Fusisporella* state of *Nectria pallidula.*

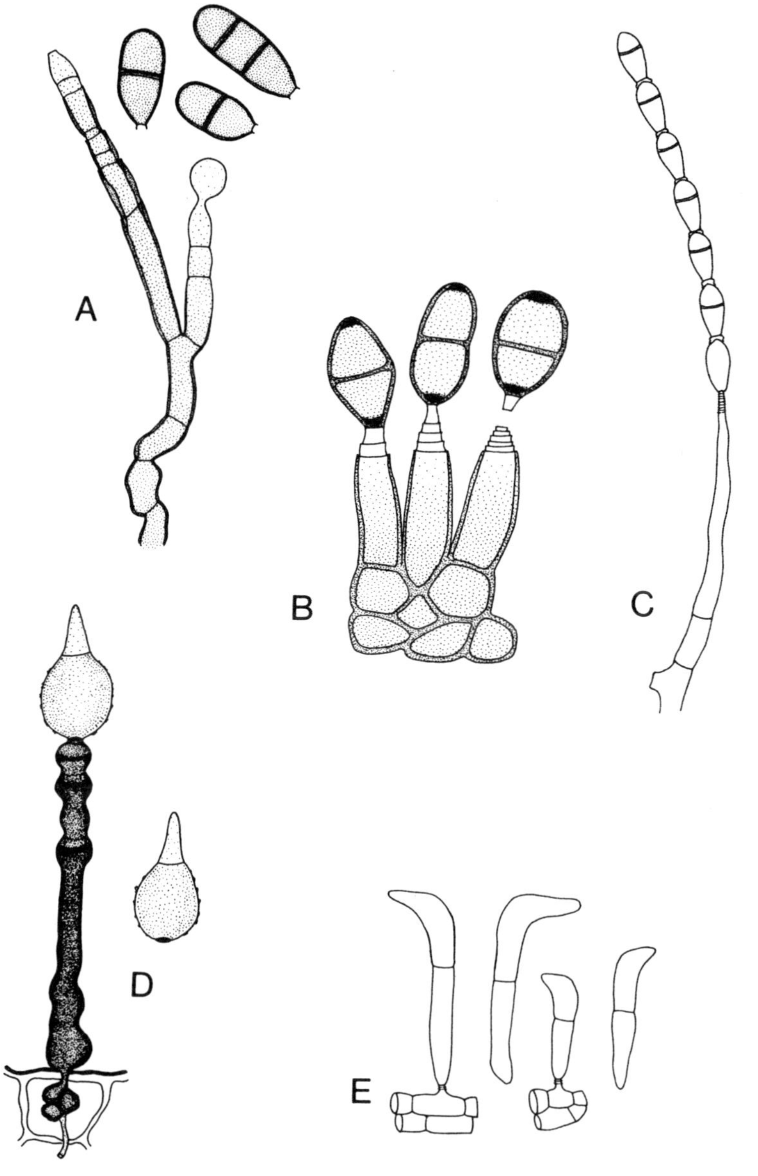

Plate 59: A *Endophragmiella pallescens;* B *Oedothea vismiae;* C *Gamsia dimera;* D *Deightoniella africana;* E *Rhynchosporium secalis.*

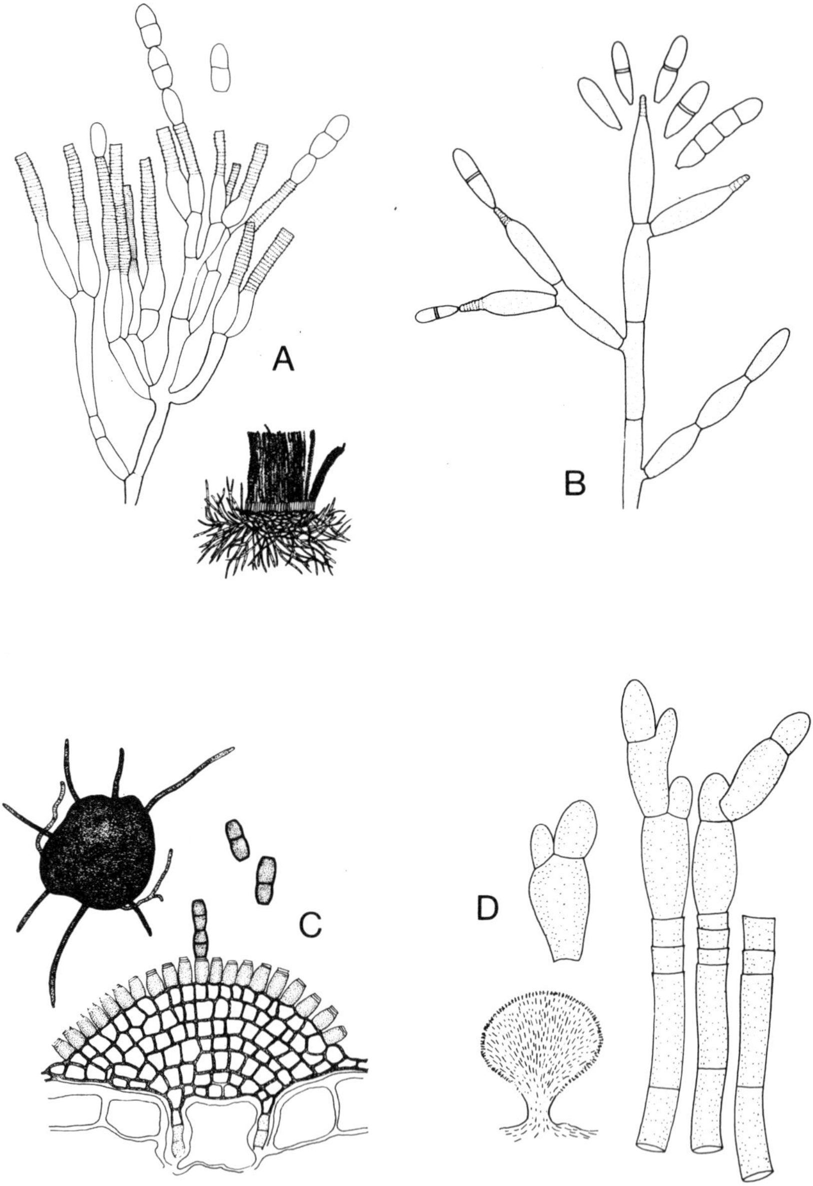

Plate 60: A *Gymnodochium* sp.; B *Exophiala salmonis;* C *Trichodochium disseminatum;* D *Annellodochium ramulisporum.*

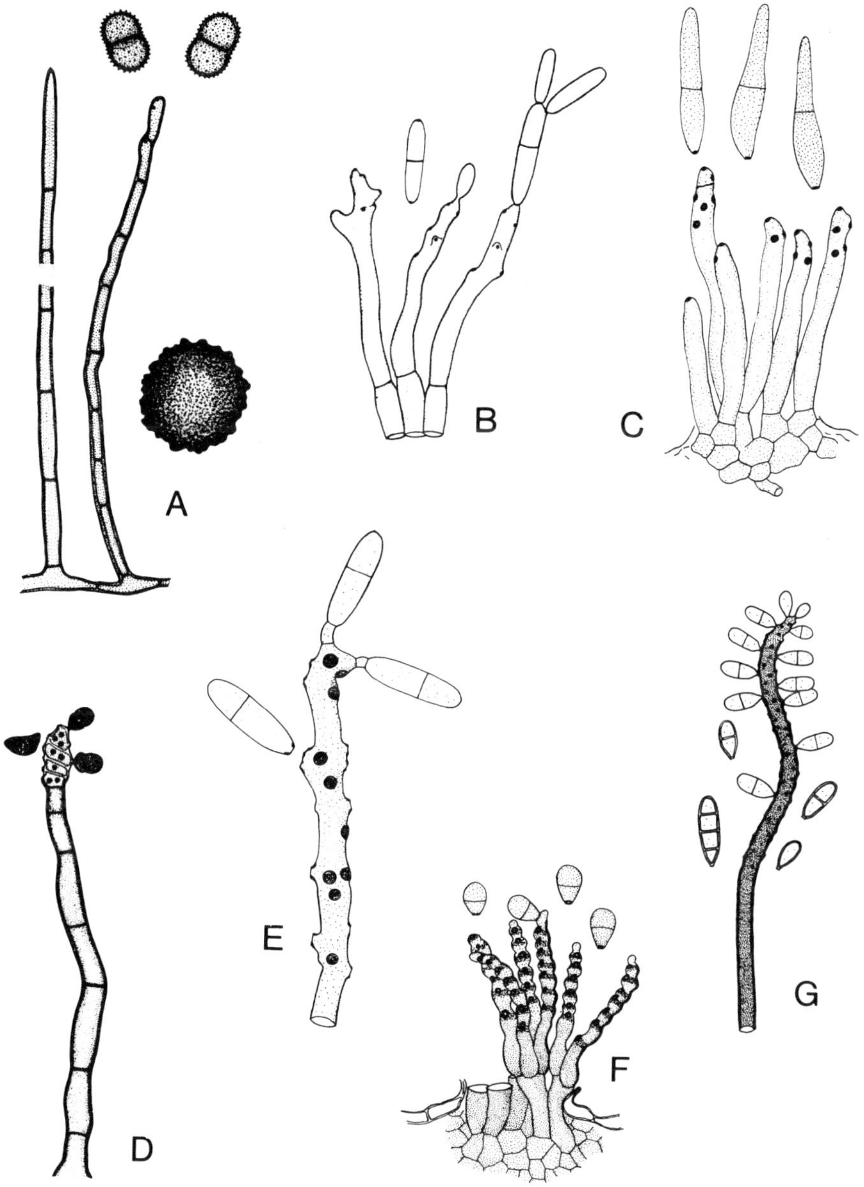

Plate 61: A *Acroconidiellina loudetiae* (with sclerotium); B *Ramularia gei;* C *Passalora depressa;* D *Martinellisia indica;* E *Sceptrifera pulchra;* F *Polythrincium trifolii;* G *Veronaea botryosa.*

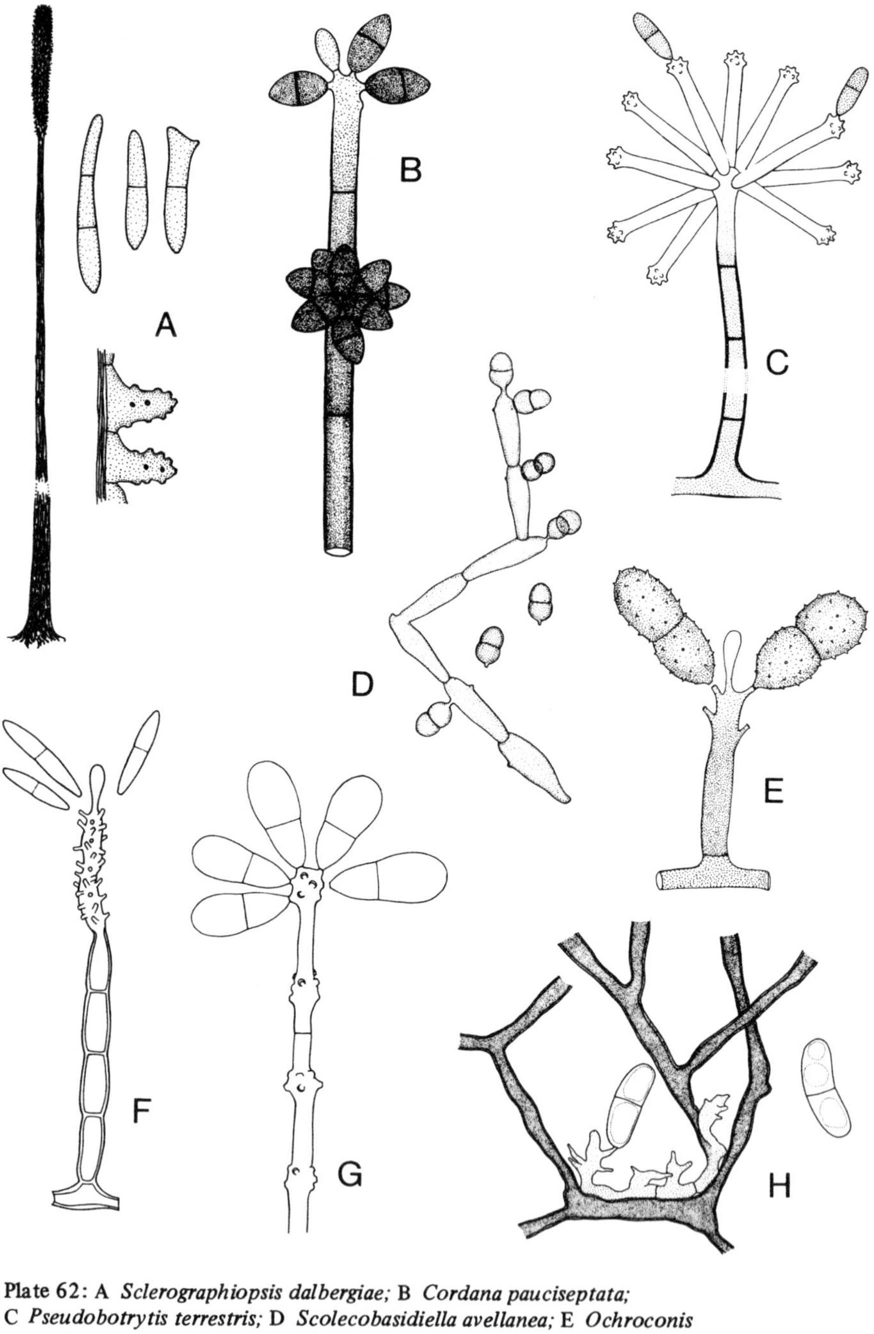

Plate 62: A *Sclerographiopsis dalbergiae;* B *Cordana pauciseptata;* C *Pseudobotrytis terrestris;* D *Scolecobasidiella avellanea;* E *Ochroconis constricta;* F *Dactylaria candidula;* G *Arthrobotrys oligospora;* H *Neta patuxentica.*

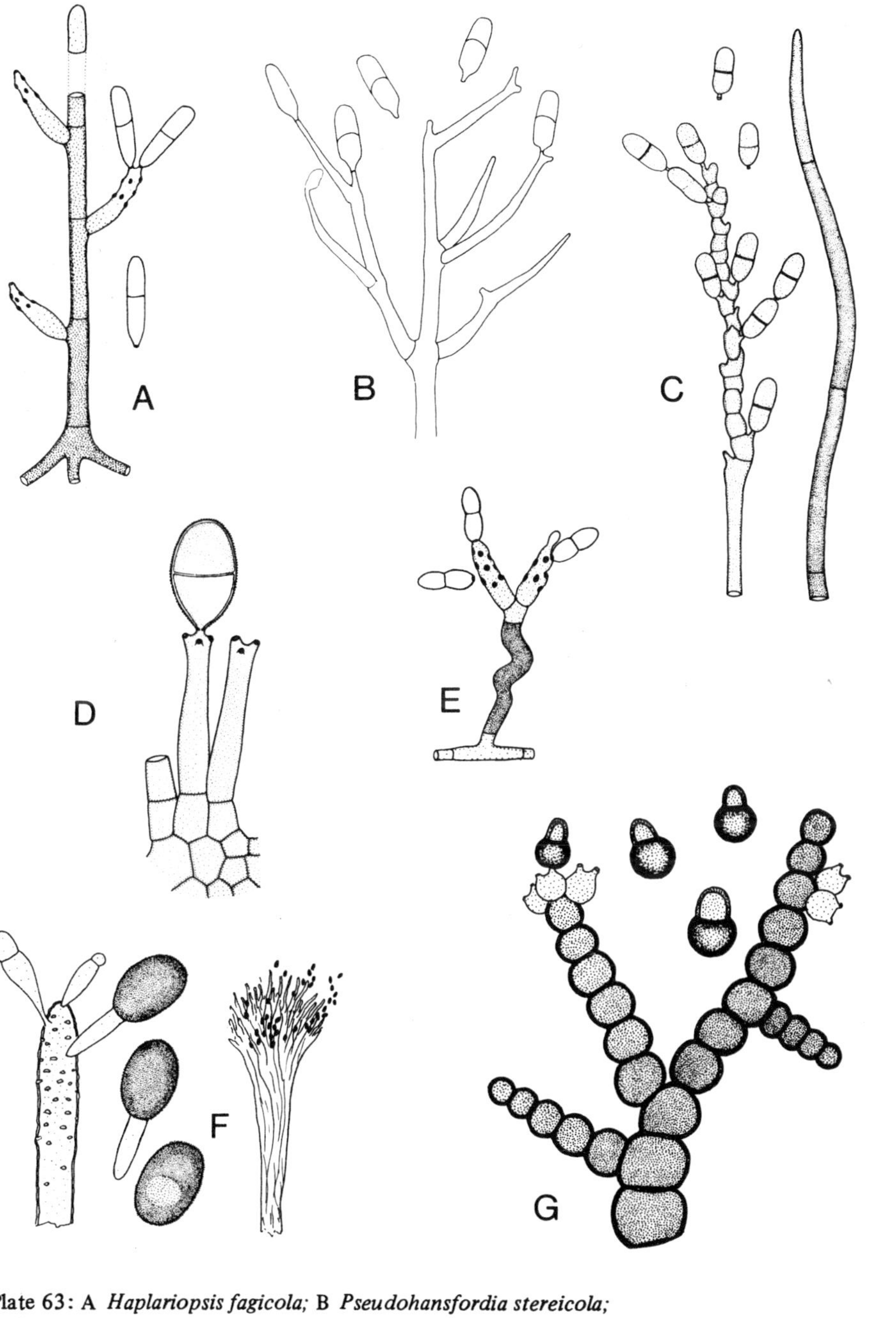

Plate 63: A *Haplariopsis fagicola;* B *Pseudohansfordia stereicola;* C *Paspalomyces aureus;* D *Asperisporium caricae;* E *Zygophiala jamaicensis;* F *Panchanania jaipurensis;* G *Capnobotrys neesii.*

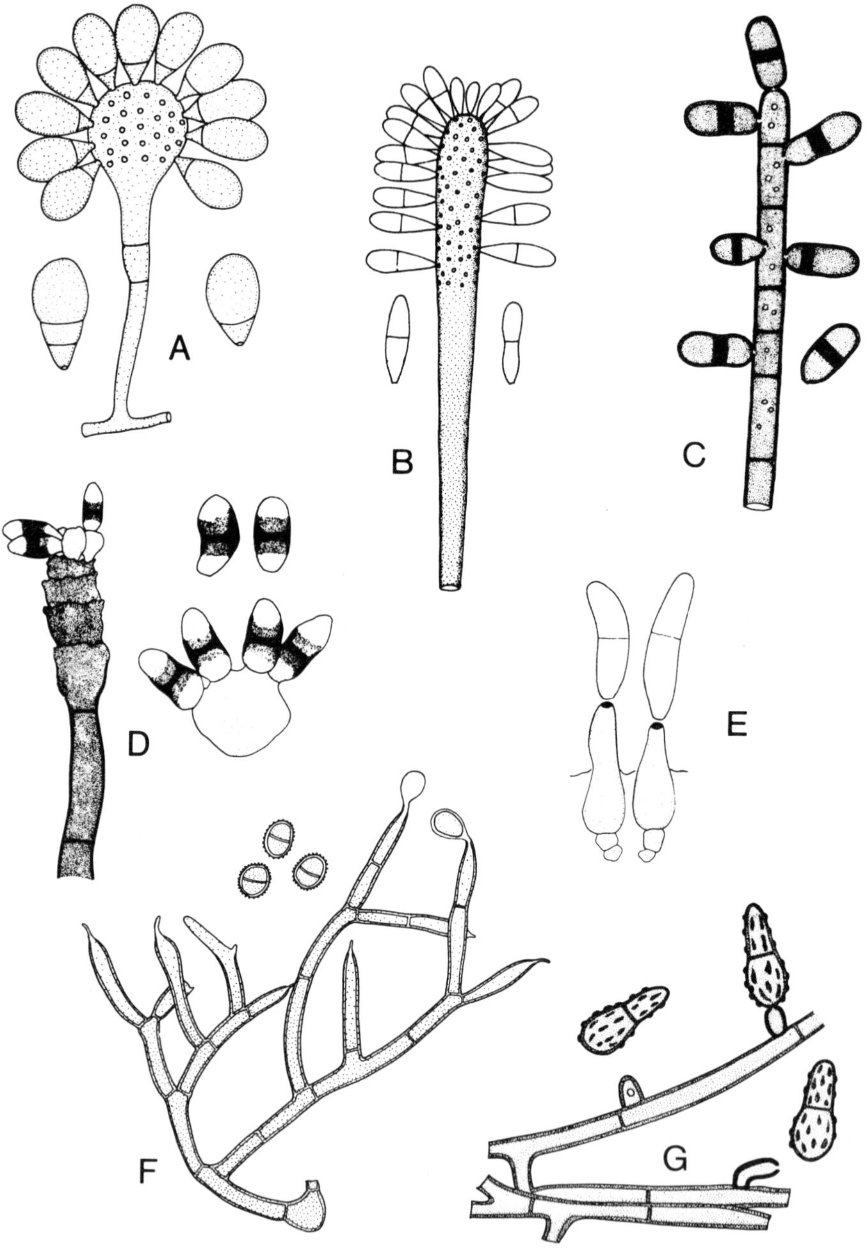

Plate 64: A *Cephaliophora nigricans;* B *Blastophorella smithii;* C *Spadicoides bina;* D *Zopheromyces fasciatus;* E *Fusicladiella pimpinellae;* F *Johnstonia colocasiae;* G *Pirozynskia farriae.*

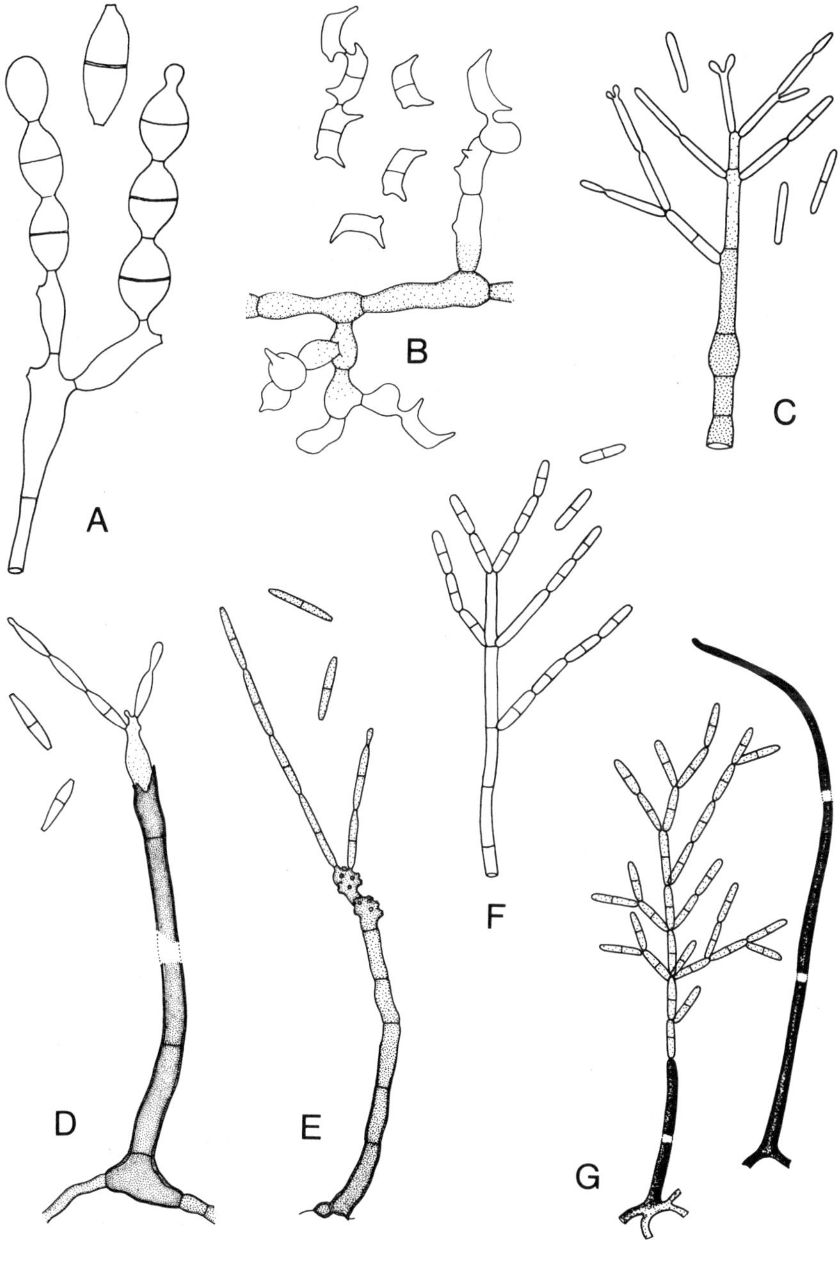

Plate 65: A *Diploospora rosae;* B *Arxiella terrestris;* C *Polyscytalum foecundissimum;* D *Pleurotheciopsis pusilla;* E *Anungitea fragilis;* F *Hormiactis alba;* G *Hormiactella fusca.*

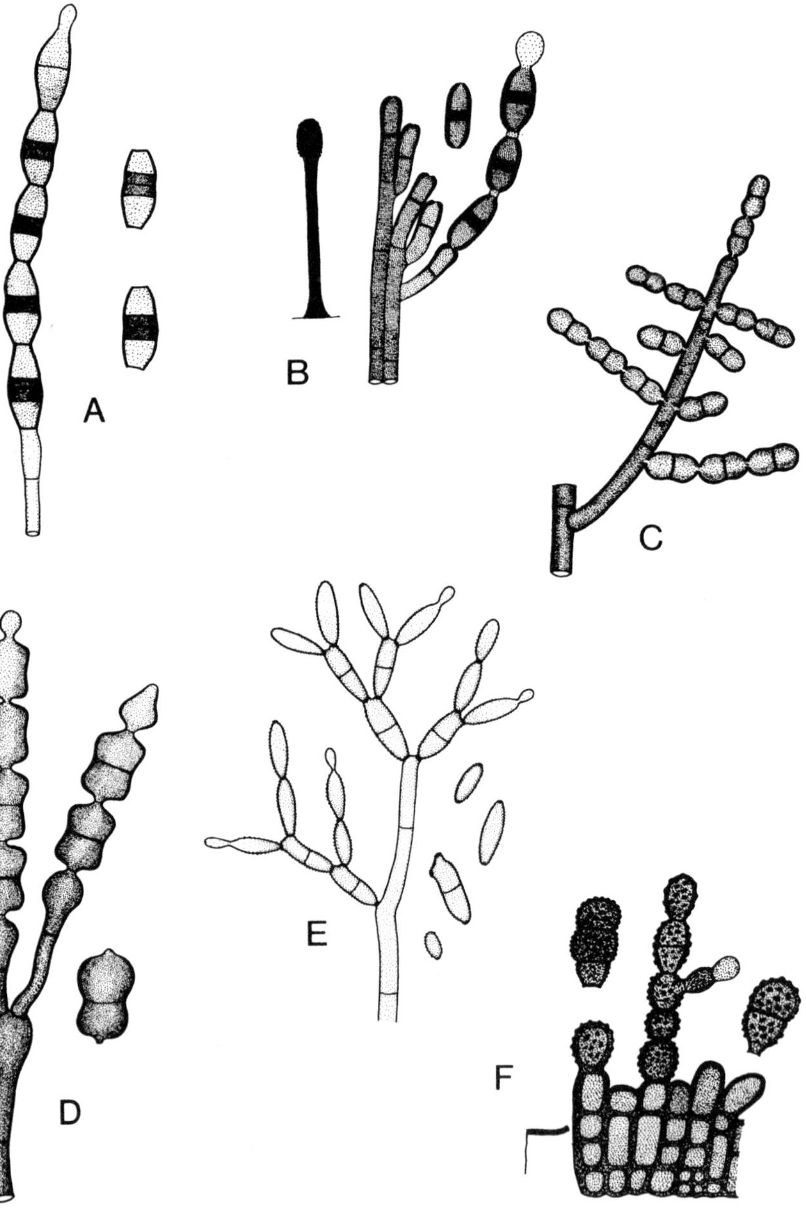

Plate 66: A *Bispora antennata;* B *Didymobotryum rigidum;* C *Diplococcium spicatum;* D *Oedemium didymum;* E *Cladosporium* sp.; F *Spilodochium vernoniae.*

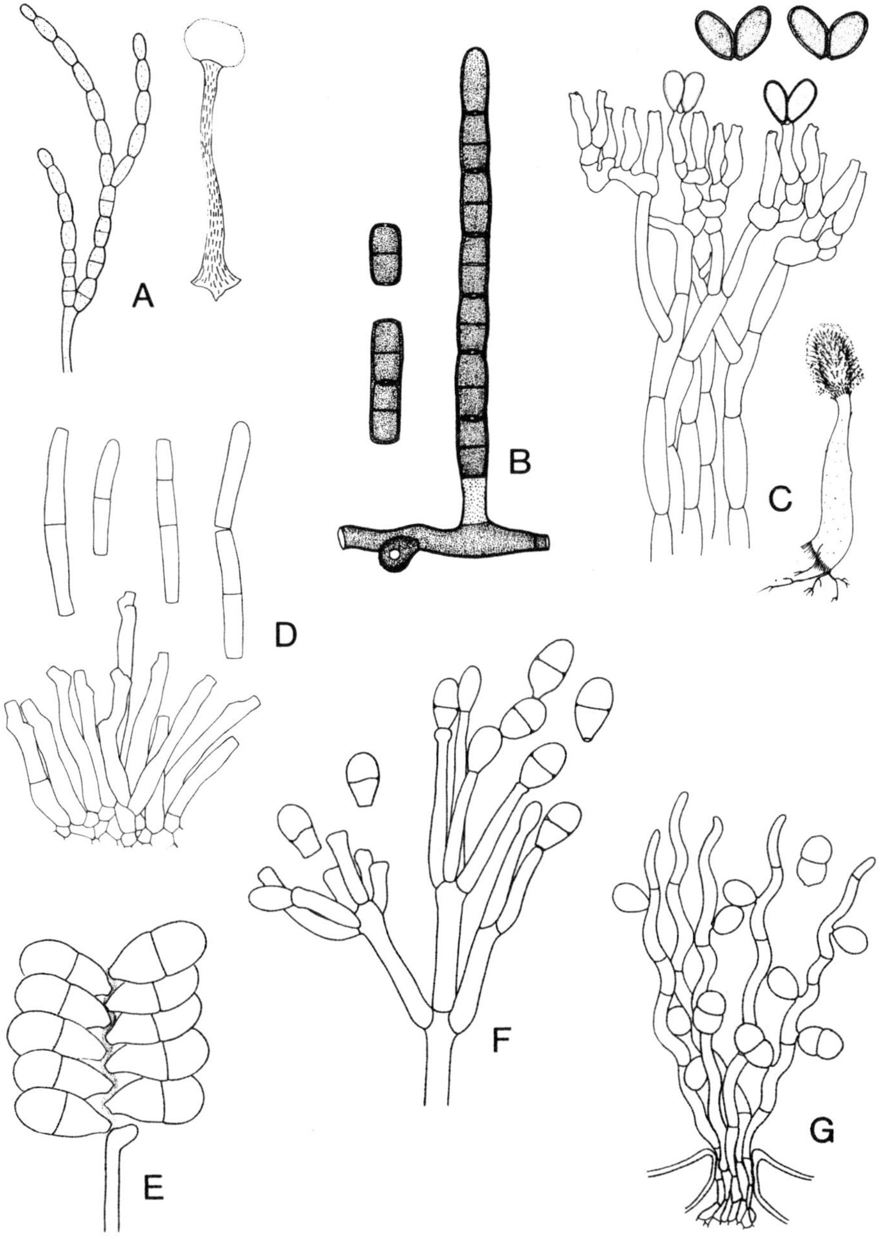

Plate 67: A *Antromyces copridis;* B *Ampulliferina persimplex;* C *Papilionospora aspergilloides;* D *Thedgonia ligustrina;* E *Trichothecium roseum;* F *Cladobotryum* state of *Hypomyces aurantius;* G *Bostrichonema alpestre.*

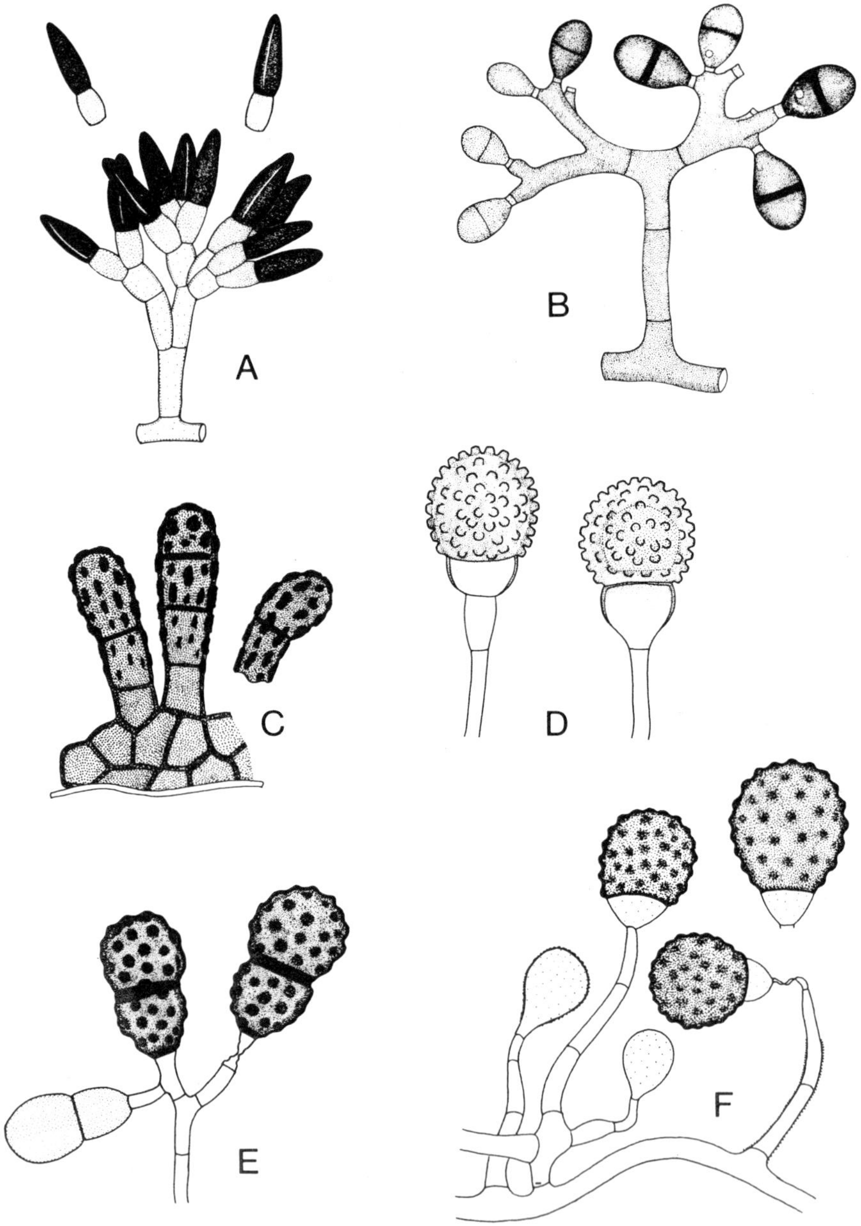

Plate 68: A *Wardomyces humicola;* B *Balanium stygium;* C *Hadronema orbiculare;* D *Mycogone perniciosa;* E *Trichocladium asperum;* F *Chlamydomyces palmarum.*

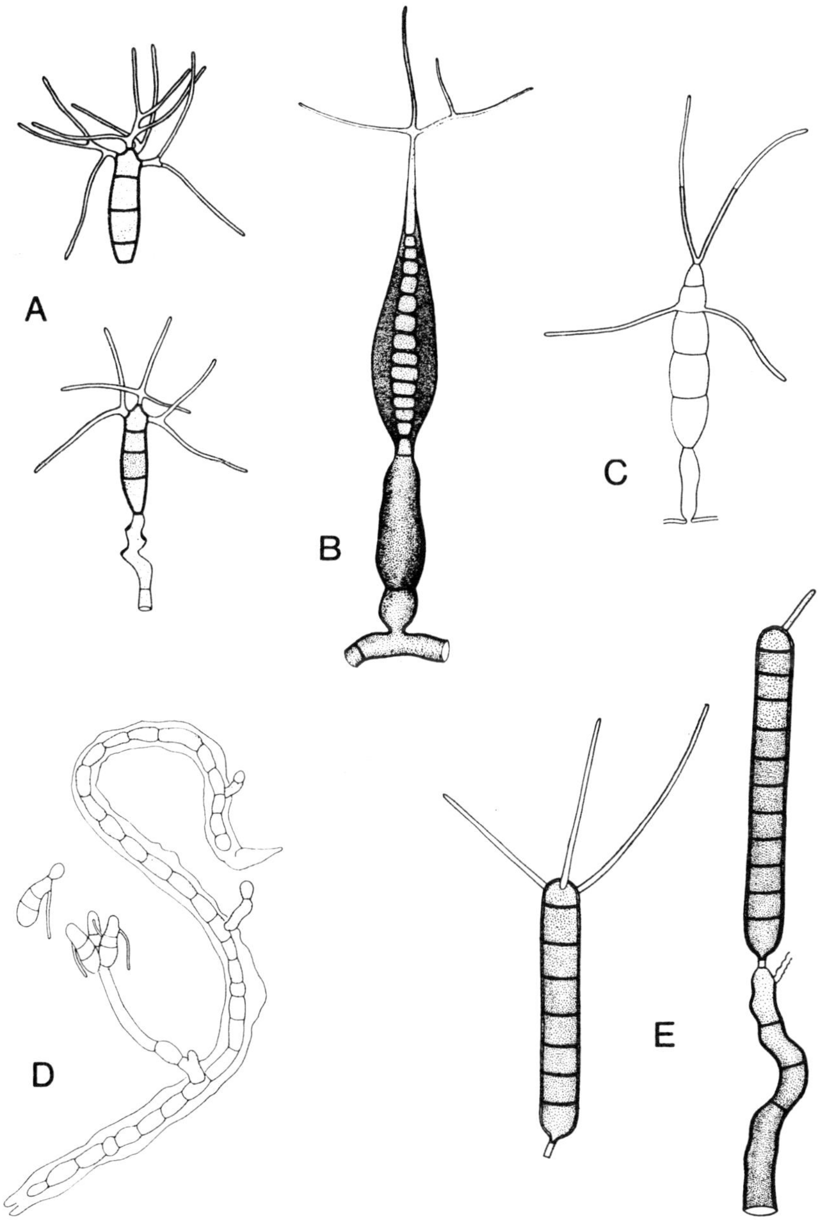

Plate 69: A *Pleiochaeta albizziae;* B *Lomaantha pooga;* C *Mastigosporium album;* D *Haptocara latirostrum;* E *Camposporium antennatum.*

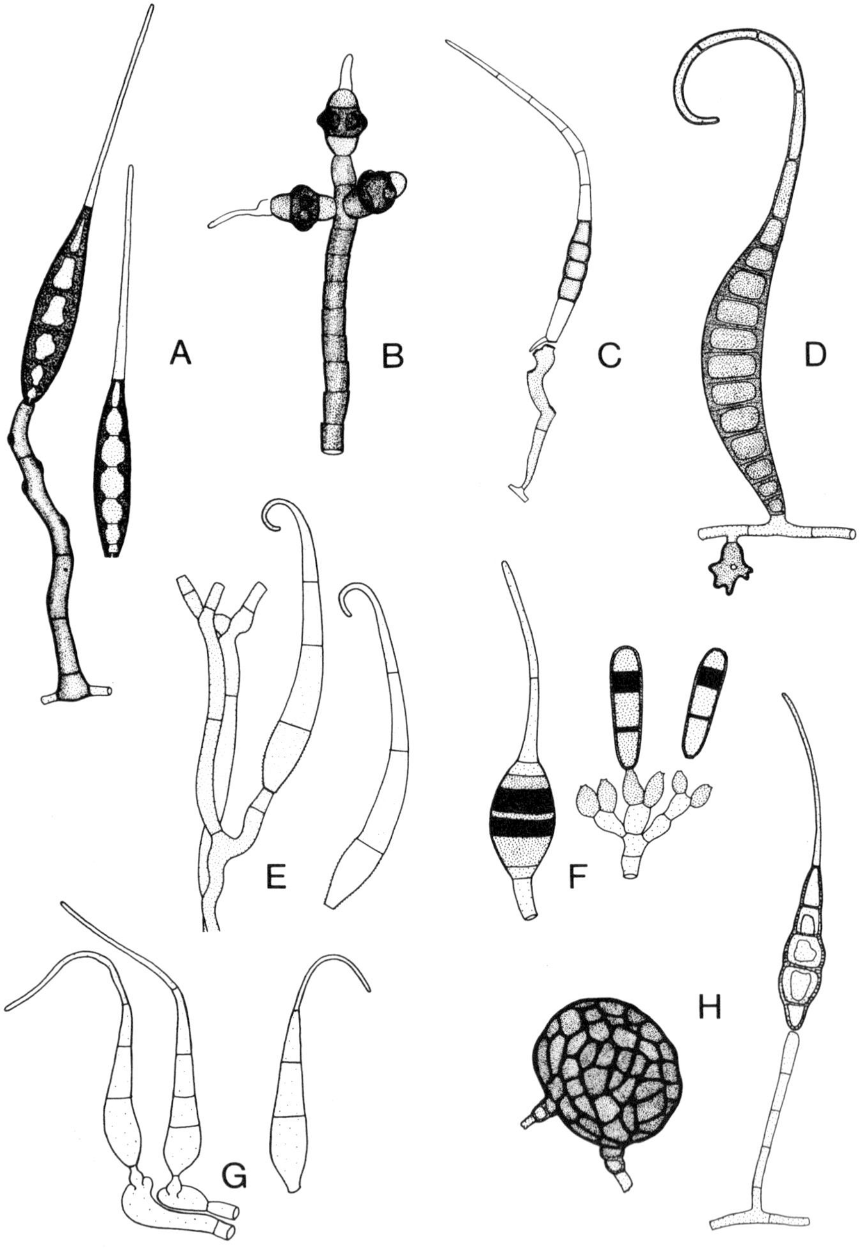

Plate 70: A *Phaeotrichoconis crotalariae;* B *Acrophragmis coronata;* C *Mycocentrospora acerina;* D *Ceratophorum helicosporum;* E *Uncispora harroldii;* F *Triadelphia heterospora;* G *Spermospora subulata;* H *Trichoconiella padwickii.*

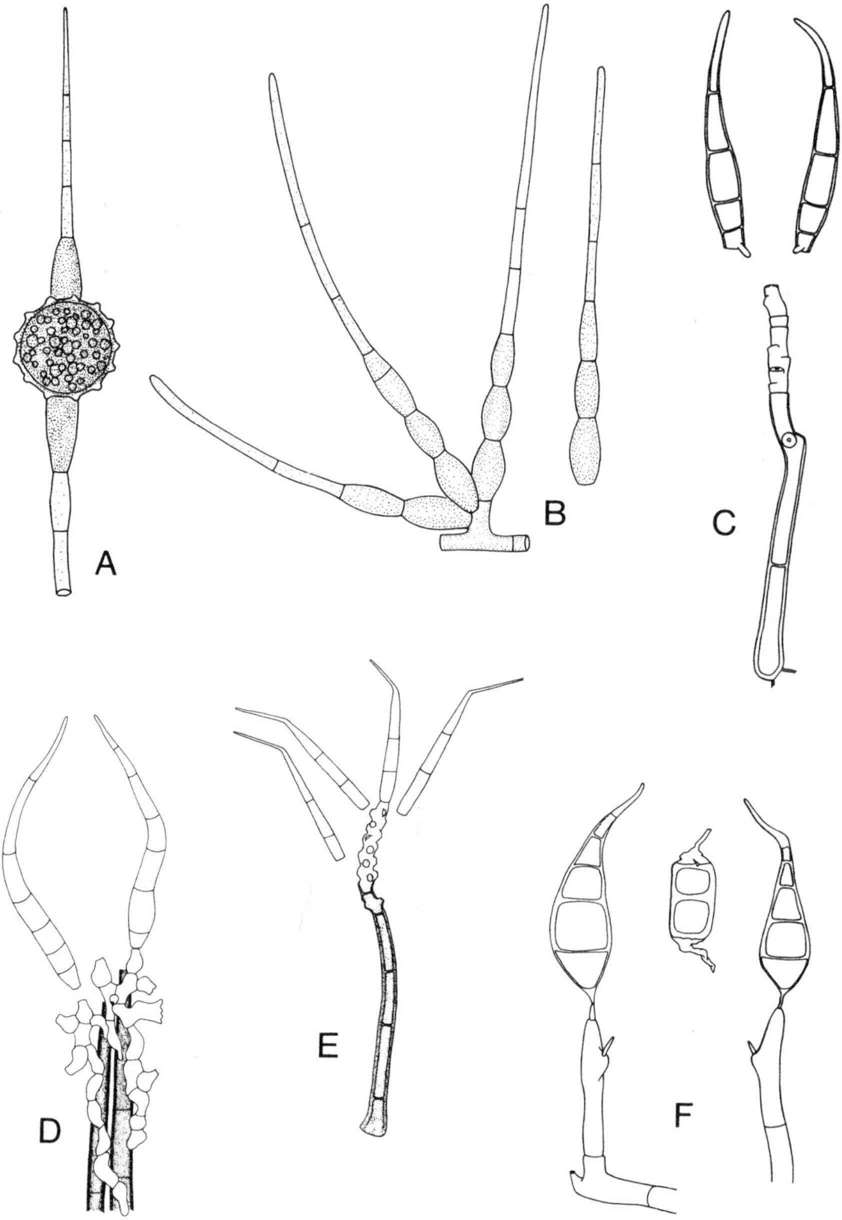

Plate 71: A *Chaetoconidium arachnoideum;* B *Arthrocladium caudatum;* C *Chionomyces meliolicola;* D *Spermosporella elegans;* E *Subulispora procurvata;* F *Trichoconis caudata.*

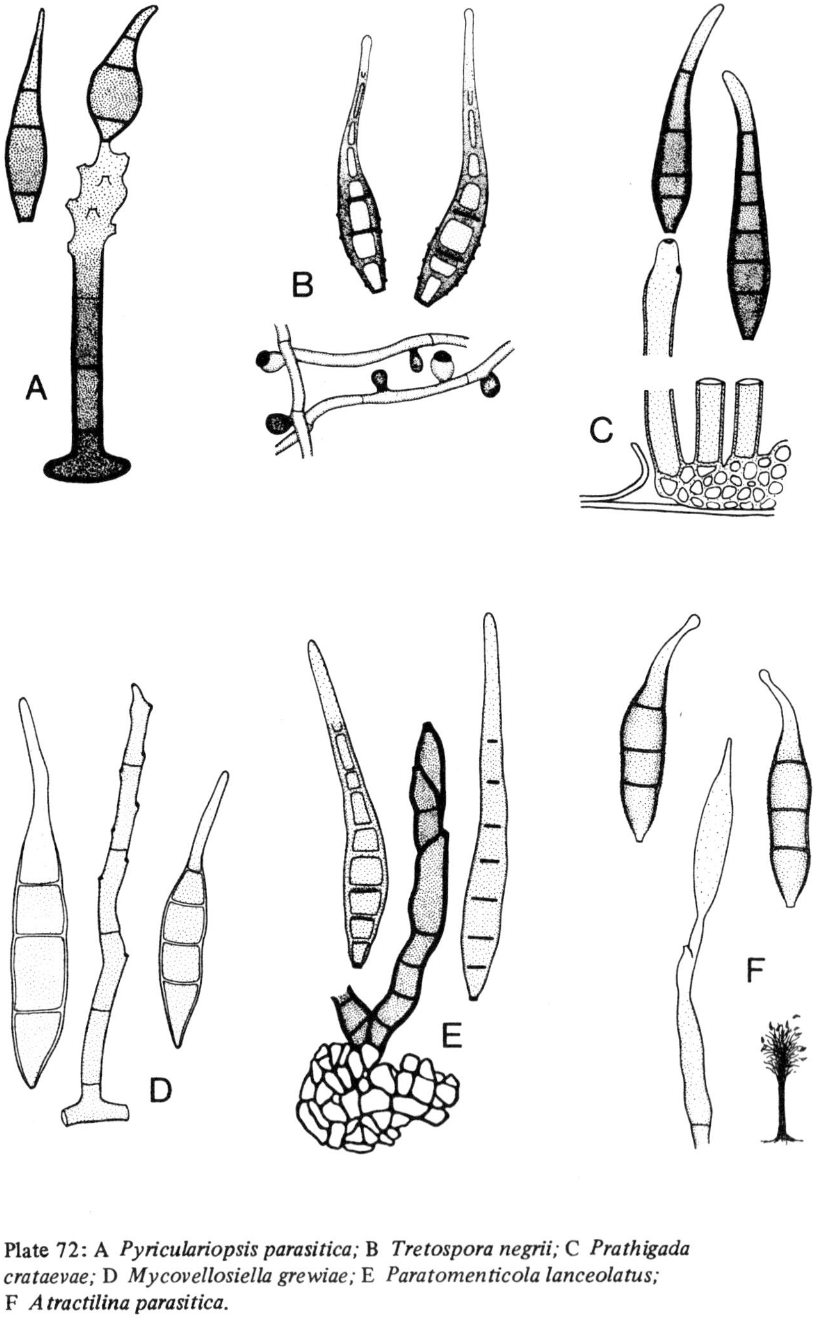

Plate 72: A *Pyriculariopsis parasitica;* B *Tretospora negrii;* C *Prathigada crataevae;* D *Mycovellosiella grewiae;* E *Paratomenticola lanceolatus;* F *Atractilina parasitica.*

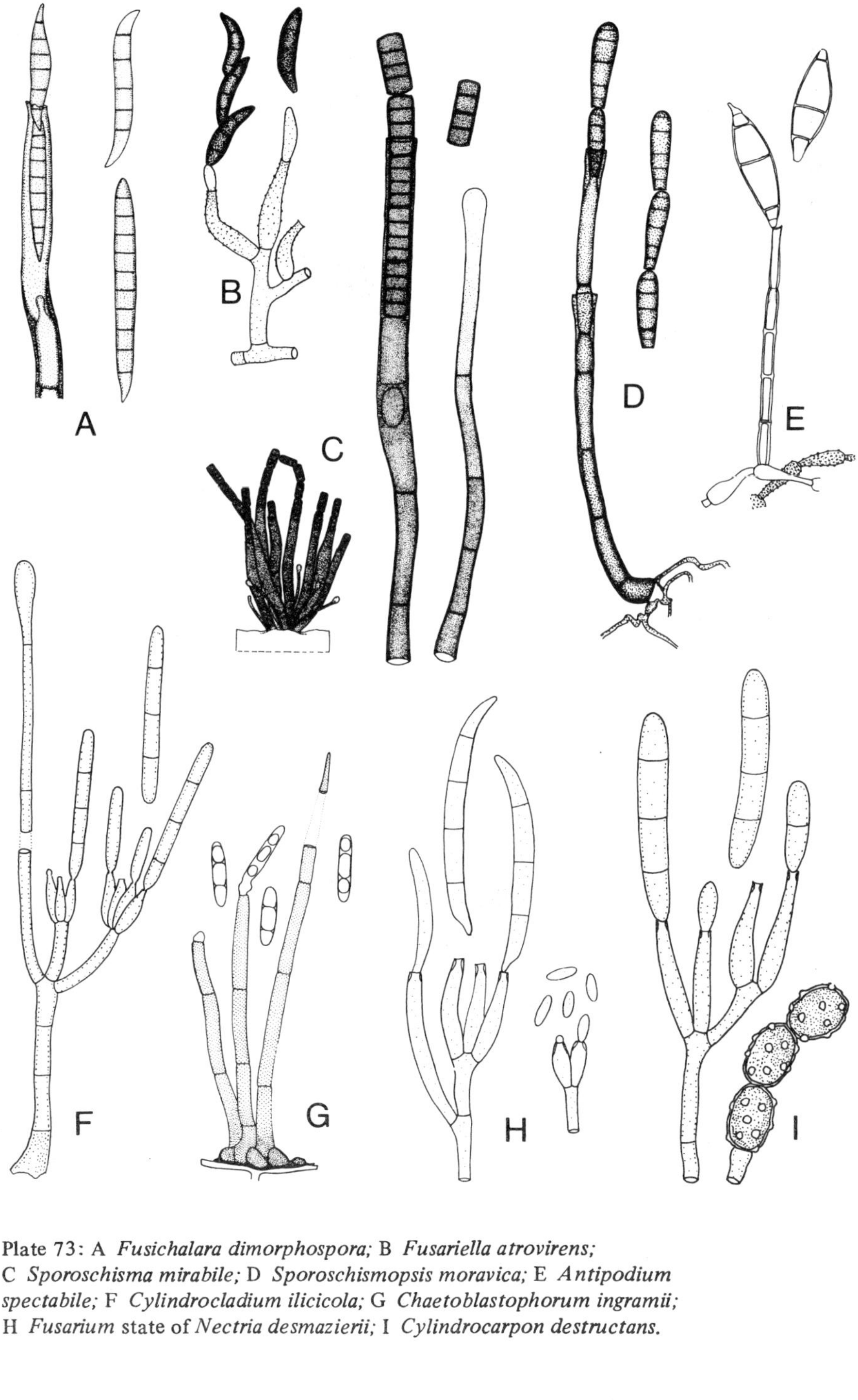

Plate 73: A *Fusichalara dimorphospora;* B *Fusariella atrovirens;* C *Sporoschisma mirabile;* D *Sporoschismopsis moravica;* E *Antipodium spectabile;* F *Cylindrocladium ilicicola;* G *Chaetoblastophorum ingramii;* H *Fusarium* state of *Nectria desmazierii;* I *Cylindrocarpon destructans.*

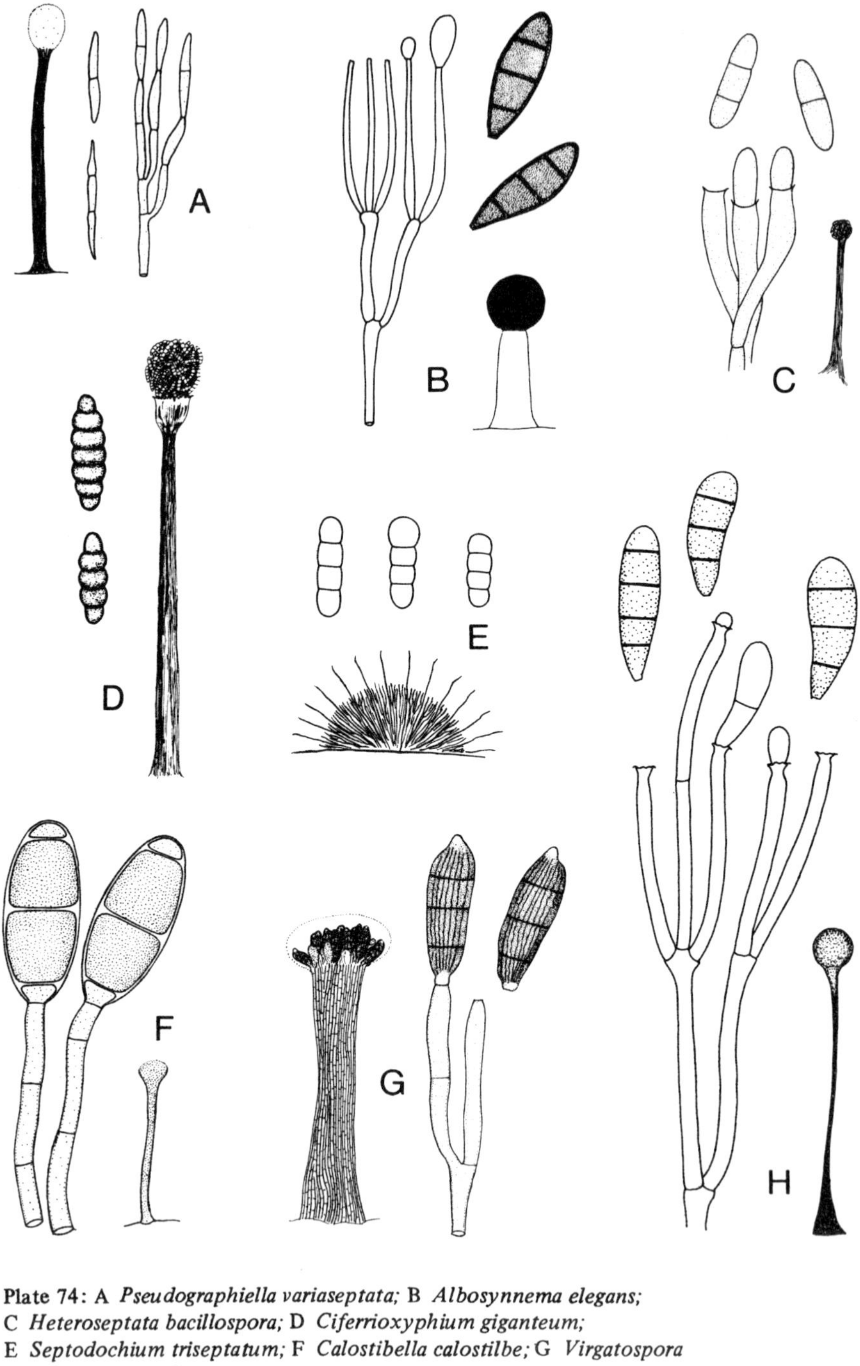

Plate 74: A *Pseudographiella variaseptata;* B *Albosynnema elegans;* C *Heteroseptata bacillospora;* D *Ciferrioxyphium giganteum;* E *Septodochium triseptatum;* F *Calostibella calostilbe;* G *Virgatospora echinofibrosa;* H *Phialoarthrobotryum triseptatum.*

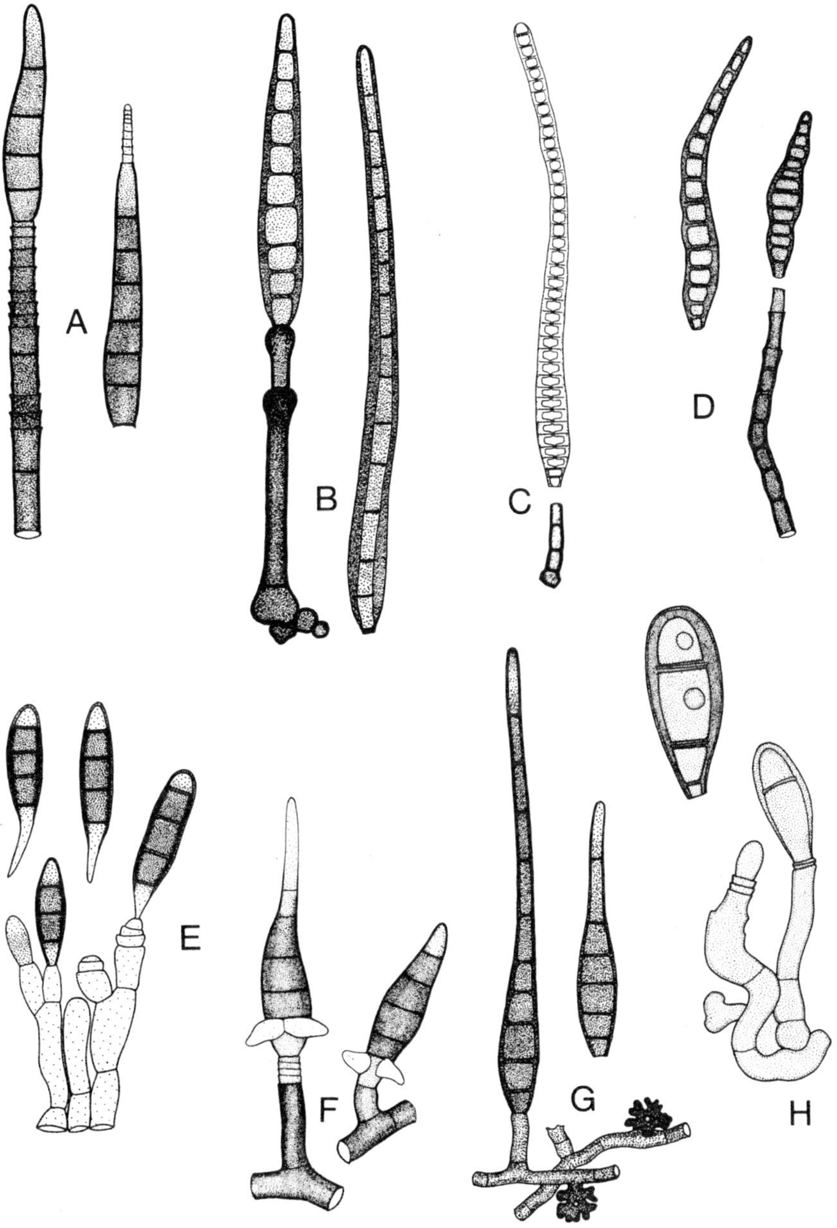

Plate 75: A *Annellophora africana;* B *Corynespora cassiicola;* C *Sporidesmium adscendens;* D *Sporidesmium bumbusicola;* E *Exosporiella fungorum;* F *Teratosperma singulare;* G *Clasterosporium flagellatum;* H *Septoidium* state of *Parodiopsis hurae.*

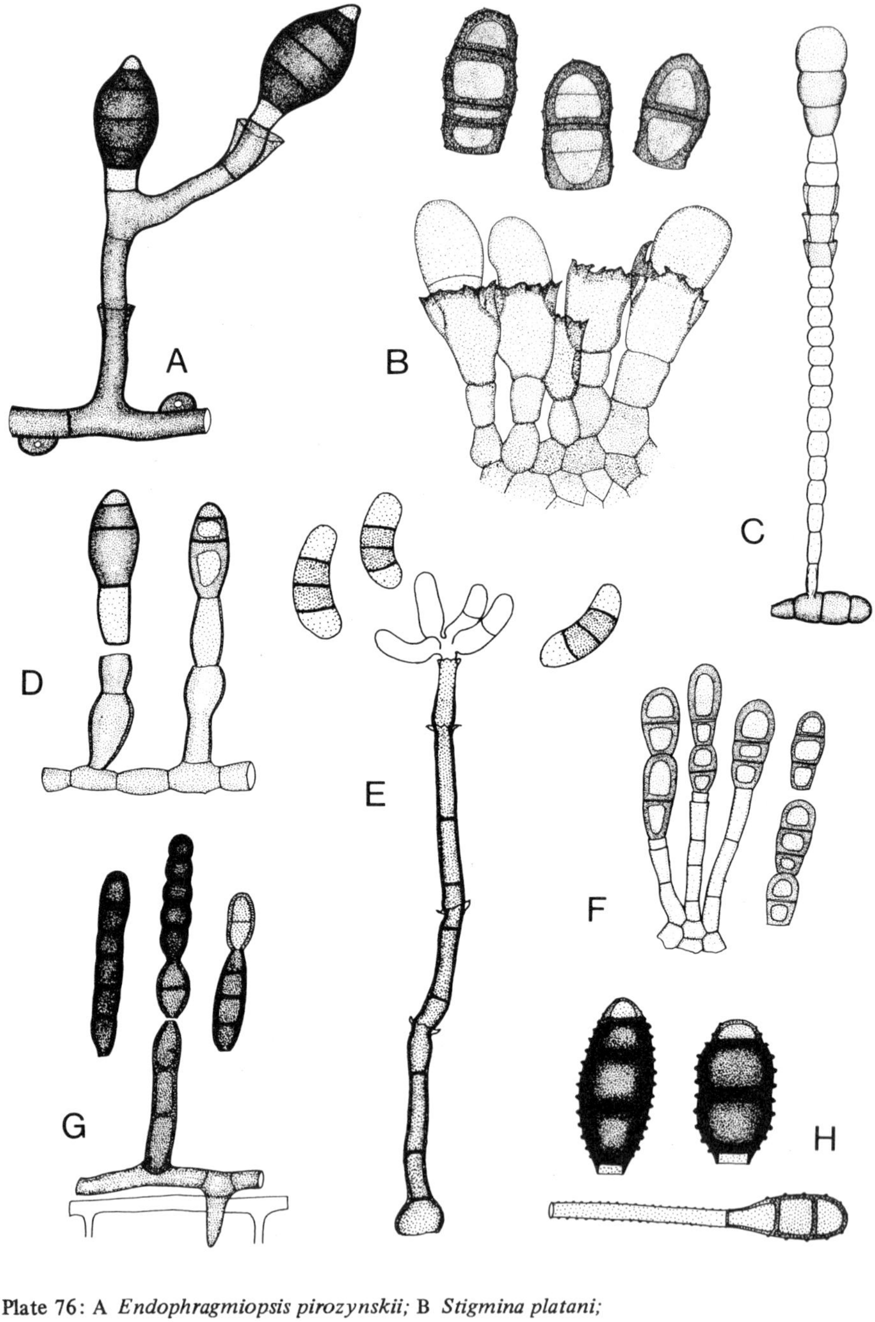

Plate 76: A *Endophragmiopsis pirozynskii;* B *Stigmina platani;* C *Antennopsis gallica;* D *Phragmospathula phoenicis;* E *Cacumisporium capitulatum;* F *Bactrodesmiella masonii;* G *Heteroconium citharexyli;* H *Stemphyliomma valparadisiacum.*

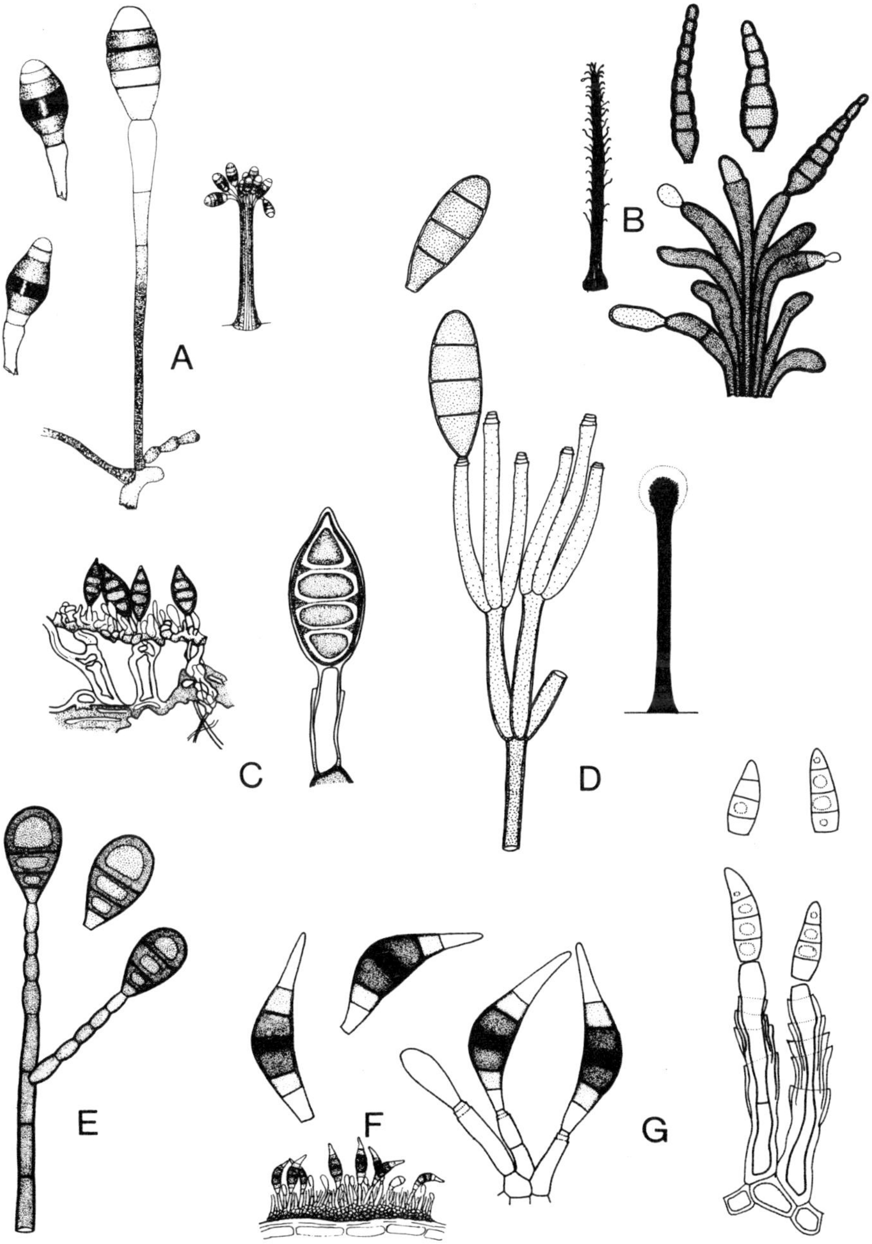

Plate 77: A *Endophragmia atra;* B *Podosporium rigidum;* C *Cooksonomyces banksiae;* D *Arthrobotryum stilboideum;* E *Brachysporiella gayana;* F *Toxosporium camptospermum;* G *Refractohilum achromaticum.*

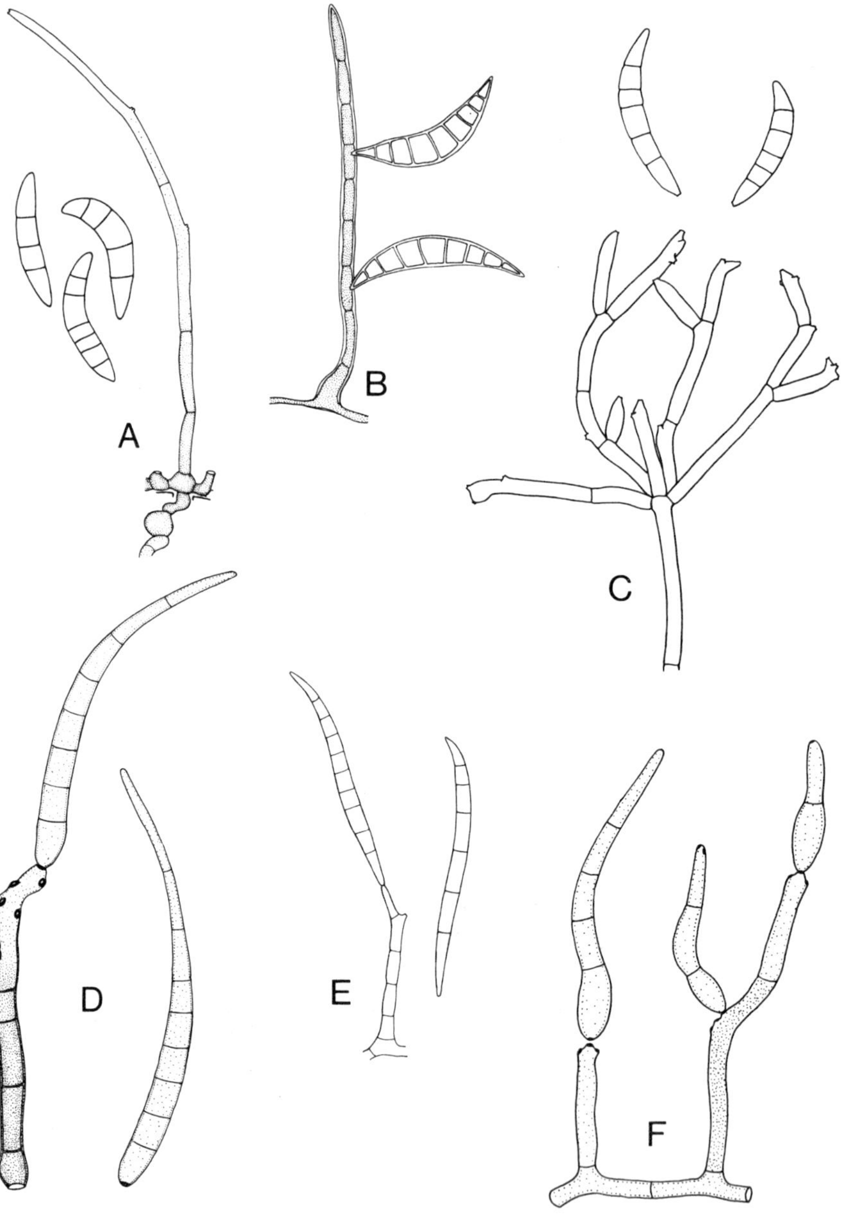

Plate 78: A *Bidenticula cannae;* B *Monodisma fragilis;* C *Pseudofusarium fusarioideum;* D *Cercospora elaeidis;* E *Vermispora grandispora;* F *Cladosporiella uredinicola.*

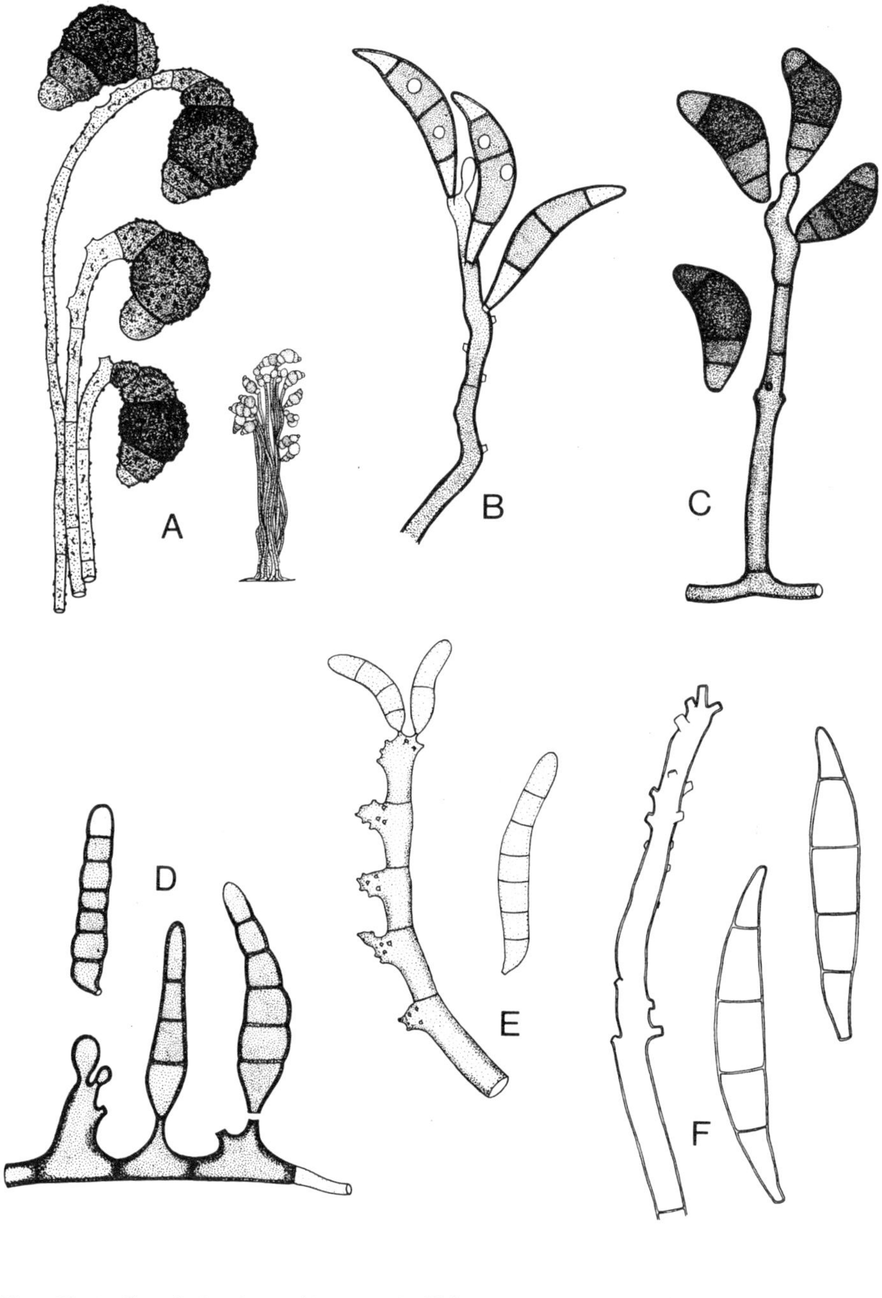

Plate 79: A *Curvulariopsis cymbisperma;* B *Nakataea* state of *Leptosphaeria salvinii;* C *Curvularia lunata;* D *Janetia euphorbiae;* E *Gonatophragmium mori;* F *Eriomycopsis bonplandii.*

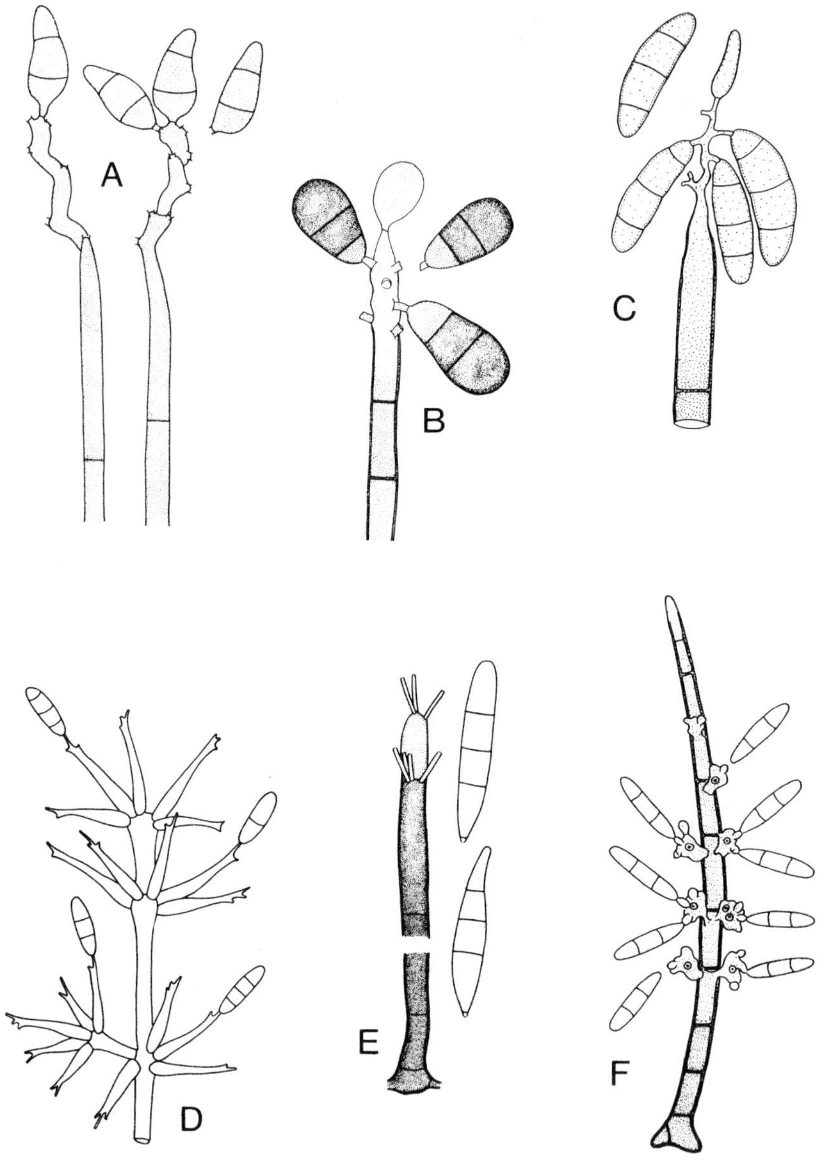

Plate 80: A *Pyricularia oryzae;* B *Brachysporium obovatum;* C *Pleurothecium recurvatum;* D *Helminthophora sphaerocephala;* E *Paratrichoconis chinensis;* F *Spondylocladiopsis cupulicola.*

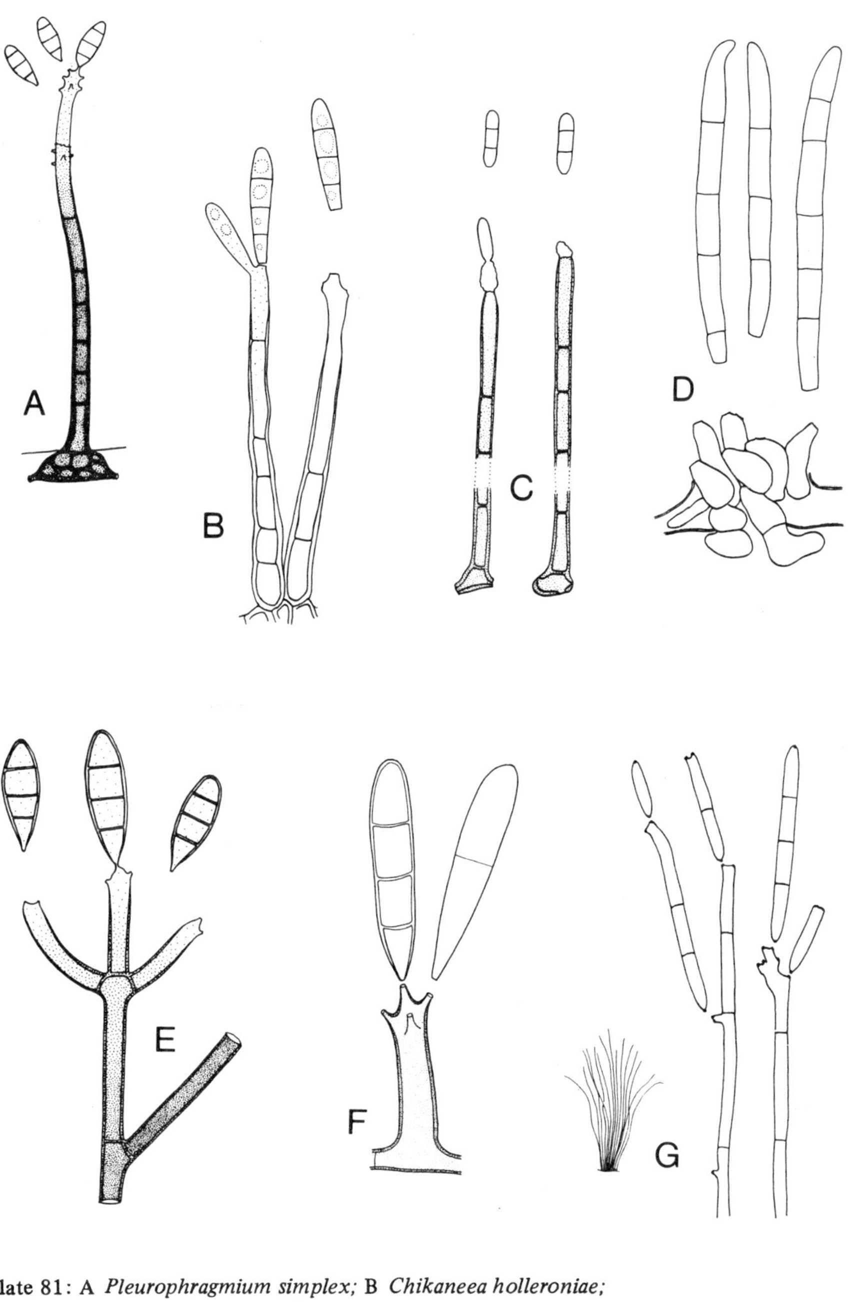

Plate 81: A *Pleurophragmium simplex;* B *Chikaneea holleroniae;* C *Blastophorum truncatum;* D *Pseudocercosporella ipomoeae;* E *Phaeodactylium alpiniae;* F *Dactylaria purpurella;* G *Ramulariopsis glycines.*

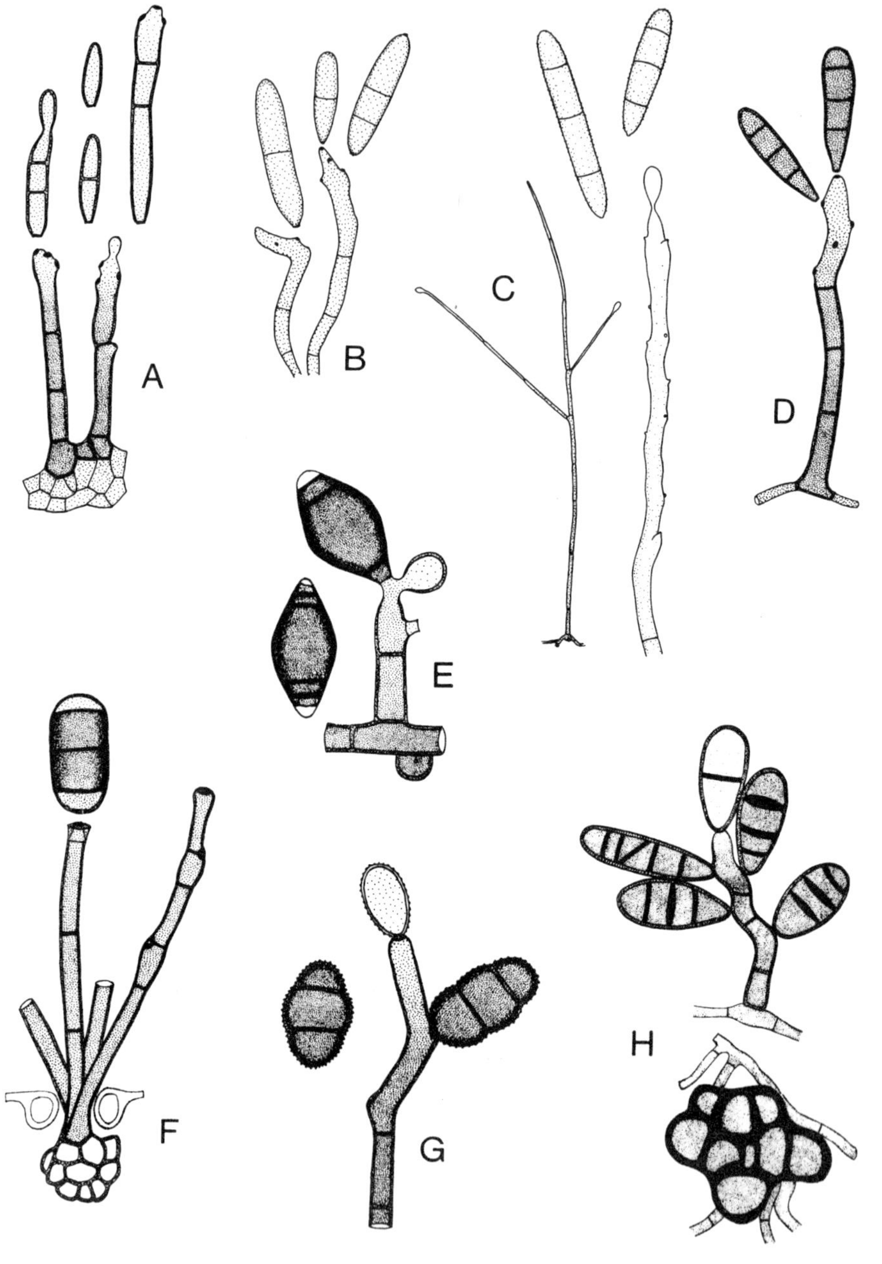

Plate 82: A *Stenella araguata;* B *Mycovellosiella cajani;* C *Redbia elegans;* D *Eriocercospora balladynae;* E *Mitteriella ziziphina;* F *Duosporium cyperi;* G *Acroconidiella tropaeoli;* H *Embellisia allii.*

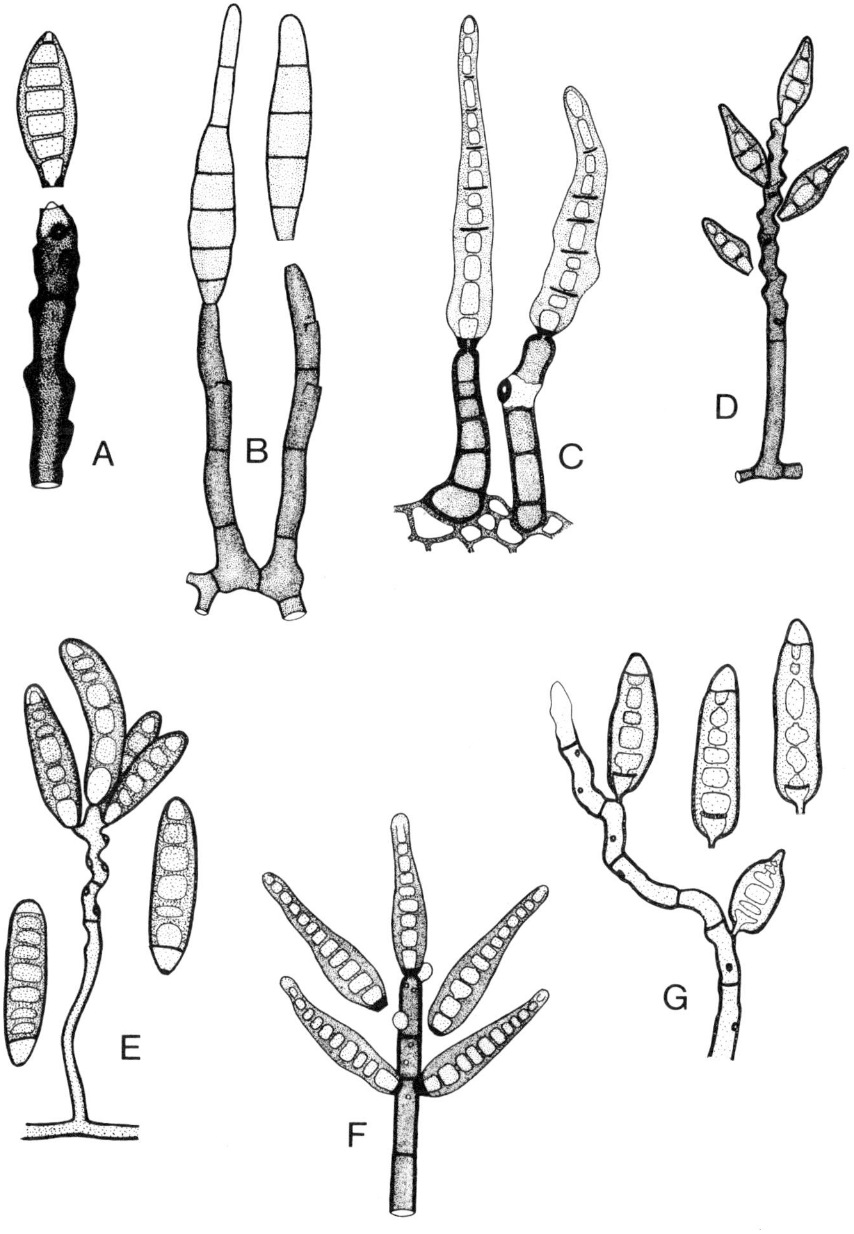

Plate 83: A *Pseudospiropes nodosus;* B *Tomenticola trematis;* C *Exosporium tiliae;* D *Spiropes guareicola;* E *Drechslera* state of *Cochliobolus bicolor;* F *Helminthosporium velutinum;* G *Exserohilum halodes.*

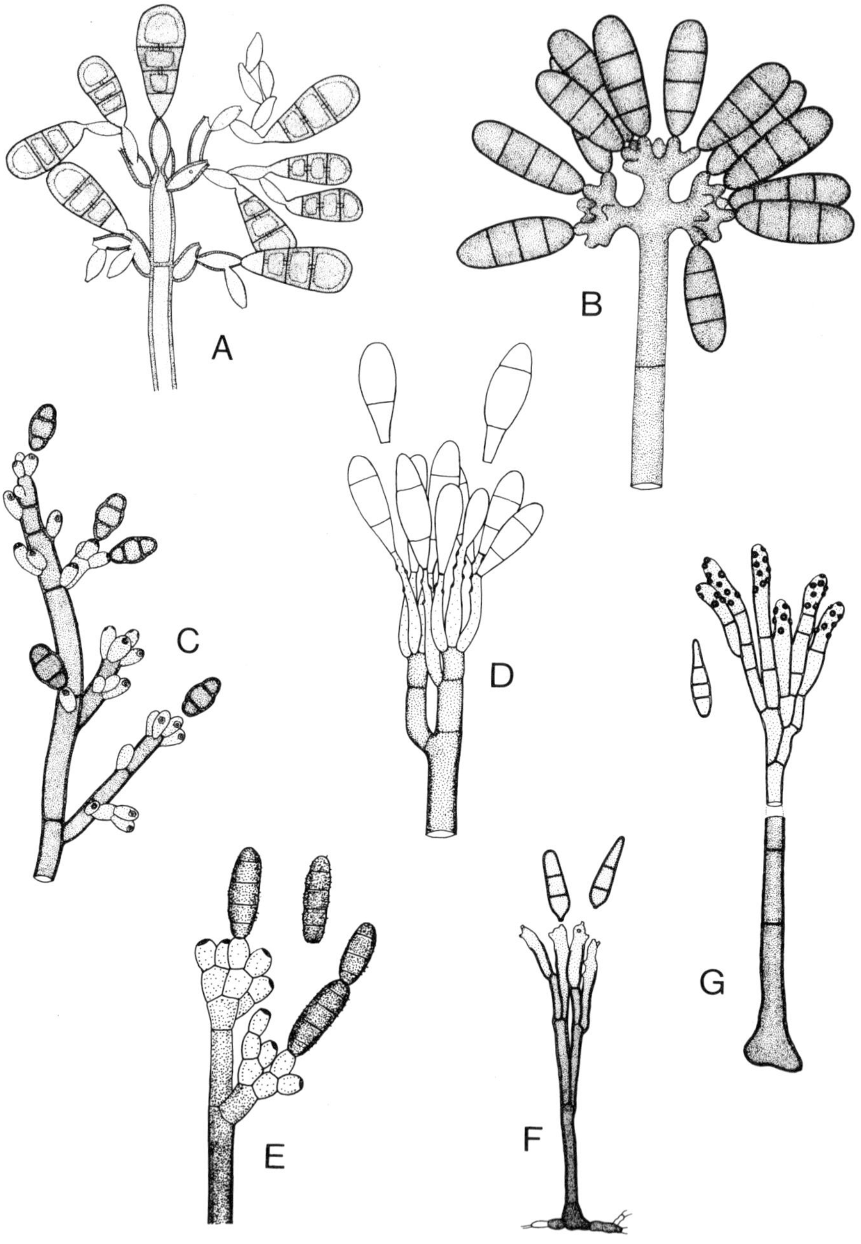

Plate 84: A *Edmundmasonia pulchra;* B *Dichotomophthora indica;*
C *Spondylocladiella botrytioides;* D *Sterigmatobotrys macrocarpa;*
E *Dendryphion comosum;* F *Parapyricularia musae;* G *Periconiella cyatheae.*

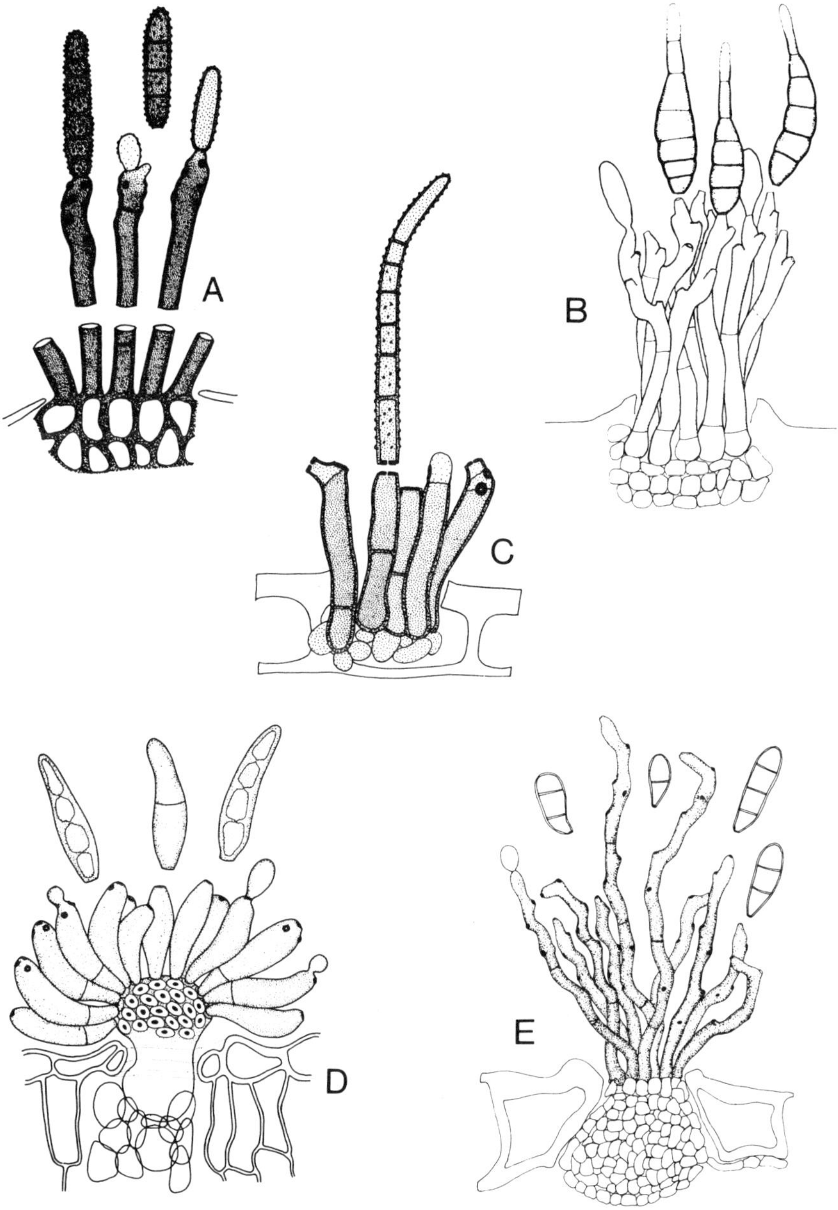

Plate 85: A *Verrucispora proteacearum;* B *Pseudocercospora vitis;* C *Stenellopsis fagraeae;* D *Camptomeris albizziae;* E *Cercosporidium chaetomium.*

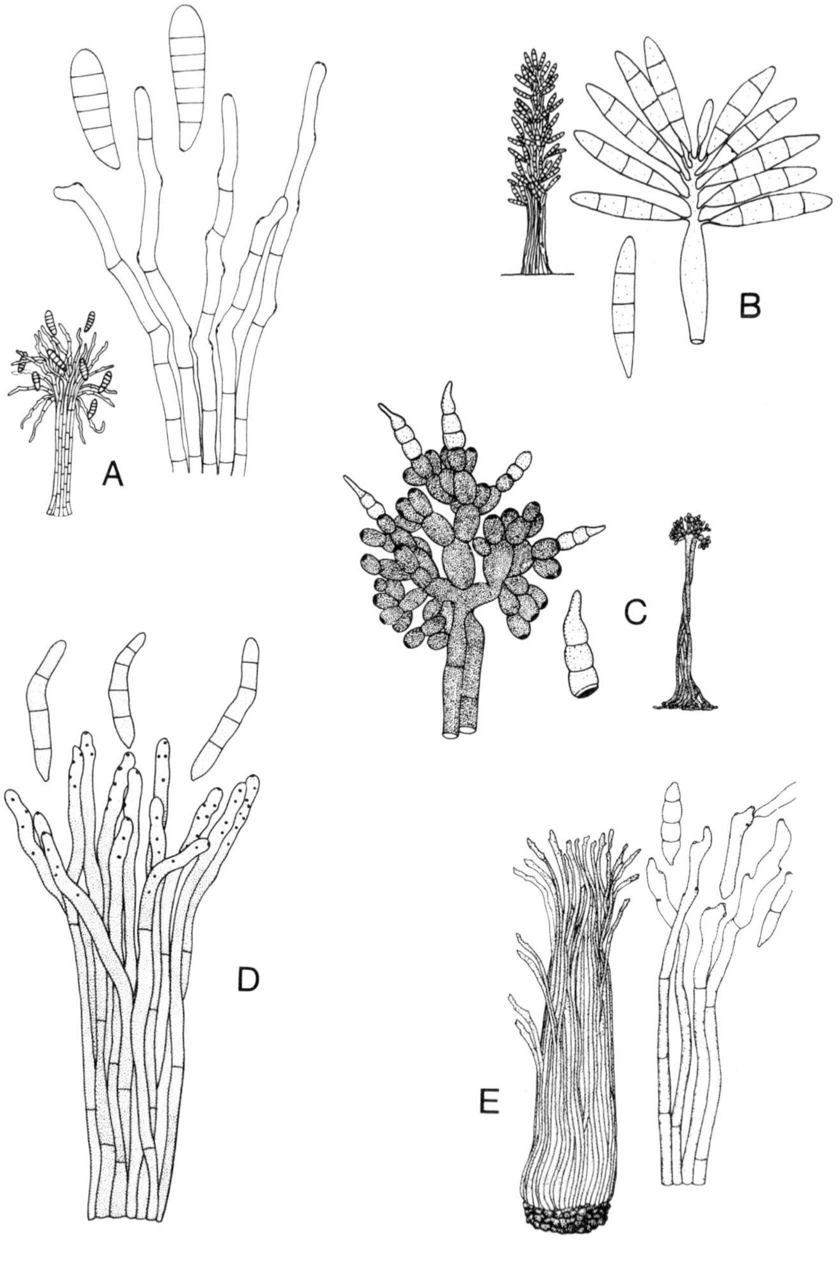

Plate 86: A *Hyalosynnema multiseptatum;* B *Arthrosporium candidum;* C *Acarocybe hansfordii;* D *Phaeoisariopsis griseola;* E *Tandonella ziziphi.*

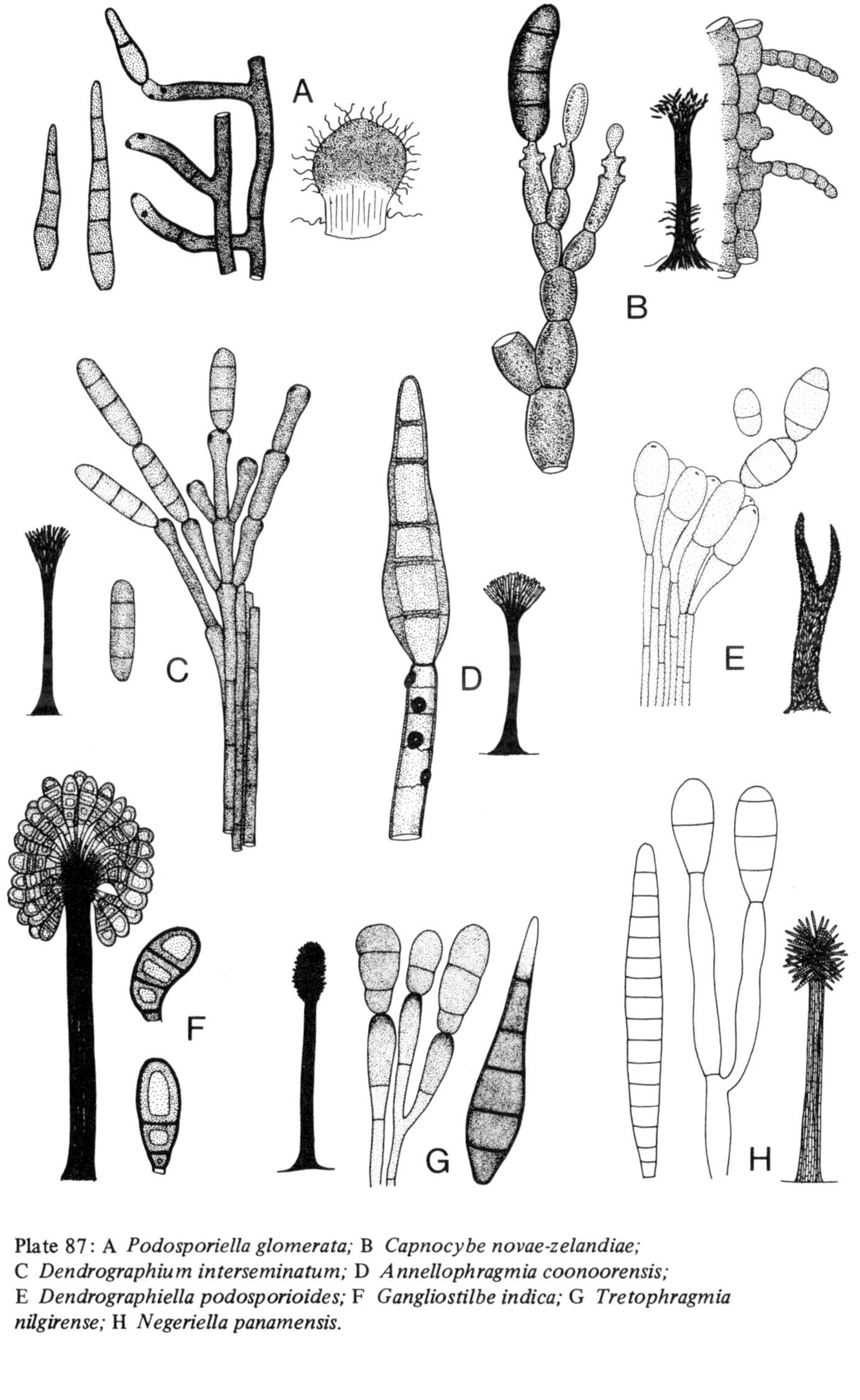

Plate 87: A *Podosporiella glomerata;* B *Capnocybe novae-zelandiae;* C *Dendrographium interseminatum;* D *Annellophragmia coonoorensis;* E *Dendrographiella podosporioides;* F *Gangliostilbe indica;* G *Tretophragmia nilgirense;* H *Negeriella panamensis.*

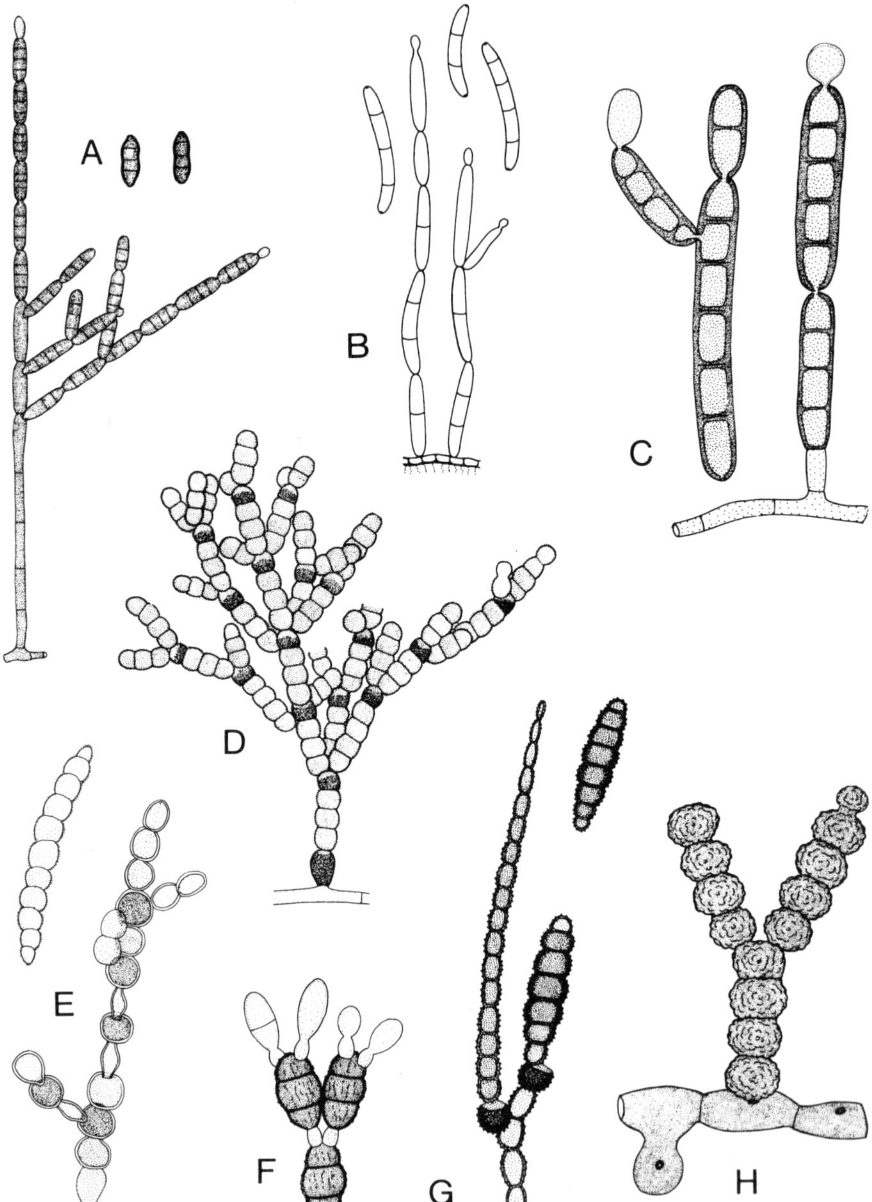

Plate 88: A *Septonema secedens;* B *Septocylindrium virens;* C *Lylea catenulata;* D *Torula herbarum;* E *Bahusaganda indica;* F *Bahusandhika indica;* G *Dwayabeeja sundara;* H *Rutola graminis.*

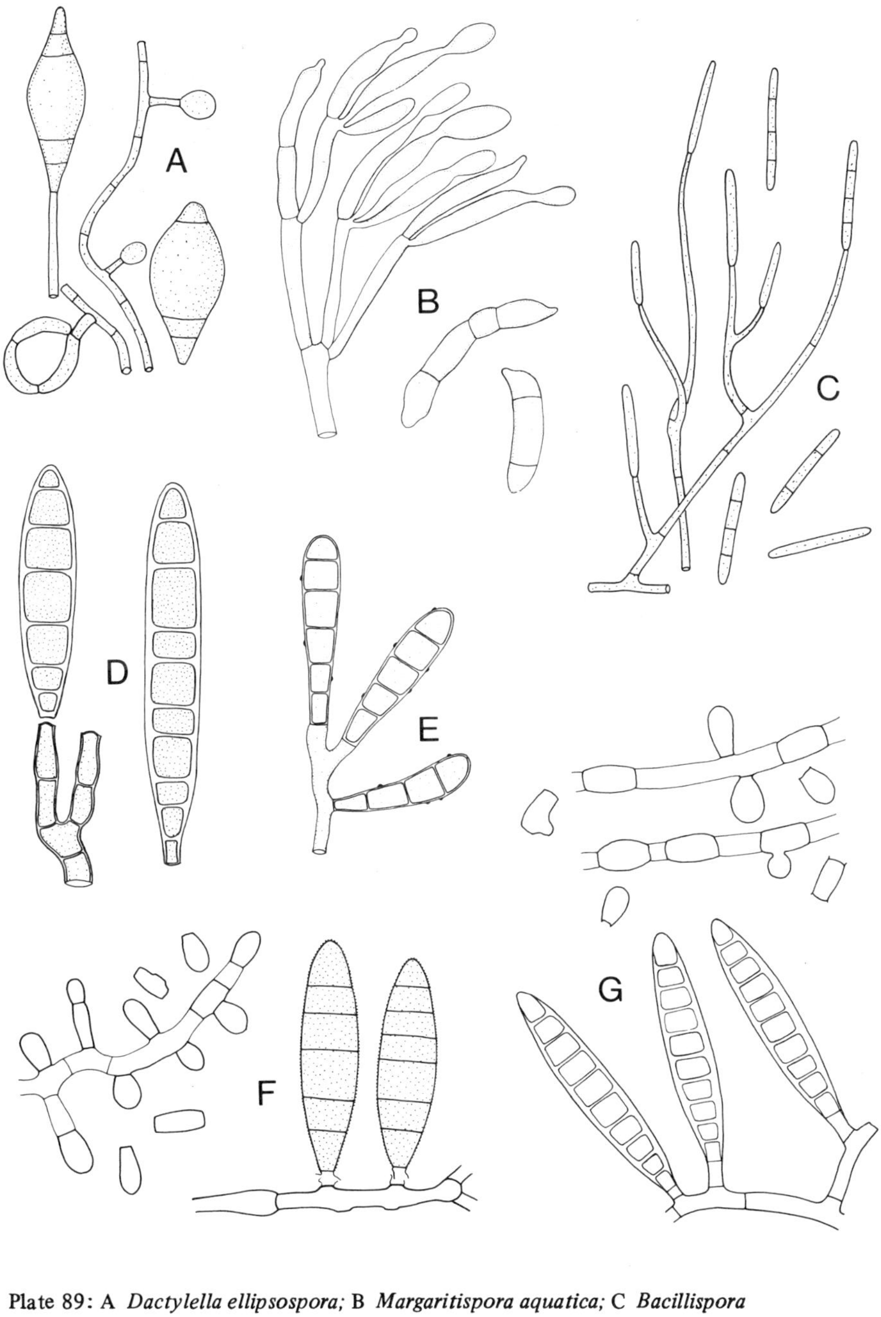

Plate 89: A *Dactylella ellipsospora;* B *Margaritispora aquatica;* C *Bacillispora aquatica;* D *Bactridium flavum;* E *Epidermophyton floccosum;* F *Microsporum gypseum;* G *Trichophyton ajelloi.*

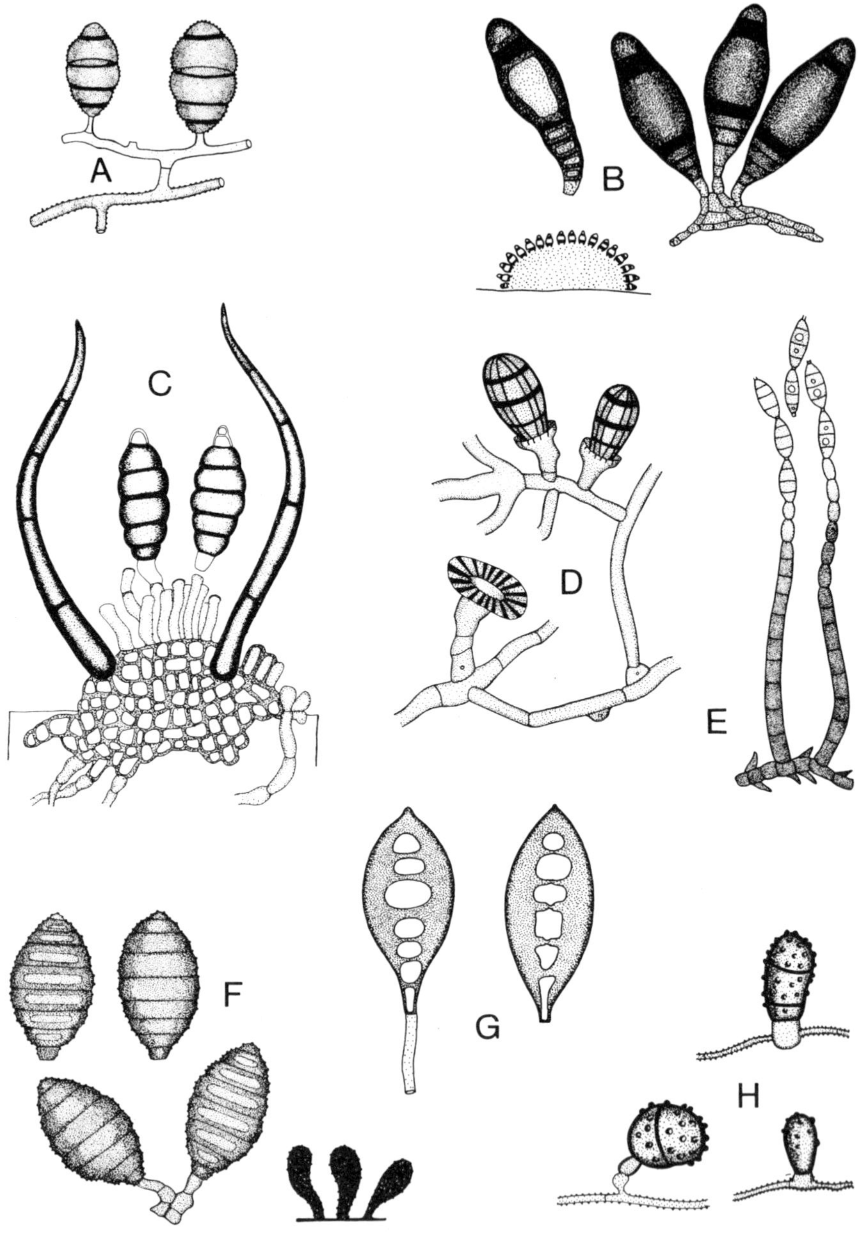

Plate 90: A *Pithomyces flavus;* B *Bactrodesmium abruptum;* C *Excipularia narsapurensis;* D *Colemaniella ossoorii;* E *Ampulliferopsis myriapoda;* F *Listeromyces insignis;* G *Murogenella terrophila;* H *Polyschema terricola.*

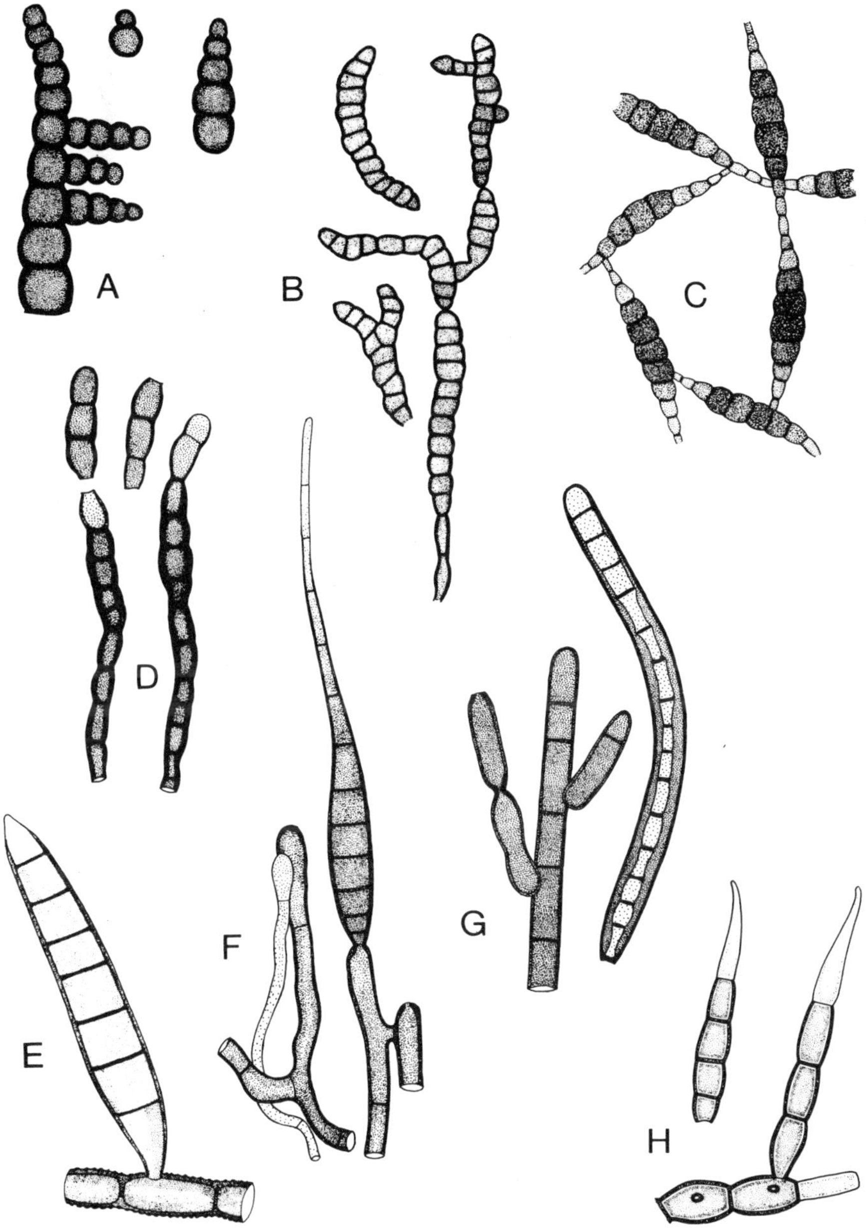

Plate 91: A *Torulopsiella* state of *Metacapnodium* sp.; B *Trimmatostroma salicis;* C *Xylomyces chlamydosporis;* D *Taeniolella exilis;* E *Antennatula atra;* F *Acarocybella jasminicola;* G *Corynesporella urticae;* H *Sessiliospora bicolor.*

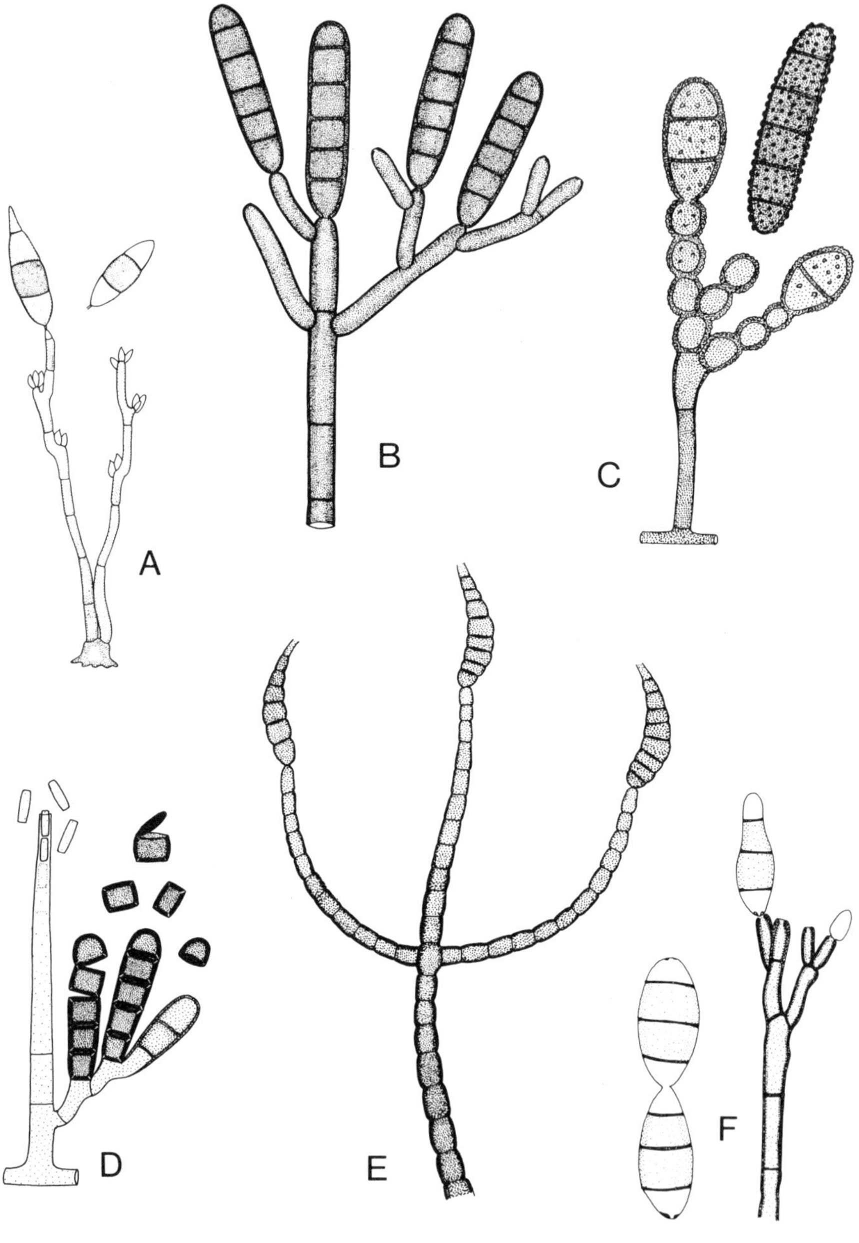

Plate 92: A *Kramasamuha sibika;* B *Dendryphiopsis atra;* C *Hormocephalum ecuadorense;* D *Trichocladium basicola;* E *Capnokyma corticola;* F *Paradendryphiopsis cambrensis.*

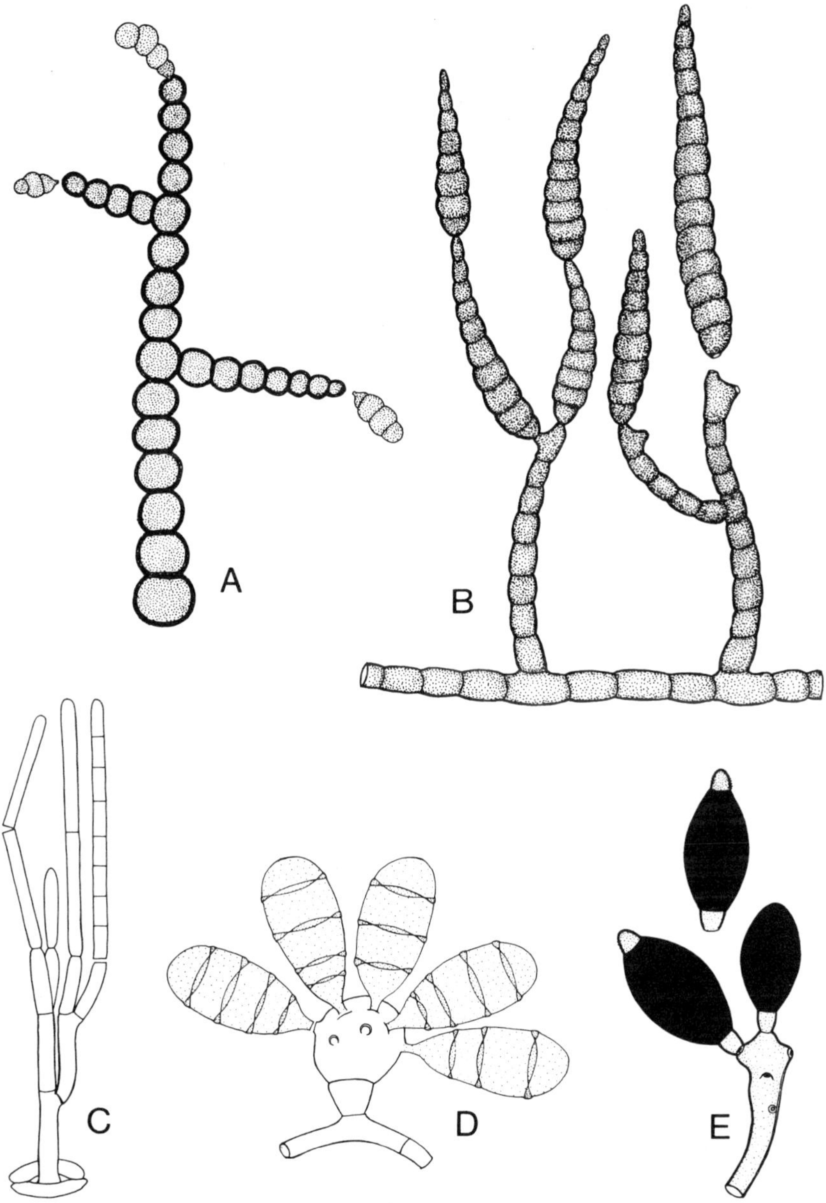

Plate 93: A *Capnosporium* sp.; B *Capnodendron trichomericola;* C *Cylindrocladiopsis lagerstraemiae;* D *Cephaliophora tropica;* E *Brachydesmiella biseptata.*

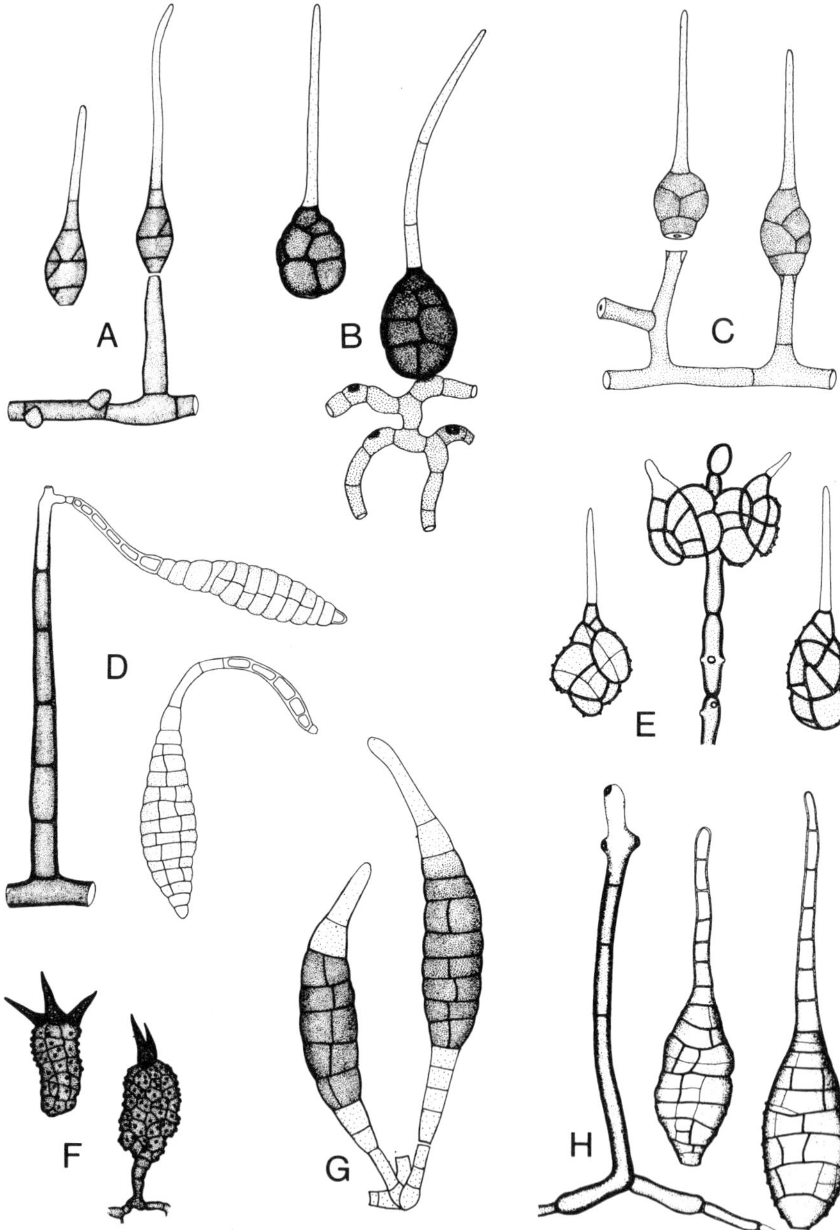

Plate 94: A *Hansfordiellopsis aburiensis;* B *Piricauda paraguayensis;* C *Hansfordiella meliolae;* D *Pendulispora venezuelanica;* E *Blastodictys hibisci;* F *Pseudopetrakia kambakkamensis;* G *Dictyodesmium ulmicola;* H *Dictyospiropes ziziphi.*

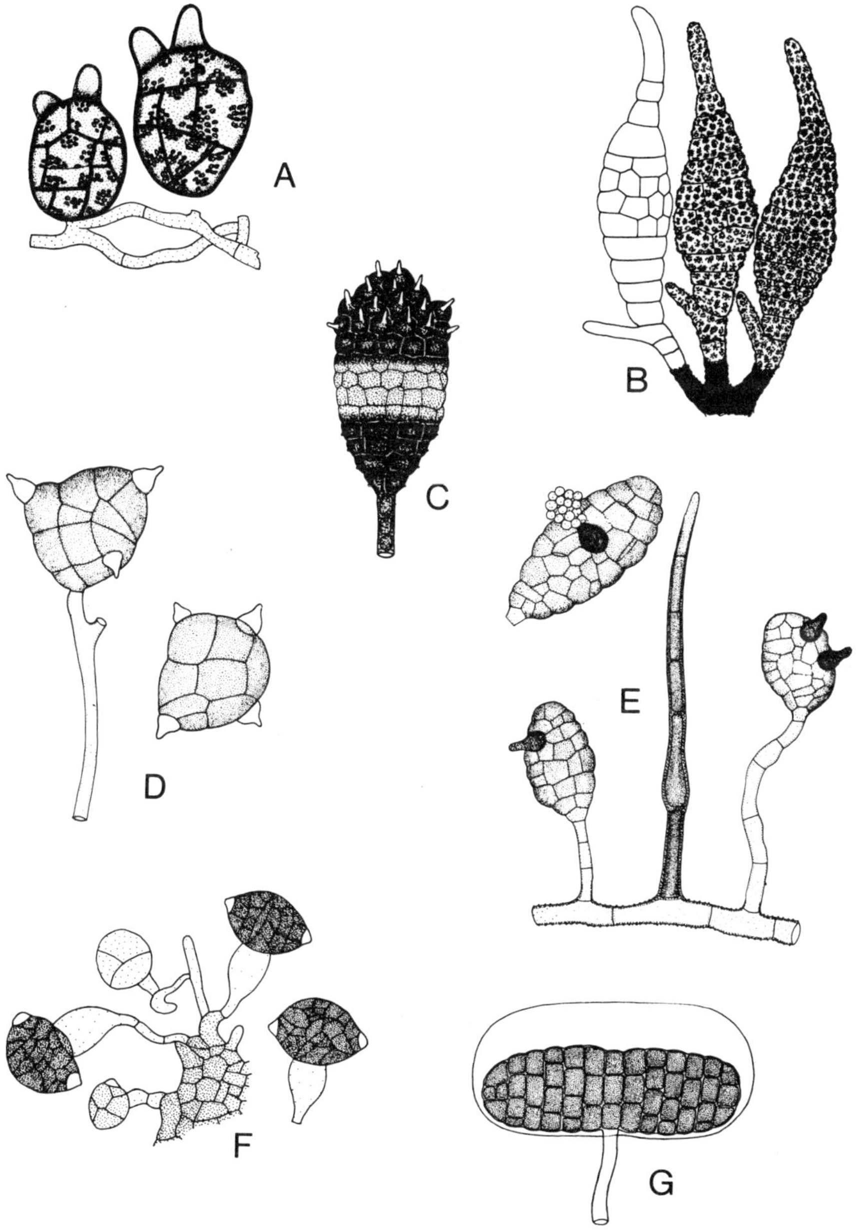

Plate 95: A *Bioconiosporium baccharidis;* B *Muiaria repens;* C *Muiogone chromopteri;* D *Oncopodiella trigonella;* E *Septosporium bulbotrichum;* F *Oncopodium antoniae;* G *Coleodictyospora cubensis.*

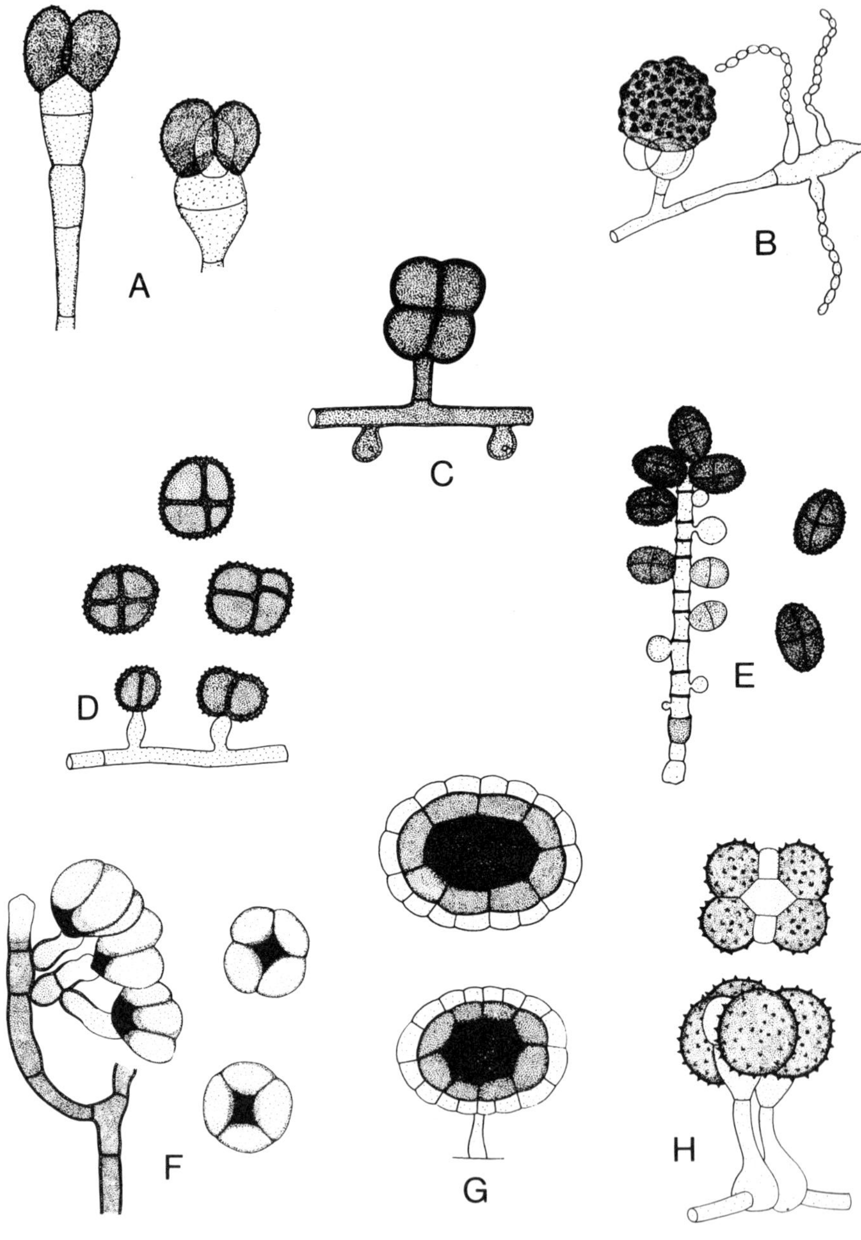

Plate 96: A *Microclava miconiae;* B *Acrospeira mirabilis;* C *Sarcinella* state of *Schiffnerula pulchra;* D *Tetracoccosporium paxianum;* E *Dictyoarthrinium quadratum;* F *Flosculomyces floridaensis;* G *Hermatomyces tucumanensis;* H *Isthmospora* state of *Trichothyrium asterophorum.*

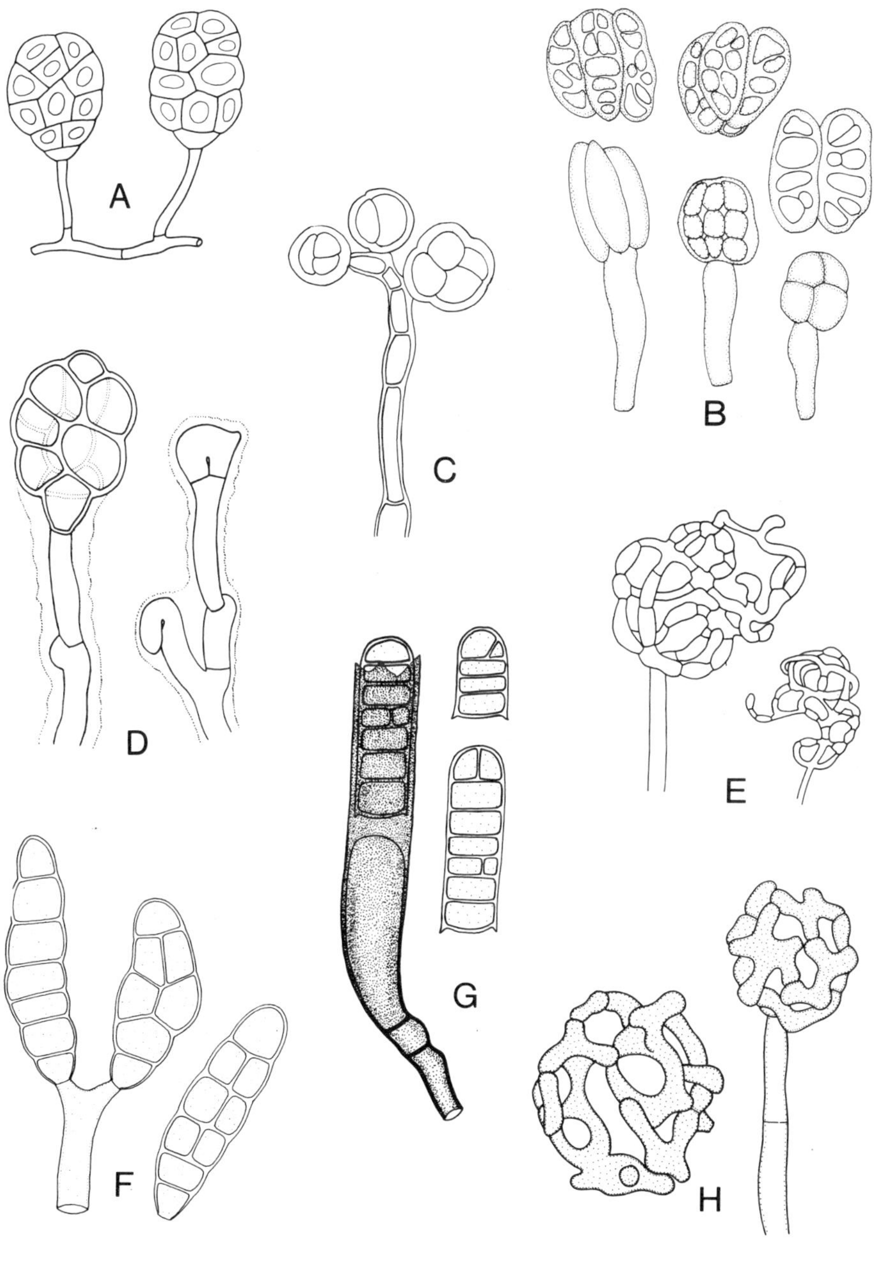

Plate 97: A *Stemphyliopsis* sp.; B *Coniodictyum chevalieri;* C *Hyalocladium moubasheri;* D *Thyrsidina carneo-miniata;* E *Strumella olivatra;* F *Miuraea degenerans;* G *Ascoconidium tsugae;* H *Clathrosphaerina zalewskii.*

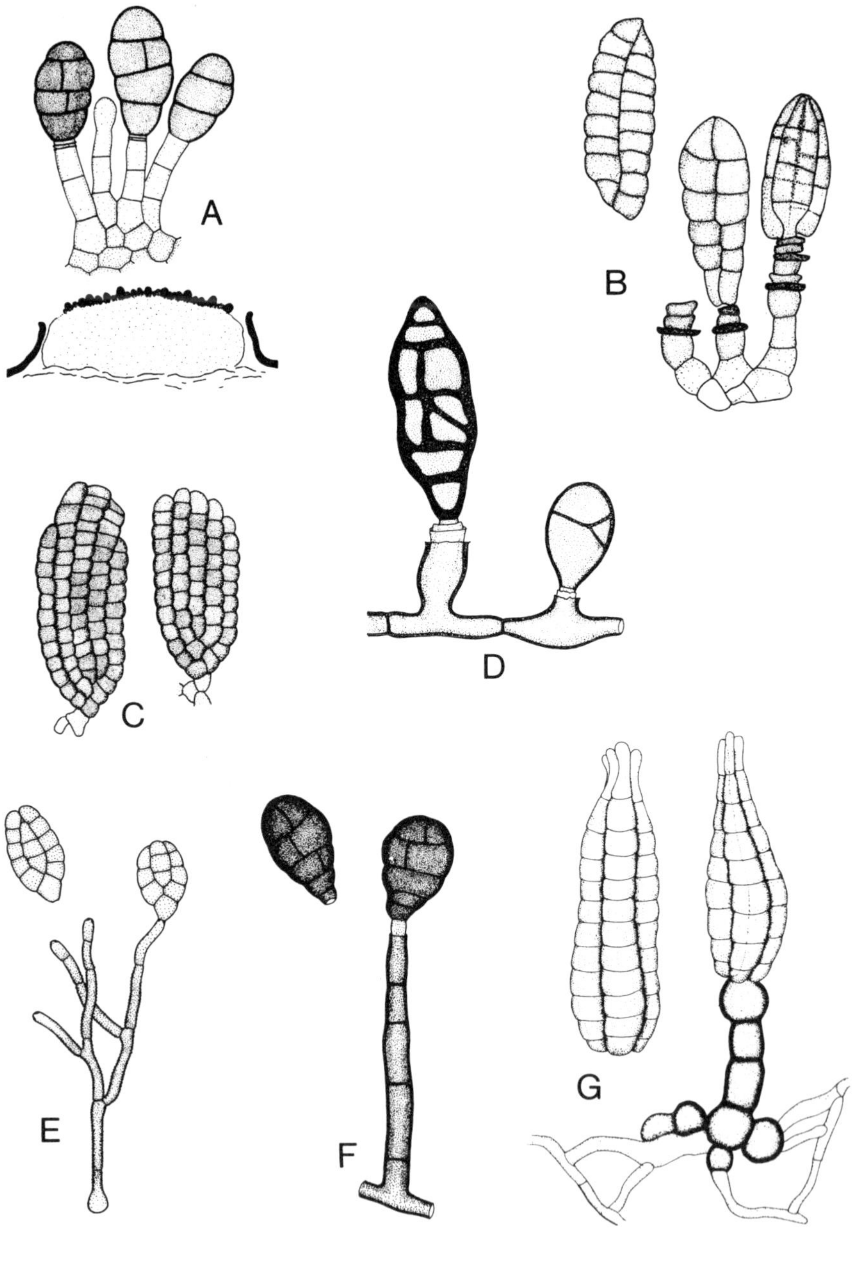

Plate 98: A *Stigmina lycii;* B *Eversia subopaca;* C *Dictyosporium elegans;* D *Annellophorella faureae;* E *Pseudodictyosporium wauense;* F *Acrodictys bambusicola;* G *Tretospeira ugandensis.*

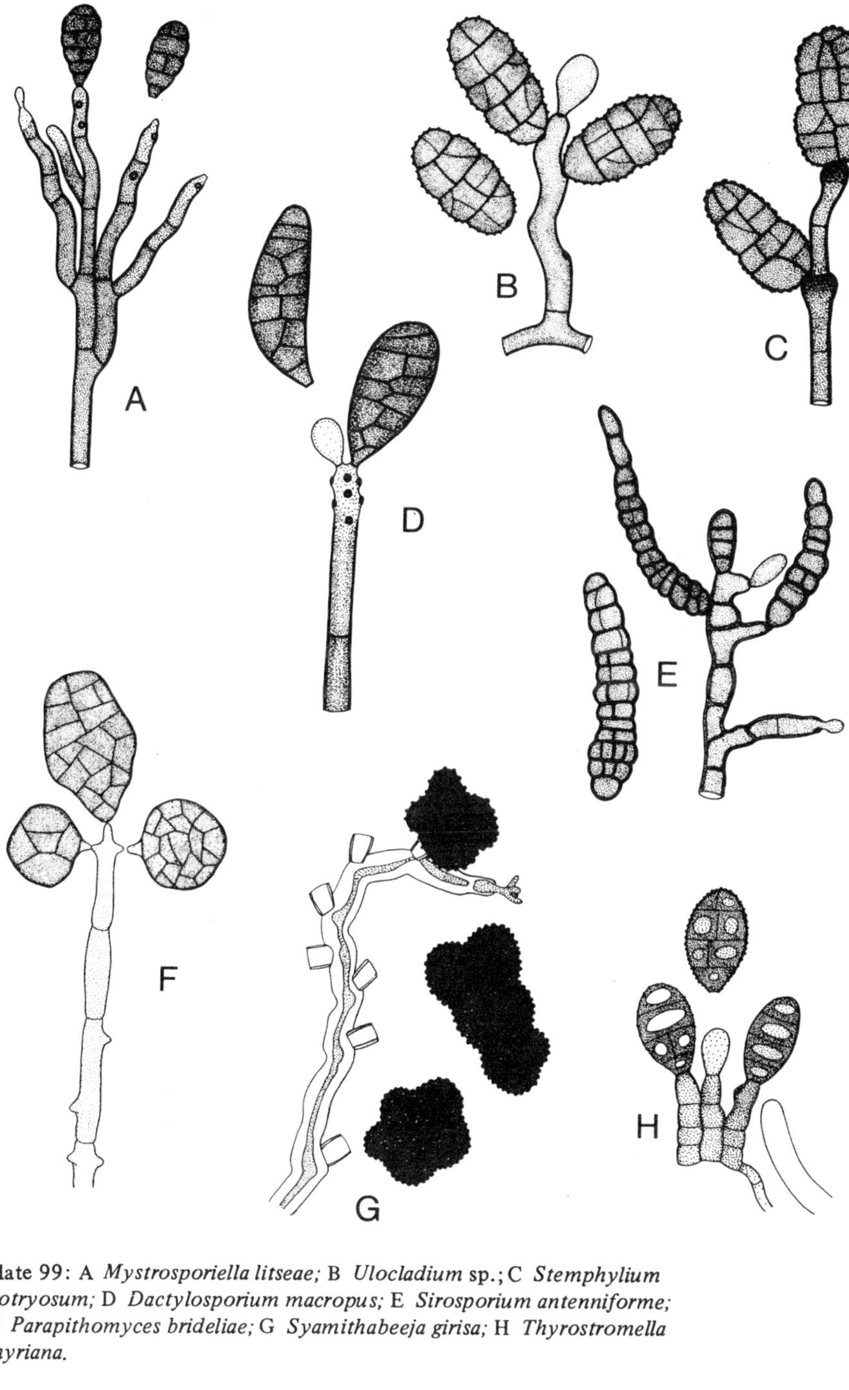

Plate 99: A *Mystrosporiella litseae;* B *Ulocladium* sp.; C *Stemphylium botryosum;* D *Dactylosporium macropus;* E *Sirosporium antenniforme;* F *Parapithomyces brideliae;* G *Syamithabeeja girisa;* H *Thyrostromella myriana.*

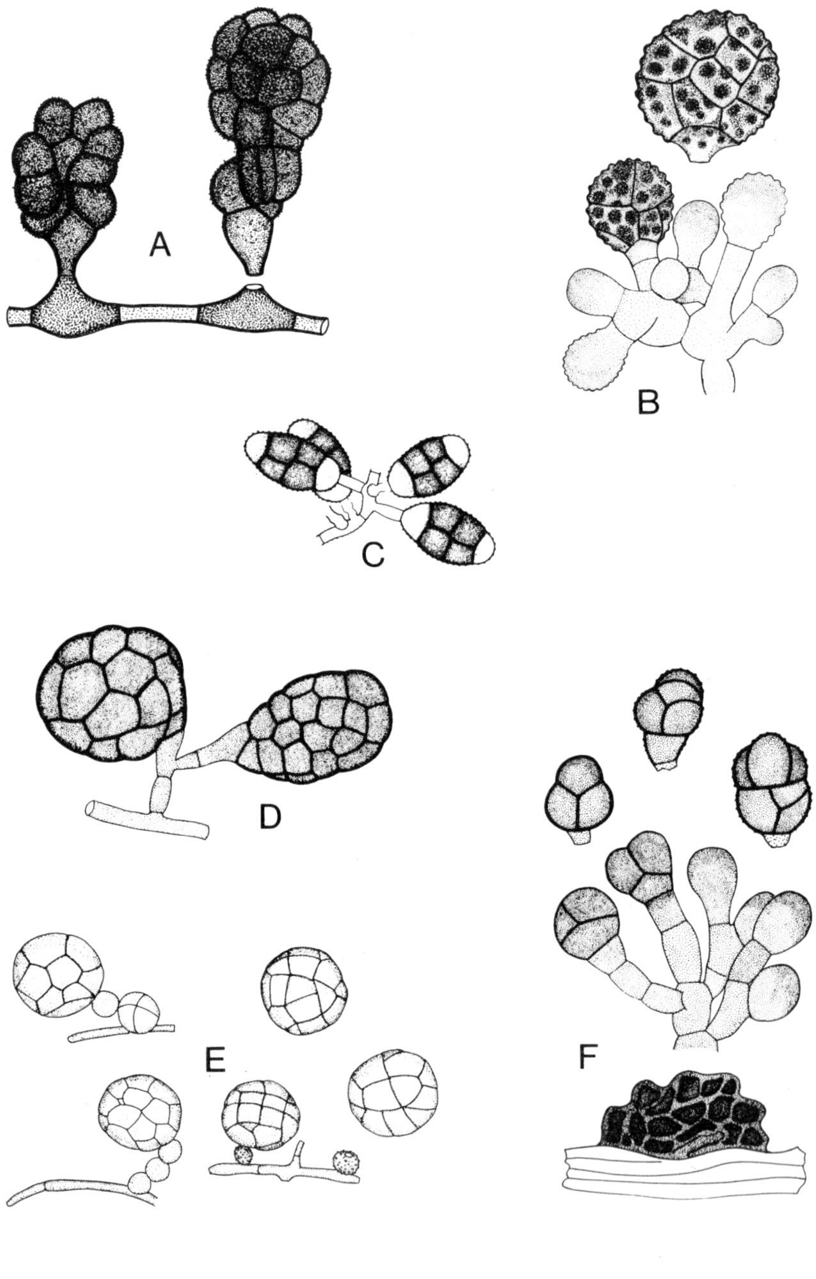

Plate 100: A *Chuppia sarcinifera;* B *Epicoccum purpurascens;* C *Pithomyces chartarum;* D *Monodictys putredinis;* E *Dictyopolyschema pirozynskii;* F *Cerebella andropogonis.*

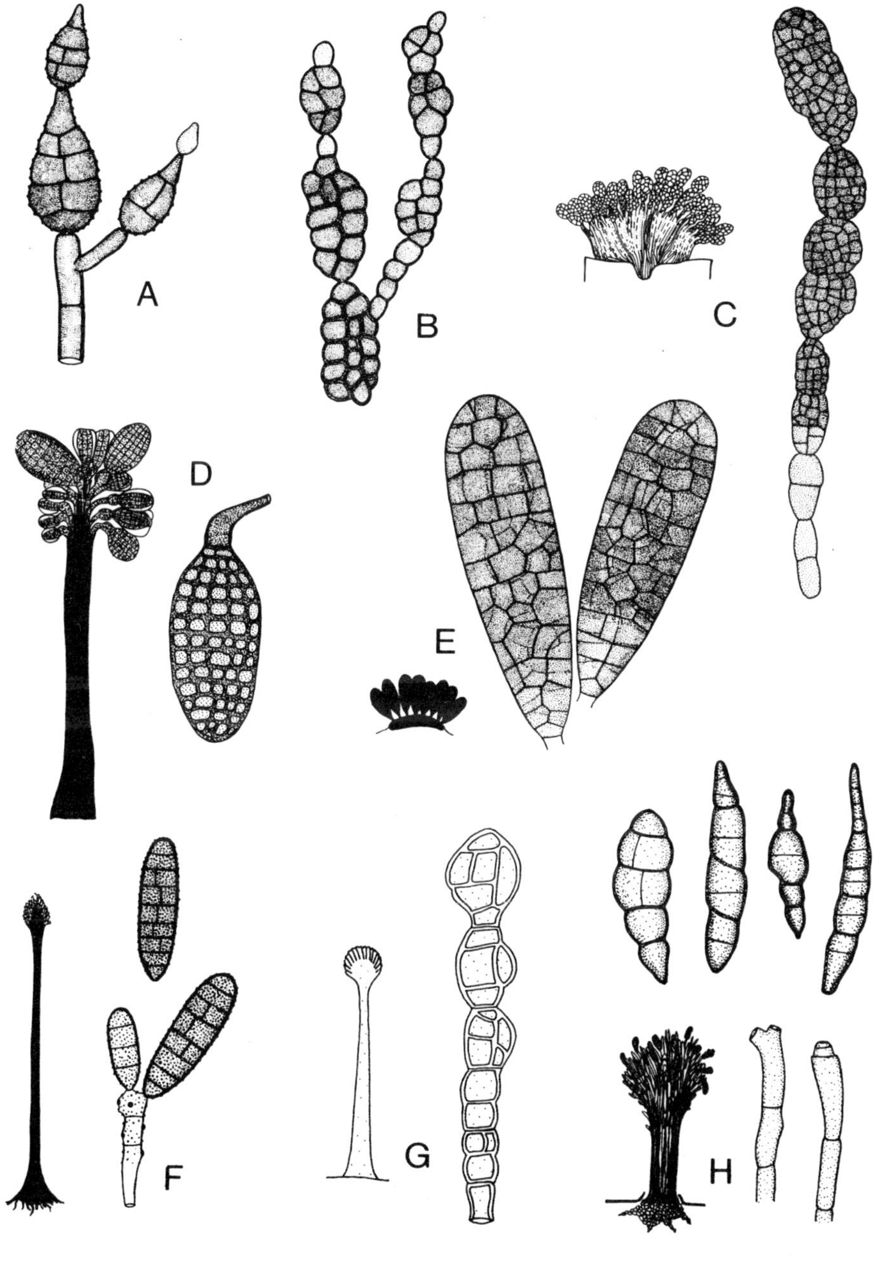

Plate 101: A *Alternaria alternata;* B *Peyronelia arbuscula;* C *Coniosporium olivaceum;* D *Kostermansinda magna;* E *Berkleasmium concinnum;* F *Sclerographium aterrimum;* G *Dictyocatenulata alba;* H *Pantospora guazumae.*

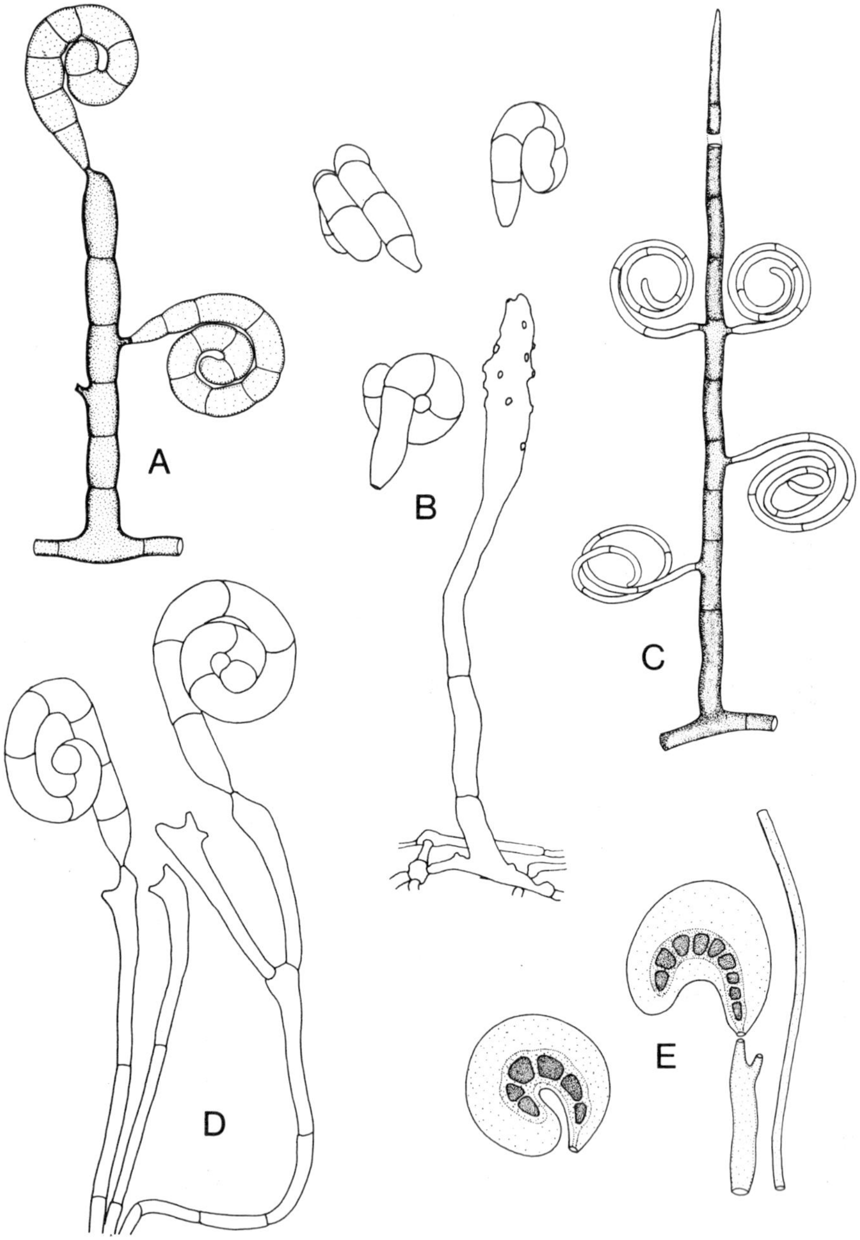

Plate 102: A *Helicoma muelleri;* B *Helicominopsis fici;* C *Helicosporium vegetum;* D *Hyalohelicomina deutziae;* E *Drepanoconis larvaeformis.*

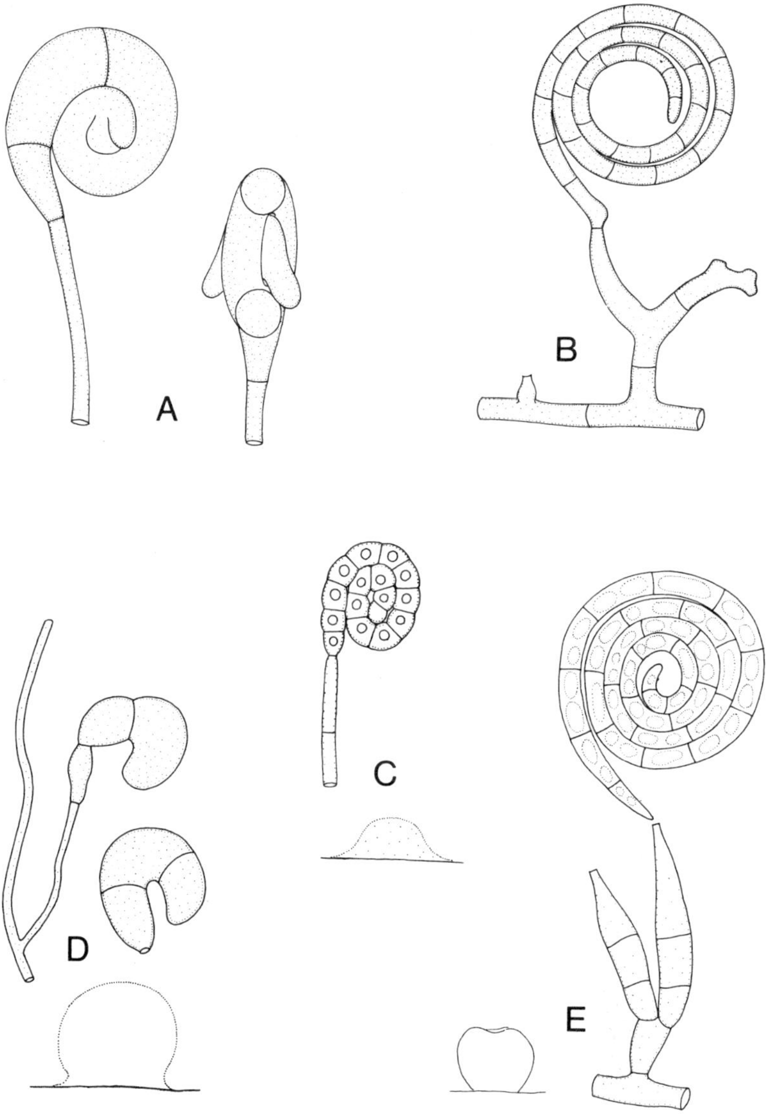

Plate 103: A *Diplorhynchus biloba;* B *Helicomyces roseus;* C *Everhartia hymenuloides;* D *Delortia palmicola;* E *Vanbeverwijkia spirospora.*

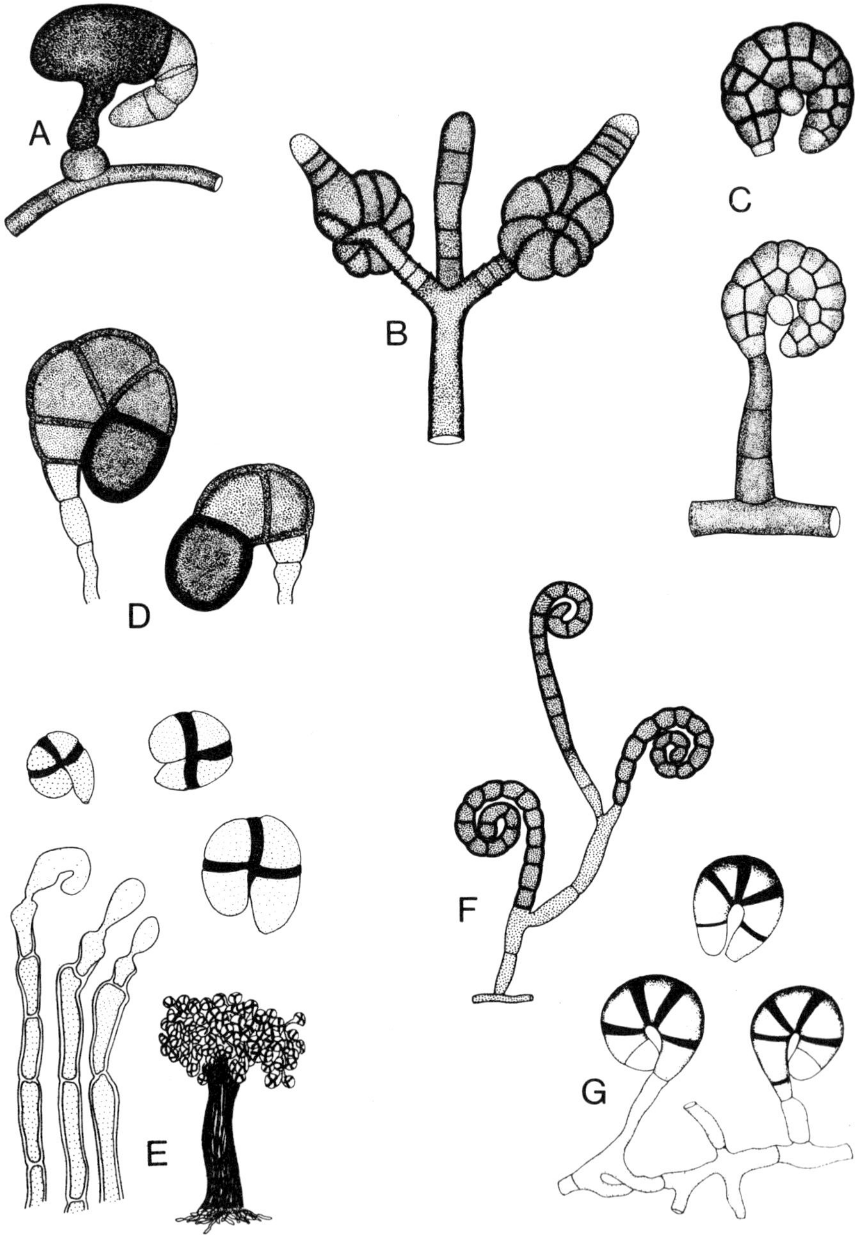

Plate 104: A *Curculiospora sydowii;* B *Circinoconis paradoxa;* C *Xenosporium berkeleyi;* D *Cirrenalia macrocephala;* E *Trochophora simplex;* F *Troposporella fumosa;* G *Troposporella monospora.*

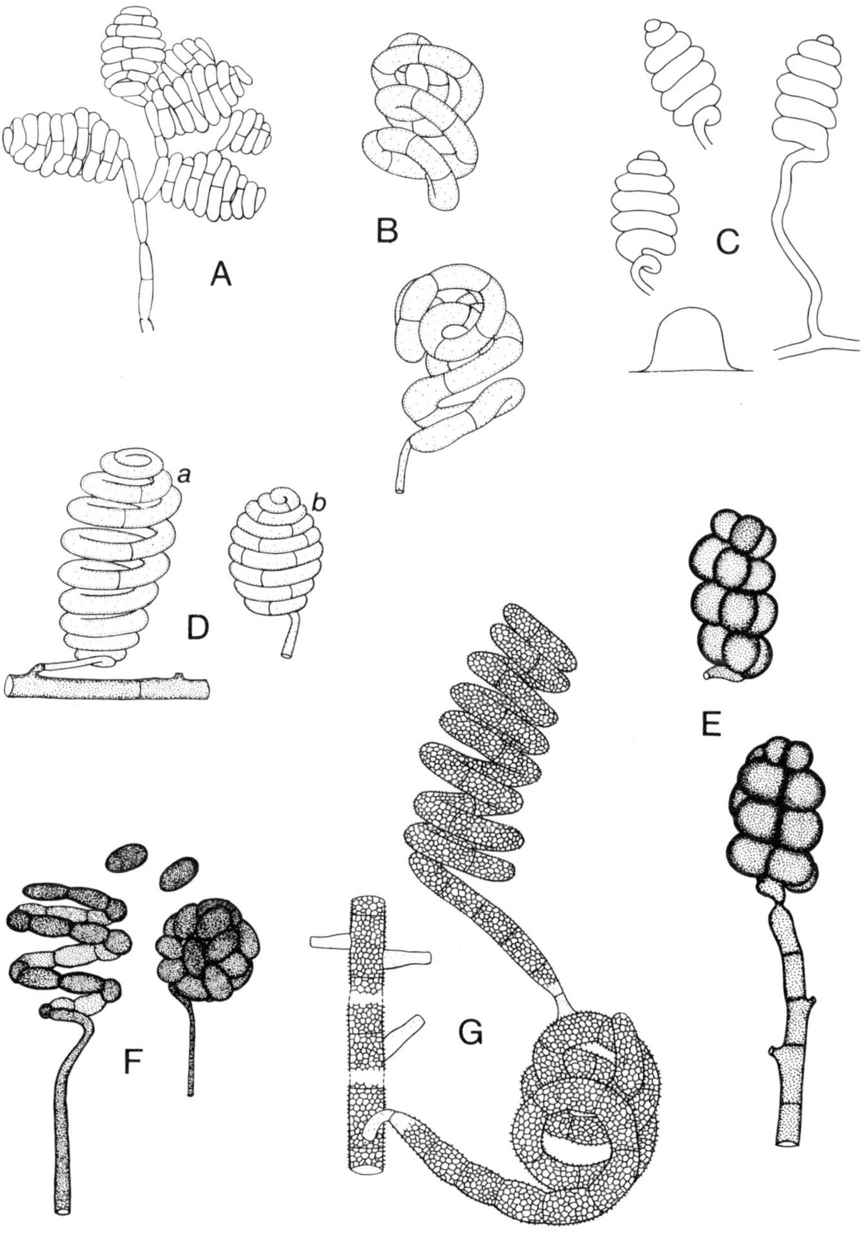

Plate 105: A *Helicodendron tubulosum;* B *Hobsonia mirabilis;* C *Troposporium album;* D *Helicoon* (a) *sessile* (b) *ellipticum;* E *Helicorhoidion botryoideum;* F *Helicocephalum sarcophilum;* G *Hiospira* state of *Brooksia tropicalis.*

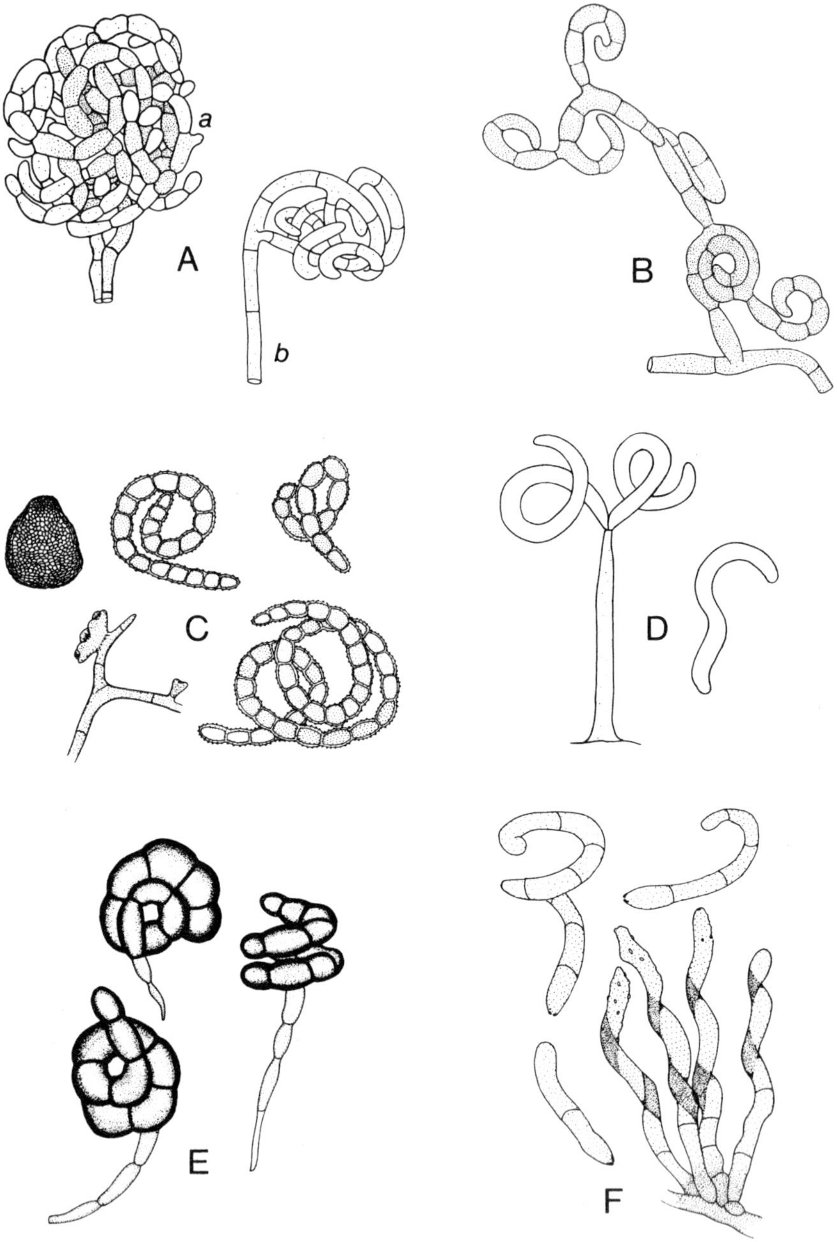

Plate 106: A *Spirosphaera* (a) *floriformis* (b) *beverwijkiana;* B *Helicodendron triglitziensis;* C *Dichotomophthoropsis nymphaearum;* D *Brachyhelicoon xylogenum;* E *Zalerion maritima;* F *Spiralum helicosporum.*

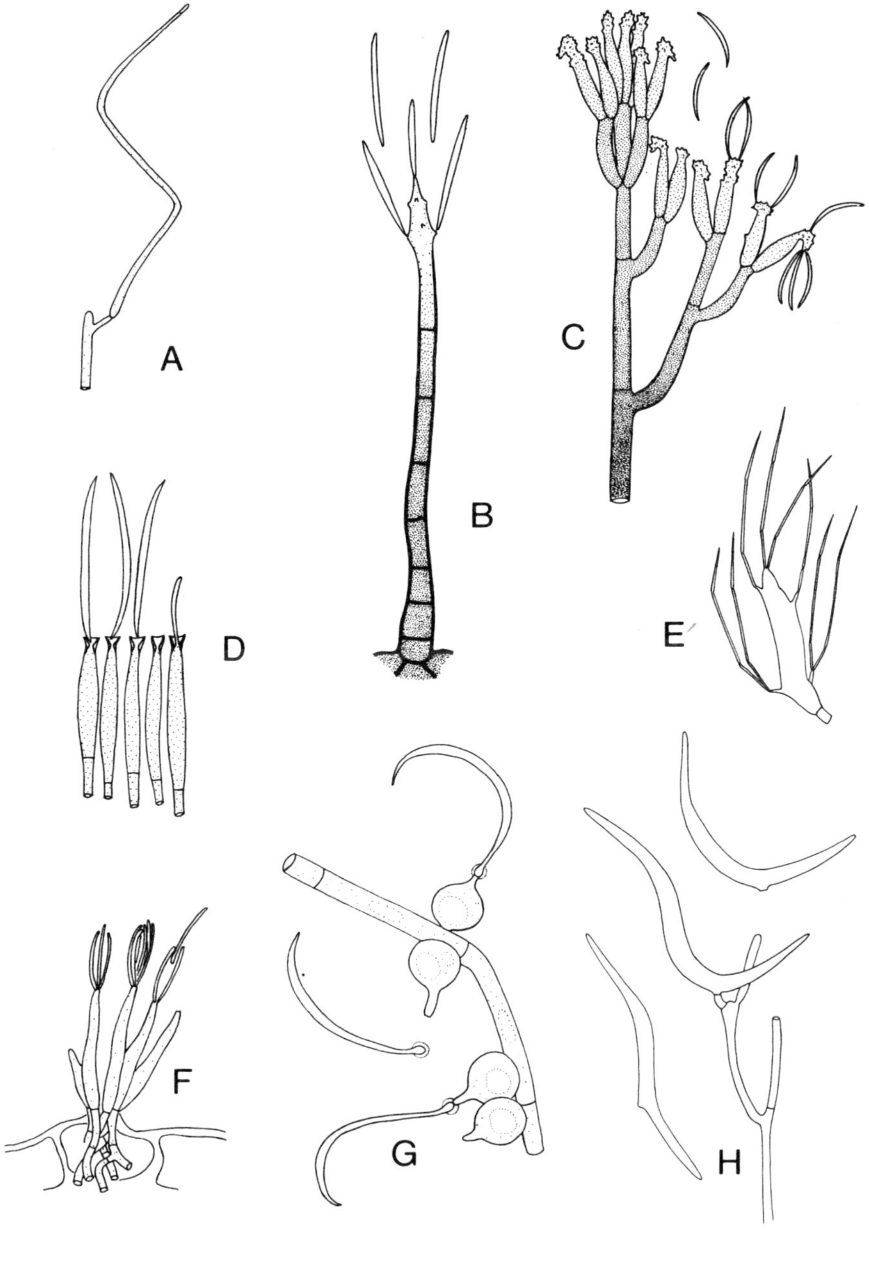

Plate 107: A *Angulospora aquatica;* B *Cercosporula corticola;* C *Harpographium* sp.; D *Kmetia exigua;* E *Ephelis mexicana;* F *Kmetiopsis hymenaeae;* G *Harposporium oxycoracum;* H *Lunulospora curvula.*

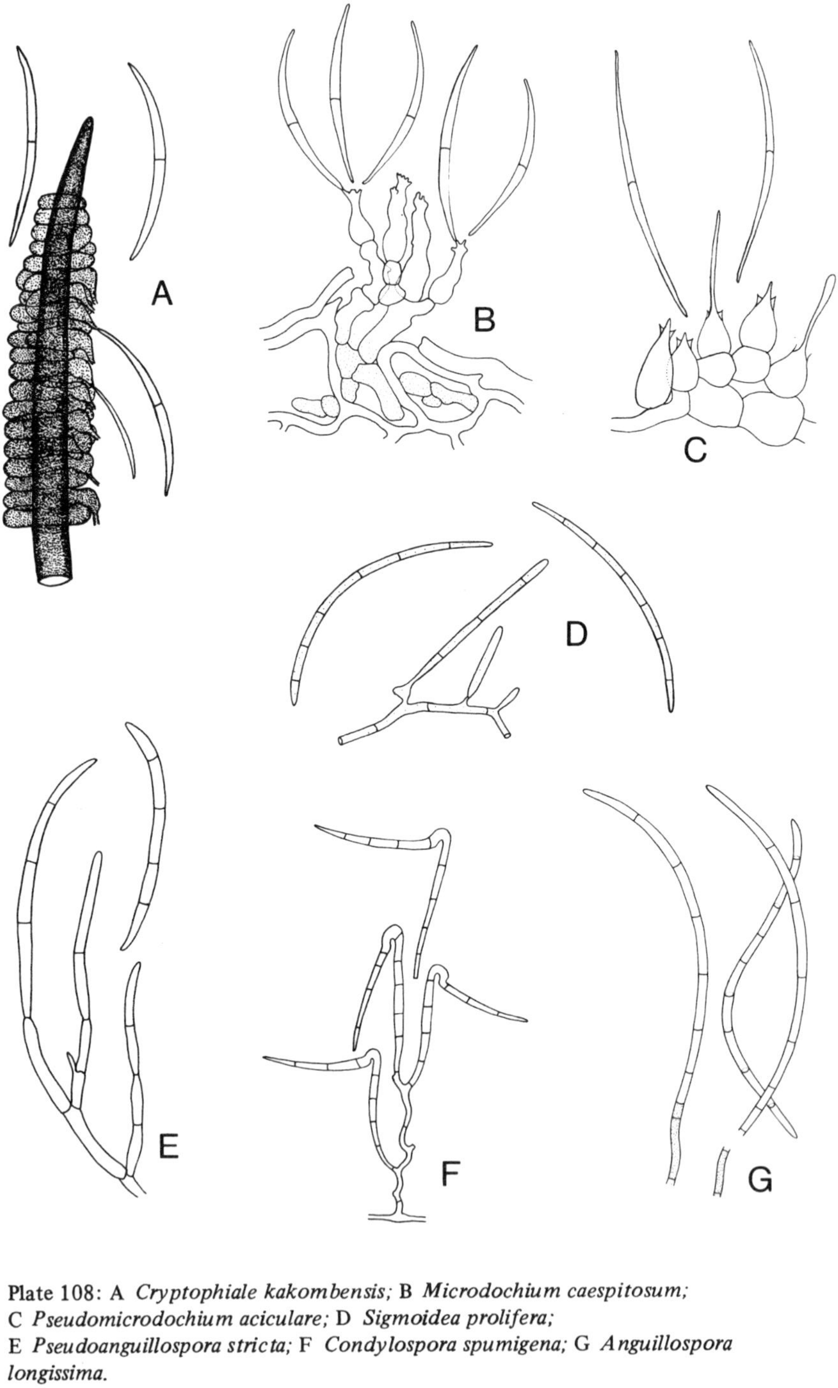

Plate 108: A *Cryptophiale kakombensis;* B *Microdochium caespitosum;* C *Pseudomicrodochium aciculare;* D *Sigmoidea prolifera;* E *Pseudoanguillospora stricta;* F *Condylospora spumigena;* G *Anguillospora longissima.*

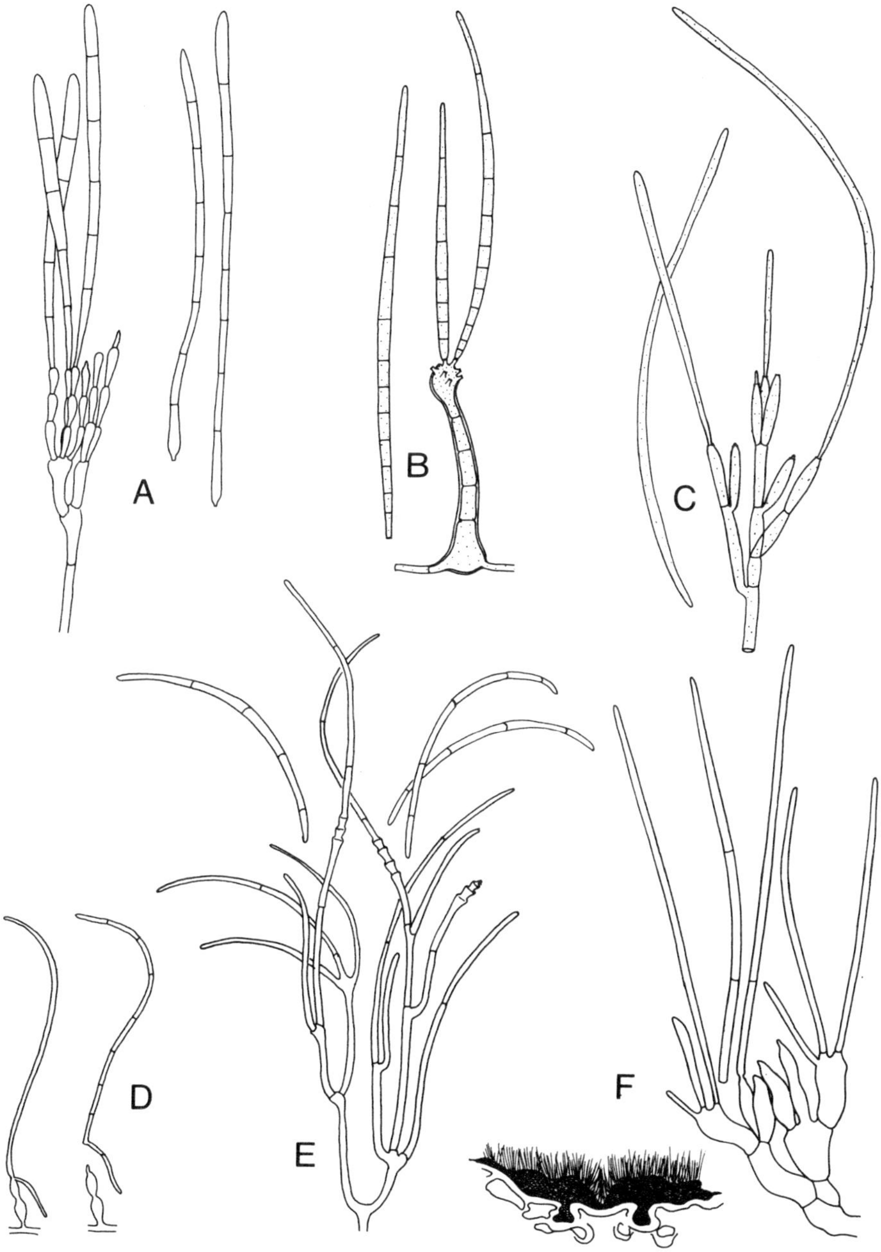

Plate 109: A *Filosporella aquatica;* B *Mirandina corticola;* C *Flagellospora curvula;* D *Calcarispora hiemalis;* E *Rogersia annelidica;* F *Linodochium hyalinum.*

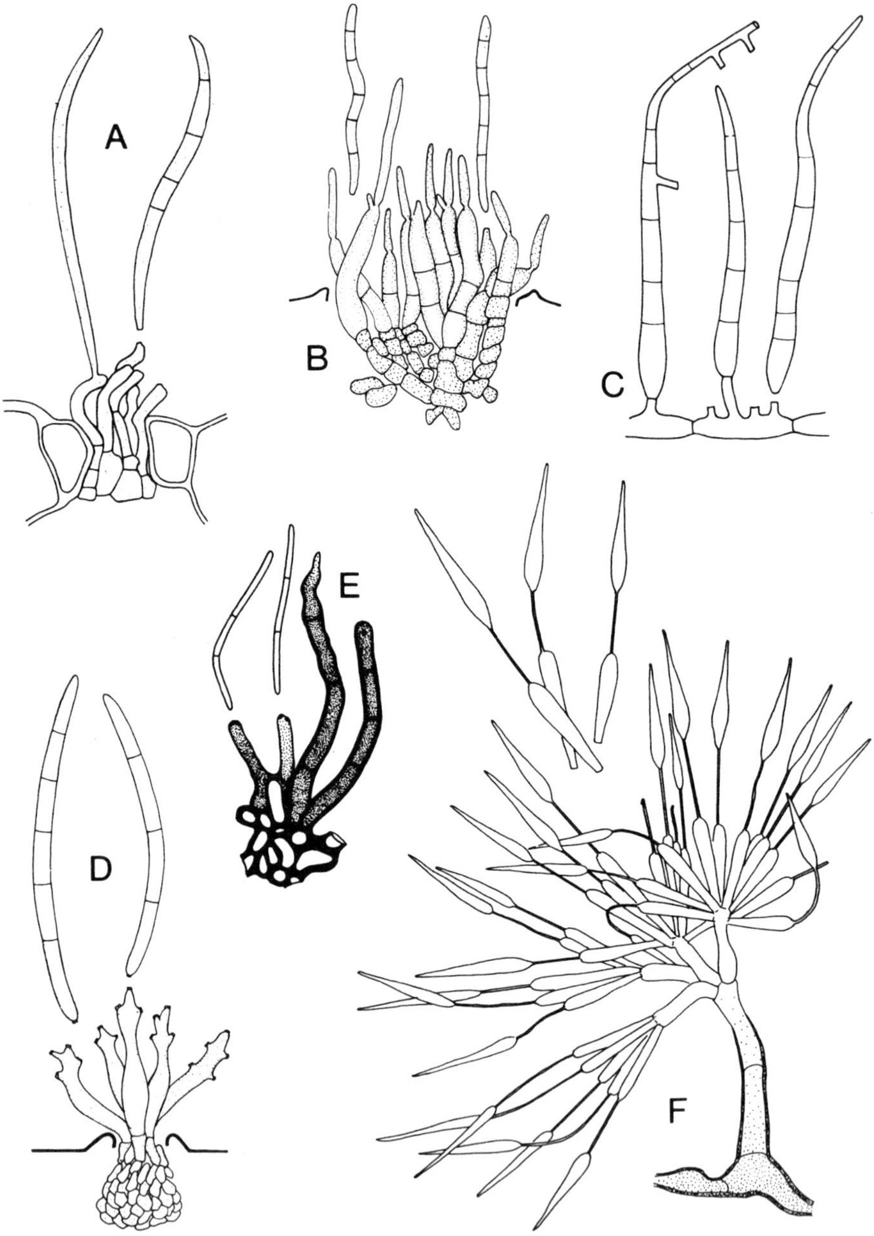

Plate 110: A *Ramulispora sorghi;* B *Cercoseptoria chamaesyceae;* C *Irpicomyces schiffnerulae;* D *Cercosporella virgaureae;* E *Schizotrichum lobeliae;* F *Pappimyces hastatus.*

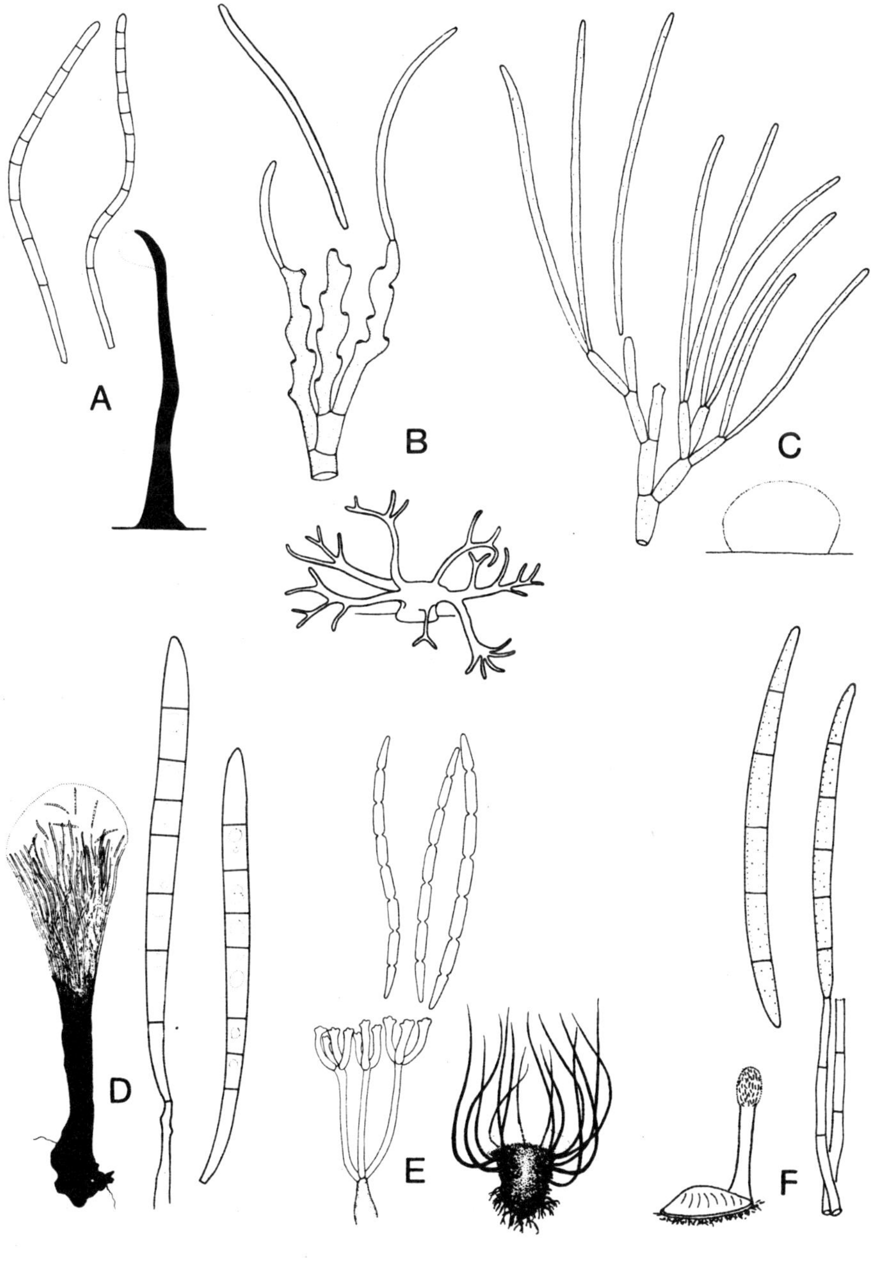

Plate 111: A *Microlychnus epicorticis;* B *Pterulopsis dummeri;* C *Linodochium hyalinum;* D *Morrisographium pilosum;* E *Wiesneriomyces javanicus;* F *Atractium flammeum.*

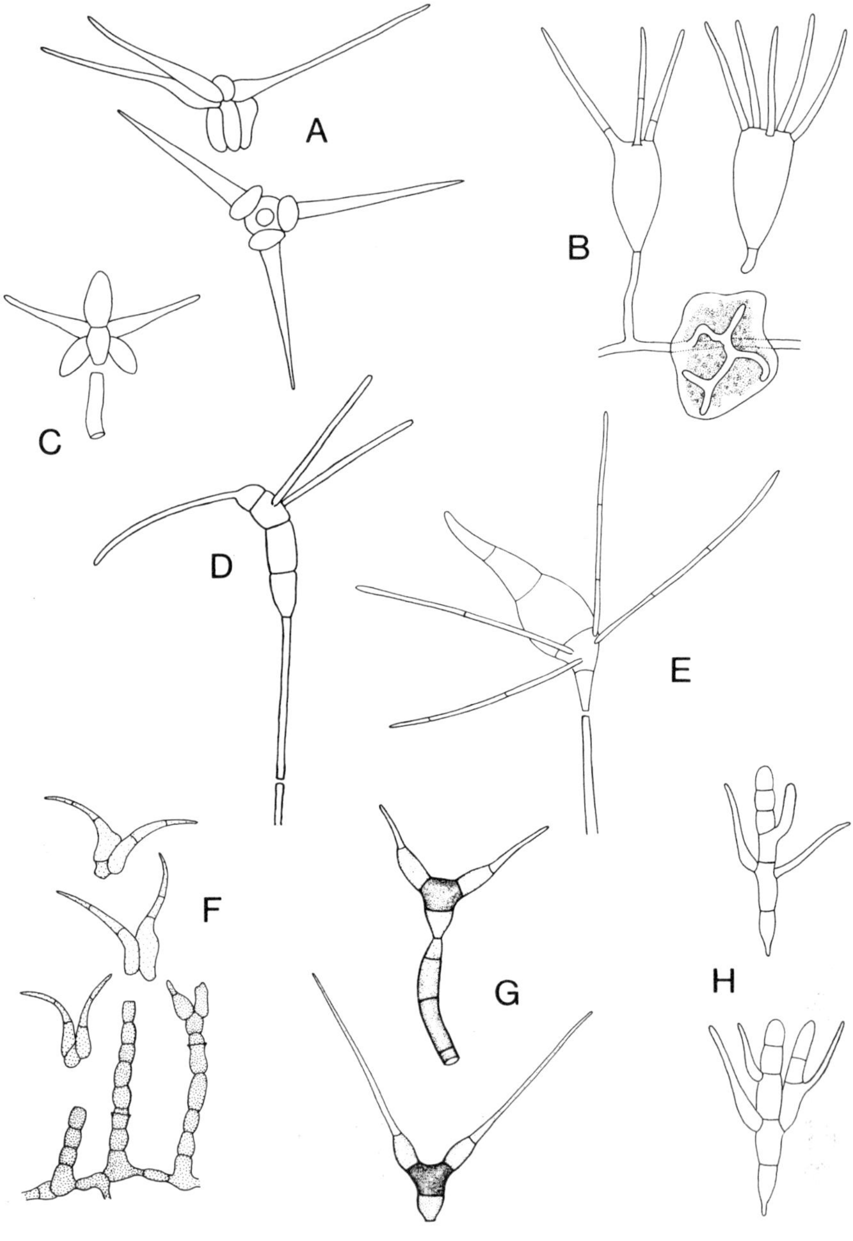

Plate 112: A *Titaea triradiata;* B *Acaulopage tetraceros;* C *Titaea hemileiae;* D *Culicidospora aquatica;* E *Laridospora appendiculata;* F *Biflagellospora japonica;* G *Iyengarina elegans;* H *Tetracladium setigerum.*

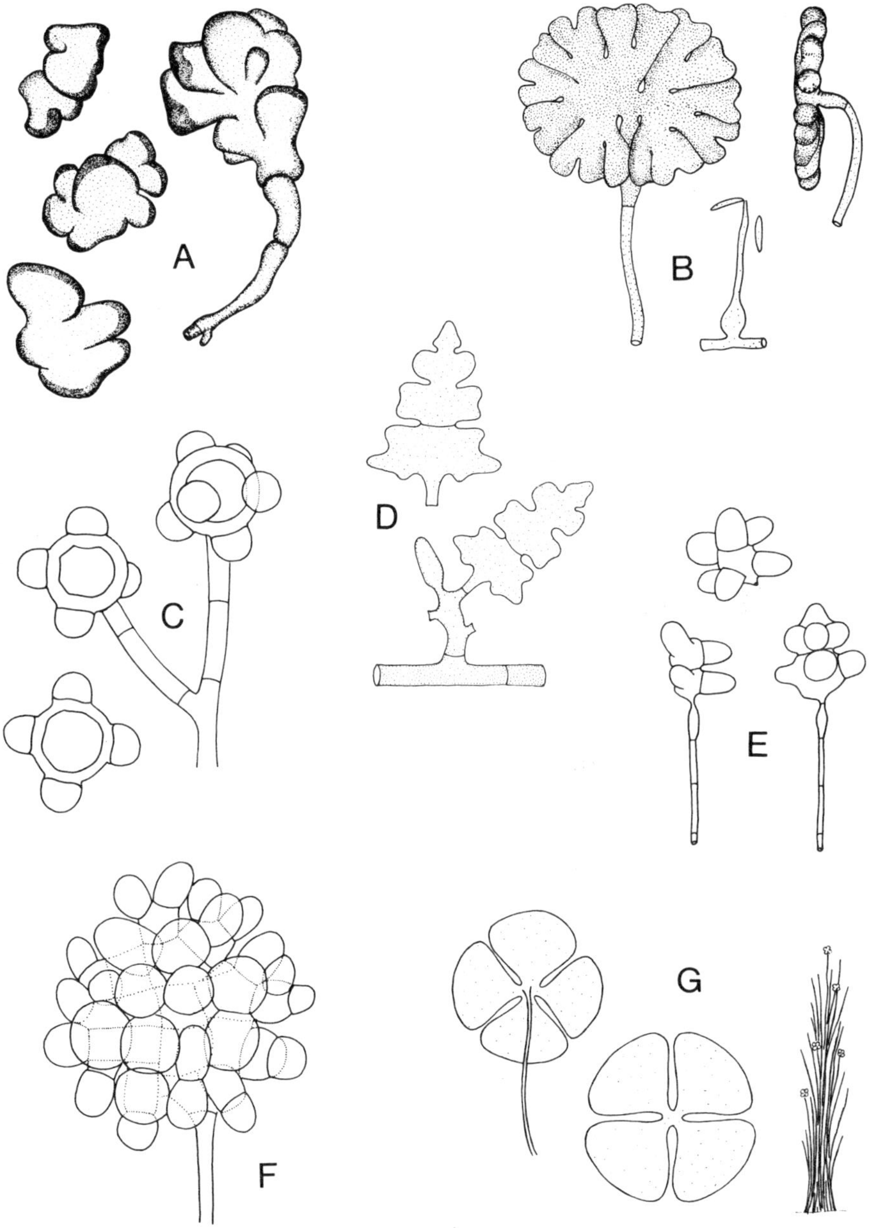

Plate 113: A *Milospium graphideorum;* B *Desmidiospora myrmecophila;* C *Stephanoma strigosum;* D *Dendrosporium lobatum;* E *Pyramidospora casuarinae;* F *Uvarispora lignicola;* G *Riessia semiophora.*

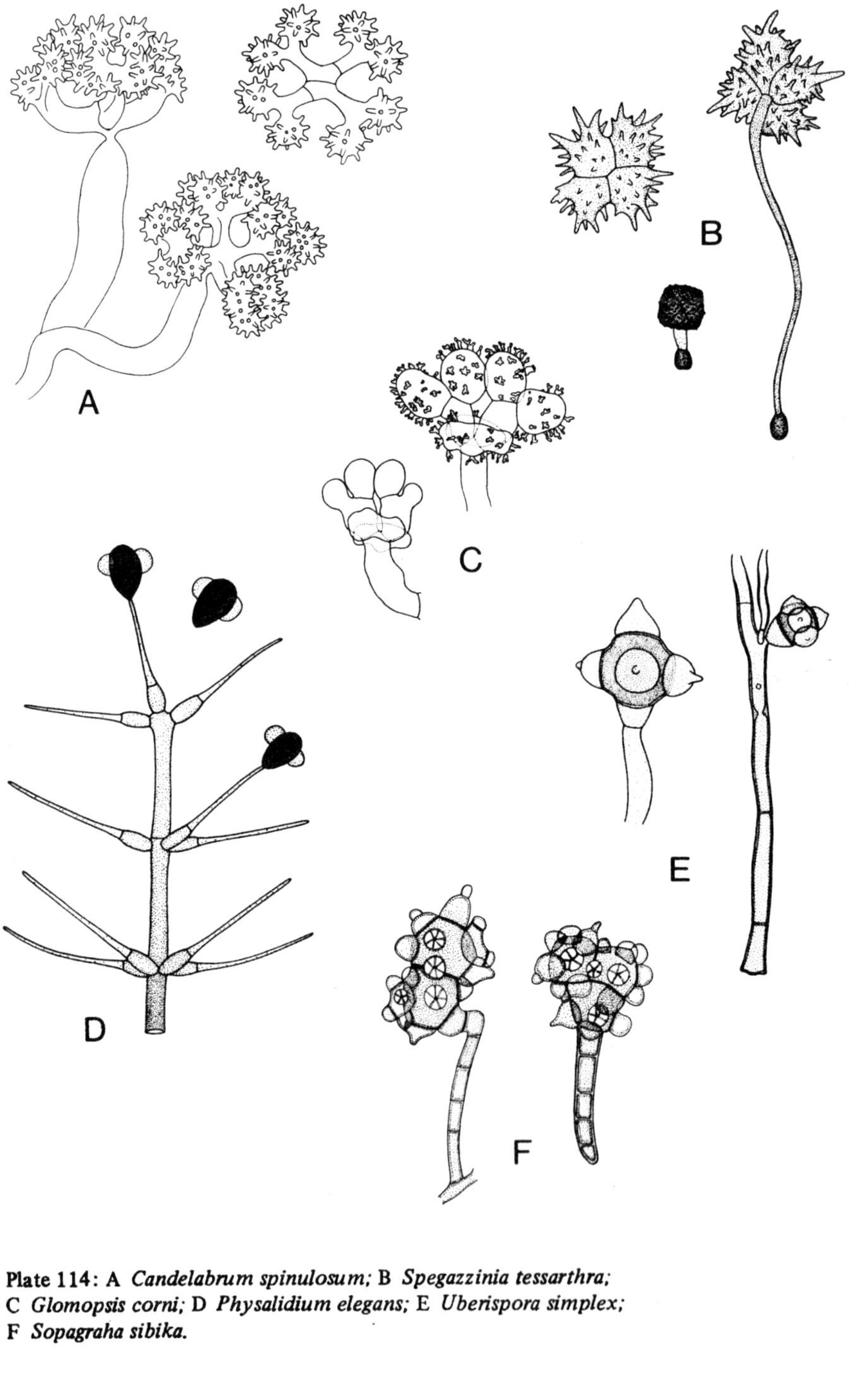

Plate 114: A *Candelabrum spinulosum;* B *Spegazzinia tessarthra;* C *Glomopsis corni;* D *Physalidium elegans;* E *Uberispora simplex;* F *Sopagraha sibika.*

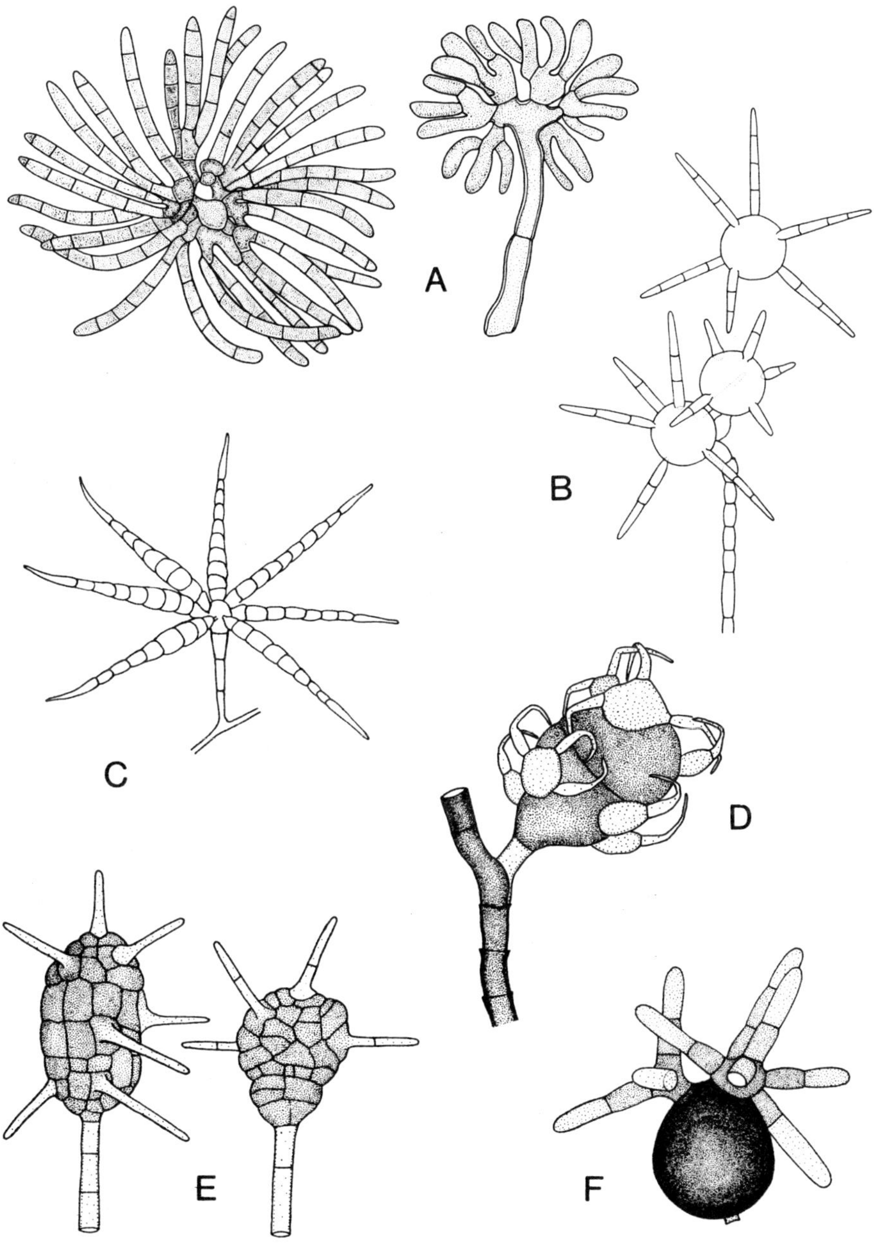

Plate 115: A *Psammina stipitata;* B *Brachiosphaera tropicalis;* C *Flabellospora verticillata;* D *Arachnophora fagicola;* E *Petrakia echinata;* F *Orbimyces spectabilis.*

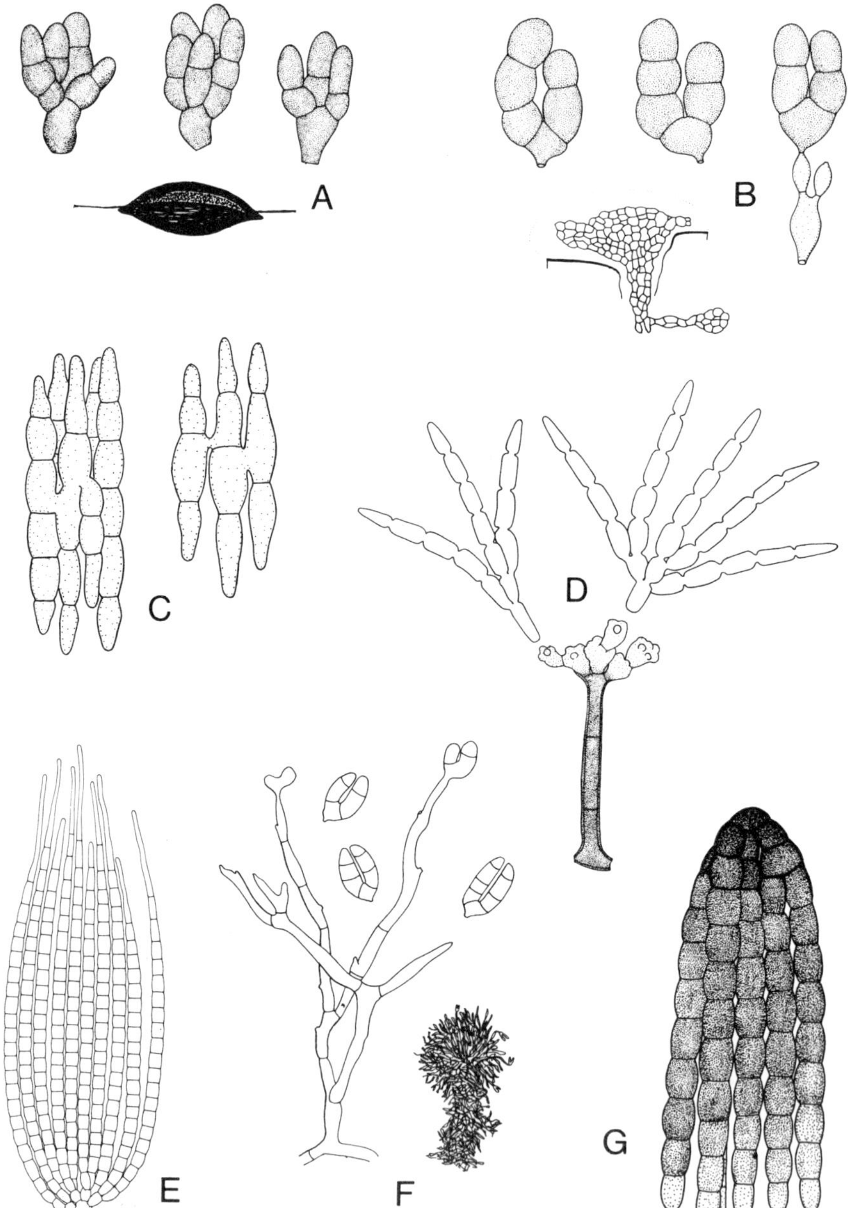

Plate 116: A *Cheiromyces stellatus;* B *Cheiromycella microscopica;* C *Amallospora dacrydion;* D *Speiropsis pedatospora;* E *Petrakiopsis elegans;* F *Dicranidion fragile;* G *Cryptocoryneum condensatum.*

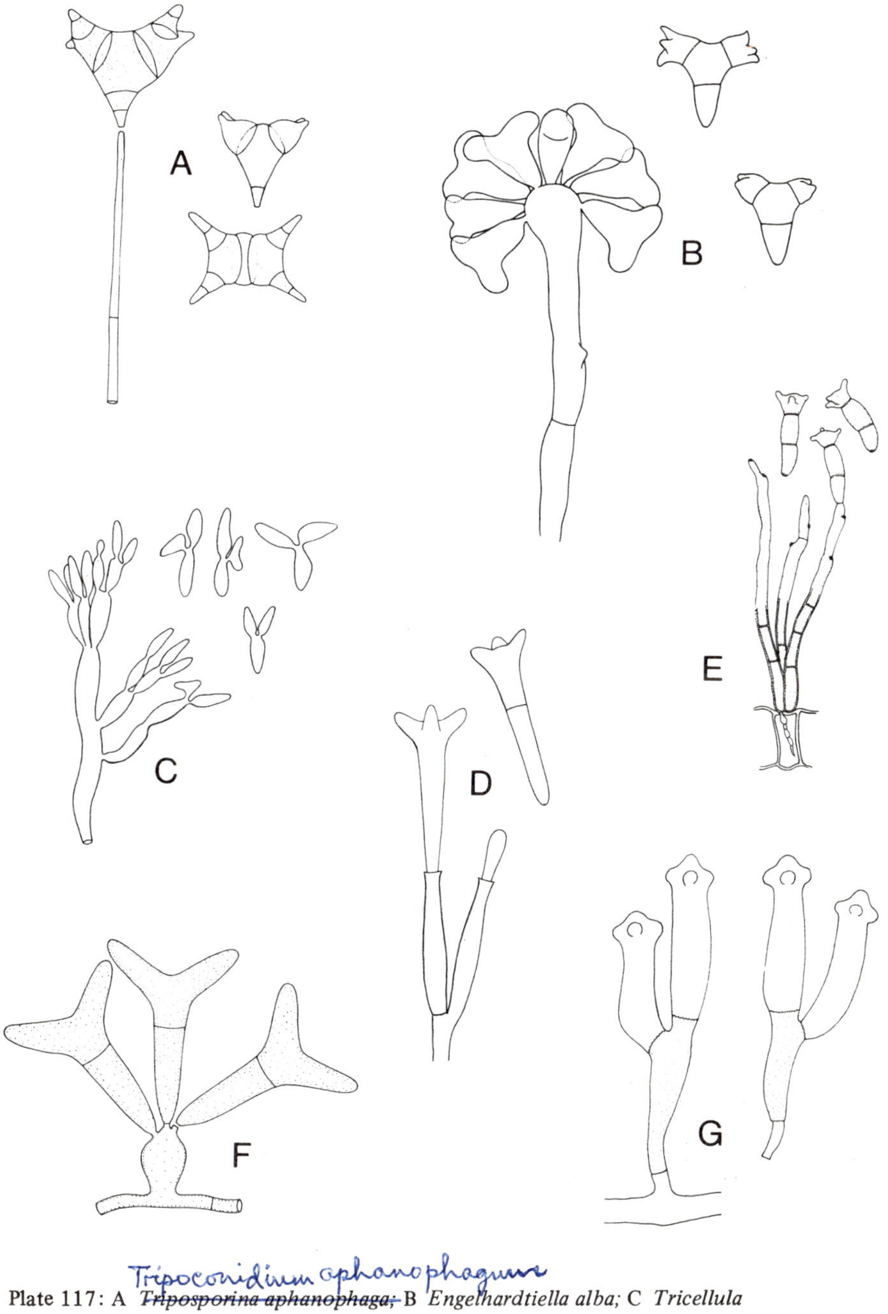

Plate 117: A ~~*Triposporina aphanophaga;*~~ [*Tripoconidium aphanophagum*] B *Engelhardtiella alba;* C *Tricellula aquatica;* D *Heliscus lugdunensis;* E *Coronospora dendrocalami;* F *Scolecobasidium terreum;* G *Diademospora ramigera.*

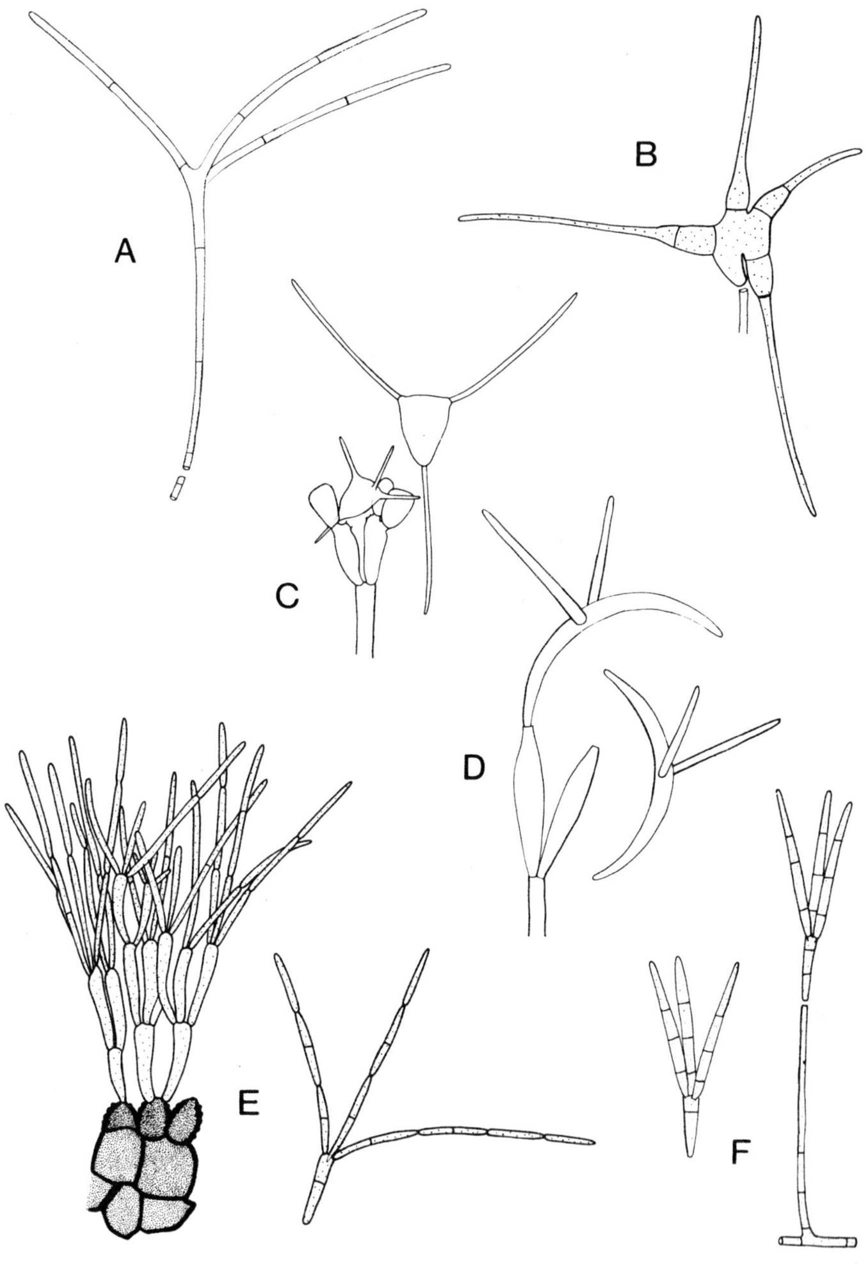

Plate 118: A *Tetrachaetum elegans;* B *Campylospora chaetocladia;* C *Clavariana aquatica;* D *Alatospora acuminata;* E *Kazulia vagans;* F *Tridentaria carnivora.*

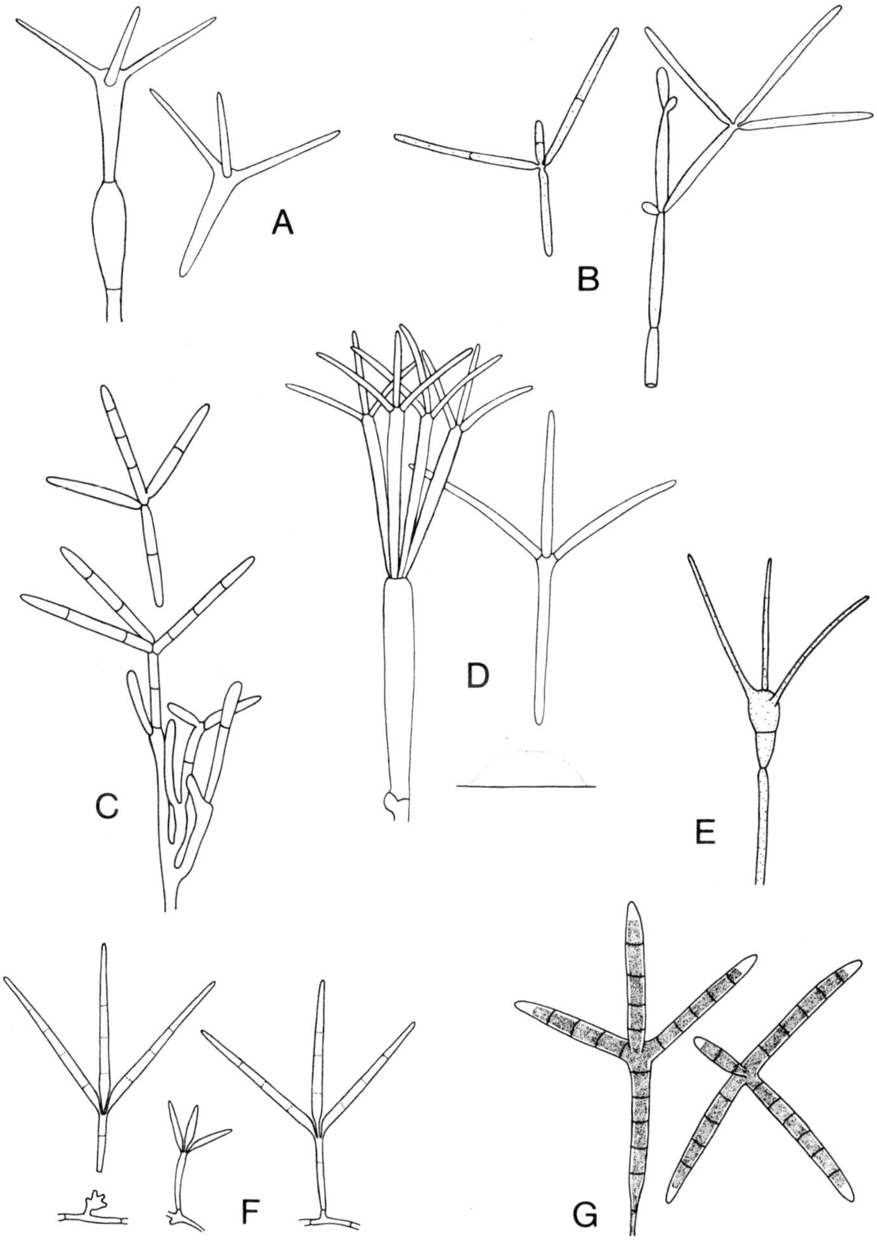

Plate 119: A *Clavatospora longibrachiata;* B *Articulospora tetracladia;* C *Fontanospora eccentrica;* D *Digitatispora marina;* E *Clavariopsis aquatica;* F *Isthmotricladia laeensis;* G *Tetrabrunneospora ellisii.*

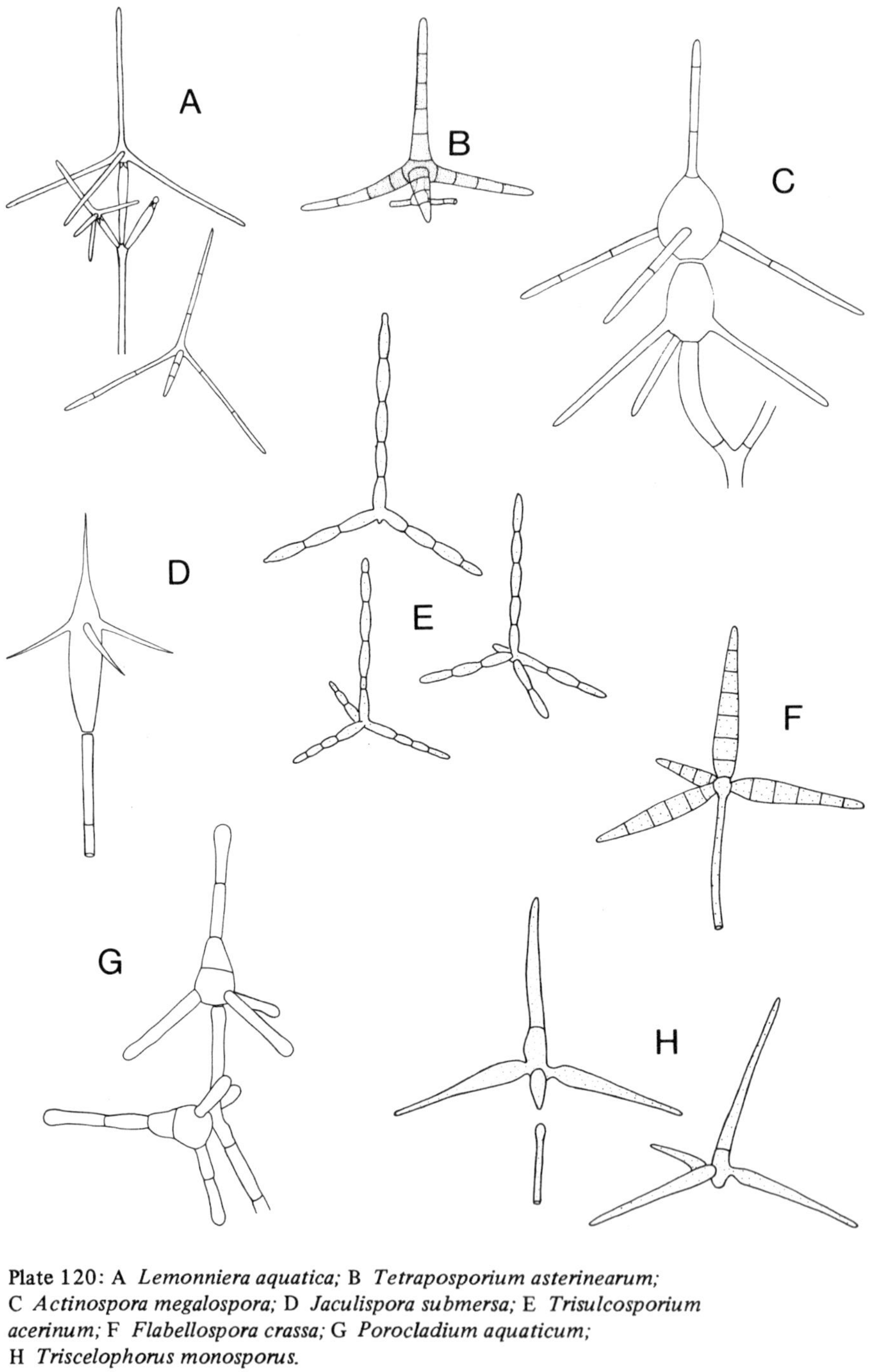

Plate 120: A *Lemonniera aquatica;* B *Tetraposporium asterinearum;* C *Actinospora megalospora;* D *Jaculispora submersa;* E *Trisulcosporium acerinum;* F *Flabellospora crassa;* G *Porocladium aquaticum;* H *Triscelophorus monosporus.*

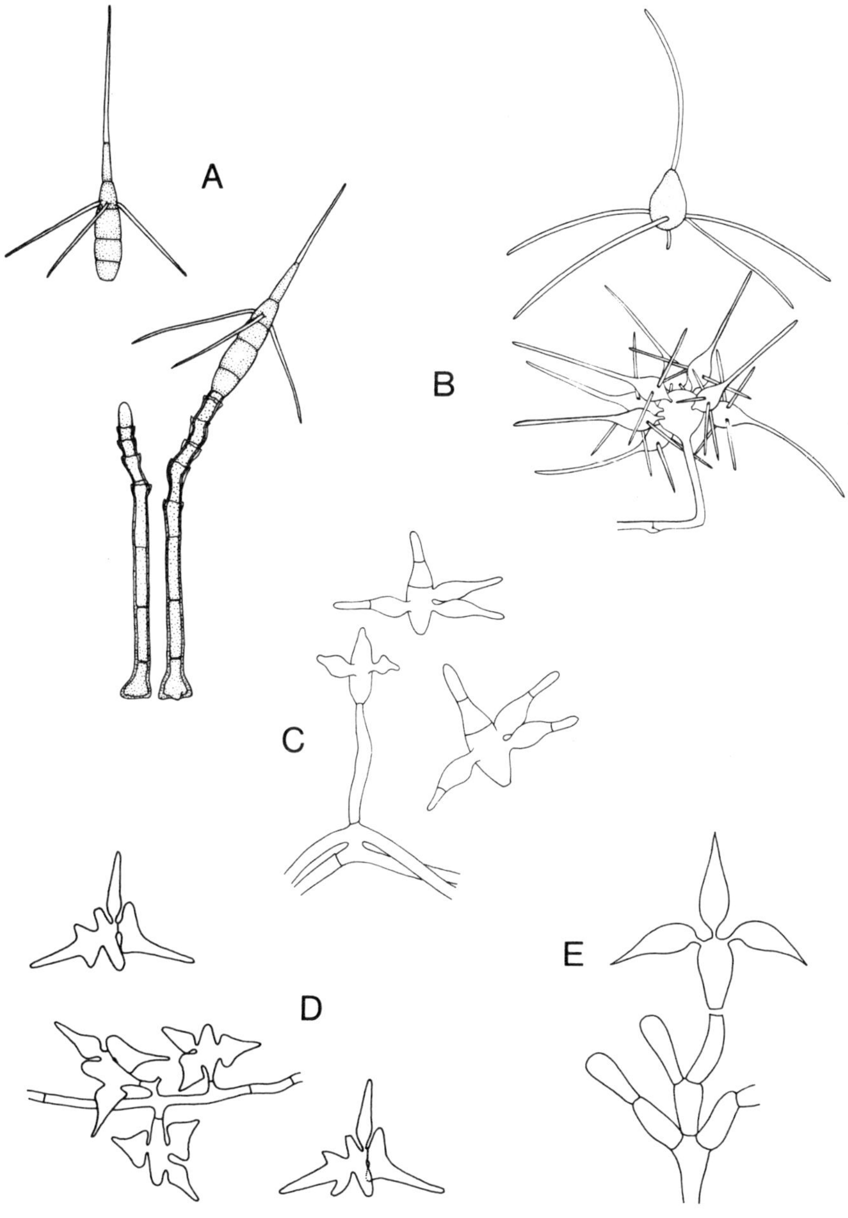

Plate 121: A *Chaetendophragmia triangularia;* B *Nia vibrissa;* C *Triramulispora obclavata;* D *Lateriramulosa uniinflata;* E *Titaea ornithomorpha.*

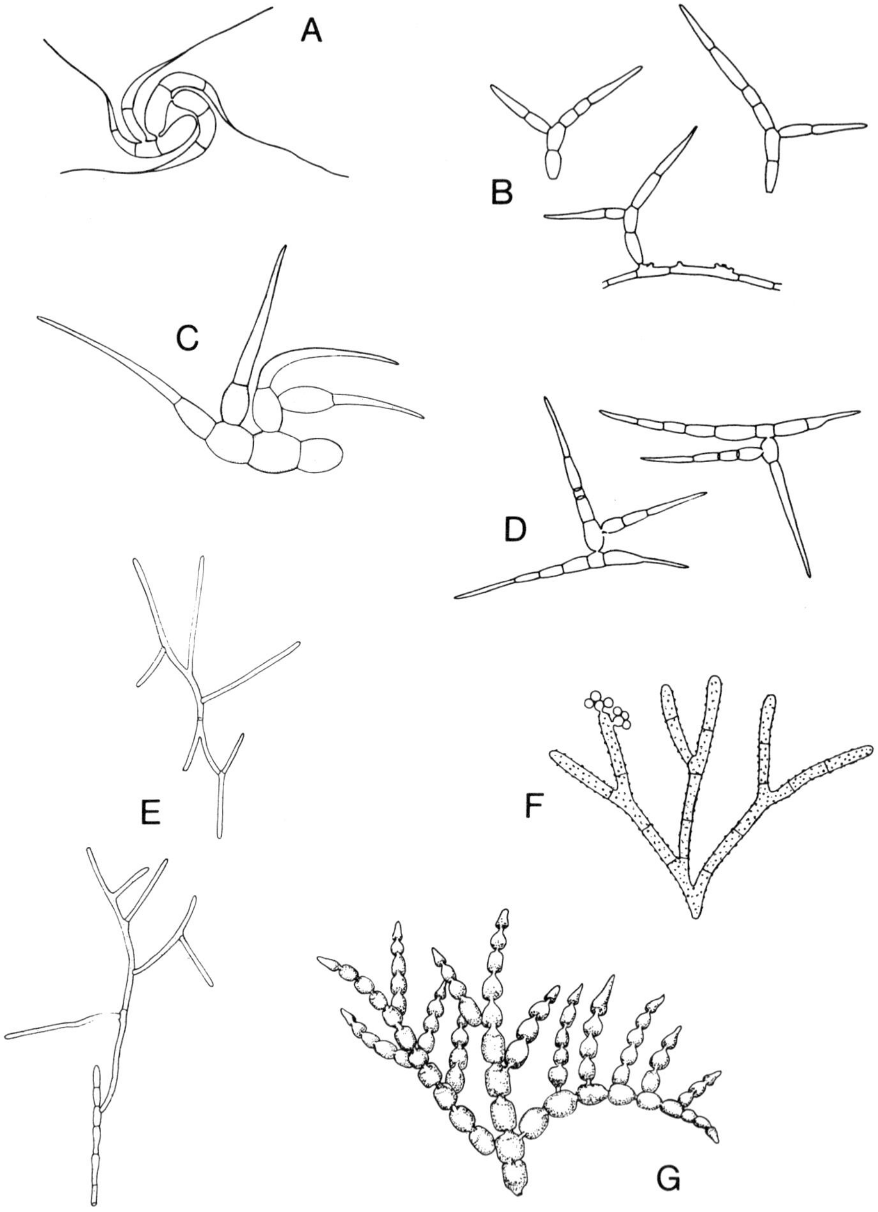

Plate 122: A *Gyoerffiella rotula;* B *Lambdasporium wauense;* C *Gyoerffiella craginiformis;* D *Pleuropedium tricladioides;* E *Polycladium equiseti;* F *Thallospora aspera;* G *Atichia* state of *Seuratia millardetii.*

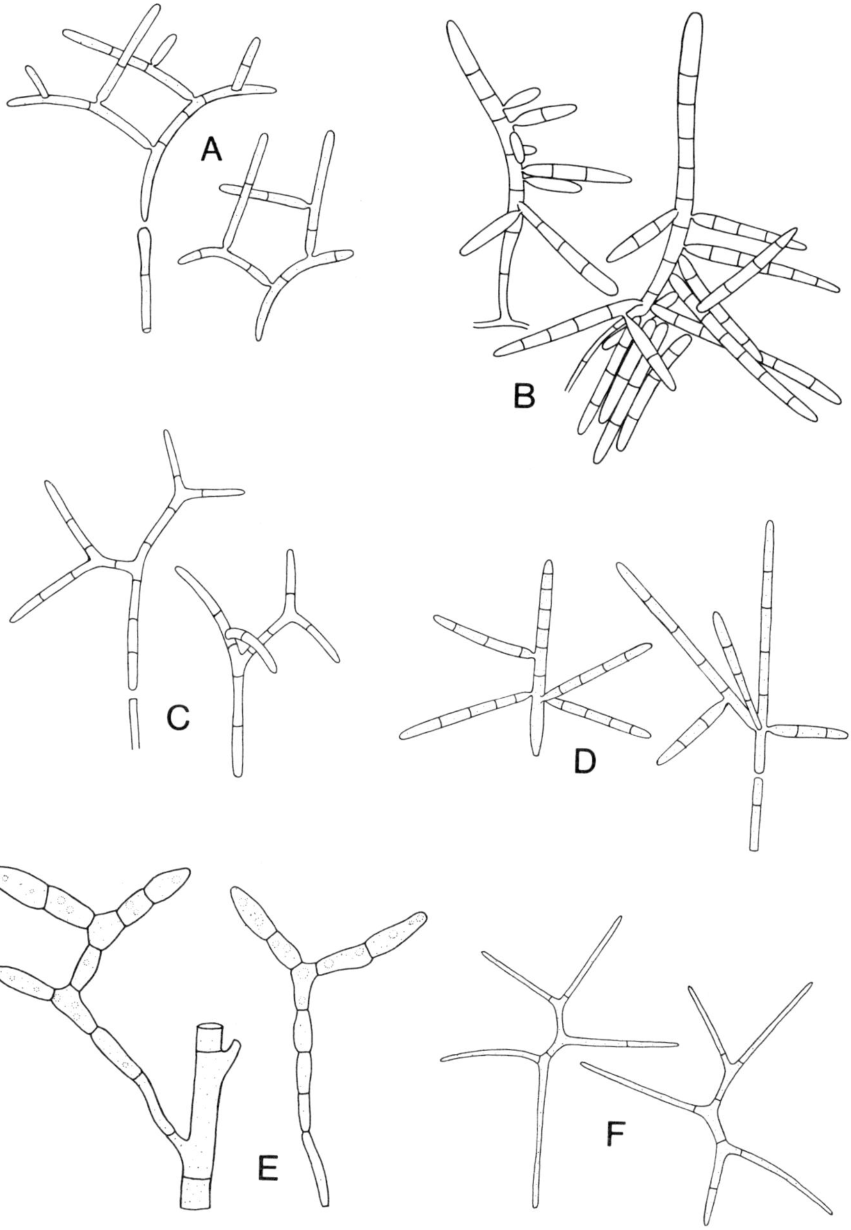

Plate 123: A *Varicosporium elodeae;* B *Dendrosporomyces prolifer;* C *Scorpiosporium minutum;* D *Dendrospora erecta;* E *Varicosporina ramulosa;* F *Tricladium anomalum.*

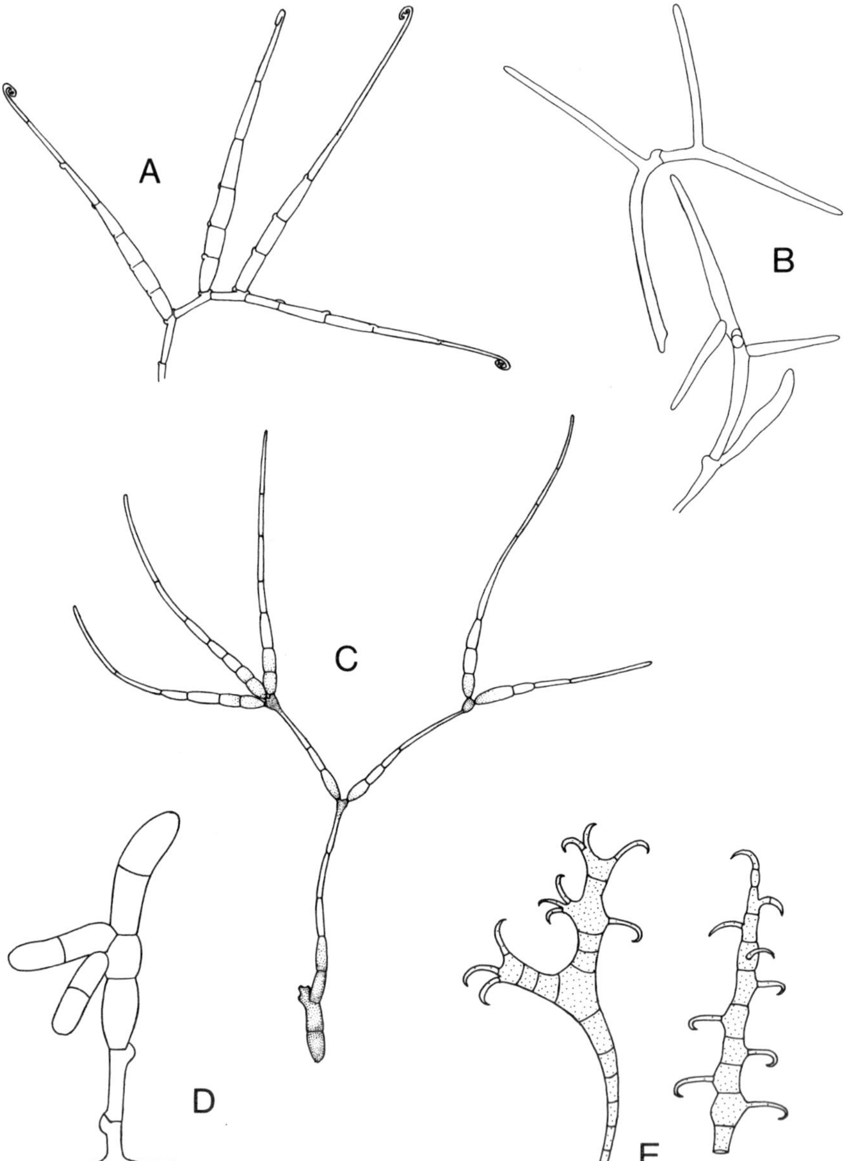

Plate 124: A *Ingoldiella hamata;* B *Taeniospora gracilis;* C *Catenosubulispora honiaraensis;* D *Titaeella capnophila;* E *Harpagomyces lomnicki.*

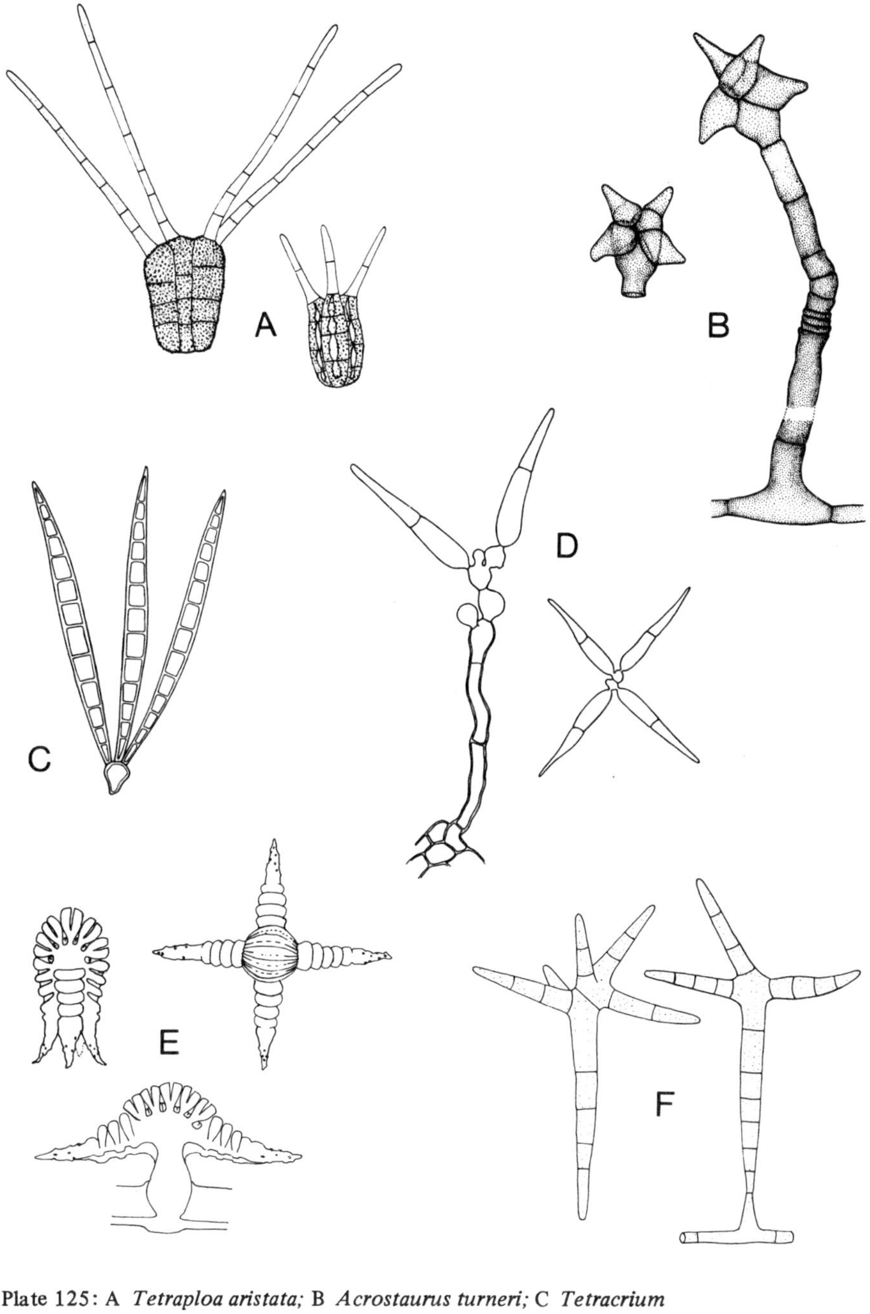

Plate 125: A *Tetraploa aristata;* B *Acrostaurus turneri;* C *Tetracrium aurantii;* D *Spermatoloncha maticola;* E *Valdensia heterodoxa;* F *Trinacrium subtile.*

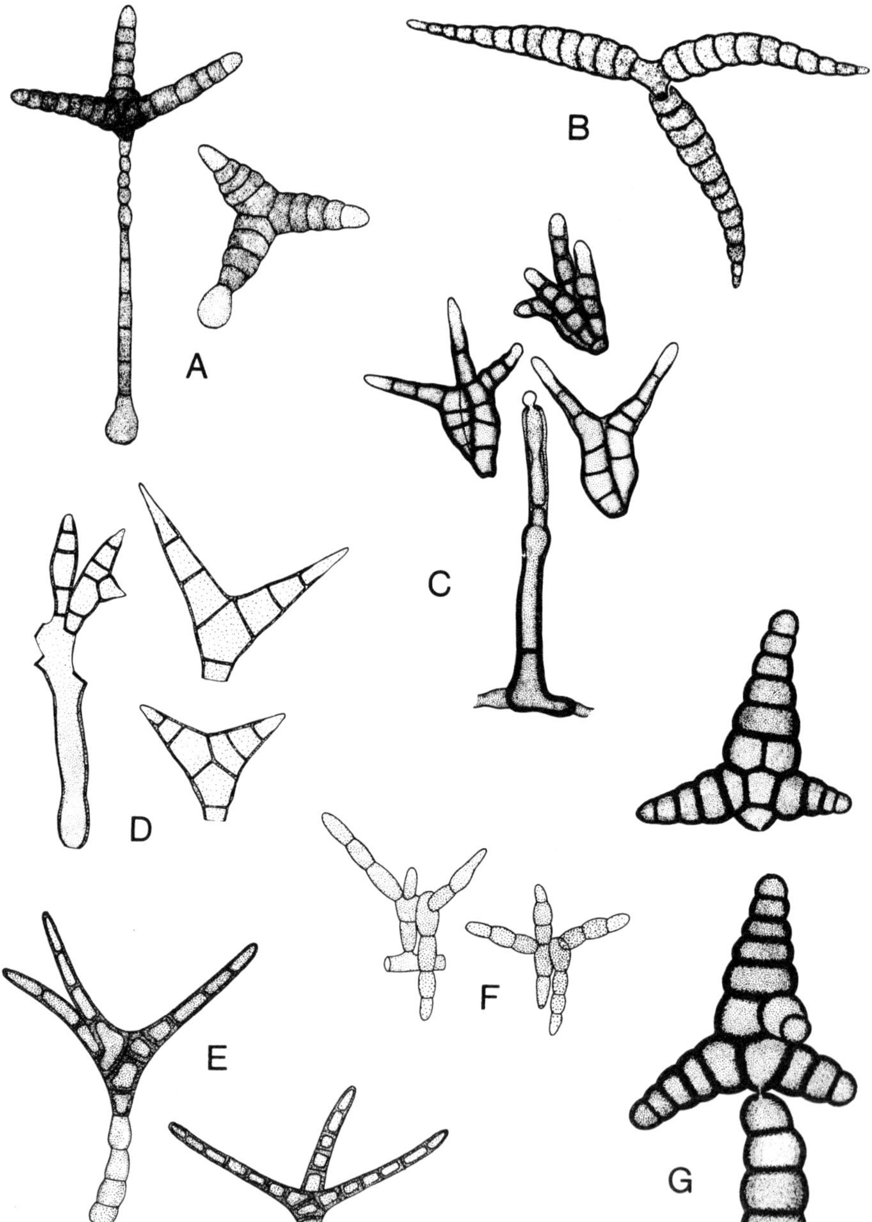

Plate 126: A *Triposporium elegans;* B *Ceratosporium fuscescens;* C *Hughesinia chusqueae;* D *Diplocladiella scalaroides;* E *Actinocladium rhodosporum;* F *Tripospermum myrti;* G *Hormiokrypsis* state of *Ophiocapnocoma phloiophilia.*

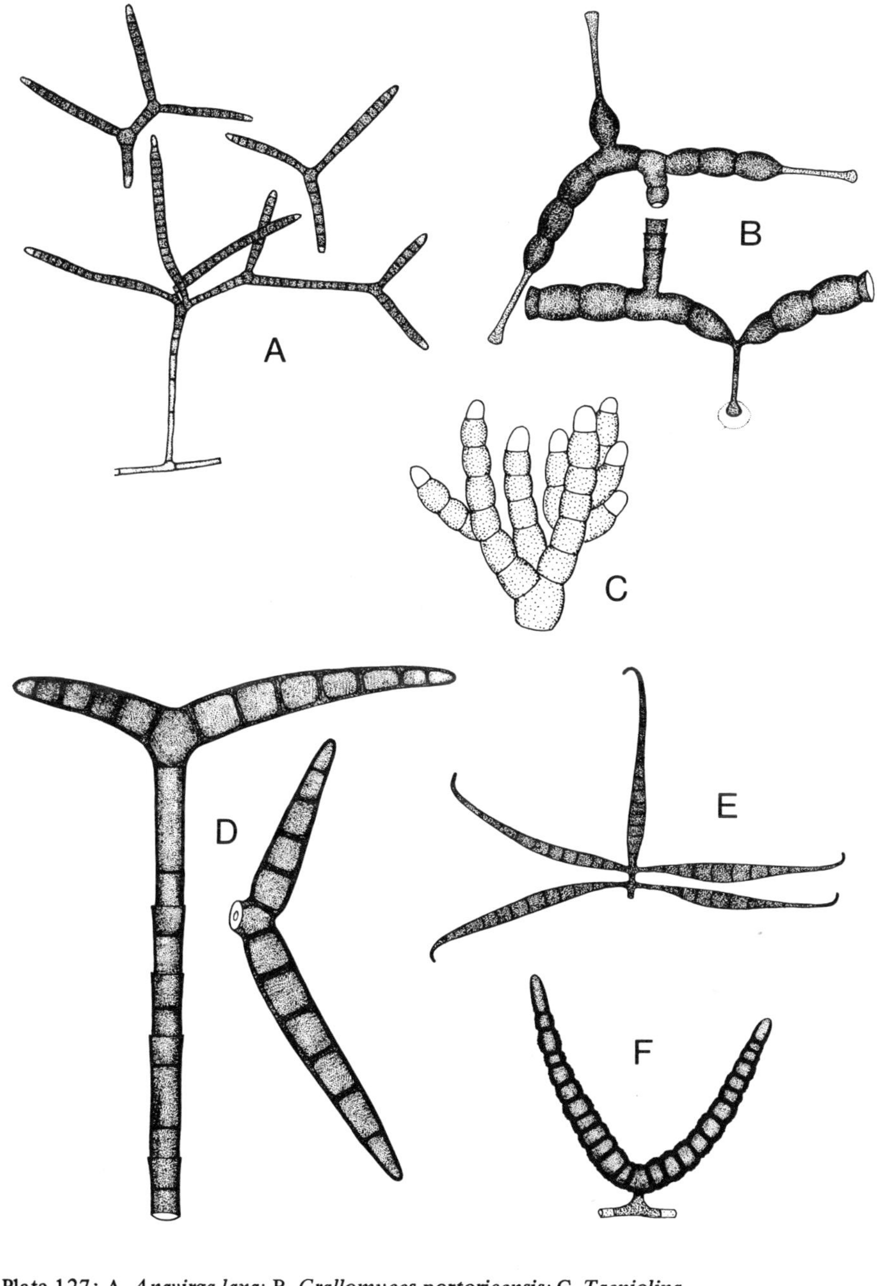

Plate 127: A *Anavirga laxa;* B *Grallomyces portoricensis;* C *Taeniolina centaurii;* D *Ceratosporella bicornis;* E *Casaresia sphagnorum;* F *Hirudinaria macrospora.*

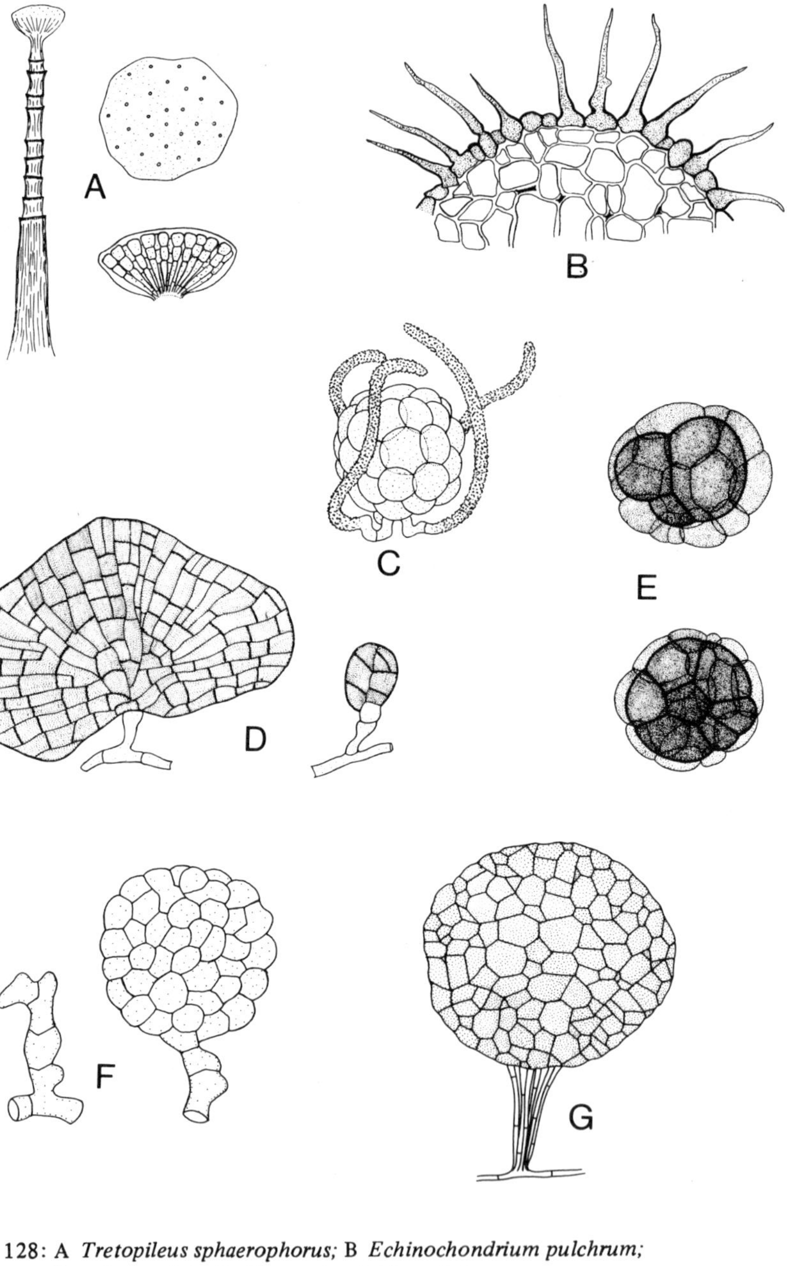

Plate 128: A *Tretopileus sphaerophorus;* B *Echinochondrium pulchrum;* C *Peyronelina glomerulata;* D *Mycoenterolobium platysporum;* E *Papulaspora sepedonioides;* F *Burgoa verzuoldiana;* G *Minimedusa polyspora.*

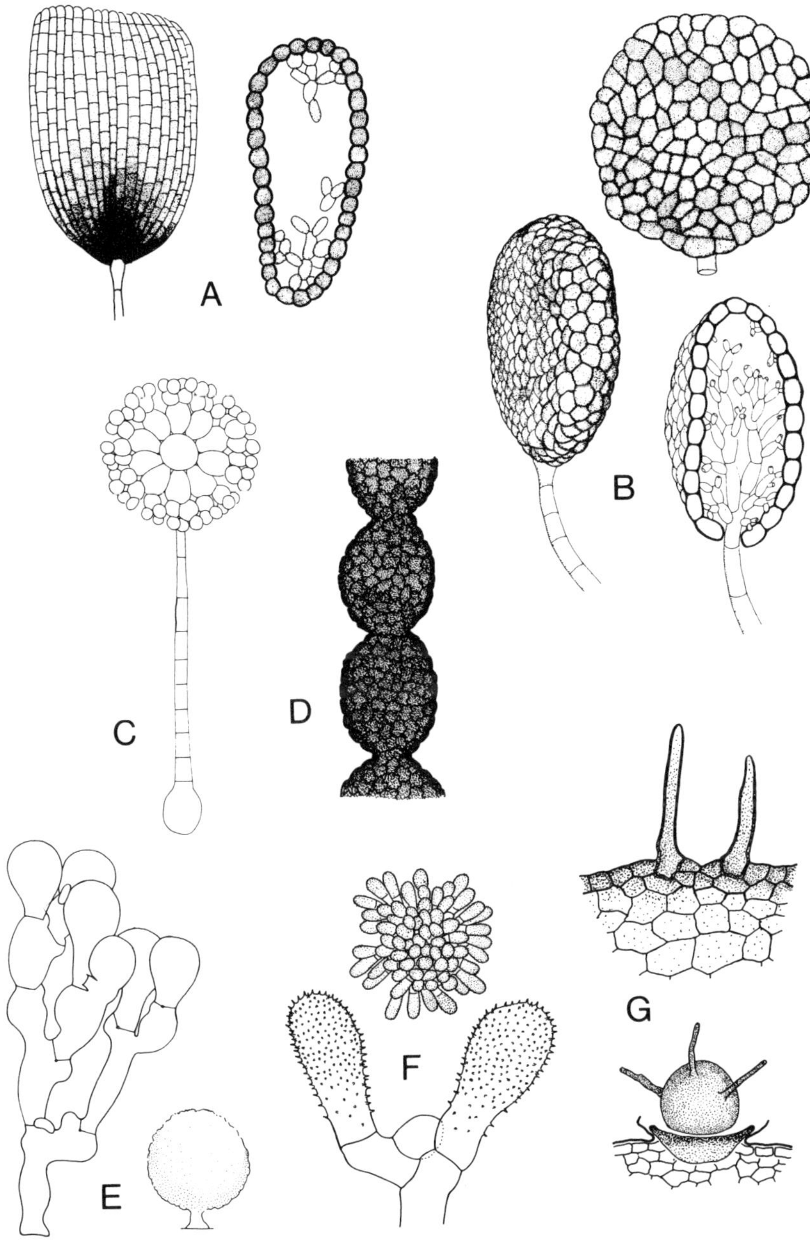

Plate 129: A *Cancellidium applanatum;* B *Beverwykella pulmonaria;* C *Cristulariella depraedans;* D *Arbuscula eugeniae;* E *Aegerita candida;* F *Spinulospora pucciniiphila;* G *Dactuliophora tarrii.*

Key-lists

GENERA WITH AMEROCONIDIA	Plate
ARTHRO HYALO AMPUL	
LINDQUISTIA (SYNNEMATA)	—
ARTHRO HYALO MISC	
WALLEMIA (SPORODOCHIA)	42 A
ARTHRO HYALO MISC?	
MONILIOPHTHORA	40 A
ARTHRO HYALO NONSP	
AMBLYOSPORIUM (ALEURI)	44 C
ANTROMYCES (SYNNEMATA)	67 A
ARTHROGRAPHIS	42 C
BOTRYOMONILIA (ALEURI?)	44 B
CYLINDROCOLLA (SPORODOCHIA)	42 F
GEOMYCES (ALEURI)	41 C
GEOTRICHUM	42 G
MALBRANCHEA (ALEURI)	41 B
MAUGINIELLA	43 A
MONILIA	38 B
MONILIELLA	43 G
MONILIOPHTHORA	40 A
NECATOR (SPORODOCHIA)	45 C
NOTHOCLAVULINA (SYNNEMATA)	—
OIDIODENDRON (ALEURI)	43 B
ONCOCLADIUM (ALEURI)	43 E
OSTEOMORPHA (CLAMPS)	45 E
OVADENDRON (ALEURI)	41 A
ROSULOMYCES	42 H
SCLEROSTILBUM (SYNNEMATA, CLAMPS)	45 D
SPHAEROSPORIUM (SPORODOCHIA)	40 D
SPORENDONEMA (ALEURI)	41 E
ARTHRO HYALO PHIAL	
WALLEMIA (SPORODOCHIA)	42 A
ARTHRO? HYALO PHIAL?	
APOSPORELLA	55 B
ARTHRO HYALO RACHI	
PARASYMPODIELLA	—
SYMPODIELLA	43 D
ARTHRO HYALO RADUL	
ROSULOMYCES	42 H
SYMPODIELLA	43 D
ARTHRO PHAEO MISC	
HELICOCEPHALUM	105 F
OJIBWAYA (SPORODOCHIA)	40 E

GENERA WITH AMEROCONIDIA	Plate
ARTHRO PHAEO NONSP	
COREMIELLA (ALEURI, SYNNEMATA)	43 F
MONILIA	38 B
OIDIODENDRON (ALEURI)	43 B
SCYTALIDIUM	42 E
STAHELIELLA	44 A
STEPHANOSPORIUM (ALEURI)	43 C
BASO HYALO ANNEL	
DENNISOGRAPHIUM (SYNNEMATA)	11 B
REFRACTOHILUM	77 G
SCOPULARIOPSIS	18 E
TYMPANOSPORIUM (SPORODOCHIA)	5 E
BASO HYALO ANNEL?	
PHIALOSTELE (SYNNEMATA)	9 D
POLYPAECILUM	20 A
BASO HYALO MISC	
BASIPETOSPORA	40 G
CLADOBOTRYUM	40 F
	67 F
FILOBASIDIELLA (CLAMPS)	35 F
OIDIUM	40 C
WALLEMIA (SPORODOCHIA)	42 A
BASO HYALO MISC?	
MONILIOPHTHORA	40 A
BASO HYALO NONSP	
AMBROSIELLA	47 B
MONILIOPHTHORA	40 A
BASO? HYALO NONSP	
NECATOR (SPORODOCHIA)	45 C
BASO HYALO PHIAL	
ACREMONIUM	2 A
ACROPHIALOPHORA	4 B
AKANTHOMYCES (SYNNEMATA)	10 F
ANGULIMAYA (SETAE)	15 C
ASPERGILLUS	8 C
BLOXAMIA (SPORODOCHIA, SYNNEMATA)	12 D
CHAETOCHALARA (SETAE)	2 H
CHALARA	2 C
CLONOSTACHYS	6 A
EXOCHALARA	3 C
FILOBASIDIELLA (CLAMPS)	35 F
GEOSMITHIA	—
GIBELLULA (SYNNEMATA)	10 D
MARIANNAEA	6 B
MERIMBLA	—
METARHIZIUM (SPORODOCHIA)	13 B
MONOCILLIUM	3 E

GENERA WITH AMEROCONIDIA	Plate
MONOSTACHYS	—
MYRIOCONIUM (SPORODOCHIA)	5 D
NALANTHAMALA (SPORODOCHIA)	14 A
NOMURAEA (SYNNEMATA)	4 D
PAECILOMYCES (SYNNEMATA)	3 B
	5 A
PENICILLIUM (SYNNEMATA)	7 B
PHIALOCEPHALA	7 D
PHIALOCLADUS	16 C
PHIALOPHORA	15 B
PHIALOPHOROPSIS	17 B
PHIALOSTELE (SYNNEMATA)	9 D
PHIALOTUBUS	5 C
SATCHMOPSIS (SPORODOCHIA)	—
SESQUICILLIUM	15 A
STACHYBOTRYS (SYNNEMATA)	7 C
TERMITARIA (SPORODOCHIA)	—
THYSANOPHORA	7 F
WALLEMIA (SPORODOCHIA)	42 A
BASO HYALO PHIAL?	
AMBROSIELLA	47 B
ENDOCONIDIUM (SPORODOCHIA)	—
MERIA	16 F
PLESIOSPORA	2 F
VOLUTINA	—
BASO? HYALO PHIAL?	
PSEUDOFUSIDIUM	36 D
BASO PHAEO ANNEL	
ANNELLODOCHIUM (SPORODOCHIA)	60 D
BATCHELOROMYCES (SPORODOCHIA)	20 C
CEPHALOTRICHUM (SYNNEMATA)	19 E
LEIGHTONIOMYCES (SYNNEMATA)	19 C
TRICHURUS (SETAE, SYNNEMATA)	19 A
WARDOMYCOPSIS	18 D
BASO PHAEO ANNEL?	
PHIALOSTELE (SYNNEMATA)	9 D
BASO PHAEO MISC	
BRIOSIA (SYNNEMATA)	43 H
CONIOSCYPHA	2 G
GILMANIELLA	49 B
OJIBWAYA (SPORODOCHIA)	40 E
THIELAVIOPSIS	50 C
BASO PHAEO PHIAL	
ACROPHIALOPHORA	4 B
ASPERGILLUS	8 C
CATENULARIA (SETAE)	3 F

GENERA WITH AMEROCONIDIA	Plate
CHAETOCHALARA (SETAE)	2 H
CHALARA	2 C
CRYPTOSTROMA (SETAE?, STROMA)	17 A
ELADIA	6 C
GLIOMASTIX	2 B
GRAPHIOTHECIUM (SYNNEMATA)	—
PAECILOMYCES (SYNNEMATA)	3 B
	5 A
PENICILLIUM (SYNNEMATA)	7 B
PHIALOCLADUS	16 C
PHIALOMYCES	7 E
PHIALOPHORA	15 B
PHIALOSTELE (SYNNEMATA)	9 D
PHIALOTUBUS	5 C
STACHYBOTRYS (SYNNEMATA)	7 C
STROMATOCREA (SPORODOCHIA)	13 A
TORULOMYCES	3 D
XANTHORIICOLA	6 G
BASO PHAEO PHIAL?	
CONIOSCYPHA	2 G
THIELAVIOPSIS	50 C
BLASTO HYALO AMPUL	
NEMATOGONUM	32 C
BLASTO HYALO MISC	
MYCELIOPHTHORA (ALEURI)	48 C
	58 B
PACHNODIUM	53 F
SPHAERIDIUM (SYNNEMATA)	38 D
BLASTO? HYALO MISC?	
SPHAERULOMYCES (AQUATIC)	36 G
BLASTO HYALO NONSP	
ALYSIDIUM	36 E
AMBROSIELLA	47 B
ARCUADENDRON (ALEURI)	41 D
BENIOWSKIA (SPORODOCHIA)	53 C
ECHINODIA (SYNNEMATA, CLAMPS)	45 A
FUSIDIUM	38 A
HORMIACTELLA (SETAE)	38 E
	65 G
HORMOMYCES (SPORODOCHIA)	36 F
MONILIA	38 B
MONILIELLA	43 G
MYCOSYLVA (SYNNEMATA)	53 E
PACHNODIUM	53 F
PTYCHOGASTER (ALEURI, CLAMPS)	51 A
SPHAEROSPORIUM (SPORODOCHIA)	40 D

GENERA WITH AMEROCONIDIA	Plate
BLASTO HYALO PHIAL?	
AMBROSIELLA	47 B
BLASTO? HYALO PHIAL?	
PSEUDOFUSIDIUM	36 D
BLASTO HYALO RADUL	
ACLADIUM	23 H
ANUNGITEA	65 E
DENTICULARIA (SPORODOCHIA)	36 B
HYALODENDRON	38 C
POLYSCYTALUM (SPORODOCHIA)	65 C
RAMULARIA	61 B
BLASTO PHAEO AMPUL	
GONATOBOTRYUM	32 E
PERICONIA (SETAE, STROMA)	37 E
BLASTO PHAEO ANNEL	
ANNELLODOCHIUM (SPORODOCHIA)	60 D
BLASTO PHAEO MISC	
ECHINOCATENA	37 B
HAPLOBASIDION	37 A
LACELLINA (SETAE, STROMA)	37 D
LACELLINOPSIS (SETAE, STROMA)	37 C
PARAPERICONIA (SETAE, SPORODOCHIA)	50 F
BLASTO PHAEO NONSP	
AGARWALIA	39 B
ALYSIDIOPSIS	39 C
ALYSIDIUM	36 E
AMPULLIFERA (PODIA)	36 A
MONILIA	38 B
PERICONIA (SETAE, STROMA)	37 E
RUTOLA	88 H
SOROCYBE (SYNNEMATA)	39 G
SPILODOCHIUM (SPORODOCHIA)	66 F
TRICHOBOTRYS (SETAE)	39 D
XYLOHYPHA	36 C
BLASTO PHAEO PHIAL	
FONSECAEA	—
BLASTO PHAEO RADUL	
ANUNGITEA	65 E
CLADOSPORIUM (STROMA)	66 E
DENTICULARIA (SPORODOCHIA)	36 B
HORMOCONIS	39 E
PERICONIA (SETAE, STROMA)	37 E
PERICONIELLA (SYNNEMATA)	84 G

GENERA WITH AMEROCONIDIA	Plate
CETERI HYALO AMPUL	
AMBIVINA	35 A
ARTHROBOTRYS	62 G
BLASTOBOTRYS (SETAE)	31 G
BOTRYOSPORIUM	32 B
CHROMELOSPORIUM (SYNNEMATA)	33 F
CLINOCONIDIUM (SPORODOCHIA)	35 D
DICHOBOTRYS (ALEURI?)	33 C
GONATOBOTRYS	32 A
GRAPHIOLA	46 E
LINDQUISTIA (SYNNEMATA)	—
MICROSTROMA (SYNNEMATA)	32 D
NIA (APPEND, SPORODOCHIA, CLAMPS)	121 B
NODULISPORIUM (SYNNEMATA)	27 D
OEDOCEPHALUM	35 G
OSTRACODERMA (PYCNIDIA)	—
OSTRACODERMIDIUM (SETAE)	32 G
PACHNOCYBE (SYNNEMATA)	32 F
PADIXONIA	34 F
PHYMATOTRICHOPSIS (ALEURI)	33 A
PULCHROMYCES	33 E
SIGMOIDEOMYCES (SETAE)	31 D
SPHONDYLOCEPHALUM	31 A
SPINIGER (CLAMPS)	35 B
	35 C
XYLOCLADIUM	28 E
CETERI HYALO AMPUL?	
GLOMOSPORA (ALEURI?, CLAMPS)	52 E
SPOROCEPHALUM	31 E
CETERI HYALO ANNEL	
CERATOCLADIUM (SETAE)	20 E
INTRAPES (SPORODOCHIA)	—
CETERI HYALO ANNEL?	
STEVENSOMYCES (SYNNEMATA)	19 F
CETERI HYALO MISC	
APHANOCLADIUM (ALEURI)	54 D
CERATOCLADIUM (SETAE)	20 E
CHAETOBLASTOPHORUM (SETAE)	73 G
CLADOBOTRYUM	40 F
	67 F
CLINOCONIDIUM (SPORODOCHIA)	35 D
COCCIDIOIDES	54 A
COSTANTINELLA (SETAE)	26 B
DESERTELLA	49 I
DICHOBOTRYS (ALEURI?)	33 C
GYROTHRIX (SETAE, STROMA)	54 G
HEMIBELTRANIA	24 D
ITERSONILIA (CLAMPS)	52 B

GENERA WITH AMEROCONIDIA	Plate
MYCELIOPHTHORA (ALEURI)	48 C
	53 B
PADIXONIA	34 F
PULCHROMYCES	33 E
STEVENSOMYCES (SYNNEMATA)	19 F
SYZYGOSPORA (CLAMPS)	51 G
VERTICIMONOSPORIUM	4 F
ZYGOSPORIUM (SETAE)	54 C
CETERI HYALO MISC?	
SPHAERULOMYCES (AQUATIC)	36 G
CETERI HYALO NONSP	
AMBLYOSPORIUM (ALEURI)	44 D
AMPHICHAETELLA (APPEND, SPORODOCHIA)	54 E
ASTEROPHORA (APPEND, SYNNEMATA, CLAMPS)	51 E
BENIOWSKIA (SPORODOCHIA)	53 C
BOTRYOTRICHUM (ALEURI, SETAE)	47 G
CHAETOCONIDIUM (APPEND)	71 A
CHANTRANSIOPSIS	55 E
CHLAMYDOMYCES (ALEURI)	68 F
CHRYSOSPORIUM (ALEURI)	54 B
DEXHOWARDIA	34 D
JACULISPORA (APPEND, AQUATIC)	120 D
MARTINDALIA (SETAE, SYNNEMATA, CLAMPS)	51 F
MYCOSYLVA (SYNNEMATA)	53 E
MYRIODONTIUM (ALEURI)	—
NEMATOCTONUS (CLAMPS)	52 C
NIA (APPEND, SPORODOCHIA, CLAMPS)	121 B
OLPITRICHUM	31 F
OVULITIS (ALEURI)	53 D
PAGIDOSPORA (CLAMPS)	52 A
SEPEDONIUM (ALEURI)	47 D
SPOROTRICHUM (ALEURI, CLAMPS)	52 D
STAPHYLOTRICHUM (ALEURI)	47 F
STEPHANOMA (ALEURI, APPEND)	113 C
TRICHOSPORIELLA (ALEURI)	53 A
TUBERCULISPORA	24 H
UMBELOPSIS	47 E
CETERI HYALO PHIAL	
COREOMYCETOPSIS	55 F
ENDOSPORELLA	55 A
HIRSUTELLA (SYNNEMATA)	10 C
INTRAPES (SPORODOCHIA)	—
MARGARITISPORA (APPEND, AQUATIC)	89 B
THAXTERIOLA	55 C

GENERA WITH AMEROCONIDIA	Plate
CETERI HYALO PHIAL?	
APOSPORELLA	55 B
CETERI HYALO RACHI	
ACRODONTIUM	23 G
BEAUVERIA (SPORODOCHIA, SYNNEMATA)	30 A
BEAUVERIPHORA (SYNNEMATA)	29 C
BLASTOBOTRYS (SETAE)	31 G
CALCARISPORIELLA	24 F
CALCARISPORIUM	26 A
COSTANTINELLA (SETAE)	26 B
DRUMOPAMA (SYNNEMATA)	29 F
GENICULOSPORIUM	25 G
HYALORHINOCLADIELLA	23 B
LEPTODONTIDIUM	23 A
MICRODOCHIUM (SPORODOCHIA)	14 E
	108 B
PSEUDOHANSFORDIA	63 B
PSEUDOMERIA	—
RHOMBOSTILBELLA (SYNNEMATA)	29 E
SOLOSYMPODIELLA	23 C
SPOROTHRIX	23 E
TRITIRACHIUM	26 C
CETERI HYALO RADUL	
ACLADIUM	23 H
AGARICOSTILBUM (SETAE, SYNNEMATA)	29 D
ARTHROBOTRYS	62 G
BASIFIMBRIA	24 A
CALCARISPORIUM	26 A
DIMORPHOSPORA (AQUATIC)	47 A
GENICULODENDRON	26 D
HEMIBELTRANIA	24 D
HYMENOSTILBE (SYNNEMATA)	28 B
MICRODOCHIUM (SPORODOCHIA)	14 E
	108 B
NODULISPORIUM (SYNNEMATA)	27 D
OLPITRICHUM	31 F
OVULARIA	24 E
PAPPIMYCES	110 F
PARAPHAEOISARIA	23 D
PHAEOISARIA (SYNNEMATA)	28 A
PSEUDOGIBELLULA (SYNNEMATA)	28 F
RAFFAELEA (SPORODOCHIA)	23 F
RAMICHLORIDIUM	25 F
RHINOCLADIELLA	25 E
RHINOTRICHELLA	25 A
SOLOSYMPODIELLA	23 C
UMBELLIDION	21 D
VERONAEA	61 G
VERTICICLADIUM	27 E
XYLOCLADIUM	28 E

GENERA WITH AMEROCONIDIA	Plate	
CETERI HYALO RADUL?		
PACHNOCYBE (SYNNEMATA)	32	F
CETERI PHAEO AMPUL		
AMPHOBOTRYS (ALEURI)	33	B
BALANIUM	68	B
BOTRYTIS (ALEURI)	34	E
NODULISPORIUM (SYNNEMATA)	27	D
STREPTOBOTRYS (ALEURI)	34	A
TRICHOSPORONOIDES	31	B
VERRUCOBOTRYS (ALEURI, STROMA)	34	C
CETERI PHAEO ANNEL		
ACROGENOSPORA	18	A
DOMINGOELLA	49	A
HEMICORYNESPORA	18	B
SPILOCAEA (SPORODOCHIA)	18	C
CETERI PHAEO ANNEL?		
PHYSALIDIUM (APPEND)	114	D
CETERI PHAEO MISC		
ARTHRINIUM (ALEURI)	46	A
	46	B
BELTRANIA (APPEND, SETAE, STROMA)	22	F
BELTRANIOPSIS (SETAE, STROMA)	22	C
BOTRYOCONIS (SPORODOCHIA)	35	E
CONIOSCYPHA	2	G
CORDELLA (ALEURI, SETAE)	46	F
DEMATOPHORA (ALEURI?, SYNNEMATA)	29	G
DESERTELLA	49	I
ECHINOCATENA	37	B
GILMANIELLA	49	B
HEMIBELTRANIA	24	D
HERPOSIRA (SETAE, STROMA)	48	D
NIGROSPORA	49	C
PAATHRAMAYA (SYNNEMATA)	9	C
PTEROCONIUM (ALEURI, SPORODOCHIA)	46	D
SADASIVANIA	37	F
WARDOMYCES (ALEURI)	68	A
ZYGOSPORIUM (SETAE)	54	C
CETERI PHAEO NONSP		
ACREMONIULA (ALEURI?)	49	H
ALLESCHERIELLA (ALEURI, SPORODOCHIA?)	51	B
AMPULLIFERA (PODIA)	36	A
BACTRIDIOPSIS	47	C
BALANIUM	68	B
BOTRYODERMA (ALEURI, SETAE)	48	A
ECHINOBOTRYUM (ALEURI, SYNNEMATA)	48	E
HARZIA	49	F

GENERA WITH AMEROCONIDIA	Plate	
HUMICOLA (ALEURI)	49	D
MAMMARIA (ALEURI)	48	B
MEGALODOCHIUM (ALEURI?, SPORODOCHIA)	50	B
MICROCLAVA	96	A
MILOSPIUM	113	A
PHYSALIDIUM (APPEND)	114	D
RHINOCLADIUM	49	G
SPOROTRICHUM (ALEURI, CLAMPS)	52	D
STREPTOBOTRYS (ALEURI)	34	A
THERMOMYCES (ALEURI)	49	E
USTILAGINOIDEA (ALEURI?, SPORODOCHIA)	50	D
VERRUCOBOTRYS (ALEURI, STROMA)	34	C
CETERI PHAEO NONSP?		
ENDOCALYX (SETAE, SPORODOCHIA, SYNNEMATA)	46	C
CETERI PHAEO PHIAL		
FONSECAEA	—	
TUBERCULINA (ACERVULI)	14	D
XANTHORIICOLA	6	G
CETERI PHAEO PHIAL?		
CONIOSCYPHA	2	G
CRASPEDODIDYMUM	16	B
CETERI PHAEO RACHI		
ACRODONTIUM	23	G
DEMATOPHORA (ALEURI?, SYNNEMATA)	29	G
HARZIA	49	F
LEPTODONTIDIUM	23	A
LOMENTOSPORA	30	D
PUCIOLA (SETAE)	25	D
SPOROTHRIX	23	E
TRITIRACHIUM	26	C
CETERI PHAEO RADUL		
ARDHACHANDRA	21	C
ASTEROMYCES	22	D
BELTRANIA (APPEND, SETAE, STROMA)	22	F
BELTRANIELLA (SETAE, STROMA)	22	E
BELTRANIOPSIS (SETAE, STROMA)	22	C
BRACHYDESMIELLA (APPEND)	93	E
CONOPLEA (STROMA)	25	C
DICYMA (SETAE)	25	B
ECHINOBOTRYUM (ALEURI, SYNNEMATA)	48	E
FUSICLADIUM (STROMA)	22	B
HADROTRICHUM (SPORODOCHIA)	30	G
HANSFORDIA (SETAE)	27	B

GENERA WITH AMEROCONIDIA	Plate
HEMIBELTRANIA	24 D
KUMANASAMUHA	30 F
LAPPODOCHIUM (APPEND, SETAE, SPORODOCHIA)	30 H
MAMMARIA (ALEURI)	48 B
MELANOGRAPHIUM (SPORODOCHIA, SYNNEMATA)	29 A
NODULISPORIUM (SYNNEMATA)	27 D
PAATHRAMAYA (SYNNEMATA)	9 C
PANCHANANIA (APPEND, SYNNEMATA)	63 F
PAPILIONOSPORA (SYNNEMATA)	67 C
PERICONIELLA (SYNNEMATA)	84 G
PHAEODAEDALEA (CLAMPS)	24 G
PSEUDOBELTRANIA	22 A
PSEUDOBOTRYTIS	62 C
PSEUDOEPICOCCUM (SPORODOCHIA)	30 E
RACODIUM	24 C
RAMICHLORIDIUM	25 F
RHINOCLADIELLA	25 E
RHINOTRICHELLA	25 A
SCEPTRIFERA	61 E
SPADICOIDES	64 C
VERONAEA	61 G
VIRGARIA	21 E
VIRGARIELLA	24 B
GLOIO HYALO AMPUL	
AUREOBASIDIUM	31 C
COLLECEPHALUS	—
HEIMIODORA (SETAE)	34 B
KABATIELLA (ACERVULI)	—
GLOIO HYALO AMPUL?	
ARTICULARIA (SYNNEMATA)	—
JACOBIA (CLAMPS)	35 B
GLOIO HYALO ANNEL	
GRAPHILBUM	19 D
LEPTOGRAPHIUM	18 F
GLOIO HYALO MISC	
BISPOROMYCES	3 J
BLASTOPHORUM	81 C
CATENOPHORA (ACERVULI)	—
CHAETOBLASTOPHORUM (SETAE)	73 G
PACHNODIUM	53 F
SYZYGOSPORA (CLAMPS)	51 G
GLOIO HYALO NONSP	
GEOTRICHUM	42 G
PACHNODIUM	53 F
SCLEROSTILBUM (SYNNEMATA, CLAMPS)	45 D

GENERA WITH AMEROCONIDIA	Plate
GLOIO HYALO NONSP?	
MATULA (SPORODOCHIA)	51 D
MICHENERA (SPORODOCHIA)	51 C
GLOIO HYALO PHIAL	
ACREMONIUM	2 A
AGYRIELLA (SPORODOCHIA)	14 B
BAHUSUTRABEEJA (APPEND)	1 E
BISPOROMYCES	3 J
BLISTUM (SYNNEMATA)	11 F
CALCEISPORA (APPEND)	2 D
CAPNOPHIALOPHORA	17 D
CHAETOCHALARA (SETAE)	2 H
CHAETOPSINA (SETAE)	17 G
CHAETOPSIS (SETAE)	57 B
CHALARA	2 C
CHLORIDIUM	3 H
CILICIOPODIUM (SYNNEMATA)	9 B
CLADORRHINUM (SETAE)	15 D
CLONOSTACHYS	6 A
CODINAEA (APPEND, SETAE, STROMA)	1 A
CODINAEOPSIS (SETAE)	1 F
COLLETOTRICHUM (SETAE, SPORODOCHIA, PODIA)	—
CORYNE (SPORODOCHIA)	11 C
CRINULA (SYNNEMATA)	9 E
CULICINOMYCES	—
CYLINDROSPORIUM (ACERVULI)	—
CYSTODENDRON	6 F
DENDRODOCHIUM (SPORODOCHIA)	12 A
DENDROSTILBELLA (SYNNEMATA)	9 G
DISCHLORIDIUM (STROMA)	3 I
GLIOCEPHALIS	8 B
GLIOCEPHALOTRICHUM (SETAE)	8 E
GLIOCLADIUM	7 A
GLOIOSPHAERA	8 D
GOIDANICHIELLA	8 F
GONYTRICHUM (SETAE)	4 A
HAINESIA (SPORODOCHIA)	—
HARZIELLA	8 A
HIRSUTELLA (SYNNEMATA)	10 C
HORMISCIOMYCES	17 C
HYALOCYLINDROPHORA (SYNNEMATA)	11 E
KABATINA (SPORODOCHIA)	—
KMETIOPSIS	107 F
KOORCHALOMA (APPEND, SETAE, SPORODOCHIA)	12 B
LEPTOXYPHIUM (SYNNEMATA)	58 E
MAHABALELLA (APPEND, SETAE)	1 B
MARIANNAEA	6 B
MENISPORA (APPEND, SETAE)	1 C
MENISPOROPSIS (APPEND, SETAE, SYNNEMATA)	56 G

GENERA WITH AMEROCONIDIA	Plate
MICROXIPHIUM (SETAE, SYNNEMATA)	17 F
MINIMIDOCHIUM (APPEND, SETAE, SPORODOCHIA)	14 F
MONOCILLIUM	3 E
MYCOLEPTODISCUS (APPEND, SPORODOCHIA)	14 C
MYRIOCONIUM (SPORODOCHIA)	5 D
MYROTHECIUM (SETAE, SPORODOCHIA, SYNNEMATA)	12 F
NEOTTIOSPORELLA (APPEND, SETAE, SPORODOCHIA)	9 H
ORAMASIA (SETAE, SPORODOCHIA)	—
PHAEOSTALAGMUS	6 D
PHIALOCEPHALA	7 D
PHIALOGENICULATA	57 G
PHIALOGRAPHIUM (SYNNEMATA)	11 A
PHIALOPHORA	15 B
RHYNCHOSPORINA	2 E
SARCOPODIUM (SETAE, SPORODOCHIA)	13 C
SAROCLADIUM	5 F
SESQUICILLIUM	15 A
SPHACELIA (SPORODOCHIA)	5 B
SPHAEROMYCES	3 G
STACHYBOTRYS (SYNNEMATA)	7 C
STACHYLIDIUM (STROMA)	4 G
STILBELLA (SYNNEMATA)	11 D
STILBUM (SYNNEMATA)	—
SYNNEMATIUM (SYNNEMATA)	10 A
THOZETELLA (APPEND, SETAE, SYNNEMATA)	1 D
	9 A
TILACHLIDIOPSIS (SYNNEMATA)	10 E
TILACHLIDIUM (SYNNEMATA)	10 B
TOLYPOCLADIUM	16 A
TRICHODERMA	16 E
TUBERCULARIA (SPORODOCHIA)	12 C
UNCIGERA	4 E
VERTICILLIUM	4 C
VOLUTELLA (SETAE, SPORODOCHIA)	12 E
WANGIELLA	15 E
ZAKATOSHIA	16 D
ZANCLOSPORA (SETAE)	4 H
GLOIO? HYALO PHIAL	
ENDOCONOSPORA (SPORODOCHIA)	13 D
HYPHOPOLYNEMA (APPEND, SPORODOCHIA)	56 F
GLOIO HYALO PHIAL?	
HARPOGRAPHIUM (SYNNEMATA)	107 C
HORMONEMA	20 D
MERIA	16 F

GENERA WITH AMEROCONIDIA	Plate
GLOIO HYALO RACHI	
BEAUVERIPHORA (SYNNEMATA)	29 C
HYALOPESOTUM	—
HYALORHINOCLADIELLA	23 B
PESOTUM (SYNNEMATA)	29 B
GLOIO HYALO RADUL	
BLASTOPHORUM	81 C
DIMORPHOSPORA (AQUATIC)	47 A
FLAHAULTIA	30 B
HAPLOGRAPHIUM	27 C
PESOTUM (SYNNEMATA)	29 B
PESTALOZZIELLA (APPEND, ACERVULI)	—
SELENOSPORELLA	21 A
VERTICICLADIELLA	27 A
GLOIO HYALO RADUL?	
DITANGIUM (SPORODOCHIA)	26 E
HETEROCEPHALUM (SYNNEMATA)	28 C
JACOBIA (CLAMPS)	35 H
GLOIO PHAEO AMPUL	
AUREOBASIDIUM	31 C
GLOIO PHAEO ANNEL	
EXOPHIALA	20 B
	60 B
GRAPHIUM (SYNNEMATA)	19 B
MELANCONIUM (ACERVULI)	—
SCEDOSPORIUM	—
GLOIO PHAEO MISC	
BLASTOPHORUM	81 C
CATENOPHORA (ACERVULI)	—
GILMANIELLA	49 B
HELICOCEPHALUM	105 F
SARCINOMYCES	—
GLOIO PHAEO MISC?	
ANTROMYCOPSIS (SYNNEMATA, CLAMPS)	45 B
GLOIO PHAEO NONSP	
PHAEOCOCCOMYCES	50 A
GLOIO PHAEO PHIAL	
ANTHOPSIS	6 E
CATENULARIA (SETAE)	3 F
CHAETOCHALARA (SETAE)	2 H
CHALARA	2 C
CHLORIDIUM	3 H
CUSTINGOPHORA	8 G
CYSTODENDRON	6 F
FONSECAEA	—
GLIOCLADIUM	7 A
GLIOMASTIX	2 B

GENERA WITH AMEROCONIDIA — Plate

GONYTRICHUM (SETAE)	4 A
LEPTOXYPHIUM (SYNNEMATA)	58 E
MYROTHECIUM (SETAE, SPORODOCHIA, SYNNEMATA)	12 F
PHIALOPHORA	15 B
SOLHEIMIA (SYNNEMATA)	9 F
STACHYBOTRYS (SYNNEMATA)	7 C
STACHYLIDIUM (STROMA)	4 G
GLOIO PHAEO PHIAL?	
HORMONEMA	20 D
GLOIO PHAEO RADUL	
BLASTOPHORUM	81 C
GLOIO? PHAEO RADUL	
LAPPODOCHIUM (APPEND, SETAE, SPORODOCHIA)	30 H

GENERA WITH CHEIROCONIDIA — Plate

BLASTO PHAEO NONSP	
TAENIOLINA	127 C
BLASTO PHAEO RACHI	
DENDRYPHION (STROMA)	84 E
BLASTO PHAEO RADUL	
DENDRYPHION (STROMA)	84 E
CETERI HYALO NONSP	
DIADEMOSPORA	117 G
PETRAKIOPSIS	116 E
PSEUDODICTYOSPORIUM	98 E
CETERI HYALO RACHI	
DICRANIDION (SYNNEMATA)	116 F
CETERI HYALO RADUL	
DICRANIDION (SYNNEMATA)	116 F
CETERI PHAEO ANNEL	
CERATOSPORELLA (STROMA)	127 D
EVERSIA	98 B
CETERI PHAEO NONSP	
CERATOSPORELLA (STROMA)	127 D
CRYPTOCORYNEUM (SPORODOCHIA)	116 G
DICTYOSPORIUM (ALEURI)	98 C
MATSUSHIMAEA	—
PSEUDODICTYOSPORIUM	98 E
TAENIOLINA	127 C
TRETOSPEIRA	98 G

GENERA WITH CHEIROCONIDIA — Plate

CETERI PHAEO NONSP?	
CHEIROMYCES (SPORODOCHIA)	116 A
CETERI PHAEO RACHI	
DENDRYPHION (STROMA)	84 E
CETERI PHAEO RADUL	
DENDRYPHION (STROMA)	84 E
GLOIO PHAEO NONSP	
CHEIROMYCELLA (ALEURI?, SPORODOCHIA)	116 B

GENERA WITH DICTYOCONIDIA — Plate

ARTHRO PHAEO MISC	
TRIMMATOSTROMA (SPORODOCHIA)	91 B
BASO HYALO MISC	
DICTYOCATENULATA (SYNNEMATA)	101 G
BASO PHAEO MISC	
CONIOSPORIUM (SPORODOCHIA)	101 C
BASO PHAEO MISC?	
BARNETTELLA (SPORODOCHIA)	—
BLASTO PHAEO NONSP	
PEYRONELIA	101 B
SCLEROCOCCUM (SPORODOCHIA)	50 E
BLASTO PHAEO RACHI	
ALTERNARIA (APPEND, SYNNEMATA)	101 A
BLASTO PHAEO RADUL	
PARAPITHOMYCES	99 F
CETERI HYALO MISC	
CONIODICTYUM (SPORODOCHIA)	97 B
RIESSIA (CLAMPS)	113 G
CETERI HYALO NONSP	
CLATHROSPHAERINA (AQUATIC)	97 H
DIHETEROSPORA (ALEURI)	—
MIURAEA	97 F
PSEUDODICTYOSPORIUM	98 E
RIESSIA (CLAMPS)	113 G
STEMPHYLIOPSIS (ALEURI)	97 A

GENERA WITH DICTYOCONIDIA — Plate

CETERI HYALO PHIAL
ASCOCONIDIUM — 97 G

CETERI HYALO RACHI
HYALOCLADIUM — 97 C

CETERI PHAEO AMPUL
BLASTODICTYS (APPEND) — 94 E

CETERI PHAEO ANNEL
ACRODICTYS (APPEND) — 98 F
ANNELLOPHORELLA — 98 D
ARACHNOPHORA (APPEND) — 115 D
EVERSIA — 98 B
PANTOSPORA (SYNNEMATA) — 101 H
SEPTOSPORIUM (APPEND, SETAE) — 95 E
STEMPHYLIUM — 99 C
STIGMINA (SPORODOCHIA) — 76 B, 98 A
TRETOPILEUS (SYNNEMATA) — 128 A
XENOSPORIUM — 104 C

CETERI PHAEO MISC
COLEMANIELLA (PODIA) — 90 D
DICTYOARTHRINIUM (ALEURI) — 96 E
DICTYOPOLYSCHEMA — 100 E
FLOSCULOMYCES (ALEURI) — 96 F
ISTHMOSPORA — 96 H
ONCOPODIUM (APPEND, SPORODOCHIA) — 95 F
SPEGAZZINIA (APPEND) — 114 B
SYAMITHABEEJA — 99 G
TRETOPILEUS (SYNNEMATA) — 128 A

CETERI PHAEO NONSP
ACRODICTYS (APPEND) — 98 F
ACROSPEIRA (ALEURI) — 96 B
BERKLEASMIUM (SPORODOCHIA) — 101 E
BEVERWYKELLA (AQUATIC) — 129 B
BIOCONIOSPORIUM (APPEND) — 95 A
CANCELLIDIUM — 129 A
CHEIROSPORA (SPORODOCHIA) — —
CHUPPIA — 100 A
COLEODICTYOSPORA — 95 G
DICTYODESMIUM (APPEND, SPORODOCHIA) — 94 G
DICTYOSPORIUM (ALEURI) — 98 C
EPICOCCUM (ALEURI, SPORODOCHIA) — 100 B
FLOSCULOMYCES (ALEURI) — 96 F
HANSFORDIELLA (APPEND) — 94 C
HANSFORDIELLOPSIS (APPEND, PODIA?) — 94 A
HERMATOMYCES (SETAE) — 96 G
KOSTERMANSINDA (SYNNEMATA) — 101 D
MICROCLAVA — 96 A
MONODICTYS (ALEURI) — 100 D
MUIARIA (APPEND) — 95 B
MUIOGONE (APPEND) — 95 C
MYCOENTEROLOBIUM — 128 D
MYXOCYCLUS (SPORODOCHIA, ACERVULI) — —
ONCOPODIELLA (APPEND) — 95 D
ONCOPODIUM (APPEND, SPORODOCHIA) — 95 F
PETRAKIA (APPEND, SPORODOCHIA) — 115 E
PIRICAUDA (APPEND) — 94 B
PITHOMYCES (ALEURI) — 90 A, 100 C
PSEUDODICTYOSPORIUM — 98 E
PSEUDOPETRAKIA (APPEND) — 94 F
SARCINELLA (PODIA) — 96 C
SEPTOSPORIUM (APPEND, SETAE) — 95 E
TETRACOCCOSPORIUM (ALEURI?) — 96 D
TETRAPLOA (APPEND) — 125 E
TRETOSPEIRA — 98 G
XENOSPORIUM — 104 C

CETERI PHAEO RACHI
EMBELLISIA — 82 H
PENDULISPORA — 94 D
SYAMITHABEEJA — 99 G
THYROSTROMELLA (SPORODOCHIA) — 99 H
ULOCLADIUM — 99 B

CETERI PHAEO RADUL
DACTYLOSPORIUM (STROMA) — 99 D
DICTYOSPIROPES — 94 H
MYSTROSPORIELLA — 99 A
PARAPITHOMYCES — 99 F
SCLEROGRAPHIUM (SYNNEMATA) — 101 F
SIROSPORIUM (STROMA) — 99 E

GLOIO HYALO NONSP
THYRSIDINA (ACERVULI, CLAMPS) — 97 D

GENERA WITH DIDYMOCONIDIA — Plate

ARTHRO HYALO NONSP
ANTROMYCES (SYNNEMATA) — 67 A

ARTHRO HYALO RACHI
PARASYMPODIELLA — —
SYMPODIELLA — 43 D

GENERA WITH DIDYMOCONIDIA	Plate
ARTHRO HYALO RADUL	
SYMPODIELLA	43 D
ARTHRO PHAEO NONSP	
AMPULLIFERINA (PODIA)	67 B
BAHUSAKALA	42 B
SCYTALIDIUM	42 E
BASO HYALO ANNEL	
GAMSIA	59 C
GYMNODOCHIUM (SPORODOCHIA)	60 A
BASO HYALO MISC	
CLADOBOTRYUM	40 F
	67 F
TRICHOTHECIUM	67 E
BASO HYALO PHIAL	
CHALARA	2 C
FUSISPORELLA (SPORODOCHIA)	58 F
PENICILLIFER	57 E
BASO PHAEO ANNEL	
ANNELLODOCHIUM (SPORODOCHIA)	60 D
BATCHELOROMYCES (SPORODOCHIA)	20 C
BISPOROSTILBELLA (SYNNEMATA)	—
GYMNODOCHIUM (SPORODOCHIA)	60 A
TRICHODOCHIUM (SETAE, STROMA, SPORODOCHIA)	60 C
BASO PHAEO PHIAL	
CHALARA	2 C
BLASTO HYALO NONSP	
DIPLOOSPORA	65 A
HORMIACTELLA (SETAE)	38 E
	65 G
HORMIACTIS	65 F
BLASTO HYALO RACHI	
PASPALOMYCES (SETAE)	63 C
BLASTO HYALO RADUL	
ANUNGITEA	65 E
ARXIELLA	65 B
DENTICULARIA (SPORODOCHIA)	36 B
DIPLOOSPORA	65 A
PLEUROTHECIOPSIS	65 D
POLYSCYTALUM (SPORODOCHIA)	65 C
RAMULARIA	61 B

GENERA WITH DIDYMOCONIDIA	Plate
BLASTO PHAEO AMPUL	
OEDEMIUM (SETAE)	66 D
BLASTO PHAEO ANNEL	
ANNELLODOCHIUM (SPORODOCHIA)	60 D
BLASTO PHAEO NONSP	
ALYSIDIOPSIS	39 C
AMPULLIFERA (PODIA)	36 A
BISPORA	66 A
DIDYMOBOTRYUM (STROMA, SYNNEMATA)	66 B
DIPLOCOCCIUM	66 C
RUTOLA	88 H
SPILODOCHIUM (SPORODOCHIA)	66 F
XYLOHYPHA	36 C
BLASTO PHAEO RACHI	
DENDRYPHION (STROMA)	84 E
BLASTO PHAEO RADUL	
ANUNGITEA	65 E
CLADOSPORIUM (STROMA)	66 E
DENDRYPHION (STROMA)	84 E
DENTICULARIA (SPORODOCHIA)	36 B
DIPLOCOCCIUM	66 C
PERICONIELLA (SYNNEMATA)	84 G
TANDONELLA (SYNNEMATA)	86 E
CETERI HYALO AMPUL	
ARTHROBOTRYS	62 G
BLASTOPHORELLA	64 B
CETERI HYALO MISC	
CLADOBOTRYUM	40 F
	67 F
FUSICLADIELLA (STROMA)	64 E
SYZYGOSPORA (CLAMPS)	51 G
CETERI HYALO NONSP	
CHLAMYDOMYCES (ALEURI)	68 F
DACTYLELLA (APPEND)	89 A
MONOTRICHUM (APPEND)	56 E
MYCOGONE (ALEURI)	68 D
CETERI HYALO NONSP?	
DWAYALOMA (APPEND, SETAE)	56 D
CETERI HYALO RACHI	
CERCOSPORIDIUM (STROMA)	85 E
DACTYLELLA (APPEND)	89 A
DENDROSPORIUM (AQUATIC)	113 D
MICRODOCHIUM (SPORODOCHIA)	14 E
	108 B
NETA (SETAE)	62 H
PASPALOMYCES (SETAE)	63 C

GENERA WITH DIDYMOCONIDIA	Plate
PSEUDOHANSFORDIA	63 B
PYRICULARIA	80 A
SOLOSYMPODIELLA	23 C
CETERI HYALO RADUL	
ARTHROBOTRYS	62 G
BOSTRICHONEMA	67 G
DACTYLARIA	62 F
	81 F
DENDROSPORIUM (AQUATIC)	113 D
DISSOACREMONIELLA	—
ERIOMYCOPSIS (APPEND)	79 F
HAPLARIOPSIS	63 A
HYPHODISCOSIA (APPEND)	56 A
ISTHMOLONGISPORA	30 C
MICRODOCHIUM (SPORODOCHIA)	14 E
	108 B
NETA (SETAE)	62 H
OVULARIA	24 E
PAPPIMYCES	110 F
PHAEOISARIA (SYNNEMATA)	28 A
PYRICULARIA	80 A
RAMICHLORIDIUM	25 F
SCOLECOBASIDIELLA	62 D
SOLOSYMPODIELLA	23 C
VERONAEA	61 G
ZYGOPHIALA	63 E
CETERI PHAEO AMPUL	
BALANIUM	68 B
CEPHALIOPHORA	64 A
	93 D
OEDEMIUM (SETAE)	66 D
ZOPHEROMYCES	64 D
CETERI PHAEO ANNEL	
DEIGHTONIELLA	59 D
ENDOPHRAGMIELLA	59 A
HEMICORYNESPORA	18 B
OEDOTHEA (SPORODOCHIA)	59 B
PHRAGMOSPATHULA (STROMA)	76 D
SPILOCAEA (SPORODOCHIA)	18 C
CETERI PHAEO MISC	
GARNAUDIA	—
WARDOMYCES (ALEURI)	68 A
ZOPHEROMYCES	64 D
CETERI PHAEO MISC?	
TRIADELPHIA (SPORODOCHIA)	70 F
CETERI PHAEO NONSP	
AMPULLIFERA (PODIA)	36 A
BALANIUM	68 B
DIPLOCOCCIUM	66 C
HADRONEMA (SPORODOCHIA)	68 C
JOHNSTONIA	64 F

GENERA WITH DIDYMOCONIDIA	Plate
PIROZYNSKIA (PODIA)	64 G
STEVENSONULA (APPEND, SETAE, SPORODOCHIA)	—
TRIADELPHIA (SPORODOCHIA)	70 F
TRICHOCLADIUM (ALEURI)	68 E
	92 D
CETERI PHAEO RACHI	
ACROCONIDIELLINA (SETAE)	61 A
CERCOSPORIDIUM (STROMA)	85 E
DENDRYPHION (STROMA)	84 E
HADRONEMA (SPORODOCHIA)	68 C
POLYTHRINCIUM (STROMA)	61 F
PYRICULARIA	80 A
CETERI PHAEO RADUL	
ACROCONIDIELLINA (SETAE)	61 A
ASPERISPORIUM (SPORODOCHIA)	63 D
BEEJADWAYA	—
CAMPTOMERIS (SPORODOCHIA)	85 D
CAPNOBOTRYS	63 G
CORDANA	62 B
DACTYLARIA	62 F
	81 F
DENDRYPHION (STROMA)	84 E
DIPLOCOCCIUM	66 C
ERIOMYCOPSIS (APPEND)	79 F
FUSICLADIUM (STROMA)	22 B
MARTINELLISIA	61 D
OCHROCONIS	62 E
PANCHANANIA (APPEND, SYNNEMATA)	63 F
PAPILIONOSPORA (SYNNEMATA)	67 C
PASSALORA	61 C
PERICONIELLA (SYNNEMATA)	84 G
PHAEOISARIOPSIS (STROMA, SYNNEMATA)	86 D
PSEUDOBOTRYTIS	62 C
PYRICULARIA	80 A
RAMICHLORIDIUM	25 F
SCEPTRIFERA	61 E
SCOLECOBASIDIELLA	62 D
SCOLECOBASIDIUM	117 F
SPADICOIDES	64 C
VERONAEA	61 G
GLOIO HYALO ANNEL	
MARSSONINA (SPORODOCHIA)	—
RHYNCHOSPORIUM (SPORODOCHIA)	59 E
GLOIO HYALO MISC	
BLASTOPHORUM	81 C
SIBIRINA	58 B
SYZYGOSPORA (CLAMPS)	51 G

GENERA WITH DIDYMOCONIDIA	Plate
GLOIO HYALO PHIAL	
CEPHALOSPORIOPSIS	57 C
CHAETOPSIS (SETAE)	57 B
CHALARA	2 C
CODINAEA (APPEND, SETAE, STROMA)	1 A
CRYPTOPHIALE (SETAE)	108 A
CYLINDROCARPON (SPORODOCHIA)	73 I
CYLINDROCLADIUM (SETAE)	73 F
CYLINDRODENDRUM	57 D
CYLINDROTRICHUM (SETAE)	57 A
DIDYMOSTILBE (SYNNEMATA)	58 D
DWAYALOMELLA (ACERVULI)	—
KUTILAKESOPSIS (SETAE, SPORODOCHIA)	58 A
LEPTOXYPHIUM (SYNNEMATA)	58 E
MARSSONINA (SPORODOCHIA)	—
MENISPORA (APPEND, SETAE)	1 C
MENISPOROPSIS (APPEND, SETAE, SYNNEMATA)	56 G
MYCOLEPTODISCUS (APPEND, SPORODOCHIA)	14 C
PHIALOGENICULATA	57 G
PSEUDOMICRODOCHIUM (SPORODOCHIA)	108 C
PUTAGRAIVAM (SYNNEMATA)	—
SEPTOMYROTHECIUM (SPORODOCHIA)	58 C
UNCIGERA	4 E
GLOIO HYALO PHIAL?	
PSEUDOGRAPHIELLA (SYNNEMATA)	74 A
GLOIO HYALO RACHI	
NETA (SETAE)	62 H
GLOIO HYALO RADUL	
BLASTOPHORUM	81 C
HYPHODISCOSIA (APPEND)	56 A
LEPTODISCELLA (APPEND, SPORODOCHIA)	56 C
NETA (SETAE)	62 H
GLOIO PHAEO ANNEL	
EXOPHIALA	20 B
	60 B
GLOIO PHAEO MISC	
BLASTOPHORUM	81 C
GLOIO PHAEO PHIAL	
CHALARA	2 C
CYPHELLOPHORA	57 F
LEPTOXYPHIUM (SYNNEMATA)	58 E

GENERA WITH DIDYMOCONIDIA	Plate
GLOIO PHAEO RADUL	
BLASTOPHORUM	81 C

GENERA WITH HELICOCONIDIA	Plate
ARTHRO PHAEO MISC	
HELICOCEPHALUM	105 F
BLASTO HYALO NONSP	
HELICODENDRON (AQUATIC)	105 A
	106 B
BLASTO PHAEO NONSP	
HELICODENDRON (AQUATIC)	105 A
	106 B
HIOSPIRA (SETAE?)	105 G
CETERI HYALO ANNEL	
FILOSPORELLA (AQUATIC)	109 A
PSEUDOCERCOSPORA (STROMA)	85 B
CETERI HYALO MISC	
DREPANOCONIS (SETAE)	102 E
CETERI HYALO NONSP	
ANGULOSPORA (AQUATIC)	107 A
DIPLORHYNCHUS	103 A
HELICOMA	102 A
HELICOMYCES	103 B
HELICOON	105 D
HELICOSPORIUM (APPEND, SETAE)	102 C
SPIROSPHAERA (AQUATIC)	106 A
TROPOSPORIUM (SPORODOCHIA)	105 C
CETERI HYALO RACHI	
HELICOMYCES	103 B
PSEUDOCERCOSPORA (STROMA)	85 B
CETERI HYALO RADUL	
HELICOMA	102 A
HELICOSPORIUM (APPEND, SETAE)	102 C
HYALOHELICOMINA	102 D
PSEUDOCERCOSPORA (STROMA)	85 B
SPIRALUM	106 F
CETERI PHAEO ANNEL	
CIRCINOCONIS	104 B
PSEUDOCERCOSPORA (STROMA)	85 B
XENOSPORIUM	104 C

GENERA WITH HELICOCONIDIA	Plate
CETERI PHAEO NONSP	
ACROSPEIRA (ALEURI)	96 B
CERATOPHORUM (APPEND, PODIA)	70 D
CIRRENALIA (SPORODOCHIA, AQUATIC)	104 D
CURCULIOSPORA	104 A
HELICOMA	102 A
HELICOON	105 D
HIOSPIRA (SETAE?)	105 G
SPIROSPHAERA (AQUATIC)	106 A
STRUMELLA	97 E
TROPOSPORELLA (SPORODOCHIA)	104 F
	104 G
XENOSPORIUM	104 C
ZALERION	106 E
CETERI PHAEO RACHI	
CURVULARIOPSIS (STROMA, SYNNEMATA, PODIA)	79 A
HELICORHOIDION	105 E
PSEUDOCERCOSPORA (STROMA)	85 B
CETERI PHAEO RADUL	
DICHOTOMOPHTHOROPSIS	106 C
HELICOMA	102 A
PSEUDOCERCOSPORA (STROMA)	85 B
SPIRALUM	106 F
TROCHOPHORA (SYNNEMATA)	104 E
GLOIO HYALO MISC?	
HOBSONIA (SPORODOCHIA)	105 B
GLOIO HYALO NONSP	
DELORTIA (SPORODOCHIA)	103 D
EVERHARTIA (SPORODOCHIA)	103 C
GLOIO HYALO PHIAL	
CRYPTOPHIALE (SETAE)	108 A
HARPOSPORIUM	107 G
VANBEVERWIJKIA (SPORODOCHIA)	103 E
GLOIO HYALO PHIAL?	
BRACHYHELICOON	106 D
GLOIO PHAEO MISC	
HELICOCEPHALUM	105 F

GENERA WITH PHRAGMOCONIDIA	Plate
ARTHRO? HYALO ANNEL?	
HORMISCIOIDEUS	—
ARTHRO HYALO NONSP	
CYLINDROCLADIOPSIS	93 C
SEPTOTRULLÚLA (SPORODOCHIA)	—
ARTHRO HYALO RACHI	
PARASYMPODIELLA	—
SYMPODIELLA	43 D
THEDGONIA (SPORODOCHIA)	67 D
ARTHRO HYALO RADUL	
CHIKANEEA	81 B
SYMPODIELLA	43 D
ARTHRO PHAEO MISC	
TRIMMATOSTROMA (SPORODOCHIA)	91 B
ARTHRO PHAEO NONSP	
AMPULLIFERINA (PODIA)	67 B
BAHUSAKALA	42 B
SCYTALIDIUM	42 E
SEPTOTRULLULA (SPORODOCHIA)	—
TORULOPSIELLA	91 A
BASO HYALO ANNEL	
REFRACTOHILUM	77 G
BASO HYALO MISC	
CLADOBOTRYUM	40 F
	67 F
BASO HYALO PHIAL	
ANTIPODIUM	73 E
CHALARA	2 C
FUSICHALARA	73 A
PENICILLIFER	57 E
SPOROSCHISMA (SETAE, STROMA)	73 C
BASO PHAEO ANNEL	
BACTRODESMIELLA	76 F
TRICHODOCHIUM (SETAE, STROMA, SPORODOCHIA)	60 C
BASO PHAEO PHIAL	
CHALARA	2 C
FUSARIELLA	73 B
FUSICHALARA	73 A
SPOROSCHISMA (SETAE, STROMA)	73 C
SPOROSCHISMOPSIS (STROMA)	73 D

GENERA WITH PHRAGMOCONIDIA	Plate
BLASTO HYALO NONSP	
SEPTOCYLINDRIUM	88 B
BLASTO HYALO RADUL	
ANUNGITEA	65 E
MYCOVELLOSIELLA (STROMA)	72 D
	82 B
PLEUROTHECIOPSIS	65 D
POLYSCYTALUM (SPORODOCHIA)	65 C
RAMULARIOPSIS	81 G
CHIKANEEA	81 B
BLASTO PHAEO ANNEL	
ANNELLOPHORA	75 A
CORYNESPORA (STROMA, SYNNEMATA)	75 B
HETEROCONIUM	76 G
BLASTO PHAEO MISC	
TORULA	88 D
BLASTO PHAEO NONSP	
AMPULLIFERA (PODIA)	36 A
CORYNESPORELLA (STROMA)	91 G
DENDROGRAPHIELLA (SYNNEMATA)	87 E
DIPLOCOCCIUM	66 C
HETEROCONIUM	76 G
LOBATOPEDIS (ALEURI?)	—
LYLEA	88 C
PARADENDRYPHIOPSIS	92 F
RUTOLA	88 H
SEPTONEMA	88 A
TAENIOLELLA	91 D
XYLOMYCES	91 C
BLASTO PHAEO RACHI	
ALTERNARIA (APPEND, SYNNEMATA)	101 A
CURVULARIA (STROMA)	79 C
DENDRYPHION (STROMA)	84 E
BLASTO PHAEO RADUL	
ANUNGITEA	65 E
CAPNODENDRON	93 B
CLADOSPORIUM (STROMA)	66 E
CURVULARIA (STROMA)	79 C
DENDROGRAPHIUM (STROMA, SYNNEMATA)	87 C
DENDRYPHION (STROMA)	84 E
DIPLOCOCCIUM	66 C
FULVIA (STROMA)	39 A
MYCOVELLOSIELLA (STROMA)	72 D
	82 B
PERICONIELLA (SYNNEMATA)	84 G
SCLEROGRAPHIOPSIS (SYNNEMATA)	62 A
STENELLA (STROMA)	82 A
TANDONELLA (SYNNEMATA)	86 E
CETERI HYALO AMPUL	
ARTHROBOTRYS	62 G
CETERI HYALO AMPUL?	
HAPTOCARA (APPEND)	69 D
CETERI HYALO ANNEL	
PSEUDOCERCOSPORA (STROMA)	85 B
SEPTOIDIUM (PODIA)	75 H
CETERI HYALO MISC	
CLADOBOTRYUM	40 F
	67 F
PARATRICHOCONIS	80 E
CETERI HYALO NONSP	
BACILLISPORA (AQUATIC)	89 C
BACTRIDIUM (SPORODOCHIA)	89 D
CHAETOCONIDIUM (APPEND)	71 A
CYLINDROCLADIOPSIS	93 C
DACTYLELLA (APPEND)	89 A
EPIDERMOPHYTON (ALEURI)	89 E
HARPAGOMYCES (APPEND)	124 E
LARIDOSPORA (APPEND)	112 E
MASTIGOSPORIUM (APPEND, SPORODOCHIA)	69 C
MICROSPORUM (ALEURI)	89 F
TRICHOCONIELLA (APPEND)	70 H
TRICHOCONIS (APPEND)	71 F
TRICHOPHYTON (ALEURI)	89 G
CETERI HYALO NONSP?	
SPERMOSPORA (APPEND)	70 G
CETERI HYALO PHIAL	
ASCOCONIDIUM	97 G
HELISCUS (AQUATIC)	117 D
MARGARITISPORA (APPEND, AQUATIC)	89 B
CETERI HYALO RACHI	
ARTHROSPORIUM (SYNNEMATA)	86 B
ATRACTILINA (SYNNEMATA)	72 F
BIDENTICULA	78 A
CERCOSPORA (STROMA)	78 D
CERCOSPORELLA	110 D
CERCOSPORIDIUM (STROMA)	85 E
CHIONOMYCES (APPEND)	71 C
DACTYLELLA (APPEND)	89 A
DENDROSPORIUM (AQUATIC)	113 D
HYALOSYNNEMA (SYNNEMATA)	86 A
MYCOCENTROSPORA (APPEND, SPORODOCHIA, AQUATIC)	70 C
PLEIOCHAETA (APPEND)	69 A

GENERA WITH PHRAGMOCONIDIA

	Plate
PLEUROTHECIUM	80 C
PSEUDOCERCOSPORA (STROMA)	85 B
PSEUDOCERCOSPORELLA	81 D
PSEUDOHANSFORDIA	63 B
PYRICULARIA	80 A
TRICHOCONIS (APPEND)	71 F
CETERI HYALO RADUL	
ARTHROBOTRYS	62 G
CAMPOSPORIUM (APPEND)	69 E
CERCOSPORA (STROMA)	78 D
DACTYLARIA	62 F
	81 F
DENDROSPORIUM (AQUATIC)	113 D
ERIOMYCOPSIS (APPEND)	79 F
EXSERTICLAVA (STROMA)	—
ISTHMOLONGISPORA	30 C
MONODISMA	78 B
PARACHINOMYCES	—
PARATRICHOCONIS	80 E
PHAEODACTYLIUM	81 E
PSEUDOCERCOSPORA (STROMA)	85 B
PYRICULARIA	80 A
REDBIA	83 C
SPERMOSPORELLA (APPEND)	71 D
SPIRALUM	106 F
SPONDYLOCLADIOPSIS (SETAE)	80 F
CETERI HYALO RADUL?	
HELMINTHOPHORA	80 D
CETERI PHAEO AMPUL	
CEPHALIOPHORA	64 A
	93 D
DICHOTOMOPHTHORA	84 B
CETERI PHAEO ANNEL	
ACROPHRAGMIS (APPEND)	70 B
ANNELLOPHORA	75 A
ANNELLOPHORELLA	98 D
ANNELLOPHRAGMIA (SPORODOCHIA, SYNNEMATA)	87 D
ANTENNOPSIS	76 C
BACTRODESMIELLA	76 F
BRACHYSPORIELLA	77 E
CHAETENDOPHRAGMIA (APPEND)	121 A
CHAETENDOPHRAGMIOPSIS (APPEND)	—
CLASTEROSPORIUM (SETAE, PODIA)	75 G
COOKSONOMYCES (SPORODOCHIA)	77 C
CORYNESPORA (STROMA, SYNNEMATA)	75 B
CORYNEUM (ACERVULI)	—

GENERA WITH PHRAGMOCONIDIA

	Plate
DEIGHTONIELLA	59 D
ENDOPHRAGMIA (STROMA, SYNNEMATA)	77 A
ENDOPHRAGMIELLA	59 A
ENDOPHRAGMIOPSIS (PODIA)	76 A
EXOSPORIELLA (APPEND, STROMA)	75 E
PHRAGMOSPATHULA (STROMA)	76 D
PODOSPORIUM (STROMA, SYNNEMATA)	77 B
PSEUDOCERCOSPORA (STROMA)	85 B
SPORIDESMIUM (STROMA)	75 C
	75 D
STIGMINA (SPORODOCHIA)	76 B
	98 A
STILBOSPORA (ACERVULI)	—
TERATOSPERMA (APPEND)	75 F
TOXOSPORIUM (SPORODOCHIA)	77 F
CETERI PHAEO MISC	
ACAROCYBE (APPEND, SYNNEMATA)	86 C
BRACHYSPORIUM	80 B
DENDRYPHIOSPHAERA	—
EDMUNDMASONIA	84 A
HORMOCEPHALUM	92 C
KRAMASAMUHA	92 A
POLYSCHEMA	90 H
SESSILIOSPORA	91 H
SPONDYLOCLADIELLA	84 C
CETERI PHAEO MISC?	
TRIADELPHIA (SPORODOCHIA)	70 F
CETERI PHAEO NONSP	
ACAROCYBELLA (APPEND)	91 F
AMPULLIFERA (PODIA)	36 A
ANTENNATULA	91 E
ARTHROCLADIUM	71 B
BACTRODESMIUM (ALEURI, SPORODOCHIA)	90 B
CAPNOKYMA	92 E
CAPNOSPORIUM	93 A
CERATOPHORUM (APPEND, PODIA)	70 D
CHAETENDOPHRAGMIOPSIS (APPEND)	—
CLASTEROSPORIUM (SETAE, PODIA)	75 G
CORYNESPORELLA (STROMA)	91 G
DENDRYPHIOPSIS	92 B
DIPLOCOCCIUM	66 C
EXCIPULARIA (SETAE, SPORODOCHIA)	90 C
HANSFORDIELLA (APPEND)	94 C
HANSFORDIELLOPSIS (APPEND, PODIA?)	94 A

GENERA WITH PHRAGMOCONIDIA Plate

HORMOCEPHALUM	92	C
LISTEROMYCES (ALEURI, SPORODOCHIA)	90	F
LOBATOPEDIS (ALEURI?)	—	
LOMAANTHA (APPEND)	69	B
MUROGENELLA (ALEURI)	90	G
PITHOMYCES (ALEURI)	90	A
	100	C
PODOSPORIUM (STROMA, SYNNEMATA)	77	B
STEMPHYLIOMMA (ALEURI?)	76	H
TRETOPHRAGMIA (SYNNEMATA)	87	G
TRETOSPORA (APPEND, PODIA)	72	B
TRIADELPHIA (SPORODOCHIA)	70	F
TRICHOCLADIUM (ALEURI)	68	E
	92	D

CETERI PHAEO NONSP?

NEGERIELLA (SYNNEMATA)	87	H

CETERI? PHAEO NONSP?

GANGLIOSTILBE (SYNNEMATA)	87	F

CETERI PHAEO RACHI

ACROCONIDIELLA	82	G
ACROCONIDIELLINA (SETAE)	61	A
ATRACTILINA (SYNNEMATA)	72	F
CERCOSPORA (STROMA)	78	D
CERCOSPORIDIUM (STROMA)	85	E
CURVULARIA (STROMA)	79	C
CURVULARIOPSIS (STROMA, SYNNEMATA, PODIA)	79	A
DENDRYPHION (STROMA)	84	E
DENDRYPHIOPSIS	92	B
DRECHSLERA (STROMA)	83	E
DUOSPORIUM (STROMA)	82	F
EDMUNDMASONIA	84	A
EMBELLISIA	82	H
EXOSPORIUM (SPORODOCHIA, SYNNEMATA)	83	C
MYCOCENTROSPORA (APPEND, SPORODOCHIA, AQUATIC)	70	C
PARATOMENTICOLA	72	E
PHAEOTRICHOCONIS (APPEND)	70	A
PLEIOCHAETA (APPEND)	69	A
PLEUROTHECIUM	80	C
PSEUDOCERCOSPORA (STROMA)	85	B
PYRICULARIA	80	A
SPIROPES (SYNNEMATA)	83	D
STENELLOPSIS (STROMA)	85	C
TOMENTICOLA	83	B

CETERI PHAEO RADUL

ACROCONIDIELLINA (SETAE)	61	A
ANNELLOPHRAGMIA (SPORODOCHIA, SYNNEMATA)	87	D
BRACHYDESMIELLA (APPEND)	93	E

GENERA WITH PHRAGMOCONIDIA Plate

BRACHYSPORIUM	80	B
CAMPOSPORIUM (APPEND)	69	E
CAMPTOMERIS (SPORODOCHIA)	85	D
CAPNOBOTRYS	63	G
CERCOSPORA (STROMA)	78	D
CURVULARIA (STROMA)	79	C
DACTYLARIA	62	F
	81	F
DENDROGRAPHIUM (STROMA, SYNNEMATA)	87	C
DENDRYPHION (STROMA)	84	E
DICHOTOMOPHTHORA	84	B
DIPLOCOCCIUM	66	C
DRECHSLERA (STROMA)	83	E
ERIOCERCOSPORA	82	D
ERIOMYCOPSIS (APPEND)	79	F
EXSEROHILUM	83	G
GONATOPHRAGMIUM	79	E
GUEDEA (SETAE)	—	
HELMINTHOSPORIUM (STROMA)	83	F
JANETIA	79	D
MITTERIELLA (PODIA)	82	E
MYSTROSPORIELLA	99	A
NAKATAEA (PODIA)	79	B
OCHROCONIS	62	E
PARACERCOSPORA (STROMA)	—	
PARAPYRICULARIA	84	F
PASSALORA	61	C
PERICONIELLA (SYNNEMATA)	84	G
PHAEOISARIOPSIS (STROMA, SYNNEMATA)	86	D
PLEUROPHRAGMIUM	81	A
PODOSPORIELLA (SPORODOCHIA, SYNNEMATA)	87	A
PRATHIGADA (APPEND, SPORODOCHIA)	72	C
PSEUDOCERCOSPORA (STROMA)	85	B
PSEUDOSPIROPES (STROMA)	83	A
PYRICULARIA	80	A
PYRICULARIOPSIS	72	A
SCOLECOBASIDIUM	117	F
SIROSPORIUM (STROMA)	99	E
SPADICOIDES	64	C
SPIRALUM	106	F
SPIROPES (SYNNEMATA)	83	D
VERRUCISPORA (STROMA)	85	A

GLOIO HYALO ANNEL

ARTHROBOTRYUM (SYNNEMATA)	77	D

GLOIO HYALO MISC

BLASTOPHORUM	81	C

GENERA WITH PHRAGMOCONIDIA Plate

GLOIO HYALO NONSP
- *MASTIGOSPORIUM* (APPEND, SPORODOCHIA) 69 C
- *SEPTOTRULLULA* (SPORODOCHIA) —

GLOIO HYALO NONSP?
- *SPERMOSPORA* (APPEND) 70 G

GLOIO HYALO PHIAL
- *ATRACTIUM* (SYNNEMATA) 111 F
- *CHALARA* 2 C
- *CIFERRIOXYPHIUM* (SYNNEMATA) 74 D
- *CODINAEA* (APPEND, SETAE, STROMA) 1 A
- *CYLINDROCARPON* (SPORODOCHIA) 73 I
- *CYLINDROCLADIUM* (SETAE) 73 F
- *CYLINDROTRICHUM* (SETAE) 57 A
- *FUSARIUM* (SPORODOCHIA) 73 H
- *MENISPORA* (APPEND, SETAE) 1 C
- *SEPTODOCHIUM* (SPORODOCHIA) 74 E

GLOIO HYALO PHIAL?
- *CALOSTILBELLA* (SYNNEMATA) 74 F
- *MORRISOGRAPHIUM* (SYNNEMATA) 111 D
- *PSEUDOGRAPHIELLA* (SYNNEMATA) 74 A

GLOIO HYALO RACHI
- *MORRISOGRAPHIUM* (SYNNEMATA) 111 D
- *PLEUROTHECIUM* 80 C
- *STERIGMATOBOTRYS* (STROMA) 84 D

GLOIO HYALO RADUL
- *BLASTOPHORUM* 81 C
- *SEPTOTIS* (SPORODOCHIA) —
- *WIESNERIOMYCES* (SETAE, SPORODOCHIA) 111 E

GLOIO PHAEO ANNEL
- *ARTHROBOTRYUM* (SYNNEMATA) 77 D
- *CACUMISPORIUM* 76 E
- *COOKSONOMYCES* (SPORODOCHIA) 77 C
- *MONOCHAETIA* (APPEND) —
- *SEIMATOSPORIUM* (ACERVULI) —

GLOIO PHAEO MISC
- *BLASTOPHORUM* 81 C
- *CACUMISPORIUM* 76 E

GENERA WITH PHRAGMOCONIDIA Plate

GLOIO PHAEO NONSP
- *SEPTOTRULLULA* (SPORODOCHIA) —

GLOIO PHAEO PHIAL
- *CHALARA* 2 C
- *CIFERRIOXYPHIUM* (SYNNEMATA) 74 D
- *CYPHELLOPHORA* 57 F
- *HETEROSEPTATA* (SYNNEMATA) 74 C
- *PESTALOTIA* (APPEND, ACERVULI) —
- *PHIALOARTHROBOTRYUM* (SYNNEMATA) 74 H
- *SPOROSCHISMOPSIS* (STROMA) 73 D
- *VIRGATOSPORA* (SYNNEMATA) 74 G

GLOIO PHAEO PHIAL?
- *ALBOSYNNEMA* (SYNNEMATA) 74 B
- *CALOSTILBELLA* (SYNNEMATA) 74 F

GLOIO PHAEO RACHI
- *CAPNOCYBE* (SYNNEMATA) 87 B
- *PLEUROTHECIUM* 80 C
- *STERIGMATOBOTRYS* (STROMA) 84 D

GLOIO PHAEO RADUL
- *BLASTOPHORUM* 81 C
- *CAPNOCYBE* (SYNNEMATA) 87 B

GENERA WITH SCOLECOCONIDIA Plate

BASO HYALO PHIAL
- *RICLARETIA* (SPORODOCHIA) 17 C

BASO? HYALO PHIAL
- *GRANULOMANUS* —

BLASTO HYALO RADUL
- *MYCOVELLOSIELLA* (STROMA) 72 D, 82 B
- *RAMULARIOPSIS* 81 G

BLASTO PHAEO ANNEL
- *CORYNESPORA* (STROMA, SYNNEMATA) 75 B

BLASTO PHAEO NONSP
- *CORYNESPORELLA* (STROMA) 91 G

BLASTO PHAEO RADUL
- *CLADOSPORIELLA* 78 F
- *MYCOVELLOSIELLA* (STROMA) 72 D, 82 B

GENERA WITH SCOLECOCONIDIA	Plate
CETERI HYALO ANNEL	
FILOSPORELLA (AQUATIC)	109 A
PSEUDOCERCOSPORA (STROMA)	85 B
CETERI HYALO MISC	
CIRCINOTRICHUM (SETAE, STROMA)	54 F
GYROTHRIX (SETAE, STROMA)	54 G
CETERI HYALO NONSP	
ANGUILLOSPORA (AQUATIC)	108 G
ANGULOSPORA (AQUATIC)	107 A
BACILLISPORA (AQUATIC)	89 C
DACTYLELLA (APPEND)	89 A
PETRAKIOPSIS	116 E
VERMISPORA	78 E
CETERI HYALO NONSP?	
LUNULOSPORA (AQUATIC)	107 H
SIGMOIDEA (AQUATIC)	108 D
CETERI HYALO PHIAL	
CALCARISPORA (AQUATIC)	109 D
FLAGELLOSPORA (AQUATIC)	109 C
CETERI HYALO RACHI	
CERCOSEPTORIA	110 B
CERCOSPORA (STROMA)	78 D
CERCOSPORELLA	110 D
CHIONOMYCES (APPEND)	71 C
CONDYLOSPORA (AQUATIC)	108 F
DACTYLELLA (APPEND)	89 A
MICRODOCHIUM (SPORODOCHIA)	14 E
	108 B
MYCOCENTROSPORA (APPEND, SPORODOCHIA, AQUATIC)	70 C
PSEUDOANGUILLOSPORA (AQUATIC)	108 E
PSEUDOCERCOSPORA (STROMA)	85 B
PSEUDOCERCOSPORELLA	81 D
PTERULOPSIS (SYNNEMATA)	111 B
SUBULISPORA (APPEND)	71 E
CETERI HYALO RADUL	
CERCOSPORA (STROMA)	78 D
CIRCINOTRICHUM (SETAE, STROMA)	54 F
DACTYLARIA	62 F
	81 F
ERIOMYCOPSIS (APPEND)	79 F
IDRIELLA	21 B
ISTHMOLONGISPORA	30 C
MICRODOCHIUM (SPORODOCHIA)	14 E
	108 B

GENERA WITH SCOLECOCONIDIA	Plate
MIRANDINA	109 B
PAPPIMYCES	110 F
PSEUDOCERCOSPORA (STROMA)	85 B
SCHIZOTRICHUM (SETAE, STROMA)	110 E
CETERI PHAEO ANNEL	
CORYNESPORA (STROMA, SYNNEMATA)	75 B
PSEUDOCERCOSPORA (STROMA)	85 B
SPORIDESMIUM (STROMA)	75 D
CETERI PHAEO MISC	
DWAYABEEJA	88 G
CETERI PHAEO NONSP	
ACAROCYBELLA (APPEND)	91 F
ARTHROCLADIUM	71 B
CORYNESPORELLA (STROMA)	91 G
IRPICOMYCES	110 C
UNCISPORA (SYNNEMATA)	70 E
CETERI PHAEO RACHI	
CERCOSEPTORIA	110 B
CERCOSPORA (STROMA)	78 D
MYCOCENTROSPORA (APPEND, SPORODOCHIA, AQUATIC)	70 C
PSEUDOCERCOSPORA (STROMA)	85 B
CETERI PHAEO RADUL	
CERCOSPORA (STROMA)	78 D
DACTYLARIA	62 F
	81 F
ERIOCERCOSPORA	82 D
ERIOMYCOPSIS (APPEND)	79 F
PSEUDOCERCOSPORA (STROMA)	85 B
GLOIO HYALO MISC	
CIRCINOTRICHUM (SETAE, STROMA)	54 F
GLOIO HYALO PHIAL	
ATRACTIUM (SYNNEMATA)	111 F
CRYPTOPHIALE (SETAE)	108 A
HARPOSPORIUM	107 G
KMETIA	107 D
KMETIOPSIS	107 F
LIBERTELLA (ACERVULI)	—
MICROLYCHNUS (SYNNEMATA)	111 A
PSEUDOMICRODOCHIUM (SPORODOCHIA)	108 C

GENERA WITH SCOLECOCONIDIA Plate

GLOIO HYALO PHIAL?		
EPHELIS (SETAE, SPORODOCHIA)	107	E
HARPOGRAPHIUM (SYNNEMATA)	107	C
MORRISOGRAPHIUM (SYNNEMATA)	111	D
GLOIO HYALO RACHI		
MORRISOGRAPHIUM (SYNNEMATA)	111	D
GLOIO HYALO RADUL		
CIRCINOTRICHUM (SETAE, STROMA)	54	F
IDRIELLA	21	B
LINODOCHIUM (SPORODOCHIA)	109	F
	110	C
RAMULISPORA (SPORODOCHIA)	110	A
SELENOSPORELLA	21	A
WIESNERIOMYCES (SETAE, SPORODOCHIA)	111	E

GENERA WITH STAUROCONIDIA Plate

BLASTO HYALO NONSP		
VARICOSPORIUM (AQUATIC)	123	A
BLASTO HYALO RADUL		
CATENOSUBULISPORA	124	C
BLASTO PHAEO NONSP		
TAENIOLINA	127	C
CETERI HYALO AMPUL		
ENGELHARDTIELLA (SPORODOCHIA)	117	B
NIA (APPEND, SPORODOCHIA, CLAMPS)	121	B
CETERI HYALO ANNEL		
BIFLAGELLOSPORA	112	F
TRIPOSPORINA	117	A
CETERI HYALO MISC		
DIGITATISPORA (CLAMPS)	119	D
RIESSIA (CLAMPS)	113	G
CETERI HYALO NONSP		
ACAULOPAGE (APPEND)	112	B
ACTINOSPORA (APPEND, AQUATIC)	120	C
ARTICULOSPORA (AQUATIC)	119	B
ASTEROPHORA (APPEND, SYNNEMATA, CLAMPS)	51	E
ATICHIA (SPORODOCHIA)	122	G
BIFLAGELLOSPORA	112	F
CAMPYLOSPORA (APPEND, AQUATIC)	118	B

GENERA WITH STAUROCONIDIA Plate

CANDELABRUM (APPEND, AQUATIC)	114	A
CLAVARIOPSIS (APPEND, AQUATIC)	119	E
CULICIDOSPORA (APPEND, AQUATIC)	112	D
DENDROSPORA (AQUATIC)	123	D
DENDROSPOROMYCES (AQUATIC)	123	B
FLABELLOSPORA (AQUATIC)	115	C
	120	F
FONTANOSPORA (AQUATIC)	119	C
GLOMOPSIS (APPEND)	114	C
GYOERFFIELLA (APPEND, AQUATIC)	122	A
	122	C
HARPAGOMYCES (APPEND)	124	E
INGOLDIELLA (CLAMPS, AQUATIC)	124	A
ISTHMOTRICLADIA (AQUATIC)	119	F
JACULISPORA (APPEND, AQUATIC)	120	D
LAMBDASPORIUM	122	B
LARIDOSPORA (APPEND)	112	E
LATERIRAMULOSA (AQUATIC)	121	D
NIA (APPEND, SPORODOCHIA, CLAMPS)	121	B
PETRAKIOPSIS	116	E
POLYCLADIUM (AQUATIC)	122	E
PSAMMINA (ACERVULI?)	115	A
PYRAMIDOSPORA (AQUATIC)	113	E
RIESSIA (CLAMPS)	113	G
SCORPIOSPORIUM (AQUATIC)	123	C
SPERMATOLONCHA (STROMA)	125	D
STEPHANOMA (ALEURI, APPEND)	113	C
TAENIOSPORA (CLAMPS, AQUATIC)	124	B
TETRACHAETUM (AQUATIC)	118	A
TETRACLADIUM (AQUATIC)	112	H
TETRACRIUM (SPORODOCHIA)	125	C
THALLOSPORA	122	F
TITAEA	112	A
	112	C
	121	E
TITAEELLA (CLAMPS)	124	D
TRICELLULA (AQUATIC)	117	C
TRICLADIUM (AQUATIC)	123	F
TRIDENTARIA (AQUATIC)	118	F
TRINACRIUM (AQUATIC)	125	F
TRIPOCONIDIUM	—	
TRISCELOPHORUS (AQUATIC)	120	H
TRISULCOSPORIUM (AQUATIC)	120	E
VARICOSPORINA (AQUATIC)	123	E
VARICOSPORIUM (AQUATIC)	123	A

GENERA WITH STAUROCONIDIA	Plate
CETERI HYALO PHIAL	
ALATOSPORA (APPEND, AQUATIC)	118 D
CLAVATOSPORA (APPEND, AQUATIC)	119 A
HELISCUS (AQUATIC)	117 D
LEMONNIERA (AQUATIC)	120 A
MARGARITISPORA (APPEND, AQUATIC)	89 B
CETERI HYALO RACHI	
BRACHIOSPHAERA (AQUATIC)	115 B
CLAVARIANA (APPEND, AQUATIC)	118 C
DENDROSPORIUM (AQUATIC)	113 D
DICRANIDION (SYNNEMATA)	116 F
FONTANOSPORA (AQUATIC)	119 C
POROCLADIUM (AQUATIC)	120 G
TRINACRIUM (AQUATIC)	125 F
CETERI HYALO RADUL	
CATENOSUBULISPORA	124 C
DENDROSPORIUM (AQUATIC)	113 D
DICRANIDION (SYNNEMATA)	116 F
KAZULIA (SETAE, SPORODOCHIA)	118 E
POLYCLADIUM (AQUATIC)	122 E
RETIARIUS	—
SPEIROPSIS (AQUATIC)	116 D
CETERI HYALO RADUL?	
TRICELLULA (AQUATIC)	117 C
CETERI PHAEO ANNEL	
ACROPHRAGMIS (APPEND)	70 B
ACROSTAURUS	125 B
ARACHNOPHORA (APPEND)	115 D
BIFLAGELLOSPORA	112 F
CERATOSPORELLA (STROMA)	127 D
CHAETENDOPHRAGMIA (APPEND)	121 A
GRALLOMYCES (APPEND, PODIA?)	127 B
IYENGARINA (APPEND)	112 G
TERATOSPERMA (APPEND)	75 F
TETRABRUNNEOSPORA (AQUATIC)	119 G
TRIPOSPORIUM	126 A
UBERISPORA (APPEND)	114 E
CETERI PHAEO ANNEL?	
PHYSALIDIUM (APPEND)	114 D
CETERI PHAEO MISC	
ISTHMOSPORA	96 H
SPEGAZZINIA (APPEND)	114 B

GENERA WITH STAUROCONIDIA	Plate
CETERI PHAEO NONSP	
ACTINOCLADIUM	126 E
ANAVIRGA	127 A
ATICHIA (SPORODOCHIA)	122 G
BIFLAGELLOSPORA	112 F
BIOCONIOSPORIUM (APPEND)	95 A
CASARESIA (APPEND, AQUATIC)	127 E
CERATOSPORELLA (STROMA)	127 D
CERATOSPORIUM	126 B
CRYPTOCORYNEUM (SPORODOCHIA)	116 G
DESMIDIOSPORA	113 B
HIRUDINARIA	127 F
HORMIOKRYPSIS	126 G
HUGHESINIA (PODIA)	126 C
LAMBDASPORIUM	122 B
MILOSPIUM	113 A
ONCOPODIELLA (APPEND)	95 D
PETRAKIA (APPEND, SPORODOCHIA)	115 E
PHYSALIDIUM (APPEND)	114 D
PSAMMINA (ACERVULI?)	115 A
PSEUDOPETRAKIA (APPEND)	94 F
SOPAGRAHA	114 F
TAENIOLINA	127 C
TETRAPLOA (APPEND)	125 A
TETRAPOSPORIUM	120 B
TRIPOSPERMUM (AQUATIC)	126 F
TRIRAMULISPORA	121 C
VALDENSIA	125 E
CETERI PHAEO NONSP?	
CHEIROMYCES (SPORODOCHIA)	116 A
ORBIMYCES (APPEND)	115 F
CETERI PHAEO RACHI	
ANAVIRGA	127 A
BRACHIOSPHAERA (AQUATIC)	115 B
CORONOSPORA	117 E
DIPLOCLADIELLA (APPEND)	126 D
CETERI PHAEO RADUL	
ANAVIRGA	127 A
SCOLECOBASIDIUM	117 F
SPEIROPSIS (AQUATIC)	116 D
GLOIO HYALO NONSP	
ATICHIA (SPORODOCHIA)	122 G
GLOEOSPORIELLA (APPEND, SPORODOCHIA)	56 B
UVARISPORA (SPORODOCHIA)	113 F
GLOIO HYALO NONSP?	
AMALLOSPORA	116 C
FURCASPORA (ACERVULI)	—

GENERA WITH STAUROCONIDIA

	Plate
GLOIO HYALO PHIAL	
BAHUSUTRABEEJA (APPEND)	1 E
GLOIO HYALO RADUL	
RAMULISPORA (SPORODOCHIA)	110 A
GLOIO PHAEO NONSP	
ATICHIA (SPORODOCHIA)	122 G

GENERA WITH BULBILS

	Plate
AEGERITA	129 E
ARBUSCULA	129 D
BEVERWYKELLA	129 B
BURGOA	128 F
CANCELLIDIUM	129 A
CHEIROSPORA	—
CLATHROSPHAERINA	97 H
CRISTULARIELLA	129 C
DACTULIOPHORA	129 G
ECHINOCHONDRIUM	128 B
GLOMOPSIS	114 C
MILOSPIUM	113 A
MINIMEDUSA	128 G
PAPULASPORA	128 E
PEYRONELINA	128 C
RHACOPHYLLUS	—
RHIZOCTONIA	—
SOPAGRAHA	114 F
SPINULOSPORA	129 F
SPIROSPHAERA	106 A
STRUMELLA	97 E
SYAMITHABEEJA	99 G
TRETOPILEUS	128 A
UVARISPORA	113 F

GENERA WITH SCLEROTIA

	Plate
AKENOMYCES	—
COCCOBOTRYS	—
SCLEROTIUM	—

STERILE GENERA

	Plate
ATTAMYCES	—
MADURELLA	—
OVEREEMIA	—
RHIZOCTONIA	—
SAPROCHAETE	—

Index of Perfect State Connections

A

ACANTHOSTIGMELLA	(A)	*XENOSPORIUM*
ACAULOPAGE	(?Z)	*ACAULOPAGE*
ACROSPERMUM	(A)	*DACTYLARIA*
ACTINODENDRON	(A)	*ONCOCLADIUM*
AITHALODERMA	(A)	*CIFERRIOXYPHIUM*
AJELLOMYCES	(A)	*BLASTOMYCES, CHRYSOSPORIUM*
AKAROPELTELLA	(A)	*SEPTONEMA, SPORIDESMIUM*
ALEURODISCUS	(B)	*MATULA, MICHENERA*
ALNIA	(A)	*SEPTOIDIUM*
AMBIVINA	(B)	*AMBIVINA*
AMORPHOTHECA	(A)	*HORMOCONIS, SOROCYBE*
AMYLOSTEREUM	(B)	"ARTHROCONIDIA"
ANIXIOPSIS	(A)	*CHRYSOSPORIUM*
ANTHOSTOMA	(A)	*?AGARICOSTILBUM*
ANTHRACOBIA	(A)	*?SCYTALIDIUM*
ANTHRACOIDEA	(B)	*CROTALIA*
APHANOASCUS	(A)	*PAECILOMYCES*
APINISIA	(A)	*CHRYSOSPORIUM*
APIOCREA	(A)	*SEPEDONIUM*
APIOSORDARIA	(A)	*CLADORRHINUM, PHIALOPHORA*
APIOSPORA	(A)	*ARTHRINIUM, NIGROSPORA, PTEROCONIUM*
APIOSPORINA	(A)	*CLADOSPORIUM, FUSICLADIUM*
APODOSPORA	(A)	"PHIALOCONIDIA"
ARACHNIOTUS	(A)	"ALEURIOCONIDIA", "ARTHROCONIDIA", *CHRYSOSPORIUM*
ARACHNOTHECA	(A)	*?ARTHROGRAPHIS*
ARMILLARIA	(B)	*RHIZOMORPHA*
ARNIUM	(A)	*CLADORRHINUM*
ARTHRODERMA	(A)	*CHRYSOSPORIUM, KERATINOMYCES, TRICHOPHYTON*
ARTHROSPORELLA	(B)	*NOTHOCLAVULINA*
ASCOBOLUS	(A)	*MONILIA, PAPULASPORA*
ASCOCALVATIA	(A)	"ARTHROCONIDIA"
ASCOCORTICIUM	(A)	*ACRODONTIUM*
ASCOCORYNE	(A)	*CORYNE, PHIALOPHORA*
ASCOPHANUS	(A)	*OEDOCEPHALUM*
ASCOTRICHA	(A)	*DICYMA*
ASPERGILLUS	(A)	*RAPERIA*
ASPERODOTHIS	(A)	*CLASTEROSPORIUM*

ASTEROMASSARIA	(A)	CORYNEUM
ASTERIDIELLA	(A)	"PHIALOCONIDIA"
ASTEROSTROMELLA	(B)	SPINIGER
ATHELIA	(B)	RHIZOCTONIA
AULOGRAPHINA	(A)	BAHUSAKALA
AUXARTHRON	(A)	MALBRANCHEA

B

BALANSIA	(A)	EPHELIS
BALLADYNOPSIS	(A)	TRETOSPORA
BATISTINULA	(A)	TRIPOSPORIUM
BETTSIA	(A)	CHRYSOSPORIUM
BIOSTICTUS	(A)	FUSIDIUM
BISPORELLA	(A)	CYSTODENDRON (?=PHIALOPHORA)
BJERKANDERA	(B)	"ARTHROCONIDIA"
BONDARZEWIA	(B)	SPINIGER
BOTRYOASCUS	(A)	RAFFAELEA
BOTRYOBASIDIUM	(B)	ACLADIUM, ALLESCHERIELLA, ALYSIDIUM, RHIZOCTONIA
BOTRYOCONIS	(B)	BOTRYOCONIS
BOTRYOTINIA	(A)	AMPHOBOTRYS, BOTRYTIS
BROOKSIA	(A)	HIOSPIRA, OVEREEMIA
BROOMELLA	(A)	MONOTRICHUM, PESTALOTIA
BUERGENERULA	(A)	RHYNCHOSPORINA-LIKE
BULBILLOMYCES	(B)	AEGERITA-LIKE
BYSSOASCUS	(A)	OIDIODENDRON
BYSSOCHLAMYS	(A)	PAECILOMYCES
BYSSOSTILBE	(A)	STILBELLA, TILACHLIDIUM

C

CALLORIA	(A)	CYLINDROCOLLA
CALLORIOPSIS	(A)	ERIOMYCOPSIS
CALONECTRIA	(A)	ACREMONIUM, CYLINDROCARPON, CYLINDROCLADIUM, FUSARIUM, VERTICILLIUM
CALOSCYPHA	(A)	GENICULODENDRON
CALYCELLINA	(A)	CHAETOCHALARA
CEPHALOASCUS	(A)	AUREOMYCES, HYALODENDRON, MONILIELLA
CEPHALOTHECA	(A)	ACREMONIUM, TRITIRACHIUM
CERATOBASIDIUM	(B)	RHIZOCTONIA
CERATOCYSTIOPSIS	(A)	HYALORHINOCLADIELLA, SPOROTHRIX, VERTICICLADIELLA
CERATOCYSTIS	(A)	ACREMONIUM, CHALARA, GRAPHILBUM, HUMICOLA, HYALODENDRON, HYALOPESOTUM, HYALORHINOCLADIELLA, LEPTOGRAPHIUM, PACHNODIUM, PESOTUM, PHIALOCEPHALA, PHIALOGRAPHIUM, SPOROTHRIX, THIELAVIOPSIS, VERTICICLADIELLA, SEE ALSO OPHIOSTOMA (A)
CHAETOMIUM	(A)	BOTRYOTRICHUM, HUMICOLA, PAPULASPORA, "PHIALOCONIDIA", SCOPULARIOPSIS, THERMOMYCES, TRICHOCLADIUM
CHAETOSARTORYA	(A)	ASPERGILLUS
CHAETOSPHAERELLA	(A)	OEDEMIUM, PHIALOCEPHALA
CHAETOSPHAERIA	(A)	BISPOROMYCES, CATENULARIA, CHLORIDIUM, CODINAEA, GONYTRICHUM, MENISPORA, SPOROSCHISMA, STACHYBOTRYS, ZANCLOSPORA
CHRISTIANSENIA	(B)	SYZYGOSPORA
CIBORIA	(A)	MYRIOCONIUM
CLATHRIDIUM	(A)	SEIMATOSPORIUM

CLATHROSPORA	(A)	*ALTERNARIA*
CLAVICEPS	(A)	*SPHACELIA*
CLEISTOIODOPHANUS	(A)	*OEDOCEPHALUM*
CLINOCONIDIUM	(B)	*CLINOCONIDIUM*
CLYPEOLELLA	(A)	*CLASTEROSPORIUM, MITTERIELLA, SARCINELLA*
COCHLIOBOLUS	(A)	*CURVULARIA, DRECHSLERA*
COLLYBIA	(B)	*SCLEROSTILBUM, SPORENDONEMA*-LIKE
COMOCLATHRIS	(A)	*ALTERNARIA*
CONIDIASCUS	(A)	*AUREOMYCES*
CONIOCHAETA	(A)	*ACREMONIUM, HORMONEMA, NODULISPORIUM, PHIALOPHORA*
CONIOCHAETIDIUM	(A)	"ALEURIOCONIDIA", *?GRAPHIUM*
CONIODICTYUM	(B)	*CONIODICTYUM, HYALODENDRON*
CONIOPHORA	(B)	*SPORENDONEMA*-LIKE
COPRINUS	(B)	"ARTHROCONIDIA", *OZONIUM, RHACOPHYLLUS*
CORDYCEPS	(A)	*ACREMONIUM, AKANTHOMYCES, CERATONEMA, GIBELLULA, HIRSUTELLA, HYMENOSTILBE, PAECILOMYCES, SPHACELIA, VERTICILLIUM*
CORIOLOPSIS	(B)	"ARTHROCONIDIA"
COROLLOSPORA	(A)	*CLAVARIOPSIS*
CORTICIUM	(B)	*HIMANTIA, MATULA, NECATOR, PAPULASPORA, SCLEROTIUM, SPINIGER, SPOROTRICHUM*
CORYNASCUS	(A)	*CHRYSOSPORIUM, MYCELIOPHTHORA, SEPEDONIUM*
CORYNELLA	(A)	*DENDROSTILBELLA*
CRATEROCOLLA	(B)	*DITANGIUM*
CRYPTENDOXYLA	(A)	*CHALARA*
CTENOMYCES	(A)	*CHRYSOSPORIUM*
CUBONIA	(A)	*PAPULASPORA*
CYMADOTHEA	(A)	*POLYTHRINCIUM*

D

DACTYLOMYCES	(A)	*PAECILOMYCES, POLYPAECILUM*
DAEDALEA	(B)	*PHAEODAEDALEA*
DALDINIA	(A)	*NODULISPORIUM, VIRGARIELLA*
DENNISIELLIA	(A)	*MICROXIPHIUM*
DESMAZIERELLA	(A)	*VERTICICLADIUM*
DEXHOWARDIA	(?B)	*DEXHOWARDIA*
DIACANTHODES	(B)	*BORNETINA*
DIAPORTHE	(A)	*LIBERTELLA*
DIATRYPE	(A)	*LIBERTELLA, SELENOSPORELLA*
DIATRYPELLA	(A)	*LIBERTELLA*
DIBELONIELLA	(A)	*MYRIOCONIUM*
DICHLAENA	(A)	*ASPERGILLUS, PENICILLIUM*
DICHOMITUS	(B)	"ARTHROCONIDIA"
DICHOSTEREUM	(B)	*SPINIGER*
DICHOTOMOMYCES	(A)	*POLYPAECILUM, SCOPULARIOPSIS*
DICTYOTRICHELLA	(A)	*RHINOCLADIELLA*
DIDYMOSPHAERIA	(A)	*FUSICLADIELLA, PERICONIA*
DIGITATISPORA	(?B)	*DIGITATISPORA, VERTICILLIUM*
DIMORPHOCYSTIS	(B)	*ACTINICEPS*
DIPLOCARPON	(A)	*MARSONINA, ?BOSTRICHONEMA*
DIPLOGELASINOSPORA	(A)	*TRICHOSPORIELLA, GEOTRICHUM*
DIPODASCUS	(A)	*GEOTRICHUM*
DISCOHAINESIA	(A)	*HAINESIA*
DISCOSTROMA	(A)	*SEIMATOSPORIUM*
DOASSANSIA	(B)	*SAVULESCUELLA*
DOASSANSIOPSIS	(B)	*DOASSANSIELLA*
DOTHIDEA	(A)	*AUREOBASIDIUM*

DOTHIORA	(A)	*HORMONEMA*
DREPANOCONIS	(B)	*DREPANOCONIS*
DREPANOPEZIZA	(A)	*MARSONINA*

E

ECHINOPODOSPORA	(A)	*CHRYSOSPORIUM, CLADORRHINUM*
EDYUILLIA	(A)	*ASPERGILLUS*
ELEUTHERASCUS	(A)	*NODULISPORIUM*
EMERICELLA	(A)	*ASPERGILLUS*
EMERICELLOPSIS	(A)	*ACREMONIUM, ?STILBUM*
EMMONSIELLA	(A)	*HISTOPLASMA, CHRYSOSPORIUM*
ENDOMYCES	(A)	*HYALODENDRON*
ENTOLOMA	(B)	*ACURTIS*
ENTYLOMA	(B)	*ENTYLOMELLA*
EPHEMEROASCUS	(A)	*VERTICILLIUM*
EPICHLOE	(A)	*SPHACELIA*
ERYSIPHE	(A)	*OIDIUM*
EUANTENNARIA	(A)	*ANTENNATULA, HORMISCIOMYCES*
EUPELTE	(A)	*CLASTEROSPORIUM, PIROZYNSKIA, SEPTOIDIUM, SPORIDESMIUM*
EUPENICILLIUM	(A)	*PENICILLIUM*
EUROPHIUM	(A)	*LEPTOGRAPHIUM, VERTICICLADIELLA*
EUROTIUM	(A)	*ASPERGILLUS*
EUTYPELLA	(A)	*LIBERTELLA*

F

FARLOWIA	(A)	*BOTRYOTRICHUM*
FARLOWIELLA	(A)	*ACROGENOSPORA*
FENNELLIA	(A)	*ASPERGILLUS*
FILOBASIDIELLA	(B)	*CRYPTOCOCCUS, FILOBASIDIELLA*
FISTULINA	(B)	*?SPOROTRICHUM*
FOMITOPSIS	(B)	*SPINIGER*
FRAGOSPHAERIA	(A)	*?SPOROTHRIX*

G

GAEUMANNOMYCES	(A)	*PHIALOPHORA*
GALACTOMYCES	(A)	*GEOTRICHUM*
GALZINIA	(B)	*BENIOWSKIA*-LIKE
GASSARIA	(A)	*ASTEROSPORIUM*
GEOPYXIS	(A)	*HANSFORDIA*
GIBBERELLA	(A)	*FUSARIUM, STAGONOSTROMA*
GLOBIFOMES	(B)	"ARTHROCONIDIA"
GLOEOCYSTIDIELLUM	(B)	"ARTHROCONIDIA"
GLOEOPHYLLUM	(B)	"ARTHROCONIDIA"
GLOEOTINIA	(A)	*ENDOCONIDIUM*
GLOMERELLA	(A)	*COLLETOTRICHUM*
GLYPHIUM	(A)	*PEYRONELIA*
GNOMONIA	(A)	*ASTEROMA, CYLINDROSPORELLA, SESQUICILLIUM*
GRANDINIA	(B)	*PAPULASPORA*
GRAPHIOLA	(?B)	*GRAPHIOLA*
GRAPHOSTROMA	(A)	*NODULISPORIUM*
GRIPHOSPHAERIA	(A)	*FUSARIUM, SEIMATOSPORIUM*
GUIGNARDIA	(A)	*HORMONEMA*
GYMNASCELLA	(A)	*MALBRANCHEA*

GYMNOASCOIDEUS	(A)	*MALBRANCHEA*
GYMNOASCUS	(A)	*CHRYSOSPORIUM, MALBRANCHEA*
GYMNOEUROTIUM	(A)	*=EDYUILLIA, Q.V.*
GYMNOSPORANGIUM	(B)	"PHIALOCONIDIA"

H

HALOSPHAERIA	(A)	*PERICONIA, TRICHOCLADIUM*
HAMIGERA	(A)	*PENICILLIUM*
HAPALOPILUS	(B)	*BENIOWSKIA*-LIKE
HAPSIDOSPORA	(A)	*ACREMONIUM*
HARPEZOMYCES	(A)	*=CHAETOSARTORYA, Q.V.*
HELICOBASIDIUM	(B)	*RHIZOCTONIA, TUBERCULINA*
HELICOCEPHALUM	(Z)	*HELICOCEPHALUM*
HELMINTHOSPHAERIA	(A)	*DIPLOCOCCIUM*
HEMICARPENTELES	(A)	*ASPERGILLUS*
HERPOBASIDIUM	(B)	*GLOMOPSIS*
HETEROBASIDION	(B)	*SPINIGER*
HOHENBUEHLIA	(B)	*NEMATOCTONUS*
HOLWAYA	(A)	*CRINULA*
HYALOSCYPHA	(A)	*CHAETOCHALARA, CLATHROSPHAERINA, HAPLOGRAPHIUM*
HYMENOPLEELLA	(A)	*MONOCHAETIA*
HYMENOSCYPHUS	(A)	*IDRIELLA, VARICOSPORIUM*
HYPHODERMA	(B)	*SPINIGER*
HYPHODONTIA	(B)	"PHIALOCONIDIA"
HYPOCREA	(A)	*ACREMONIUM, GLIOCLADIUM, TRICHODERMA, VERTICILLIUM*
HYPOCREOPSIS	(A)	*STILBUM, STROMATOCREA*
HYPOMYCES	(A)	*CLADOBOTRYUM, GLIOCLADIUM, MYCOGONE, SEPEDONIUM, SIBIRINA, STEPHANOMA, SYMPODIOPHORA, TRICHOTHECIUM, VERTICILLIUM*
HYPOXYLON	(A)	*GENICULOSPORIUM, HANSFORDIA, NODULISPORIUM, ?PERICONIELLA, TRIPLICARIA, XYLOCLADIUM*
HYSTERIUM	(A)	*CONIOSPORIUM*

I

INONOTUS	(B)	*PTYCHOGASTER*
IODOPHANUS	(A)	*OEDOCEPHALUM*
ITERSONILIA	(B)	*ENTYLOMELLA*

J

JACOBIA	(?B)	*JACOBIA*

K

KERATINOPHYTON	(A)	*CHRYSOSPORIUM*
KERNIA	(A)	*GRAPHIUM, SCOPULARIOPSIS*
KORFIELLA	(A)	*CONOPLEA*

L

LAETIPORUS	(B)	*CHRYSOSPORIUM, SPOROTRICHUM*
LAETISARIA	(B)	*?RHIZOMORPHA*
LASIOBOTRYS	(A)	*ULOCLADIUM*
LASIOSPHAERIA	(A)	*HELICOMA, PHIALOPHORA*
LAURILIA	(B)	*SPINIGER*
LENTINUS	(B)	*DIGITELLUS*
LEPIOTA	(B)	*COCCOBOTRYS*
LEPTOSPHAERIA	(A)	*ALTERNARIA, CAMPOSPORIUM, CERCOSPORA, NAKATAEA* (SEE *MAGNAPORTHE*), *PESTALOTIA*
LEPTOSPOROMYCES	(B)	*TAENIOSPORA*
LEUCOAGARICUS	(B)	*ATTAMYCES*
LEUCOGYROPHANA	(B)	*SCLEROTIUM*
LEUCOSPHAERIA	(A)	*SPOROTHRIX*
LEVEILLULA	(A)	*OIDIUM*
LEVISPORA	(A)	*ACREMONIUM*
LILLIPUTIA	(A)	*PHIALOCEPHALA*
LOPHOTRICHUS	(A)	*HUMICOLA*
LYCROSTROMA	(B)	*MICHENERA*

M

MACRONODUS	(A)	*MALBRANCHEA*
MAGNAPORTHE	(A)	*NAKATAEA, PYRICULARIA, SCLEROTIUM*
MAGNUSIOMYCES	(A)	*GEOTRICHUM*
MASSARINA	(A)	*ANGUILLOSPORA, DACTYLELLA, PYRICULARIA*
MATRUCHOTIA	(?B)	*MATRUCHOTIA*
MAURODOTHINA	(A)	*PIROZYNSKIA*
MELANCONIS	(A)	*CORYNEUM, MELANCONIUM, STILBOSPORA*
MELANOCHAETA	(A)	*SPOROSCHISMA*
MELANODOTHIS	(A)	*RAMULARIA*
MELANOMMA	(A)	*PSEUDOSPIROPES*
MELANOPSAMMA	(A)	*STACHYBOTRYS*
MELANOPSAMMELLA	(A)	*GONYTRICHUM*
MELANOSPORA	(A)	*HARZIA, HARZIELLA, PAPULASPORA, PROTEOPHIALA*
MELANOTUS	(B)	"ARTHROCONIDIA"
MELIOLA	(A)	*JAINESIA*
MELIOLINA	(A)	*CATENULARIA*-LIKE
MELIOLIPHILA	(A)	*CHIONOMYCES, ERIOMYCOPSIS*
MERISTACRUM	(Z)	*MERISTACRUM*
METACAPNODIUM	(A)	*CAPNOBOTRYS, CAPNOCYBE, CAPNOPHIALOPHORA, CAPNOSPORIUM, TORULOPSIELLA*
METULODONTIA	(B)	*AEGERITA*
MICROASCUS	(A)	*SCOPULARIOPSIS, WARDOMYCES, WARDOMYCOPSIS*
MICROCOLLYBIA	(B)	*SCLEROSTILBUM*
MICROCYCLUS	(A)	*CERCOSPORA, FUSICLADIUM, PASSALORA*
MICRONECTRIELLA	(A)	*FUSARIUM*
MICROSPHAERA	(A)	*OIDIUM*
MICROTHECIUM	(A)	*HARZIELLA, PAPULASPORA, PHIALOPHORA*
MICROTHELIA	(A)	*DENDRYPHIOPSIS*
MILADINA	(A)	*ACTINOSPORA*
MOLLISIA	(A)	*ANGUILLOSPORA, PHIALOPHORA (?=MYRIOCONIUM)*
MONASCUS	(A)	*BASIPETOSPORA*
MONILINIA	(A)	*MONILIA, MYRIOCONIUM*
MORCHELLA	(A)	*COSTANTINELLA*
MONOGRAPHELLA	(A)	*FUSARIUM*
MUELLERITES	(A)	*AUREOBASIDIUM*-LIKE
MYCOACIA	(B)	"ARTHROCONIDIA"

MYCOARACHIS	(A)	*ACREMONIUM*
MYCOCITRUS	(A)	*ACREMONIUM*
MYCOSPHAERELLA	(A)	*CERCOSEPTORIA, CERCOSPORA, CERCOSPORELLA, CERCOSPORIDIUM, CLADOSPORIUM, FUSICLADIELLA, MIURAEA, OVULARIA, PARACERCOSPORA, PASSALORA, PHAEOISARIOPSIS, POLYTHRINCIUM, PSEUDOCERCOSPORA, RAMULARIA, STIGMINA*
MYRIOSCLEROTINIA	(A)	*MYRIOCONIUM*
MYTILIDION	(A)	*SEPTONEMA*
MYXOTRICHUM	(A)	*MALBRANCHEA, MYXOTRICHELLA, OIDIODENDRON*

N

NANNIZZIA	(A)	*MICROSPORIUM*
NECTRIA	(A)	*ACREMONIUM, ATRACTIUM, CALOSTILBELLA, CILICIOPODIUM, CYLINDROCARPON, DENDRODOCHIUM, DIDYMOSTILBE, FLAGELLOSPORA, FUSARIUM, FUSISPORELLA, GLIOCLADIUM, HELISCUS, MYROTHECIUM, STILBELLA, TUBERCULARIA, TUBERCULISPORA, VERTICILLIUM, VOLUTELLA*
NECTRIOPSIS	(A)	*FUSARIUM, STACHYLIDIUM*
NEOBULGARIA	(A)	*MYRIOCONIUM*
NEOCOSMOSPORA	(A)	*ACREMONIUM (STATE OF FUSARIUM)*
NEOPARODIA	(A)	*CHUPPIA, SARCINELLA*
NEOSARTORYA	(A)	*ASPERGILLUS*
NEUROSPORA	(A)	*MONILIA, PAPULASPORA*
NIA	(B)	*NIA*
NIESSLIA	(A)	*MONOCILLIUM*
NIGROSABULUM	(A)	*ACREMONIUM*
NUMMULARIA	(A)	*NODULISPORIUM, XYLOCLADIUM*
NYCTALIS	(B)	*ASTEROPHORA*

O

OLIGOPORUS	(B)	*PTYCHOGASTER*
ONYGENA	(A)	*SPORENDONEMA*-LIKE
OPHIOCAPNOCOMA	(A)	*CAPNOPHIALOPHORA, HORMIOKRYPSIS, TORULOPSIELLA*
OPHIOCORDYCEPS	(A)	*HIRSUTELLA*
OPHIONECTRIA	(A)	*ANTIPODIUM, HELICOMYCES, HELICOSPORIUM*
OPHIOPARODIA	(A)	*SEPTOIDIUM*
OPHIOSTOMA	(A)	(SEE ALSO *CERATOCYSTIS*), *GRAPHIUM, HYALODENDRON, PACHNODIUM, PESOTUM, RAFFAELEA, SPOROTHRIX, VERTICICLADIELLA*
ORBILIA	(A)	*DICRANIDION*
OTTHIA	(A)	*STIGMINA*
OVULINA	(A)	*OVULITIS*

P

PANUS	(B)	*SCLEROMA*
PARADIDYMELLA	(A)	*SEIMATOSPORUM*
PARKERELLA	(A)	*CYPHELLOPHORA*-LIKE
PARODIELLINA	(A)	*ARNAUDIA, SEPTOIDIUM*
PARODIOPSIS	(A)	*SEPTOIDIUM*
PECKIELLA	(A)	*ACREMONIUM, SEPEDONIUM*
PELLICULARIA	(B)	*RHIZOCTONIA, SCLEROTIUM*

PELORONECTRIELLA	(A)	ACREMONIUM
PENICILLIOPSIS	(A)	PENICILLIUM
PENIOPHORA	(B)	AEGERITA, ANTHINA, "ARTHROCONIDIA", SPINIGER
PERENNIPORIA	(B)	"ARTHROCONIDIA"
PERISPORIOPSIS	(A)	SEPTOIDIUM
PERONEUTYPELLA	(A)	PHAEOISARIA
PESTALOPEZIZA	(A)	PESTALOTIA
PESTALOSPHAERIA	(A)	PESTALOTIA
PETRIELLA	(A)	GRAPHIUM, SPOROTHRIX
PETRIELLIDIUM	(A)	GRAPHIUM, SCEDOSPORIUM
PETROMYCES	(A)	ASPERGILLUS
PEZIZA	(A)	CHROMELOSPORIUM, DICHOBOTRYS, OEDOCEPHALUM
PHALLUS	(B)	RHIZOSTROMA
PHANEROCHAETE	(B)	CHRYSOSPORIUM, SPOROTRICHUM
PHELLINUS	(B)	"ARTHROCONIDIA"
PHILLIPSIA	(A)	BOTRYTIS-LIKE
PHLEBIA	(B)	"ARTHROCONIDIA"
PHLEBOPUS	(B)	"ARTHROCONIDIA"
PHLEOGENA	(B)	HEYDENIA, MARTINDALIA
PHYLLACTINIA	(A)	OIDIUM
PHYSALOSPORA	(A)	ARTHRINIUM
PILGERIELLA	(A)	SEPTOIDIUM
PLACOSOMA	(A)	?SPORIDESMIUM
PLECTANIA	(A)	CONOPLEA
PLECTOSPHAERELLA	(A)	FUSARIUM
PLEOCHAETA	(A)	OIDIUM
PLEOSPORA	(A)	ALTERNARIA, DENDRYPHION, STEMPHYLIUM
PLEUROTUS	(B)	ANTROMYCOPSIS, GEMMULARIA
PLUNKETTOMYCES	(A)	MALBRANCHEA
PODONECTRIA	(A)	TETRACRIUM
PODOSORDARIA	(A)	?CHRYSOSPORIUM
PODOSPHAERA	(A)	OIDIUM
PODOSPORA	(A)	PHIALOPHORA
PODOSTROMA	(A)	TRICHODERMA
POLYPORUS	(B)	ECHINODIA, MYLITTA, PIETRAIA, SPOROTRICHUM
PORIA	(B)	"ARTHROCONIDIA", GEMMULARIA
POROCONIOCHAETA	(A)	HORMONEMA
PORONIA	(A)	?BLASTOBOTRYS, ?PACHNOCYBE, XYLOCLADIUM
POTEBNIAMYCES	(A)	AUREOBASIDIUM, HORMONEMA
PRINGSHEIMIA	(A)	HORMONEMA
PSEUDEUROTIUM	(A)	ACREMONIUM, SPOROTHRIX
PSEUDOCOCHLIOBOLUS	(A)	CURVULARIA, HELMINTHOSPORIUM
PSEUDOGUIGNARDIA	(A)	ARTHRINIUM
PSEUDOGYMNOASCUS	(A)	GEOMYCES
PTERIDIOSPERMA	(A)	PHIALOPHORA
PSEUDOMASSARIA	(A)	BELTRANIELLA
PSEUDONECTRIA	(A)	SESQUICILLIUM, TILACHLIDIUM, VOLUTELLA
PSEUDOVALSA	(A)	CORYNEUM
PSEUGOGYMNOASCUS	(A)	GEOMYCES
PTERIDIOSPERMA	(A)	PHIALOPHORA
PUNCTULARIA	(B)	PTYCHOGASTER, SPORENDONEMA-LIKE
PUTTEMANSIA	(A)	TETRACRIUM
PYRENOPEZIZA	(A)	CYLINDROSPORIUM, PHIALOPHORA
PYRENOPHORA	(A)	CORYNESPORA, DRECHSLERA
PYRONEMA	(A)	OEDOCEPHALUM
PYTHIUM	(O)	ARTOTROGUS
PYXIDIOPHORA	(A)	CHALARA

Q

QUATERNARIA	(A)	*LIBERTELLA*
QUESTIERIA	(A)	*QUESTERIELLA*

R

REMISPORA	(A)	*PERICONIA*
RENISPORA	(A)	*CHRYSOSPORIUM*
RESINICIUM	(B)	*SPINIGER*
RHOPALOMYCES	(?Z)	*RHOPALOMYCES*
RHYNCHOSPHAERIA	(A)	*MONOCHAETIA*
RIGIDOPORUS	(B)	"ARTHROCONIDIA"
ROUMEGUERIELLA	(A)	*PHIALOCEPHALA*
ROSELLINIA	(A)	*DEMATOPHORA, NODULISPORIUM*
ROZITES	(B)	*ATTAMYCES*
RUTSTROEMIA	(A)	*MYRIOCONIUM*

S

SAGENOMA	(A)	*ACREMONIUM*
SAGERIA	(A)	*ASCOCONIDIUM*
SARCOSCYPHA	(A)	*?HYALODENDRON*
SARTORYA	(A)	*ASPERGILLUS*
SCHIFFNERULA	(A)	*MITTERIELLA, QUESTERIELLA, SARCINELLA*
SCIRRHIA	(A)	*HADROTRICHUM*
SCLEROCLEISTA	(A)	*ASPERGILLUS*
SCLEROTINIA	(A)	*AMPHOBOTRYS, BOTRYTIS, MYRIOCONIUM, SCLEROTIUM, VERRUCOBOTRYS*
SEAVERINA	(A)	*VERRUCOBOTRYS*
SEBACINA	(B)	*FLAHAULTIA*
SEPTOBASIDIUM	(B)	*ALEURODOMYCES, HARPOGRAPHIUM*
SEPTOTINIA	(A)	*SEPTOTIS*
SETOSPHAERIA	(A)	*DRECHSLERA, EXSEROHILUM*
SEURATIA	(A)	*ATICHIA*
SISTOTREMA	(B)	*BURGOA*
SORDARIA	(A)	*PAPULASPORA*
SPHAERELLA	(A)	*ASTERISPORIUM*
SPHAERONAEMELLA	(A)	*CHALARA, GABARNAUDIA*
SPHAEROSPORELLA	(A)	*DICHOBOTRYS*
SPHAEROSTILBE	(A)	*ATRACTIUM*
SPHAEROTHECA	(A)	*OIDIUM*
SPHAERULINA	(A)	*CERCOSPORA*
SPONGIPELLIS	(B)	"ARTHROCONIDIA"
SPOROPHORMIS	(A)	= *WARCUPIELLA, Q.V.*
STEREUM	(B)	*SPINIGER*
STREPTOTINIA	(A)	*STREPTOBOTRYS*
STRIGOPODIA	(A)	*CAPNOPHIALOPHORA*
STROMATINIA	(A)	*SCLEROTIUM*
SYDOWIA	(A)	*HORMONEMA*
SYMPHYOSIRINIA	(A)	*SYMPHYOSIRA*
SYNCLEISTOSTROMA	(A)	= *PETROMYCES, Q.V.*

T

TALAROMYCES	(A)	*MERIMBLA, PAECILOMYCES, PENICILLIUM*
TERMITOMYCES	(B)	*TERMITOSPHAERA*

THANATEPHORUS	(B)	*RHIZOCTONIA*
THAXTERIA	(A)	*OEDEMIUM*
THAXTERIELLA	(A)	*HELICOMA*
THECOTHEUS	(A)	*RHINOCLADIELLA*
THERMOASCUS	(A)	*PAECILOMYCES, POLYPAECILUM*
THIELAVIA	(A)	*ACREMONIUM, CHRYSOSPORIUM, MYCELIOPHTHORA, SEPEDONIUM, TRICHOSPORIELLA*
THUEMENELLA	(A)	*TRICHODERMA*-LIKE
THYRONECTRIA	(A)	*TUBERCULARIA*
TILLETIA	(B)	*TILLETIELLA*
TOMENTELLA	(B)	*ALYTOSPORIUM*
TORRUBIELLA	(A)	*GIBELLULA, ?GRANULOMANUS, HIRSUTELLA, VERTICILLIUM*
TRACYA	(B)	*SAVULESCUELLA, TRACYELLA*
TRAMETES	(B)	"ARTHROCONIDIA", *SPINIGER*
TRECHISPORA	(B)	*DEXHOWARDIA, ?OSTEOMORPHA, PAPULASPORA*
TREMELLA	(B)	*HORMOMYCES*
TRIANGULARIA	(A)	*PHIALOPHORA*
TRICHOCOMA	(A)	*PENICILLIUM*
TRICHOLOMOPSIS	(B)	*RHIZOCTONIA*
TRICHOMETASPHAERIA	(A)	*DRECHSLERA*
TRICHOPELTHECA	(A)	*HORMISCIOMYCES, PLOKAMIDOMYCES, TRICHOTHALLUS*
TRICHOPHAEA	(A)	*DICHOBOTRYS*
TRICHOSKYTALE	(A)	*PENICILLIUM*
TRICHOSPHAERELLA	(A)	*ACREMONIUM, GONYTRICHUM*
TRICHOTHYRIUM	(A)	*ISTHMOSPORA*
TRIPOSPORIOPSIS	(A)	*TRIPOSPERMUM*
TUBERCINIA	(B)	*PAIPALOPSIS*
TUBEUFIA	(A)	*HELICOSPORIUM, PENDULISPORA*
TULASNELLA	(B)	*RHIZOCTONIA*
TYPHULA	(B)	*SCLEROTIUM*
TYROMYCES	(B)	"ARTHROCONIDIA", *PTYCHOGASTER*

U

UNCINOCARPUS	(A)	*MALBRANCHEA*
UNCINULA	(A)	*OIDIUM*
URNULA	(A)	*CONOPLEA, STRUMELLA*
USTILAGO	(B)	*SPOROPHORA*
UTHATOBASIDIUM	(B)	*RHIZOCTONIA*

V

VALDENSINIA	(A)	*VALDENSIA*
VALSONECTRIA	(A)	*SPOROTHRIX*
VARARIA	(B)	*DEXHOWARDIA*-LIKE, *SPINIGER*
VENTURIA	(A)	*CLADOSPORIUM, FUSICLADIUM, POLLACCIA, SPILOCAEA*

W

WAITEA	(B)	*RHIZOCTONIA*
WALLROTHIELLA	(A)	*ACREMONIUM, GLIOMASTIX*
WARCUPIELLIA	(A)	*ASPERGILLUS*

X

XENODIUM	(A)	*XENODIELLA*
XENOMERIS	(A)	*HORMONEMA*
XENOPUS	(B)	*XENOPUS*
XEROMYCES	(A)	*FRASERIELLA*
XYLARIA	(A)	*NODULISPORIUM, PADIXONIA*
XYLOGONE	(A)	*BAHUSAKALA*
XYNOPHILA	(A)	"ARTHROCONIDIA"

Y

YAMAMOTOA	(A)	*CLASTEROSPORIUM, MITTERIELLA, SARCINELLA*

Z

ZENDERA	(A)	*GEOTRICHUM*
ZOPFIA	(A)	*CLADOSPORIUM*
ZOPFIELLA	(A)	*CHRYSOSPORIUM, HUMICOLA,* "PHIALOCONIDIA"
ZUKALIA	(A)	*?KAZULIA*
ZYGOMYCETES	(?Z)	*ACAULOPAGE, RHOPALOMYCES*

References

Ainsworth, G. C. 1971. "Ainsworth and Bisby's Dictionary of the Fungi.", 6th ed., Commonwealth Mycological Institute, Kew. (5th ed., 1961)

Ainsworth, G. C. 1973. Introduction and keys to higher taxa. *In* "The Fungi", Vol. IVA (Ainsworth, *et al.*, eds.), pp. 1-7, Academic Press, New York.

Ainsworth, G. C., F. K. Sparrow and A. S. Sussman (eds.). 1973. "The Fungi an Advanced Treatise." Vols. IVA and IVB, Academic Press, New York.

Arx, J. A. von. 1970. "The Genera of Fungi Sporulating in Pure Culture." Cramer, Lehre.

Arx, J. A. von. 1974. "The Genera of Fungi Sporulating in Pure Culture." 2nd ed. Cramer, Vaduz.

Bandoni, R. J. 1972. Terrestrial occurrence of some aquatic Hyphomycetes. *Can. J. Bot.* **50**:2283-2288.

Barnett, H. L. 1955. "Illustrated Genera of Imperfect Fungi." Burgess, Minneapolis.

Barron, G. L. 1968. "The Genera of Hyphomycetes from Soil." Williams and Wilkins, Baltimore, Maryland.

Bary, A. de 1858. Quoted in Whittaker, 1969, ref. 84.

Bender, H. B. 1931. "The Genera of Fungi Imperfecti." Thesis, Yale University.

Carmichael, J. W. 1962. *Chrysosporium* and some other aleuriosporic Hyphomycetes. *Can. J. Bot.* **40**:1137-1173 and Pls. I-VIII.

Carmichael, J. W. 1971. Blastospores, aleuriospores and chlamydospores. *In* "Taxonomy of Fungi Imperfecti" (W. B. Kendrick, ed.), pp. 50-70. Univ. of Toronto Press, Toronto.

Carmichael, J. W. 1979. Cross-reference names for pleomorphic fungi. *In* "The Whole Fungus" (Bryce Kendrick ed.), **1**:31-41. National Museums of Canada, Ottawa.

Clements, F. E. and C. L. Shear. 1931. "The Genera of Fungi." Hafner, New York.

Cohn, F. 1875. Quoted in Whittaker, 1969, ref. 71.

Cole, G. T. and R. A. Samson, 1979. "Patterns of Development in Conidial Fungi." Pittman, London.

Donk, M. A. 1962. The generic names proposed for Hymenomycetes—XII. Deuteromycetes. *Taxon* **11**(3):75-104.

Ellis, M. B. 1971. "Dematiaceous Hyphomycetes." Commonwealth Mycological Institute, Kew.

Ellis, M. B. 1976. "More Dematiaceous Hyphomycetes." Commonwealth Mycological Institute, Kew.

Fries, E. M. 1821-32. "Systema Mycologicum." Greifswald.

Gams, W. 197l. "*Cepholosporium*-artige Schimmelpilze (Hyphomycetes)." Gustav Fischer Verlag, Stuttgart.

Haeckel, E. 1866. Quoted in Whittaker, 1969, ref. 29.

Hennebert, G. L. 1971. Pleomorphism in Fungi Imperfecti. *In* "Taxonomy of Fungi Imperfecti." (W. B. Kendrick, ed.), pp 202-203. University of Toronto Press, Toronto.

Hennebert, G. L. and L. K. Weresub. 1979. Terms for states and forms of fungi, their names and types. *In* "The Whole Fungus." (W. B. Kendrick, ed.), **1**:27-30. National Museums of Canada, Ottawa.

Heslop-Harrison, J. 1962. Purposes and procedures in the taxonomic treatment of higher organisms. *In* "Microbial Classification" (G. C. Ainsworth and P. H. A. Sneath, eds.), pp. 14-36. Cambridge University Press, London and New York.

Hogg, J. 1860. Quoted in Whittaker, 1969, ref. 26.

Hughes, S. J. 1953. Conidia, conidiophores and classification. *Can. J. Bot.* **31**:577-659.

Hughes, S. J. 1958. Revisiones Hyphomycetum aliquot. *Can. J. Bot.* **36**:727-836.

Hughes, S. J. 1959. Starting point of nomenclature of Hyphomycetes. *Taxon* **8**:96-103.

Ingold, C. T. 1975. "Guide to Aquatic Hyphomycetes." Freshwater Biol. Assoc. Scientific Publication No. 30, Ambleside, Cumbria, England.

Ingold, C. T. 1979. Advances in the study of so-called aquatic Hyphomycetes. *Amer. J. Bot.* **66**:218-226.

Kendrick, W. B., (ed.) 1971. "Taxonomy of Fungi Imperfecti." University of Toronto Press, Toronto.

Kendrick, Bryce, (ed). 1979. "The Whole Fungus." Vols. 1 and 2. National Museums of Canada, Ottawa.

Kendrick, W. B. and J. W. Carmichael, 1973. Hyphomycetes. Chapter l0. *In* "The Fungi", Vol. IVA, (Ainsworth, *et al.*, eds). pp. 323-509. Academic Press, New York.

Kendrick, B., F. DiCosmo and committees. 1979. Teleomorph-anamorph connections in Ascomycetes. *In* "The Whole Fungus" (Bryce Kendrick, ed.) pp. 283-359. National Museums of Canada, Ottawa.

Kendrick, B. and T. R. Nag Raj. 1979. Morphological terms in Fungi Imperfecti. *In* "The Whole Fungus" (Bryce Kendrick ed.), **1**:43-61. National Museums of Canada, Ottawa.

Korf, R. P. and L. M. Kohn, 1979. Later starting point blues. I. *Monilia fructigena*. *Mycotaxon* **9**:521-522.

Kudo, R. R. 1946. "Protozoology." 3rd ed., Thomas, Springfield, Illinois.

Lindau, G. 1907. Hyphomycetes (erste Halfte) *In* "Rabenhorst's Kryptogamem-Flora von Deutschland, Oesterrich and der Schweiz." 2 ed., Vol. 1, Sect. VIII, Eduard Kummer, Leipzig.

Lindau, G. 1910. Hyphomycetes (zweite Halfte) *In* "Rabenhorst's Kryptogamem-Flora von Deutschland, Oesterrich and der Schweiz." 2 ed., Vol. 1, Sect. IX, Eduard Kummer, Leipzig.

Link, H.F. 1809. Observationes in ordines plantarum naturales. *Mag. Ges. naturf. Freunde, Berlin,* **3**:3-42.

Luttrell, E. S. 1979. Deuteromycetes and their relationships. *In* "The Whole Fungus", (Bryce Kendrick, ed.), **1**:241-264. National Museums of Canada, Ottawa.

Mason, E.W. 1933. Annotated account of fungi received at the Imperial Mycological Institute. List II, (Fascicle 2). *Mycol. papers C.M.I.* **3**:1-67.

Matsushima, T. 1971. "Microfungi of the Solomon Islands and Papua-New Guinea." Matsushima, Kobe.

Matsushima, T. 1975. "Icones Microfungorum a Matsushima Lectorum". Matsushima, Kobe.

Morgan-Jones, G., T. R. Nag Raj and Bryce Kendrick. 1972. Icones generum Coelomycetum I. *Univ. Waterloo Biol. Series* **3**:1-42. (The Icones continues with various authors to part X and index in 1979.)

Nag Raj, T. R. and Bryce Kendrick 1975. "A Monograph of *Chalara* and Allied Genera." Wilfred Laurier Univ. Press, Waterloo.

Nicot, J. 1977. La systematique des champignons imperfaits: problemes et perspectives. *Revue de Mycol.* **4**:381-395.

Persoon, C. H. 1801. "Synopsis Methodica Fungorum." Gottingen.

Pirozynski, K. A. and L. K. Weresub. 1979. The classification and nomenclature of fossil fungi. *In* "The Whole Fungus." (Bryce Kendrick ed.), **2**:653-688. National Museums of Canada, Ottawa.

Saccardo, P. A. 1882-193l. "Sylloge Fungorum." Padua.

Sigler, L. and J. W. Carmichael, 1976. Taxonomy of *Malbranchea* and some other Hyphomycetes with arthroconidia. *Mycotaxon* **4**:349-488.

Simmons, E. G. 1952. Culture studies in the genera *Pleospora, Clathrospora,* and *Leptosphaeria. Mycologia* **44**:330-365.

Smith, G. 1962. The morphological approach to the taxonomy of microfungi *In* "Microbial Classification" (G. C. Ainsworth and P. H. A. Sneath, eds.), pp. 111-118. Cambridge University Press, London and New York.

Stafleu, F.A. *et al.* (eds.) 1972. "International Code of Botanical Nomenclature." Oosthoek's Uitgeversmaatschappij N. V., Utrecht.

Stafleu, F.A. *et al.* (eds.) 1978. "International Code of Botanical Nomenclature." Bohn, Schetema & Holkema, Utrecht.

Stanier, R. Y., M. Doudoroff and E. A. Adelberg. 1957. "The Microbial World." Prentice-Hall, Englewood Cliffs, New Jersey.

Sutton, B. C. 1977. Coelomycetes VI. Nomenclature of generic names proposed for Coelomycetes. *Mycol. papers C.M.I* **141**:1-253.

Tode, H. J. 1790. "Fungi Mecklenburgensis selecti."

Vuillemin, P. 1910 Les conidiospores. *Bull. Soc. Sci. Nancy* III, **11**:129-172.

Vuillemin P. 1911. Les aleuriospores. *Bull. Soc. Sci. Nancy* III, **12**:151-175.

Wakefield, E. M. and G. R. Bisby 1941. List of Hyphomycetes recorded for Britain. *Trans. Brit. Mycol. Soc.* **25**:49-126.

Weresub, L. K. 1979. On the question of naming pleomorphic anamorphic fungi. *In* "The Whole Fungus" (Bryce Kendrick, ed.), **2**:689-709. National Museums of Canada, Ottawa.

Weresub, L. K. and K. A. Pirozynski. 1979a. Pleomorphism of fungi as treated in the history of mycology and nomenclature. *In* "The Whole Fungus" (Bryce Kendrick, ed), **1**:17-25. National Museums of Canada, Ottawa.

Weresub, L. K. and K. A. Pirozynski, 1979b. The classification and nomenclature of fossil fungi. *In* "The Whole Fungus" (Bryce Kendrick, ed.) **2**:653-688. National Museums of Canada, Ottawa.

Whittaker, R. H. 1969. New concepts of kingdoms of organisms. *Science* **163**:150-160.

Wilson, C. L., W. E. Loomis and T. A. Steeves. 1971. "Botany." 5th ed. Holt, Reinhart & Winston, New York.

Abbreviations

(A)	-Ascomycota
ALEURI	-keyword for aleuric dehiscence see part IVB2
AMERO	-keyword for amerosporae see part IVB2
&	-ampersand for “and” or “et”
AMPUL	-keyword for ampulliform see part IVB2
ANNEL	-keyword for annelliform see part IVB2
APPEND	-keyword for appendages on conidia see part IVB2
AQUATIC	-keyword for aquatic habitat see part IVB2
ARTHRO	-keyword for arthrocatenate see part IVB2
ASCO	-division Ascomycota
(B)	-Basidiomycota
BASO	-keyword for basocatenate see part IVB2
BLASTO	-keyword for blastocatenate see part IVB2
BASIDIO	-division Basidiomycota
(C)	-Coelomycetes
CBS STUDIES	-Studies in Mycology, Centraalbureau voor Schimmelcultures
CEPHALO. SCHIMMEL.	-see Gams 1971 in References
CETERI	-keyword for “the rest” see part IVB2
CHALARA	-see Nag Raj & Kendrick 1975 in References
CHEIRO	-keyword for cheirosporae see part IVB2
CLAMPS	-keyword for clamp connections see part IVB2
CMI DESCRIPTIONS	-CMI Descriptions of Pathogenic Fungi and Bacteria
CMI PAPERS	-Mycological Papers of the Commonwealth Mycological Institute
COELO	-class Coelomycetes
CONIDIAL FUNGI	-see Cole & Samson 1979 in References
DESCR	-description

DH	-see Ellis 1971 in References
DICTYO	-keyword for dictyosporae see part IVB2
DIDYMO	-keyword for didymosporae see part IVB2
ECOL	-ecology
EMEND	-emendation
FIG	-figure(s)
GENERA	-see von Arx 1970 or 1974 in References
GLOIO	-keyword for gloiosporae see part IVB2
GUIDE	-see Ingold 1975 in References
(H)	-Hyphomycetes
HELICO	-keyword for helicosporae see part IVB2
HYALO	-keyword for hyalosporae see part IVB2
HYPHO	-class Hyphomycetes
HYPHOMYCETES	-see Barron 1968 in References
ICBN	-International Code of Botanical Nomenclature
ICONES	-see Matsushima 1975 in References
ILLUS	-illustration(s)
INED	-*ineditus* -not published
IN LIT.	-*in littera* -in a letter
MDH	-see Ellis 1976 in References
MICROFUNGI	-see Matsushima 1971 in References
MISC	-keyword for miscellaneous type of conidiogenous cells see part IVB2
MONOGR	-monograph
NEW COMB(S)	-new combination(s)
NEW SP (SPP)	-new species
NOM. CONS.	-*nomen conservandum*
NOM. NOV.	-*nomen novum*
NOM. REJ. PROP.	-*nomen rejiciendum propositum*
NONSP	-keyword for nonspecialized conidiogenous cells see part IVB2
ORIG	-original
ORTHO	-orthographic
P	-page(s)
PHAEO	-keyword for phaeosporae see part IVB2
PHIAL	-keyword for phialiform see part IVB2
PHRAGMO	-keyword for phragmosporae see part IVB2
PHYSIOL	-physiology
PL	-plate(s)
PODIA	-keyword for hyphopodia or stromatopodia see part IVB2
RACHI	-keyword for rachiform see part IVB2
RADUL	-keyword for raduliform see part IVB2
REFS	-references

X

Y

Z

REVIS	-revision
SCOLECO	-keyword for scolecosporae see part IVB2
SER	-series
SP (SPP)	-species
STAURO	-keyword for staurosporae see part IVB2
TAB	-table(s), used for illustrations in some older publications
TBMS	-Transactions of the British Mycological Society
VAR	-variant
(Y)	-class Blastomycetes (yeasts)
(Z)	-division Zygomycota

Index to Plates